# Principles of
# Supply Chain Management

## SECOND EDITION

# Series on Resource Management

## RECENT TITLES

# Principles of
# Supply Chain Management

## SECOND EDITION

Richard E. Crandall
William R. Crandall
Charlie C. Chen

**CRC Press**
Taylor & Francis Group
Boca Raton  London  New York

CRC Press is an imprint of the
Taylor & Francis Group, an **informa** business

CRC Press
Taylor & Francis Group
6000 Broken Sound Parkway NW, Suite 300
Boca Raton, FL 33487-2742

© 2015 by Taylor & Francis Group, LLC
CRC Press is an imprint of Taylor & Francis Group, an Informa business

No claim to original U.S. Government works

Printed on acid-free paper
Version Date: 20140721

International Standard Book Number-13: 978-1-4822-1202-0 (Hardback)

---

**Library of Congress Cataloging-in-Publication Data**

---

Crandall, Richard E., 1930-
    Principles of supply chain management / Richard E. Crandall, William R. Crandall, Charlie Chen. -- Second edition.
        pages cm
    Includes bibliographical references and index.
    ISBN 978-1-4822-1202-0
    1. Business logistics. 2.  Physical distribution of goods--Management. 3.  Marketing channels--Management.  I. Crandall, William, 1956- II. Chen, Charlie C., (Associate professor) III. Title.

HD38.5.C73 2015
658.5--dc23                                                                    2014027314

---

Visit the Taylor & Francis Web site at
http://www.taylorandfrancis.com

and the CRC Press Web site at
http://www.crcpress.com

*To Jean—yesterday, today, and tomorrow*

*To Sue, my wife and friend forever*

*To Fu-Mei, my mother and mentor forever*

# Contents

## Section II   Demand Perspective

## Section III Supply Perspective—Distribution, Production, Procurement, and Logistics

## Section IV    Need for Integration

# Section V   Financial and Information Technology Perspectives

## Section VI   The Future

# *Preface to the Second Edition*

As we look back at the preface to the first edition of this book, we recall our optimism that it would provide a good understanding of first, supply chains, and second, supply chain management. As we pointed out, adding the term *management* to *supply chain* opened up a different world. While supply chains could be physically assembled in a straightforward manner, managing them required skills and approaches that were not needed when businesses operated more autonomously. Building communication, collaboration, and trust into supply chains presented new challenges, especially as supply chains became more globally dispersed and complex. We like to think we did a good job of spelling out the fundamentals of supply chain management in our first edition. However, just as supply chains and their management have evolved, our presentation in this second edition has also evolved.

Some companies have built noteworthy supply chain systems and are managing them effectively and efficiently. We want to give those companies the credit they deserve and feature them as representative of the way in which industrial firms are coping with the greater challenges of global competition, increasing consumer demands, and increasing volatility in the business environment throughout the world. While the opportunities for growth in markets abound, so do the risks. Even in the best managed companies, supply chain disruptions are becoming of greater concern.

In this second edition, we have added two major sections to each chapter. At the beginning of the chapter, we include a company profile in which we describe a real company that illustrates the main theme of the chapter. For example, in Chapter 1, we describe how a company like Proctor & Gamble must build and manage multiple supply chains to obtain all of the products they sell. In Chapter 2, we show how a company like Zara manages a supply chain that enables them to rapidly introduce new clothing models within an unparalleled quickness. Other company profiles offer an introduction to the topics covered within the chapter.

We have also added a Hot Topics section at the end of each chapter. In these, we introduce situations that can cause disruptions in supply chains and may involve ethical or environmental considerations. It could be the result of risky supply chain design or failure to adequately monitor the sources of supply. In some cases, it could be the result of natural disasters beyond the control of supply chain members, although not necessarily beyond their responsibility to anticipate.

Table P.1 lists the companies featured in the company profiles and the subject matter of the hot topics.

We have kept the organization of the book consistent with the first edition, except for the addition of the new Chapter 9 on logistics. Please see the following introduction for a fuller description of the organization of the book. There are some issues or topics that are in the forefront of supply chain managers' concerns. Rather than add separate chapters, we have included extended discussions of these topics within chapters in which we believe they most logically fit. Some of the most evident are the issues of outsourcing and reshoring (Chapter 9), sustainability (Chapter 10), risk management (Chapter 17), and data analytics and knowledge management (Chapter 18).

There are a number of other additions to this edition to make the material as current as possible. While we recognize the need to get the book published, we are constantly trying to get one more new idea in before the editors finally shut the door on us. We hope you find our efforts worth your time.

**TABLE P.1**

List of Company Profiles and Hot Topics

| Chapter Title | Company Profile | Hot Topic |
|---|---|---|
| 1. Evolution of supply chains | Proctor & Gamble | Why offshore outsourcing is nothing to be toyed with |
| 2. Supply chains as a system | Inditex (Zara) | How a natural disaster can cripple a supply chain |
| 3. Determining customer needs | McDonald's | Is there human trafficking in your supply chain? |
| 4. A system to meet customer needs | GE Aviation | Sweatshops: two sides of the same coin |
| 5. Demand management | NextEra Energy | The problem of cheap |
| 6. Distribution and retailing | Lowe's | The problem of the middleman and a milk crisis in China |
| 7. Production and service processes | Caterpillar | Clothing, the Achilles heel of the supply chain |
| 8. Procurement/purchasing | Nestlé | Why apple juice should be made from … apples |
| 9. Logistics: The glue that holds the supply chain together | Transportation Insight | Container shipping and its risk points |
| 10. Reverse supply chains | Genco | Reshoring: revisiting the make or buy decision |
| 11. The need to integrate | Cisco | AECL encounters problems with its cancer fighting machine—Part 1 |
| 12. Why integration is difficult | Boeing | AECL encounters problems with its cancer fighting machine—Part 2 |
| 13. How to build an integrated supply chain | Interface | AECL encounters problems with its cancer fighting machine—Part 3 |
| 14. Information flow along the supply chain | SAP | Boeing 787: a new supply chain model in the commercial aircraft industry—Part 1 |
| 15. Funds flow along the supply chain | Wells Fargo | The Boeing 787: problems in the supply chain—Part 2 |
| 16. ROI for supply chains and other issues | Apple | The Boeing 787: pushing the limits of outsourcing—Part 3 |
| 17. Trends in supply chain management | Amazon | Finding solutions to the sweatshop problem |
| 18. Preparation for the future | Google | How social media knocked down the lean, finely textured beef industry |

# *Preface to the First Edition*

Supply chains affect individuals and organizations, both directly and indirectly. *Supply chain* is a term found daily in both trade and academic publications. Whether you are a customer or a supplier, you depend on supply chains for your necessities and luxury goods, as well as various other services. Why are supply chains so pervasive in our everyday lives?

In his book, *The World Is Flat, A Brief History of the Twenty-First Century*, Thomas Friedman listed supply chaining as one of the ten flatteners that would make the world more competitive on a global scale. Supply chains are the mechanisms by which products move from the raw materials stage through a series of transformative processes to the ultimate consumer. As a concept, supply chains are intuitively obvious. We recognize that the cartons of orange juice we buy at the local supermarket originated in an orange grove, or that the automobile we bought at the dealer required a number of supply chains to furnish all of the features included in its assembly. Even when supply chains extend across the world, we understand something about how they operate and why they are important.

When we add the word *management* to *supply chain*, we enter a new conceptual arena. Peter Drucker, one of the most prolific writers about management, offers the following statement in his book, *Management*:

> While management is a discipline—that is, an organized body of knowledge and as such applicable everywhere—it is also 'culture.' It is not value-free science. Management is a social function and embedded in a culture—a society—a tradition of values, customs, and beliefs, and in governmental and political systems. Management is—and should be—culture-conditioned; but, in turn, management and managers shape culture and society.

This statement suggests that supply chain management is more than understanding where products come from and how they get to the consumer. In the days when humans lived in caves and never traveled more than a few miles during their entire lives, supply chains were simple and could be easily managed. In present times, when even the smallest of companies may be dealing with suppliers and markets on a worldwide basis, supply chain management attempts to deal with a myriad of technical, organizational, and cultural challenges. These challenges can be overwhelming and often defy the rational management practices that worked so well in the past.

In this book, we explain why supply chain management is so different, and more difficult, than managing an individual business or some other form of organization. We examine the various components of a supply chain in detail; we then show how the pieces must fit together if the supply chain is to be effective. We then outline an approach to achieve supply chain integration that leads to effective management. We conclude that Peter Drucker was right. Supply chain management is more than managing technology and the supporting infrastructure. It requires a blending of organizational cultures within a specific company and among all members of a supply chain.

# *Acknowledgments*

We thank a number of people for helping us to prepare the second edition of this book. Much of the credit goes to William T. Walker, or Bill to most of us who have had the good fortune to call him a colleague and friend. Bill helped us get our latest book published with CRC Press, *Vanishing Boundaries, How Integrating Manufacturing and Services Creates Customer Value*. Bill read every chapter and offered a number of insightful suggestions to clarify and organize the topics in a meaningful way. It is a book that deals with the vanishing boundaries between manufacturing and services sectors, and how knowledge is being transferred between these two major types of businesses. A portion of that book examined the role of supply chains in facilitating knowledge transfer. It was thus a natural evolution to extend a portion of that work to a full-blown textbook on supply chain management. We want to acknowledge Bill's willingness to share his ideas and insights to help us make this second edition even better. Thank you Bill for your continued support. We also thank Dr. Al Harris for his insightful comments in many areas, but especially in the IT sections.

We are especially grateful to Jennifer Proctor, editor-in-chief, and Elizabeth Rennie, managing editor, of the *APICS* magazine. With their permission, we drew heavily from the following Relevant Research columns written by Dick.

Chapter 10—Crandall, R.E., Opportunity or threat? Uncovering the reality of reverse logistics, *APICS Magazine*, 16(4), 24, 2006. Section on "Benefits of Reverse Logistics" was adapted from this article.

Chapter 11—Crandall, R.E., Dream or reality? Achieving lean and agile integrated supply chains, *APICS Magazine*, 15, 10, 20, 2005. Section on "Setting the Stage" was adapted from this article.

Chapter 13—Managerial Comment 1 was adapted from Source: Adapted from Crandall, R.E., *APICS Magazine*, 16(6), 20, 2006.

Chapter 15—Crandall, R.E., Let's talk, better communication through IOS, *APICS Magazine*, 17(6), 20, 2007. Section on "Interorganizational Systems" was adapted from this article.

Chapter 15—Crandall, R.E., Exploring Internet EDI, *APICS Magazine*, 15(8), 18, 2005. Section on "Electronic Data Interchange" was adapted from this article.

Chapter 15—Crandall, R.E. and Main, K.T., Cash is king, *APICS—The Performance Advantage*, 12(1), 36, 2002. Section on "Supply Chain Funds Flow" and "A Comprehensive Example" was adapted from this article.

Chapter 16—Crandall, R.E., Beating impossible deadlines: A variety of methods can help, *APICS Magazine*, 16(6), 20, 2006. Section on "Programs Requiring Close Supply Chain Relationships" was adapted from this article.

Chapter 17—Crandall, R.E., Productive renovations, converting your business from a transaction-oriented to a process-oriented model, *APICS Magazine*, 17(2), 16, 2007. Section on "Process Evolution was adapted from this article.

Chapter 17—Crandall, R.E. (2009) Riding the green wave, more businesspeople catch on to the sustainability movement, *APICS Magazine*, 19 (1), 24–27. The section on Sustainability was adapted from this article.

Chapter 17—Crandall, R.E. Perceptions of peril, evaluating risk management in supply chains, *APICS Magazine*, 22(3) 21–25, 2012. Section on "Risk Management" was adapted from this article.

Chapter 18—The section "Expand Knowledge Management" was adapted from Crandall, R.E., From data to wisdom, Is knowledge management the key to the future? *APICS Magazine*, 17(9), 30, 2007.

Chapter 18—The section "Acquire Data Analytics Capabilities" was adapted from Crandall, R.E., The big data revolution, investigating recent developments in collection and analytics, *APICS Magazine*, 23(2), 20–23, 2013.

Thanks Jennifer and Beth for your help and encouragement.

We further thank APICS (The Association for Operations Management) for selecting the first edition of this book as one of four recommended reading books in preparation for their CSCP certification in supply chain management. This recognition by our peers made us feel we have something worth saying.

We thank all of the companies we worked for during our time in industry prior to joining the academic ranks. Dick spent over 25 years in industry working for manufacturing companies in the electronic industries, service companies in wholesale and retail, and an international consulting company, where he worked with a variety of manufacturing and service companies. Rick spent over 10 years in the food service industry, most of which was with ARAMARK in contract food services. Charlie's area of expertise is information technology and he has extensive international experience, especially in Taiwan, Japan, and China. We believe that our varied backgrounds have enabled us to take a panoramic view of global supply chains.

To supplement our own experience, we dug deeply into the extensive research by practitioners, consultants, and academics. We reviewed hundreds of articles and dozens of books to uncover the concepts, principles, and techniques that are included in this book. We hope this information will be helpful in your business career.

Amy Blalock, the project coordinator for our book at CRC Press, has been helpful in providing guidance about all of the things we needed to do and has answered any number of questions that we came up with. Amy, we appreciate your understanding and patience in our stumbles. Thanks for covering for us, and for helping us through all the steps in preparing the book for publication.

We would like to thank Lara Zoble, editor, Taylor & Francis Group. Lara provided the encouragement to write the second edition to this book and helped us focus our thinking as we evaluated different ways to make this second edition better. Lara also gave ideas about ways to make the tables and figures more interesting.

We are also thankful to all of those who had the unenviable job of trying to make our words sound enlightened or at least less obtuse, as well as worrying about grammar, punctuation, pagination, location of exhibits, and all those other things we were very willing to outsource to experts like them. We thank Richard Tressider at Taylor & Francis and Christine Selvan at SPi Global, and the editorial staff, for reviewing and editing of the book. We are sure there are a number of other people at CRC Press who helped make this book a reality. However, in this age of electronic communication, we do not always have a chance to meet everyone personally. Thank you; we hope you enjoyed it as much as we did.

Finally, we specially acknowledge the contributions of Jean, wife of Dick and mother of Rick. She encouraged us, but, even more importantly, she read every word (several times) and offered many helpful suggestions. Despite her best efforts, there may still be errors. If you find them, blame us, not her, or any of the people mentioned earlier.

**Richard E. (Dick) Crandall**
**William R. (Rick) Crandall**
**Charlie C. Chen**

# Authors

**Richard E. "Dick" Crandall** is a professor in the College of Business at Appalachian State University (ASU), Boone, North Carolina. He is certified in production and inventory management (CFPIM) and supply chain professional (CSCP) by the APICS-The Association for Operations Management. He earned his PhD in production/operations management from the University of South Carolina, Columbia, and is a registered professional engineer and a certified public accountant. Prior to joining ASU, Dick worked as an industrial engineer and in management positions for manufacturing and service companies. He was a consultant with a major consulting firm, installing systems for both operations and financial applications. With Rick Crandall, he coauthored the book *Vanishing Boundaries, How Integrating Manufacturing and Services Creates Customer Value*, by CRC Press, Taylor & Francis Group.

**William "Rick" Crandall** currently serves as a professor of management at the University of North Carolina at Pembroke. He received his PhD in business administration with a focus on organizational behavior and human resource management from the University of Memphis, Tennessee. His primary research interest is in the area of crisis management, helping organizations cope with catastrophic events. He is the author of the book, *Crisis Management in the New Strategy Landscape* (coauthored with John Parnell and John Spillan, also of the University of North Carolina at Pembroke), released by Sage Publications. He is also active in researching issues related to supply chain management. Prior to entering higher education, Dr. Crandall worked in management for ARA Services (now Aramark), a service management firm based in Philadelphia.

**Dr. Charlie C. Chen** was educated at Claremont Graduate University, California, and earned his PhD in management information systems. He is a professor in the Department of Computer Information Systems at Appalachian State University, Boone, North Carolina. His research interests include project management and supply chain management. He is a member of the Association for Information Systems and Decision Sciences Institute and is certified by the Project Management Institute as a project management professional (PMP). Dr. Chen has published in journals such as *Communications of Association for Information Systems*, *Behaviour and Information Technology*, *Journal of Knowledge Management Research & Practice*, and the *Journal of Information Systems Education*. Dr. Chen is a dedicated transnational scholar and a trip leader for study-abroad programs in Asia (Japan and Taiwan).

# Introduction

This is a book about supply chain development and management. Supply chains are made up of a complex variety of businesses and support organizations. In this book, we first describe the component parts and then explain how they are put together to make an integrated entity—the supply chain—that provides products and services the consumer wants, effectively and efficiently.

We have added a new chapter (Chapter 9) to include a more comprehensive coverage of logistics. As supply chains become more complex and dispersed around the world, the movement of goods (logistics) becomes more important, even critical. As we point out, logistics is the glue that holds a supply chain together.

In each chapter, we provide descriptions and examples to adequately explain the topics. In addition, we describe some items in more detail because they are newer or important enough to merit the added coverage. A few of these related topics include

- Outsourcing and reshoring—Chapter 9
- Sustainability—Chapter 10
- Supply chain risk management—Chapter 17
- Knowledge management—Chapter 18

The book is divided into six major sections, as shown in Table I.1. We explain why supply chains are important and how they are presently evolving. We describe the major components of a supply chain and how they need to be closely integrated if the supply chain is to be effective. In addition to the flow of goods and services, supply chains depend on information and funds flow for completeness. Finally, we look at the current issues and trends in supply chains, and consider possible additional developments in the future.

## Chapter Outline

### Section I: Overview of Supply Chain Management

This section describes supply chains and explains why supply chain implementation and management became imperatives for most businesses during the last decade of the twentieth century.

1. Evolution of Supply Chains

This chapter describes the evolution of supply chains from a disconnected series of steps, in moving products from the point of origin to the point of consumption, to a seamlessly integrated flow of goods.

**TABLE I.1**

Principles of Supply Chain Management Overview

| Core Area | Chapter Title |
| --- | --- |
| Section I: Overview of Supply Chain Management<br>Overviews the history of supply chains and the need to use a systems approach. | 1. Evolution of Supply Chains<br>2. Supply Chains as a System |
| Section II: Demand Perspective<br>Examines the supply chain from the customers' point of view. Customers are the demand side of the supply chain. | 3. Determining Customer Needs<br>4. A system to Meet Customer Needs<br>5. Demand Management |
| Section III: Supply Perspective—Distribution, Production, Procurement and Logistics<br>Examines the supply chain from the suppliers' point of view. Suppliers are the means of meeting customer demands. | 6. Distribution and Retailing<br>7. Production and Service Processes<br>8. Procurement/Purchasing<br>9. Logistics—The Glue That Holds the Supply Chain Together<br>10. Reverse Supply Chains |
| Section IV: Need for Integration<br>Describes why and how to integrate the supply chain components. | 11. The Need to Integrate<br>12. Why Integration Is Difficult<br>13. How to Build an Integrated Supply Chain |
| Section V: Financial and Information Technology Perspectives<br>Describes how supply chains require funds and information flows. | 14. Information Flow along the Supply Chain<br>15. Funds Flow along the Supply Chain<br>16. ROI for Supply Chains and Other Issues |
| Section VI: The Future<br>Discusses major business trends and their effect on future supply chain needs. | 17. Trends in Supply Chain Management<br>18. Preparation for the Future |

### 2. Supply Chains as a System

This chapter describes the input–transformation–output (ITO) model and how these elements are linked together to become functioning supply chains for both manufacturing and service businesses. The ITO model is the DNA of the supply chain. The inputs are things and people; the economic inputs, such as employees, equipment, facilities, and systems, are classified as resources employed in the transformation processes.

## Section II: Demand Perspective

This section expands on the introduction in Chapter 2 about customers. Customers are the recipients and evaluators of the goods and services provided by the supply chain. Customers include the ultimate consumer of the final product. Customers also include the internal participants along the supply chain.

### 3. Determining Customer Needs

This chapter describes the major sections of the supply chain, beginning with the customer. It deals with the topic at both the strategic and the operational levels.

### 4. A System to Meet Customer Needs

This chapter takes the information developed in Chapter 3 and transforms it into an operational system to meet the identified needs and wants. The design process will be described for a typical supply chain with some variations noted for different supply chain configurations.

### 5. Demand Management

This chapter considers the demand and supply capabilities required to satisfy volume, timing, and other physical requirements. As in Chapter 4, a typical supply chain is described with some explanation of the potential variations.

## Section III: Supply Perspective—Distribution, Production, Procurement, and Logistics

This series of chapters in this section describes the supply side of the supply chain. It will include the three major functional areas of distribution, production, and procurement. It also includes a description of how to design and manage the service side of the product/service bundle.

### 6. Distribution and Retailing

This chapter describes the functional areas closest to the consumer—the wholesale and retail activities—and their role in getting the products and services to the consumer.

### 7. Production and Service Processes

This chapter describes the manufacturing processes and the added services necessary to provide a complete package for the customer.

### 8. Procurement/Purchasing

This chapter addresses the supply side to production. Purchased materials and services represent over half of the product costs in most industries. As more companies move to a *core competency* strategy, they will outsource more of the component manufacturing to suppliers and become primarily an assembly operation.

### 9. Logistics: The Glue That Holds the Supply Chain Together

This chapter describes how the logistics function, using various modes of transportation, moves products and services along the supply chain and helps connect the multiple participants together to assure effective and efficient flows of goods. We also provide a comprehensive look at the outsourcing movement and its implications for supply chain management.

### 10. Reverse Supply Chains

The earlier chapters described the activities involved in the supply chain that moves goods and services toward the ultimate consumer. This chapter will describe the activities that facilitate the flow of goods, and the accompanying services, in the reverse direction, commonly termed reverse logistics.

## Section IV: Need for Integration

In earlier chapters, we described individual components of a supply chain. In this section, we describe the need to integrate those components into an integrated supply chain. It is first necessary to integrate the functions within one company and then to integrate that company with other members of the supply chain.

11. The Need to Integrate

This chapter stresses the need to move to an integrated supply chain. It is a major decision for a company and requires top management support and participation because a major commitment of resources will be needed. Very few companies have moved completely to integrated supply chains.

12. Why Integration Is Difficult

This chapter explores why it is difficult to implement an integrated supply chain. Aside from the complexity of the task, it requires more than making positive moves with technology; it also requires overcoming natural resistance within the organization.

13. How to Build an Integrated Supply Chain

This chapter outlines an approach to building an integrated supply chain. It includes case studies to illustrate the steps in the process.

## Section V: Financial and Information Technology Perspectives

The first 13 chapters described the physical aspects of the supply chain and the flow of goods and services. In this section, we will describe the flow of information and funds among supply chain participants.

14. Information Flow along the Supply Chain

This chapter describes the information flow along the supply chain. While some aspects of information flow have been covered in earlier chapters, this chapter describes in more depth the applications and technologies used in facilitating effective information flow.

15. Funds Flow along the Supply Chain

This chapter describes funds flow along the supply chain. It describes some of the technology used and the entities involved in electronically moving funds among supply chain partners.

16. ROI for Supply Chains and Other Issues

From a financial perspective, supply chain participants support each other. Hence, there is a need to find a way to equitably distribute the benefits and costs incurred in the supply chain among these participants.

## Section VI: The Future

At the present time, few companies have implemented completely effective supply chains. While the topic is of current interest, will there be a time when companies lose interest in supply chain management and move on to some new program? If so, what will it be?

17. Trends in Supply Chain Management

Changes in the business environment have triggered the need for integrated supply chains. Will changes in the future reinforce the need for supply chains or make them less beneficial?

18. Preparation for the Future

While the future will not be like the past, it is difficult, if not impossible, to accurately forecast all of the possibilities that could arise in the future. Thus, a company has to be prepared to provide some structure in the supply chain and to be able to adapt to changing conditions.

# Section I

# Overview of Supply Chain Management

# 1

## Evolution of Supply Chains

**Learning Outcomes**

After reading this chapter, you should be able to

- Define what a supply chain is
- Discuss the importance of supply chain management (SCM)
- Describe the evolution of supply chains
- Discuss how different government policies affect the supply chain
- Identify the current trends that link supply chain participants
- Identify the current developments in SCM
- Discuss the obstacles to supply chain integration
- Identify companies with successful supply chains

Chapter 1 is an introduction to supply chains and why they are important to today's businesses. Products have to be moved from the point of origin—the farm or the mine—to the point of use—another organization or the ultimate individual consumer. As businesses expand into global markets, they have also expanded their sources of supply throughout the world. Supply chains have become longer and more complex. Most companies have more than one product line. This means they probably have multiple supply chains, serving different customers and requiring different suppliers. Managing a supply chain is more difficult than managing the same operations internally because of differences in policies, procedures, and, most of all, cultures.

As you read this company profile and the chapter in the book, think about some of the decisions a manager would have to make in organizing and managing the supply chains necessary to assure good customer service and profitable operations. These decisions could require answers to some of the following questions:

- What is important to our customers? Is it price, quality, fast deliveries, or all of these?
- Which operations should we do ourselves? Which should we outsource? Why?
- Where should our suppliers be located? Why?
- How do we evaluate the effectiveness of our supply chain(s)?
- What additional questions do you think are relevant to SCM?

## Company Profile: Procter & Gamble (P&G)

In Table 1.1, you will see a list of the top 25 supply chains for U.S. companies, as compiled by Gartner Research. The table shows the top 25 for 2008 and 2013. During that 5-year period, 15 of the 25 companies in 2008 were not in the top 25 for 2013, indicating the need to be consistently good to remain at the top. Apple was ranked no. 1 for all of these years and much has been written about the valuable contribution of their supply chains during that period, especially in meeting the tremendous demand for their iPod, iPhone, and iPad. Tim Cook, the present CEO, is given a lot of the credit for developing Apple's supply chains during his tenure as COO.

Procter & Gamble (P&G) is another company that was ranked among the top companies every year. P&G's supply chain story may be as impressive as Apple's when you consider the number of different supply chains the company needs to support their array of products. In 2012, the company had revenues of nearly $84 billion, of which over 90% came from their "Billion-Dollar Brands," those 25 brands that had revenues of over $1 billion each. P&G has five reportable product segments: beauty (24% of net sales with $5 billion brands), grooming (10% of net sales with $4 billion brands), health care (15% of net sales with $4 billion brands), fabric care and home care (32% of net sales with $9 billion brands), and baby care and family

**TABLE 1.1**

A Comparison of Top 25 Supply Chain Companies: 2008 and 2013

| 2008 | 2013 |
|------|------|
| 1. Apple | 1. Apple |
| 2. Nokia | 2. McDonald's |
| 3. Dell | 3. Amazon.com |
| 4. Procter & Gamble | 4. Unilever |
| 5. IBM | 5. Intel |
| 6. Wal-Mart Stores | 6. Procter & Gamble |
| 7. Toyota Motor | 7. Cisco Systems |
| 8. Cisco Systems | 8. Samsung |
| 9. Samsung Electronics | 9. Coca-Cola Company |
| 10. Anheuser-Busch | 10. Colgate-Palmolive |
| 11. PepsiCo | 11. Dell |
| 12. Tesco | 12. Inditex |
| 13. Coca-Cola Company | 13. Wal-Mart Stores |
| 14. Best Buy | 14. Nike |
| 15. Nike | 15. Starbucks |
| 16. SonyEricsson | 16. PapsiCo |
| 17. Walt Disney | 17. H&M |
| 18. Hewlett-Packard | 18. Caterpillar |
| 19. Johnson & Johnson | 19. 3M |
| 20. Schlumberger | 20. Lenovo Group |
| 21. Texas Instruments | 21. Nestlé |
| 22. Lockheed Martin | 22. Ford Motor |
| 23. Johnson Controls | 23. Cummins |
| 24. Royal Ahold | 24. Qualcomm |
| 25. Publix Super Markets | 25. Johnson & Johnson |

*Sources:* AMR Research, 2008; Gartner Press Release, May 23, 2013.

care (19% of net sales with $3 billion brands). In 2012, Pampers became their first $10 billion brand. All told, they have over 80 different brands and, within each brand, there are multiple products. It becomes apparent they need a variety of different supply chains to obtain the materials and products to manufacture and distribute their many products.

P&G is a global company. They sell products in more than 180 countries, with Pampers, the leading disposable diaper product, serving 25 million babies in over 100 countries. Their 2012 sales are distributed among the following geographic areas: North America (the United States and Canada) 39%, Western Europe 19%, Asia 18%, Latin America 10%, and Central and Eastern Europe, Middle East and Africa (CEEMEA) 14%. They have on-the-ground operations in approximately 75 countries and manufacturing operations in 40 countries throughout the world. Their supply chains must deal with diversity and complexity.

P&G sells through mass merchandisers, grocery stores, membership club stores, drug stores, department stores, salons, and in high-frequency stores. Their sales to Wal-Mart Stores and its affiliates represented approximately 14% of 2012 sales. No other customer represented more than 10% of net sales and the top 10 customers accounted for 31% of total unit volume. They sell in a highly competitive environment against other branded or private label products, in which price is often the differentiating factor. Consequently, their supply chains have to produce low-cost products.

Product quality is also a prime consideration because so many of the products are used for personal care or for home care; consequently, they have a high profile in the consumer market. This means P&G must provide products that meet a high standard, both in content and packaging. The product must look good on the retailer's shelf as well as perform well in its use.

P&G's supply chains must be responsive. Retailers must have high inventory turns to achieve their targeted return on investment. As a supplier, P&G must help their customers have enough products to avoid stockouts, but not too much, to avoid excess inventory. The retailer groups that P&G supply offer numerous product promotions and sales; therefore, P&G must be able to meet the peaks and valleys of demand from their customers. Their supply chains must be responsive to both time and volume fluctuations.

P&G is a sustainable company. As part of their vision for the future, they have some ambitious goals that are as follows: use 100% renewable or recycled materials for all products and packaging; design products that delight consumers while maximizing conservation of resources; have zero consumer or manufacturing waste going to landfills, and power all of plants with 100% renewable energy. One of their major contributions was to develop Cold Water Tide to reduce the need to heat water for washing clothes.

P&G is a socially responsible company. It was founded in 1837. They have been successful over the intervening years by making and selling branded products that offer value at competitive prices. As part of their Purpose statement, they say: "At P&G, we touch lives in small but meaningful ways. Billions of them. Every day. The simple, inspiring way to think about this is that P&G brands serve about 4.6 billion of the nearly seven billion people on the planet today. Before P&G can serve the world's remaining consumers profitably, we can reach them altruistically. We can improve their lives in ways that enable them to thrive, to increase their quality of living and, over time, to join the population of consumers we serve with P&G brands."

P&G's supply chains must deal with risks that can arise throughout the world. The sheer magnitude and complexity of their supply chains mean they face daily disruptions at some time and place in their supply chains. The fact that customers rarely see the effects of these disruptions is a credit to the reliability and adaptability of P&G's supply chains.

This profile of P&G should give you a good idea of what a supply chain means to a company. Supply chains must respond to the demands in the marketplace for lower costs, higher quality, responsiveness, and flexibility. At the same time, they must be financially strong, a good corporate citizen, and an active leader in achieving improvements in sustainability. We will spend the rest of this book explaining how supply chains can be designed and managed to achieve these challenging objectives.

*Source:* Adapted from the P&G website, http://www.pg.com/en_US/index.shtml.

## What Is a Supply Chain?

A supply chain involves various participants who perform a sequence of activities in moving physical goods or services from a point of origin to a point of consumption. To avoid redundancy, we will use *products* to include both products and services throughout the book. A simple supply chain involves participants in the following order from upstream (toward the origin) to downstream (toward the ultimate consumer): source supplier, manufacturer, distributor, retailer, and customer. Information flows along the supply chain toward the customer to aid in tracking the flow of products and flows upstream from the customer to suppliers to aid in identifying the characteristics of the demand for the product. At each stage of the supply chain, the customers pay their suppliers for the goods and services received; consequently, funds flow upstream from the ultimate consumer to the original supplier. The *APICS Dictionary* defines a supply chain as "the global network used to deliver products and services from raw materials to end customers through an engineered flow of information, physical distribution, and cash" (Blackstone 2013).

At first, supply chains were closely associated with the supply side, or the upstream part of the process. As supply chain importance increased, the customer or the demand side of the process received more attention. To recognize this expansion of scope, some used the term value chain. However, the consensus at present is that supply chain is the accepted term to cover both the demand and the supply sides of the entire process. We will use supply chain throughout this book.

SCM describes the functions used to manage the activities of delivering products, producing information, and generating increased revenue for stakeholders involved at different stages of a supply chain. The Council of Supply Chain Management Professionals defines SCM as follows:

> Supply chain management encompasses the planning and management of all activities involved in sourcing and procurement, conversion, and all logistics management activities. Importantly, it also includes coordination and collaboration with channel partners, which can be suppliers, intermediaries, third party service providers, and customers. In essence, supply chain management integrates supply and demand management within and across companies. (CSCMP 2008)

The supply chain flow begins when suppliers, such as farms or mines, deliver raw materials to manufacturers for producing goods based on purchase orders (POs). A network of suppliers in multiple tiers is usually involved, depending on the complexity of the final product. For instance, a car is a sophisticated product made of thousands of components ranging from the engine, brake system, wheels, transmission, suspension, lamps, and electronics. A competitive manufacturer in the automobile industry needs to seamlessly manage a network of suppliers to ensure car parts ordered get delivered on time. After receiving raw materials, a manufacturer begins the production process according to a production plan that is based on demand forecasts or customer orders. Manufacturers can also be arranged in tiers if the final finished products need to proceed through multiple assembly processes. Manufacturers that produce intermediate products are called component manufacturers, or fabricators. All intermediate products flow to the final product manufacturer (e.g., Apple, Nike, Samsung, Ford Motor, Toyota, and Boeing) for the final assembly. The finished goods continue the journey to the wholesalers or distributors, then to retailers, and finally into the hands of customers.

If customers are not satisfied with the products or services, they may return the purchased goods to retailers as part of the reverse logistics process. The returned goods then journey back to the distributors, wholesalers, manufacturers, and perhaps even to the raw material suppliers. Since supply chain activities take place in sequence, problems with any one of these activities can slow or stop the flows of goods, information, and funds. Supply chain problems (e.g., back orders, excessive inventory, insufficient funds, customer complaints, and product defects) can result in an inefficient supply chain that will be plagued with disruptions. On the other hand, an orchestrated collaboration among suppliers, manufacturers, distributors, wholesalers, and retailers can result in an efficient supply chain. All these supply chain partners can profit from the efficiency of their supply chain because the smooth flows of goods, information, and funds can enable all participants to receive goods, correct information, and revenues on a timely basis. As a result, it is critical to understand how to manage supply chains in different formats and improve their efficiency to benefit all supply chain participants. Figure 1.1 shows the supply chain participants in a variety of industries and how products flow from upstream suppliers to downstream customers. The blocks at the bottom of each activity represent the resources required to enable the transformation process.

## Importance of SCM

As global competition increases and businesses extend their reach into foreign markets, they become more dependent on their supply chains. A supply chain depends on the continued profitability of each of these participants. Therefore, it is desirable to optimize the total profit—the difference between revenue and costs—of a supply chain and fairly distribute the proceeds to all participants.

Sustainability of any business depends on its ability to continuously improve its profits, that is, the financial bottom line. This universal rule is applicable to old and new businesses as well as to small and large enterprises. The means to achieve this goal varies according to the strategies of the business. Some companies pursue product innovations and improve customer services, whereas others reduce operating costs and streamline delivery processes. Regardless of the means taken to achieve business goals, all firms need to synchronize the flows of products, information, and funds. The goal of constructing an efficient supply chain is to help generate revenues throughout the value chain (Porter 1996). Thus, the revenues can flow back to business partners to reward their share of contributions.

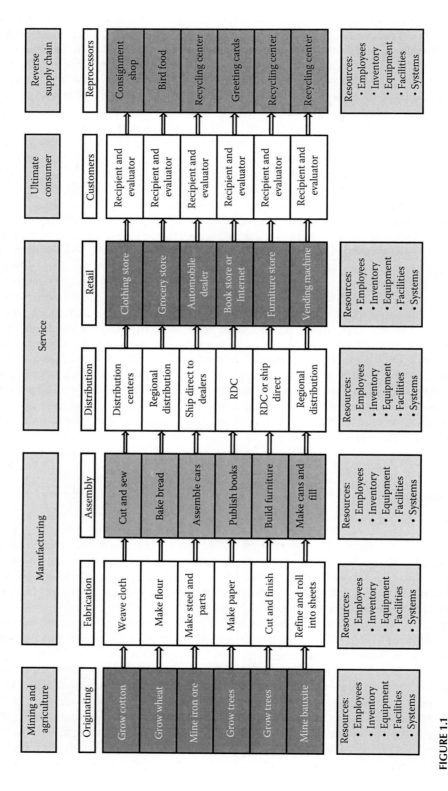

**FIGURE 1.1**
Examples of supply chains for various industries.

In addition to economic, or financial, success, companies are recognizing the need to meet their social responsibilities and their environmental responsibilities. Coupled with the need for financial success, these three areas constitute what some are calling the *triple bottom line* (Elkington 1999). We will describe this emerging concept more fully in Chapter 10.

## Evolution of Supply Chains

Supply chains are not new; they evolved over time. Supply chains were formed differently to cope with the assorted challenges that existed at different times in history. A historical tour of supply chains can help to understand those earlier challenges and show new ways to resolve contemporary challenges. The early supply chains included a discrete series of steps that moved products from the point of origin to the point of consumption. Those early forms are being transformed by progressive companies into seamlessly integrated supply chains that move goods, information, and funds across different companies and countries.

### Early Supply Chains

Supply chains were formed when humans first began trading with each other. Traders obtained goods from merchants and delivered the requested goods to buyers. The process of locating the right suppliers to provide goods for customers was a very basic form of a supply chain. Since most trading was limited to local buyers and sellers, it posed fewer challenges with regard to communication among trading parties, delivery of goods to buyers, and collection of payments from buyers.

An Arabian tribe, the Nabataeans, monopolized trade in the Mediterranean continent over 2000 years ago. The tribe's successful dominance of trade depended on the construction of a hub-and-spoke supply chain network (Hull 2008). Many value-added services were available to sustain this network, including partnerships, secured transportation of goods, and innovative packaging.

The early form of supply chains continued to evolve when the Venetian merchant Marco Polo traveled through the Black Sea into China via the Silk Road from AD 1275 to 1295. This journey took 20 years to accomplish and was probably the first time an expedition crossed the European and Asian continents (Silkroad Foundation 2008). As such, Europeans were able to obtain scarce goods, such as gunpowder, compasses, printing supplies, and noodles from China.

The global supply chain to link the world took another step forward when Christopher Columbus discovered the Americas in 1492. Later, the invention of the steam engine and other technologies lowered transportation costs and increased the distance and speed of moving goods across the world. Today, while many supply chains remain local in scope, many others represent a global perspective.

### Local Supply Chains

In the first half of the twentieth century, vertical integration was a popular form of business structure. This strategy meant that a business attempted to be largely self-sufficient. If its primary business was the assembly of finished products, they also housed their own fabrication departments to make component parts. On the downstream side, they may have had their own distribution centers and, in some cases, their own retail stores. One of the classic examples of this arrangement was the River Rouge plant of Ford Motors. They brought in iron ore from company-owned mines by way of Lake Michigan, processed it through the foundries and metal working shops and on to the final moving assembly line, all within the span of 2–3 days (Hounshell 1985). The advantages of vertical integration were direct control of all operations, reduced work-in-process inventories, and reduced cycle time from the raw material stage to the finished goods stage.

Companies liked to have their suppliers nearby to reduce the lead times required to obtain goods and to have them readily accessible should problems arise. In some industries, suppliers tended to cluster around their main customers. In automobiles, Detroit became the center of such a cluster. Textile and furniture industries also attracted such arrangements in the Carolinas, and farm equipment manufacturers tended to operate in the Midwest. These local clusters matured into tightly connected supply chains.

Initially, companies confined their business to local markets. As markets expanded geographically and customers became more diverse, products became more complex to satisfy the greater diversity in the marketplace. It was increasingly difficult for one business to do everything for itself, at least, as well as outside suppliers could do it for them. Consequently, many businesses began to shift from a *make* to a *buy* arrangement, which led to what is now known as outsourcing. The advantages were a greater variety of products at lower costs. The disadvantages were loss of direct control and longer lead times to obtain products.

Over the last half of the twentieth century, many U.S. companies expanded their markets into other countries. They also found that competition increased as foreign companies moved into the United States. In order to effectively compete, many businesses turned to global supply chains.

## Global Supply Chains

Expansion of local supply chains into global supply chains opened access to new markets and new sources of supply. Offshore outsourcing—buying from foreign suppliers—was initially attractive because of lower purchase costs. However, it also extended the supply chain, reduced the level of control over suppliers, and increased lead times from days or weeks to weeks or months. In general, global supply chains are now the approach of choice for most companies.

Maintaining a constant flow of products, information, and funds remains the primary objective for businesses today. This requires a number of facilitating activities to move the goods, information, and funds through the supply chain. These facilitating activities depend upon a firm's infrastructure, human resource management, financial management, accounting management, technology development, and procurement management. These activities play important supportive roles for the normal operations of core business activities.

As supply chains grew in size, they involved more business partners. Wal-Mart, a mega retailer, currently has hundreds of suppliers in its first tier as well as many more suppliers in other tiers working for them. Given these circumstances, communication problems alone can be overwhelming. Therefore, reducing the number of suppliers is a natural pursuit of manufacturers to optimize the buyer–supplier relationship. The success of many of the best SCM practices, including lean production and quick response systems, depends on establishing a close relationship among supply chain participants.

As the trade volume among a limited number of business partners increases, the need emerges for an information system to automate the transactional processes and to minimize errors. Wal-Mart and Cisco Systems are two examples of companies that share real-time inventory and shipping information with their top suppliers via an enterprise information system. As such, these two retailers and their suppliers can jointly make decisions based upon real-time, accurate information. Traditional electronic data interchange (EDI) is an information system that emerged in the 1990s to meet the growing needs of large enterprises to process voluminous data. EDI is a proprietary value-added network (VAN) designed to meet the specialized needs of large enterprises through facilitating more efficient transactions with major business partners. We will describe more about supply chain information systems in Chapter 14.

With advances in transportation and communication technology, goods and services can now be potentially sourced from anywhere around the world. Global supply chains are formed when supply chain participants are spread across country boundaries. Although many supply chain opportunities are readily available, many more challenges lie ahead for multinational firms. The challenges of managing supply chains on a worldwide basis fall into three categories: governmental, geoeconomic, and cultural (Jessup and Valacich 2008). Governmental challenges include working within different political systems, meeting regulatory requirements, achieving data-sharing directives, and working with Internet capabilities. Geoeconomic challenges include working under time zone differences, managing infrastructure reliability, interacting with participants of various demographic profiles, and working with a limited availability of expertise. Cultural challenges consist of working with participants who have different ethics and attitudes about ways of doing business. All these challenges can substantially increase the uncertainty of sourcing goods on a global basis.

The proliferation of Internet usage enables supply chain participants to engage in transactions without knowing each other on a personal basis. However, geographical distance, language, time zones, cultural differences, irreconcilable customs, and incompatible information systems all become barriers to the construction of a smooth supply chain. These barriers increase the difficulty of managing a supply chain and affect the planning, sourcing, making, delivering, and returning processes. The lack of an orchestrated effort to coordinate these processes can lead to a disordered, inefficient, and ineffective supply chain. The rapidly falling costs of information and telecommunication technologies and their ubiquity enable the creation of digital supply chain networks. This improvement further accelerates the pace of globalization and the creation of today's global supply networks in digital formats.

A global supply chain crosses country boundaries and needs to address the local complexity caused by governmental, geoeconomic, and cultural differences. Take the purchasing function, for instance. Many country-specific factors can influence transactional costs in the purchasing function; some of these factors are internal and others are external (den Butter and Linse 2008). Internal factors, such as intellectual property costs, transportation costs, and quality control practices, can increase purchasing costs. External factors that can also inflate purchasing costs include currency exchange rates, local labor costs and safety standards, labor availability, and cultural differences. Many of these local factors can lower the efficiency of a global supply chain. In practice, the global operation of many supply chains remains disintegrated, disconnected, and not spontaneous (Singh 2005). It is important to understand the relevance of these factors and to assess their impact on the operations of a global supply chain. Among cultural factors, the form of government affects the common way of conducting business within a country and is one major obstacle to the improvement of the supply chain's operational performance (Yaibuathet et al. 2008).

Expansion to global supply chains has necessitated dramatic improvements in electronic communications. Information visibility is fundamental to the success of today's supply chain networks. Information flow drives the movement of products and funds. The advances in information systems technology increase information transparency and make it easier to manage supply chain participants.

The importance of information systems support can be seen from the communication perspective. Think about the number of channels that need to be in place to facilitate the communication of business partners. It takes at least $N \times (N - 1)/2$ channels for $N$ number of business participants to communicate with each other. For three business partners to communicate with each other, it takes at least three channels. Imagine having 1000 suppliers for Wal-Mart trying to communicate. The number of channels that are needed for all 1000 suppliers to communicate with each other would be 499,500! If it takes about 1 min to finish each conversation, it would take 8325 h before all supply chain participants could communicate with each other one time. This communication inefficiency calls for an improved process.

Information systems combine hardware, software, and telecommunication networks to automate the data collection, creation, and distribution process. The advances in information systems technology solve the aforementioned communication issue. According to Moore's law, computer processing performance doubles every 18 months (Mollick 2006). This proposition has been generally accurate in predicting the velocity of information systems' progress since the 1970s. As a result, increased information processing ability has gradually overcome the information overload problem.

Supply chain participants realize the potential benefits of using information systems to streamline their supply chain networks. Of the types of information systems, Internet EDI, enterprise systems, and radio-frequency identification (RFID) have played the most important roles in managing global supply chains. More recently, companies are beginning to use social media sites and other unstructured data to capture data. Each of these technologies will be discussed later in this book.

A recent survey focused on the progress being made across three capabilities that are essential to supply chain success today—visibility, analytics, and flexibility. As in past surveys, clear levels of competency emerged among the respondents. The key findings included the following:

- Leaders tend to have broader and more global views of SCM, and they are typically early adopters of new technologies and planning systems.
- Supply chain leaders influence revenue gain at double the rate of laggards.
- Companies with wider scopes of visibility and higher-quality data are outperforming their industry rivals.
- A firm with combined visibility and analytics capabilities is almost twice as likely to be a fast grower (top 20% of growth) than a firm that is below average in both skill areas.
- The combination of visibility and flexibility enables the greatest business growth and profitability (CSC 2013).

The Silk Road was one of the early supply chains. In a comprehensive review of today's supply chains, the *new* Silk Road is described as reemerging as a primary conduit of goods and services, especially from the Far East to Europe. Goods are moved with multimodal means, from camels to high-speed rail. Communication and legal barriers are overcome through ingenuity and determination. Modern concepts such as cross docking facilitate the movement of goods at natural transfer points. As in earlier centuries, the Silk Road is still alive and well (Haksöz et al. 2012).

Variations in supply chain performance among Europe, North America, and the Asia Pacific region could be due to the differences in government forms. We will discuss how two major forms of government—state-controlled government and market-driven government—can affect the operations of global supply chains differently.

## Changing Government Orientation

Although there are numerous variations on the basic form of government, we will consider only the two predominant forms: state controlled, such as communism, and market driven, such as democracies.

### State-Controlled Governments

A state-controlled government can be viewed as the opposite form of a market-driven government. Corporations in the state-controlled government are often owned by the local governments and need to undertake business activities on behalf of its owner government. Although many state-owned enterprises are not domestically or internationally competitive, they are allowed to exist as long as they follow the directed strategies from the owner government. Too much control from government can create many obstacles to the optimization of SCM. One scenario is that a state-controlled government may mandate that all state-owned enterprises stop producing orders for their international customers because of a national crisis, such as a flood or an earthquake. Such a mandate would direct all production efforts be allocated to disaster relief efforts. In this type of a situation, all transportation vehicles would become the custody of the government in order to support the movement of relief supplies to the affected areas of the country.

When the government intervenes on business matters, the market is institutionalized and could potentially disrupt the supply of goods scheduled to be delivered to international customers on time. Supply chains are difficult to manage under state-controlled governments because national policies can supersede private interests. International firms interested in conducting business in state-owned government need to be aware of the constraints of operating supply chains. Examples of state-owned governments monopolizing different industry sectors can potentially disrupt the smooth flow of supply chains.

### Market-Driven Governments

A market-driven government is a laissez-faire system in which customers dictate business direction. In order to maximize their profits, supply chain partners need to collaborate to meet the needs of their customers. For instance, in order to attract customers via personalized service, Dell created the direct-sales business model. Customers can order computers directly from Dell and have their product delivered to them from one of its manufacturers. Dell created a direct link with its manufacturers in order to avoid extra tiers in the supply chain. This strategy resulted in increased customer satisfaction and cost savings.

Since there are fewer supply chain participants, more profits are available for Dell and its manufacturers. Amazon.com is also demonstrating a new business model that is changing the face of retailing, as we will discuss more in Chapter 6.

## Current Trends That Link Supply Chain Participants More Closely

Several current trends are contributing to the creation of global supply chains. These trends are (1) relationship building in supply chains, (2) proliferation of electronic business, (3) rise of developing economies, (4) offshore outsourcing, (5) need for quality improvement, and (6) changing customer demands. These trends are forcing supply chain participants around the world to work more closely together than ever before.

### Relationship Building

Supply chains require that companies build relationships with one another. While there are numerous programs aimed at doing this, we will discuss three of the currently popular programs: customer relationship management (CRM), supplier relationship management (SRM), and product life cycle management (PLM).

#### *Customer Relationship Management*

The ultimate objective of CRM is to create a long-lasting relationship with customers. Good relationships with profitable customers pay off in the long run because the cost of retaining customers is less expensive than recruiting new customers. A company needs to perform a number of customer-related functions in order to satisfy their customers and earn their trust. CRM is comprised of operational, analytical, and collaborative activities. Operational CRM activities include order taking, invoice/billing, and offering help desks, call centers, field services, recall, and customer services. Analytical CRM consists of capturing, storing, filtering, cleansing, and analyzing customer data. These activities provide intelligence for the company in order to run promotional events and commercials and to provide personalized services to customers. For instance, a customer can receive a personalized greeting card from a car dealership on his/her birthday. Collaborative CRM addresses communication, collaboration, and coordination issues resulting from the transactional process between vendors and customers. A number of technologies are available to support operational, analytical, and collaborative CRM. We will describe them more fully in Chapter 4.

Although CRM technologies are useful, they do not guarantee customer satisfaction and long-lasting customer relationships. Customer building requires an ongoing personal touch and a subsequent business process transformation. For instance, workflow management and call centers may be able to provide information about customers. However, if functional departments are not willing to share information about customers with each other, then these technologies will not be able to achieve optimal performance. Customers may still not be able to enjoy a one-stop solution when the finance department is not willing to share billing information with the technical support department. As a result, it is critical that a company change its way of conducting business to a customer-centric approach to capitalize on the technological capabilities of CRM solutions.

### Supplier Relationship Management

SRM addresses business issues associated with the procurement or sourcing process. Noteworthy SRM issues include spending analysis, catalog management, sourcing implementation, sourcing execution, payment, settlement, contracting, and monitoring of supplier performance. The goals in resolving these issues are to build close relationships with key suppliers and to flexibly manage the network of suppliers. A close supplier relationship can create benefits, such as cost reductions on goods and services, streamlined procurement processes, and regulatory compliance.

SRM activities are not directly related to a transaction process, but indirectly support the transactional process. Various technologies are available to augment the success of SRM activities. One major software provider addresses several major areas: procure-to-pay optimization, catalog management, self-service procurement, operational sourcing, centralized contract management, and operational reporting (SAP 2013). We will describe SRM more fully in Chapter 6.

### Product Life Cycle Management

A company is competitive if the lead time for introducing products and services to the marketplace is shorter than that of its competitors. A new product poses threats to the established marketplace and invites intense competition. Many new products may not even gain acceptance by customers or succeed in penetrating the existing marketplace. The challenge is to ensure that new products and services are successfully introduced and accepted in the marketplace.

The product life cycle (PLC) concept states that all products proceed through five stages: development, introduction, growth, maturity, and decline. The revenues of a company increase over time as new products and services successfully enter the market and increase in sales volume. In order to maintain the momentum of growing a new product or service, a company needs to integrate all of the supply chain functions, from designing, supplying, manufacturing, marketing, to aftersales service. Successfully managing the PLC offers several business benefits including the following:

- Shortens the time to market
- Reduces research and development (R&D) costs
- Optimizes product designs
- Improves product quality
- Reduces waste
- Increases the success rate of new products

Several technologies are emerging to help companies deliver these PLC benefits. We will describe PLM systems in more depth in Chapter 4.

### Electronic Business

Commercialization of the Internet led to the proliferation of electronic business. Electronic business, or e-business, is the use of electronic means, primarily the Internet, to conduct business. Another term that is also similar to e-business is e-commerce. "Electronic commerce (EC or e-commerce) describes the process of buying, selling, transferring, or exchanging products, services, or information via computer networks, including the Internet. E-business is a somewhat broader concept. In addition to the buying and selling of goods and services, e-business also refers to servicing customers, collaborating with business partners, and performing electronic transactions within an organization" (Rainer and Cegielski 2011, p. 201). Because of the similarity of terms, we will use the broader term of e-business in this book.

E-business helps remove the location and time barriers in a business transaction. It also empowers individual and business customers, manufacturers, logistics providers, and suppliers to collaborate with each other electronically. E-business fueled a revolution in the manner in which supply chains operate; it allows supply chain partners to fulfill unique needs at different stages of the supply chain.

For example, many companies experience problems with sourcing and procurement. These problems include manual and inefficient purchase processes, ineffectiveness in managing the purchasing budget, low levels of compliance with internal and external regulations, and poor purchase decision-making procedures. Many e-procurement solutions, such as procurement portals, trading exchanges, and e-procurement software, have risen to resolve some of these problems.

E-business can aid in managing supply chain procurement problems through the use of e-distributors, e-procurements, trade exchanges, and industry consortia.

A tightly integrated and relatively small number of trading partners is a prerequisite for the creation of proprietary supply chain networks. The openness of e-business can largely extend benefits to supply chain partners not having a proprietary relationship with each other (Turban et al. 2007). We will describe e-business in more detail in Chapter 6 as it relates to the retailing function.

## Developing Economies

The economic growth of emerging markets in Eastern Europe and Asia has been one of the major forces flattening the world (Friedman 2006). In the face of stiff competition and maturing domestic markets, many firms in developed countries have exploited business opportunities in these emerging markets. At the same time, businesses in developed countries found they could lower their operating costs by using suppliers in the emerging countries. This strategy had the added benefit of providing economic growth to the emerging countries. After over two decades of development in the outsourcing areas, many suppliers in those emerging markets acquired skills and knowledge in delivering products to meet and, sometimes, to exceed the requirements of their customers in the developed countries. Suppliers in emerging markets mushroomed as they advanced along the learning curve. Many multinational firms are now enjoying the benefits brought on by the increased availability of suppliers in this growing market. Yet, new managerial challenges are arising. Several of the most salient challenges include maintaining quality control, fostering intercultural communications, and meeting all relevant regulatory requirements. The movement to use foreign suppliers is an extension of the make-or-buy decision to today's offshore outsourcing.

## Outsourcing

Offshore outsourcing is a natural response to the increasing worldwide competition for lowering costs within global supply chains. Like any strategic decision, offshore outsourcing is a vehicle for an organization to stay profitable and remain competitive both domestically and internationally. Service companies like IBM outsources to transform its own business from materials processing to information processing (Dalesio 2004). Dell outsources to multiple tiers of overseas suppliers to support its mass customization business model. Banks establish call centers in India to provide cost-effective CRM solutions to local customers in the United States. Countries with developing economies are popular locations for outsourcing venues. These include major economies such as China, India, Russia, and Brazil and a host of emerging countries.

The stress of maintaining a smooth supply chain intensifies when the geographical distance between outsourcing clients and providers increases. As offshore outsourcing continues to evolve, supply chain participants need to continuously improve in at least five areas: (1) supplier management, (2) overall supplier performance, (3) quality of work, (4) comparative efficiency, and (5) comparative costs (Rottman and Lacity 2004). All areas of offshore outsourcing improvements should align with the vision and strategy of a business in order to derive long-term, strategic benefits.

Offshore outsourcing may present a public relations problem for the company that chooses to outsource. When domestic jobs are *lost* to a firm overseas, the citizens in the home country often resent the decision to outsource. This remains a difficult situation however, since the same citizens typically desire lower prices at the local retailer. We will discuss the outsourcing movement more fully in Chapter 9.

## Need for Quality Improvement

While low cost may be the most desired objective for products, quality is not far behind. Costs can be determined with specific measures; quality has a variety of measures, as seen in the following definition from the International Organization for Standardization (ISO 2013).

> The *quality* of something can be determined by comparing a set of inherent characteristics with a set of requirements. If those inherent characteristics meet all requirements, high or excellent quality is achieved. If those characteristics do not meet all requirements, a low or poor level of quality is achieved.

*Quality* is, therefore, a question of degree. As a result, the central quality question is: How well does this set of inherent characteristics comply with this set of requirements? In short, the *quality* of something depends on a set of inherent characteristics and a set of requirements and how well the former complies with the latter.

According to this definition, *quality* is a relative concept. By linking quality to requirements, ISO 9000 argues that the *quality* of something cannot be established in a vacuum. *Quality* is always *relative to* a set of requirements. Product quality can be measured by the operating characteristics, performance, and reliability of the product. Service quality is often the result of individual preferences and is more difficult to quantify. In general, quality is said to be achieved if it meets or exceeds customer expectations or requirements.

One problem encountered in offshore outsourcing is that the products manufactured may be of unacceptable quality. Recently, several foreign-made exports such as toys, milk, tires, and pet food have been cited with poor quality problems. Several of these issues will be discussed in more depth in *Hot Topics* that are included at the end of each chapter. Various factors can contribute to poor quality across the supply chain, including the following:

- Pressure on suppliers to reduce costs can lead to lower quality.
- Business practices in some countries do not place a high value on quality.

- Outsourcing company may fail to communicate quality requirements to suppliers.
- Suppliers may not have the capability in their processes to maintain a desired level of quality.

These factors indicate the goal of having an efficient global supply chain with suppliers, especially those in emerging countries, is far from being achieved. It is difficult to achieve close quality control when reducing operation costs is the central concern of SCM. All supply chain participants need to engage in a concerted effort toward continuous quality improvement as part of the business mission.

## Changing Customer Demands

Customers define the needs in today's affluent society. In order to compete for a customer's attention, companies focus on improving their forecasting accuracy for customer demand. The ability to improve forecasting can decrease response time to the needs of customers. By facilitating collaboration between marketing and operations, the sales and operations planning (S&OP) process seeks to balance supply and demand and is one of the effective SCM solutions.

### *Decreasing Response Time*

Decreasing response times is one of the many strategic benefits of a well-managed global supply chain. With response time decreased, a global supply chain improves its organizational agility—the ability to be flexible and to effectively compete in the global marketplace. An agile participant in the global supply chain can make effective use of flexible information systems, have high sensitivity to customer needs, integrate business processes, and collaborate better with each other. Dell's make-to-order business models and Procter & Gamble's fast response to the needs of customers are examples of how decreasing response time helps gain a competitive advantage. Some firms adopt the practice of cross docking by reorganizing the distribution center layout so that products can arrive, be regrouped, and depart for destinations on the same day. This cross-docking solution can result in the reduction of lead times, inventories, and response time.

We will describe the growing importance of logistics in Chapter 9. This is the glue that holds a supply chain together and helps to manage the response time, especially in the face of increasing offshore outsourcing.

### *Lean Supply Chains to Reduce Waste*

The *bullwhip effect* has been widely cited to describe how a somewhat stable demand at the consumer level becomes an increasingly variable demand as the orders move upstream along the supply chain. Four major contributors to the bullwhip effect are demand signal processing, a loosely controlled return policy, order batching, and price variations (Lee et al. 2004). The bullwhip effect creates waste in the supply chain, including non-value-added administrative processes and inventory.

One possible solution to the bullwhip effect problem is to communicate demand at the consumer level quickly upstream to all suppliers. While suppliers still have to wait for orders from their customers, they have a better idea of what the actual demand is at the end of the supply chain. Taking action upon receiving the demand signal can help reduce waste and contribute to an increased profit margin among parties across the supply chain

(Hung et al. 2006). Tight collaboration among business partners can create a close-loop system, which is a prerequisite to the lean supply chain (LSC).

Large-scale original equipment manufacturers (OEMs) across industries, such as Hyundai, Dow Corning Corp., Procter & Gamble Co., Land O'Lakes Inc., Lucent Technology, and Pfizer, have successfully made inroads into LSC management. These OEMs are striving to further streamline their supply chains by pressuring or enticing their domestic and international suppliers to follow suit. Waste reduction in the supply chain is an ongoing battle for firms to stay lean and efficient (Manrodt et al. 2008).

## Current Developments in SCM

While the trends discussed earlier are macro in nature, that is, occurring on a broad economic level over a number of countries, the current developments discussed next are more micro in nature in that they occur at the industry level. These developments signal that supply chains must become more integrated and efficient in their operations.

### Power Has Shifted from Manufacturers to Retailers

A firm needs a clear value offering to stay competitive. After all, customers decide to purchase products and services if the value offered to them is clear and attractive. Cost leadership and differentiation are two commonly employed generic strategies to offer value propositions (Porter 1980). A cost leadership strategy, often referred to as a low-cost strategy, seeks to lower costs throughout the entire business process. The savings achieved are then passed on to the customer. For example, Wal-Mart's passion is to keep its operating costs as low as possible and to offer *everyday low prices* to their customers (Fishman 2006). Southwest Airlines minimizes turnaround time to offer customers the lowest price per passenger mile. McDonald's constructs a global sourcing network in order to offer competitive meals on a worldwide basis. Strength in miniaturization and interface design gives Apple a dominating position in the iPhone and iPad marketplace. The Kanban system allows Toyota to deliver products to customers just in time. Sophisticated supply chain networks enable Dell to offer a make-to-order promise to its customers. These best practices have one thing in common: a close integration across the supply chain to deliver distinct value propositions.

The value proposition delivery process begins with understanding the requirements of various stakeholders, including customers, regulatory institutions, and the market. The R&D department begins by analyzing these requirements and proposes new products and services to meet customer demands. New products are tested via prototyping to understand their feasibility and marketability. Once products pass the prototype test, the company begins the mass production process to introduce the newly developed products into the marketplace. The success of a product itself does not guarantee the success of the business. Many services, such as marketing and sales, need to play active roles in supporting the production and introduction process. Hence, a successful product launch can attract customers and profits so that the company can continue to fine-tune the product and to develop newer products.

In order to acquire and retain customers, retailers sometimes squeeze profit margins from their upstream supply chain partners. Thus, in order to support a customer-driven business, manufacturers have no choice other than to gradually shift its power to retailers. As the integration among business partners across the supply chain becomes tighter, the bargaining power of suppliers diminishes and shifts to the customer. This process occurs because the price of the product is being driven down, and customers, who are typically price sensitive, choose lower-cost venues (Porter 1980). Advanced information technologies, such as search engines, artificial intelligence, online communities, and data analytics, are pushing information to customers according to their needs. As a result, the increased information accessibility, availability, and transparency give customers bargaining power to negotiate with retailers.

## Consolidation of Small, Local, or Regional Retailers into National Chains

Mega retailers, such as Wal-Mart and Target, may be profitable businesses because of their economies of scale and scope. However, their presence has taken away the diminishing profits of small retailers and many of these have shut down. A few competitive small retailers consolidated with local or regional retailers and emerged into national chains. In the automobile industry, Parts Plus acquired Independent Auto Parts of America (IAPA) to increase network size and bargaining power against its suppliers. Agrium–UAP consolidated with the purpose of increasing its operational efficiency. Such strategic moves created a domino effect as other retailers did the same (Sfiligoj 2008). In the grocery industry, several retailers consolidated, resulting in the absorption of more than 4000 grocery stores (Kaufman and Waves 2007). In the department store arena, Macy's is an example of a company that has combined regional department store chains into a mega department store company.

When a national chain uses the consolidation strategy, many tangible and intangible benefits can be delivered. They can attract volume purchase discounts, realize economies of scale and scope, increase store traffic, enhance customer loyalty, and reduce the time to market. Despite these potential benefits, consolidated retailers also face many challenges related to aligning different corporate cultures and management practices. They need to make these alignments before they can attain the economic synergies.

## Emergence of "Killer Category" Retailers

*Killer category* or *big box* retailers, such as Home Depot, Lowe's, and IKEA, often acquired smaller retailers in the same business sector. In some of these acquisitions, the degree of commonality in corporate culture was high between the acquiring company and the company acquired. This strategy allowed *killer category* retailers to minimize the impact of different corporate cultures and practices in their consolidation efforts.

Most importantly, the bottom line in each retail sector has substantially improved. A longitudinal study examining multiple sectors of *killer category* retailers over a 17-year period found these retailers increased both sales revenue and per-capita spending (Sampson 2008). The industry sectors studied included the book publishing, sporting goods, home center, electronic, toy, home furnishing, and grocery sectors.

## From a Make-and-Sell Mentality to a Sense-and-Respond Orientation

When manufacturing companies were the predominant entities, they often adopted a make-and-sell strategy. They developed products they thought would be accepted in the marketplace and which they were able to produce profitably. After that, they figured out how to market those products. This worked well when the products were oil, steel, automobiles, and passenger train travel. These were generic products with limited variety and few suppliers. The manufacturers could own or control the retail outlets for their products and forecast the demand well enough to match supply and demand.

Today, however, markets are saturated with suppliers, and consumers have the option of being more selective. Suppliers have to continue to expand their product lines in the hopes of offering something new and different enough to attract buyers. This situation makes it more difficult to forecast demand—both the amount and the variety. Consequently, there is a need for organizations to move from a *make-and-sell* model of operating to a *sense-and-respond* model. Only by making this transition can they hope to be successful (Haeckel 1999).

In the make-and-sell model, manufacturers planned production to meet expected demand and then purchased goods from suppliers. Sales and marketing activities did not always correlate with manufacturers and suppliers. Manufacturing units sought efficient production runs on single model products, while marketing departments sought out products with different model offerings. The linkage between upstream and downstream activities was missing from the make-and-sell supply chain model. This absence of an orchestrated effort in managing the linear supply chain resulted in long cash-to-cash cycles, out-of-stock or overstock problems, and long accounts payable and/or accounts receivable cycles. The linear model, then, was acceptable only when product proliferation was not an issue.

However, the make-and-sell model is eroding in today's customer-driven business. New developments include brand proliferation and shortened PLCs, intensifying competition for consumer attention, globalization, media fragmentation, e-business, and mobile commerce. These developments resulted in the increased importance of sensing and responding to the needs of customers. Thus, it is important for a company to construct a demand-driven supply chain (DDSC) to cope with these new challenges. An effective DDSC includes three major overlapping areas (Hofman et al. 2011):

- *Supply management*: Manufacturing, logistics, and sourcing
- *Demand management*: Marketing, sales, and service
- *Product management*: R&D, engineering, and product development

DDSC is a sense-and-response model for fixing supply chain problems, derived from the make-and-sell model.

## Obstacles to Supply Chain Integration

Although companies want to participate in smooth-running supply chains, they must overcome certain obstacles to achieve this objective. Some of the most common obstacles include the need for globalization, the need to establish common interests, the need for an interorganizational system (IOS), the need for multiple supply chains within a company, and last, but far from least, the need for trust among participants.

## Need for Globalization

Countries vary in their degree of logistics friendliness. The national differences and the trend of product proliferation can create operational and integration problems when the supply chain expands internationally. For example, an international business must deal with product proliferation issues, such as mass customization and multiple product configurations to meet regulatory requirements of different countries. In addition, technical, social, and communication challenges are encountered as information sharing between customers and suppliers spreads throughout the geographic regions of the world.

In recent years, international companies devised different ways to cope with these SCM challenges. For example, customers and suppliers have formed interorganizational teams to increase cooperativeness in international alliances and international competitiveness. In the American automotive industry, carmakers adopted a modularity model, a modified version of the Japanese keiretsu model, in order to achieve the ideal build-to-order practice (Ro et al. 2007).

Supply chains encounter rising social and environmental requirements that vary with different countries. The ability of an international enterprise to meet those requirements can have a positive short-term and long-term contribution to both profitability and competitiveness.

## Complexity of Arranging Entities with Common Interests

Supply chain partners with common interests and interdependence can help promote trust and interorganizational learning throughout the supply chain. However, roadblocks exist in building a learning supply chain. These roadblocks originate from the lack of common interests shared by supply chain partners and the complexity of managing these interests. Bessant et al. (2003) identified several barriers to the interorganizational learning process for different industries:

- Supply chain partners are self-centric and do not see the bigger picture of realizing the potential synergies.
- Supply chain partners do not trust each other.
- Supply chain partners are not involved in the changes needed to transform the chain.
- Supply chain partners do not have adequate knowledge and skilled resources to enact changes in the chain.
- Supply chain partners have an inability to disseminate learning further than the first tier of suppliers.
- Learning support for supply chain partners is not available or consistent.
- There exists a low-cost culture that creates destructive preconceptions.
- Supply chain partners are not open about their fears and concerns.

In practice, common interests exist not only in the cooperative supply chain but also in the competing supply chain. Cooperative supply chain partners seek long-term relationships in order to mutually benefit each other and to increase profitability. In contrast, competing channels of distribution have common interests in controlling intrachannel opportunism. Striving for a good reputation is a useful mechanism to control opportunistic behavior of supply chains. However, this intangible asset is hard to acquire, since goodwill takes time to build. Both cooperative and competing supply chains need to continuously evolve to address the complexity of having business partners reach common interests.

## Lack of Effective Interorganizational Systems

IOS, such as enterprise systems, EDI, and extranets, transcend organizational boundaries and link supply chain participants closely so that constant information flows can be maintained. Streamlined information flows can help derive many business benefits, ranging from cost reduction, inventory reduction, mass customization, global communication, minimization of response times, to improved information visibility. However, a company cannot develop an effective IOS alone. Rather, a substantial coordination effort between business partners is indispensable for the success of an IOS deployment. Business partners have different concerns and react to the ideas of IOS implementation in different manners. It is commonly agreed that a costly IOS investment can only be justified if a long-term, close relationship between business partners is committed or is already in place. Measures of effectiveness such as return on investment (ROI) are key determinants in convincing supply chain partners to adopt a costly IOS. The technology competence of suppliers also varies greatly; consequently, technological incompatibility can lead to the failure of an IOS implementation.

## Need for Multiple Supply Chains within Companies

Shortening PLCs, varying expectations for differentiated products and services, and offshore outsourcing trends have resulted in the need for multiple supply chains for companies. Collaborative research efforts among business partners and improvements in computer-aided design (CAD) technologies resulted in the increased product innovations, thereby shortening product life. The rapid increase of product lines can result in additional management complexities, since each product line has unique supply chain requirements. The process of sourcing auto parts for a company's main business is different from the process of sourcing travel services. In the auto parts example, products require a purchaser with specialized knowledge, whereas providing travel services does not. When introducing a new product, companies will need to change the process of planning, sourcing, making, and delivering. Incongruities can cause a delay in introducing a new product. For example, problems with the management of multiple suppliers caused Boeing to delay the delivery of the Dreamliner 787 (Sanders 2009).

A company typically manages three types of supply chains: new products, service parts, and indirect purchases (Fickle 2006). Supply chains managing new products have an internal focus, whereas the other two types have an external focus. Maintaining a balance of internal and external supply chains can increase customer satisfaction and improve the bottom line of a business. In order to achieve this balance, it is important to address the challenges associated with multiple supply chains. These challenges result from the specific needs of different products, customers, value chain segments, and industries.

## Lack of Trust between Participants

Trust among participants is an indispensable factor for the success of an effective global supply chain. Trusting organizations are more willing to invest in relationship building and supply chain activities, such as R&D, forecasting, planning, training, information systems investment, quality improvement practices, and information sharing with their business partners.

Recalls of imported goods are examples of adverse outcomes that result when interorganizational trust among supply chain participants is missing. The importers forced suppliers to compete with each other on price in order to win an offer. In response, suppliers attempted to make sure they underbid each other and, in the process, sacrificed product quality. A vicious cycle across the supply chain can be created without the presence of trust; hence, trust is a necessary step toward a healthy supply chain.

Moreover, trust can stop opportunist behavior from appearing. When opportunist behavior declines, the opportunity to share information with business partners improves. When business partners propose that suppliers increase supply chain activities to facilitate their transactions, an organization trusting its partners is more likely to reach a consensus for the proposed activities. An organization may also think it worthwhile to invest in the proposed activities because of the trust with its partners.

When Wal-Mart mandated their suppliers would have to implement RFID initiatives, it brought the interorganizational trust issue to the test (Webster 2008). Some suppliers trusting Wal-Mart as a reliable buyer perceived the proposed investment as an opportunity to improve information visibility, thereby increasing their strategic agility. On the other hand, those distrustful organizations suspected Wal-Mart was trying to increase its bargaining power against them. It is easier to intercept confidential information when the information exchange process between the sender and the recipient is widely distributed, instead of being centrally managed. A supplier also needs to face the risk of leaking confidential information to its competitors by sharing confidential information with their same customers. Lack of trust among organizations may outweigh the benefits of any proposed supply chain activities.

Trust is an important factor that influences an interorganizational relationship. When a firm is willing to outsource its information technology (IT) functions, it becomes more involved with its sourcing partners in the implementation of outsourcing projects. In an outsourcing relationship, the degree of trust can improve the level of cooperation between partners.

There are several benefits of maintaining trust in an outsourcing relationship. Trust serves as an enabler for organizations to share important (or even confidential) information. Trust in business partners can increase the willingness of a firm to be more open in negotiations and to share information. Some firms behave opportunistically and take advantage of an outsourcing relationship to maximize one-time financial gains. But in a trusting relationship, open discussions can take place, and partners can jointly formulate an outsourcing strategy that benefits both parties in the long term. A firm is also more likely to form long-term alliances with a partner that it trusts. Furthermore, a firm that trusts its partners is not only more likely to share information but also more likely to invest capital and labor in improving and maintaining the outsourcing relationship. This trust occurs because one or both partners realize that a successful outsourcing project is beneficial for both the outsourcer and the provider.

In addition to building trust within a partnership, building public trust is also an important issue from the perspective of the outsourcing provider. For example, an outsourcing provider that mishandles confidential information about consumers is likely to lose the trust of its existing partners. A provider that subcontracts work to offshore firms that are known to disregard intellectual property rights is also likely to lose trust as well as future business. In this regard, an outsourcing provider can build public trust by continuously demonstrating its commitment to security and privacy.

## Examples of Companies with Successful Supply Chains

Gartner publishes an annual *Top 25* manufacturers and retailers with respect to their supply chain capability. Table 1.1 shows the Top 25 publicly known companies based on performance indicators including return on assets, inventory turns, revenue growth, and expert opinions from within Gartner and from peer groups. Although the study reviews a number of manufacturing and retail companies, it does not include specialized service industries, such as airlines, banks, utilities, insurance, and transportation. Although the researchers go to great lengths to select performance indicators that truly reflect supply chain performance, they continue to search for more comprehensive measures that would better indicate a company's supply chain performance. However, the information they seek is not readily available in most company reports (Hofman 2013).

Table 1.1 compares the Top 25 list for 2008 and 2013. Fifteen of the twenty-five companies listed in 2008 are not on the 2013 list, indicating an active turnover of companies. Of the 10 that survived over the 5-year period, it is obvious they have demonstrated consistent attention to a variety of best practices. How these best practices have been implemented in the Top 25 companies varies. However, these best practices have some things in common such as strong brands, a balanced and consistent supply chain presence, superior financial performance, and recognition by industry experts.

Supply chain leaders can achieve economies of scale with less complexity via the building of strong brands. This manifests itself in fewer items and more volumes shipped per item. In the companies listed in Table 1.1, the Coca-Cola Company stands out as one that has built a strong brand image throughout the world.

Supply chain leaders tightly align business strategy with operational strategy. Metrics used to measure supply chain effectiveness are interrelated. These leading companies are good at managing the balance of metrics among cash-to-cash performance, inventory level, accounts receivable, and accounts payable.

Supply chain leaders are active in managing downstream data, out-of-stock problems, and customer demand. Information transparency in customer data is high.

Supply chain leaders collaborate with their upstream suppliers in executing product and supply chain innovations as well as corporate social responsibility.

We will describe these practices, along with other worthwhile ideas, and how companies use them in the remaining chapters of this book.

## Summary

A supply chain consists of various participants who perform a sequence of activities in moving goods, capital, and information from a point of origin to a point of consumption and vice versa. Without supply chains, business commerce would not exist. Early supply chains were simple, basically involving a party that made goods and a customer that bought them. As supply chains expanded, they became more formal and included suppliers, manufacturers, distributors, retailers, and, finally, the end user customer. Today, global supply chains consist of multiple supply chains working together throughout a number of different countries.

Several current trends are causing supply chain partners to work more closely with each other. These trends include (1) technological improvements in supply chains, (2) the proliferation of electronic business, (3) the rising of developing economies, and (4) new supply chain strategy formulations. These four trends are more macro in nature in that they are occurring on a broad level over a number of countries. At the microlevel, several developments have occurred that also signal that supply chains need to operate in a more integrated and efficient manner. These developments include (1) a power shift from manufacturers to retailers; (2) the consolidation of small, local, or regional retailers into national chains; (3) the emergence of *killer category* retailers; and (4) the movement from a make-and-sell mentality to a sense-and-respond orientation.

While there has been considerable progress in making supply chains more effective and efficient, a number of obstacles exist that impede the full supply chain integration. Four of these obstacles include (1) the complexity of arranging supply chain partners with common interests, (2) the lack of effective IOS, (3) a need for multiple supply chains within companies, and (4) the lack of trust between participants.

This chapter has been an overview of supply chains and SCM. We will cover all of these topics in more depth in later chapters. In Chapter 2, we describe how supply chains must be organized using a systems approach. In Section II, we view supply chains from the demand perspective. Chapter 3 describes how to determine customers' needs and wants. Chapter 4 describes how to design a system to meet customer needs, and Chapter 5 explains how companies develop strategies to meet demands.

Section III focuses on the supply side of the supply chain. Chapter 6 covers the distribution of products to customers through distributors and retailers. Chapter 7 describes the production, or manufacturing, functions. Chapter 8 explains the role of purchasing in supply chains, and Chapter 9 describes how the logistics functions link all of the pieces of the supply chain together. Chapter 10 describes the role of reverse logistics in supply chains.

In Section IV, we state the case for integrated supply chains. Chapter 11 explains the need for integrated supply chains, Chapter 12 uncovers the obstacles encountered in integrating supply chains, and Chapter 13 describes some steps necessary in building integrated supply chains.

The chapters in Section V cover information and financial flows within supply chains. Chapter 14 covers information flows and Chapter 15 describes funds flow. In Chapter 16, we address the difficult question of how to share the costs and benefits equitably among supply chain participants.

Finally, in Section VI, we look forward to the future. In Chapter 17, we highlight reasons the future will not be like the past and the need for preparation. In Chapter 18, we describe ways in which companies can prepare for the future by implementing sound SCM practices.

At the end of each chapter, we will include a section called *Hot Topics*. They will be short summaries of issues related to SCM. The first one is about problems that can occur from outsourcing.

### Hot Topic: Outsourcing to Low Wage Countries

The evolution of supply chains has seen a gradual movement of manufacturing moved to offshore locations. For manufacturing industries in the United States, outsourcing to companies overseas has been in practice for several decades. Initially, much of the business went to Mexico after the passage of NAFTA (North American Free Trade Agreement) in 1994. However, within a decade of its passing, a number of U.S. companies began to outsource to countries in Asia, in particular China and Vietnam. The attraction was that products could be produced at lower costs because of the lower wages paid to factory workers in these countries.

## MAKE VERSUS BUY

This situation is an example of the make-or-buy decision that all businesses must grapple with. Making the product means the business has total control over costs and quality. Outsourcing the making of the product gives up a great deal of that control. Of course, management will argue that it writes specifications (specs) for exactly how the product should look when it is finished. Write demanding specifications and you will get a high-end product, or at least, that is the goal. In many cases, that scenario works out fine. Many high-quality products are made in Asia, successfully.

However, there are times when the quality of the product delivered is deficient, and, in some cases, defective. One area that has been particularly hard hit is the toy manufacturing industry. Outsourcing the making of toys in China is not an unusual practice as the largest toymakers have been licensing their manufacturing in Asia for decades (Barboza and Story 2007). It is estimated that 80% of all toys manufactured in the world are made in China by migrant workers (Shell 2009). However, in recent years, two companies, Mattel, Inc. and RC2, encountered difficulties in their relationship with supply chain partners in China.

## TOY PROBLEMS

Mattel had a 15-year relationship with its Chinese supplier, Early Light Industrial Company, before encountering a major crisis in 2006/2007 with the discovery of lead paint on certain toy products. Early Light had subcontracted the painting of toys to another subcontractor, Lee Der Industrial Company, which used an unauthorized paint containing lead (Wisner 2011). Herein lies one of the problems of overseas sourcing; suppliers use subcontractors who may or may not abide by the rules of the contract.

RC2 makes the popular Thomas the Train wooden sets. In 2007, lead paint was found on its toys originating from Dongguan, China, a factory town that is the base of much of the world's toy manufacturing. The discovery was made by RC2 in one of its regular quality checks. Once the lead was identified, the company made a voluntary recall of its affected toys. Unfortunately, the scope of the recall covered toys made from January 2005 to June 2007, indicating that RC2 went for over 2 years unaware that lead was on its toys (Barboza and Story 2007).

Of course, it is possible that quality problems could emerge even with companies that source their products in the home country. However, when the products are made overseas, a different type of negative stigma emerges. When products are made overseas, jobs are lost in the home country. When a displaced worker learns that a defective product that he or she once made was made with defects, in say, China, then it is actually a double blow to the psyche of American (or other home country) labor. The company that receives that product also suffers. In the case of Mattel, it is estimated the 2007 related recalls cost the company $40 million (Wisner 2011).

## QUESTIONS FOR DISCUSSION

1. Quality problems can occur with a supplier in both domestic and globally outsourced companies. How then can a company ensure that the quality of its supplier-provided products remains high?

2. What communication strategies can a company use when it is criticized for outsourcing to an overseas company?

## REFERENCES

Barboza, D. and Story, L., RC2's train wreck, *The New York Times*, June 19, 2007. http://www. nytimes.com/2007/06/19/business/19thomas.html?pagewanted=all&_r=0, accessed June 20, 2013.

Shell, E., *Cheap: The High Cost of Discount Culture*. New York: Penguin, 2006.

Wisner, J., The Chinese-made toy recalls at Mattel, Inc., *Business Case Journal*, 18(1), 16–30, 2011.

## Discussion Questions

1. What are the major benefits of effective SCM?
2. What potential problems result from poor SCM?
3. How have supply chains evolved in their scope and complexity?
4. What key steps should companies take to achieve the greatest benefits from supply chains?
5. A close integration among participants is indispensable to the creation of an efficient supply chain. What major obstacles could stop participants from forming a close relationship?
6. How is electronic business affecting SCM practices?
7. How do leading companies cope with the current challenges in SCM?
8. Why is trust so important in effective SCM?
9. How would you measure the performance of a supply chain?
10. Pick a company on the Top 25 list (Table 1.1). What can you find out about their supply chain?

## References

Bessant, J., Kaplinsky, R., and Lamming, R., Putting supply chain learning into practice, *International Journal of Operations & Production Management*, 23(2), 167, 2003.

Blackstone, J.H., *APICS Dictionary*, 14th edn., APICS—The Association for Operations Management, Chicago, IL, 2013.

Council of Supply Chain Management Professionals, *CSCMP Supply Chain Management Definitions*, 2008. http://cscmp.org/aboutcscmp/definitions.asp, retrieved October 22, 2008.

CSC, *The Ninth Annual Global Survey of Supply Chain Progress*, 2013. http://assets1.csc.com/insights/downloads/Supply_Chain_2012FinalReport3.7a, accessed June 10, 2013.

Dalesio, E.P., IBM in N.C. hums at center of outsourcing debate, *USA Today*, March 17, 2004, http://usatoday30.usatoday.com/tech/news/2004-03-17-ibm-outsource_x.htm, accessed August 4, 2014.

den Butter, F.A. and Linse, K.A., Rethinking procurement in the era of globalization, *MIT Sloan Management Review*, 50(1), 76, 2008.

Elkington, J., *Cannibals with Forks: Triple Bottom Line of 21st Century Business*, Capstone Publishing Ltd., Oxford, U.K., 1999.

Fickle, K., How to manage multiple supply chains, *Logistics Today*, 47(1), 26, 2006.

Fishman, C., *The Wal-Mart Effect: How the World's Most Powerful Company Really Works—And How It's Transforming the American Economy*, The Penguin Press, New York, 2006.

Friedman, T.L., *The World Is Flat: A Brief History of the Twenty-First Century*, Farrar, Straus and Giroux, New York, 2006.

Haeckel, S.H., *Adaptive Enterprise, Creating and Leading Sense-and-Respond Organizations*, Harvard Business School Press, Boston, MA, 1999.

Haksöz, Ç., Seshadri, S., and Iyer, A.V., *Managing Supply Chains on the Silk Road*, CRC Press, Taylor & Francis Group, Boca Raton, FL, 2012.

Hofman, D., *The Gartner Supply Chain Top 25 for 2013*, 2013, Gartner Research. http://www.gartner.com/technology/supply-chain/top25.jsp, accessed June 10, 2013.

Hofman, D., O'Marah, K., and Elvy, C., *The Gartner Supply Chain Top 25 for 2011*, 2011, Gartner. http://www.gartner.com/id=1709016, accessed June 1, 2011.

Hounshell, D.A., *From the American System to Mass Production, 1800–1932: The Development of Manufacturing Technology in the United States*, The John Hopkins University Press, Baltimore, MD, 1985.

Hull, B.Z., Frankincense, myrrh, and spices: The oldest global supply chain? *Journal of Macromarketing*, 28(3), 275, 2008.

Hung, W., Chu, J., and Lee, C.C., Strategic information sharing in a supply chain, *European Journal of Operational Research*, 174(3), 1567, 2006.

International Organization for Standardization (ISO), 2013. http://www.praxiom.com/iso-definition.htm#Quality, accessed June 11, 2013.

Jessup, L.M. and Valacich, J.S., *Information Systems Today: Managing in the Digital World*, 3rd edn., Pearson Prentice Hall, Upper Saddle River, NJ, 2008.

Kaufman, P. and Waves, A., Strong competition in food retailing despite consolidation, *Amber Waves*, 5(1), 5, 2007.

Lee, H.L., Padmanabhan, V., and Whang, S., Information distortion in a supply chain: The bullwhip effect, *Management Science*, 50(12), 1875, 2004.

Manrodt, K.B., Vitasek, K., and Thompson, R., The lean journey/part II: Putting lean to work, *Logistics Management*, 47(8), 74, 2008.

Mollick, E., Establishing Moore's law, *IEEE Annals of the History of Computing*, 28(3), 62, 2006.

Porter, M.E., *Competitive Strategy*, Free Press, New York, 1980.

Porter, M.E., What is strategy? *Harvard Business Review*, 74(6), 61, 1996.

Rainer, R.K., Jr. and Cegielski, C.G., *Introduction to Information Systems*, 3rd edn., John Wiley & Sons, Inc., Hoboken, NJ, 2011.

Ro, Y.K., Liker, J.K., and Fixson, S.K., Modularity as a strategy for supply chain coordination: The case of U.S. auto, *IEEE Transactions on Engineering Management*, 54(1), 172, 2007.

Rottman, J.W. and Lacity, M.C., Proven practices for IT offshore outsourcing, *Cutter Consortium*, 5(12), 1, 2004.

Sanders, P., Boeing tightens its grip on Dreamliner production—Company is in talks to buy fuselage factory from supplier; supply-chain woes have dogged 787 program, *Wall Street Journal*, July 2, 2009, B.1.

SAP Supplier Relationship Management, 2013. http://www54.sap.com/solution/lob/procurement/software/srm/index.html, accessed June 10, 2013.

Sfiligoj, E., The road not yet taken, *Croplife*, 171(2), 6, 2008.

Silkroad Foundation, Marco Polo and his travels, 2008. http://www.silk-road.com/artl/marcopolo.shtml, accessed April 20, 2014.

Sampson, S.D., Category killers and big-box retailing: Their historical impact on retailing in the USA, *International Journal of Retail & Distribution Management*, 36(1), 17, 2008.

Singh, M., Supply chain reality check, *MIT Sloan Management Review*, 46(3), 96, 2005.

Turban, E., Lee, J.K., King, D., and McKay, J., *Electronic Commerce, 2008*, Prentice Hall, Upper Saddle River, NJ, 2007.

Webster, J.S., Wal-Mart's RFID revolution a tough sell, *Network Word*, 25(36), 34, 2008.

Yaibuathet, K., Enkawa, T., and Suzuki, S., Influences of institutional environment toward the development of supply chain management, *International Journal of Production Economics*, 115(2), 262, 2008.

# 2

## Supply Chains as a System

**Learning Outcomes**

After reading this chapter, you should be able to

- Describe the components of the input–transformation–output (ITO) model.
- Describe the four types of flows in a supply chain.
- Provide examples of supply chains in different industries.
- Identify the difference between internal and external customers.
- Explain the difference between open and closed systems.
- Describe the effects of external influences on supply chains.
- Discuss the obstacles and enablers of supply chain integration.
- Explain how supply chain performance can be measured.
- Explain why costs, resources, and benefits must be fairly allocated along the supply chain.
- Describe the ways companies create value along the supply chain.

**Company Profile: Zara's**

This chapter describes the basic components of a supply chain. The basic unit is the ITO model. Inputs are transformed into outputs at numerous stages along the supply chain. The ITO processes are designed to facilitate the physical flow of goods and services. In addition, they also depend on information flow and funds flow. Supply chains are affected by both internal (within the organization) and external (the business environment) influences. There are benefits that result from good supply chain systems; however, there are obstacles that make supply chain system design and implementation difficult.

This company profile is about Zara, a division of Inditex, one of the leading apparel companies in the world. Zara has designed a supply chain system that works extremely well. Some of the questions you should think about as you read the company profile and the chapter in the book include:

- What are some of the key elements in the Zara system?
- Which parts of the Zara supply chain are most important? The least important?

- What are the benefits of Zara's supply chain system?
- What part of Zara's supply chain gives them the best competitive advantage?
- What additional questions do you think relate to supply chain systems?

### Inditex: Zara

This chapter is about a supply chain as a system. A system is a regularly interacting or interdependent group of items forming a unified whole toward the achievement of a goal (Blackstone 2013). A supply chain is a system that moves product from the point of origin at a mine or farm through a series of value-adding activities such as manufacturing and distribution, until the product reaches the ultimate consumer. The supply chain also moves information upstream from the customer back to the supplier and also downstream from suppliers to customers. The supply chain assures the flow of funds from customer to supplier. In order to achieve these flows effectively, the supply chain is composed of links between individual participants (members) of the supply chain, forming a system to move goods, information, and funds.

Zara shows its revamped image based on four guiding principles: beauty, clarity, functionality, and sustainability. Simplicity is the byword throughout the store as part of Zara's mission to bring customers into direct contact with fashion. Inditex opens its first stores in Armenia, Bosnia-Herzegovina, Ecuador, Georgia, and the Former Yugoslav Republic of Macedonia and reaches 6000 stores worldwide (Inditex Web Site, Store Locations, http://www.inditex.com/en/who_we_are/timeline).

Inditex is a company that has built a supply chain system that is extremely successful. They were first listed in the Gartner Top 25 Supply Chains in 2012 when they were ranked No. 19. In 2013, they moved up to No. 15 (Gilmore 2013). They have reached this level of recognition by developing a supply chain that emphasizes fast response times and agility in meeting variability in their marketplace—high fashion clothing. Their growth confirms their success. From their beginning in 1975, they have grown to over 6000 stores throughout the world.

Zara is their leading brand, with nearly 2000 stores, serving 86 markets and generating net sales of €10,541 million, or approximately $13.7 billion (Inditex 2013). They do this by managing a supply chain with the almost unbelievable response time of 2 weeks (instead of an industry average of 6 months) from the time they begin to design the new product until that new product is available in the store. They use a pull system that produces what customers want instead of a push system that designs a product and then tries to figure out how to sell it. Their somewhat unorthodox supply chain system is designed to be agile and react quickly to changes in the marketplace.

They are able to achieve a fast response through developing and maintaining "shared situation awareness, a team's ability to recognize a pattern in a fluid situation and use it to anticipate what might happen next" (Sull and Turconi 2008, p. 8). Teams develop shared situation awareness in three steps: (1) observe the raw data, (2) spot patterns to form hypotheses about how the situation might unfold, and (3) test the hypotheses (Sull and Turconi 2008).

The employees at Zara achieve their results through a combination of cultivating customer awareness, communicating between stores and corporate, collaborating among organization functions, and committing to meet rigid design, production, and delivery time schedules.

## INFORMATION GATHERING

At the store level, the store manager is responsible for finding out what customers buy and what they do not buy. They do this without a plethora of written reports, but by talking directly with salespersons and observing the racks to see what is selling. The manager and employees even look at the clothes tried on but not bought to see if they can detect patterns—colors, styles, etc. The store manager is responsible for deciding what items to carry in the store, not a centralized computer. They communicate upward through regional and country managers, most of whom have been store managers, to the central design teams. At the same time, corporate-level designers are also reviewing daily sales to determine trends or patterns. The information gathering combines hard data with informal inputs to get a sense of the changes taking place among consumers. One of the strengths of the system is that new designs come quickly and on a regular schedule.

## PRODUCT DESIGN

Ideas for new items flow from the stores to headquarters at La Coruna, Spain. There, teams consisting of designers, marketing specialists, and production planners collaborate to convert the idea into reality. The layout of the building is designed to facilitate the *togetherness*. There are three long building areas—one each for men's, women's, and children's clothing. Within each, there are no offices, only open spaces with desks and tables to encourage cross-functional discussions. Employees are hired to be team players, not prima donna designers. The team comes up with a design concept and has a prototype sewn, and employees model the clothing item. If the item does not meet approval, it is discarded; if it does, it goes on to the production area. Only about 25% of the prototypes finally reach the stores; however, the teams generate over 10,000 new designs each year (Petro 2012). Team members are changed often to encourage the flow of new ideas. They also test their prototypes by displaying them in an area called *Fashion Street*, located in the same building as the designers. *Fashion Street* is a long parade of stores, with existing and proposed shop formats. This makes it possible for architects, designers, and visual merchandisers to visualize not only the clothing but the store format and layouts to see how they work together as a system (Sull and Turconi 2008).

## MANUFACTURING

Zara is able to move the new designs through the manufacturing phase because they use a number of local subcontractors in the La Coruna area. Approximately 50% of their manufacturing is done close by in Spain, Portugal, and Morocco, for the fashion items that have short life cycles and which they want to move into stores quickly, with minimum inventory. Some of these factories are highly automated; hundreds of others are small subcontractors who can respond quickly to Zara's time schedules. The factories also operate with just-in-time (JIT) production systems developed in cooperation with logistics experts from Toyota Motor. More staple items, such as T-shirts, are outsourced to low-cost countries such as Eastern Europe, Africa, and Asia (Capell 2008). Zara also produces in small batches with high variety. They keep some excess capacity in the factories to be able to move quickly when new designs are needed (Ferdows et al. 2004). In 2012, they purchased from a total of 1434 suppliers with 446 in the European Union (EU), 136 in non-EU Europe, 48 in America, 112 in Africa, and 672 in Asia (Inditex 2012 Annual Report, p. 63).

## DISTRIBUTION

Once the clothing is produced, it is shipped to the stores, twice each week, to keep the stock fresh and the inventory low. Their logistics system provides continual replacement and guarantees that stores will receive new products in no more than 2 days (24–36 h in Europe and 48 h in the rest of the world) (Inditex 2012 Annual Report, p. 43). Many of the items are already priced and on racks to reduce the time required at the stores to make them ready for sale. The distribution of goods is on a rigid schedule to assure new items arrive in the stores quickly and on a definite schedule. Stockouts offer an opportunity to introduce new items and encourage customers to visit the store often.

## STORES

In 2012, their 1763 stores were located throughout the world, with 1121 in Europe, 218 in America, and 424 in Asia and the rest of the world, including 140 in China and 88 in Japan (Inditex Store Location 2013). Their stores are spacious, 10,000 ft$^2$ or so, but are not crowded with endless racks of clothes. In fact, there may be a limited number of items, conveying the impression that a customer should buy now, because it may not be there later (Ferdows et al. 2004). Stores are located in fashionable, high-traffic areas. The company describes their new store concept, introduced in 2012 this way:

> This innovative interior design was presented at the brand's 666 Fifth Avenue store in New York. The new image revolves around the principles of beauty, clarity, functionality and sustainability. The store turns around two long axes, hallways, or "walkways," leading to individualized spaces or "cubes" on each side, in which the various collections are shown in an individualized way. The wood of the furniture has been finished in textures that are reminiscent of fabrics such as linen or silk, in elegant and neutral colors. In addition, with regard to sustainability, the store is in line with the guidelines governing the Inditex Group's eco-efficient stores.
>
> **Inditex 2012 Annual Report**

In addition to the Fifth Avenue location, in 2012, they opened stores in the Park House area of London at 460-490 Oxford Street and on No. 92, Champs-Elysées, Paris.

## RESULTS

Does this tightly integrated supply chain produce results for Inditex and Zara? There are a number of indicators that it does. Inditex SA is now the world's largest clothing retailer with revenue of over $19 billion in 2012. It has higher inventory turns, fewer markdowns, and a higher average gross margin on sales than its competitors. Although they operate contrary to some of the conventional practices of clothing retailers—less low country outsourcing, more spare production capacity, air shipments instead of ocean carrier, and a higher concern for supplier working conditions—they end up with a higher return on investment (ROI).

## SYSTEM

They have a self-reinforcing system built on three principles: close the communication loop, stick to a rhythm across the supply chain, and leverage the capital assets to increase supply chain flexibility (Ferdows et al. 2004). Zara puts all of the pieces together into a system that gets the job done. Sanford C. Bernstein offers advice to rivals who hope to mimic Inditex's results: "Don't follow the Zara Pattern halfheartedly. The Inditex way is an all-or-nothing proposition that has to be fully embraced to yield results" (Capell 2008).

*Source:* Adapted from Crandall, R.E., *APICS Mag.*, 24(1), 25–27.

## REFERENCES

Capell, K., Zara thrives by breaking all the rules, *Business Week*, October 20, 66, 2008.

Crandall, R.E., The race for style, evaluating the breakneck speed of Zara's integrated supply chain, *APICS Magazine*, 24(1), 25–27.

Ferdows, K., Lewis, M.A., and Machuca, J.A.D., Rapid-fire fulfillment, *Harvard Business Review*, 82, 104–110, 2004.

Gilmore, D., Understanding the Gartner top 25 supply chain rankings, *Supply Chain Digest*, March 1, 2013. http://www.scdigest.com/assets/FirstThoughts/13-03-01.php?cid=6791, accessed June 19, 2013.

Inditex Annual Report, 2012. http://www.inditex.com/en/shareholders_and_investors/investor_relations/share, accessed June 19, 2013.

Inditex Web Site, Store Locations. http://www.inditex.com/en/who_we_are/stores?zone=0, accessed June 20, 2013.

Petro, D., The future of fashion retailing: The Zara approach (part 2 of 3), *Forbes*, October 25, 2012.

Sull, D. and Turconi, S., Fast fashion lessons, *Business Strategy Review*, 19(2), 4–11, 2008.

## Introduction

In Chapter 1, we described the evolution of supply chains and their importance in today's business environment. In this chapter, we describe the basic components of supply chains and how they fit together to form the supply chain system. Supply chain management (SCM) is the design, planning, execution, control, and monitoring of supply chain activities with the objective of creating net value, building a competitive infrastructure, leveraging worldwide logistics, synchronizing supply with demand, and measuring performance globally (Blackstone 2013). Viewed another way, it is the result of a series of coordinated steps necessary to transform raw materials into finished products and services that the consumer needs or wants. A company manages a series of business processes in order to transform inputs into outputs that have value to a customer. Customers value a pan of corn bread higher than a bag of unprocessed cornmeal. A well-furnished apartment costs more than a nonfurnished one to a tenant. In the context of SCM, the inputs consist of things and people. The outputs are goods or services consumed by customers in manufacturing or service businesses. The ITO model is the DNA of the supply chain. Both manufacturing and service businesses employ the ITO model to create a systematic, functioning supply chain to sustain and become competitive. Figure 2.1 shows a basic ITO model.

In this chapter, we describe the major business processes and how to employ these processes to transform inputs into outputs. Resources are consumed during the transformation process and companies must manage these resources wisely in order to improve their supply chain systems. These resources include employees, equipment, facilities, inventory,

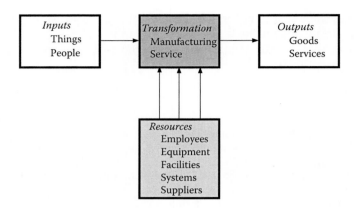

**FIGURE 2.1**
Basic ITO model.

systems, and supply chain partners. All of these resources require periodic replenishment and upgrading to maintain a competitive edge for the supply chain.

The role of the customer is important, since a customer's perceived value determines the actual value of the outputs (products or services). All products and services evolve through a life cycle. Managerial challenges exist at each stage of the product life cycle. Internal and external customers are involved through the product life cycle. Internal customers transform the received inputs into valuable outputs. External customers exchange financial resources for the goods and services they order. All customers are recipients of goods and services as well as evaluators of the product received. Customers' knowledge about a product can influence their product preference. Incorporating a customer's preference into the development of a product can help a company develop marketable new products faster, with a higher success rate.

As a system, supply chain participants need to continuously interact with their environments, so they can develop products more effectively. Rapid internal and external changes can disrupt the normal operations of a supply chain. A change in organization structure is an internal change that can be disruptive. Uncertainties in customer demand or lead time are external, uncontrollable changes. Hence, SCM requires a close collaboration among participants. At different stages in the supply chain, participants introduce various interests and management philosophies. Conflicts of interests among supply chain participants are examples of changes that can potentially disrupt the normal operation of the supply chain.

Integration problems arise when supply chain participants are connected. All supply chains have different compositions. When different corporate cultures, information systems, and business practices merge into a supply chain, obstacles develop that can stop or slow the integration process. Identifying the major obstacles to supply chain integration and knowing how to address these obstacles are important topics for a business to learn.

Performance measurement systems are needed to assess the performance of a supply chain. A supply chain derives benefits and incurs costs. Operational efficiency, shortened product lead time, and lower inventory costs are examples of tangible benefits and costs. Intangible benefits include customer satisfaction, information visibility, and faster decision-making processes. Direct costs include implementation of new technology, meetings between customer and supplier, and retraining of employees. Participants of a supply chain can incur intangible costs as well. These include the loss of customer goodwill when a product is defective or the loss of a business order when a supplier is out of stock. These intangible benefits and costs create a situation whereby measuring supply chain effectiveness is difficult. A balanced, integrative approach to measuring tangible and intangible benefits and costs is essential for the continued growth of supply chains.

## Supply Chain Systems

A holistic view of supply chain systems can provide insights on how supply chain participants operate independently and collectively toward achieving a real-time, global supply chain. The ITO model is a concept that is applicable not only to SCM but also to other systems. The ITO model can assist in understanding the complexity of supply chain systems. Bertalanffy (1969) first proposed the *general systems theory*. This theory asserts that a real system continuously evolves by interacting with external environments and integrating new properties into the system. A holistic perspective of viewing the continually evolving process of a system is essential to understand how system parts function both independently and collectively toward a common goal. Other disciplines, including physics, biology, information systems, technology, and sociology, have grounded their works on the *general systems theory*.

The ITO model organizes a supply chain system in a sequence of inputs, transformation, and outputs. These three components vary with industry sectors. Supply chain systems can be constructed for both the manufacturing and the service sectors.

### Inputs

The inputs for a manufacturing process usually include raw materials, such as wood, steel, water, plastic, or chemicals. Production processes, such as carving wood, processing steel, and blending chemicals, begin after raw materials arrive. The output of the production process includes finished goods, such as furniture, toys, or automobiles.

In contrast, inputs for service sectors include people, such as patients, travelers, and vacationers to satisfy customer demand for services, such as medical care, in-flight services, and recreational entertainment. Hospitals, airlines, and vacation destinations offer services (transformation process) to satisfy customers' needs. After receiving services, these same people are now outputs—patients recover from illness, travelers arrive at a destination, and vacationers have a great recreational experience.

The basic ITO model is applicable to a general understanding of supply chain systems for both manufacturing and service sectors. A closer look at the operations of both sectors indicates that differences do exist in all three essential elements. Manufacturing inputs are things, whereas service inputs are primarily people. The manufacturing process is tangible and standardized, whereas service processes are often intangible and vary greatly among service providers. For example, it is easier to see how a table is built than to learn how to please a customer at a restaurant. Measuring outputs of manufacturing sectors is often easier than measuring outputs of service sectors. A finished good needs to meet particular product specifications and can be measured consistently. A group of tourists can have divergent experiences after using the same customer service at the same tourism destination. A higher degree of variations exists in the output of service sectors than in the output of manufacturing sectors.

## Transformation

Much of the research in management has focused on the transformation process. The transformation process in manufacturing is usually obvious, because it involves a physical change in the product. Some variables in the transformation of services are tangible, such as determining the number of store sites for a retail business. Other variables are intangible, such as a caregiver's attention to a patient or the level of judgment required by a bouquet arranger in a custom flower shop.

The transformation process can be viewed as a technical function, capable of standardizing or customizing the product. Companies adapt the transformation process to any of the variations that may be encountered in the operating environment. For example, the service transformation can be rigid, the use of ATMs, or flexible, providing human bank tellers.

Automation, the substitution of technology for people, is an example of the active use of technology in providing services. Automation usually increases over time as the service becomes more standardized, as in the case of travel insurance policies sold through vending machines. The amount of automation also increases as the technology capability increases, such as in airline reservation systems now available to every travel agent and online flight selection for the Internet user.

The transformation process must be adaptable to variations in tasks and rates of service. The service can be continuous, an electric utility, or discrete, a lawyer preparing a house purchase agreement. Companies can service customers at their place of business, as in a bakery; in the customer's house, as by an interior decorator; or over a communication system, as in the home video shopping channel. The presence of many variables in the transformation process provides a uniqueness and identity on the part of the service provider.

## Outputs

Outputs take the form of goods, or things, which customers can measure and evaluate to see if they conform to specifications. A thin plastic paper is an example of an output that is used to wrap afterdinner mints. It must be attractive and meet other physical specifications; simultaneously, it provides part of the service package at a restaurant. Measuring the output for a customer may be a complex undertaking, because each customer is unique. The measure of a completed output may be as indefinite as the perception of the customer. For example, attending a training course on the use of a word processing software package transforms a person with little knowledge to a person with more knowledge. However, the knowledge gained varies with each person. Clearly, defining the output for a pure service is difficult, and the achievement of a satisfactory output requires variations in the process used. An understanding of the expectations of the customer before providing the service may preclude subsequent customer dissatisfaction or rework of the service.

The amount of customization required of a service varies among outputs, especially when the output is a person instead of a thing. Another variation is the rate of demand by the customer. Demand is rarely at a constant rate; therefore, the transformation process has to be able to accommodate both customizing the service for the customer and adjusting the demand rate.

The success of any business depends on how well the outputs reflect the critical success factors of the business. Critical success factors are those items a business must do well in order to succeed. Examples include the cost of service, the timeliness of service, and the quality of service. Careful planning is required to provide the service package that maximizes the benefits of critical success factors.

**TABLE 2.1**

Examples of ITO Processes in Different Industries

| Business Types | Inputs | Transformation Process | Outputs |
|---|---|---|---|
| *Manufacturing and other producing sectors* | | | |
| Electric utility | Coal, oil, gas, solar | Power generation and distribution | Power for residential house |
| Building construction | Land, steel, concrete | Architecture design and construction | An office building |
| Toy firms | Plastic, chemicals, and paint | Manufacture of final products | Toys |
| *Service sectors* | | | |
| Restaurant | Hungry people | Food preparation and service | Regular patron |
| Hotel | Arriving travelers | Room preparation and services | Happy customer |
| Hospital emergency room | Untreated patients | Diagnosis and treatment | Treated patients |
| Public accounting firm | Unaudited companies | Auditing and accounting services | Audited company |

## Manufacturing versus Services

Table 2.1 compares ITO elements between manufacturing and service sectors. Within the manufacturing sector, an electric utility company constructs power plants to maximize its capacity to serve residential areas. The high-voltage power at the power plant is redistributed as low-voltage power for each residential house. In the service sector, food preparation at a restaurant begins with the customer's order. The chef in the kitchen receives the order and begins cooking the food to order. The server delivers the food to the customer to eat. When the transformation process is successful, the customer is full and satisfied with the overall food service.

In looking at the differences between the two sectors, an interesting phenomenon is developing. The boundary between manufacturing and services blurs as these two processes become closely related and inseparable in contemporary business practice (Crandall and Crandall 2014). Dell's assemble-to-order business model promises that customers can customize orders and have products delivered to them directly from manufacturers in a few days. A customer contacts a local accountant for tax reporting services and has a tax report generated and submitted electronically to an outsourcing partner in India. The information system of a car dealer and automobile manufacturer is synchronized and integrated, so the service representative can provide real-time information to a customer on the availability of the car they desire. These examples demonstrate that the ITO model provides a lens to uncover the complexity of supply chains in both manufacturing and service sectors. We will continue to examine each element in detail in the following sections.

## Characteristics of Supply Chains

Supply chains can be structured in different forms to improve business performance in areas such as operational efficiency, agility, lean management, customer satisfaction, inventory levels, and response time to market. When business partners operate without seeking to integrate their supply chains, they will not communicate, cooperate, or collaborate with each other. As a result, disruptions to daily operations, such as having excess inventory or backorders, often result. When companies recognize these issues, they can take actions to coordinate with each other. However, before the synergy of a partnership is realized, the supply chain must be reconstructed and integrated.

A supply chain can change in focus during its lifetime. In its early stages, supply chain managers seek to improve operational performance, by increasing efficiency, improving resource allocation, accurately predicting customer demand, increasing cash flow speed, and eliminating waste. As the supply chain advances in complexity, its focus turns to the improvement of strategic business benefits, such as fostering innovation, achieving competitiveness in information, sustaining market growth, and constructing higher entry barriers to potential competitors.

Once a supply chain is completed and integrated, the supply chain partners need to evaluate how they are performing in terms of its major functions: the physical, financial, and informational flows and the supply chain relationships (Cavinato 2005). The first three describe the flows along a supply chain; the relational supply chain describes the linkages between entities that compose the supply chain. William T. Walker provides examples of how the flow of materials, information, and cash depends on the linkages between customers and suppliers (Walker 2005). Understanding these elements can help a business assess its strengths and weaknesses, a necessary endeavor in strategic analysis and planning. From this understanding, the business can increase its ability to perform well in these functions and, hence, sustain itself and grow.

## Physical Flow

Physical flow is the actual movement of goods or the delivery of services across the supply chain. Financial, informational, and funds flows play supporting roles to ensure the core supply chain functions smoothly and efficiently from one business partner to another. Therefore, all supply chain partners attempt to optimize the physical flow to ensure that customers receive goods on time and at a reasonable price. In order to accomplish these goals, operational costs, such as inventory, transportation, and manufacturing, need to be optimized. At the same time, business partners must closely collaborate with each other and streamline the physical flow to reduce waste. Success at moving physical flow can lower costs and increase revenues. For instance, manufacturers obtain materials from multiple suppliers at an e-trading site, such as Alibaba.com, in order to lower the purchasing costs. Suppliers from Alibaba.com work with the Li & Fung Group, a major trading company in Hong Kong, to minimize logistics costs and ensure the timely delivery of goods to retailers (Cho 2006). Retailers promote goods to customers in an effective manner so that revenues increase and the quality of customer service improves. If customers are satisfied with the goods purchased, retailers will continue to order from manufacturers. Moving physical goods from upstream to downstream supply chain seamlessly is indispensable to the sustainability of a supply chain.

However, many problems can stop the smooth movement of goods, including demand forecasting errors, market volatility, and capacity bottlenecks at a given logistics point. According to the Boston Consulting Group, demand forecasts for the market acceptance of new drugs can be in the range of −75% to +300% in the product introduction stage (Stalk 2008). Many factors can increase market volatility. The recent subprime market crisis affected customer demand for houses, which created ripple effects on the demand for home improvement goods and financial investment products.

The threat of a labor strike can also affect the supply chain. The ports of Los Angeles and Long Beach are reaching peak capacity due to the surge of arriving containers from countries that manufacture the many goods that are offshore outsourced by American companies. Past labor strikes at these ports created huge disruptions in the U.S. supply chains. The threat of strikes in the future indicates that the vulnerability of these supply chains may continue to be fragile.

Considering these challenges associated with the flow of goods, supply chain managers are striving to optimize every decision to move physical goods smoothly through the supply chain.

## Information Flow

When goods move from one location to another, information requires updating and dissemination to supply chain partners. The absence of information synchronicity can result in overstocking, backorders, poor decision-making processes, distrust between supply chain partners, and slow responses to market changes. A smooth flow of information enables the supply chain manager to address common questions that surface in the day-to-day running of the organization:

- When receiving an order, should an organization accept or reject it?
- Should a manufacturer order more raw materials when the inventory reaches the reorder point?
- Is there enough market demand to justify the volume of an order?
- How soon can a retailer promise the delivery of a special order?

Any supply chain manager would have trouble answering these questions if information were not synchronized and integrated across the supply chain.

Another challenge to information flow is the difficulty of synchronizing information across the supply chain. The difficulty derives from the existence of information in both explicit and implicit forms (Yuva 2002). Explicit SCM information includes transactional data, procedures, policies, lessons learned, performance criteria, and revenue reports. Explicit supply management information can be created, captured, refined, stored, managed, and disseminated in paper-based or digital formats. Moving the explicit information across the supply chain is indispensable to support the flow of goods and funds. After receiving explicit information, an effective supply chain manager should have the ability to reorganize and transform information into useful knowledge.

Implicit SCM information, such as insights, intuition, corporate values, organizational culture, and SCM experience, are difficult to document. Implicit information often resides in the minds of information holders, such as supply chain managers. It is a challenge to capture and transfer the implicit information to business partners to improve supply chain effectiveness.

A learning organization needs to capture both explicit and implicit information and convert them into actionable knowledge (Walczak 2008). A learning supply chain needs to utilize implicit and explicit information in different ways to realize different business benefits. For example, the objective of the cash-to-cash cycle in supply chains is to increase the speed of cash flow. Therefore, explicit information about accounts receivable, inventory, and accounts payable is critical information to capture. An innovative supply chain has the goal of shortening the lead time in the development of new products and services in order to have the first-mover advantage. Sharing implicit information across business partners is also important but more difficult to achieve. Making efforts to build interorganizational trust is indispensable to the willingness of business partners to share implicit information.

## Funds Flow

When goods move or services are provided, business partners expect monetary compensation from their customers. Funds need to flow in order to support the movement of goods and services from their origins to their final delivery to the end user and vice versa. The Center for Strategic Supply Leadership conference asserts that capital flows are important inside a supply chain (Roberts 2002). The following points illustrate the importance of funds flow to sustain the operation of a supply chain:

- An internal physical supply chain typically contains more than 70% of organizational assets.
- An average of 55% of total revenue in a company is spent on purchased materials.
- Maintenance, repair, and operations (MRO) activities account for 7% of the total revenue in a company.
- Services account for 18% of the total revenue.
- A total of 80% of revenue dollars is spent on supply chain activities.

All of these costs are direct or indirect costs associated with the movement of products and information. As the supply chain extends further, hidden costs, including overstocks, stockouts, defective inventory, and market fluctuations, escalate the total cost of ownership (TCO). Consider Mattel's initiative in recalling toys due to quality control problems with its Chinese supplier. Mattel needed not only to spend money in recalling toys but most importantly to save its corporate image and reputation (Stalk 2008).

Most supply chain managers are aware that financial measures—including direct and indirect costs as well as sales revenues—cannot capture all information related to the business operation. A need exists to portray supply chain efficiency and effectiveness in a more comprehensive manner. The balanced scorecard offers managers and executives a more comprehensive view of organizational performance. This approach includes both tangible and intangible benefits and costs. Bolstorff and Rosenbaum (2007) offer the example of a food product group that utilized the balanced scorecard to measure the internal, external, and shareholder performance. Metrics to measure the internal performance of this company included supply chain cost and supply chain asset management efficiency. External performance measures consisted of monitoring the reliability, responsiveness, and flexibility of the supply chain. Profitability and effectiveness of return were used to measure shareholder values derived from the improvement of the supply chain.

## Relational Flow

Supply chain business partners need to both compete and cooperate with each other in order to streamline the flow of goods, capital, and information. For example, Wal-Mart mandated that its top suppliers adopt radio-frequency identification (RFID). In return, Wal-Mart shares real-time sales and inventory information with those suppliers. Improvement in delivery and product quality can increase customer purchases over time and result from both collaboration and competition among supply chain partners. Retailers can provide better customer service on-site, thereby reducing the number of returned goods. The savings in returned goods can then be shared between retailers and suppliers without having suppliers solely absorb the returned goods.

These examples show a balanced relationship among business partners, thereby providingbenefits to all in the supply chain.

These balanced relationships are preceded by foundational relationships formed by key supply chain participants. Foundational relationships require supplier–buyer relationships, unique buyer–supplier linkages, customer–supplier relationship management, and supplier rationalization (Roberts 2002). If the business goal is to accelerate the rate of product innovation, the marketing partner in the supply chain needs to share market information with the R&D partners in the chain. If the business goal is to increase cash flow, a policy to reward suppliers who utilize cash payments can be established. In contrast, the success of a green—environmentally friendly—supply chain depends on the collaboration of participants to cut waste and control inventory levels. These foundational relationships later lead to the establishment of balanced relationships among other business partners in the supply chain.

Foundational relationships generally fall into four basic relational styles: adversarial arm's length, nonadversarial arm's length, adversarial collaborative, and nonadversarial collaborative relationships. The arm's-length relationship has only a short-term focus. Suppliers or buyers can easily dictate the terms of each transaction that results in conflicts between chain partners, since the suppliers have a high degree of bargaining power. On the other hand, mutual benefits could be achieved if the bargaining power of supply chain partners were equal and each partner willing to consider the interests of the other partners without compromising their own individual interests. In contrast to the arm's length relationships, collaborative relationships have a long-term focus. Business partners in the adversarial collaborative relationship are less willing to share confidential information than those in the nonadversarial collaborative relationships, but a willingness to work toward a long-term relationship is still present. In the nonadversarial collaborative relationship, both sides share information with each other and work together toward the accomplishment of long-term goals.

The relational supply chain is also dynamic. When market conditions change, business partners respond by rethinking their relationships. A natural disaster can force partners to change their focus from product innovation to logistics efficiency. As such, the relationship between business partners is restructured to meet the urgent needs of providing goods to customers in disaster-affected areas. In general, the buyer is primarily concerned with purchasing capability, while the seller is focused on customer management. Many supply chain concepts have been proposed to address challenges associated with the relational supply chain. On the buyer side, the most often discussed topics include lean management, agile SCM, and transaction and economic costs (Cox et al. 2004). On the seller side, topics such as customer portfolio management, relationship portfolio mapping, and resource-based strategy are receiving much attention from both academia and practitioners.

## Examples of Supply Chains in Different Industries

Measuring supply chain performance varies with different industries. In the fashion industry, the product life cycle is short and the market demand is seasonal. Therefore, measures such as cash-to-cash cycle time, asset turns, and inventory days of supply are utilized. In contrast, PC manufacturers follow mass customization strategies and utilize measures

**TABLE 2.2**

Supply Chains for Different Industries

| SC Process or Industry | Supplying | Manufacturing | Distributing | Retailing | Consuming | Recycling |
|---|---|---|---|---|---|---|
| Pharmaceutical | Plastics, chemicals | Compounding, packaging | Wholesalers, regional distribution centers (RDCs) | CVS pharmacy, Walgreen's, Wal-Mart Pharma | Clinics, hospitals, physician offices | Recalls, disposal |
| Automobile | Steel sheets, aluminum ingots, polymer pellets | Original equipment manufacturers (OEMs) | RDCs, wholesalers | Car dealerships, auction markets, online sellers | Individual and corporate users | Recalls, salvage yards, reverse engineering |
| Furniture | Imported woods, unfinished parts, finishers | Cutting, assembly, and finishing | Furniture shippers, UPS | Thomasville, Broyhill, and others | Home owners, offices | Defects, damaged, worn out |
| Food | Corn or wheat seed, animal livestock | Flour mills, food processors | Regional DCs, general wholesalers | Grocery stores, restaurants, caterers | Individual customers | Expired (out of date), excess |
| Clothing and textiles | Cotton, wool, plastics | Spun yarn, greige goods, dyed cloth | Clothing intermediaries | Belk, Macy's, Gap | Individual consumers, military | Worn out, out of fashion |

such as short-order fulfillment cycle time and upside supply chain flexibility. Table 2.2 shows examples of how the outputs of supply chains vary with different industries.

Suppliers in different industries provide different raw materials to manufacturers. Cotton and wool are supplies needed to manufacture clothing materials. Seed, wheat, and livestock are inputs for food materials. Automakers use steel blanks, aluminum ingots, and polymer pellets. Imported woods and unfinished parts are needed to make furniture. Manufacturers then convert these materials into unfinished or finished goods, such as spun yarn, flour, or auto parts. Tiers of manufacturers need to be in place to help convert unfinished goods to finished goods. Some of these manufacturers locate domestically to be close to their markets; others are located overseas, usually to take advantage of lower-cost labor. Once goods are finished in the production process, regional and local distribution centers pick up, transport, and offload these goods to wholesalers or directly to the retailer for sale to the final customer. Customers consume these products and discard them whenever they are defective, expired, or reaching the end of their life cycle.

## Internal and External Customers

Manufacturing and services alike deliver goods and services to internal and external customers in order to receive monetary and/or intangible returns. Internal customers are functional departments or other divisions within a single company. All transformation activities are under the control of the company management. Ideally, a cooperative mode of operation exists among internal supply chain participants. An external customer is one outside the company. Suppliers have less control over external customers and rely on a collaborative relationship to achieve the best results.

Customers vary in their needs, likes, and product knowledge. After receiving outputs, internal and external customers play three important roles: recipients, evaluators, and participants. In some instances, the customer can be an input to the transformation process, for example, a student at a university; a participant, a shopper using a self-checkout terminal; or a designer, a customer who decides on a workout schedule at a fitness center. Customers first evaluate if the delivered goods and services meet their needs and likes. For example, the ordering of a broadband connection service enables a small business owner to conduct marketing research and sell toys on the Internet. The subscribed Internet services could be reliable or unreliable. The reliability of the Internet connection is critical to the daily operation of this small business. The owner is an evaluator in determining if he or she should continue subscribing to the broadband service from the same Internet service provider (ISP). The owner can also play the role of a participant by offering feedback to the ISP's customer service representatives to help improve their service offerings. Customer participation becomes a popular practice along with the growing acceptance of built-to-order and mass customization business models. Vacation packages, personal computers, and dinner entrees can all be customized to the needs of customers. Figure 2.2 displays the multiple roles of customers.

Customers possess varying levels of knowledge about the products and services they receive. A sophisticated customer is more demanding for product features than a novice customer. A novice is probably happy with a standardized sound system. On the other

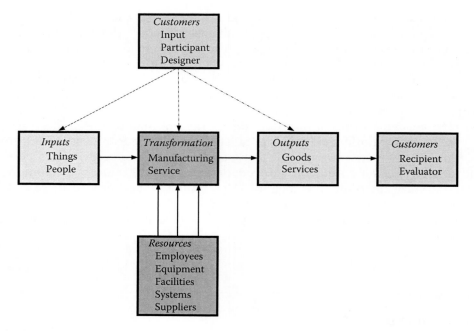

**FIGURE 2.2**
Role of customers in the ITO model.

hand, an expert in sound systems is interested in customizing these systems in order to have an optimal entertaining experience. Some members of a tourist group may have more knowledge about civil war history, while others in the group may possess little knowledge. When touring the Gettysburg battleground in Pennsylvania, the former group may not be as pleased with and interested in the basic information presentation as the latter group. These scenarios suggest that product and service offerings need to be constructed differently to best satisfy the varying needs of internal and external customers.

## Open Systems versus Closed Systems

A closed system is a system that is isolated from external influences. While such a condition can exist in carefully controlled scientific experiments, it rarely exists in the business environment. Closed systems do not interact with external influences, unlike open systems, which must learn to interact and adapt to outside influences. These external influences have different effects on the operations of a supply chain. For instance, passing a federal law to restrict the use of asbestos insulation could potentially put an insulation installer out of business overnight. External customers can dictate business operations and direction. In such an open system environment, the relationship between external customers and business is in both competitive and cooperative modes.

General systems theory has many variations. For instance, holism seeks to view the whole system, not just the parts. Goal seeking works at recognizing the importance of a predetermined goal for the system to achieve. Entropy theory states that a system has a tendency to be in a chaotic order over time. Systems have components and each component contains subcomponents. All systems differ from each other because they are created to fit their applications. Of all these system variations, the distinction between open systems and closed systems can help us better understand the complexity of supply chain systems. This classification rests upon the basis of resource availability (Schoderbek et al. 1990). A closed supply chain system includes the three essential elements of an ITO system. Outputs of an ITO system become inputs of another ITO system. All transformation activities are under the control of supply chain participants within the closed system. As such, no additional resources will enter the system to disrupt the normal operation of an ITO system. This assumption allows an ITO system to be closely examined with respect to internal interactions. In practice, a company never operates in a closed system because internal and external factors are constantly affecting operations. Figure 2.3 shows how, in an open system, competitors, economy, technology, government, environment, and society are examples of external forces influencing business operations.

In an open system, a supply chain participant's decision can affect decisions of other participants. For example, the length of a sales promotion at a retailer can affect customer sensitivity, the wholesale equilibrium price, the number of product orders, and product prices under fluctuating demand conditions (Kogan and Herbaorn 2008). All business partners in the supply chain should consider the difference between closed and open systems when making individual decisions.

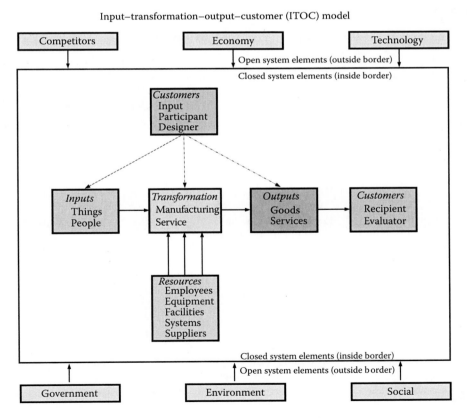

**FIGURE 2.3**
Closed versus open systems.

## Effect of External Influences on Supply Chains

The presence of external factors poses many uncertainties. The degree and velocity of changes to external factors can disrupt the normal operations of supply chains. For example, offshore outsourcing is a vehicle for an organization to stay profitable and competitive domestically and internationally. External uncertainties increase when trade regulations in India change, scope of agreements increases, suppliers go out of business, or quality problems cannot be fixed. All these changes can lead to higher operational costs, regulatory penalties, longer information and financial flows, and increased inventory. Supply chain participants need to address these increased operational risks. For instance, less clearly defined procedures and processes may be a problem with an offshore outsourcing relationship, perhaps more so than with traditional domestic contractual relationships. Supply chain managers must be aware that all of these external risks can potentially disrupt supply chains.

In addition to the uncertainties involved with offshore outsourcing, other external factors include government policies, legislation, the rise of emerging economies, competitive opportunities, public interests, advances in information systems, and innovative business models. Table 2.3 offers a representative list of primary external factors and their potential influence on supply chain systems.

## Obstacles and Enablers of Supply Chain Integration

A supply chain is most effective when its functions are closely integrated into a complete system. While this is a highly desirable objective, many obstacles to supply chain integration still exist at different stages of the value chain. We will describe some of these obstacles and enablers briefly in the following and in more detail in Chapter 12.

**TABLE 2.3**

Potential Impact of External Factors on Supply Chain Systems

| External Factors | Potential Disruptions to Supply Chain Systems |
|---|---|
| Offshore outsourcing | Short-term, flexible contractual relationships may create distrust between supply chain participants. |
| New government legislation | Plant operations in home or host countries may have to change to meet new government requirements. |
| Emerging economies | Reliance on multilayer suppliers causes difficulty in controlling quality (Mattel's lead-contained toys case). |
| Competitive opportunities | Establishing foreign subsidiaries in other countries creates communication problems. |
| Public interests | Increased public awareness of information privacy can restrict information flow between countries. |
| Information systems advances | Enterprise systems cause integration issues; data collection advances create information overload. |
| Proliferation of e-business | Open systems and cloud computing replace closed systems characterized by long-term relationships. |
| Innovative business models | Mass customization strategies increase coordination costs. |

## Obstacles

Developing new products and services by collaborating with suppliers yields many benefits, such as a reduction of costs and product development time as well as improvement of product design. The effective integration with suppliers in new product development (NPD) activities requires shared training, education, management, and trust among NPD project team members (Ragatz et al. 1997). However, most companies lack the skills, competencies, and shared mindset needed to implement the extended enterprise integration (Davis and Spekman 2004). These roadblocks exist in the upstream activity and need to be removed.

However, shortening the product life cycle brings great uncertainty to NPD. The life span of products, such as high-tech electronics, toys, and fashion apparel, is seasonal and may last only a few months. Changing customer preferences creates first-mover advantages and late-mover disadvantages. Goods not sold fast enough become excess inventory, creating pressures for retailers to downsize order and liquidate inventory.

Independent scheduling between supply chain participants can create excess inventory holding costs and high frequency of schedule revisions. Integrated production schedules minimize the negative effects of schedule revisions, such as increased assembly rate adjustment costs, safety stockholding costs, and coordination costs to communicate schedule changes to all supply chain members. Achieving integrated production schedules comes with a price; collaborative forecasting methods are necessary as well as an information sharing policy. Supply chain participants should agree on adopting the same forecasting method. However, this is not an easy task since many external variables, such as climate, population, and local customer demand, can influence forecasting accuracy. Optimal forecasting accuracy requires consideration of both internal and external factors.

Supply chain participants also need a standard information sharing policy. For example, high-tech products tend to have a shorter product life cycle than other products. Consequently, it is common to see that multiple generations of products coexist in the same market at the same time. Customers in different markets, for example, urban versus rural markets, may have different sensitivities to the different versions of products. Sharing this market-sensitive information across partners enables them to adjust production schedules accordingly. A major obstacle at the production stage occurs when supply chain participants are not convinced that standardizing information sharing is actually to their advantage.

Countries vary in their degrees of logistics friendliness. These national differences and the trend toward product proliferation can create operational and integration problems when a supply chain is expanded internationally. An international business must deal with product proliferation issues, such as mass customization and product configurations that meet regulatory requirements of different countries (Sahling 2006). Information sharing between customers and suppliers spreads throughout the geographic regions of the world and further poses technical, social, and communication challenges. Interorganizational systems (IOS), such as extranets and EDI, transcend organizational boundaries by automating the information-sharing process. However, managerial problems, such as trust and power structure, can limit the willingness of supply chain participants to share information with each other. Table 2.4 summarizes other obstacles to supply chain integration.

**TABLE 2.4**

Obstacles to Supply Chain Integration for Functional Areas

| Supply Chain Activities | Major Obstacles |
| --- | --- |
| NPD | Shorter product life span creates problems for late movers, including increased development costs and lost sales. |
| Operations: production schedule | Independent scheduling between supply chain participants can create excess inventory holding costs and a high frequency of schedule revisions. |
| Logistics: movement of goods and services | A global supply chain is faced with supply variability and inconsistent adherence to quality and human resource practices. |
| Marketing: CRM | CRM effectiveness requires corresponding changes to policies, procedures, performance measurement, and information visibility. |
| Reverse logistics: aftersales service | Increased litigation and regulation from government; increased pressure from special interest groups. |

## Enablers

While there are obstacles to supply chain integration into a system, there are also forces that enable systems integration. Web technology is a key enabler for integrated supply chains because of its ability to increase information visibility and capital flows, as well as to reduce coordination costs; therefore, a growing number of companies are assimilating web technologies into internal and external supply chains.

Customer relationship management (CRM) integrates customer information across the supply chain and tracks customer interactions throughout the organization, regardless of where, when, or how the interaction occurs. An effective CRM strategy can derive such benefits as 24/7/365 operation, individualized service, improved information sharing, planning integration, customer satisfaction, and enhanced product development. CRM systems have grown in complexity to help (1) automate the fundamental business processes for interacting with the customer, (2) analyze customer behavior and perception to provide business intelligence, and (3) provide effective and efficient communication with customers. However, the deployment of CRM systems does not guarantee the success of a CRM strategy. The long-lasting problems with interfunctional and interorganizational trust, power structure among supply chain participants, and the tendency of local optimization pose challenges to the implementation of a successful CRM strategy. It is indispensable to enact corresponding changes to organizational policies, procedures, performance measurement, and information-sharing attitude (Jessup and Valacich 2008).

## Performance Measurement

If supply chain activities are not measured, they cannot be managed. Integrated, extended supply chains can derive direct and measurable benefits—reduced production costs, increased number of customers served, lowered inventory levels, higher quality, faster response times, fewer stockouts, and higher on-time deliveries. Other benefits that are more difficult to measure and verify include business process transformation, organizational learning capacity, process consistency, innovation capacity, market share, customer satisfaction, customer retention, and overall competitive position. Some of these benefits are long-term, strategic benefits, whereas others are short-term, operational benefits. One recurring challenge for firms is the inability to articulate the indirect, intangible strategic costs and benefits. This point is a valid concern since supply chain professionals constantly face the challenge of demonstrating the long-term value of supply chain initiatives to their organizations. To create a defensible long-term competitive position through strong supply chain integration, all supply chain participants should effectively measure direct and indirect benefits and costs.

A global supply chain consists of multiple participants performing business activities for each other. The geographical locations of supply chain participants vested with different performance metrics create performance measurement problems. These problems have become more critical to the operation of supply chain systems in order to maintain its visibility, viability, and vitality.

Global sourcing is a prevalent supply chain practice. A benchmarking study showed the presence of an integrative performance measurement system is one of the best practices to support global sourcing (Trent and Monczka 2005). An effective performance measurement helps supply chain participants track the performance of global sourcing initiatives. This study found that performance measurement has largely improved for the studied companies in such areas as (1) capturing company-wide savings realized and expected to be realized from global sourcing, (2) assessing the impacts of global sourcing initiatives on corporate financial measures (e.g., return on assets, quick ratio, and return on inventory), (3) learning about returns on investment for global sourcing projects, and (4) measuring the buying location performance of suppliers.

TCO is a proposed performance measurement to manage the extended supply chain. TCO analysis considers the incremental effects of tangible and intangible costs on alternative solutions (Fawcett et al. 2007).

A successful outsourcing project can generate tangible benefits including reduced operating costs, reduced error rates in the production process, and lower cycle time. Other benefits include increased market share, new customers, and reduced customer dissatisfaction. Indicators commonly used in the field of project management are also useful in measuring the benefits of an outsourcing project. These indicators include ROI, internal rate of return (IRR), net present value (NPV), and the payback period.

One of the major benefits of outsourcing is the overhaul of a firm's cost structure, which is a function of production costs, transaction costs, and maintenance costs (Williamson 1996). Production cost is the cost to provide products and services and includes the cost of capital, labor, and materials. Because these three items are easy to measure, production costs can be readily measured as well.

The transaction cost is the cost of transferring goods and services between suppliers and customers. In outsourcing relationships, it is the cost of transferring goods and services from the provider to the outsourcer. This cost must be measured, because in an outsourcing relationship, organizations need to monitor, control, and manage activities related to the transaction. Two components of the transaction cost are the contractual cost and the operational cost. The contractual cost includes the costs associated with underwriting, policing, and enforcing a contract, whereas the operational cost includes those related to searching, holding inventory, and transmitting/processing information.

In addition, the assessment of transaction costs depends on three factors: asset specificity, uncertainty, and transaction frequency. Understandably, the lower the specificity of the asset, the greater the risks associated with the outsourcing project. The more uncertain the transaction, the more difficult it is to assess the tangible benefits of an outsourcing project.

Lastly, the higher the frequency of outsourcing transactions, the higher the transactional cost. A major goal of an outsourcer is not to "create monopoly, but to economize on transaction costs" (Williamson 1996, p. 3).

Maintenance costs are essentially the coordination costs that are expended in the delivery of goods and services. In information technology (IT) outsourcing projects, after-sales services represent a major component of maintenance costs; these services include system training, maintenance, support, and upgrade. Factors such as structured formal contracts, frequent and effective interorganizational communication, and supply chain partner trust can lower maintenance costs. In addition, adopting standardized exchange formats and structuring business environments via standards or e-business portals can lower maintenance costs. Maintenance costs would also be lower if there are few outsourcing providers.

Examining the tangible benefits by assessing the reduction in production, transaction, and maintenance costs is an economically rational approach of making decisions regarding outsourcing projects. Kalakota and Robinson (2003) took this approach to assess the tangible benefits of cost savings and found that typically in offshore outsourcing, there is a 70% reduction in costs due to the lower cost of labor. However, there is an additional 20% hike in transaction costs and another 20% addition in monitoring costs. Thus, the overall cost savings of offshore outsourcing is typically 30%. The reduction of production, transaction, and maintenance costs can result in the increase of business agility. In determining the TCO, it is important to consider transaction and monitoring costs.

Technology is changing rapidly; as a result, the life cycle of products and services is shrinking. Companies feel pressure to increase the speed at which they bring new products and services to market. This time-based competition is now the norm for conducting business in many industries and markets. Whenever a company can quickly bring innovative products and services to market, it possesses a competitive advantage over its competitors, as Apple has demonstrated in recent years. However, such an advantage cannot be sustained unless the company continues its innovative efforts. Agile organizations carry on innovative efforts and adapt quickly to changing market conditions. These organizations often discover that investing in production capacity is a countermeasure because the life span of a particular technology may outlive the life of the production facility. To an agile organization, there are intangible benefits of outsourcing. These benefits include divesting noncore activities and allowing the firm to focus on its core competencies as well as securing needed resources for contingencies. Strategically, these benefits include acquiring new skills, knowledge, and expertise for future growth and improving business flexibility via partnerships.

Given the benefits and inevitable costs of outsourcing, an unambiguous and consistent measurement approach is required for a productive outsourcing partnership. In devising such a system, it is important to incorporate tangible and intangible benefits/costs into the metrics. One such system has five interrelated metrics. Service levels achievement (SLA) consists of a set of contractually obligated high-level performance benchmarks for the outsourcing partnership. Quality and standards are the measurements related to the reduction in errors and the improvement in operational performance. Process deals with the manner in which partners reduce the cycle time in their manufacturing process. The number of new ideas presented to the executive board quarterly shows innovation. Cultural fit is assessed using a survey of employees from the partnering companies. These metrics are also consistent with the reward and incentive system by being the basis of bonus calculations (Linder 2004).

Often, organizations find it necessary to develop their measurement system gradually instead of specifying all performance metrics at the outset of the partnership relationship. The evolutionary approach of adopting measurement systems has the advantage of taking into account the invariable growth in the relationship over time, especially in today's fast-changing business environments. Total cost of outsourcing (TCO) analysis is one of many performance measurements that can help assess long-term strategic benefits of supply chain initiatives. If an integrative performance measurement is not in place, a firm will risk assigning minimal importance to supply chain activities. Many organizations treat proposed SCM initiatives as an *expense* rather than an *investment* in the business. This mindset is usually due to the lack of measuring the strategic success of SCM initiatives in support of corporate strategy. Thus, it is vitally important that organizations begin to validate SCM initiatives by developing integrative performance measurements.

## Allocation of Costs, Resources, and Benefits along the Supply Chain

Supply chain networks are comprised of activities and relationships that need to be managed in order to maximize customer value and achieve sustainable competitive advantage. For instance, NPD initiatives need to integrate upstream and downstream activities in order to yield the benefits of reduced costs and improved quality of purchased materials, reduced product development time, and improved access to technology. Supply chain participants operate both independently and collaboratively via different business processes, activities, and decisions. As such, all supply chain partners accrue some costs and resources, whereas others are allocated to individual participants. Therefore, it is important to ensure that costs, resources, and benefits are fairly allocated along the supply chain.

Strategic cost management aligns business values with cost drivers. All supply chain participants need to analyze that business processes, activities, and decisions actually create tangible and intangible costs and benefits. This cost driver analysis can help distinguish between value-added and non-value-added activities. Southwest Airlines aimed to become a low-cost, friendly, and reliable airliner. In order to support this goal, Southwest Airlines redesigned its supply chain so that value-added activities, such as utilizing hub locations and standardizing equipment, were implemented. Strategic cost management allows Southwest Airlines to customize the design of its supply chain based on the value–cost alignment. As such, other supply chain participants perform only the value-added activities needed to justify operational costs. Table 2.5 summarizes these changes (Parnell 2008).

Supply chain participants should assign key metrics to different supply chain activities when allocating profits, costs, and resources across the supply chain. When embarking on lean Six Sigma projects, participants at different stages of the supply chain receive benefits, absorb costs, and use resources at varying degrees. A fair allocation of benefits, costs, and resources across the supply chain can ensure that business values and cost drivers are aligned equitably. Thus, supply chain participants can focus on delivering value-added cost activities. Various metrics can be used to measure these projects, such as inventory investment, profit/loss, forecast accuracy, lead time, and unplanned orders. We will probe more deeply in this area in Chapter 16.

**TABLE 2.5**

Southwest Airlines' Allocation of Costs, Resources, and Benefits

| Strategic Operations and Benefits |
| --- |

Utilizes secondary airports whenever possible in major cities; flies to smaller airports in less populated markets

    Lowers cost by reducing the amount of time the aircraft is on the ground.

    Less air traffic and congestion improves on-time reliability.

    Better matches aircraft availability to customer demand.

    Improves efficiency of transferring passengers between regions.

    Serves more markets with fewer airplanes.

Standardizes aircraft (Boeing 737)

    Improves efficiency of pilots and crew

    Promotes faster maintenance and flight turnaround

    Better able to take advantage of interchangeability of parts and supplies

Implements performance-enhancing upgrades to the equipment

    Lower fuel costs, maintenance costs, noise level.

    Improved mileage range.

    Increased customer satisfaction.

    Leather passenger seats increase comfort as well as durability.

Does not use preassigned seating, different classes of service, or meal service

    Lowers cost

    peeds turnaround of the aircraft

Implements careful hiring procedures and cross-training of associates

    Employees support the mission of friendliness and customer service.

    Flexible to meet the needs of whatever job is required, within reason.

Promotes customer self-ticketing, early adoption of ticketless travel, and no use of travel agencies to reduce costs and promote efficiency

*Source:* Adapted from Parnell, J., *Strategic Management: Theory and Practice*, 3rd edn., Atomic Dog Publishing, Cincinnati, OH, 2008. With permission.

## Value Creation as the Ultimate Objective

The complexity of managing a business in today's global environment is increasing. Managing in this environment goes beyond customizing a known product for an individual customer. Instead, it involves addressing all of the internal and external issues that may exist, while at the same time, making the product at a competitive cost and lead time. Operating in a complex, chaotic environment tests the capabilities of any company.

Value creation is an essential element of business strategy and dictates what core competencies are required and how to compete using these competencies. A sustainable business needs to have a clear value proposition to attract customers willing to do business with you. Wal-Mart's "Everyday low price!" is a low-cost strategy. Kia Motors' "10 years, 100,000 miles warranty" emphasizes customer service. Dell's assemble to order is a mass customization strategy. Toyota's "Shorter time to make" is a product innovation strategy. Southwest Airlines' value proposition is based on low-cost, safety, reliability, and friendliness, according to the company's mission statement. These value propositions are supported with supply chains—a network of manufacturers and service providers that work together to convert and move goods, capital, and information from inputs, through transformation, to outputs.

In the process of using supply chain networks to deliver these value propositions, organizations face the complexity of forces that are external to the organization's control. These forces require businesses to develop global strategies that address product design (products must be designed to fit the market they serve), processes (to reflect the best blend of resource costs, manufacturing protocols, and local preferences), and locations (blending a myriad of decision criteria in the face of potentially dramatic change).

An organization should modify its value propositions to cope with these changes in external forces. Take low-cost country sourcing, for instance. The advantages deteriorate as labor costs and quality problems increase or employee human rights and safety are violated. Customers will demand change. Many companies use flexible supply chains to meet variable demands based on the new value proposition of nearshore strategies. Shorter product life cycles, increases in mergers and acquisitions, and emergence of product life cycle management (PLM) have pressured supply chain participants to increase the degree of openness of their supply chain networks (SanFillipo 2007). Amazon.com's initial value proposition of becoming a pure online bookstore was supported with a tight collaboration with other book warehouses. As an increased number of customers demanded a wider variety of products and timely delivery, Amazon.com repositioned its value proposition to be "The largest online retailer" by building huge warehouses and carrying larger inventories. Amazon.com's current mission statement is "To build a place where people can come to find and discover anything they might want to buy online" (Amazon.com 2009).

The complexity of managing external changes along with new value propositions is difficult and even introduces paradoxes that increase the confusion. According to a special report by Deloitte Touche Tohmatsu, some major paradoxes of complexity include the following:

- *The optimization paradox*: Despite the potentially huge economies from designing supply chains from a global view, most manufacturers optimize locally.
- *The customer collaboration paradox*: Despite the need to be much more responsive to customers, few manufacturers are collaborating closely with them.
- *The innovation paradox*: Product innovation is continuing to accelerate, yet few manufacturers are preparing their supply chains for faster new product introduction.
- *The flexibility paradox*: Flexibility is a key priority, but it is being sacrificed in the drive to cut unit cost.
- *The risk paradox*: Keeping supply chain quality high is critical, yet manufacturers' risk of supply chain failures keeps growing (Deliotte 2003).

Traditional decision making relied on linear thinking, an assumption that the future will be largely like the past. Today, however, many businesses are finding that using the past as a basis for planning the future may not be very useful; in fact, it may be downright misleading. But how can organizations apply nonlinear thinking approaches? It takes a completely new theory, a theory appropriately called complexity theory. William Frederick describes complexity theory as a variant of chaos theory that explains the organizational and evolutionary dynamics that occur as complex living systems interact with each other and their environments. He suggests that both the corporation and community are natural systems interacting and coevolving in response to environmental factors. The corporation is hypothesized as a complex adaptive system that must adapt itself to the environment in which it operates (Frederick 1998). In the future, dealing with complexity will be a necessary core competency for any successful organization.

Many companies may find complexity, or its variant, chaos theory, as too *far out* to be seriously considered. However, a few companies are already trying to decide how to incorporate this theory into their thinking. Managing the complexity of external changes poses risks and opportunities to business operations. A competitive company needs to continuously seek and test new value propositions in the increasingly complex and chaotic environment.

## Summary

In this chapter, we described the individual components that make up a supply chain by using the ITO model. The model delineates important considerations at each of those three processes in a supply chain system. Manufacturing and service industries are compared to understand how these three processes differ from each other. Each supply chain is comprised of four types of flows: goods, funds, information, and relational flows. These four flows are often not synchronized with each other; thus, business partners across the supply chain often fail to reach optimal decisions. Internal and external forces, such as market volatility, business policy, interorganizational trust, and differences in information systems, further weaken the ability of supply chain managers to make optimal decisions.

An effective supply chain manager recognizes these obstacles to supply chain system integration. Improving the bottom line is the universal goal to any business manager, including the supply chain manager. However, simply seeking an improved bottom line at the expense of other performance indicators can be a major mistake. This shortsighted view does not consider the indirect costs and benefits that account for a major share of supply chain performance. The balanced scorecard is an effective system to help measure these tangible and intangible performance metrics in the supply chain. While the individual components are important, an effective supply chain depends on the integration of those components into an integrated system. In addition, businesses are beginning to recognize their extended responsibilities to society and the environment, as part of the sustainability umbrella. In the following chapters, we will describe the varied elements of the supply chain in greater depth.

The basic theme of the chapter is that a supply chain is a system. We described how supply chains must be designed to operate in both closed and open system environments, if they are to be fully effective.

## Hot Topic

The hot topic for this chapter is about the effect that natural disasters can have on the supply chain system.

## Hot Topics in the Supply Chain: How a Natural Disaster Can Cripple a Supply Chain

In this chapter, you learned how supply chains are actually a series of ITO processes. An interruption at any ITO point in the supply chain can stop the movement of goods within that chain. For example, an earthquake in Asia can cause factories in other parts of the world to shut down.

For supply chain managers, a natural disaster can lead to a large-scale disruption to a company's supply resources. As a result, the company is then unable to meet the commitments it has made to its customers (Zsidisin et al. 2005). Examples of natural disasters that can disrupt the supply chain include major weather events, earthquakes, floods, and even volcanoes.

### BACKGROUND

One of the most visible examples in recent years of a natural disaster was the trilogy of events that occurred in Japan during March of 2011. An earthquake, tsunami, and nuclear meltdown at the Fukushima Nuclear Power Plant left the country reeling in human suffering with thousands dead and even more homeless. The impact on supply chains was massive, causing disruptions in the automobile industry in need of electrical components and certain shades of paint (Shappell 2012).

In regard to vulnerability, this particular automobile supply chain was extended and yet concentrated in one geographic area. In terms of the extended supply chain, components made in Japan were needed all over the world. However, in terms of concentration, many companies had single-sourced their suppliers in this region of Japan, which left them open to vulnerability since no backup was readily available for certain types of goods.

Supply chain disruptions can reduce revenue, decrease market share, and threaten production and distribution activities (Healings 2012). Ultimately, a natural disaster can inflict significant damage not only on the supply chain but also on a company's annual financial report. As a result of the earthquake and tsunami in Japan, some companies disclosed they had missed earnings targets in 2011 (Dempsey 2012).

Supply chain disruptions can also have a negative effect on stock prices. As Kumar (2009, p. 37) notes, "Supply chain disruptions have been found to impact negatively shareholder value by as much as 8–10% and are amplified in 'time sensitive' environments where early market introduction is critical to success." Furthermore, Hendricks and Singhal (2008) estimate that supply chain disruptions can cause a shareholder value decrease of as much as 10%.

### FACTORS THAT INTENSIFY THE EFFECT OF A NATURAL DISASTER

On a good day, managing supply chains is challenging, but throw in a natural disaster and even the best laid plans can be upset. Furthermore, some of the practices of current supply chain managers actually intensify the effect of a natural disaster. Consider the following:

#### CARRYING SMALLER INVENTORIES

One of the main practices in contemporary supply chains is the application of JIT/lean manufacturing principles. However, carrying lower inventories brings with it vulnerability. "Specifically, today's lean supply chains are becoming increasingly – 'fragile' – that is, less able to deal with shocks and disruptions that can have a significant, if not catastrophic, impact on the firm" (Zsidisin et al. 2005, p. 46). Less inventory does lead to cost savings when operations are normal. However, because of a lack of buffers, a natural disaster can easily stop the production line quickly, which brings its own set of costs.

### SINGLE SOURCING

Single sourcing is another common practice in contemporary supply chains. However, when the main vender is hit with a crisis, the companies it supplies will be impacted as well. An interruption in the delivery schedule can cause production to grind to a standstill at affected factories.

### DAILY DELIVERIES

Daily deliveries are used at many facilities that operate with lean operating systems. However, daily deliveries also imply there is little reserve stock in place. Even a minor weather event can cause a stoppage or delay in deliveries, which could lead to the shutdown of the production line in a lean system.

### AUTOMATION

Another practice many companies strive for is automation in the manufacturing environment. Automation helps buffer the impact of wage increases among workers, especially when new contracts are negotiated in union environments. Automation also leads to a reduction in workplace accidents and injuries. However, as machinery and technology become more complex, the potential for equipment breakdown increases. When labor cannot be substituted for capital, breakdowns become more serious as the production process will be highly dependent on the technology working correctly.

Hurricane Ike illustrated how a natural disaster impacted the highly automated oil refinery industry in the Gulf of Mexico. In 2008, the gulf supplied about 20% of the nation's oil producing capacity (Lee and Thurman 2008). When the storm hit in August 2008, refineries *shut in* operations to minimize physical damage to oil producing facilities. After the storm passed, production slowly resumed, but not fast enough to offset gas shortages in major cities such as Nashville, TN; Atlanta, GA; and Charlotte, NC.

Natural disasters are part of the supply chain manager's risk portfolio. With supply chains extending worldwide, it needs to be remembered that what goes on in one part of the world will affect operations in another part.

## QUESTIONS FOR RESEARCH AND DISCUSSION

1. How can companies that follow JIT/lean management practices hedge against interruptions in their supply chains?
2. How has a natural disaster affected the supply chain where you work? What provisions did your company make to address the disaster?

## REFERENCES

Dempsey, J., Consider your supply chain risk, *The Corporate Board*, 33(194), 21–25, 2012.
Hendricks, K.B. and Singhal, V.R., The effect of supply chain disruptions on shareholder value, *Total Quality Management*, 19(7–8), 777–791, 2008.
Kumar, S., Risk management in supply chains, *Advances in Management*, 2(11), 36–39, 2009.
Lee, G. and Thurman, E., Southeast retail deals with gas shortage, *WWD: Women's Wear Daily*, 196(68), 17, 2008.
Shappell, B., Falling sun, *Business Credit*, 114(3), 10–12, 2012.
Zsidisin, G., Ragatz, G., and Melnyk, S., The dark side of supply chain management, *Supply Chain Management Review*, 9, 46–52, 2005.

## Discussion Questions

1. Explain why the ITO model is called the DNA of the supply chain.
2. Describe the four types of flows inside a supply chain. Discuss how these different flows complement each other.
3. Using Table 2.2 as a guide, describe other types of supply chains not listed in the table.
4. In today's customer-driven market, an effective supply chain needs to be responsive to requests from internal and external customers. Who are these internal and external customers? How are their needs different from each other?
5. What types of external influences impact the supply chain? What types of influences have you seen at places you may have worked?
6. Supply chain integration is a desirable goal that is faced with both obstacles and enablers. An effective supply chain manager is capable of removing these obstacles and capitalizing on enablers to help integrate supply chains with its business partners. How can a supply chain manager address these obstacles and enablers of supply chain integration?
7. Business processes that cannot be measured cannot be managed. Managing supply chain performance requires the measurement of both tangible and intangible benefits and costs. How can supply chain performance be measured?
8. Describe strategies that companies use to create value for their customers through effective management of their supply chains.
9. How does outsourcing increase or decrease the risk of supply chain disruptions?
10. Discuss the potential use of chaos theory in SCM.

## References

Amazon.com, *Amazon mission statement*. http://www.samples-help.org.uk/mission-statements/amazon-mission-statement.htm, retrieved from January 20, 2009.

Bertalanffy, L.V., *General System Theory*, George Braziller, New York, 1969.

Blackstone, J.H., *APICS Dictionary*, 14th edn., APICS—The Association for Operations Management, Chicago, IL, 2013.

Bolstorff, P. and Rosenbaum R., *Supply Chain Excellence: A Handbook for Dramatic Improvement Using the SCOR Model*, 2nd edn., American Management Association, New York, 2007.

Cavinato, J., Supply chain logistics initiatives: Research implications, *International Journal of Physical Distribution & Logistics Management*, 35(3/4), 148, 2005.

Cho, V., Factors in the adoption of third-party B2B portals in the textile industry, *The Journal of Computer Information Systems*, 46(3), 18–31, 2006.

Cox, A., Lonsdale, C., Sanderson, J., and Watson, G., *Business Relationships for Competitive Advantage: Managing Alignment and Misalignment in Buyer and Supplier Transactions*, Palgrave Macmillan, New York, 2004.

Crandall, R.E. and Crandall, W.R., *Vanishing Boundaries, How Integrating Manufacturing and Services Creates Customer Value*, CRC Press, Boca Raton, FL, 2014.

Davis, E.W. and Spekman, R.E., *The Extended Enterprise: Gaining Competitive Advantage through Collaborative Supply Chains*, Prentice Hall, Upper Saddle River, NJ, 2004.

Deloitte, The challenge of complexity in global manufacturing, Deloitte Touche Tohmatsu, Special Report, 2003.

Fawcett, S.E., Ellram, L.M., and Ogden, J.A., *Supply Chain Management: From Vision to Implementation*, Prentice Hall, Upper Saddle River, NJ, 2007.

Frederick, W.C., Creatures, corporations, communities, chaos, complexity, *Business and Society*, 37(3), 358, 1998.

Healings, S., Planning for supply chain disruption, *Supply Chain Europe*, Jan–Feb, 28–29, 2012.

Jessup, L. and Valacich, J., *Information Systems Today: Managing in the Digital World*, 3rd edn., Prentice Hall, Upper Saddle River, NJ, 2008.

Kalakota, R. and Robinson, M., *Service Blueprint: Roadmap for Execution*, Addison-Wesley Professional, Boston, MA, 2003.

Kogan, K. and Herbon, A., A supply chain under limited-time promotion: The effect of customer sensitivity, *European Journal of Operational Research*, 188(1), 273, 2008.

Linder, J.C., *Outsourcing for Radical Change: A Bold Approach to Enterprise Transformation*, American Management Association, New York, 2004.

Parnell, J., *Strategic Management: Theory and Practice*, 3rd edn., Atomic Dog Publishing, Cincinnati, OH, 2008.

Ragatz, G.L., Handfield, R.B., and Scannell, T.V., Success factors for integrating suppliers into new product development, *The Journal of Product Innovation Management*, 14(3), 190, 1997.

Roberts, J.S., Value X 4 equals excellence, *Inside Supply Management*, 13(8), 38, 2002.

Sahling, L., Scourge of the lean supply chain, *Journal of Commerce*, 1, 1–2, October 9, 2006.

SanFilippo, E., More ways to collaborate as PLM market consolidates, *Manufacturing Business Technology*, 25(7), 46, 2007.

Schoderbek, P.P., Schoderbek, C.G., and Kefalas, A.G., *Management Systems—Conceptual Considerations*, Richard D. Irwin Inc., Homewood, Chicago, IL, 1990.

Stalk, G., *5 Future Strategies You Need Right Now*, Harvard Business School Press, Boston, MA, 2008.

Trent, R.J. and Monczka, R.M., Achieving excellence in global sourcing, *MIT Sloan Management Review*, 47(1), 24, 2005.

Walczak, S., Knowledge management and organizational learning: An international research perspective, *The Learning Organization*, 15(6), 486, 2008.

Walker, W.T., *Supply Chain Architecture: A Blueprint for Networking the Flow of Material, Information, and Cash*, CRC Press, Boca Raton, FL, 2005.

Williamson, O., *The Mechanism of Governance*, Oxford University Press, New York, 1996, p. 3.

Yuva, J., Knowledge management: The supply chain nerve center, *Inside Supply Management*, 13(7), 34, 2002.

# Section II

# Demand Perspective

# 3

## Determining Customer Needs

### Learning Outcomes

After reading this chapter, you should be able to

- Explain why it is important to group customers into homogeneous segments
- Identify the needs of the ultimate customer
- Identify factors that determine customers' needs
- Describe the methods to forecast demand
- Describe the six attributes of a well-designed product
- Describe alternative approaches to product design
- Determine how many supply chains a company needs
- Explain why serving internal customers is important

### Company Profile: McDonald's

To be successful, a company must find out what the customer wants and then develop the supply chains to meet those wants. Sometimes, this requires that the company make major changes in their product lines to meet ever-changing customer preferences. In addition, the company must forecast the level of demand to be able to provide enough capacity to meet the demand. McDonald's is such a company. Over the years, the company has made major changes in their products and methods of servicing customers. As you read this company profile and the chapter in the book, think about some of these questions:

- How did McDonald's find out what customers wanted?
- Where did the ideas for some of their most successful products come from?
- How important was the introduction of the drive-through to customer service?
- What additional changes should they make?

What additional questions do you think relate to designing supply chains to meet customer needs?

## Company Profile: McDonald's

McDonald's first appeared in the Gartner Top 25 supply chain listing in 2010, when they were ranked 11th. They moved to 8th in 2011 and 2012 and to 2nd in 2013. McDonald's has a global supply chain that moves product to over 34,000 restaurants in 118 countries. The company has grown steadily since its 1940 beginning in San Bernardino, California (McDonald's 2013).

To achieve its position as the leader in the fast-food industry, McDonald's has had to adapt to changing consumer tastes. They have morphed from a company with a limited menu of hamburgers and drinks to one with multiplatform offerings designed to appeal to customers of all ages and tastes. They have done this by recognizing that the customer plays a large role in determining how a fast-food restaurant must operate to be successful.

Their mission statement leads with this sentence: "McDonald's brand mission is to be our customers' favorite place and way to eat and drink." They expand on this in their value statements, the first of which is "We place the customer experience at the core of all we do. Our customers are the reason for our existence. We demonstrate our appreciation by providing them with high quality food and superior service in a clean, welcoming environment, at a great value. Our goal is quality, service, cleanliness and value (QSC&V) for each and every customer, each and every time" (McDonald's 2013).

Has the company demonstrated their commitment to their customers with their actions? What have they done to keep their customers coming back? In this profile, we outline a number of changes that resulted from the company's recognition they had to change to keep pace with transitions in the business environment, especially customer preferences.

### BEGINNINGS

*Starting point*: McDonald's was started in 1940 by the McDonald brothers, Maurice (Mac) and Richard (Dick). It opened as *McDonald's Bar-B-Q* as a carhop drive-in with 25 barbecued items on the menu. The brothers soon recognized most of their sales came from selling hamburgers so they closed the drive-in and, in 1948, moved to a streamlined self-serve restaurant with a limited menu of hamburgers, cheeseburgers, French fries, soft drinks, shakes, and apple pie. There was not even a place to sit down to eat. The name was changed to simply *McDonald's*. In 1954, Ray Kroc, a milk shake equipment salesman, visited the restaurant in San Bernardino, California, and was convinced the concept could be expanded and arranged with the brothers to let him set up franchises throughout the country. In 1955, Kroc's first McDonald's opened in Des Plaines, Illinois (near Chicago). Kroc emphasized the values later captured as quality, service, cleanliness, and value (QSC&V) in the company's value statement (Wikipedia 2013).

The 1960s and 1970s were periods of growth for the company. Ray Kroc completed the acquisition of the company in 1961 from the McDonald brothers. He then began to introduce changes in menu items, store layouts, and advertising:

- 1962: Opened the first restaurant with seating in Denver, Colorado.
- 1963: The menu has expanded from hamburgers (beef) to fish (Filet-O-Fish) in Cincinnati, Ohio, in a neighborhood heavily populated with Catholics who practiced abstinence (the avoidance of meat) on Fridays.
- 1967: Opened its first foreign restaurant in British Columbia, Canada.

- 1968: Introduced the Big Mac, developed by Jim Delligatti, a franchisee in Pittsburgh. (Note: several of the new *signature* sandwiches have been developed by franchisees who recognized their appeal to customers.)
- 1972: Started serving breakfast with the introduction of the Egg McMuffin and later added a full line of breakfast items.
- 1975: Opened its first drive-through window in Sierra Vista, Arizona, with a goal to provide service in 50 s or less.

During this period, the advertising was targeted toward children and families. The belief was that if the children want to come, the parents would have to bring them. The company also recognized their social commitment by opening the first Ronald McDonald house in 1974 in Philadelphia (Wikipedia 2013).

## GROWTH AND TURMOIL

The 1980s was a decade of intense competition, principally with Burger King and Wendy's. Despite the competition, McDonald's continued to grow, both domestically and globally. They also expanded from the suburbs to urban areas and continued to diversify their menus:

- 1980: Introduced the McChicken sandwich (not successful) and McNuggets (so intensely successful the company had to develop a completely new supply chain to meet the demand).
- 1985: Introduced ready-to-eat salads to lure more health-conscious consumers.

During the 1980s, the company successfully expanded its markets in Japan, Canada, Germany, Great Britain, Australia, and France.

The 1990s was an unsettled period in the company's history. Its successes came largely from outside the United States, where they showed increasing flexibility with respect to local food preferences and customs. They opened the first kosher McDonald's in Israel, developed a chain in Arab countries that used *halal* menus that complied with Islamic laws for food preparation, entered India with a Big Mac made with lamb called the Maharaja Mac, and opened the first McSki-Thru in Lindvallen, Sweden. By 1997, about 50% of sales came from foreign countries (Wikipedia 2013).

The 1990s also saw some blunders. On the menu side, such items as fried chicken, pasta, fajitas, and pizza failed. Consumers appeared to be confused by the uncertainty in menu items and especially the advertising and promotions. The Leo Burnett ad agency was terminated as a result. Some of the sales promotions were successful, such as the Teenie Beanie Baby promotion and an alliance with Disney/Pixar in 1998. This, along with the addition of DDB Needham, a new ad agency, enabled the company to keep moving forward. In the late 1990s, they also announced a "Made for You" campaign to introduce special-order sandwiches, as a counter to competitors such as Wendy's who made sandwiches to order. In the late 1990s and early 2000s, the company made a number of acquisitions to diversify their offerings:

- 1998: Bought a minority interest in Chipotle Mexican Grill
- 1999: Bought Aroma Café, a UK chain of 23 upscale coffee and sandwich shops
- 1999: Added Donatos Pizza, a Midwestern chain of 143 pizzerias based in Columbus, Ohio

- 2000: Acquired Boston Market, a bankrupt company with 850 outlets
- 2001: Bought a one-third interest in Pret A Manager, an upscale urban-based chain specializing in ready-to-eat sandwiches made on the premises (Wikipedia 2013)

## RECENT CHANGES

By the 2000s, the company needed a change. The "Made for You" system failed. Although it improved the quality of the food, it also increased service times and proved labor intensive. In addition, McDonald's had to battle its image as a purveyor of fatty, unhealthful food. Sales declined and there was even a loss in the fourth quarter of 2002. The actions that turned the company around included the following:

- 2002: Began to divest the companies acquired in the late 1990s to concentrate on their core business.
- 2003: Started a global marketing campaign to promote a new healthier and higher-quality image. It was called "I'm lovin' it" and began simultaneously in more than 100 countries around the world.
- 2003: Introduced premium salads, the McGriddles and chicken selects.
- 2004: Dropped supersize options.
- 2005: Opened a Wi-Fi service in selected restaurants.
- 2006: Began a *forever young* branding by redesigning restaurants to appeal to young adults. The new model includes three area zones—a linger zone to accommodate those who might dawdle and socialize while sitting comfortably on armchairs or sofas using free Wi-Fi, a zone with counters for patrons in a hurry to grab and go, and a third zone for families or groups with movable seating arrangements. The harsh colors and hard plastics were replaced with custom earth tones and flexible padded fabrics. Considering the first McDonald's had no seating, the current configuration is designed to encourage customers to enjoy the newer and hopefully inviting environment.
- 2007: Introduced snack wraps.
- 2008: Introduced the Chicken Biscuit and burritos.
- 2009: Introduced McCafe coffees, including lattes, cappuccinos, and mochas.
- 2010: Introduced real fruit smoothies.
- 2012: Started posting calorie counts for menu items.

During all of this period of transitions, the company continued to grow and change. Change is difficult in a smaller chain of company-owned stores; it is extremely taxing in a company the size of McDonald's with 80% of the stores owned by franchisees, who must agree with the changes (Wikipedia 2013).

One of the notable achievements by McDonald's is their position in Russia. They opened the first restaurant in 1990 with the aim of building an independent supply chain within the country. They now have over 200 company-owned restaurants that cover 15% of the landmass but 80% of the Russian people and economy. As in other countries, they have added special menu items such as forest berry and cherry mushroom pies and the Beef-a-la-Russ sandwich. They developed their entire supply chain locally, even growing the potatoes to get the higher quality they wanted. They expect to have sources for 100% of their needs soon to minimize the distance products must move and reduce the delays from moving across countries with different duties, taxes, customs, and the like (Tulip 2010).

During much of the first decade of the twenty-first century, the CEO was Jim Skinner, a long-time employee. Skinner became CEO in 2004 when Jim Cantalupo, a long-serving CEO, died and was succeeded by Charlie Bell, who was diagnosed with colorectal cancer and resigned within 6 months. Skinner was relatively unknown to the outside world but was considered an operations whiz internal to the company. Under Skinner, McDonald's reached new levels of dominance in the fast-food industry with sales in 2012 of $77.4 billion, followed well back by KFC with sales of $17.9 billion, Subway at $15.2 billion, Burger King at $14.7 billion, and Starbuck's at $14.2 billion. McDonald's sales were greater than their next four competitors, combined. Skinner believed that hospitality and friendly relationships are important, but he never lost sight of the fact that McDonald's customers want fast response. In killing a product of deli sandwiches because they could not be prepared in 55–60 s, he said: "McDonald's customers at the drive-through don't want to be chatted with, and they don't want to wait two minutes for a turkey sandwich" (Kowitt 2011, p. 77).

McDonald's will change menus, store design, color schemes, advertising slogans, and whatever else to provide what their customers want.

## REFERENCES

Kowitt, B., Why McDonald's wins in any economy, *Fortune*, 164(4), 47, September 5, 2011.
McDonald's website, 2013.
Picture of Russian Olympic restaurant. http://www.telegraph.co.uk/finance/newsbysector/retailandconsumer/8650272/McDonalds-Olympic-restaurant-will-be-biggest-in-the-world.html.
Tulip, Feeding the people, an interview with Khamzat Khasbulatov, CEO of McDonald's Russia, January/February 2010. www.scemagazine.com.
Wikipedia, History of McDonald's, 2013. http://en.wikipedia.org/wiki/History_of_McDonald%27s, accessed June 20, 2013.

## Introduction

Supply chains not only need to lower operating costs but also need to deliver value-added products and services to meet the needs of internal and external consumers. When customers are satisfied with products and services received, they respond in a positive way. Otherwise, the flows of goods, information, and funds will not be smooth, thereby lowering the financial and sustainability benefits among the supply chain participants.

Consumer behavior is difficult to predict, because uncontrollable factors such as changing market conditions and evolving technology trends also affect business. Although personalized customer service is desirable for any business, it is more practical for most businesses to group customers into segments and develop products to meet the needs of each homogeneous customer segment. Consumers within the same segment will have similar needs. Only after a business develops an in-depth understanding of each customer segment can managers take further steps to explore the possibility of delivering personalized services to each individual customer.

An internal customer is the recipient (person or department) of another person's or department's output (good, service, or information) within an organization (Blackstone 2013). An external customer is a person or organization that receives a good, a service, or information but is not part of the organization supplying it. The ultimate consumer is the individual who buys the good or service for one's own use, that is, a homemaker who buys a vacuum cleaner for use at home or a company that buys an industrial vacuum cleaner for its use in the office or plant.

Internal and external customers are involved at different stages of the supply chain and have differing needs and wants. Classifying these needs based on product/service features, economic needs, location needs, and service needs enhances an understanding of their relative importance. There are direct and indirect methods for identifying customer's needs. Learning how to use these methods is the first step toward achieving success in managing demand-driven supply chains.

Traditional metrics that were used to measure supply chain performance in meeting customer's needs have limited usefulness in today's affluent and customer-driven marketplace. Various contemporary measures are emerging to cope with this challenge; therefore, it is important to learn about these measures, along with their strengths and limitations. This chapter expands on the introduction presented in Chapter 2 concerning customer's needs, by focusing on the demand side of the supply chain. The specific topics to be discussed include the following:

- Group customers into homogeneous segments.
- Determine the needs of the ultimate consumer.
- Accurately determine customer needs.
- Develop reliable demand forecasts.
- Determine the attributes of a well-designed product.
- Consider alternative approaches to product design.
- Determine the number of supply chains needed within a company.
- Develop meaningful performance measurement systems.
- Respond to the needs of internal customers.

## Group Customers into Homogeneous Segments

This overall strategy allows a firm to group customers into homogeneous segments where customers within the same segment share common attributes and respond to similar market stimulus. In contrast, heterogeneous market segments behave differently and need a different marketing strategy. Customers are concerned with the utility value and price of their products/services as well as the convenience of purchasing them. Supply chain partners need to collaborate with each other to effectively address these concerns. Multichannel distribution, that is, selling a product through various outlets such as retail stores, the Internet, or grocery stores, is an effective means to serve multiple customer segments. This arrangement creates synergies and exploits the economies of scale. The presence of a multichannel distribution system is one of the many effective means to support segmentation efforts in marketing strategy.

All business activities across the supply chain, including sales, operations, delivery, and aftersales service, need to be structured differently and tightly intertwined for different market segments. As such, a company may have to construct multiple supply chains in order to meet the needs of its different customer segments. In order to support its market segmentation strategy, Gap Inc. constructed three supply chains in order to deliver three distinct product brands—Old Navy for price-sensitive customers, GAP for trend-conscious customers, and Banana Republic for high-quality, luxury apparels (Buckley 2004).

Companies like GAP Inc. grouped customers into homogeneous segments so value offerings could become a differentiator to increase the overall business and business in all segments. At the same time, they wanted to minimize the negative effects of marketing cannibalization. Companies of all industries have used the following value offerings (Kotler 1999):

*Convenience*: The 7-Eleven convenience chain in Asia has become a 24/7/365 store to offer shoppers convenient hours for buying standard food items and for paying utility bills and traffic tickets.

*Customization*: Levi's sales clerks can measure a person's exact dimensions and store this information into a database. Customers can customize Levi's products based on the information on the Internet.

*Faster service*: Wells Fargo Bank offers its commitment to provide service in 5 min or they will credit a customer's account $5 as a penalty for not paying immediate attention to the customer's needs.

*More and/or better service*: Saks, an upscale department store chain, builds customer loyalty by providing personal shopping experiences, such as showing clothes to customers who make appointments, home delivery services, and a birthday gift for VIP customers.

The number of supply chains required by a company varies according to the number of customer segments identified. An effective market segmentation strategy can differentiate one market segment of service offerings from another. When the degree of homogeneity within each market segment is improved, a company can increase its marketing effectiveness.

As data collection and analysis improves, the size of customer segments decreases, to the point where some authorities believe companies will have to eventually deal with individual consumers as the target market, or *markets of one*. Needless to say, this requires the capability not only to determine customer needs at this level but also to deliver the products and services to meet these individual demands.

## Determine Needs of the Ultimate Consumer

What are the needs of the ultimate consumers? How are their needs different from those of internal customers within the same company? For simplicity, we will use the term *needs* to represent *needs and wants*.

Ultimate consumers are the final recipient of goods and services as well as the originator of positive revenue and customer information flow. In the context of business-to-customer (B2C) trading, individual customers purchasing goods at a retailer, such as Wal-Mart or Target, are ultimate consumers. In the context of business-to-business (B2B) trading, ultimate customers are business partners sourcing goods from another business partner. For instance, Wal-Mart and Sam's Club are external customers of Procter & Gamble and Panasonic. In turn, Procter & Gamble and Panasonic are external customers of electronic component manufacturers.

The goal of supply chain management (SCM) is to ensure that the ultimate customers receive the right products and services at the right place, time, and price. Advances in various technologies have resulted in shorter product life cycles, increased product proliferation, sophisticated customer requirements, and, not surprisingly, increased demand for newer products in a shorter time. The importance of these new developments has put an emphasis on the use of supply chains to deliver quality products and services in a shorter lead time to the ultimate customers (Vokurka 1998). A satisfied consumer can create customer loyalty for product and service offerings, thereby building brand loyalty.

## Product Needs

The primary product attributes desired by most consumers include low, or at least competitive, prices, high quality, fast response times, and product lines with new features that roll out on a regular basis. At one time, suppliers viewed some of these attributes as trade-offs. For example, low costs did not always correlate with high quality. Because many products were made to stock in stores, shorter response times were not considered a high priority. Today, with the emphasis on maintaining low levels of inventory, fast response times are becoming essential.

## Service Needs

Service needs are also changing. Products can no longer stand alone; they must be supported with a service package that includes the following:

- *Easy availability of product information*: After making a purchase, customers want more information that enables them to optimally operate and maintain their purchases.
- *An attractive buying environment (either in store or online)*: Customers want attractive as well as functional places to shop.
- *Competent and helpful salespersons*: Customers are better informed today and while they may need less help in making their selection, they want knowledgeable sales personnel available when they do need help.
- *Lifelong product support*: Customers are more aware of total product life cycle costs. They are inclined to be more demanding of sellers to provide support for the products, even when those products are at the later stages of their life cycles.

Most businesses are classified as either manufacturing or service; however, almost all businesses today sell a combination of goods and services.

## Marketing Mix

A company constantly faces the challenge of increased market uncertainties where customer needs are changing. This inability to address market uncertainties can lead to uncertain estimates of market attractiveness, prototype failures, and inappropriate timing of product launches (Rice et al. 2008). In order to cope with this challenge, a company must first understand its customer's typical needs and then determine how to best meet those needs by improving SCM functions of sales, distribution, revenue models, and customer interaction.

Typical customer needs correspond with the 4Ps of the traditional marketing mix— *product, price, placement, and promotion*. Products are tangible goods, while services are intangible. Sometimes, it is hard to separate products from services because customers perceive them as integrated product/service bundles. Google is an example of such a bundle because of its ability to benefit both the tangible needs of customers and the service needs of customers (Gawer and Cusumano 2008). FedEx and UPS attract customers, not only because of their ability to deliver products to destinations but also because customers can receive real-time updates about the status of product delivery via the tracking service. Poor service may turn a customer to another source, even if they are satisfied with the original product. Therefore, when customers determine a product choice, the product/service bundle provides a more complete picture of customer needs with respect to the value offerings of the product.

In the customer's eyes as well as from the manufacturer's perspective, price is often used to reflect the relative quality of products or services. Price and quality are two economic costs to consumers. Most consumers have perceptions about the trade-off between these two economic costs, and their willingness to make the trade-off depends on their economic needs. Improved quality of products can often increase a customer's willingness to pay more for a product.

Placement addresses the delivery timing and geographical location of the product or service. Some degree of latitude is involved in product delivery. A restaurant requires frequent deliveries of perishable products such as dairy and produce items. However, for frozen and canned goods items, deliveries can be less frequent since these products can be stored for longer periods without perishing. In this example, the location of the companies that supply these items is less important to the restaurant manager as long as the supplier can make consistent deliveries. In a manufacturing setting, the suppliers may need to be located close to the plant making the product since more frequent deliveries are required. For example, just-in-time (JIT) is the concept of delivering goods and services to these manufacturers so that in-process inventory can be minimized, stockouts can be avoided, and return on investment of a business can be improved.

Promotion involves making the merits of the product or service known to the potential customer. In the retail sector, promotion occurs commonly through traditional advertising venues, such as newspapers, magazines, the Internet, television, radio, and mail. Industrial suppliers use more selective means of promotion, such as trade shows, company websites, and field sales representatives. With the increased interest by consumers in buying online products, company websites are becoming an essential part of the promotion efforts.

## Manufacturing and Service Supply Chains

Manufacturing supply chains differ from service supply chains in product and service characteristics. Table 3.1 lists the major differences between pure products and pure services. Customers buy tangible goods in order to use them for their satisfaction. These customers may not be interested in learning about all the back-end activities, including channel efficiencies, distribution, and quality management, needed to support the delivery of goods. In contrast to products, the process of delivering services is very important. Each process counts. Take a theme park tour, for instance. The process of obtaining tickets, asking for directions within the theme park, dining at the food services outlets, and asking questions of the customer service representatives is an important part of the total enjoyment experience. Therefore, the typical needs of customers vary greatly with the consumption of products or services.

**TABLE 3.1**

Product and Service Characteristics

| Pure Products | Pure Services |
| --- | --- |
| Tangible goods—they can be tasted, touched, seen, smelled, or listened to. | Intangible services—they can be *experienced* and enjoyed. |
| Inventory management—how much inventory should we keep on hand? | Capacity management—how large should our service facility be? |
| Product is made to order (MTO), assembled to order, or made to stock. | The service is provided in collaboration with the consumer. |
| Production and distribution are centralized. | Service locations decentralized closer to demand points. |
| Utilizes economies of scale—lower price through longer production runs. | Utilizes economies of scope; expanded services without increased overhead costs. |
| Transactional marketing. | Relationship marketing. |
| Control of channels. | Internal marketing. |
| Channel efficiency. | Channel agility. |

## Accurately Determine Customer Needs

Satisfying customer needs is a fundamental goal of doing business. An increased number of satisfied customers can result in increased sales, improved customer loyalty, lower attrition rates, cross- and up-selling opportunities, and lowered customer acquisition costs. The effectiveness of industrial and personal sales largely depends on the ability of a firm to determine accurately the needs of its ultimate and internal customers.

However, customer needs are objective, and wants are subjective. Many times, customers do not even know their needs until someone creates them. As Apple's CEO Steve Jobs stated, "You can't just ask customers what they want and then try to give that to them. By the time you get it built, they'll want something new" (Bilimoria 2007). Customer needs are dynamic and have to be discovered, identified, and satisfied.

### Approaches

Companies use direct and indirect approaches to better understand the needs of their customers. The most straightforward approach is to interact directly with customers. Surveys, focus groups, and product sampling are three effective methods in directly contacting customers and identifying their needs. All data collected from these direct approaches are primary data. In the absence of direct customer interaction, many creative avenues are readily available for companies to conduct market research. Affinity marketing studies can provide insights on customers' buying patterns and trends. Market segmentation studies can help the company understand customer needs in different market segments. Statistical tools such as analysis of variance (ANOVA) can support the analysis of market segmentation as it seeks to discover differences in tastes among the various segments. R&D personnel can brainstorm new product concepts and then test them out with small samples of customers. The indirect approach relies on tools, such as prototyping, mindmapping, and group support systems, to increase productivity in idea generation and customer wants identification. All these indirect market research approaches rely on secondary data, but can also effectively provide insights on customer needs. Consequently, primary and secondary data together are useful in determining customer needs.

A new method of looking at past history involves the *big data* movement. This involves collecting vast amounts of data from seemingly unrelated sources and then using sophisticated data analytics to sift out patterns that can zero in on customer preferences (Crandall 2013).

## Market Research

Market research is a systematic approach to collecting and analyzing data about target markets, competition, and potential customers. The goal of market research is to increase a company's intelligence in discovering customer needs. If a company can accurately determine what the customer wants, follow-up sales and marketing methods can be more cost effective and relationships with customers can be improved.

Market intelligence is an important feature in analytical CRM software that has high business impact (Doyle 2005). Market intelligence can improve sales performance and maximize revenues. An effective market intelligence approach can address the following questions:

1. Who are our customers?
2. Where are our customers located?
3. How do our customers want to be serviced?
4. What exactly do our customers want in their products and services?
5. How much demand exists for our products?
6. What are our customers' likes and dislikes of our products?
7. What product features are particularly attractive to our customers?

A company can undertake market research in-house or rely on a market research agency to provide the service. Market research utilizes different research techniques and methodologies to provide insights on customer needs:

- *Questionnaires*: A list of questions that are asked of potential consumers concerning their likes and dislikes about products and/or services. Questionnaires can be face to face, by mail, by telephone, and, increasingly, online.
- *Focus group*: A small group led by a facilitator that seeks to learn customer preferences.
- *In-depth interviews*: An extension of the questionnaire mentioned earlier, only more detailed and conducted in a face-to-face format.
- *Analyzing secondary market data*: That is, previously published materials.
- *Market segmentation analysis*: This method seeks to find commonalities in various groups or segments of consumers.
- *Statistical analysis*: Various statistical approaches can be used to identify both differences and commonalities in consumer groups.

When applying these methodologies to understand customer behavior, it is important to recognize that consumer behavior is susceptible to dynamic changes in the external business environment. Therefore, it is critical to scan those changes as well and consider them when conducting market research. Variables that affect consumer behavior include competitor offerings, changes in technology, fluctuations in the economy, government decisions, social trends, and changes in the business environment. Market intelligence also needs to include quantitative and qualitative demand forecasting methods. Information accuracy about customer needs and its subsequent demand can help a company effectively design products and services.

While many of the research methods described earlier are still valid, some practitioners are not completely satisfied with the results. Consequently, they are advocating the use of direct observation of consumers to see how they use the products and what problems they are having in *coping* with the product's inadequacies. This helps to identify product design fixes and additional product features needed.

## Open System Scanning

In Chapter 2, we described closed and open systems. In this section, we describe how these open system variables can affect customer needs.

### Competition

Most companies are experiencing active, sometimes severe, competition. It is not enough to just recognize that competitors are in the marketplace; to be successful, management must understand what the competition is doing and how to effectively compete against them. Management must study the competition to learn about their strategies, strengths, and weaknesses and project their likely courses of action. Information about competitors is available from a variety of sources, but it is up to the management to decide how to organize that information in a way that can help in its own strategic response.

New methods of serving customers abound. One of the most dynamic is in the use of the Internet by customers to find out more about the product they are considering for purchase, either in traditional retail stores or through a company's website. The ubiquity of the Internet has lowered the bargaining power of suppliers, but increased that of buyers by offering better information symmetry (Parnell 2008). Tens of thousands of organizations are exchanging millions of messages monthly. The Internet EDI solution to increase information transparency is exerting enormous pressure on both buyers and sellers to lower prices and increase product quality. The barriers to entry are lower. Efforts to establishing a web presence are minimal when using web hosting companies and current technologies. Substitute products or services have mushroomed. Amazon was originally a book wholesaler but has evolved into an entertainment wholesaler. The entrance of Amazon into toy and consumer electronics industries has created threats to many companies, such as ToysRUs, Circuit City, and Best Buy. The second largest electronic chain store, Circuit City, filed for bankruptcy in the year 2008 due to not only economic contraction but also the intense competition and shrinking profits at retailers. Recently, Best Buy has had to dramatically revise its position as a major retailer. Effective market research should examine closely these industry forces and assess how they affect the demand and supply of goods and services for customers.

### Technology

The innovative use of information technology can bring forth additional ways of servicing customers. Many companies are utilizing technologies, such as social networking tools, to understand group and individual behavior. About 80% of today's information is unstructured and in text formats, such as email and various word-processed documents.

This situation lends itself well to various types of data mining uses. The emergence of data analytics is an example of how new technologies are enhancing the efforts to determine more precisely customer needs.

Some of the industries that use data analytics include airlines for determining what price it charges on each flight; banks, for deciding how to best provide customer service across its multiple offerings; and retailers, in deciding how much and where to place their inventories for optimum effectiveness (Court 2012).

Another study reports applications in identifying fraud in real time; evaluating patients for health risks, changes in consumer sentiment, and need for service on a jet engine; and exploring network relationships in social media sites (Davenport et al. 2012).

Real-time information about product and inventory can be shared among supply chain partners. Advances in information technology enable companies to offer creative value propositions, such as make-to-order options, personalized customer services, comparison shopping venues, reverse auctions, competitive pricing, and user-friendly return policies.

Companies that know how to make effective use of these new technologies have first-mover advantages. However, the short life span of many of these technologies can often be duplicated by competitors. MySpace was the innovator of the online social networking community movement. Facebook was the follower and eventually surpassed the world traffic of MySpace and is currently dominating the market. Yahoo started the first information portal. Yet, Google invented the search engine based on a proprietary algorithm to deliver search results with a higher accuracy rate than Yahoo's search engine. As a result, Google now dominates the search market.

When conducting market research, a company needs to carefully assess the direct effect of technology innovation on customer needs. Failure to keep pace with technology can be disastrous to a business.

The innovation diffusion model asserts the adoption or acceptance of new technology progresses through five growth stages: (1) innovators, (2) early adopters, (3) early majority, (4) late majority, and (5) laggards (Rogers 1962). Many innovations do not succeed in completing these five processes and are discontinued at any one of those five stages. One study indicated that, for many discontinuous or disruptive innovations, a chasm or gap exists between the first two adopter groups (innovators and early adopters) and the early majority group (Moore 1991). The life span of a technology varies greatly and it is hard to predict how far a new technology will progress through these five stages.

### Economy

Economic conditions can have a direct effect on customer needs. One way to assess this impact is by looking at the price elasticity of demand (PED). Customers increase or decrease the quantity of their purchases in response to changes in the price of different goods and services. When customer sensitivity to price changes is high, the demand is said to be elastic. Customers will increase their purchases if the price of goods and service decreases. On the other hand, customers will reduce their purchases if the price of goods increases.

A number of factors can influence the degree of elasticity: (1) substitution, (2) income level, (3) necessity, and (4) duration. The elasticity is higher when

1. More substitute products or services are available for the consumer to pick from
2. The price of the product rises higher than what the consumer's income level can support

3. The good or service is no longer considered a necessity to the consumer when the price goes up
4. The price stays at higher levels and when this occurs, consumers often look for alternate products at lower prices

Whenever one of these four conditions is met, customers may react more aggressively to price change. As a result, total revenue for a company will fall. On the other hand, the total revenue will rise if the aforementioned conditions were not met. Since PED can directly affect the total revenues of products and services, a company needs to be aware of its dynamics.

In addition to the variables described, three additional variables can affect the needs of customers: government decisions, social trends, and changes in the business environment. Although somewhat indirect, the effect of these relationships should not be underestimated.

### Government

Governments regulate the business environment through institutional policies. As a response to the financial crisis in 2008, the U.S. government decided on a *bailout* plan designed to resolve the subprime market crisis and stimulate the U.S. economy. Using tax incentives to promote the use of hybrid cars or other alternative energy programs is another example to influence the needs of customers via government action.

Offshoring options have substantially reduced the cost of component parts that are eventually assembled into U.S. manufactured products such as automobiles. Government decisions can often increase the enthusiasm for offshoring ventures. China established special economic areas where foreign investors can waive taxes during their start-up years. The North American Free Trade Agreement (NAFTA) was created in 1994 and contributed to the rise of apparel and textile factories around the U.S.–Mexico border. Such factories, also known as maquiladoras, have been controversial since many U.S.-based jobs were lost because of them.

These types of government decisions can influence consumer demand in various ways. Offshoring and offshore outsourcing initiatives tend to lower product costs, which can increase consumption. However, when jobs are lost due to outsourcing, spending power is decreased for those groups that lose their jobs. This can be a particularly delicate situation for some industries, such as aircraft manufacturing, where contracts are often granted with political strings attached.

Government policy can dictate the direction of the national and global economy, which can then affect local economic activities and, ultimately, customer needs and spending patterns. Because of the effect of government policy on the needs of customers, companies need to carefully assess these causal relationships for both existing and potential government actions.

### Social Trends

Social trends emerge when individuals are doing similar things across a local or national culture. Many goods are susceptible to the influence of social trends. The proliferation of iPods was the direct result of the digital culture. iPods became a medium to share music with others. Sharing videos at YouTube and posting and sharing information about mutual friends via Facebook or Twitter are also a social trend. Double income families and the growing acceptance of late marriages are other social trends related to the family unit. Changing family structures are even affecting music use and consumption (Nuttall and Tinson 2007). We are moving into the personalized *Starbucks economy* from the one-choice *Ford economy* (Schulman 2007). Market research analysts have also discovered that more families are demanding a second master bedroom. This social trend has affected the purchase of furniture (Gunin 2008).

Sometimes, social trends can even appear to be in conflict with each other. For example, there is a growing interest in leading healthier lifestyles in the United States, and yet, fast-food restaurants continue to be favorite eating spots among Americans. Sport utility vehicles, notorious for their gas guzzling capabilities, are still popular in the United States, despite the fact that a green revolution is also a growing national social trend. Even though social trends may not always make rational sense, they need to be watched by businesses. All these social trends are important catalysts and can affect the behavior and attitude of people in consuming goods and services. How social trends affect the needs of customers is an important market research issue to address.

### Business Environment

New product development cannot be effectively executed without taking into consideration the impact of the business environment. Many types of risk are present in this environment and can potentially disrupt the normal operations of supply chains. Therefore, it is important to identify the major risks and learn how they can potentially affect the results of demand forecasting. Environmental risks range from price fluctuations of component materials and services, to the changing security requirements of global logistics.

The Internet has removed entry barriers to many industry sectors, resulting in intensified competition. Moreover, many new competitors have developed new products in order to enter a profitable business arena.

Sometimes, a market crisis in the business environment can reap devastating effects. The subprime market crisis exhausted the cash of investment firms and put some of them out of existence. Political and economic crises can disrupt the normal operations of supply chains. As a result, the demand for goods will sharply change.

Many domestic and international regulations are also increasing the cost of supply chain operations. Ignorance or underestimation of business environmental risks can lead a firm to extinction.

A sort of corporate radar is necessary to identify potential environmental risks that can affect customer demands. Such market intelligence needs to continuously scan at least eight basic environments and take actions to avoid these risks. Questions to be answered in each basic environment scanning are as follows (Albrecht 2000):

- *Customer environment*: What geographic factors, social values, and psychographic changes are occurring? How will these changes affect customer demands?
- *Economic environment*: What are the current and future dynamics of markets, capital sources, the state of the national economy, and the state of the international economy? How sensitive are the demands of our products to these factors?
- *Technological environment*: What technological breakthroughs and solutions are available to improve the capability of the enterprise and competitors? How will these technological innovations affect the way business is conducted in the future?

- *Social environment*: How do social values, trends, styles, preferences, and cultural patterns affect the receptivity of consumers to different kinds of products?

- *Political environment*: How do government decisions at the local, national, regional, and global levels affect the business environment? How do these decisions affect managing the local enterprise?

- *Legal environment*: How do legal considerations, such as antitrust, intellectual property, and trade protectionism constrain or relax the ways of doing business? Some countries are more litigious than others. How will legal regulations in different countries affect business operations differently?

- *Geophysical environment*: The availability and accessibility of resources vary greatly among countries. Geophysical resources include road and telecommunication infrastructure, abundance of national resources, and sources of skilled talents. How will geophysical resources affect the options of doing business?

A thorough scanning of these basic environments can help minimize operational risks.

## Develop Reliable Demand Forecasts

Developing forecasts for customer demand is an important SCM function.

Good forecasting requires forecasting the right things and then forecasting things right (Blanchard 2007). It is important to distinguish between demand and supply. A company must first forecast its expected demand; then, it must plan how to supply product to meet the demand.

The presence of direct and indirect variables can affect the accuracy of demand forecasting results. When demand is relatively stable, these variables can often be programmed into a decision-making model to help forecast customer demand. When demand has greater variation in it, these models become less reliable and must be supplemented with adaptive techniques. In addition to quantitative forecasting methods, a variety of qualitative forecasting methods are also available for use. In the methods described in the following sections, actual results are compared with the forecast and the differences shown as forecast errors. A good rule to follow is to use the simplest forecast method that results in accurately predicting actual demand, or small forecast errors. Examining large forecast errors sometimes leads to the recognition that trends and seasonal factors contribute to forecast errors. This is a signal to use more advanced forecasting methods to isolate the effects of trends and seasonal factors (Walker 2005).

### Quantitative Forecasting Methods

Quantitative forecasting methods fall into two broad categories: time series and explanatory, also known as causal methods. Time-series methods build upon historical data. The pattern in historical data becomes a baseline to forecast the pattern of future customer demand. Data collected in past years can help predict customer demand in the next year. On the other hand, explanatory methods collect data about all variables that may have potential effects on customer demand. For instance, brand awareness (a proxy for demand) can be dependent upon the consumer's perception of the product, the celebrity

**TABLE 3.2**

Summary of Quantitative Forecasting Methods

| | | Forecasts | | | | | Forecast Errors | | | |
|---|---|---|---|---|---|---|---|---|---|---|
| Week | Actual | 3 Month MA | WMA 3,2,1 | Expo Low Alpha | Expo Trend Adj. | Expo with Trend Adj. | 3 Month MA | WMA 3,2,1 | Expo Low Alpha | Expo with Trend Adj. |
| 1 | 25.0 | | | 25.0 | | 25.0 | | | | |
| 2 | 29.0 | | | 25.0 | | 25.0 | | | | |
| 3 | 35.0 | | | 25.4 | 0.1 | 25.5 | | | | |
| 4 | 47.0 | 29.7 | 31.3 | 26.4 | 0.3 | 26.7 | 17.3 | 15.7 | 20.6 | 20.3 |
| 5 | 30.0 | 37.0 | 40.0 | 28.4 | 0.7 | 29.5 | −7.0 | −10.0 | 1.6 | 0.5 |
| 6 | 27.0 | 37.3 | 36.5 | 28.6 | 0.7 | 30.3 | −10.3 | −9.5 | −1.6 | −3.3 |
| 7 | 22.0 | 34.7 | 31.3 | 28.4 | 0.7 | 30.7 | −12.7 | −9.3 | −6.4 | −8.7 |
| 8 | 30.0 | 26.3 | 25.0 | 27.8 | 0.5 | 30.4 | 3.7 | 5.0 | 2.2 | −0.4 |
| 9 | 35.0 | 26.3 | 26.8 | 28.0 | 0.5 | 30.9 | 8.7 | 8.2 | 7.0 | 4.1 |
| 10 | 40.0 | 29.0 | 31.2 | 28.7 | 0.6 | 31.9 | 11.0 | 8.8 | 11.3 | 8.1 |
| 11 | 41.0 | 35.0 | 36.7 | 29.8 | 0.8 | 33.6 | 6.0 | 4.3 | 11.2 | 7.4 |
| 12 | 30.0 | 38.7 | 39.7 | 30.9 | 0.9 | 35.3 | −8.7 | −9.7 | −0.9 | −5.3 |
| 13 | | 37.0 | 35.3 | 30.9 | 0.8 | 35.7 | | | | |
| Avg. | 32.6 | Running sum of forecast errors | | | | RSFE | 8 | 3.5 | 44.9 | 22.8 |
| | | Mean absolute deviation | | | | MAD | 9.5 | 8.9 | 7 | 6.5 |
| | | Tracking signal (RSFE/MAD) | | | | TS | 0.8 | 0.4 | 6.4 | 3.5 |

who promotes the product, consumer income, consumer age, and promotional events. Explanatory methods can help managers understand how these factors interrelate and influence brand awareness.

Several techniques and tools are available within each category of quantitative forecasting methods. Each technique has specific applications for different forecasting situations. Managers should understand the strengths and weaknesses of each technique and decide how to leverage them to the advantage of their company.

Table 3.2 provides an overview of these quantitative methods. All of these methods provide good forecasts if the past demand history is relatively stable with small variations. On the other hand, large fluctuations in past history, except for trend or seasonal causes, can result in erratic forecasts.

We will describe four time-series methods: simple moving average (SMA), weighted moving average (WMA), exponential smoothing, and simple regression. Table 3.2 summarizes the results from forecasts, using each of these methods. Figure 3.1 displays the patterns formed by each method.

## Simple Moving Average

The moving average is a time-series technique to smooth out short-term fluctuations. A 3-month moving average uses the past three periods of data, a 4-month moving average uses the past four periods of data, and so on. The longer the averaging period used, the smoother the forecast becomes. In addition, moving average forecasts always lag actual results; the longer the averaging period, the longer the lag. In addition to smoothing and lagging, the moving average also is limited in that it can provide only short-term forecasts.

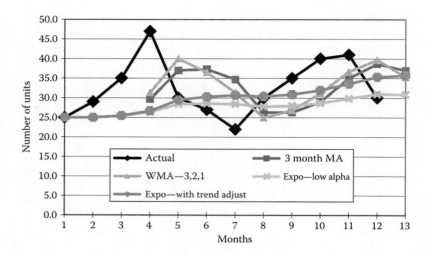

**FIGURE 3.1**
Graph of forecast methods.

In Table 3.2, a 3-month moving average is presented. The average for the first three periods is 29.7 (25 + 29 + 35) divided by 3; this becomes the forecast for the fourth period. The actual in the fourth period is 47.0. The difference between the forecast and the actual is 17.3 (Actual − Forecast), and this is shown under the Forecast Error section of the table.

By adding the forecast errors over a period of time—nine periods, in this example—it is possible to gauge the accuracy of the forecast. Three measures are commonly used: running sum of forecast errors (RSFE), mean absolute deviation (MAD), and tracking signal (TS). The RSFE is the arithmetic sum of the forecast errors. A positive result (12.7) indicates the actual is consistently higher than the forecasts; viewed another way, the forecast is consistently underestimating the actual. The MAD (10.5) is the average of the absolute forecast errors and indicates the amount of average variation, both high and low. The TS (1.2) is the ratio of the RSFE divided by the MAD. The TS can be either positive or negative; the smaller, the better.

### Weighted Moving Average

The WMA is a variation of the SMA. In Table 3.2, the WMA forecasts are shown alongside the 3-month moving average. The results are different as the WMA assigns more weight to the most recent periods. The assumption is that recent observations have a stronger influence on future changes than less recent observations. Table 3.2 shows the weights assigned to a data series decrease from a weight of 3 at period $n − 1$, to 2 for period $n − 2$, and to 1 for period $n − 3$. In Figure 3.1, the WMA is slightly more responsive to the actual results than the SMA.

The formula for calculating the WMA is as follows:

$$\text{Forecast}_n = \frac{\text{Actual}_{n-1}x_3 + \text{Actual}_{n-2}x_2 + \text{Actual}_{n-3}x_1}{6}$$

For Period 4, Forecast = (35 × 3 + 29 × 2 + 25 × 1)/6 = 31.3.

In the area of SCM, SMA and WMA methods are used to forecast customer demand for products over short period; however, they are not useful in forecasting for extended periods.

## Exponential Smoothing

Exponential smoothing is another time-series method of forecasting. While the moving average method requires using multiple periods of past data, exponential smoothing uses only the previous period's forecast and actual data. It adjusts the previous period forecast by multiplying an alpha ($\alpha$) factor of less than 1.0 by the difference between the previous actual and forecast amounts and by adding the result to the previous forecast. The formula is

$$\text{Forecast}_n = \text{Forecast}_{n-1} + (\text{Actual}_{n-1} - \text{Forecast}_{n-1})$$

For Period 3 (alpha = 0.1), Forecast = 25 + (29 − 25) × 0.1 = 25.4.

The formula shows that the higher the alpha factor, the greater the weight assigned to actual results. While exponential smoothing considers prior period results, it does so without the need to retain those results. Figure 3.1 shows that the exponential smoothing curve does not track the actual results very well; it lags badly. This is confirmed in the RSFE calculation of 44.9, which is much higher than for the other methods.

## Exponential Smoothing with Trend Adjustment

Actual results often include trends, either increasing or decreasing. SMAs and simple exponential smoothing do not capture those trends quickly. To compensate for this, a trend adjustment factor, called beta ($\beta$), can be added to the exponential smoothing formula to add an increment to the forecast that reflects the trend amount. The following are the formulas employed:

Forecast without trend adjustment

$$\text{Forecast}_n = \text{Forecast}_{n-1} + (\text{Actual}_{n-1} - \text{Forecast}_{n-1})\alpha$$

Trend adjustment (using a beta factor of 0.2)

$$\text{Trend adjustment}_n = (\text{Forecast}_n - \text{Forecast}_{n-1}) + \text{Trend adjustment}_{n-1}(1-\beta)$$

Example: Trend adjustment$_4$ = 0.2 × (26.4 − 25.4) + 0.1 × (0.8) = 0.28, or 0.3 rounded.

Forecast including trend adjustment

$$\text{Forecast with trend adjustment}_n = \text{Forecast}_n + \text{Trend adjustment}_n$$

Example: Forecast with trend adjustment$_4$ = 26.4 + 0.3 = 26.7.

We can see in Table 3.3 that a trend adjusted exponential smoothing forecast is better than exponential smoothing without the trend adjustment, but it still lags the actual. Using a higher alpha factor would help to track wide changes in the actual demand.

## Seasonal Factor Forecasting

Demand forecasting may also need to account for seasonal variables, and none of the methods described earlier detect seasonal patterns directly. To accomplish this, it requires several steps:

1. Display the actual results, as shown in the left side of Table 3.3. This example is for demand in each quarter of 3 years.

**TABLE 3.3**

Developing a Seasonal Forecast

| Quarter | Seasonalized Data Demand (Units) | | | | Seasonal Factors Ratio of Each Quarter to Average Quarter | | | |
|---|---|---|---|---|---|---|---|---|
| | Actual Year 1 | Actual Year 2 | Actual Year 3 | Forecast Year 4 | Year 1 | Year 2 | Year 3 | Average for Years 1–3 |
| 1 | 24 | 25 | 26 | 27 | 1 | 1 | 1 | 1 |
| 2 | 14 | 15 | 16 | 16 | 0.58 | 0.6 | 0.62 | 0.6 |
| 3 | 34 | 35 | 36 | 38 | 1.42 | 1.4 | 1.38 | 1.4 |
| 4 | 24 | 25 | 26 | 27 | 1 | 1 | 1 | 1 |
| Total for year | 96 | 100 | 104 | 108 | 4 | 4 | 4 | 4 |
| Average quarter | 24 | 25 | 26 | 27 | | | | |
| % increase | | 4 | 4 | 4 | | | | |

2. Calculate the relative weight of each quarter's demand, shown on the right side of Table 3.3. For example, the first and fourth quarters are equal to an average quarter, the second quarter is approximately 0.6 times an average quarter, and the third quarter is approximately 1.4 times the average quarter.

3. Project a forecast for the total year, using a regression line. In this example, the total for Year 4 is forecast to be 108 units. The data in the table show an increasing demand trend for each year.

4. Multiply the total year's forecast by the quarterly weights to get a seasonalized forecast for each quarter.

### Regression Analysis

The regression analysis is called a causal method of forecasting. The causal method attempts to predict demand by identifying other variables that can influence demand. For example, demand for a menu item in a restaurant can be influenced by the customers' income, the price of the menu item, their concern for eating healthy, and whether the meal being consumed is lunch or dinner. Sometimes, even the weather outside can influence demand. For example, if the temperature is very cold, demand for soups or chili may go up!

This forecast method makes use of a regression model to increase the forecast accuracy. For instance, if experts think the weather and consumer income could significantly affect the sales volume of apparels, a model can be constructed to predict the sales volume of apparels based on these two predefined variables. This method constructs a causal model based on a rigid judgmental process. A linear dependence relationship between a series of independent variables and a dependent variable is formulated as follows:

$$Y \text{ (dependent variable)} = \text{Constant} + R_1^2 X_1 + R_2^2 X_2 + R_3^2 X_3 + R_4^2 X_4 + \cdots$$

where

$R$ is a coefficient

$X$ is an independent variable

Depending on the topic researched, independent variables could vary from weather conditions to income levels. Assume such variables as economic growth, income levels, unemployment rates, and government policies can influence the sales volume of cars. All these predictive variables would be considered independent variables and would appear on the right of the formula to help forecast the sales volume of cars (the dependent variable).

Regression analysis has two major limitations. First, unlike exponential smoothing that considers all past observations, the regression model only incorporates a predefined set of data based on the judgment of experts. Second, independent variables that are highly related to each other should not all be included in the regression model. This condition, known as multicollinearity, can cause the predicted demand to be inaccurate.

For a more complete description of the methods described earlier, see any of a number of production and operations management textbooks (Heizer and Render 2011; Stevenson 2012).

## Qualitative Forecasting Methods

When historical data are not available or not applicable to the studied subjects, an effective quantitative forecasting model cannot be constructed. In these cases, qualitative forecasting methods can be utilized. A major approach central to qualitative forecasting methods is to rely on the insights of expert opinions. The collective intelligence and experience of experts can help derive insights on customer demand for unique products or services. Two major qualitative forecasting methods are the survey and the Delphi method.

### Survey Method

The survey process begins when a company identifies a reference population or a representative group in order to help understand the buying patterns of its potential customers. The responses collected from the survey group are called primary data. Primary data can be useful in understanding consumer characteristics, thoughts, behavior, and attitudes of customers toward products and services. Conducting a survey can be time consuming, yet, when done correctly, can yield valuable consumer information.

When interpreting survey data, several factors should be considered. First, the *self-reporting* process may reflect personal bias, errors, omissions, and untrue responses. Second, surveys are often discarded, particularly if they appear in the mail or e-mail formats. Thus, the lack of survey data and the total amount of surveys submitted should be considered as well. A low response rate may not be a reliable indicator of how a potential consumer market really feels about a particular product or service offering.

### Delphi Method

The Delphi method overcomes some of the major limitations of the survey method by adopting anonymity, controlled feedback, and statistical responses (Fowles 1978). The Delphi method relies on a panel of experts to engage in a series of systematic, interactive discussions. Experts are chosen based on the fit of their expertise to the subjects under study. If the study involves consumer electronics, experts in the area of electronics and related fields are solicited. Nutritionists would be invited if the researched products were vitamins or supplements. Questions are then posed to these experts and after rounds of discussions, ideas and opinions are narrowed down, until a consensus is reached. Since the responses are solicited in an anonymous manner, the experts are less likely to feel self-conscious about their ideas if they appear to go against the trend of other responses in the group. This advantage can contribute to an increase in the number of ideas generated. Ultimately, it is hoped that this technique will yield a better forecast prediction than if all of the ideas were generated in an open environment where each expert knows everyone else's input.

## Collaborative Forecasting

Forecasting methods that use past history are applicable only to products that have relatively stable demand patterns. As demand fluctuation increases, demand forecasts need to reflect these fluctuations. Often, this can best be done by using the most current information about ultimate demand at the downstream end of the supply chain. Companies look to collect as much feedback from their customers as they can to be better informed about the changing demand patterns. Even better, supply chain participants need to join efforts in improving collaborative planning, forecasting, and replenishment (CPFR) activities. CPFR effectiveness can help a company better understand consumer behavior and forecast product and consumer demand.

## Determine the Attributes of a Well-Designed Product

Six attributes are often found in products with a sound design: (1) functionality, (2) validity, (3) manufacturability, (4) reliability, (5) serviceability, and (6) recyclability. A product with good functionality works to satisfy customer's needs. The validity of a product generates higher values to customers than the cost spent to acquire the product. The manufacturability attribute focuses on the efficiency of making the product. A well-designed product can simplify the manufacturing process and reduce the time to market. The quality of the product often relates to how desirable the product is to the final customer. Serviceability focuses on the ease of performing maintenance or repairs on the product during its effective life span. Recyclability considers how the products can be recycled back through the reverse logistics supply chain. The presence of these six attributes in the product design can increase not only the efficiency of the SCM but also customer satisfaction.

### Functionality (Product Works to Satisfy Customers' Needs)

Product functionality pertains to the ability of a product to operate as it promises. For instance, a customer who purchases a sports car is interested in the functions of engine performance, sporty styling, and aggressive handling capabilities. Safety-concerned customers would be more interested in buying a car with safety features, such as side air bags, impact absorbing interiors, crash-resistant door pillars, and antilock braking systems. Customers form expectations about the functional benefits they desire from the products they purchase. Customers are satisfied when their functional expectations and usage experience match.

### Validity (Product Has Value and Functions at a Reasonable Cost)

Although the product attribute of functionality is important, how much functionality is sufficient to meet customer needs and wants? The answer seems to vary. Microsoft Word has dozens of features. How many of those features are commonly used by regular end users? You can find a similar phenomenon when using consumer products. The functionalities and performance of today's consumer products are entirely adequate for most consumers. However, further engineering efforts to increase functionality may not substantially increase a customer's perception of product usefulness (Israelsohn 2008). Therefore, the key question to ask is "How much value is ultimately necessary for the customer?" Product validity can be improved by analyzing and removing unnecessary attributes in products. Only the functional capability the customer truly seeks in a product should be engineered into the product design. When customers receive the final product, it should have the highest perceived value the customers desire at the price they are willing to pay (Mohan 1991). Product validity is essential to the market success of a product.

## Manufacturability (Product Can Be Efficiently Produced)

Manufacturability is the degree of ease in making products during the production process. This is an important product attribute because a great, innovative product may not be easily producible. This situation occurs when there is a disconnect or communication problem between product design teams and manufacturing personnel. An effective designer should recognize that, while the optimization of product design is essential, it is not always sufficient by itself.

Decisions made at the product design stage determine to a large extent the product's final cost and quality. Failure to take into consideration the attribute of manufacturability when conducting product design can greatly increase production costs, create poor quality products, and slow the time to market. Therefore, the ultimate goal of good product design is to optimize both the design and the entire production system, including managing of raw materials, suppliers, manufacturing processes, labor force capabilities, and distribution procedures (Shukor and Axinte 2009).

## Reliability (Product Has a Variety of Quality Attributes)

Reliability, or quality, is a multifaceted attribute consisting of availability, compatibility, manufacturability, expandability, flexibility, functionality, maintainability, usability, portability, reliability, efficiency, and security. A quality product does not necessarily meet all of these attributes at the same time. Therefore, it is important to know the implied needs demanded by consumers and then match these needs with the corresponding quality attributes. For instance, disposable contact lenses are designed to meet the convenience and sanitary needs of busy consumers. Having customers use the product for a year is never the intended purpose. Toyota's Camry is one of its bestselling cars because customers love its quality attributes of fuel economy, spacious interiors, and reliability. Redesigning that particular car model into a sportier car would create a mismatch between the needs of targeted customers and their corresponding quality requirements.

The quality concept is also applicable to the service industry. Customers desire both expedient and courteous service from those with whom they conduct business. The quality of the call center experience is one service contact area that often receives low ratings from customers. A frequent complaint is one must wait a long time before they can discuss their problem with the right person. Competing on the basis of time is an important quality consideration in the service industry.

## Serviceability (Product Can Be Serviced during Its Effective Life)

Serviceability is the ease in which servicing can be performed on a product during its useful life. The importance of serviceability increases over the life span of products. This is particularly true for industrial products such as automobiles, airplanes, and locomotives. It is necessary that either the original manufacturer or a third-party provider carry replacement parts. Product designs that put cost savings ahead of serviceability have resulted in decreased profitability and increased maintenance costs (Barkai 2005). In the end, poor serviceability not only increases total cost of ownership (TCO) but also decreases customer satisfaction and product loyalty.

### Recyclability (Product Can Be Recycled along the Reverse Logistics Supply Chain)

Recyclability is the capability within the supply chain to return unusable products, defective items, items to be repaired or recalibrated, and environment-friendly goods back to the original manufacturer or a third-party provider. In order to improve recyclability, a company needs to address various reverse logistics supply chain activities. The reverse product flow begins with the returning of goods to some contact point, such as the customer service center. After verifying the product warranty and payment records, the customer service representative credits the customer account. The returned product will continue its journey toward its upstream origins by traveling through distribution, product disassembly, repair, calibration, and component separation, until reaching a point where it is finally recycled or destroyed.

In summary, functionality, validity, and quality account for a portion of a well-designed product. The product design also needs manufacturability, serviceability, and recyclability. These six attributes indicate that the construction of an effective supply chain is indispensable in developing well-designed products.

---

## Consider Alternative Product Design Approaches

Supply chains can be physically efficient or market responsive (Yucesan 2007). When customer demand is stable and products are commodity or functional goods with low margins, the supply chain needs to lower operational costs and maximize its resource utilization. In contrast, customer demand for fashion and information technology goods is usually uncertain. When such goods are in great demand, increases in profit margin can result within a short time. The success of the iPod captured more than 90% of the market share from the digital music industry. On the other hand, Toshiba's failure in marketing the HD DVD format created a disaster to its product development projects. The challenge in creating a market-responsive supply chain is the ability to accelerate product development and production and to put on the brakes when the market is not favorable toward new products (Handfield 2008). Apple's later successes with the iPhone and iPad have confirmed its reputation for developing innovative products, at least under the leadership of Steve Jobs. In recent months, some impatience has developed in the marketplace with Apple's what some construe a slower pace of innovation. Therefore, market-responsive supply chains are needed because they emphasize product innovations, modular designs, and shortened time to market.

Aligning the interests of business partners across the supply chain can maximize performance of individual business partners and the entire supply chain. It is also important to align product design methods with the needs of the supply chain partners. When this alignment is achieved, the agility and adaptability of supply chains can be largely increased (Lee 2006). Physically efficient supply chains can employ quality function deployment (QFD) and design for manufacturability (DFM) methods to maximize product performance at minimum cost. In contrast, market-responsive supply chains can utilize concurrent engineering (CE) and design for sustainability (DFS) to reduce the time to market and increase responsiveness to market demand for greener products. The following section details some of these product design methods and explains how each method can support the varying needs of the supply chain.

### Quality Function Deployment

QFD is a "method to transform user demands into design quality, to deploy the functions forming quality, and to deploy methods for achieving the design quality into subsystems and component parts, and ultimately to specific elements of the manufacturing process" (Akao 1994). It is an effective method to understand customer needs. Methods used to exploit QFD include the voice of customer (VOC) and prioritization of customer needs.

Direct customer interactions can reveal customer needs. This information can be converted into product and service requirements important to customers, after which R&D teams can translate customer requirements into the product design. It is more likely that customers will be satisfied with the developed products and services, because the entire product development process has been constructed to increase customer satisfaction. The focal point of the QFD process is a matrix called the *house of quality (HOQ)*. The APICS Dictionary (2013) describes the HOQ as a structured process that relates customer-defined attributes to the product's technical features needed to support and generate these attributes. This technique achieves this mapping by means of a six-step process:

1. Identify customer requirements. This describes *what* is to be done.
2. Identify supporting technical features to satisfy the requirements. This describes *how* it can be done.
3. Correlate the customer requirements with the supporting technical features. This describes how well the *hows* satisfy the *whats*.
4. Identify the relationship among the technical features. This describes how well the *hows* interact.
5. Assign priorities to the customer requirements and technical features. This describes which of the *hows* to evaluate first.
6. Evaluate competitive stances and competitive products. This describes how competing products are satisfying the customer *whats*.
7. Determine which technical requirements to deploy in the product design. This describes the *hows* to be included in the final product.

For a fuller discussion of QFD, see Crandall (2010).

### Concurrent Engineering

CE is another technique to develop new products in response to customer expectations based on the consensus of cross-functional team members. Team members need to collaborate, trust, and share information about customer expectations with each other throughout the product development process. CE expedites the product development process in the parallel rather than the sequential mode. This cycle consists of many stages, including planning, implementation, reviewing, and modifying, in order to continuously improve the product development process. This strategy can shorten the time to market and help to optimize an effective and efficient product development process. A continuous improvement and refinement cycle is a core element of a CE strategy. Companies, such as NASA, Boeing, and General Electric, are currently adopting CE to gain competitiveness in shortening the time to market.

## Design for Manufacturability

DFM is a methodology to engineer products for the best manufacturability. The best manufacturability costs less, reduces lead times, and strives for the highest quality in design. Most importantly, effective utilization of DFM can support business strategies, including standardization, mass customization, build to order, and product line rationalization. In general, product designs should reduce variability, confusion, and complexity in the production process by following rules such as (1) reducing the number of parts, (2) making assembly foolproof, (3) simplifying the assembly process, (4) making the product easy to test, and (5) avoiding excessively tight tolerances (Walleigh 1989).

Standardizing design and manufacturing processes can increase flexibility in developing the product line and reduce R&D costs. With decreased product life cycles, customer demand is changing dynamically. Standardization may appear to run counter to being able to satisfy dynamic changes in customer demands. However, a hybrid approach, such as mass customization, is an effective method to customize customer needs at the mass level. Assemble-to-order or build-to-order strategies help minimize inventory and the need for detailed forecasting. Product line rationalization allows a firm to optimize the productivity of product lines by eliminating or outsourcing unproductive products to a third party. This strategy can free up a strangled production line and optimize its capacity.

## Design for Sustainability

Social and environmental problems can easily damage a company's reputation. BP's oil spill off the Gulf coast of the United States remains one of the most talked-about environmental disasters in history. For years, Nike endured accusations of promoting sweatshops when using overseas contractors, despite the fact that they regularly monitor their contractors (Zadek 2004). More recently, Apple and other companies have incurred negative press reports because of their use of Foxconn, a China company with reported violations in employee working conditions (Dou 2013). These types of problems become corporate liabilities, if not prevented and managed properly. A number of external stakeholders will exert pressure on the organization to act in a socially responsible manner. Activist organizations and government regulators exert pressure on companies to stop controversial practices. Customers demand safer, healthier, environmentally friendly products produced in a socially responsible manner. Certain legislative requirements require companies to adopt environmental safeguards and disclose their internal business practices concerning the meeting of environmental regulations. Many businesses are beginning to look at these requirements seriously and are seeking to develop profitable and sustainable opportunities out of them.

The ability to meet the increasing customer demand for environmentally friendly goods can help sustain a business. Offering these goods does more than just meet a growing social trend in society; it also contributes to the well-being of human beings in general. Practicing energy conservation, utilizing renewable energy sources, and recycling contaminated by-products are examples of *doing good* for less and can help a business to stay profitable. Companies must realize that their adverse impact on the environment can also be largely reduced and even eliminated in the long term.

Many business sectors, such as chemical, agricultural, auto manufacturing, and the film-developing industry, are embracing the concept of DFS. Another popular term becoming more widely used is the *triple bottom line*. This is a concept that encourages businesses to aim for profitability, positive social involvement, and improved environmental operation as an integrated approach to their business strategies (Elkington 1998).

Environmental concern is a key consideration when designing products for sustainability. A good ecodesign approach helps ensure new products do not pollute the environment and retired products can be recycled back to minimize the waste. A sustainable product needs to meet these three requirements in order to create a sustainable business. Products designed with the welfare of the environment in mind benefit both domestic and international customers. Such *green* product innovations do not guarantee profit; however, it is important to improve manufacturing efficiency and product quality as well as to exploit the market opportunities that desire these types of products.

## Determine the Number of Supply Chains Needed by a Company

A supply chain involves multiple buyers and suppliers in the transactional process. Buyer–supplier relationships progress from a basic partnership between a single buyer and a single seller to a more complex supply chain, where various customers and suppliers with different interests transact with each other (Mentzer 2001). The challenge of servicing customers grows in complexity as the number of relationships within the supply chain increases. The number of *handoffs* between the buyer and supplier increases when each transaction crosses the boundaries between participants in the supply chain. With this increasing complexity, a firm must continuously redesign and optimize the value-added processes of the supply chains in which they participate.

One important decision for supply chain professionals is to determine the number of supply chains needed to service their customers. This decision can involve aligning a supply chain with a customer segment, a product/service bundle, or a supplier category. Understanding this alignment helps the supply chain professional better understand how many supply chains are necessary to best service customers vested with different interests.

### Align with Customer Segment

Market-driven supply chains are constructed to meet customer demands, whereas manufacturing-centered supply chains are designed to lower internal operational costs. In order to meet the requirements of shorter product life cycles and increased changes in customer demand, suppliers and customers need to work closely to create a market-driven supply chain. This kind of supply chain is effective at understanding customers' requirements and reducing the time needed to respond to their needs.

The first step in the construction of a market-driven supply chain is to identify the value segments based on customer requirements. To accomplish this, companies must market their products and services to groups of customers with similar profiles, such as spending habits, behaviors, gender, geographical locations, and age. The selection of market channels varies with the customer segment chosen. For instance, customers in an urban location may favor online sites and local retail outlets, whereas customers in the rural area may prefer regional malls for one-stop shopping. An online home delivery grocery business is particularly attractive to customers who have scarcity of time (Boyer and Frohlich 2006). Customer segmentation addresses differing customer needs by categorizing them. The practice of customer segmentation enables a firm to increase marketing effectiveness and improve the use of limited marketing resources.

Supply chain design affects how different customer segments are serviced. Effective sales and operation decisions must consider purchasing, production, and distribution processes. Some customers value delivery reliability because they operate in an unpredictable environment, whereas others are more sensitive to price. A firm needs to customize the design of their supply chains to meet the differing needs of these two customer segments. Agile supply chains are more effective to support the customer segment favoring delivery reliability because they often order in low volume, but in high variety. In contrast, lean supply chains rely on large volume orders but with low variety. Lean supply chains are more fit for a customer segment that is sensitive to price. The placement of a multichannel distribution system to make a constant trade-off between process integration and channels' separation cannot only serve customer segments but also exploit economies of scale (Agatz et al. 2008).

The alignment between the market segment and the number of supply chains is part of the design of market-driven supply chains. It is important to manage the linkage between supply chain processes and customer segments (Jüttner et al. 2006).

## Align with Product/Service Bundle

Providing excellent services is an indispensable component to the successful operation of a supply chain in producing and delivering physical goods to customers. The services provided by McDonald's go beyond just providing the food to the customer. The consistency of food quality, sanitation, and speedy service are successful factors in McDonald's business. Customer needs can be determined based on (1) key elements of transactional processes, (2) product–process matrix, and (3) service intensity matrix.

The delivery of products/services into the hands of customers is not the end to customer service. Customer service consists of three categories of transaction elements: pretransaction, transaction, and posttransaction elements (Carter et al. 2007). All these elements together contribute to a complete customer service experience. For example, a product return policy can affect the perceptions of customers about a company's commitment to customer service. Out-of-stock problems affect the order fill rate and the order cycle time. Informing customers about their order status helps increase the trust of customers in placing orders. A firm can improve its customer service by aligning product and service offers with each transactional process. Critical success elements, such as the customer service policy, accessibility, the organization structure, and system flexibility, need to be integrated in the pretransaction phase. When goods and services move into the transaction phases, business partners need to collaborate with each other to integrate the order cycle time, inventory availability, and the order fill rate.

Aftertransaction service elements that need to be integrated consist of availability of spares, warranty, customer complaints, and claims.

As customer needs change, production processes also need to change to most effectively satisfy those needs. For instance, customers who want low-volume, high-variety products need customized service, best provided by a flexible process, such as in a make-to-order process. As the degree of product standardization increases, products can be mass-produced and serviced via the automation of the production process.

## Align with Supplier Category

A company can also align its supply chain with supplier categories. Scale, uncertainty, and the specificity or uniqueness of the assets are three critical elements to help determine the needs of customers, thereby improving profits for companies (Chopra and Meindl 2007). When a firm is large in scale, servicing the needs of internal customers can achieve economies of scale and make profits possible. In contrast, using a third party is a better alternative when a firm's size is small and achieving economies of scale is not possible. A company can own and operate its own service centers and expect to make a continuous profit if the demand for its products and services is predictable. When asset specificity is high, it is difficult for a third party to provide services to internal customers. Therefore, a company needs to provide its own internal services in order to make a profit. In contrast, a third party can help deliver services to ultimate customers and make a profit if the asset specificity is low.

## Align with Common Incentives

Supply chain partners with different interests, costs, risks, and rewards of doing business will encounter conflicts from time to time. In order to minimize these conflicts, it is important to align the performance measurements of supply chain partners with their marketing and management efforts. However, this alignment may be difficult to accomplish if the business partners are not motivated to reveal actions and information to each other (Narayanan and Raman 2006). Companies are often reluctant to share too much information since this can potentially put business partners in a disadvantageous position. As such, partners are less willing to share information unless the participants can develop a win–win scenario.

Furthermore, the decision of using one incentive to promote the interests of one business partner can discourage the goals of another business partner. Offering sales discounts for high-volume purchases may increase sales volume for the supplier, yet increase inventory costs for the retailer.

In addition to improving the trust level, supply chain partners can improve their relationships by redesigning the performance measurement as follows:

- Acknowledge that problems with incentives can exist.
- Identify poorly designed incentives.
- Create or redesign incentives that will induce partners to behave in ways that maximize the supply chain's profits.

Performance measurement along the supply chain is still largely in the early stages of development.

## Respond to Needs of Internal Customers

When an organization is structured in terms of functional departments (sometimes referred to as silos), conflicting goals may develop between departments. Purchasing managers have the goal of buying products at a low price while not sacrificing quality. Marketing managers have the goal of introducing new products with features, while production managers desire long production runs of standardized products. These same managers are also the internal customers of the supply chain. Internal customers have varying needs; sometimes, their needs conflict with each other. Determining and meeting the needs of internal customers are prerequisites for SCM success.

The marketplace is changing rapidly. Companies that can swiftly adjust their strategies are often the ones that succeed in their industries. To improve agility to meet changing conditions, organizations are shifting from managing all aspects of the business in-house to engaging outside partners to run those activities that are not part of their core competencies. This process of engaging outside partners, or outsourcing, involves contracting processes and functions to outside organizations that specialize in the contracted activities. In doing so, such organizations use outsourcing as part of their overall strategy to stay flexible or agile. For example, more companies are making the strategic decision to outsource some or even all of their information systems (IS) functions. The same is true in the area of human resources (HR), as many companies outsource the payroll and applicant screening functions.

The actual practice of outsourcing internal functions evolved in the last few decades. Outsourced functions are both operational oriented (e.g., system administration/support, IT infrastructure, help desk, and call center) and strategically oriented (e.g., business process, system and architecture planning, R&D) (Ware 2003). The average value of offshore outsourcing contracts runs into millions of dollars, and the present trend is to increase outsourcing to leading countries such as India, China, Philippines, Ireland, and Russia. Companies are increasingly outsourcing internal functions to partners who are geographically dispersed. This is disruptive to the flow of goods and information among internal customers within an organization. Consequently, it is important for an agile organization to understand and meet internal customer requirements.

## Conclusion

This chapter discussed the needs of external customers for products and services. Delivering these customer needs requires supply chains with a combination of manufacturing and service processes. Market research is the first step in increasing business intelligence in an open system environment. Direct effect variables include competition, technology, and economy. Indirect effect variables include government, social trends, and environment. All these variables can influence demand forecasting. Qualitative and quantitative forecasting methods are available to assist supply chain professionals in predicting customer demands. However, as demand variability increases, demand forecasts require more collaboration with customers to develop better forecasts. Products and services can be customized in their designs via many approaches, including QFD, CE, DFM, and DFS. Supply chains can be customized in order to align with the customer segment, the product/service bundle, and the supplier category. Performance measurements can be constructed to better understand the effectiveness of the customized supply chains to meet customer needs. It is also important to remember the needs of internal customers when developing global supply chains.

## Hot Topic: Human Trafficking

The hot topic for this chapter is about human trafficking among suppliers in the supply chain.

### Hot Topics in the Supply Chain: Is There Human Trafficking in Your Supply Chain?

Human trafficking, or modern-day slavery, is not a topic most of us are comfortable talking about. Unfortunately, with the extension of global business, the reality is that this trend is part of the supply chain for many transnational companies (Eckes 2011). Human trafficking leads to forced labor and there is a good chance that some of the products you use might have been made by someone who was working against their will.

#### BACKGROUND

According to the International Labor Office, there are approximately 20.9 million victims of forced labor in the world. Forced labor involves three conditions. First, there is work or services provided by an employee for a third party. Second, the work is performed under the threat of a penalty. Such penalties can be physical, financial, or psychological. Finally, the person doing the work or providing the service cannot freely leave their job (International Labor Office 2013).

The Asia Pacific region of the world accounts for the highest number of forced laborers, at 11.7 million people, or 56% of the total. Africa is second at 3.7 million people or 18% of the total. The main industries affected include agriculture; food processing; fishing; manufacturing of textiles, garments, and electronics; mining; and domestic work (International Labor Office 2013).

A key factor involved in human trafficking is the presence of migration patterns, which causes citizens in one country to move to another in search of better economic opportunities. Migrant workers are among the most abused class of citizens in the world because of their low skill levels and desperation to find work. Migrants often become targets of criminal gangs, who capture them and subject them to forced labor or prostitution. The victims may end up in textile sweatshops in China, the sex industry in South America, or in agricultural jobs in East Europe (Eckes 2011).

Of particular concern to supply chain managers is the possibility that an outsourced, offshored product may have been made by someone in a forced labor situation in Asia. The risk is so real that one state, California, has enacted its own form of supply chain monitoring.

#### CALIFORNIA TRANSPARENCY IN SUPPLY CHAINS ACT OF 2010

The state of California passed its own law in human trafficking as it applies to supply chains. The law requires retailers and manufacturers to state what they are doing to eliminate human trafficking from their supply chains. The act became effective on

January 1, 2012, and applies to all companies doing business in California that have sales over $100 million.

The specifics of the act require companies to disclose on their websites the details of what they are doing in evaluating and addressing the potential for slavery and human trafficking in their supply chains (Deschenaux 2011). The law is more informational in nature, meaning that companies must state what they are doing to ensure the integrity of their supply chains; however, they do not have to cease operations with a supplier if it is found to be involved in human trafficking (Jaeger 2012). Still, the pressure will be on companies in the future to ensure their supply chains are clean of human trafficking, as the risks associated with such practices can lead to negative publicity, business interruptions, public outcry, loss of consumer trust, and, ultimately, shareholder wealth (Interfaith Center of Corporate Responsibility 2011).

## QUESTIONS FOR RESEARCH AND DISCUSSION

1. Do an Internet search and look at the websites of some major firms doing business in California. What information did you find in regard to the California Transparency in the Supply Chain Act?
2. What do you think the future will be in regard to this type of monitoring of supply chains? Do you think other states will enact similar legislation?

## REFERENCES

Deschenaux, J., California law focuses on human trafficking, *HR Magazine*, 56(11), 22, November 2011.
Eckes, A. Jr., The seamy side of the global economy, *Global Economy Journal*, 11(3), 1–26, 2011.
Interfaith Center of Corporate Responsibility, Effective supply chain accountability: Investor guidance on implementation of the California Transparency in Supply Chains Act and beyond, November 2011. http://www.iccr.org/resources/2012/2011SupplyChainGuide.pdf (retrieved January 7, 2014).
International Labor Office, Stopping forced labour and slavery-like practices: The ILO strategy, 2013. http://www.ilo.org/wcmsp5/groups/public/---ed_norm/---declaration/documents/publication/wcms_203447.pdf (retrieved January 6, 2014).
Jaeger, J., California passes new supply chain transparency law, *Compliance Week*, 24, 63, March 2012.

## Discussion Questions

1. Why is it important to group customers into homogeneous segments?
2. How do you determine the needs of the ultimate consumer?
3. How do you decide the needs of the internal customers?
4. What are some of the typical customer needs?
5. How do you segment customers by product/service bundle features, economic needs, location needs, and service levels?
6. There are direct and indirect methods to determine customer needs. Name two primary methods within each category.
7. Using a product that you have in your house or apartment, discuss how well you think it is designed.

8. What factors determine how many supply chains are needed within a company to meet customer needs?

9. What are the traditional measures for meeting customer needs? What are the major limitations to these measures?

10. What are the contemporary measures for meeting customer needs? What are the major strengths in these new measures?

---

## References

Agatz, N.A.H., Fleischmann, M., and van Nunen, J.A.E.E., E-fulfillment and multi-channel distribution: A review, *European Journal of Operational Research*, 187(2), 339, 2008.

Albrecht, K., *Corporate Radar*, American Management Association, New York, 2000.

Barkai, J., Design for serviceability, *Machine Design*, 77(12), 134, 2005.

Bilimoria, K., In my opinion, *Management Today*, May, 10, 2007.

Blackstone, J.H., *APICS Dictionary*, 14th edn., APICS—The Association for Operations Management, Chicago, IL, 2013.

Blanchard, D., *Supply Chain Management Best Practices*, John Wiley & Sons, Inc., Hoboken, NJ, 2007.

Boyer, K.K. and Frohlich, M.T., Analysis of effects of operational execution on repeat purchasing for heterogeneous customer segments, *Production & Operations Management*, 15(2), 229, 2006.

Buckley, N., Shoppers must wait for quote-free prices to hit the streets, *Financial Times*, July 21, 8, 2004.

Carter, J.R., Slaight, T.H., and Blascovich, J.D., The future of supply management—Part 2: Technology, collaboration, supply chain design, *Supply Chain Management Review*, 11(7), 44, 2007.

Chopra, S. and Meindl, P., *Supply Chain Management: Strategy, Planning and Operation*, Pearson Prentice Hall, Upper Saddle River, NJ, 2007.

Court, D., Putting big data and advanced analytics to work, McKinsey, 2012. http://www.mckinsey.com/features/advanced_analytics (includes video), accessed May 15, 2013.

Crandall, R.E., House plans, quality function deployment helps design a better product, *APICS Magazine*, 20(4), 28–31, 2010.

Crandall, R.E., The big data revolution, investigating recent developments in collection and analytics, *APICS Magazine*, 23(2), 20–23, March–April 2013.

Davenport, T.H., Paul, B., and Randy, B., How 'big data' is different, *MIT Sloan Management Review*, Fall, 54(1), 43–46, 2012.

Dou, E., Apple shifts supply chain away from Foxconn to Pegatron, *Wall Street Journal*, May 29, 2013. http://online.wsj.com/article/SB10001424127887323855804578511122734340726.html, accessed June 17, 2013.

Doyle, S., A sample road map for analytical CRM, *Journal of Database Marketing & Customer Strategy Management*, 12(4), 362, 2005.

Elkington, J., *Cannibals with Forks: The Triple Bottom Line of 21st-Century Business*, Capstone Publishing, Oxford, U.K., 1998.

Fowles, J., *Handbook of Future Research*, Greenwood Press, Westport, CT, 1978.

Gawer, A. and Cusumano, M.A., How companies become platform leaders, *MIT Sloan Management Review*, 49(2), 28, 2008.

Gunin, J., With IT panel: Social trends impact furniture marketing, *Furniture Today*, 32, 45, August 25, 2008.

Handfield, R.B. (ed.), *Consumers of Supply Chain Risk Data*, Auerbach Publications/Taylor & Francis Group, Boca Raton, FL, 2008.

Heizer, J. and Render, B., *Operations Management*, 10th edn., Pearson Prentice Hall, New York, 2011.

Israelsohn, J., What does "better" mean? *EDN*, 53(4), 26, 2008.

Jüttner, U., Christopher, M., and Baker, S., Demand chain management—Integrating marketing and supply chain management, *Industrial Marketing Management*, 35(8), 989, 2006.

Kotler, P., *Kotler on Marketing: How to Create, Win, and Dominate Markets*, Free Press, New York, 1999.

Lee, H.L., The triple—A supply chain, *Harvard Business Review*, 82(10), 102, 2006.

Mentzer, J.T. (ed.), Managing the supply chain: Managerial and research implications, in *Supply Chain Management*, Sage Publications, Thousand Oaks, CA, 2001.

Mohan, R.N., Defining product value in industrial market, *Management Decision*, 29(1), 14, 1991.

Moore, G.A., *Crossing the Chasm: Marketing and Selling Technology Products to Mainstream Customers*, Harper Business, New York, 1991.

Narayanan, V.G. and Raman, A., *Aligning Incentives in Supply Chains*, Harvard Business School Press, Cambridge, MA, 2006.

Nuttall, P. and Tinson, J., Keeping it in the family: How teenagers use music to bond, build bridges and seek autonomy, *Advances in Consumer Research*, 35(1), 450, 2007.

Parnell, J., *Strategic Management: Theory and Practice*, 3rd edn., Atomic Dog Publishing, Cincinnati, OH, 2008.

Rice, M.P., O'Connor, G.C., and Pierantozzi, R., Implementing a learning plan to counter project uncertainty, *MIT Sloan Management Review*, 49(2), 54, 2008

Rogers, E.M., *Diffusion of Innovations*, Free Press, Glencoe, IL, 1962.

Schulman, S., Taste: The trend in trends, *Wall Street Journal*, September 21, W13, 2007.

Shukor, S.A. and Axinte, D.A., Manufacturability analysis system: Issues and future trends, *International Journal of Production Research*, 47(5), 1369, 2009.

Stevenson, W.J., *Operations Management*, 11th edn., McGraw-Hill, New York, 2012.

Vokurka, R.J., Supplier partnerships: A case study, *Production and Inventory Management Journal*, 39(1), 30, 1998.

Walker, W.T., *Supply Chain Architecture: A Blueprint for Networking the Flow of Material, Information, and Cash*, CRC Press, Boca Raton, FL, 2005.

Walleigh, R., Product design for low-cost manufacturing, *The Journal of Business Strategy*, 10(4), 37, 1989.

Ware, L.C., Weighing the benefits of offshore outsourcing [electronic version], September 2, 2003. http://www2.cio.com/research/surveyreport.cfm?ID = 62 (retrieved December 1, 2005).

Yucesan, E., *Competitive Supply Chains: A Value-Based Management Perspective*, Palgrave Macmillan, New York, 2007.

Zadek, S., The path to corporate responsibility, *Harvard Business Review*, 82(12), 125, 2004.

# 4

## A System to Meet Customer Needs

### Learning Outcomes

After reading this chapter, you should be able to

- Identify the five major components of the supply-chain operations reference (SCOR) model
- Discuss the differences between the Global Supply Chain Forum (GSCF) model and the SCOR model
- Describe the major features and benefits of customer relationship management (CRM)
- Discuss the importance of product life cycle management (PLM) within the context of supply chain management (SCM)
- Describe the ways supply chains can be visually depicted
- Discuss how designing production processes will vary according to the type of product made and the volume needed
- Discuss the various resource requirements needed in a customer-driven supply chain
- Describe the concepts of core competency and the total cost of ownership (TCO)
- Describe the responsibility and authority of entities operating in the supply chain
- Evaluate the design of supply chains with respect to their abilities to achieve business objectives

This chapter will take the information developed in Chapter 3 and transform it into an operational system to meet the identified needs and wants of the customer. The design process will be described for a typical supply chain with some variations noted for different supply chain requirements. As an introduction, we will briefly outline two of the most widely referenced supply chain models—the SCOR model and the GSCF model. Then we will describe two major programs used as important components of supply chain systems—CRM and PLM. With that background, we will identify the characteristics of an effective supply chain system.

### Company Profile: GE Aviation

This chapter describes how companies build relationships with customers. The objective is to retain the customer over an extended period of time by providing not only products but also related services that the customer values. While the product may be standard, the services are often designed for a particular customer's needs, thereby creating a lasting relationship between the customer and supplier. General Electric (GE) Aviation has done this with their jet engine business. As you read this company profile and the chapter in the book, think about these questions:

- Why are services so important in the jet engine business?
- How would you describe CRM?
- How does a good supply chain help in a CRM program?
- What is the role of information in building customer relationships?

What additional questions do you think relate to building customer relationships?

### Company Profile: General Electric

Chapter 4 describes systems that enable companies to provide customers with products and services that strengthen the relationship between supplier and customer. The customer gets better products and the supplier benefits from increased customer retention. Two of the programs that will be described are CRM and PLM. As part of their strategic plan, GE is focusing on five choices, one of which is to build deep customer relationships based on an outcome-oriented model (GE Annual Report 2012, p. 7).

In this company profile, GE will be featured to show they use both programs in dealing with their customers in the GE Aviation segment of their business. In 2012, the Aviation segment generated $19,994 million in revenues, or 13.6% of total company revenues. The segment provided $3,747 million in profit, or 16.4% of total segment profit (GE Annual Report 2012, p. 42). It is a growing business in a highly competitive market, in which GE is one of the major players.

### CUSTOMER RELATIONSHIP MANAGEMENT

It is obvious that GE must sell jet engines that perform will in service. However, it is the added services the Aviation segment provides that helps to build continuing relationships. They are the world's leading producer of large and small jet engines for commercial and military aircraft. They have products that directly support the engines such as avionic systems, mechanical systems, integrated power systems, and components parts, electrical and mechanical. They also have support systems that include air traffic optimization services, integrated logistics management, and aftermarket support services (geaviation.com 2013).

GE Aviation also offers a variety of additional services that include the following:

- *Maintenance*: Overhauls with original equipment manufacturer (OEM) materials and repair methods
- *Material*: New parts, used parts, repair services, and technology upgrades
- *Asset management*: Engine leasing and related financial services, such as rentals and engine exchanges
- *TRUEngine™ program*: Certification of an engine that has been maintained with OEM parts and overhaul practices
- Optimization programs to reduce costs and increase reliability
- *myEngines™*: Information source to guide decision making
- *Diagnostics*: Provides information to help improve reliability, operations, and cost
- *Fuel and carbon solutions*: Enables airlines to save fuel with support of consultation and technology from GE

- *ClearCore™ engine wash*: An innovative *core* wash and collection system that improves engine efficiency and lowers overall operational expenses
- *MRO network*: A worldwide network of repair and parts facilities

In addition, the organization offers consulting services to help customers decide which of the services would provide the best combination package for the customer (geaviation.com 2013).

GE Aviation offers a full range of services. As companies increase their use of these services, GE can strengthen their relationships with customers. GE has built a part of their customer relationship program in collaboration with Salesforce, a leading CRM system provider. According to Salesforce, GE Aviation started by using Salesforce to connect with commercial customers and exchange data. However, they soon found that the system could increase collaboration with customers by making the data more meaningful. In addition to conventional data collection, social networks are helping GE Aviation update its brand image and become more approachable to consumers. A video describing this program can be found at http://www.salesforce.com/customers/stories/ge.jsp (Salesforce 2013).

## PRODUCT LIFE CYCLE MANAGEMENT

Another program that can be used to strengthen customer relationships is PLM. PLM is a system designed to collect information about a product during its design and preserve that information throughout the life of a product. The availability of that information can be used to assure proper use, maintenance, repair, and disposal of the equipment. Siemens, one of the leading PLM software companies, offers the following description.

*PLM* can be defined as an information strategy: it builds a coherent data structure by consolidating systems. You can also call PLM an enterprise strategy: it lets global organizations work as a single team to design, produce, support and retire products, while capturing best practices and lessons learned along the way. Siemens PLM Software views PLM as an information strategy, an enterprise strategy and, ultimately, a transformational business strategy. They see it as a comprehensive approach to innovation built on enterprise-wide access to a common repository of product information and processes.

PLM software allows companies to manage the entire life cycle of a product efficiently and cost effectively, from ideation, design, and manufacture through service and disposal. Computer-aided design (CAD), computer-aided manufacturing (CAM), computer-aided engineering (CAE), product data management (PDM), and digital manufacturing converge through PLM.

PLM is unique from other enterprise software solutions because it drives top-line revenue from repeatable processes. By providing the application depth and breadth needed to digitally author, validate, and manage the detailed product and process data, PLM supports continuous innovation (Siemens 2013).

GE Intelligent Platforms (GE-IP) is a division of GE, headquartered in Charlottesville, VA. It provides industrial software, control systems, and embedded computing platforms to optimize customers' assets and equipment. "Our goal is to help our customers' grow the profitability of their businesses through high performance solutions for today's connected world. We work across industries including power, manufacturing, water, mining, oil & gas, defense and aerospace" (GE Intelligent Platforms 2013).

GE-IP uses PLM for products that may have a life of years or even decades. During that time, there is a need to monitor the system to improve its performance and ensure that it is being serviced and supported at a minimum cost and with minimum disruption.

GE-IP has the knowledge, the infrastructure, and the facilities to provide that level of service. The PLM system is designed to provide customer care for as long as a product is in service, the key element of which is long-term, and continuous, service (GE Intelligent Platforms 2013).

GE-IP operates in two closely related markets: automation and embedded computing. In automation, they provide computer boards, systems, and software packages to control and report on factory production lines and production tools. In the embedded market, they produce boards and systems used in defense, telecommunications, aviation, and other types of similar applications. Through the use of sensors, they are able to monitor on a real-time basis the operation of a piece of equipment or system. If operating problems are detected, they are able to provide corrective or repair services, including replacement parts.

## OTHER PROGRAMS

While CRM and PLM have been developed as separate programs, it is apparent they are complementary. Both are concerned with the creation of information that can be shared by users throughout the life of a product. "GE's Fuel & Carbon Solutions (FCS) helps airlines reduce fuel consumption by combining advanced data analytics and industry expertise. FCS can integrate and analyze vast amounts of data – about 1.5 terabytes per customer per year – from flight operations, airline systems and flight data recorders. It then uses the resulting operational intelligence to identify, implement and monitor changes in the way flights are planned and flown, saving fuel and improving efficiency" (GE Annual Report 2012, p. 14).

In addition to CRM and PLM, GE Aviation is also counting on another innovation in manufacturing—additive manufacturing or 3D manufacturing. Over the next 8 years, they expect to produce over 100,000 3D printed parts for its LEAP and GE9X engines. Each of these engines will have 19 3D printed fuel nozzles, which will be 25% lighter and five times more efficient than conventional nozzles, which should lead to fuel savings in aircraft with the engine (Ivas 2013). Additive manufacturing will make availability of replacement parts faster and reduce the level of inventory needed. See the following excerpt from GE's website about the use of 3D printing in developing new appliance parts.

> Louisville, Ky (July 11, 2013) (NYSE:GE)—In the Rapid Prototyping Center (RPC) that just opened at GE's Appliance Park in Louisville, Ky., a 3D printing machine quickly lays down layer after layer of plastic resin to create a fabric softener dispenser for a new top load washing machine model. Another machine makes a prototype of a new design of grates for gas ranges that will be used with consumer focus group research. In the past, designing the right part could be very costly and take months to complete. In every aspect of product design and engineering, the RPC accelerates the development process and reduces costs. (GE News 2013)

GE is a company that is moving heavily into services to support the products they manufacture. As Jeffrey Immelt, CEO, says in his Letter to Shareowners, "We have grown our service revenue from $21 billion to $43 billion over the past decade. Services represent about 75% of our industrial earnings. With $157 billion of service backlog, we have the momentum to grow in the future. We believe in a solutions-oriented selling model, one that can deliver outcomes for customers. In Aviation, we create value through the performance of our new technologies. We only win when our customers win" (GE Annual Report 2012, p. 7).

## REFERENCES

GE Annual Report, 2012, http://www.ge.com/ar2012/, accessed July 1, 2013.

GE Aviation, Products and services, 2013, www.geaviation.com/systems/products-and-services/, accessed July 1, 2013.

GE Intelligent Platforms, PLM, product lifecycle management for mission critical electronics, 2013, http://www.ge-ip.com/about/companyinfo, accessed July 1, 2013.

GE News, http://www.genewscenter.com/Press-Releases/At-a-Rapid-Pace-GE-Changes-the-Way-it-Develops-Appliances-40f0.aspx, accessed July 1, 2013.

Ivas, P., GE aviation looks to speed up additive manufacturing process, 2013, www.caddedge.com/...d-design-and-3d-cad-software-news/bid/181064/, accessed June 28, 2013.

Salesforce, GE's social transformation, 2013, www.salesforce.com/customers/stories/ge.jsp, accessed June 26, 2013.

Siemens, What is PLM software?, 2013, www.plm.automation.siemens.com/en_us/plm/, accessed June 28, 2013.

## Supply Chain Models

The Supply Chain Council (SCC), an independent, not-for-profit, global organization, formulated the SCOR model as a process reference model to guide its corporation members, in order to cope with the growing challenges associated with SCM. The SCOR model spans from suppliers' suppliers to customers' customers and helps address challenges faced by parties within the extended supply chain. The SCOR model, shown in Figure 4.1, guides companies in executing five processes according to a series of SCM standards: (1) planning, (2) sourcing, (3) making, (4) delivering, and (5) returning (Supply Chain Council 2013). Conversely, the SCOR model does not describe such related business practices of sales and marketing (demand generation), research and technology development, product development, and some elements of postdelivery customer support. The model assumes but does not explicitly address training, quality, information technology (IT), and non-SCM administration.

Each process dictates a definite set of SCM activities for a company to follow. Details about those activities within each process are listed in Table 4.1. The SCC has announced

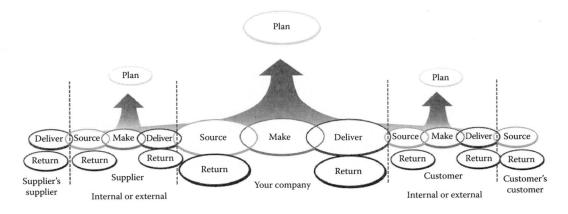

**FIGURE 4.1**
SCOR model scope and structure. (From Supply Chain Council, SCOR®, Supply Chain Operations References model: SCOR Version 10.0, Washington, DC, 2013, pp. 3–5, http://www.supply-chain.org/resources/scor.)

**TABLE 4.1**

Supply Chain Activities of Five SCOR Processes

| SCOR Process | Definitions |
| --- | --- |
| Plan | Processes that balance aggregate demand and supply to develop a course of action that best meets sourcing, production, and delivery requirements. |
| Source | Processes that procure goods and services to meet planned or actual demand. |
| Make | Processes that transform a product to a finished state to meet planned or actual demand. |
| Deliver | Processes that provide finished goods and services to meet planned or actual demand, typically including order management, transportation management, and distribution management. |
| Return | Processes associated with returning or receiving returned products for any reason. These processes extend into postdelivery customer support. |

*Source:* Supply Chain Council, SCOR®, Supply Chain Operations References model: SCOR Version 10.0, Washington, DC, 2013, pp. 3–5, http://www.supply-chain.org/resources/scor. With permission.

Version 11 of their model, which adds a sixth activity, ENABLE, to their model, which will be incorporated into all SCOR training and certification programs beginning in 2013.

## Global Supply Chain Forum Model

SCM is "the integration of key business processes from end user through suppliers that provides products, services, and information that add value for customers and other stakeholders," according to the GSCF model (Lambert et al. 1998). This model takes cross-functional and cross-business perspectives to examine the effectiveness of supply chain operations.

The GSCF model comprises three primary elements: the supply chain network structure, the supply chain business processes, and the associated management components. The supply chain network structure consists of the network relationships among business partners across the supply chain. The supply chain business processes are the lifeblood to sustain these network relationships. The GSCF model addresses eight key business processes as follows:

- *CRM*: Develop and maintain relationships with customers so that cross-functional customer teams can tailor product and service offerings to meet customer needs and wants.
- *Customer service management*: Establish a one-stop contact center where information across functions and business is aggregated to service customers.
- *Demand management*: Reduce demand variability and increase supply chain flexibility by balancing customers' requirements with supply chain capabilities.
- *Order fulfillment*: Streamline all supply chain activities from product design, customer requirements acquisition, production, and distribution to accepting returned goods.
- *Manufacturing flow management*: Move products across the production units throughout the entire supply chain to improve manufacturing flexibility.
- *Supplier relationship management (SRM)*: Develop and maintain relationships with networks of suppliers.

- *Product development and commercialization*: Increase the speed in bringing new products to market via collaboration with suppliers and customers.
- *Returns management*: Streamline reverse logistics activities, including returns, recalls, and maintenance (Lambert et al. 1998).

Each business process is further composed of strategic and operational subprocesses where supply chain activities are actually performed. These supply chain activities include the key elements of planning and control, organization structure, information flows, power structure, leadership, corporate culture, and risk/reward structures.

The SCOR and GSCF models have several commonalities. First, both models use a multitier approach to create alignment between strategic planning and operational implementations. Second, both models adopt cross-functional and cross-company perspectives to create a linkage between business partners across the supply chain. In spite of these two commonalities, the two models vary in four criteria: scope, intracompany connectedness, intercompany connectedness, and drivers of value generation (Lambert et al. 2005). Scope refers to the extent to which the adopted model can support corporate strategy. Intracompany connectedness refers to the degree of formal and informal contacts between employees of different functional departments. Intercompany connectivity refers to the degree of collaboration between business partners across the supply chain. The drivers of value generation refer to approaches used to achieve supply chain excellence, such as reducing costs, increasing revenues, and improving asset efficiency. Table 4.2 summarizes the differences between the SCOR and GSCF models with respect to these four criteria. The GSCF framework has advantages over the SCOR model in providing implications on improving CRM functions. The GSCF model is primarily based on the interconnectedness between CRM and SRM. CRM is an important aspect to the continuous support of the execution of corporate and functional strategies in the GSCF model.

**TABLE 4.2**

Differences between SCOR and GSCF

| Area of Difference | SCOR | GSCF |
|---|---|---|
| Scope | Has a direct impact on the operations strategy, but an indirect impact on corporate strategy | Has an impact on corporate and functional strategies primarily through CRM and SRM processes |
| | Prescribes the activities that are directly related to the product flow | Focus is on relationship-related supply chain activities |
| Intracompany connectedness | Modeled after the workflow reengineering process to ensure that three primary functions are integrated: purchasing, manufacturing, and logistics | Seeks to determine the level of involvement of all corporate functions for each business process |
| Intercompany connectedness | Connects business partners via transactional activities related to the plan, source, make, deliver, and return functions | Driven by CRM and SRM processes to identify improvement opportunities in supply chain efficiency and product/service offerings |
| Drivers of value generation | Focuses on the improvement of operational efficiency, cost reductions, and improvement in asset utilization | Focuses on the revenue implications of all activities performed by supply chain partners |

*Source:* Adapted from Lambert, D.M. et al., *J. Bus. Logist.*, 26(1), 25, 2005. With permission.

## Customer Relationship Management

CRM refers to business processes executed by a company with its business partners to service customers and sustain relationships with them. CRM includes four categories of business processes: (1) the customer contact process, (2) the back-office process, (3) the business-to-business (B2B) relationship building process, and (4) the marketing/sales data analysis. Customers directly contact the company via accessible communication channels, including phone, e-mail, instant message, fax, or a face-to-face meeting with a sales representative. Operational efficiency is an important performance measure in the front-office process. The back-office process focuses on functional aspects of the business, including planning, billing, advertising, finance, accounting, and manufacturing. For instance, suppose a number of customers bring a coupon to a retailer and expect it to be honored during an event promotion period. If the stock planning process is not accurate, overstock or back order problems could occur to upset a customer's shopping experience. These problems can result in a loss of customer goodwill and lower sales revenues. Customer relationships can also be extended to business partners, industry consortiums, trade associations, R&D centers, distributors, wholesalers, and retailers. Maintaining a positive relationship with business customers is also indispensable to the success of CRM. Sharing and exchanging transactional data between business partners' internal functions are critical to support the flow of goods and information in a successful supply chain. When transactional data are received, they need to be analyzed and converted into useful information that is beneficial to supply chain partners. For example, a marketing campaign relies heavily on useful information (e.g., market share and sales volume by regions, product categories, and promotion periods).

CRM implementations need to integrate business processes within and across companies. The success of CRM depends on the reliable delivery of not only marketing and sales activities but also other business activities. An example is Walmart's return policy, "All merchandise purchased from Walmart.com may be returned either to a store or by mail within 90 days of receiving it, unless otherwise noted below. Additional guidelines are also noted below" (Walmart 2013). In order to honor the return policy and streamline the reverse logistics, Wal-Mart needs to integrate its return centers with distributors and suppliers. Only when these functions are integrated across the company can CRM be performed to satisfy customers.

CRM must be implemented from an interorganizational process perspective. Many process-oriented frameworks are available to examine SCM and are also applicable to examining CRM. Among all frameworks, the SCOR and the GSCF provide detailed guidelines to enable implementation (Lambert et al. 2005).

CRM is "a marketing philosophy based on putting the customer first. The collection and analysis of information designed for sales and marketing decision support (as contrasted to enterprise resources planning information) is required to understand and support existing and potential customer needs. It includes account management, catalog and order entry, payment processing, credits and adjustments, and other functions" (Blackstone 2013).

### Origins of CRM

How exactly did CRM get started? One driver was global competition, which forced many companies to become more customer oriented as a means of securing a competitive advantage. Progressive businesses wanted to become *customer centric*. Many expressed this as simply formalizing their ever-present mantra of *caring for the customer*; however, a number of managers viewed it as a chance to blend capabilities not heretofore available. A host of information technologies made it possible to collect and analyze data about customers and then to translate that knowledge into meaningful marketing strategies.

Another driver was the need for enterprise resource planning (ERP) vendors to find new products to sell. After the dizzying burst of ERP implementations leading up to Y2K (Y2K was the acronym used to designate the time when computer clocks would turn from December 31, 1999–January 1, 2000. Because many software programs used only two digits for the year, there was widespread concern that when the clocks turned from 99 to 00, data would be lost or distorted. ERP vendors assured potential customers their software used four digits for the year and would not have a problem; consequently, they were able to sell ERP systems as a precautionary step for many companies), it became apparent that the next wave of software implementations would occur in SCM. SCM offered a bonanza of applications that included CRM, along with other processes that needed software and implementation consulting help. While specialist vendors Salesforce.com developed CRM software, the major ERP vendors (SAP and Oracle) were selling extensions of their ERP packages that included CRM software. Even Microsoft wanted a share of the market, with the introduction of its Microsoft Dynamic CRM Live in 2007 (Bois 2006).

### What Can CRM Do?

CRM programs can accomplish the following activities:

- CRM collects information about customers, primarily from sales transactions, but also from a number of other *touch points*, such as complaints or inquiries. E-CRM is a popular means of recording website contacts, searches, and other nonsales activities. For example, Amazon.com recommends books based on your previous searches and requests that you provide reviews of books that you have purchased.

- From the collected data, a CRM program organizes the customer base into segments, or groups, of similar customers. The groups may be organized around age, income level, location, books searched, or whatever the marketing group deems useful. The ultimate objective is to get to a group of one (an individual customer), if that is practical.

- The marketing organization designs a program to appeal to the groups described earlier. While the primary emphasis of a CRM program is customer retention, marketers are not above designing sales programs that will attract new customers. As an example, Winer (2001) describes several types of programs including customer service, frequency/loyalty, customization, rewards, and community building programs.

- The programs designed to enhance the relationship with existing customers are implemented. Sometimes, salespeople may feel they are asked to change from a *hunt and kill* mode to a *tend the farm* mode (Nairn 2002). This terminology means that progress in these programs will be monitored closely and results will be measured.

- The CRM program develops a set of metrics to measure the results and to revise, modify, discontinue, and reverse the marketing initiatives that have been introduced.

### Benefits

The primary benefit of a CRM program is increased customer retention. Customer retention has greater value than customer acquisition. Businesses should expect to see improved financial results from their CRM program (Kennedy and Alfred 2004; Lambert 2006).

In addition to the tangible benefits, there are intangible benefits. The relationships with customers should be more open and effective. Hopefully, things will go smoothly; however, if there are incidents, the closer relationship should help in their resolution.

Internally, the need to develop cross-functional programs should increase collaboration among internal functions. Even without considering operations functions such as purchasing, production, distribution, and inventory management, there is a need to get sales, marketing, and IT people more comfortable with one another. Nairn (2002) highlights the difficulty in communication between the "emotion-driven sales force and the clinical binary-driven IT expert" as difficult unless intelligently managed.

## CRM Processes

As indicated by the SCOR and GSCF models, SCM is comprised of business processes across functions and business partners. As one of the key business processes, CRM focuses on developing and managing relationships with internal and external customers. CRM business processes consist of strategic and operational subprocesses as follows (Croxton et al. 2001):

*Strategic subprocesses*:

- Review corporate and marketing strategy.
- Identify criteria for categorizing customers.
- Provide guidelines for the degree of differentiation in the product/service agreement (PSA).
- Develop a framework of metrics.
- Develop guidelines for sharing process improvement benefits with customers.

*Operational subprocesses*:

- Differentiate customers.
- Prepare the account/segment management team.
- Internally review the accounts.
- Identify opportunities with the accounts.
- Develop the PSA.
- Implement the PSA.
- Measure performance and generate profitability reports.

CRM is part of a demand chain solution to address challenges in attracting potential customers and creating customer loyalty. Demand chains address issues related to customer demand, including product information, the product itself, services related to the purchased products, the product's ease of use and performance, the purchasing experience, and aftersales support (Fawcett et al. 2007). For example, Dell offers business partners web-based services 24/7/365 days a year. Amazon.com individualizes services for customers based on their shopping history. FedEx and UPS allow customers to track the status of their package. Airlines provide real-time updates of flight schedule changes to the mobile devices of passengers. The ability to address these issues can help companies gain a competitive advantage in the marketplace.

A comprehensive CRM solution is equipped with three primary areas of functions: operational, collaborative, and analytical functions. Operational functions can automate the fundamental downstream supply chain processes, including marketing, sales, and aftersales services. Collaborative CRM functions help the entire organization efficiently and effectively communicate with internal and external customers across the supply chain. Analytical CRM functions provide insights on customer behaviors and perceptions of product features, price, and service. Business intelligence can be improved with the support of analytical CRM functions.

The success of CRM practices depends on a company's capability to leverage customer information in B2B and B2C channel relationships (Christiaanse and Venkatraman 2002). Companies can customize these channels in two ways—relation/transactional and high tech/low touch—to improve the relationship between customers and suppliers (Ellis-Chadwick et al. 2001). Relational channels provide emotional appeal to customers via the four marketing Ps (price, place, promotion, and placement). In contrast, the transactional relationship downplays the emotional aspects and emphasizes the IT capability to streamline a transactional process. For both channel-building approaches, human and technology emphasis can have different degrees of involvement. Although IT supports are indispensable for the success of CRM, too much reliance on IT applications can bring diminishing returns to sales performance (Ahearne et al. 2004). The ultimate goal of IT usage is to optimize the performance of marketing tasks in achieving a mass customization strategy. How to effectively use IT to support a CRM strategy is both a talent and an art.

In the face of managing multiple sales teams after acquiring large organizations, Siemens Medical Solutions standardized its marketing and sales processes to integrate their sales teams (Camaratta 2005). As a result, the operational efficiency of the sales teams was substantially improved and the company sales objectives were achieved. This example shows that business processes and systems need to be flexible and amenable to integration across the supply chain in order to cope with changes in the business and technological environments. A company needs to incessantly improve their ability to synchronize information and processes across various customer touch points to build up a close connection among analytical, operational, and collaborative functions of CRM.

## Problems

Despite the benefits of CRM, there have been setbacks in its implementation. Although many companies anticipated CRM as a strategic practice, some IT-enabled CRM projects have been limited in their strategic focus (Payne and Pennie 2005). There are a number of reputable companies providing software and consulting services. However, a lack of strategic alignment and integration within the user company has caused many CRM projects to fail. Due to organizational variations, CRM technologies need to complement with other technologies in order to help a company compete strategically.

Other users of CRM have not been ecstatic about their results. Many have failed to realize the *perfect order* status heralded by CRM proponents. A Meta Group study reported a 55%–75% implementation failure rate (Chan 2005). According to Gartner Group, 55% of all CRM (software solutions) projects did not produce results (Lambert 2006). In a survey of senior executives, 25% reported that these software tools failed to deliver profitable growth and, in many cases, damaged long-standing customer relationships. Clearly, CRM has ways to go before its proposed benefits are fully realized.

Why have CRM programs not been more successful? It would be easy to blame the software vendors for *overhyping* their product, but that does not excuse the companies that bought the software from doing their part. In a very readable book, Keiningham et al.

describe 53 myths they attribute to customer loyalty programs (Keiningham et al. 2005). These myths include a number of promises that have not been met satisfactorily in practice:

- One of the leading problems has been the myopic view of considering CRM as solely an IT technology and not a strategic process. It is not enough to create a database of customers, no matter how cleverly designed.

- Another limited-scope perspective has been to consider CRM as just a marketing program. While marketing is the driver, they need cross functional support from the rest of the organization.

- Sometimes, businesses design the program around what they think they can do (their capability), rather than around the needs of the customer. As a result, the business may become efficient at doing things that the customer does not care about or respond to.

- Some companies fail to get widespread support within their organization for the CRM program. Sales personnel may not like the closer monitoring, finance managers may not agree with the deployment of resources, operations managers may feel left out, and IT staff may resent the cavalier attitude toward their innovative systems design.

- Companies may try to do too much too soon. Most researchers advocate a selective approach to the implementation of CRM.

- Customers may be *turned off* instead of *turned on* because of the excessive attention they receive. While subtlety is not generally associated with marketing efforts, it may be a trait to cultivate in CRM.

- There may be a lack of communication among interested parties. Another study examined the relationship among IT technology and infrastructure capabilities (technology that underpins the availability, quality, and depth of customer information), human analytic-based capabilities (the diverse skills and experience of employees necessary to interpret and use CRM data effectively), and business architecture and structural capabilities (the incentives and controls for employee behavior that support CRM). The study found that all three capabilities, not just technology alone, are necessary for a successful CRM program. The authors also concluded that these capabilities are not easy to acquire and maintain; therefore, many companies do not have these capabilities and this contributes to the lack of success for some CRM programs. As they put it, "CRM suffers when It is poorly understood, improperly applied, and incorrectly measured and managed" (Coltman et al. 2011, p. 217).

## CRM's Future

Should we conclude that CRM is ready to join a host of other *management fads* and fade away? Just the opposite. From its origin in the mid-1990s, CRM had a burst of popularity until 2000, when it fell into decline because of limited success and tighter IT budgets. After a low point around 2004, it appears to be in vogue again. Compton (2004) predicts renewed growth, particularly in nontraditional service areas, such as government and education. One reason for the renewed interest is that many supporters conclude that CRM is not just a marketing program; it is an essential part of the modern supply chain.

CRM research supports the conclusion that it is a management *fashion* and not a *fad*. It should have a great appeal in service industries, such as banks (Giltner and Richard 2000), public accounting (Snyder 2013), and insurance (West 2001). Electronic CRM has great promise. Feinberg et al. (2002) report on the progress in retailing. Shah and Mirza (2005) provide, in great depth, the use of web services to achieve effective CRM.

The rapid increase of social media sites and user activity offers opportunities in CRM program to take advantage of the access to customers and prospective customers. While the opportunities are great, the strategies to reach individuals in an effective way are still being studied and developed (Ang 2011; Klie 2011).

Companies may be misusing CRM programs by operating in ways that are in the interest of the provider but not in the interest of customers and other parties. Frow et al. (2011) call this the dark side of CRM. They identified numerous examples of this type of behavior and point out their list may not be exhaustive. Their examples include the following:

- *Communication-based dark side behavior*: Information misuse (selling or making available user information to other groups), customer confusion (complex pricing or confusing information about products), dishonesty (overcharges or pressure to buy unneeded items), and privacy invasion (collection of more information than needed)
- *Manipulating alternatives*: Customer favoritism (preferred treatment), customer *lock-in* (high switching costs in contracts), relationship neglect (especially in long-term relationships where customers are taken for granted), and financial exploitation (excessive late payment fees or gaps in insurance policies)
- *Side effects*: *Spillover* effects (pop-up ads or unsolicited telephone calls at dinner time) and ecological impacts (promotion of high sugar content drinks among the young or unneeded medications)

The authors caution that companies need to be sure their CRM programs follow an *enlightened* path.

CRM is a long way from achieving its potential. Some companies will get it all together and reap the benefits by assimilating the principles of CRM into their normal business practices. Others will go through the motions, stall, and move into a newer program that promises to be easier to implement and provide even greater benefits.

## Product Life Cycle Management

Another program that provides a system framework for supply chains is called PLM. It is concerned with managing a product, and its accompanying information, throughout the life of the product.

### What Is PLM?

PLM is a relatively new acronym in the management literature. Articles began appearing in 2000 and have steadily increased during the past few years. Sussman (2002) defined PLM as "the marketplace name for the comprehensive framework of technology and services that permits extended product teams—inside and outside of an enterprise—to collaboratively conceptualize, design, build, and manage products throughout their entire lifecycles." This definition pointed out two primary objectives of PLM—it is both a technology to manage information and a process to promote collaboration among departments within a single company or between separate companies.

Swink (2006) expands the scope of PLM and distinguishes between PLM as a management process and a PLM system. The system is a collection of hardware and software technologies that support PLM and is generally available for design and manufacturing engineers. The PLM management process extends the functionality to include CRM, SCM, and ERP. This process links product data with customer data, processing data, cost data, and resource planning data and makes these data accessible to everyone within the firm as well as to customers and suppliers.

### How Did PLM Evolve?

PLM started out as an extension of CAD to include electronic data–sharing capabilities among users, both internal and external to a company. Since these origins, PLM has grown rapidly to become a strategic initiative that parallels SCM as a global management concept.

CAD has been a well-accepted technology for several decades. It captures a vast amount of data about products and makes these data available to downstream participants in the product design and production processes. CAM enabled product design to be connected with the production process and uses applications of computer numeric control (CNC) to reduce product development times and production costs.

CAD was also linked with CAE, which is the "process of generating and testing engineering specifications on a computer workstation" (Blackstone 2013). CAD is linked with PDM systems used to track bills of materials and product revisions.

As the amount of data about products accumulated, new computing and data transmission capabilities made engineering data available in a variety of formats to downstream users. Thus, the concept of PLM was born.

### What Does PLM Include?

As the name implies, PLM includes the use of product information not only during its research and design stage but also during its production, postsale service, and recycling stages.

Gould (2005, p. 58) described it this way. "What's in a PLM system? Some authoring tools, computer-aided design (CAD); large dollops of simulation and visualization; lots of manufacturing data systems (e.g., computer-aided process planning [CAPP] and configuration management); heavy-duty infrastructure stuff (database management systems [DBMS] and data communications); and plenty of behind-the-scenes infrastructure utilities, such as web-based user interfaces and application programming interfaces (API)."

PLM life cycle information can include

- Design data and material lists from suppliers and their suppliers
- Bills of material and their many derivatives and revisions
- Design-for-manufacturing input, revisions history, notes, and source data
- Marketing and sales input
- Cost accounting data

- Test and performance criteria
- Manufacturing process requirements
- As-built information, including quality assurance data and regulatory compliance confirmation and certification
- Service and use history (McClellan and Harrison-Broninski 2006)

PLM has three core components: authoring, visualization, and business process support. Authoring includes the mechanical, electrical, and software design aspects. Visualization lets users collaborate across a wide range of participants in real time. Business process support provides a wide range of data about the product unit or batch during and after manufacturing that lead back to authoring. "CAD systems and other authoring tools have been available for many years. It's the extended cradle-to-grave view of the product that makes PLM a useful business tool, expanding horizontally to include wider participation and vertically to include data from more of the processes and events that make up the product lifecycle" (McClellan and Harrison-Broninski 2006, p. 37). They go on to say PLM handles the first two components well, but will need additions of business process management (BPM) and service-oriented architecture (SOA) to fully support the third component.

Companies face pressure to bring new models to market quickly. "Most major aerospace/ defense and automotive manufacturers depend on PLM systems, which include CAD/ CAM design and manufacturing tools, 3-D visualization software, CAE simulations and product data management (PDM), to direct the many highly complex and interwoven elements of modern manufacturing designs" (Waurzyniak 2012, p. 67).

## What Does PLM Not Include?

Since its inception, PLM, like most management programs, has become more holistic. From its beginning as a technology extending CAD, it is now a strategic initiative that some consider even broader than other global programs, such as SCM. One author lists the following phases as within the scope of PLM: requirements, portfolio management, planning, conceptual design, product engineering, manufacturing engineering, simulation validation, build and produce, test and quality, sales and distribution, in-service operation, maintenance and repair, and disposal and recycling (Gould 2005).

PLM does not generally include the planning and execution of the strategies and tactics to maximize market presence; the commercial or transactional activities related to ordering, shipping, and fulfilling orders or payments (i.e., ERP); the assigning of factory floor resources, managing material flows, task scheduling, line balancing, or equipment maintenance scheduling (i.e., MES); and the automation of customer or prospect databases (Gould 2005).

For our purposes, we will exclude the information that deals with the actual sale and production of a product's flow within the supply chain, such as in sales and operations planning, production planning and scheduling, purchasing, and distribution. This is a parallel information flow and, like PLM, under the umbrella of SCM.

## What Are the Benefits of PLM?

PLM started in the aerospace and automotive industries. The automotive industry wanted faster product development, and the aerospace industry wanted improved planning of repair and maintenance needs for spare parts. Boeing is an effective user of PLM with their 787 Dreamliner, and Toyota is the most progressive user of PLM in the automotive industry (Waurzyniak 2006).

An early study by AMR Research found benefits in three areas: (1) infrastructure savings, through the elimination of redundant manual data entry and deliveries of hard copies through couriers; (2) reduction of operating costs, usually within 6–12 months after companies go live with PLM; and (3) strategic competitiveness impacts, such as in time to market, market share, and gross margins, usually in 3–5 years (O'Marah 2003).

Another industry that embraced PLM is the consumer-packaged goods sector because of pressures to increase revenues and improve operating efficiencies. As supplements to PLM, they adopted product data and pack management, portfolio management, product requirement management, collaborative product design and visualization, and functional design integration. As a result, they achieved the following benefits:

- Faster time to market
- Longer, more profitable product life
- Reduced product development and production cost
- Increased product life cycle margins
- Improved product quality
- Increased development capacity
- Improved customer service and satisfaction
- Rapid quoting/responses
- Increased component reuse
- Increased enterprise use of PLM data in marketing and sales
- Better supply chain relationships (Nelson 2006)

There are reports of more novel PLM applications. Jimmie Johnson, winner of five consecutive NASCAR championships (six in total), owes much of his success to his team's use of PLM to track where and how parts are used and how to adapt to manufacturer's design changes (Bartholomew 2007). The NASCAR teams for Evernham Motor Sports, Joe Gibbs Racing, and Hendrick Motor Sports found PLM to be a valuable addition to their set of tools. One benefit is that simulation techniques reduce the need for expensive and time-consuming aerodynamic testing in wind tunnels (Sutton 2007).

PLM will aid in meeting FDA compliance requirements for medical device manufacturers. ArthroCare Corp., a manufacturer of minimally invasive surgical products, implemented a PLM program in 2003 to help streamline its product development process. They report that the implementation reduced record keeping and improved document control by making data available in real time across all the company's sites (Jusko 2006).

The conclusion by most authors is that cost-reduction benefits come faster, but may not be sufficient to justify the expense. However, the real payoffs come later, in 3–5 years, by producing revenue increases that would not materialize without the use of PLM.

## What Are the Obstacles?

If PLM is so promising, why is not everyone using it? As with many new programs, a number of obstacles prevent quick adoptions. PLM is complex because it consists of a variety of functional experts, large numbers of suppliers, vast quantities of data, and high rates of product introductions. Many PLM initiatives fail because they do not avoid six common traps: (1) lack of a complete vision, (2) not having a business case, (3) failure to construct a business release approach, (4) making PLM a departmental initiative, (5) placing too much burden on application software, and (6) no owner of the product life cycle (Conner 2004).

A study of collaboration building found that 80% of companies were dissatisfied with their collaborative development efforts for new products. Four barriers prevented effective relationships:

- *Physical and temporal barriers*: These impede real-time, rich communication relationships among team members. While technology is available, there is a need to design the timing and makeup of team structures to foster balance among the product design objectives of the team members.
- *Organizational and hierarchical barriers*: Standard operating procedures often create functional barriers that must be broken down. It may be necessary to realign reporting relationships and reward structures. Collaboration is more difficult if it includes multiple companies.
- *Relational barriers*: Individuals may be unwilling to collaborate due to a preconceived loss of power or status. It may be difficult to find employees who have both the technical and personal skills to work effectively in a collaborative environment.
- *Knowledge barriers*: Although the technical means to communicate may exist, and the team organization may be adequate, there is a final need to develop and classify knowledge and knowledge retrieval systems to make collaboration a reality (Swink 2006).

PLM has evolved into an engineering concept, and, in many cases, supply chain professionals have not been included in product life cycle initiatives nor have representatives from sales and marketing. If PLM is to succeed, it must expand its scope by including all functional areas of the business (Conner 2004).

Companies are finding that PLM technology is coming along quickly, although it is still a long way from being a completely integrated package. While the technology exists, there is a need to alter the infrastructures and cultures of companies to make PLM a complete success.

## What Is the Present Status of PLM?

The status of PLM implementations is mixed. Many initiatives have fallen short of their original expectations (Conner 2004). Swink (2006) concluded that a comprehensive PLM system has not been developed, although steady progress is being made. Teresko reports "PLM is not just a set of technologies, but a strategic business approach that integrates people, processes, business systems, and information" (Teresko 2007). Thilmany found that companies still have not adequately linked PLM to their manufacturing execution systems (MES) so information can easily move back and forth from engineering to manufacturing (Thilmany 2007).

PLM is being extended to the shop floor to connect with MES by such software developers as Siemens and Dassault (Waurzyniak 2012).

PLM is also being used by retailers to quickly and easily distribute final product specifications in different languages and with the latest changes. Suppliers also benefit because they spend less time and resources interpreting or modifying production directions (Goyal 2013).

In many technology innovations such as PLM, the potential benefits have not been fully realized. As new applications for PLM emerge, the realization of all the possibilities by any one system or company recedes. No single company can implement all of the applications that PLM designers envision. While this does not diminish the potential, it does mean it is more difficult to choose the specific components or applications of PLM that will be the most beneficial for a company.

The pieces that make up a PLM system exist; the task is putting them together into an integrated system and then introducing that system throughout a company's network of supply chains.

## What Does the Future Hold?

PLM has come a long way in a short time as a business tool. While there has been growth in the marketplace, there has also been consolidation among providers. The wave of mergers, acquisitions, and alliances between traditional PLM vendors and suppliers of enterprise software provides greater opportunities for data sharing and eventual collaboration. Examples of this merger trend include Oracle buying Agile Software and Siemens buying UGS. Microsoft also has been involved in several alliances. SAP is extending their strong position as an ERP provider with forays into additional areas such as CRM, SCM, and PLM. It appears that further consolidation is likely, much as in the CRM market (SanFilippo 2007). The scope of PLM continues to expand, from a technology to a system to a management concept to a strategic philosophy, with the evolution from CAD to PDM to PLM to PLM 2.0 (Brown 2008).

There is also the question of where PLM fits into the hierarchy of management programs. Is it going to be the umbrella that covers all other management systems such as CRM and SCM, or will it be a component of a larger concept? In order for companies to achieve increased profitability with PLM, they must tap into supply chain principles and expertise (Conner 2004). PLM can be thought of as the master system that includes PDM, CRM, SCM, and ERP (Swink 2006).

The use of PLM systems is being extended from manufacturing to a variety of industries, such as consumer goods—razors—and high fashion clothing. Where manufacturers worry about a large number of parts, retailers worry about a large number of stock keeping units (SKUs) involving colors, sizes, fabrics, styles, and packaging (Thilmany 2013).

In the early stages of PLM development, the high cost and difficulty in implementing it were barriers to entry for many small- and medium-sized companies. However, lower-cost PLM alternatives are coming available, which are also making it possible to serve a variety of customers (Waurzyniak 2012).

A number of companies are developing cloud-based applications not only for the traditional PDM of PLM but also for the digital manufacturing tools that form the core of engineering programs (Waurzyniak 2013).

One of the latest studies advocates the use of closed-loop PLM to denote an integrated approach to product life cycles that include three phases:

1. Beginning of life (BOL) during which the product is conceptualized, defined, manufactured, and sold.
2. Middle of life (MOL), when the product is used, serviced, and maintained.
3. End of life (EOL), during which the product can be refurbished and reused; the components can be reassembled and refurbished; the materials can be reclaimed with or without reassembly; and/or the product can be disposed with or without incineration (Haydaya and Marchildon 2012; Kiritsis 2011).

There is little doubt that PLM, as a managerial concept, is important and appropriate to most businesses. However, the fact that it is a desirable objective does not make it any easier to implement successfully.

## Supply Chain Configuration

Supply chains can be visually depicted in the form of a map. A supply chain can consist of three networks: manufacturing, service, and a reverse network (McCormack and Johnson 2002). Customers buy products and services from manufacturing and service networks and return products in various forms in reverse networks. In this supply chain map, customers are the drivers of these three networks of activities.

### Basic (Generic) Supply Chain

Another representation of a supply chain views it in terms of four basic zones: (1) the upstream zone, (2) the downstream zone, (3) the reverse stream zone, and (4) the spot-sourcing zones (McCormack and Johnson 2002). The primary objective of the upstream zone is to source raw materials from networks of suppliers to the transformation process, where these raw materials are converted into value-added components. Many core competencies are critical to the success of this sourcing and transformation process. These include material technology, process technology, capacity management, logistics, capital investment, innovation, and forecasting.

After the transformation process, the value-added components enter the midstream zone to be manufactured into finished products. The core competencies necessary at this midstream zone focus on production. These competencies include production engineering, capacity management, inventory management, planning and forecasting, materials handling, lot tracking, procurement, materials engineering, quality management, information systems, and product development.

The final products need to be moved into the hands of end user customers for consumption. Therefore, the downstream zone is created to fulfill product orders and provide services to customers. Core competencies at this zone are marketing and sales related. These include expertise in marketing, selling, demand management, customer service, order taking, order processing, warehousing, packaging and repackaging, postponement, quality management, financial services, payment processing, and CRM information systems.

Sometimes, customers may not be satisfied with their product because of a defect or some aspect of poor customer service. In other cases, customers may decide to recycle their products (e.g., batteries, personal computers, film, and ink cartridges) to collection points that take these items. Whether it is returning a defective product, or recycling an old one, customers at this point in the supply chain are participating in the reverse stream zone. There are other reasons products may need to be returned including a need to repair or recalibrate the used product or, perhaps, a product recall is in effect. In order to support these reverse stream activities, core competencies needed include excellent customer service, a return collection process, a return credit policy, warranty tracking capabilities, real-time inventory management, disassembly capabilities, troubleshooting, calibration and repair expertise, hazardous material transport abilities, and environmentally responsible recycling.

Spot sourcing is an auction market for buyers and sellers to source goods from each other without constructing the other three basic zones. The auction market has several variations. A forward auction is formed when a buyer sources from multiple suppliers and chooses the supplier with the lowest bid. A reverse auction is formed when many buyers compete with each other to source from a sole supplier. The buyer with the highest bid can close the deal. The third variation of an auction market is when matching buyers and sellers are needed when multiple buyers and sellers are present.

### Variations for Different Industries

Although supply chains vary among industries, even among companies in the same industry, there are some common features. Figure 4.2 depicts a general supply chain map with an originating point in the extractive (mining) or agriculture industries. The product then proceeds into manufacturing, which usually has some form of preliminary fabrication to make components, which are then assembled into finished products. From there, the products move into the distribution channels, usually consisting of an intermediate distribution center and a final destination at a retail store. There, the consumer buys the product. In today's e-business environment, sometimes, the shipment may go directly to the consumer from the manufacturer or distribution center without passing through the retail store.

In each stage of the supply chain, resources are required to transform the arriving inputs into the outputs to be passed on to the next stage.

### Supply Chain Mapping

To determine the components of a specific supply chain, a process called supply chain mapping can be used. This involves identifying the major entities involved in each of the zones of the supply chain. For a given participant, such as a final assembler, their internal records can be used to identify the principal customers and suppliers. For other participants, it is necessary to consult public records to decide who their major customers and suppliers are. A comprehensive explanation of this process can be found in Walker (2005).

Figure 4.3 is an example of a supply chain map developed from public information. Similarly, Figure 4.4 shows the process of recycling a battery throughout the reverse stream zone.

---

## Determining Resource Requirements

A supply chain involves the use of resources, such as facilities, equipment, employees, inventory, and information systems, in moving products or services from suppliers to customers or vice versa. The resources required are different depending on which of the four supply chain zones are being activated.

### Facilities

Facilities are essential to the movement of goods and services across the supply chain. Unlike other resources, facilities are physically located and cannot be easily changed due to their permanence. Therefore, successful execution of facility planning has a long-term, lasting impact on the competitiveness of a supply chain.

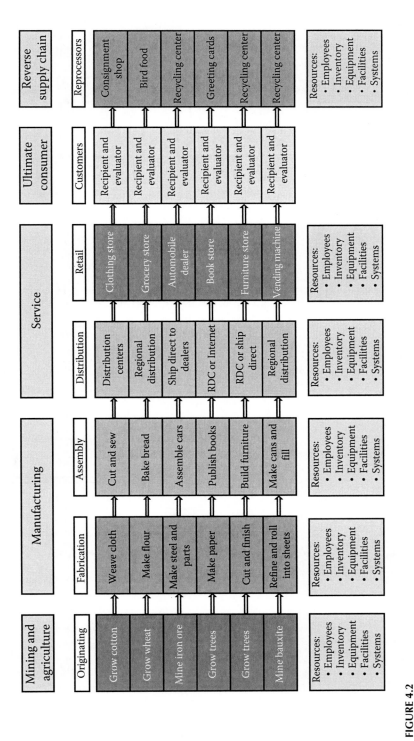

**FIGURE 4.2**
Examples of different supply chains.

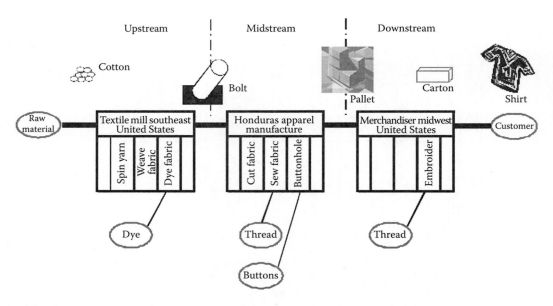

**FIGURE 4.3**
Supply chain map for apparel. (From Walker, W.T., *Supply Chain Architecture: A Blueprint for Networking the Flow of Material, Information, and Cash*, CRC Press, Boca Raton, FL, 2005. With permission.)

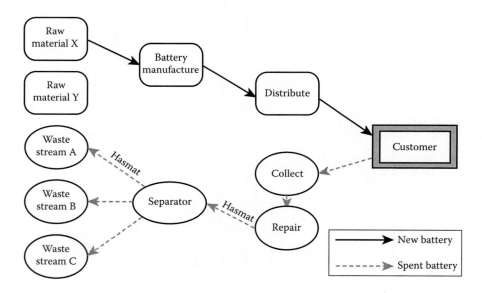

**FIGURE 4.4**
Mapping a reverse supply chain network. (From Walker, W.T., *Supply Chain Architecture: A Blueprint for Networking the Flow of Material, Information, and Cash*, CRC Press, Boca Raton, FL, 2005. With permission.)

Facilities as a resource are present in all four supply chain zones. In the upstream zone, a firm needs an R&D center to innovate new products and services and to pilot test them for mass production. In the case of sourcing raw materials from suppliers, a purchasing department would need to contact and negotiate with suppliers to determine the purchasing specifications and to agree on an acceptable purchasing cost. Suppliers store raw

**TABLE 4.3**

Facility Location Strategies and Their Characteristics

| Facility Location Strategy | Characteristics | Examples |
|---|---|---|
| Offshore factory | The offshore factory manufactures parts and finished goods and then exports them back to the main business in its home country. | Intel's factory in Penang, Malaysia, produces commodity components. |
| Source factory | The source factory involves plant management functions, such as supplier selection and production planning. | Hewlett-Packard's plant in Singapore produces calculators and keyboards. |
| Server factory | The server factory takes advantage of tax incentives and minimizes exchange risk and logistics costs. | Coca-Cola's international bottling plants serve local markets. |
| Contributor factory | The contributor factory engages in R&D, production planning, and procurement and suppliers' development. | Sony's factory in Bridgend, Wales, is an R&D center. |
| Outpost factory | The outpost factory has access to resources within the respective industry cluster. | Lego's factories in Denmark, Germany, Switzerland, and the United States are sources of innovative models and toys. |
| Lead factory | The lead factory is the most sophisticated and innovates new product and business processes throughout the entire organization. | Hewlett-Packard's factory in Singapore transformed itself into a knowledge-intensive factory. |

*Source:* Compiled from Ferdows, K., *Harv. Bus. Rev.*, 75(2), 73, 1997.

materials in a warehouse and then deliver them to the manufacturer. The manufacturer takes the raw materials and transforms them into finished goods. In the downstream zone, a distributor transports finished goods to a wholesaler or retailer for sales to the final consumer. In the reverse zone, a customer service center receives defective products and returns them to the manufacturer. A recycle center takes in goods and transforms them into reusable materials. Ineffective management of facilities and their locations can increase the cost of managing supply chains, impede the offering of good customer services, and lower the competitive advantage for the firms involved.

Optimization of facility location is another important consideration to the sustained competitiveness of a supply chain. An ideal facility location can minimize transportation costs, mitigate the influence of exchange rates, reduce labor rates and taxes, lower supply chain uncertainty, and support manufacturing practices (Bhatnagara and Sohal 2005). More importantly, the facility location should be aligned with the business strategy. Table 4.3 shows six facility location strategies growing in their strategic importance: offshore, source, server, contributor, outpost, and lead factories. On one end of the continuum is the offshore factory, which is the most primitive facility with the sole purpose of manufacturing commodity parts and exporting them to buyers. On the other end is the lead factory, which is the most advanced facility form with the prime objective of transforming a factory into a learning organization. The other facility strategies fall somewhere in between these two extremes. In choosing a location strategy, it is important that the business align its location needs with its overall business strategy (Ferdows 1997).

## Equipment

Various equipment or tools can be leveraged to support activities in different supply chain zones. Machinery to design new products and transportation vehicles to move goods are important equipment in the upstream zone. Manufacturing machinery, production assembly line configurations, and all types of transportation vehicles are equipment utilized in the downstream zone. In the reverse zone, equipment is needed to take recycled goods and disassemble them into parts and then into raw materials. Within the spot zone, computing equipment and websites help facilitate the online auction process.

Equipment can range from inexpensive general-purpose tools, such as screwdrivers and hammers, to expensive special-purpose equipment, such as automatic integrated circuit assembly equipment. Special equipment may be designed to assist workers or, in some cases, to replace the worker, especially in dirty or dangerous jobs.

## Employees

The skills and talents required to perform different functions within the supply chain zones vary greatly. In the upstream zone, intellectual capital is needed in the area of researching and developing new products and services. The developed ideas then need to be tested for ease of manufacturing. Once prototypes are accepted, they move into the manufacturing process where talent in production management is needed to plan how to meet the anticipated demand by the market. In order to accurately predict customer demand, market information is needed from the marketing and sales department. This area represents another important skill set needed by employees within the supply chain. The resulting collaboration process also needs to involve distributors in order to deliver products and services to customers.

Recycling specialists need to examine goods to be recycled back to the upstream supply chain. This process is important since improperly recycled goods can become waste and slow the recycling process, as well as cause unnecessary damage to the environment.

## Inventory

Customer demand fluctuates and is often difficult to predict. Allowing for some excess inventory is an important buffer to avoid the back order problem, especially in supply chains with volatile or variable deliveries. However, too much inventory increases holding costs, which can lead to a competitive disadvantage. Therefore, maintaining optimal inventory levels in order to control inventory costs and avoid back orders is a critical skill set within SCM.

Inventory control cannot be achieved unless the inventory level can be optimized at each step across the entire supply chain. This task means that all business partners in the supply chain need to collaborate in the process of optimizing inventory levels. Many factors can undermine the intention of achieving optimal inventory levels. Managing philosophies vary among companies and can therefore influence inventory management decisions. The relative bargaining power of buyer against seller is also a factor influencing the decision on how much inventory to keep. The physical distance needed to transport goods is another factor having an impact on inventory management decisions. Physical distance is critical since it is a key cost component of the final product as well as the lead time necessary to deliver those products.

## Information Systems

Information systems play supportive roles in moving goods and services. Within the upstream zone, the SRM software is often utilized. The SRM software has such functionalities as supplier management, spend analysis, category management, forward and reverse auctions, contract life cycle management, and cost and quotation management. The CRM software involves information systems utilized in the downstream zone. CRM supports four areas of supply chain operations: (1) front-office operations, (2) back-office operations, (3) business relationships, and (4) customer data analysis. A firm can utilize CRM to understand and interact with customers as well as build up loyal relationships with their customers. Information flow is described more fully in Chapter 14.

## Designing Processes to Match with Products

The process of delivering products and services varies with their variety and purchase volume. In addition, the degree of customization and expected development time must also be considered. Processes generally fall along a continuum from project based to rigid, as shown in Figure 4.5 (Crandall and Crandall 2013). Project-based processes are utilized when the delivery of products and services requires high customer contact, relatively high throughput time, and a high degree of process flexibility. In contrast, a rigid process is more feasible when the product is highly standardized with minimal variety and requires a structured flow, high capital intensity, low unit cost, and operational efficiency. Project-based

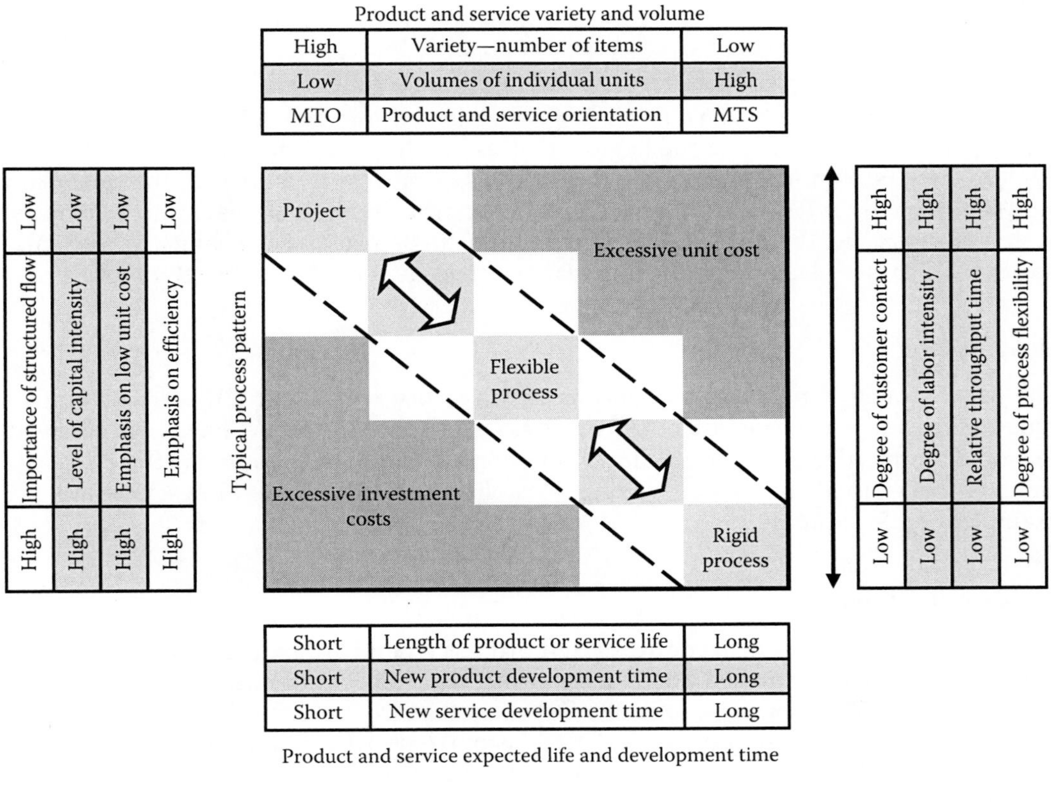

| High | Variety—number of items | Low |
| Low | Volumes of individual units | High |
| MTO | Product and service orientation | MTS |

Product and service variety and volume

| Short | Length of product or service life | Long |
| Short | New product development time | Long |
| Short | New service development time | Long |

Product and service expected life and development time

**FIGURE 4.5**
Product–process matrix. (From Crandall, R.E. and Crandall, W.R., *Vanishing Boundaries, How Integrating Manufacturing and Services Creates Customer Value*, 2nd edn., CRC Press, Boca Raton, FL, 2014. With permission.)

delivery processes are effective at dealing with situations where the variety of items is high, the volume of individual units is low, the length of product or service life is short, and new product development time is short. Rigid delivery processes are effective where production volumes are high and the product life cycles are long. Flexible processes are designed to provide the capability to produce high volumes of a variety of products. Designing processes to match the product characteristics can increase supply chain effectiveness.

There are four major SCM processes: (1) make to stock (MTS), (2) assemble to order (ATO), (3) make to order (MTO), and (4) engineer to order (ETO). We describe them in further detail in Chapter 7.

## Make to Stock

MTS is a traditional manufacturing strategy that makes products ahead of time in anticipation of potential demand. It uses inventory to buffer unpredictable customer demands and deliver products to customers at a scheduled due date. One of the problems with the MTS approach is the infamous *bullwhip effect*. This occurs when business partners do not collaborate and share information with each other. As a result, suppliers must speculate what the customer demand will be and, thus, cannot optimize their order decisions. As such, MTS often results in inventory levels that do not closely match demand. Although MTS may be able to increase the order service level and fill rate, manufacturers risk carrying products in inventory that could become outdated and have low profitability.

## Assemble to Order

The ATO practice increases the efficiency of delivering customized orders by first manufacturing parts or modular subassemblies and then assembling them into finished goods. Since the lead time needed to complete the manufacturing process is shortened, the time delay between receiving and delivering customer orders can be decreased. This result can increase customer satisfaction because of the shortened time to deliver and the ability to customize orders. ATO is also beneficial to the manufacturer because of reduced inventory levels.

## Make to Order

The MTO strategy attempts to (1) minimize inventory levels and (2) offer customized products by starting the manufacturing process after customer orders have been received. Customers are willing to trade waiting time for a customized product if the perceived value is higher than the cost of waiting. Custom furniture and wedding cakes are two examples of this strategy.

## Engineer to Order

An ETO strategy begins the product design process when customers convey unique customer specifications. Finished goods are not manufactured until customers approve the product design. Therefore, it is probable for a firm to stock raw materials and assemblies, but not finished goods simply because the customized order cannot be predetermined. Each customized order may require completely different bills of materials and manufacturing processes. As such, the lead time to deliver products and services could be lengthy and unpredictable. Customers are often heavily involved in the design and manufacturing process in the execution of the ETO manufacturing strategy.

## Determining the Mix of Make and Buy

A company can sustain competitive advantages when its resources and capabilities are valuable, rare, imperfectly imitable, and not easily substitutable (Barney 1997). These resources and capabilities include assets, business processes, information and knowledge, corporate culture, supply chain networks, and a secure infrastructure. A resource has value when it can help improve operational efficiency and effectiveness. Wal-Mart streamlines its supply chain networks to lower operational costs and delivers the lowest cost products to customers. The possession of strategic resources can directly contribute to competitive advantage, thereby improving operational performance (Newbert 2008). Apple Computer Inc. is noted for its product innovation ability. However, when a company does not possess these strategic resources, they must determine whether they should develop these resources as a core competency or purchase them from other companies.

For a long time, businesses have considered the basic decision in manufacturing of whether they should make a part or buy it from an outside supplier. In today's global business environment, the *buy* decision has become known as *outsourcing*. When the purchase is from a supplier in a foreign country, the term *offshore outsourcing* is used. This is in contrast to *offshoring*, which is used to describe establishing, or buying, and owning a business in another country.

### Core Competency Concept

A core competency is a business activity that a company can perform better than the competition or is considered so important that it must be done within the company. All companies need to continuously concentrate on their core competencies in order to distance themselves from competitors. Specialized and hard-to-copy knowledge can be used as an organizational core competency to compete in today's hypercompetitive business environment. Outsourcing practices arise because they allow a company to transfer some of the organization's noncore activities, such as production of parts or subassemblies; maintenance, repair, and operations (MRO) services; payment services; IT applications development; and legal services, to outside suppliers.

Outsourcing moves functions from inside a company to outside suppliers. Business proponents consider offshore outsourcing a strategic way to reduce operational costs, increase business agility, and expand international market share. As outsourcing forms grow in complexity and maturity, offshore outsourcing concentrations and business challenges intensify.

While modern outsourcing practices are growing in complexity, the strategic viability of offshore outsourcing needs to be properly assessed. A commonly cited reason for offshore outsourcing is to lower the cost of performing the contracted work. However, competitive advantages based solely on lowering costs have begun to deteriorate over the long run. As a result, it is important to properly assess these benefits before deciding what core competency to keep in-house and what noncore business activities to outsource.

### Total Cost of Ownership

When deciding whether to outsource or not, companies should consider the TCO. This requires adding to the direct cost of buying a good or a service the indirect costs, which can be both tangible and intangible. Minimizing TCO is important to procurement management.

Procurement professionals are exposed to two dimensions of transactional costs: (1) objective versus subjective factors and (2) internal versus external factors (den Butter and Linse 2008). Objective factors rely on facts, whereas subjective factors are subject to the perceptions of users. Objective factors are often related to financial issues, such as direct costs, quality improvement, on-time delivery, product life cycle costs, and transportation costs. Subjective issues include unethical behavior, diminished confidence in a brand, and adversarial labor relations. These subjective factors can result in an increase in operational costs and need to be assessed carefully when determining the TCO.

Internal factors relate to the organization itself and can be seen in the company's market position, brand, and the company reputation. Internal factors affect the relative bargaining power of buyers and sellers. External factors occur outside of the organization and can be seen in governmental regulations, market and industry labor costs, and currency exchange rates. When making a purchase decision, both internal and external factors can influence the objective and subjective values involved in the transaction.

## Cost Reduction versus Revenue Increase Considerations

When deciding which core competency to keep in-house, a company should assess operational and strategic benefits in addition to cost reduction. One recurring challenge for firms is the inability to articulate the actual benefits (e.g., cost savings, increased profit, or improved performance) when making an outsourcing decision. In outsourcing, there are both short-term operational benefits and long-term strategic benefits. In assessing short-term operational benefits, most firms adopt measurable metrics (e.g., production life cycles, inventory levels, and the number of customers served). However, long-term strategic benefits are often more difficult to measure. These benefits include customer satisfaction, business process transformation, organizational learning capacity, process consistency, international market expansion, and the capacity to innovate. Management constantly faces the challenge of demonstrating the long-term value of outsourcing initiatives to their organizations.

If a firm does not systematically assess long-term strategic benefits, top management risks assigning minimal importance to some obvious benefits, including labor cost reduction. Many organizations treat proposed outsourcing initiatives as an expense rather than an investment in the business. That treatment is usually due to the lack of measurement of strategic successes of outsourcing decisions in support of corporate strategy. Thus, it is vitally important that organizations begin to validate outsourcing initiatives by developing an integrative measurement system.

Trading partners need to weigh both tangible and intangible benefits and costs through the life of an offshoring relationship. Cost savings and reaching out to global customers are two of the most cited tangible benefits by offshoring advocates. Revenue growth has been shown in all geographical areas. The concept of TCO enables top offshore outsourcing performers to achieve total visibility, control, and transparency in the performance measurement process.

## Effect of Outsourcing Movement

The outsourcing movement can potentially deliver many business benefits. Tangible benefits include savings in capital investment, improvement of cash flow and cost accountability, and a reduction and controlling of operating costs. Outsourcing strategies can provide intangible benefits—access to new information technologies and scarce talents, concentration on core competencies, improved quality performance accountability, and better response to changes in the external environment.

Outsourcing decisions involve risks as well. One risk is the failure to consider hidden costs, such as vendor search and contracting, transition costs, costs of managing outsourcing projects, and the transition back to in-house production after outsourcing relationships have ceased. A second set of risks concerns the areas of shirking, poaching, and opportunistic repricing (Clemons 1991). Shirking occurs when a vendor does not perform up to expectation, but insists on charging full payment. Poaching occurs when confidential information about outsourcing projects is shared with competitors of the outsourcing firm. Opportunistic repricing occurs when vendors violate the contract and overcharge the outsourcer for unanticipated enhancements. Outsourcing involves operational risks, such as technological obsolescence, high turnover rates of professionals, and loss of strategic control (McClellan and Harrison-Broninski 2006). Outsourcing limits flexibility and requires larger cash investment in inventory because of the longer supply chains (Walker 2005).

In order to minimize these operational risks, outsourcing relationships need to be constantly maintained and updated to optimize performance that benefits both trading partners. As such, outsourcing should involve an ongoing evaluation of fixed and nonrecurring costs as well as variable and recurring costs.

Many companies face the challenge of consolidating business processes and outsourcing noncore functions in order to properly control operational costs (Mills and Sims 2008). Economic concerns are the first and foremost concern for offshoring partners to achieve sustainability in their relationships. A sustainable offshoring relationship needs to assess political and economic factors, the strength and track record of the supplier, supplier risks, and the governance structure of contracts (Bushell 2008).

## Aligning Entities along the Supply Chain

A firm's outsourcing strategy has a long-term impact on its sustainable competitiveness. The other business partners along the supply chain are also affected. Therefore, it is important to prepare a project plan to align the business interests of all partners along the supply chain. The project plan should be constantly monitored to cope with changes in the business environment. A project plan needs to address two important issues: (1) What entities should be involved in the supply chain? (2) How should authority and responsibility be allocated among these entities? Each of these is explained in further detail in the subsequent section.

In aligning participants along the supply chain, companies should try to avoid channel conflict. Channel conflict occurs when manufacturers (brands) disintermediate their channel partners, such as distributors, retailers, dealers, and sales representatives, by selling their products direct to consumers through general marketing methods and/or over the Internet through e-commerce. Another version of channel conflict occurs when a supplier is also a competitor. This situation can cause significant complications in trying to establish collaborative, knowledge-sharing relationships (Tsay 2004).

## Entities to Be Involved

In today's customer-driven economy, business partners along the supply chain have an array of business processes and systems to cope with varying customer demands. This reality has resulted in the proliferation of multiple contact points and channels to service customers. The result is that customers often have incompatible formats in working with their supply chain partners (Chan 2005). In order to create a unified view of customers across the supply chain, it is important that business partners create, exchange, share, store, and analyze information in a collaborative fashion. Not being able to do so can cause three types of disparities in the CRM ecosystem: (1) different functions and projects create silos of information, (2) information collected from different channels are not integrated, and (3) information analytic processes vary with company needs (Kelly 1999). The presence of these disparities can result in dissatisfied customers since sales and marketing activities based on this inconsistent information are not effective. Neither can upstream activities, such as new product development and the manufacturing process, be effective at meeting customer needs and wants.

## Allocation of Authority and Responsibility among Entities

Outsourcing jobs does not necessarily mean outsourcing responsibility. Mattel's lead-contained toys produced by its Chinese supplier put the company's reputation under fire. Customers may not understand or even care that upstream suppliers may be contributing to this product defect. From the customer's point of view, they just want a defect-free product, regardless of its origins. Therefore, it is important to look at the outsourcing venture as a team project. The members of the team are the business partners along the supply chain who contribute to the success of the final product. Clear roles and responsibility for all participant companies need to be specified, so that project authority can be appropriately allocated.

An outsourcing strategy can take on five formats: (1) domestic purchasing, (2) international purchasing only as needed, (3) international purchasing as part of sourcing strategy, (4) an integrative sourcing strategy across worldwide locations, and (5) an integrative sourcing strategy across worldwide locations and functional groups (Trent and Monczka 2005). Each succeeding level grows more complex in terms of functional as well as locational coordination and integration. As such, roles and responsibilities of business partners increase as the outsourcing relationship compounds.

## Collaboration Process

Rapid changes in technology make the management of an information systems outsourcing contract an increasingly important issue. Because technology changes so quickly, the duration of the contract often outlives the service offerings, and technology is often obsolete by the time the contract expires. The outsourcer's goal is for its provider to fulfill the terms of the contract, while the provider's goal is often to amend the contract to extend the terms of the contract, possibly with more expensive, updated technology. Conflicting goals may cause tensions to develop between the outsourcer and provider. Mitigating any potential conflict is important in an outsourcing relationship.

Trust is an important factor that influences an interorganizational relationship (Smith et al. 1995). When a firm is willing to outsource, it becomes more involved with its sourcing partners in the implementation of projects (Premkumar and Ramamurthy 1995). In an outsourcing relationship, the degree of trust can further the degree of cooperation between partners.

Some firms behave opportunistically and take advantage of an outsourcing relationship to maximize one-time financial gains. But in a trusting relationship, open discussions can take place, and partners can jointly formulate an outsourcing strategy that will benefit both parties in the long term. A firm is also more likely to form long-term alliances with a partner that it trusts. Furthermore, a firm that trusts its partners is not only more likely to share information but also more likely to invest capital and labor in improving and maintaining the outsourcing relationship. This may occur because one or both partners realize that a successful outsourcing project is beneficial for both the outsourcer and the provider.

In addition to building trust within a partnership, building public trust is also an important issue from the perspective of the outsourcing provider. For example, an outsourcing provider that mishandles confidential information of consumers is likely to lose the trust of its existing partners. Likewise, a provider that subcontracts work to offshore firms that are known to disregard intellectual property rights is also likely to lose trust as well as future business. In this regard, an outsourcing provider can build public trust by continuously demonstrating its commitment to security and privacy.

## Implementation Plan

Like any project management plans, an implementation plan for outsourcing projects needs to include a set of essential elements, including an overview of the project, project organization, management and technical processes utilized, work to be performed, schedule, and budget (Schwalbe 2007):

- *Project overview*: This section should include project names, a brief description of the project, sponsor names, deliverables of the project, supporting materials, and a list of definitions and acronyms.
- *Project organization*: Organizational charts, project responsibilities, and process-related information.
- *Management and technical processes*: Management objectives, project controls, risk management, project staffing, and technical processes.
- *Work to be performed*: Major work packages, key deliverables, and other work-related information.
- *Schedule*: Summary schedule, detailed schedule, and other schedule-related information.
- *Budget*: Summary budget, detailed budget, and other budget-related information.

Outsourcing requires a project approach because it involves a series of steps requiring multifunctional participation.

## Evaluating the System Design

The supply chain system design must pass several rigorous tests. Will it accomplish its objectives? Is it sustainable over a long period of time? Is it flexible enough to be adapted to changing conditions?

## Will It Accomplish Its Objectives?

Customers have been driving the movement of supply chain designs. For instance, manufacturing strategies have been shifting from MTS, ATO, MTO, and ETO toward mass customization, in order to provide personalized products and services to larger numbers of customers. The success of a CRM strategy depends on the construction of a well-orchestrated supply chain across business partners. A business can reference the SCOR model in designing their supply chain systems: planning, sourcing, making, delivering, and returning (Supply Chain Council 2013). However, it is rare to see a firm provide all business activities. As such, firms begin concentrating on what they can do the best, their core competencies, and sourcing the other activities from business partners. This resource-based view of designing supply chains has resulted in cost reductions and new market expansion.

Whether outsourcing takes place onshore or offshore, its practice is increasing because of the demands to both satisfy customer needs and lower operational costs. However, there are hidden costs that could offset the potential savings and earnings of these outsourcing initiatives. In addition, the increased complexity and integration efforts associated with more complex outsourcing strategies often offset potential cost savings. Take a typical 250-seat call center that may cost several millions of dollars to set up. Many hidden expenses, such as high turnover rates, power outages, work inefficiency, and inadequate training, all need to be taken into account in order to cover the cost savings as the direct result of cheap labor (Farrell 2006). Therefore, more complex sourcing strategies are often required if the outsourcer desires higher benefits to the firm, such as organizational agility.

A firm can increase its supply chain agility via technical and managerial solutions. Technical solutions include SOA (Ren and Lyytinen 2008), semantic web technology (Zisk 2008), and basic industrial engineering tools. Strategic management of outsourcing is the first managerial solution to strategize offshoring relationships, thereby increasing supply chain agility. Maintaining a positive social influence contributes to the establishment of long-term, trusting relationships, which can also increase supply chain agility. Utilizing integrative measurements can also increase the accountability of offshoring relationships, thereby advancing agility in supply chains.

When evaluating the design of a supply chain, it is important to first recognize customer demands and use that knowledge to formulate a supply chain strategy to meet those demands. A firm can selectively choose core supply chain functions to perform and source noncore functions to business partners. The TCO concept should be used to assess if the objective of using the supply chain to reduce cost or increase profit has been accomplished.

## Is It Sustainable?

A supply chain is dynamic in nature because its business partners need to interact with each other within a business environment that is constantly changing. Under these circumstances, a *win–win* perspective among supply chain partners is indispensable to sustain the business relationships in the supply chain network. Although some partners in the supply chain may be pleased with a loose return policy, manufacturers and suppliers may not be able to survive when retailers take advantage of the situation. Likewise, carrying too much inventory may be able to satisfy some customers in the supply chain. Yet, the excessive inventory cost can sabotage the financial health of suppliers. In the area of trust, not sharing marketing and sales information may cause product design teams to be unable to engineer and develop a product that meets the customer's requirements.

Many supply chain initiatives, such as demand management, manufacturing flow management, customer service management, SRM, and returns management, are effective at addressing today's SCM challenges. However, a lack of cooperation, collaboration, and integration among business partners can substantially lower the effectiveness of these SCM initiatives.

## Is It Flexible?

Supply chain flexibility is the ability of business partners to reconfigure and redeploy resources at different instances, usually because of some change in the external environment. For instance, in the past, a tsunami in Japan, floods in Thailand, and political insurrections in Egypt stalled airports and other transportation means, which disrupted the normal operations of the associated supply chains. How can business partners restructure themselves to bypass these disruptive events within a short period of time and quickly resume to normal operations via spot or systematic sourcing solutions? The ability to be responsive and flexible can minimize the potential loss caused by unanticipated events. Some companies develop contingency and crisis management plans to account for such shocks in the supply chain.

Outsourcing empowers an organization to be flexible and to effectively compete in the global marketplace. An agile organization is one that has a flexible information systems architecture, high sensitivity to customer needs, integrated business processes, and collaborative relationships (Christopher 2000). Thus, outsourcing has the potential to help transform a company into an agile organization. However, an outsourcing strategy is a necessary but not sufficient condition for an agile organization. This is because strategies formulated by top management take time to permeate through the organization and to the outsourcing partners. The learning curve also slows the transformation process of a company into an agile organization. A breakdown in the psychological contract can also result in the ineffectiveness of organizational learning in the outsourcing context (Schlosser et al. 2006).

To shorten the time it takes for the entire organization to adopt a particular strategic direction, the social factors of trust and power must be balanced among supply chain partners. In addition, an integrative measurement system to assess outsourcing projects is also essential. Intercompany coordination via information sharing can increase flexibility in delivery quantity and due-date requirements (Chan and Chan 2009). Therefore, to increase organizational agility through outsourcing, a strategic emphasis of outsourcing, positive social influences, information sharing, and an integrative measurement approach are all necessary.

Supply chain flexibility enables a firm to sense and respond to business opportunities at the enterprise-wide level. Although uniformity and standardization of governance structures can support organizational agility (Kevin and Worrell 2008), implementation across trading partners is often difficult to achieve. While it may be difficult to standardize an interorganizational governance structure, it is feasible to establish strategic, long-term relationships among business partners across the supply chain.

## Summary

In this chapter, we described two basic supply chain models—the SCOR model and the GSCF model. The SCOR model is based on the recommendations of the SCC, a nonprofit global organization. This model is based on the concepts of business process reengineering, benchmarking, and best practices. The SCOR model integrates five SCM processes that exist within each partner and across partners: planning, sourcing, making, delivering, and returning. The GSCF model takes a cross-functional and cross-business perspective to examine the effectiveness of supply chain operations. The GSCF model comprises three primary elements: the supply chain network structure, the supply chain business processes, and the associated management components.

We also described two current programs that are being used to develop supply chain systems—CRM and PLM. CRM refers to business processes executed by a company with its business partners to service customers and sustain relationships with them. CRM is essential in helping to build trust into the supply chain network. PLM is concerned with managing a product, and its accompanying information, throughout its sequential life phases. Because products and supply chains are dynamic, considerations must be made for product changes throughout the supply chain processes.

We then described a basic supply chain model and variations of that model across several industries. Next, we discussed the resources required in a supply chain and the types of processes that fit within a supply chain. Finally, we covered the role of outsourcing and its effect on supply chain system design. The key consideration in this *make* or *buy* decision is to determine which functions of the company are part of its core competencies. Those areas of operations that fall outside of the core competencies are candidates for outsourcing.

## Hot Topic: Sweatshops

Despite the best efforts of companies to design and build effective supply chains, they sometimes run into difficulties. The hot topic for this chapter is about one of those difficulties.

### Hot Topics in the Supply Chain: Sweatshops—Two Sides of the Same Coin

Most readers are aware that sweatshops are linked to offshore outsourcing. Typically, one hears that a company has outsourced the making of a product to a company in Asia for *cost reasons*. Usually, we are accustomed to thinking that the cost savings are mainly related to labor. However, less strict environmental regulations and relaxed to nonexistent worker safety codes also constitute cost savings, although many companies may not want to admit that up front. Savings can also be achieved by lengthening the work hours for workers, yet keeping pay low. Collectively, low pay, lax worker safety regulations, and long hours lead to what has been commonly referred to as a sweatshop.

No CEO or supply chain manager wants to admit that sweatshops are part of their supply chains. And yet, the reality is that if offshoring has gone to a developing country, there is a good chance there may be a sweatshop close by. Welcome to the world of international trade.

#### BACKGROUND

Global competition has resulted in intense price competition for certain classes of goods. Many of these can be found in the likes of Wal-Mart, Target, Dollar General, and a host of other discount savvy retailers. To keep costs low for the customer, manufacturers must meet the demands of discount retailers. It is the retailers, not the manufacturers, who have the upper hand when it comes to bargaining power. Retailers look for manufacturers who can produce products at the least cost possible. This scenario means even U.S.-based manufacturers must outsource some or all the making of the product to companies in Asia or other developing countries. Ellen Shell (2009, p. 209) describes it this way:

> When competition is mostly about price, innovation often takes a backseat to cost cutting. Laying off workers and hiring cheaper ones is one sure way to enhance the bottom line. Another is to scour the world for low-wage workers, especially those in countries with lack-luster enforcement of environmental and workers' rights regulations.

Scouring the world for low-wage workers may not seem like a noble endeavor for some, but it is the reality of some modern-day supply chains. The irony is that it is consumers who drive this process. Consumers want (make that demand) the lowest cost possible on certain goods, which means retailers must offer those goods. In a free market, consumers are free to shop where they can find the best bargain, which means cost cutting by retailers can be intense. In response, retailers must dictate to manufacturers the cost parameters in the making of a product. This often means sourcing to companies in developing countries, with low-paid workers, lax safety standards, autocratic supervision of employees, and questionable environmental standards (Eckes 2011). Not only do companies compete with each other to find low-cost sources in developing countries, but the developing countries compete with each other to attract business, a phenomenon that has been labeled the race to the bottom. In the race to the bottom, governments cut back on welfare programs for employees in order to entice businesses to set up operations or contracts in their countries (Chan and Peng 2011). This scenario results in a swing of power away from the employee and into the hands of the multinational corporation (MNC) and its contractor in the host country.

## CRITICS OF SWEATSHOPS

Not surprisingly, there are many critics of sweatshops, and there should be. Unsafe work environments with less than local-market pay and where supervisors may be coercive should not be an acceptable norm in any country. Indeed, the thought of MNCs saving money by having products made under these conditions is deplorable in the eyes of the company's stakeholders. And yet, sweatshops continue to be a problem. Bangladesh has become a symbol of unsafe building structures in the garment industry (Eckes 2011); Thailand a haven for aquatic sweatshops in the shrimp industry (Shell 2009); and China, hosting a factory environment of long-hour, military-style supervision and hot working conditions (Chan and Peng 2011).

In a nutshell, critics see sweatshops as a violation of basic worker rights and welfare. Their viewpoint is that no company should treat its employees poorly or pay substandard wages. Instead, MNCs should work with their suppliers to upgrade the factories and management practices.

## SUPPORTERS OF SWEATSHOPS

To say that one is a supporter of sweatshops probably is not the right phrase, but nonetheless, there is a school of thought that maintains that this form of offshore outsourcing does not always deserve the negative reputation it has acquired. First of all, sweatshops do create jobs for those in developing countries. In fact, without some form of work, many citizens would face unemployment altogether with no means of supporting a family (Lund-Thomsen 2008; Shell 2009). Workers in developing countries do not understand when university students call on their schools to boycott sweatshops. To these workers, boycotting sweatshops means taking away their sole means of income (Miller 2003). For women, the only alternative to working in a sweatshop may be living in prostitution or a hard life as an agricultural worker (Eckes 2011).

Second, for a developing country, sweatshops may be the main link to global supply chains. Indeed, "global supply chains matter for international development ... as the ability of countries to prosper depends on their participation in the global economy, which is largely a story about their role in global supply chains" (Gereffi and Lee 2012, p. 24). Developing countries want to make a name for themselves and show the rest of the world they can be a player in the global economy. Being part of the supply chain for an MNC can be seen as a badge of honor for these countries.

Finally, developing countries tend to have large populations that need to be employed. For these countries, finding employment for their massive populations is a goal in itself. It does not necessarily matter if the jobs are good, just so that there are jobs somewhere that can employ the multitudes. China illustrates this dilemma well. Indeed, much of China's economic reforms and striving to enter the World Trade Organization (WTO) were strategies to employ the country's massive population (Chan and Peng 2011).

## QUESTIONS FOR RESEARCH AND DISCUSSION

1. Suppose that two developing nations are competing for MNCs to contract manufacturing in their countries. What avenues are available to the political leaders of these countries to lure business to their areas? How could one country gain an advantage over the other country in gaining the MNC contracts?

2. Suppose that a family member or friend approaches you and argues that MNCs should never outsource to countries outside of the home country. Given what you have learned about global supply chains, what would you tell that person?

## REFERENCES

Chan, C. and Peng, Z., From iron rice bowl to the world's biggest sweatshop: Globalization, institutional constraints, and the rights of Chinese workers, *Social Service Review*, 85, 421–445, September 2011.

Eckes, A. Jr., The seamy side of the global economy, *Global Economy Journal*, 11(3), 1–26, 2011.

Gereffi, G. and Lee, J., Whey the world suddenly cares about global supply chains, *Journal of Supply Chain Management*, 48(3), 24–32, 2012.

Lund-Thomsen, P., The global sourcing and codes of conduct debate: Five myths and five recommendations, *Development and Change*, 39(6), 1005–1018, 2008.

Miller, J., Whey economists are wrong about sweatshops and the antisweatshop movement, *Challenge*, 46(1), 93–122, 2003.

Shell, E., *Cheap: The High Cost of Discount Culture*, Penguin Books, New York, 2009.

## Discussion Questions

1. Why is CRM important in the effectiveness of a supply chain? Identify a company that you think has a good CRM focus. What reasons would you give for identifying this company?

2. Explain the concept of PLM. Why do supply chain managers need to understand the importance of PLM?

3. What is a supply chain map? Draw out a simple supply chain map for the food service operation at your school.

4. Identify the five types of resources needed in a supply chain as discussed in this chapter. Which resources are the most scarce where you work? In what ways does your company work to sustain these resources?

5. Identify and differentiate the following production methods: MTS, MTO, ATO, and ETO. In your explanation, give examples of each.

6. What should a company consider as it decides whether to make a product or have it outsourced?

7. What products do restaurants typically make themselves? What products do they typically outsource?

8. Explain why TCO is an important consideration in the design of a supply chain.

9. Explain how offshore outsourcing affects the design of a supply chain.

10. The basic theme of Figure 4.5 is that the process used to make a product must align with the product needs. Discuss.

---

## References

Ahearne, M., Weinstein, L., and Srinivasan, N., Impact of technology on sales performance: Progressing from technology acceptance to technology usage and consequence, *Journal of Personal Selling and Sales Management: Special Issue on Customer Relationship Management: Strategy, Process and Technology*, 24(4), 297, 2004.

Ang, L., Is SCRM really a good social media strategy? *Journal of Database Marketing & Customer Strategy Management*, 18(3), 149–153, 2011.

Barney, J.B., *Gaining and Sustaining Competitive Advantage*, Addison-Wesley Publishing Company, Reading, MA, 1997.

Bartholomew, D., PLM makes Jimmie Johnson go fast; NASCAR frontrunner credits product life-cycle management for unbelievable winning season on the track, *Baseline*, November 2007.

Bhatnagara, R. and Sohal, A.S., Supply chain competitiveness: Measuring the impact of location factors, uncertainty and manufacturing practices, *Technovation*, 25(5), 443–456, May 2005.

Blackstone, J.H., *APICS Dictionary*, 14th edn., APICS—The Association for Operations Management, Chicago, IL, 2013.

Bois, R., Live, From Boston! It's Microsoft CRM, www.amrresearch.com, Boston, MA, July 12, 2006.

Brown, J., The PLM evolution of the metal benders, *Manufacturing Business Technology*, February 15, 2008.

Bushell, M., Go offshore but do your homework, *Global Telecoms Business*, 97, 84, March/April 2008.

Camaratta, J., From theory to execution, *Marketing Management*, 14(4), 16, 2005.

Chan, H.K. and Chan, F.T.S., Effect of information sharing in supply chains with flexibility, *International Journal of Production Research*, 47(1), 213, 2009.

Chan, J.O., Toward a unified view of customer relationship management, *Journal of American Academy of Business*, 6(1), 32–38, March 2005.

Christiaanse, E. and Venkatraman, N., Beyond Sabre: Expertise exploitation in electronic channels, *Management Information Systems Quarterly*, 26(1), 15, 2002.

Christopher, M., The agile supply chain: Competing in volatile markets, *Industrial Marketing Management*, 29(1), 37, 2000.

Clemons, E.K., Evaluation of strategic investments in information technology, *Communications of the ACM*, 34, 22, 1991.

Coltman, T., Devinney, T.M., and Midgley, D.F., Customer relationship management and firm performance, *Journal of Information Technology*, 26(3), 205–219, 2011.

Compton, J., It's not business as usual, *Customer Relationship Management*, 8(12), 32–37, December 2004.

Conner, M.P., The supply chain's role in leveraging PLM, *Supply Chain Management Review*, 8(2), 36, 2004.

Crandall, R.E. and Crandall, W.R., *Vanishing Boundaries, How Integrating Manufacturing and Services Creates Customer Value*, 2nd edn., CRC Press, Boca Raton, FL, 2014.

Croxton, K.L., Garcia-Dastugue, S.J., Lambert, D.M., and Rogers, D.S., The supply chain management processes, *International Journal of Logistics Management*, 12(2), 13–36, 2001.

den Butter, F.A. and Linse, K.A., Rethinking procurement in the era of globalization, *MIT Sloan Management Review*, 50(1), 76, Fall 2008.

Ellis-Chadwick, F.E., Mchardy, P., and Wiesehofer, H., On-line customer transactions: A journey from close connective relationships to anonymous transactional relationships? *A Marketing Odyssey*, Cardiff, U.K., p. 90, July 2001.

Farrell, D., Smarter offshoring, *Harvard Business Review*, 84(6), 84, June 2006.

Fawcett, S.E., Ellram, L.M., and Ogden, J.A., *Supply Chain Management: From Vision to Implementation: International Edition*, Pearson Higher Education, Upper Saddle River, NJ, 2007.

Feinberg, R.A., Kadam, R., Hokama, L., and Kim, I., The state of electronic customer relationship management in retailing, *International Journal of Retail & Distribution Management*, 30(10), 470–481, 2002.

Ferdows, K., Making the most of foreign factories, *Harvard Business Review*, 75(2), 73–88, 1997.

Frow, P., Payne, A., Wilkinson, I.F., and Young, L., Customer management and CRM: Addressing the dark side. *The Journal of Services Marketing*, 25(2), 79–89, 2011.

Giltner, R. and Richard, C., Re-think customer segmentation for CRM results, *The Journal of Bank Cost & Management Accounting*, 13(2), 3–19, 2000.

Gould, L.S., Additional ABCs about PLM, *Automotive Design & Production*, 117(12), 58, 2005.

Goyal, V., Consistent products, *Retail Merchandiser*, 54(1), 4–5, 2013.

Haydaya, P. and Marchildon, P., Understanding product lifecycle management and supporting systems, *Industrial Management + Data Systems*, 112(4), 559–583, 2012.

Jusko, J., Complexity cure, *Industry Week*, 255(12), 45, 2006.

Keiningham, T.L., Vavra, T.G., Aksoy, L., and Wallard, H., *Loyalty Myths: Hyped Strategies That Will Put You Out of Business—And Proven Tactics That Really Work*, John Wiley & Sons, Inc., Hoboken, NJ, 2005.

Kelly, D., Focus on the customer, *Sales & Marketing Management*, 151(12), 3a, 1999.

Kennedy, M.E. and Alfred, M.K., Using customer relationship management to increase profits, *Strategic Finance*, 85(9), 36–42, March 2004.

Kevin, P.G. and Worrell, J.L., Organizing IT to promote agility, *Information Technology and Management*, 9, 71–88, March 2008.

Kiritsis, D., Closed-loop PLM for intelligent products in the era of the internet of things, *Computer-Aided Design*, 43(5), 479–501, 2011.

Klie, L., Socially challenged, *Customer Relationship Management*, 15, 14, May 2011.

Lambert, D., Cooper, M., and Pagh, J., Supply chain management, implementation issues and research opportunities, *International Journal of Logistics Management*, 9(2), 1–19, 1998.

Lambert, D.M., The customer relationship management process (Chapter 2), *Supply Chain Management Processes, Partnerships, Performance*, Lambert, D.M. (Ed.), Supply Chain Management Institute, Sarasota, FL, pp. 25–40, 2006.

Lambert, D.M., Garcia-Dastugue, S.J., and Croxton, K.L., An evaluation of process-oriented supply chain management frameworks, *Journal of Business Logistics*, 26(1), 25, 2005.

McClellan, M. and Harrison-Broninski, K., Product meets process, *Intelligent Enterprise*, 9(2), 37, 2006.

McCormack, K.P. and Johnson, W.C., *Supply Chain Networks and Business Process Orientation: Advanced Strategies and Best Practices*, Vol. 27, CRC Press, Boca Raton, FL, 2002.

Mills, T. and Sims, S., Balancing electronic efficiencies and paper-processing costs, *Banking Strategies*, 84(1), 40, 2008.

Nairn, A., CRM: Helpful or full of hype? *Journal of Database Management*, 9(4), 376–382, July 2002.

Nelson, T., Innovating new product development with product lifecycle management, *Supply Chain Europe*, 15(2), 40, 2006.

Newbert, S.L., Value, rareness, competitive advantage, and performance: A conceptual- level empirical investigation of the resource-based view of the firm, *Strategic Management Journal*, 29(7), 745, July 2008.

O'Marah, K., The business case for PLM, *Supply Chain Management Review*, 7(6), 16, 2003.

Payne, A. and Pennie, F., A strategic framework for customer relationship management, *Journal of Marketing*, 69(4), 167–177, October 2005.

Premkumar, G. and Ramamurthy, K., The role of interorganizational and organizational factors on the decision mode for adoption of interorganizational systems, *Decision Sciences*, 26(3), 303, 1995.

Ren, M. and Lyytinen, K., Building enterprise architecture agility and sustenance with SOA, *Communications of the AIS*, 22, 75, 2008.

SanFilippo, E., More ways to collaborate as PLM market consolidates, *Manufacturing Business Technology*, 25(7), 46, 2007.

Schlosser, F.K., Templer, A., and Ghanam, D., An integrated agenda for understanding the impact of HR outsourcing on organisational learning orientation, *Journal of Labor Research*, Special issue on Outsourcing Management, 27(3), 291, 2006.

Schrage, M., Connect demand with supply: The CIO's next big challenge will be figuring out how to bridge the company's sometimes conflicting CRM and SCM objectives, *CIO*, 18(4), 1–3, November 15, 2004.

Schwalbe, K., *Information Technology Project Management*, 5th edn., Course Technology, Boston, MA, 2007.

Shah, J.R. and Mirza, B.M., Effective customer relationship management through web services, *The Journal of Computer Information Systems*, 46(1), 98–109, Fall 2005.

Smith, K.G., Carroll, S.J., and Ashford, S.J., Intra- and interorganizational cooperation: Toward a research agenda, *Academy of Management Journal*, 38(1), 7, 1995.

Snyder, M., Maximizing revenue opportunities with CRM systems, *The CPA Journal*, 83(1), 15, 2013.

Supply Chain Council, SCOR®, Supply Chain Operations References model: SCOR Version 10.0, pp. 3–5, Washington, DC, 2013. http://www.supply-chain.org/resources/scor.

Sussman, D., Survey says PLM gaining momentum, *MSI*, 29(9), 27, 2002.

Sutton, M., Races to virtual engineering, *Ward's Auto World*, 43(12), 20, 2007.

Swink, M., Building collaborative innovation capability, *Research Technology Management*, 49(2), 37, 2006.

Teresko, J., PLM market grows to meet integration challenges, *Industry Week*, 256(6), 19, 2007.

Thilmany, J., Engineering meets manufacturing, *Mechanical Engineering*, 129(12), 20, 2007.

Thilmany, J., Lifecycle management: It's not just for engineers anymore. PLM Chic, *Mechanical Engineering*, 135(3), 38–41, 2013.

Trent, R.J. and Monczka, R.M., Achieving excellence in global sourcing, *MIT Sloan Management Review*, 47(1), 24, 2005.

Tsay, A.A., Channel conflict and coordination in the e-commerce age, *Production and Operations Management*, 13(1), 93, 2004.

Walker, W.T., *Supply Chain Architecture: A Blueprint for Networking the Flow of Material, Information, and Cash*, CRC Press, Boca Raton, FL, 2005.

Walmart, Walmart Returns Center, 2013. http://help.walmart.com/app/answers/detail/a_id/190, accessed May 15, 2013.

Waurzyniak, P., Collaborating with PLM, *Manufacturing Engineering*, 137(4), AAC1, 2006, accessed May 15, 2013.

Waurzyniak, P., PLM systems further extend reach to the shop floor, *Manufacturing Engineering*, 149(5), 67–70, 72–75, 2012.

Waurzyniak, P., Driving digital tools to the cloud, *Manufacturing Engineering*, 150(4), 51–52, 54–58, 2013.

West, J., Customer relationship management and you, *IIE Solutions*, 33(4), 34–37, April 2001.

Winer, R.S., A framework for customer relationship management, *California Management Review*, 43(4), 89–106, Summer 2001.

Zisk, S., Business process semantics—An opportunity for convergence, *DM Review*, 18(9), 16, 2008.

# 5

## Demand Management

### Learning Outcomes

After reading this chapter, you should be able to

- Define demand management
- Describe the various approaches to accepting demand
- Identify the steps involved in implementing a demand management program
- Identify the challenges in merging supply and demand into a management process
- Contrast demand management approaches in manufacturing and service industries
- Describe the four strategies a business can use to manage demand
- Identify the factors that affect the selection of a demand management strategy
- Describe how influencing factors affect the selection of a demand management strategy
- Describe a model that integrates demand and supply management
- Provide examples of programs associated with each of the four demand management strategies
- Discuss the relationship of demand management strategies along the supply chain

### Company Profile: NextEra

Earlier chapters have stressed the importance of finding out what customers want or need. This chapter is about developing strategic programs to meet those needs effectively and efficiently. The entire supply chain, from retailers upstream through the manufacturing and originating groups, must decide how to best allocate their resources. Industries, and companies within industries, use different demand management strategies; some use multiple strategies to fit different product and market requirements. As you read this company profile and this chapter in the book, think about these questions:

- How does a company find out what the demand for its product should be?
- How much extra capacity should a company have (more than its expected demand)?

- Is a demand forecast the same as a sales forecast? Why or why not?
- How should a company plan for major spikes in seasonal demand?

What additional questions do you think relate to supply chain demand management?

---

### Company Profile: NextEra

Chapter 5 is about demand management, which may sound like an oxymoron. How do you manage demand? Do you tell your customers when they should order from you or instruct accident victims when to come to the emergency room? We understand that we should manage the supply side, but how should companies manage demand?

NextEra Energy, Inc., is a company that is continually trying to manage demand. They are an electric utility that have a responsibility to their customers to provide electric power when their customers want it. Because they have captive customers, they are subject to public scrutiny if they fail to deliver the services when needed. At the same time, they must be efficient because their price increases (rate increases) must be approved by government agencies. How do they do it? How do they do it in a sustainable way? We will use them as a representative of the electric utility industry.

#### SOURCES OF DEMAND AND SUPPLY

First, let us look at the types of demand an electric utility must deal with. Their customers include commercial office or retail buildings, commercial factories, commercial transportation vehicles, private electric automobiles, apartment buildings, single-family dwellings, schools, and IT data centers. Within each type, there may be multiple sources of different demands. For example, in a single-family dwelling, the sources of demand include lights, air conditioning, kitchen appliances, computers, television, and other power users. Demand from these users varies during the day and over seasons during the years.

When demand varies widely over time, the utility has to have additional capacity to meet total, or peak, demand. Power outages, or failure to meet demand, cause customer complaints and perhaps even danger to users, such as in hospitals or stoplights. Additional capacity is expensive and may require years to attain. Often, rate increases are based on the need for additional capacity because the need for electric power is increasing steadily through the years.

What are the sources of supply? At the present time, the conventional methods include burning coal, gas, and oil to turn turbines to generate electricity. Another way is through the use of nuclear power.

Some of the emerging alternative energy ways to generate electricity are through solar collectors, wind turbines, or geothermal wells. The general objective is to reduce the amount of electricity generated through burning fossil fuel and increase the amount generated through alternative energy sources. This poses a major decision for most companies. Which source of supply should get the capital investment money available?

The basic question seems to boil down to this: how does an electric utility meet widely varying demands while, at the same time, operating as efficiently to keep electric bills as low as possible? The answer is obvious. They must reduce the variation in demand or use some form of demand side management (DSM). As one writer puts it: "Today's electric utilities need to perform a delicate balancing act between the need to meet the dynamic load requirements of their customers and the need to improve energy efficiency and conservation" (Hedrick 2013, p. 2).

## EVOLUTION

DSM is not new. Following the energy crisis of the early 1970s, the electric utility industry began a quiet revolution in the late 1970s and 1980s by developing DSM and integrated resource planning (IRP). The premise of DSM is that there are benefits to both utilities and their customers to change energy use patterns, whether by shifting demand to different periods, reducing demand at specific times, or reducing overall energy use through energy-efficient technologies (York et al. 2007).

Two decades ago, Nadel and Geller (1996) issued a status report on the progress of DSM programs. They reported that utility DSM had gone through three major eras since the late 1970s: (1) information and loans, (2) resource acquisitions, and (3) preparation for a more competitive utility industry.

*Information and loans*: The information and loans programs were primarily designed to educate consumers and businesses to encourage them to invest in cost-effective DSM measures. During this period, load management programs began to encourage customers to reduce their use of energy during peak periods. In general, consumers were not interested in voluntarily reducing their energy consumption.

*Resource acquisitions*: In the 1980s, states and regions began implementing IRP processes that featured DSM programs designed to reduce power consumption through investments in energy-saving programs. Often the utilities subsidized consumers in making the investment through rebates or allowances. However, these programs took time and money from utilities and were still not eagerly accepted by consumers.

*Preparation for a more competitive industry*: As the utility industry moved from a highly regulated industry to one in which market forces and competition were becoming more important, utility DSM was also going through a transition, from regulated to competitive (Nadel and Geller 1996).

In a more competitive industry, utilities are looking more closely at DSM as a means of gaining a competitive advantage.

## DEMAND MANAGEMENT STRATEGIES

In Chapter 5, we describe four demand management strategies: provide, match, influence, and control.

- *Provide*: Maintain enough capacity to meet peak demand.
- *Match*: Vary capacity to meet demand fluctuations during different time periods.
- *Influence*: Encourage customers to change their demand patterns to reduce peak loads.
- *Control*: Limit available capacity to reduce peak loads and reduce capital investment required (Crandall 1993).

Maximum capacity is required when using a provide strategy; minimum capacity is required when using a control strategy. Consequently, most utilities would like to move away from a strictly provide strategy to a blend of other strategies to reduce the amount of capacity in which they must invest. This requires that utilities become more efficient by reducing energy losses while also reducing peak loads. Provide and match are reactive strategies and rely on supply management to be effective. Influence and control are proactive and reflect the trend toward increased emphasis on DSM.

Utilities must consider each of the four strategies. They must provide enough capacity over the long term to meet increasing demand and to replace those sources of capacity that have reached the end of their useful lives or have become obsolete because of new technology. It is difficult to forecast future demand, especially when new types of demand, such as electric vehicles, are emerging where there is little past history on which to base future demand. From an investment perspective, it is difficult to decide which type of energy sources warrants the investment. Should the company invest in traditional fossil fuel plants or alternative energy sources still on the downward slope of costs and upward slope of consumer acceptance?

Match sounds like a desirable strategy. But how does a utility vary capacity as demand changes? It is not like a fast-food restaurant or retail store where capacity can be changed through flexible employee scheduling. In the short term, utilities have fixed capacity as regards facilities and work force. They have limited opportunities to react to daily fluctuations in demand unless they can develop reliable data. This requires extensive data collection and analysis at a microlevel of detail. The smart grid is moving toward greater awareness about demand patterns that will make it easier to forecast demand. Another possibility for better matching of supply with demand is with electric storage. Most manufacturing companies can buffer demand through building inventory of finished products. Building an inventory of electrical energy has been a challenge that has not yet been solved, although there are numerous efforts to increase the capability of batteries to act as a storage medium.

Influence is the strategy that provides benefits for both the utility and their customers. It reduces the total capacity investment and also reduces operating costs. This results in lower energy costs for the consumer. However, it means consumers must change their demand patterns by moving consumption away from the peak periods to lower-demand periods. In the past, consumers have not been willing to do this voluntarily. There are indications that off-peak pricing can be effective, if consumers become aware of the benefits to them. The rise of new sensor technology will make it possible for consumers to understand how shifting the time for washing clothes, or other discretionary uses of energy, from daytime to nighttime will reduce peak loads. Several studies have found that demand for electricity is somewhat elastic (higher prices will reduce demand) and dynamic pricing models can reduce peak load demand (Quillinan 2011).

Control may sound like an undesirable strategy; however, it is sometimes necessary during the short term, especially if demand spikes occur as a result of weather. For example, periods of excessive heat may significantly increase air conditioning energy demands. Another cause of a spike in energy availability could be the required overhaul of an energy-producing unit. In either case, it may be necessary to limit the use of energy among consumers. Usually, the utility would have a predesigned plan to decide the sequence in which users would be denied availability of energy. Another use of control would be in the start-up of a heavy energy consumer, such as a server farm or a factory with heavy energy requirements. In these cases, the utility would work with the user to set up a start-up schedule of energy availability until the implementation is completed.

As the technology becomes more sophisticated and the smart grid concept takes hold, utilities may use the control strategy more as sensors help divert demand from nonessential uses to off-peak periods. In many cases, the customer may not even be aware of this diversion and will depend on the utility to maintain continuity where necessary. As technology has progressed, DSM has moved from voluntary to automatic.

## OBSTACLES

What are the obstacles to changing DSM strategies? They can be boiled down into three categories: technology, infrastructure, and culture. Technology involves the processes used to determine demand and techniques used to meet that demand. Infrastructure involves the ways in which demand is delivered from supplier to consumer and the organizational requirements to get that done. Culture includes the acceptance of consumers, and suppliers, to participate in demand management programs.

Technology applications are expanding rapidly, especially in collecting data about demand patterns and evaluating the operation of electric grid systems. They include new methods of generating electricity—modular nuclear facilities to reduce investment and operating risks (Wamsted 2011), clean coal, and reduced polluting gas. One of the newest and most promising technologies is the smart grid. The smart grid is a network that uses sensors and smart meters to collect data from suppliers and consumers to provide an integrated picture of what is happening within that grid system. "Clearly, intense effort is underway to build a grid with a 21$^{st}$-century brain – an electric system whose various parts can talk together, automate, self-heal and make consumers active energy managers, all with game-changing speed and scope" (Wood 2013, p. 48).

As Wood's paper suggests, the quest for a smart grid is just beginning and may never end. FPL, a division of NextEra Energy, Inc., is a company that is starting the journey. Their initial effort involved installing 4.5 million smart meters alongside more than 10,000 sensors at a cost of $800 million. Almost immediately, they realized benefits by identifying 400 malfunctioning transformers and correcting them before they caused a power outage. In some cases, they were able to solve smaller problems without sending out a repair crew (Brunner 2013). FPL believes the smart grid will help improve future reliability and customer service. "Through real-time access to data at the meter and on our lines, FPL can now better detect and prevent outages, and restore service faster when outages do occur. Data is also available to customers through an Energy Dashboard, available online and via mobile device, which is designed to help them monitor energy consumption and make more informed decisions about their usage" (NextEra Energy 2013). While technologies are still being developed, the progress to date supports the idea that demand patterns can be identified, analyzed, and managed.

Infrastructure has to do with the way the electricity is carried from the generating source—generating site, wind turbine, solar collector, or geothermal well—to the ultimate consumer. The infrastructure for the traditional fossil fuel–generating plants is well established with transmission lines and transformer stations to building wiring, although it is still not as well connected in a national grid system as many would like. Think of all the work that is needed to set up the same kind of network to carry power from the alternative energy sites. Doing this will require capital investment, land acquisition rights, production of equipment, and a host of other project elements to be successful. Consider this: in their 2012 Annual Report, FPL lists 19 conventional (oil, gas, coal) generating plants that can produce 20,696 megawatts (MW) of power, or an average of 1,089 MW each. The report also lists 89 wind facilities capable of generating 10,057 MW of power, or 113 MW each. The wind facilities have as few as 10 turbines and as many as 803 turbines. Each turbine and each facility have to be connected to transmission lines, all new infrastructures. One of the more interesting *little* tasks is to build a network of charging stations for electric cars.

Culture is an element that focuses on how well consumers—industrial, institutional, and private—accept and participate in DSM. While the technologies can produce and monitor the amount of energy used and distribution systems can move power as needed, the ultimate consumer must participate in DSM to avoid the need for governments to legislate and regulate energy management—a situation most would like to prevent. In speaking of the opportunities to incorporate new designs into building construction, Paul Torcellini, head of building programs at Department of Energy's (DOE) National

Renewable Energy Laboratory says, "You have to get it into the culture" (Wamsted 2011, p. 47). DSM will be a much rougher road without customer acceptance and participation. Because of concerns about privacy, only 20% of all consumers are likely to sign up for automated response programs, and more than a third do not want utilities to control systems in their home regardless of the savings potential (Wamsted 2011).

## SUSTAINABILITY

As a final challenge for electric utilities, they must use DSM programs to reduce the negative environmental impact of generating and distribution of power. Reducing the amount of generating capacity required will reduce the negative environmental impact. Replacing conventional fossil fuel plants with alternative energy sources will also contribute to greening the environment. By doing so, utilities will better fulfill their social responsibilities.

The U.S. federal government is playing a role in supporting the advancement of the smart grid. In 2009, the DOE awarded $3.4 billion to 30 investor-owned utilities and a like number of municipalities to pilot 100 smart grid projects. Industry is matching the federal grants for a total investment of more than $8 billion. The program provides $1 billion for the installation of smart meters to enable communication between utilities and customers. Another $2 billion will support building the networks of smart grids. A total of $400 million will be used to modernize transmission lines to reduce line losses and increase the distance power can be transmitted, and the final $25 million will help in the development of manufacturing capabilities of smart devices (Swirbul 2010).

NextEra Energy Resources, a division of NextEra Energy, Inc., and a companion company to FPL, is devoted primarily to generating electricity through alternative energy sources—wind, natural gas, nuclear, and solar. They believe they are the North American leader in wind energy generation, which represents 57% of the company's energy generation. They have over 100 wind-generating locations, primarily in the southwest United States. Their goal is to be a leader in producing electricity from clean and renewable resources (NextEra Energy, Inc., 2013). FPL offers seven residential DSM programs to increase customer energy efficiency and reduce peak load demand. These programs included home energy surveys, air conditioning audits, load management reviews, building envelope, new construction design, duct system testing and repair, and low-income weatherization. As of 2012, these programs had 278,499 participants.

Technology is going to make it possible to reduce energy consumption and, at the same time, provide better service to customers. Part of the solution depends on the consumer. Lowe's, the home improvement company, is actively promoting a *smart-house* system that will give users "the ability to remotely control and monitor systems such as thermostats, door locks, power consumption, camera and motions sensors, on a computer or mobile device" (Portillo 2012). The systems will work wirelessly and connect with meters in future smart grids. How will consumers react? A survey of experts found that just over half predicted widespread acceptance; however, almost half predicted most smart-house efforts would fail because of consumers' lack of trust and system complexity (Portillo 2012).

Greener electricity does not come easily. Think about DSM the next time you flip the switch and expect the lights to come on. Better still turn those lights out when they are not needed, or your smart meter may do it for you.

## REFERENCES

Brunner, G., Florida rolls out large-scale smart grid, puts rest of country to shame, extremetech. com, May 3, 2013, accessed July 2, 2013.

Crandall, R.E., Demand management—Today's challenge for service industries, Dissertation, University of South Carolina, Columbia, SC, 1993.

Hedrick, K., Demand side management: Why utility-directed load management programs make more sense than ever before, LandisGyr, 2013, http://www.landisgyr.com/webfoo/wp-content/uploads/2012/11/Landis-and-Gyr-DSM.pdf, accessed July 8, 2013.

Nadel, S. and Geller, H., Utility DSM, What have we learned? Where are we going? *Energy Policy*, 24(4), 289–302, 1996.

NextEra Energy, Inc., 2012 Annual Report, p. AR-2, 2013.

Portillo, E., Are you ready for a house with a brain? *Charlotte Observer*, charlotteobserver.com/business, July 22, 2012, accessed August 1, 2012.

Quillinan, J.D., Pricing for retail electricity, *Journal of Revenue and Pricing Management*, 10(6), 545–555, July 13, 2011.

Swirbul, C., Smart spending, *Transmission & Distribution World*, p. 4, November 2010.

Wamsted, D.J., Super technologies that could change the game for the electric industry, *Electric Perspectives*, 36(1), 38–49, January/February 2011.

Wood, E., More of a journey than a destination, *Power Engineering International*, 20(5), 48–53, May 2013.

York, D., Kushler, M., and Witte, P., Examining the peak demand impacts of energy efficiency, *Management Quarterly*, 48(3), 2–15, Fall 2007.

## Introduction

The onset of global competition has severely impacted American manufacturing in that customers have developed an increased expectation, of what makes up satisfactory quality, price, and service. This same trend is appearing in many service industries, as international competition is becoming more evident in retailing, airline travel, tourism, banking, and real estate. As competition increases, and customer expectations rise, the need to manage demand becomes more important. To put it simply, the future of many manufacturers may depend on how well they integrate their product offerings with their service package. This strategy requires companies to provide a complete goods–service package that can be customized, in both time and content, for individual customers (Chase and Garvin 1989; Crandall and Crandall 2014).

Both manufacturing and service industry managers recognize the need for effective demand management. Manufacturing companies often use inventory as a buffer to compensate for demand fluctuations while satisfying their desire to maintain a constant level of production. Demand management has also been a matter of concern in service industries, where perishability of the service may often preclude the use of inventory, thereby requiring a much closer matching of demand and supply capacity.

Demand management has often been portrayed as a process driven by top management that funnels through the financial planning process until it reaches the operating levels where marketing and operations must collaborate to develop an approach to use. Figure 5.1 shows this hierarchical approach.

In earlier chapters, we described how companies should consider their customers when designing a supply chain. If all goes well, the customers will respond by placing orders or creating demand on the upstream part of the supply chain. This chapter will consider the demand and supply capabilities required to satisfy the demand in volume, timing, quality, flexibility, and administrative requirements.

The *APICS Dictionary* defines demand management as "(1) the function of recognizing all demands for goods and services to support the marketplace. It involves prioritizing demand when supply is lacking. Proper demand management facilitates the planning and use of resources for profitable business results; (2) In marketing, the process of planning, executing, controlling, and monitoring the design, pricing, promotion, and distribution of products and services to bring about transactions that meet organizational and individual needs. Syn: marketing management." It goes on to define the demand management process as "a process that weighs both customer demand and a firm's output capabilities, and tries to balance the two. Demand management is made up of planning demand, communicating demand, influencing demand, and prioritizing demand" (Blackstone 2013).

The definitions convey the idea that demand management involves both the demand side and the supply side. There must be a balance between blindly accepting all forms of demand and then trying to satisfy it and accepting only demand that can be satisfied with the existing supply capability.

This chapter examines the various issues associated with demand management in both the manufacturing and service industries. We will begin by proposing an expanded description of the term, followed by a discussion on why demand management is important. We then provide a classification of demand management strategies along with the internal and external factors that may influence the choice of these strategies. We will then examine the relationship between the independent and dependent variables and conclude by suggesting ways the demand management process fits within the framework of strategic planning.

## Definition of Demand Management

In order to consider the simultaneous need to manage both demand and supply, demand management will be defined in this book as

> The total integrated effort of various functions within an organization, especially top management, marketing, and operations. The activity attempts to satisfy customer demand with an acceptable quality level of service, at a satisfactory level of effectiveness (delivery of service), and with organizational efficiency (cost of service), both during the short-term and the long-term. It involves the forecasting of demand so that resources, such as materials, workers, equipment, facilities, systems, and organizational structure can be appropriately planned and applied to meet the expected demand. (Crandall 1993)

The key components of an effective demand management program include the following:

- A strong orientation toward the customer's requirements, whether a producer, an in-house department, or a consumer. These requirements include the amount or timing of the product or service, the quality requirements, and the composition of the total service package. The term *customer* implies a more personal (behavioral) understanding of the source of demand, as compared to *market*.

- The integration of sales and marketing, operations, and top management to enable the parties involved to achieve a solution from among a set of feasible alternatives. The best solution will satisfy the customer's requirements at the lowest possible cost.

- A strategic or long-term planning focus as well as a tactical or short-term planning focus. The need for improved service quality is becoming a primary strategic issue and requires a different interpretation of *demand*.

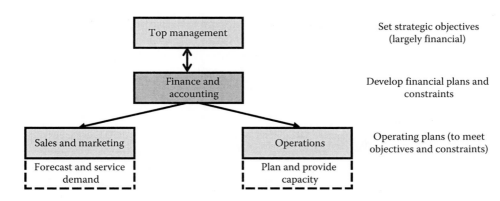

**FIGURE 5.1**
Traditional approach to demand management.

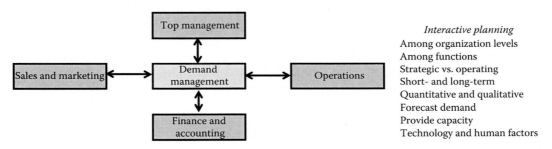

**FIGURE 5.2**
Contemporary approach to demand management.

The approach required by the aforementioned definition differs from the traditional *top-down* approach shown in Figure 5.1. Figure 5.2 illustrates how this definition requires an interactive approach, with open communications among the various functions.

The process begins with setting corporate objectives, which involves the participation of top management and the corporate staff. These objectives include planning the volume of business to be done (sales) and the cost of providing those services. The next step requires each functional area (sales and marketing, operations, finance, and accounting) to evaluate its own capabilities and formulate department strategies to meet both their functional goals and corporate objectives. They should compare their local plans to see if they mesh satisfactorily to meet all the company objectives; if not, they must resolve conflicts and inadequacies to develop a feasible solution to the corporate objectives. If they cannot develop a feasible solution, either departmental objectives or corporate objectives must be revised. The point is that this strategic process is an interactive one that may involve a number of iterations before final strategies are reached. There are several issues or trade-offs to be considered in this interactive planning process.

One issue is the amount of capacity to be maintained. Because service companies are seldom able to store services as inventories, as in manufacturing, it is usually necessary for them to have an excess of capacity over demand. Service demand patterns may have wide fluctuations unless otherwise controlled; therefore, a major decision in service operations is to decide how much excess capacity must be provided. As expected, less fluctuation in demand will require less excess capacity. This relationship presents the possibility of trade-offs between excess capacity costs and costs of controlling demand fluctuations.

While this interactive approach to planning is desirable, it is often difficult to achieve in practice. Showalter and White (1991) call this integrated approach *demand-output management* and state that "success in any business organization depends upon the conscious co-ordination of marketing strategy and production policy."

## Importance of Demand Management

Demand management, as defined earlier, is an important consideration for most businesses. Meeting customer demand effectively is essential to securing additional business; providing capacity efficiently is essential to continued economic well-being of the firm. In the services sector, utilities have been practicing a form of demand management for years (sometimes called load management), both because the practice is beneficial and because regulatory bodies require it. A number of authorities identified the need for demand management several decades ago. Sasser said: "Managing demand and supply is a key task of the service manager. To manage the shifting balance that characterizes service industries, managers need to plan rather than react" (Sasser 1976). Lovelock (1984) suggested marketing has an important role in demand management. "One solution to the demand problem that falls within the province of operations is to tailor capacity to meet variations in demand. Another solution that should logically be entrusted to marketing is tailoring demand to match available capacity."

Another study identified five dimensions of quality: tangibles, reliability, responsiveness, assurance, and empathy. Of these five, the two that correspond most closely to demand management are reliability (ability to deliver the promised service dependably and accurately) and responsiveness (willingness to help customers and provide prompt service). In their study, the researchers found that the groups studied considered reliability as the most important dimension of quality, with responsiveness as the second most important. In all groups studied, approximately two-thirds of the respondents considered these two dimensions of reliability and responsiveness the most important (Zeithaml et al. 1990). It follows that demand management is important in providing high-quality service.

In their description of companies that excel, Heskett et al. (1990) provided the following insight: "Managing demand and supply requires a different kind of understanding about the nature of demand for a service concept as well as efforts to coordinate the management both of customers and the people who serve them. In most services, it is the single greatest determinant of profitability."

Demand management is important and should be viewed as an integrated system. While Figure 5.2 conveys the need for integration, it does not place demand management within the framework of the total planning process for a company.

## Managing Demand

Managing demand involves deciding which demand a company will accept and the form in which they want to accept it. It is primarily concerned with determining the product line they should offer to best serve their customers and, at the same time, doing it profitably.

In a supply chain, demand has two principal sources—independent and dependent. The independent demand is the demand of the ultimate consumer at the end of the supply chain. Dependent demand is the demand that occurs at each junction of the supply chain. It is also referred to as derived demand and is a function ultimately of independent demand (Mentzer 2006).

There are several possible approaches to managing demand as demand varies over time, product mix, and customer mix. The demand management process must adapt to keep up with these changes and to anticipate some changes ahead of time so that demand strategies can be developed to effectively deal with them.

### Accept All Demand

Some companies believe they should accept all forms of demand. If they do, some of their demand will be on the fringe of their product or service line, which could be unprofitable or require excessive resources. Although this may be a short-term expedient strategy, especially for start-up firms, it is not likely to be a good long-term strategy. Over time, companies should adopt a more selective approach.

### Select the Types of Demand to Accept

Most companies develop guidelines for deciding what type of demand to accept. In the process, some forms of demand will be rejected. This may include imposing limits on the number of items in the product line, such as sizes in a battery manufacturer, the number of different bagels in a bakery, or the number of students in a supply chain management class. Companies may also have other criteria for rejecting orders, such as small order size, rush deliveries, or excessive customization requirements. They should design the guidelines to enable them to attract profitable or potentially profitable orders and design their product lines to fit each market niche, or even each customer. In considering how to develop a demand management approach to large customers, Hill states, "Merchandise planning and demand management tools optimize assortment mixes not only by account, but also by each door within the account" (Hill 2000).

### Solicit External Input in Forming Demand Patterns

After the company decides on how to limit their product and service offerings, they should find some way to test their demand strategy with the marketplace. They can check with some of their key customers, study what their competitors are offering, and maintain a continuing search for products to add or delete from their product line. Chapter 3 included a more comprehensive discussion on product line development.

### Design the Form in Which Demand Will Be Accepted

Once the company decides on the scope of their product and service lines, they should describe these offerings in an informative way for presentation to their customers. Customers are usually more willing to accept limitations if they know about them ahead of time than if they find out after they have gone to the trouble and expense of preparing an order and then having it rejected.

## Impose Constraints on Demand Submission

The company must realize it cannot be all things to all people. As a result, limiting the product line will be necessary, even if it means turning down a customer occasionally. Tailoring the product line requires constant attention. One way to identify new product opportunities is to spot products that customers are asking for but which are not available from the company. Rejecting orders is contrary to the natural instinct of most business people; however, it should enable the company to concentrate its resources effectively on those products that best serve its customers.

## Managing Supply

This chapter will focus on how companies manage the supply side of their operations to meet the demand. This will include a description of the demand management strategies available to companies and how they can select the strategy best for them. An outline of the steps necessary to implement the demand management program is included.

### Select a Demand Management Strategy

Companies should decide on a basic strategy to use in managing their demand. They may choose different strategies if they have multiple businesses or product lines. There are four possible strategies:

- Provide (maintain enough supply capacity to meet peak demand)
- Match (vary supply capacity to meet fluctuating demand)
- Influence (modify demand through voluntary actions of customers)
- Control (limit demand variation through policies and practices)

Selecting a strategy requires a careful analysis of a number of factors. These factors will be described, along with an explanation of how they influence the demand management strategies.

### Develop a Demand-Forecasting System

Before demand can be managed, there must be a way to identify and anticipate it. Most often, companies use a forecasting system to do this. Most quantitative forecasting systems use past history as a base demand level and supplement it with additional intelligence from the marketplace and internal sources. Often, the supplemental information contains judgment forecasts from knowledgeable persons. In progressive organizations, company representatives collaborate with their customers to develop consensus forecasts.

The forecasting system implies anticipation of demand in volumes, product mix, and response times, from which plans can be prepared. As the planning time horizon shrinks, the plans move from proactive to reactive. Lapide recommends that companies develop long-term, medium-term, and short-term demand forecasts and plans to most effectively match supply and demand (Lapide 2006).

Simple systems will not solve the tough forecasting problems that most companies face, and a more sophisticated system will require knowledgeable people to use it. Forecasting is an analytical task, not a clerical task. A quantitative forecasting system should include

- *Integrated inventory planning*: The most critical, and most often lacking, ingredient
- *Up-to-date modeling logic*: With the ability to level, trend, and account for seasonality
- *Selectable forecast calendars*: Fast moving requires different time horizons than slow moving
- *Measuring errors*: In dollars, to separate the important from the trivial
- *Separate forecast initialization and revision processes*: Adapt to a best fit as patterns emerge

Above all, the system should be flexible enough to keep pace with an organization's growth (Parker 1996).

### Determine the Resource Requirements to Meet the Demand

Once a demand forecast is available, operations managers have to prepare a plan to meet the demand. This involves production planning in manufacturing companies and employee planning and scheduling in service companies. Companies must specify the quantity of employees, equipment, facilities, and inventory required during the forecast demand period. While this can be done independently of the marketing group, most companies find it works better when the two functions—marketing and operations—work together in a collaborative manner.

---

## Merging Supply and Demand into a Demand Management Process

A company should develop strategies for managing demand and a system to forecast it. Having achieved this, the company needs a process whereby the demand and supply functions within the business work together to prepare a plan to best satisfy the demand requirements.

### Sales and Operations Planning to Match Short-Term Supply and Demand

Within a company, the most popular short- and medium-term approach to match supply and demand requirements is called sales and operations planning (S&OP). Richard Ling and others first introduced this concept over two decades ago; however, it did not gain widespread acceptance until the past few years. It is a process in which the marketing and production planning groups meet on a regular basis to reconcile differences between demand expectations and supply capabilities. It is defined as "a process to develop tactical plans that provide management the ability to strategically direct its businesses to achieve competitive advantage on a continuous basis by integrating customer-focused marketing plans for new and existing products with the management of the supply chain. The process brings together all the plans for the business (sales, marketing, product development, manufacturing, sourcing, and financial) into one integrated set of plans. It is performed at least once a month and is reviewed by management at an aggregate (product family) level. The process must reconcile all supply, demand, and new-product plans at both the detail and aggregate levels and tie to the business plan. It is the definitive statement of the company's plans for the near to intermediate term, covering a horizon sufficient to plan for resources and to support the annual business planning process. Executed properly, the sales and operations planning process links the strategic plan for the business with its execution and reviews performance measurement for continuous improvement" (Blackstone 2013).

## Collaboration among Supply Chain Participants

In addition to the need for collaboration within an organization in demand management, participants outside the organization in the supply chain should also collaborate to enhance the demand management process. "If demand information can be communicated throughout the entire supply chain, each trading partner would know how much product to have available and when. Less inventory would be needed as a hedge against uncertainty. Lead times could be shortened as less unneeded product would be made, freeing up production capacity. Sales would increase because the right amount of product would be available at the right points of consumption. As a result, all trading partners in the supply chain would reduce the cost of goods sold and increase their profits" (Taylor and Fearne 2006).

As attractive as demand collaboration along the supply chain sounds, the facts are that many companies run into obstacles in achieving true collaboration, such as the following:

- The pace of adopting new ways of doing business is slow.
- Demand information supplied by customers is not put to use in the other trading partners' own demand, supply, logistics, and corporate planning in an integrated manner.
- Demand management and supply management processes are not integrated, and S&OP is not utilized to synchronize demand and supply.
- There is a lack of trust among trading partners to share pertinent information and collaborate on decision making.
- There may be a desire to partner but not to commit to executing the communicated plans.
- There is a common view that demand collaboration is a technology solution and that the current technology is too complex (Crum and Palmatier 2004).

Achieving effective collaboration is not easy, but companies that have been successful in its implementation find it worthwhile.

Formal management improvement programs have been developed to include the use of demand management. Some of the more recent initiatives include quick response (QR), developed in the textile industry; efficient consumer response (ECR) developed for the grocery industry; and the latest collaborative planning, forecasting, and replenishment (CPFR). Although these programs were developed for the retailer–manufacturer interface, they have spread to the upstream sector of supply chains, such as in the agrifood supply chains (Taylor and Fearne 2006).

In addition to the programs described earlier, one of the newer programs is called demand-driven manufacturing (DDM). DDM is an outgrowth of lean initiatives and is a synchronized, closed-loop between customer orders, production scheduling, and manufacturing execution, all while simultaneously coordinating the flow of materials across the supply chain. The Aberdeen Group found that the leaders in DDM achieve significant improvements in on-time delivery, reduced safety stock, and faster response times (Littlefield 2007).

## Demand Management in Manufacturing

In manufacturing, demand management is a highly integrative activity through which all potential demands on manufacturing capacity are collected and coordinated (Vollman et al. 1992). The functions included forecasting, order entry, order-delivery-date promising, physical distribution, and other customer contact–related activities. This activity included the management of day-to-day activities that effectively responded to the demand of customers. Long-term considerations such as product and process design, capacity levels, and strategic plans were included in other manufacturing activities.

Some have advocated an expanded role for demand management in manufacturing. In assessing why some manufacturing companies have invested heavily in computer-based planning systems, such as enterprise resources planning (ERP), but have not achieved documented performance improvement, Ling and Palmatier suggested: "Efforts to date have been tilted too much towards managing the supply side of operating a business. Equal effort has not been expended in managing demand. To strengthen performance, many companies are helping to integrate Sales, Marketing, Manufacturing and Management with a more effective Demand Management process" (Ling and Palmatier 1987).

Andrews pointed out traditional functions of demand management are generally listed as order entry, forecasting, and order promising. While sales planning is often included, it is a too limited viewpoint. He recommends a fundamental restructuring and enhancement of demand management activities to more tightly connect the operational side of the business with the customer (Andrews 1987).

The APICS definition shown earlier recognized demand management as a distinct management function but confined its scope to the production planning part of the organization. Ling recommended a change in the definition to recognize the need to support the customer and to facilitate the management and utilization of resources (Ling 1991). He suggested that effective demand management should involve the participation of both the sales/marketing and operations functions to simultaneously plan for customer support and resource utilization. Ling expanded the S&OP process to include participants at all management levels and functions. He stressed the need to integrate the strategic and business plans and to follow up to assure compliance with these plans (Ling 1992).

Showalter and White considered the need to integrate the marketing and operations functions in *demand-output management* when they reported that the literature exhibited a pronounced lack of marketing perspective. Most previous work has been conducted by operations management researchers who included the marketing perspective simply as a constraint on the production planning process (Showalter and White 1991).

Customer requirements, as well as the resources available, should be considered when planning production. This requires that marketing and operations work closely together to plan for both short-term and long-term requirements. In manufacturing, the traditional way to provide response to the customer was to use inventories and order backlogs as buffers. Today, the increasing emphasis on shorter lead times and greater flexibility in responding to customer orders has led to the rapid increase in lean manufacturing programs throughout many manufacturing sectors. With lean manufacturing programs, there is an emphasis to reduce both order backlog and inventories; this makes the need to integrate S&OP (demand management) an even more urgent consideration.

As previously mentioned, S&OP has become the primary program used by many companies to manage demand. However, S&OP is primarily a short-term practice and there is a need for companies to view demand management as a strategic imperative. S&OP has evolved from a way to balance supply and demand to a means of linking financial goals to operating goals. The inability of finance and operations to speak the same language is a perennial problem and can lead to significant misunderstandings (Parker 2008).

One of the problems in managing demand in manufacturing is that the available capacity can change with changes in product mix. This means that planning at the product level of detail is sometimes necessary to assure that demand can be satisfied.

In Chapter 4, we described four approaches to manufacturing: make to stock (MTS), assemble to order (ATO), make to order (MTO), and engineer to order (ETO). Each approach uses a combination of inventory and lead times to manage demand in the most efficient and flexible way. If a longer lead time can be tolerated by the customer, inventory can be carried in the least finished state. This reduces the cost of carrying inventory and, at the same time, gives the manufacturer more flexibility in meeting demand variations.

There is a recognized need for the practice of demand management in both manufacturing and service industries.

## Demand Management in Services

Sasser was one of the first writers to offer examples of service companies managing both supply and demand. He first defined two basic strategies: a *chase-demand* strategy in which capacity is changed to fit demand and a *level-capacity* strategy in which capacity is kept constant and demand is controlled. He then described two basic approaches: altering demand and controlling supply. For example, demand can be altered through such measures as discount pricing in restaurants, which effectively shifts demand from peak periods to non-peak periods. Airlines also employ this measure with special tours for nonbusiness travelers. Sasser described the use of part-time employees to provide added capacity during peak periods, as in bank teller operations, or the increased involvement of customers in the service process to reduce waiting, as in salad bars at restaurants (Sasser 1976).

Lovelock later wrote of managing demand, particularly from the marketing perspective. He suggested that the problem is the worst for capacity-constrained service organizations that face significant variations in demand levels and suggested two solutions: (1) rely on the operations function to tailor *capacity* to meet variations in demand and (2) entrust the marketing function to tailor *demand* to match available capacity (Lovelock 1984). Both solutions are variations of the strategies proposed by Sasser. Lovelock showed that companies may encounter periods of excess demand, followed by periods of excess capacity; therefore, they must develop workable strategies for both situations. Pricing can be used to both reduce demand (increase prices) and increase demand (decrease prices), and when insufficient capacity is available, reservation systems or formalized queuing can be used. He concluded that balancing capacity and demand is crucial to the success of a company.

Heskett et al. devoted an entire chapter to "Managing Demand and Supply" in their book *Service Breakthroughs* (Heskett et al. 1990). They built on the works of Sasser and Lovelock and stressed the need to evaluate the relationships between the nature of demand (amount, speed, and predictability of change), the basic measure of risk (capacity utilization required to break even), and the amount of capacity needed (costs of lost sales, holding capacity, and poor service). They advocated taking a systems approach to the problem by considering multiple influencing factors in the environment as well as internal capabilities. They further suggested that service companies should recognize the need, and importance, of matching their available resources to the demand for their services.

*Demand management* has primarily been used in services to represent the marketing or customer-oriented perspective. *Managing supply* or capacity management describes the actions taken by the operating side of the business to satisfy the demand. A key issue in most capacity-constrained service industries is *yield management*, a methodology used to help companies sell the right inventory unit, to the right type of customer, at the right time, and for the right price. Yield management has gained widespread acceptance in both the airline and hotel industries. Kimes provided an analysis of appropriate situations for the effective application of yield management and emphasized the need for quick and easy-to-use methods for profitably managing capacity in capacity-constrained service firms (Kimes 1989).

Each approach to demand management described earlier implied a segregated function. Consequently, there has been considerable emphasis on developing either marketing-oriented or process-oriented service models; however, they did not always reconcile the potential conflict between these occasionally divergent views. Figure 5.1 illustrates this hierarchical approach to the problem. The implication is that top management formulates strategic objectives, the financial function converts these strategic objectives to financial objectives, and then marketing and operations use the financial objectives to formulate their functional (local) plans to satisfy the corporate objectives. Problems can occur however, when the functional plans are not compatible with one another or if there is insufficient interaction among the marketing, operations, finance, and top management functions.

There is agreement that the separation of the selling and marketing function from the operations function is much less definite in services than in manufacturing. The term *customer encounter* represents a unique feature of services and implies the coming together of demand and supply, so there is a vital interface between marketing and operations (Collier 1983). The fact that most service businesses do not have the buffers of physical inventory or order backlog intensifies the importance of managing this customer encounter. This suggests a need for an integrative model that recognizes the close relationship between the marketing of the product (demand side) and the operations of the service activity (supply side).

The fact that almost all services have the characteristics of intangibility, perishability, heterogeneity, and simultaneity of production and consumption causes the customer encounter to be a key point in the total process of providing a service. (Sasser 1976). Each customer has unique characteristics and wants. This uniqueness adds variety and complexity to the demand pattern, the service provided, and even the processes used to provide the service.

## Proposed Demand Management Strategies

There are two primary alternatives in managing demand and supply: manage the capacity to match the demand and alter the demand to better match the capacity available (Sasser 1976). Figure 5.3 shows the effects of these strategies as the alternatives of match (manage the capacity) and influence (alter demand). In these variations, the service business attempts to balance the economic trade-offs between the cost of excess capacity and the opportunity cost of lost business because of excess lead time.

Crandall offered two variations of those alternatives described by Sasser (Crandall 1993). Figure 5.3 shows them as provide (constant capacity to meet peak demand) and control (constant capacity to meet average demand) strategies. While providing sufficient capacity to meet peak demand at any time (provide) may be thought of as economically infeasible and controlling demand to a level flow (control) may be thought of as undesirable from a customer viewpoint, successful service companies actually use both strategies. The provide strategy includes quick lube businesses where *no waiting* is part of the service package; kitchen capacity in restaurants, where changes in capacity are restricted or expensive; and the use of technology to provide inexpensive sources of capacity, such as the use of vending machines to sell insurance at airports or automatic teller machines (ATMs) on college campuses to provide students with cash. The control strategy includes scheduling of return visits by doctors, turning away the overflow of customers by airlines by refusing to accept reservations for flights when there are no seats available (the practice of overbooking is an attempt to maximize resource utilization although there may be some adverse effect on customer service), and the use of regular schedules for deliveries by wholesale businesses.

A provide strategy can be thought of as an extreme form of managing demand through providing excess capacity or in deciding how much excess capacity to provide given uncertainty. Likewise, a control strategy can be thought of as an extreme form of managing

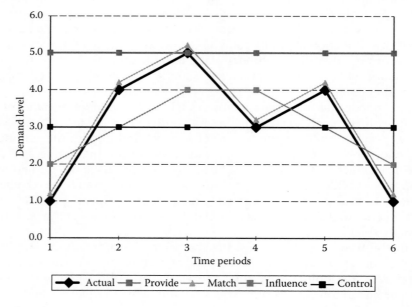

**FIGURE 5.3**
Demand management strategies.

demand with lead time or scheduled arrivals. Match and influence strategies can be viewed as combination or intermediate strategies. The real issue is when the deployment of extreme and combination strategies should take place.

To review, let us look at the four strategies in a nutshell:

*Provide*—Attempt to have sufficient capacity at all times to meet peak demand. While this means there may be excess capacity, the provider believes this is more important than taking the chance of losing business.

*Match*—Attempt to anticipate demand patterns so that capacity levels can be changed as needed. This means scheduling the work force carefully and using subcontractors or other temporary arrangements.

*Influence*—Attempt to change customer demand patterns to obtain good utilization of resources. This requires that marketing carefully plan promotions, pricing, and other marketing programs, so they are closely coordinated with operations.

*Control*—Keep demand variation to a minimum to fully utilize unique services and need high-cost resources to provide the level of customer service expected.

Provide and control strategies imply a level capacity, the former at peak demand and the latter at average demand. Match and influence strategies imply variable capacity, the former to accept demand as given and the latter to alter demand to reduce the range of fluctuations.

---

## Factors That Affect Selection of a Demand Management Strategy

The selection of a demand management strategy can be influenced by many factors. In this chapter, a number of factors will be described as the most likely to be used as drivers in the selection of demand management strategies. Thomas identified resource differences that can affect strategy selection (Thomas 1978); Lovelock described differences in the market, or demand, that affect strategies (Lovelock 1983); and Heskett et al. suggested several conditions that encourage the strategic choice between a chase-demand and a level-capacity strategy (Heskett et al. 1990). The following section contains a brief description of how certain factors could influence demand management strategy.

Demand management strategies may be influenced by many internal or external factors. Table 5.1 lists factors grouped by type of resource, marketplace variables, and strategic considerations. For a more complete description of each variable, see Crandall (1993). This tentative list can be expanded or condensed by the company when arriving at a demand management strategy.

Using the factor continuums as a scale, a profile can be developed to show where a business or industry fits. A brief summary of an empirical study using these factors as independent variables and the demand management strategies as dependent variables is included later in this chapter.

### Resources' Value

Demand management strategies can be related to the types of resources required. Table 5.2 shows the expected relationship of each strategy either with facilities and equipment or with employees. In those industries that have tangible inventories, such as retailing and

**TABLE 5.1**

Factors That Influence Demand Management Strategies

| Influencing Factors | Factor Continuums | | |
|---|---|---|---|
| *Resources employed* | | | |
| Materials | Standard, tangible | ⟷ | Custom, intangible |
| Employees | Low skill | ⟷ | High skill |
| Equipment | Standard, low cost | ⟷ | Custom, high cost |
| Facilities | Standard, low cost | ⟷ | Custom, high cost |
| Systems | Standard, low cost | ⟷ | Custom, high cost |
| Others | Not unique | ⟷ | Unique |
| *Demand factors* | | | |
| Type of service | Routine, short cycle | ⟷ | Custom, long cycle |
| Type of process | Standard, common | ⟷ | Unique, customized |
| Type of customer | Naive, shops around | ⟷ | Knowledgeable, loyal |
| Demand variability | Short term, large | ⟷ | Long term, small |
| Demand predictability | High | ⟷ | Low |
| Repeat business | A little | ⟷ | A lot |
| *Strategic or environmental factors* | | | |
| Strategic focus | Long term | ⟷ | Short term |
| Performance objectives | Customer service | ⟷ | Financial (profits) |
| Performance measures | Quality and service | ⟷ | Financial (costs) |
| Competitive situation | High, easy to enter | ⟷ | Low, difficult |
| Life cycle stage | Early stages | ⟷ | Late stages |
| Size of technical core | Small (high contact) | ⟷ | Large (low contact) |

**TABLE 5.2**

Demand Management Strategies and Resource Factors

| Demand Management Strategy | Heavy Dependence on Facilities or Equipment | Heavy Dependence on Employees |
|---|---|---|
| Provide (warehouse) | Cost of capacity is low, compared with the cost of not having capacity. | |
| Match (fast-food restaurant) | | Capacity can be adjusted easily by scheduling workers. |
| Influence (electric utility) | Cost of capacity is high and must be used efficiently. | |
| Control (doctor's office) | | High skill of workers makes it possible, even necessary, to schedule customers. |

wholesaling, seasonal changes in demand can be matched with corresponding changes in inventory levels. A provide strategy may be appropriate when excess facilities and equipment capacity are low in cost and high in effective output, especially when compared with the cost of not having capacity available. An example is the use of ATMs at banks. The match strategy can be appropriate, because it is usually possible to schedule employees and make it possible to match capacity with demand, as in fast-food restaurants. An influence strategy may be appropriate when there is heavy dependence on facilities and equipment

**TABLE 5.3**

Demand Management Strategies and Demand Factors

| Demand Management Strategy | Small Fluctuations in Demand | Large Fluctuations |
|---|---|---|
| Provide (janitorial cleaning service) | Small difference between peak and average capacity requirements; service can be improved at a small cost. | |
| Match (fruit farms at harvest time) | | Provide temporary, such as seasonal, changes in capacity without the need to maintain year-round capacity. |
| Influence (grocery stores) | Small fluctuations may be random or unexpected, requiring some adjustments. | |
| Control (hospital emergency room) | | Large, unpredictable fluctuations will preclude good service unless some effort is made to control. |

so that there is a need to limit demand fluctuations to reduce the capacity investment requirements, as for airlines. Finally, a control strategy is often appropriate when a skilled work force provides a unique service for which the customer will accept scheduling or waiting in line for the service, such as for doctors or lawyers.

## Type of Demand

Viewing the choice of demand management strategies from a marketing perspective should also be considered. For example, one possibility is to consider the size or predictability of demand fluctuation, as shown in Table 5.3. Where there is little fluctuation and the demand is predictable, the provide and match strategies can be accomplished with less risk and cost than when there are large fluctuations or unpredictable demand. On the other hand, large fluctuations with predictable demand encourage the use of influence and control strategies. There is little doubt that market conditions affect the choice of demand management strategy. When considering the variability in customers, some factors in this group become more important in deciding the most appropriate demand management strategy.

## Top Management Strategies

Besides concern about the resources employed or the market conditions served, the choice of a demand management strategy will be influenced by top management strategies. For example, businesses often want to pursue a strategy of high-quality, fast-response, customized customer service. Banks use both multiskilled employees and technology to provide a variety of services to both individuals and businesses. Another example is the wide variety of services provided by travel agencies, with online access to prices and reservation systems. These examples are in contrast to limited strategies of pursuing only financial objectives, such as profits and return on investment. In the past, many companies only looked at *the financials*. However, as the aforementioned examples suggest, many contemporary managers are likely to select high-service strategies, such as provide and match, as opposed to the lower-capacity-level strategies of influence and control. Table 5.4 shows a matrix of strategy factors versus demand management strategies.

**TABLE 5.4**

Demand Management Strategies and Strategic Factors

| Demand Management Strategy | Long-Term, Quality Service, Small Technical Core | Short-Term, Profit-Oriented, Large Technical Core |
|---|---|---|
| Provide (commercial banks) | Long-term customer relationships, with emphasis on quality and customer service that necessitates availability of capacity | |
| Match (department stores) | | Emphasis on profits and expectation of short-term customer relationships necessitates efficient utilization of capacity. |
| Influence (wellness centers) | Low competition level or early stages of service life cycle with low customer mobility | |
| Control (package delivery companies) | | High industry entry barriers, high contact with customers and mature service area. |

The factors described earlier should help companies to decide which demand management strategies are most appropriate for their business. A specific approach to using these factors will be presented later in this chapter.

## Relationship between Factors and Strategies

We now have a classification of the demand management strategies available to companies and an assortment of factors that could influence the choice of these strategies. Table 5.5 shows how demand management strategies (dependent variables) can be related to the influencing factors (independent variables).

The demand management strategies are arrayed as a continuum from provide to control, with match and influence as interim strategies. The influencing factors are shown on the left side of Table 5.5. The descriptions on the left side of the table for each factor correspond to the provide and match strategies, while the descriptions on the right side of the table correspond to the influence and control strategies. A profile of points can be prepared for an individual company for the factor groupings; this profile can then be used to suggest the demand management strategies that would fit best with their situation. An example of how this process works for two hypothetical companies is shown in Table 5.5.

The profile on the left side of the table (the X's) describes a service that does not have high-value resources, does not provide custom service with a custom process, and is more oriented to customer service than resource utilization. This company will favor provide and match strategies. The company shown on the right side of the table (the O's) has significantly different characteristics with high-value resources, custom services and processes, and an emphasis on resource utilization. This company will favor influence and control strategies.

The profiles shown earlier are hypothetical; in actual practice, the patterns may have wider variations. The amount of variation makes it difficult to clearly select demand management strategies. This wide variation could be the result of at least two different

**TABLE 5.5**

Two Examples of Factors That Influence Demand Management Strategies

| Influencing Factors | | Demand Management Strategies Provide, Match, Influence, and Control Factor Ratings | | | | | | | |
|---|---|---|---|---|---|---|---|---|---|
| | | −3 | −2 | −1 | 0 | 1 | 2 | 3 | |
| *Resources employed* | | | | | | | | | |
| Materials | Standard, tangible | | x | | | o | | | Custom, intangible |
| Employees | Low skill | x | | | | o | | | High skill |
| Equipment | Standard, low cost | | x | | | o | | | Custom, high cost |
| Facilities | Standard, low cost | | x | | | o | | | Custom, high cost |
| Systems | Standard, low cost | | | x | | | o | | Custom, high cost |
| Others | Not unique | x | | | | | o | | Unique |
| *Demand factors* | | | | | | | | | |
| Type of service | Routine, short cycle | | x | | | | o | | Custom, long cycle |
| Type of process | Standard, common | | x | | | o | | | Unique, customized |
| Type of customer | Naïve, shops around | | x | | | | o | | Knowledgeable, loyal |
| Demand variability | Short term, large | x | | | | | o | | Long term, small |
| Demand predictability | High predictability | x | | | | | o | | Low predictability |
| Amount of repeat business | Small amount | | | x | | | o | | Large amount |
| *Strategic or environmental* | | | | | | | | | |
| Strategic focus | Long term | | x | | | o | | | Short term |
| Performance objectives | Customer service | | x | | | | o | | Financial (profits) |
| Performance measures | Quality and service | | x | | | | o | | Financial (costs) |
| Competitive situation | High, easy to enter | x | | | | | o | | Low, difficult |
| Life cycle stage | Early stages | | | x | | | | o | Late stages |
| Size of technical core | Small (high contact) | | x | | | | | o | Large (low contact) |

causes: (1) the companies surveyed did not have a well-developed demand management process in place and may not have knowingly selected their demand management strategies, and (2) the factors used are still in the development stage and additional research will be necessary to establish which ones are the best drivers of demand management strategies.

The arrangement of the factors in Table 5.1, when compared with the demand management strategies in the matrix form, shown in Table 5.5, suggests that low-value responses will drive a company toward provide and match strategies, with high-value responses driving a company toward influence and control strategies. The following are examples:

- *Resources*: High-value, custom resources favor influence and control strategies; low-value, standard resources favor provide and match strategies.
- *Demand*: Custom service and process, along with low predictability, favor influence and control strategies; standard services with high predictability favor provide and match strategies.
- *Strategy*: Financial and short-term measures favor influence and control strategies; customer service and long-term measures favor provide and match strategies.

For a fuller explanation of the rationale of the relationships between the individual influencing factors and the demand management strategies, see Crandall (1993).

## Model for Integrating Demand and Supply Management

The approach suggested earlier is that marketing and operations managers should work closely together to formulate a strategy that will incorporate both capacity and demand planning, or *demand management* as defined earlier. Showalter and White addressed the need for future research in this area:

> The variability of demand coupled with the direct exposure of the productive system to demand fluctuations suggests that balancing demand rate with output rate may represent an interdependent management problem with several previously unexplored ramifications. This interdependency must be considered if management is to achieve minimum costs for the organization. Although research has focused on resolving the demand-output problem in manufacturing systems, little interest has been given to it in service organizations. (Showalter and White 1991)

They go on to develop a proposed general model that provides equations for both cost and effectiveness functions.

Another approach considers the desired, or intuitive, strategy that would be selected by marketing or operations. Table 5.6 shows provide and match as the desired strategies for marketing because both strategies use a capacity buffer to handle unexpected demand or to anticipate the need for extra capacity. With adequate capacity, the quality level of customer service should be improved and the response time decreased. Conversely, the preferred strategies for operations would be influence and control, because these strategies try to minimize the capacity required. Lower capacity levels usually imply lower investments and, assuming no decline in demand, a higher financial return.

On the other hand, these positions reverse when considering the capability of marketing and operations. Marketing is in a better position to influence and control because of their closer relationship with the customer. Operations is in a better position to provide and match, because these strategies require the change in resources or capacity levels. Table 5.6 explains this relationship. Viewed from this perspective, it becomes apparent that marketing and operations must work closely together to find the correct balance among demand management strategies.

Figure 5.4 provides a proposed model of how marketing and operations can use their independent planning as inputs to an integrated demand management strategy. The design of this model suggests that the activity in the planning process progresses from left to right and that marketing and operations move closer together to produce an integrated

**TABLE 5.6**

Comparison of Marketing and Operations Demand Management Strategies

| Management Function | Intuitive (Desired) | Capability (Able to Do) |
| --- | --- | --- |
| Marketing | Provide | Influence |
|  | Match | Control |
| Operations | Influence | Provide |
|  | Control | Match |

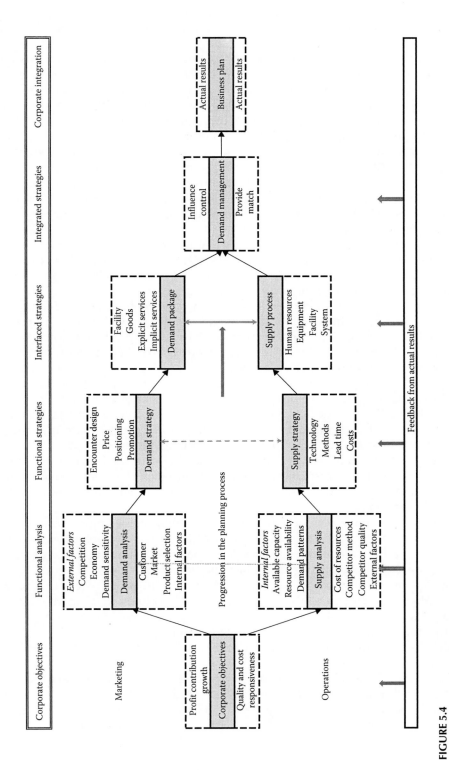

**FIGURE 5.4**
Integrated demand management planning model.

demand management plan, which then becomes a part of the total business plan for the company. Although not explicitly shown, the model implies the presence of feedback at all stages in the process.

The model in Figure 5.4 presents the idea that while marketing and operations begin their evaluation process separately, each has a responsibility to maintain contact with one another to preclude widely divergent plans. As the planning process moves to the right, marketing and operations merge closer together, signifying they maintain closer communications and coordination, until their individual strategies are fully integrated at the demand management juncture. Both marketing and operations start with common inputs as to objectives and general strategies of the company. The increased emphasis on strategic planning should help companies be aware of the need to consider long term and short term, tangible and intangible factors, and the importance of congruence between objectives and performance measures. If marketing and operations do not have common objectives, it is unlikely they will be able to develop an integrated demand management program.

Once they have specified their objectives, each functional area can assess their own internal capabilities and their external, or open system, environment. This is the stage designated as *functional analysis*, in which marketing and operations look at the strengths available to them and the threats that may severely hamper them.

After each function has assessed the situation, they can prepare their individual functional strategies, in which they decide what would provide a local optimum approach, considering only the implications for their particular function. While they formulate this strategy, they also should be sure to identify the critical areas of their strategy, that is, what must be preserved versus what can be modified.

When functional strategies are set, marketing and operations can compare their strategies to evaluate the fit. Marketing provides the service package design and operations provides the service process design. If they are compatible, the result can be quickly translated into the total demand management plan, with the areas of interface between marketing and operations clearly identified and highlighted. If the separate plans are not compatible, as often occurs in real practice, they must be reconciled. Recognition of differing approaches requires a resolution of strategies to reach the desired outcome. The coordinated demand management program then becomes part of the input for the general business plan of the company.

Table 5.7 contains a hypothetical example of how this process could work. It shows how the corporate function sets the overall organizational goals and objectives and how the functions of marketing and operations work separately to determine their capabilities and functional goals before blending these programs into a plan that best satisfies the corporate goals. It is important that each of the major functional areas understands clearly what they can do before they agree to a final strategy. Earlier, the planning process was described as a sequence of discrete steps. While this provides a logical approach, in actual practice, there will be frequent interruptions to an orderly progression, and the discrete steps will overlap and intersect with other activities in the planning process.

Table 5.7 is a guide to demand management planning; however, it is impossible to provide a rigidly structured approach that applies to even one service business, much less a general approach that can be universally used. Companies must customize the strategic and demand management planning processes to fit their own needs.

**TABLE 5.7**

Planning Example for Marketing and Operations

| | Corporate Objectives | Functional Analysis | Functional Strategies | Interfaced Strategies | Integrated Strategy | Corporate Plan |
|---|---|---|---|---|---|---|
| Sales and marketing | | External scan: finds a new competitor who emphasizes shorter service; economy is stable. Reviews internal: finds customer complaints have increased; contact level with customers is low and causes slow resolution of complaints; employee turnover has increased. | Decide primary emphasis should be on reducing order entry and shipping errors caused by inexperienced employees in the warehouse. A secondary strategy will be to handle customer complaints with personal contact from trained employees. | Establish a group to handle customer complaints through personal contact, by either telephone or face-to-face. Provide extensive training and decide on a provide strategy to assure that demand is satisfied. | | |
| Corporate | Establish objectives to (1) improve quality, (2) increase sales by a specified percentage, and (3) maintain current profit margins. | Inadequate contact among the three functions, except to clarify the objectives and the planning time schedule. | Discuss the proposed strategies. Decide the planned automation program represents a major change but should help both groups. Key factors for success include the worker training program and the automation implementation program. | Establish a multifunctional task force (marketing, human resources, and operations) to design and implement the warehouse automation program. Consider both customer service requirements and operating costs in the system design. Design the system with excess capacity for future expansion. | Monitor implementation programs closely and modify, as necessary, to assure success. The emphasis on provide and match strategies should enable the company to cope with the problems during the implementation of the improvement programs, after which the incidence of problems should decrease. | Successful implementation of these demand management programs and strategies should address the major problems that would prevent the attainment of the corporate goals. |
| Operations | | External scans: own wage rates lower than market. Internal scans: capacity is tight at peak periods; demand patterns have wide fluctuations, especially at the beginning of the week; and the cost of automating is moderate. | Select a strategy of automating the order processing system, from order entry through shipping, to reduce costs of operations and to reduce lead time. This will also reduce the number of employees needed but will increase the skill levels required. | Implement a program to develop employee teams, empowered to adjust schedules to meet fluctuations in demand. Provide training to develop multiskilled employees, enabling the use of a match demand management strategy for employees. | | |

## Programs Used to Implement Demand Management Strategies

A number of programs could be used to implement demand management strategies. Table 5.8 shows programs grouped under each demand management strategy.

### Provide Strategy Programs

The programs in this category assure that capacity will be available to meet maximum, or near-maximum, demand. Sometimes, this can be accomplished with automated processing, such as ATMs at banks. While there may be times when customers have to wait, usually, there is ample excess capacity to meet all demand. Another way to provide excess capacity is to work through intermediaries, such as when airlines use online travel websites to sell tickets or when insurance companies work through independent agents. Demand in excess of capacity becomes the domain of the travel agent and the independent insurance agent. Sometimes, only a part of the resources have excess capacity, such

**TABLE 5.8**

Demand Management Programs

---

Provide capacity to meet peak demand
    Use automation for round-the-clock operation.
    Develop intermediaries, i.e., travel agents.
    Provide excess capacity for long-lead-time facilities.
    Contract for standby equipment for peak periods.
    Hire additional employees to minimize wait time.
    Invest in multiskilled employees or cross-train existing employees.
Match capacity to demand fluctuations
    Use flexible staffing of full-time employees.
    Use part-time employees, or allow overtime of full-time employees.
    Share capacity with other firms.
    Increase customer participation in service process.
    Use alternative (sometimes less efficient) methods.
    Shift nonessential work to off-peak periods.
Influence demand to smooth capacity requirements
    Use off-peak period or some other form of dynamic pricing.
    Use advertising and other forms of promotion.
    Utilize personal contact from provider to customer.
    Create a reservation system.
    Develop complementary services.
    Limit variety of customers; implement niche positioning.
Control demand to level-capacity requirements
    Separate high and low contacts; control technical core.
    Develop distinctive competencies to retain customers.
    Obtain contractual commitments.
    Preschedule from one visit to the next.
    Inventory customers in a waiting line.
    Reduce variation among customers with conditioning.

---

as in the size of a convocation hall at a university, where it would be impossible to adjust the size to fit the demand. Companies can also contract for standby equipment for peak periods—cities contracting for help in snow removal, hiring excess employees to be available during all demand periods, investing in multiskilled workers, and cross-training existing employees.

## Match Strategy Programs

Match strategies attempt to adjust capacity up or down to closely correspond with the demand. Usually, the resource most often varied is the work force, although other resources can be altered, under certain conditions. Two frequently used approaches are (1) to schedule full-time employees on staggered shifts to match more employees with heavier demand and (2) to use part-time employees, or let regular employees work overtime, to handle peak periods. A way to match equipment capacity with demand is to share capacity with other firms, in the hope that their peak periods will occur at different times. Electric utilities use this approach very successfully as they import or export power to other utilities throughout the country over an electric power distribution network. The service process can be designed to be flexible so that the customer can participate more actively when demand is high, such as in many self-service operations, for example, building and supply retailer. Sometimes, capacity can be matched to demand by opening more units of capacity, even if it is less efficient, such as in the use of backroom clerks to serve as tellers in a bank or in the use of manual photocopy machines when the automatic sorter and collating copier are loaded. One of the simplest ways to utilize capacity during peak periods is to shift nonessential work, such as filing, to off-peak periods. Another interesting way of changing capacity is to match demand by changing the layout of retail stores to accommodate seasonal merchandise or changes in demand patterns. Even the shelf-space allocations can change as demand for different items varies (Hill 2000).

## Influence Strategy Programs

This strategy attempts to reduce the amount of demand variation, so that there is less need to adjust capacity. The customer voluntarily changes their arrival time or rate to alleviate the high fluctuations of the demand pattern. While the service provider can use a matching strategy, it requires less adjustment after the demand fluctuation has been reduced. Dynamic pricing, or using variable price schedules, can be a powerful tool to attract customers to off-peak demand periods. For example, economic incentives can be used in early-bird specials at restaurants or low weekend room rates at hotels that specialize in business travelers. A similar motivator is to advertise or use other forms of promotion to attract customers to off-peak periods, such as having a retail store send letters to customers telling them that they can get sales-promotion prices if they shop a day early. Some companies, such as a fitness center, use an informal process by having their employees tell certain customers they will not need to wait as long if they come at the suggested time. A more formal move is to create a reservation system so that demand is much more predictable. The service operation can also create complementary services so that the customer can be flexible in the service required. For example, in a supermarket, a customer who sees a long line at the deli may go to the produce section to shop instead, then return to the deli when the line is shorter.

## Control Strategy Programs

This strategy attempts to fully utilize the capacity resources by limiting demand variation. The customer accepts the need to vary their arrival patterns but not always in as cooperative a fashion as under the influence strategy. One traditional approach is to separate the high- and low-contact areas to establish the technical core where demand is controlled carefully. Processing mortgage payment checks in a bank is an example of the use of the technical core approach. Another approach is to develop distinctive competencies so that the customer is willing to endure a reservation system or long lead times to obtain service, such as in a doctor's or dentist's office. Sometimes the service provider can obtain contractual commitments, such as for a wholesaler who delivers a retail store's order every Thursday afternoon. Some professional offices also schedule return visits for repeat customers, that is, an appointment for teeth cleaning or taking piano lessons. The use of waiting lines is a form of demand control (this is probably the one a college or university uses the most), although most service businesses try to avoid this method as much as possible. A long-term approach is to reduce the variation among customers by training the customer in the service process so that most non-value-added process steps are eliminated. The self-service checkout lines in a grocery store are an example of this approach. Customers can choose to stand in lines (which are often longer) where cashiers ring up and bag the groceries, but they have the option of self-checkout if they desire.

These program classifications are not meant to be rigid definitions; they are only examples of the kinds of programs that a business can choose. Many service companies choose programs that may represent different strategies; however, they tend to concentrate their efforts in one or two strategic areas.

## Demand Management along the Supply Chain

Up to this point, we have described the demand management process as a generic approach, without regard about how it fits within the supply chain. While individual companies along the supply chain can develop their own demand management strategies, it is important for them to coordinate their efforts with their supply chain partners. Some likely scenarios will be described next; however, there are numerous possible strategy combinations.

There are two issues that sometimes interfere with the smooth matching of supply and demand—channel conflict and disintermediation. Channel conflict occurs when manufacturers disintermediate their channel partners, such as distributors, retailers, dealers, and sales representatives, by selling their products directly to consumers through general marketing methods and/or over the Internet through e-commerce. Using multiple supply chain—selling direct and selling through intermediaters—to serve the same customer market segment creates supply chain channel conflict; disintermediation removes the conflict by removing the intermediate resellers. While companies can operate with channel conflict, it introduces the potential for strained relationships with intermediate suppliers or resellers.

In general, supply chains are designed to dampen the fluctuations in demand—reducing the bullwhip effect. Accomplishing this will tend to lessen the differences among strategies; that is, the difference between maximum demand and average demand will be reduced. This means that operating with a provide or match strategy will enhance customer service and will not be as demanding on a company's resources than if there was not an effective supply chain.

## Retail

At the retail position in the supply chain, the demand management strategy will almost always be in the provide and match areas—provide for facilities and equipment and match for employees and inventory. While retailers can use the influence strategy through sales and promotions, these spikes in demand can be disrupting to upstream suppliers. It is unlikely that retailers will ever be able to use a control strategy with individual consumers.

## Wholesale

At the wholesale or distribution level, the demand management strategies will tend to move more toward influence and control because of the heavy dependence on moving goods from the distribution centers to the retail stores. Distribution centers usually have a set route that they follow to make deliveries to the stores; it is impractical to respond to individual stores on an as-ordered basis.

## Manufacturer

Manufacturers usually have a heavy investment in facilities and equipment; therefore, it is imperative that they are able to produce goods at a reasonably level rate. Their typical approach to meeting variable demand, such as seasonal, is to build inventory during the slow months and sell it down in the high-demand months. However, the move toward lean operations makes excess inventory undesirable and forces manufacturers to work harder at reducing demand variability by collaborating more closely with their downstream customers.

## Mining and Agriculture

Companies that mine raw materials or grow products have little choice when it comes to demand management strategies. They mine materials at a consistent rate because it is too expensive to modify their output rates. Agriculture companies grow products during the growing season and dispose it as quickly as they can, because of limited shelf life. While they can vary their product mix and levels of output over the long term, they have little flexibility during the short term.

---

## Summary

This chapter explained what demand management is and how it fits within the framework of strategic planning. Four demand management strategies were proposed and described. A number of factors were identified for use in deciding how their existence influences a company's selection of its primary and secondary demand management strategies.

A systematic approach to demand management planning was proposed, and examples were provided to illustrate how the approach could be used. Table 5.8 shows a number of specific programs that could be used to operationalize the chosen strategies.

The progress in demand management is still somewhat uncertain. The focus of supply chain management is changing from planning, implementing, and controlling operations in order to meet customer demand to sensing demand, shaping a market response, and driving profitable and reliable supply. In the area of demand management, companies are shifting from corporate forecasting and demand and supply matching to demand sensing and demand visibility. This means companies must continue to upgrade their manufacturing operations by reducing costs and improving quality and customer service. They must redefine manufacturing's role in the demand-driven supply chain. Unfortunately, the future constraint may be sourcing and building a reliable supply base, not manufacturing (Cecere 2006).

It is important to recognize that demand management, as a subsystem within the strategic and business planning system, is not yet a well-defined function in most companies. While it has great potential value, it requires additional refinement as a management process.

### Hot Topics in the Supply Chain: The Problem of Cheap

There is a problem that every supply chain manager should be aware of: it is called Gresham's law. The law is described in many economic textbooks, and some readers might remember learning about it in a college economics class. In its original context, Gresham's law discussed the implications of when good and bad money are circulating in the economy. Good money, such as gold-based coins, hold their value, while bad money, such as paper bills, do not. Eventually, the good coins are moved out of circulation because they are considered valuable and hence collected by hobbyists and others who appreciate the worth of the coin. What remains in circulation are low-value coins and paper money.

### GRESHAM'S LAW AND DISCOUNT GOODS

Writer Ellen Shell (2009) extended the application of Gresham's law to consumer goods. She maintains that in a discount-oriented consumer market, products will eventually sink down in quality to the lowest common denominator because shoppers are simply looking for the best price, even if the quality of the product is marginal. What results is an array of products that are low in quality and cheap in price. "It's simple economics that companies will sell the worst products that people are willing to pay for" (Segan 2008, p. 68). This can become frustrating for shoppers who really desire to spend a little more and buy a better-quality product. Gresham's law, according to Shell, dictates that only mediocre products will be left in the marketplace.

Of course, not all product lines follow this progression toward cheap quality. The ones most susceptible seem to be those sold in big box stores, dollar stores, and discount retail stores (Shell 2009). However, there can be collateral damage to those companies that truly want to sell a good product; they must be aware of the distribution and retail channels that are *out there*. For example, a high-quality product that is marketed through the discount channels will eventually sink down to the lowest common denominator in that market. This means what was once a good product will need to be *cheapened* so it can sell in the market. Cheapening the product means lowering its manufacturing cost, so it can be sold to a retailer, who will sell it to the discount conscious shoppers. Eventually, only the cheapest products will sell and those of higher quality will no longer appeal to the consumer. Hence, by Gresham's law, bad products drive out good products.

## WHERE HAVE ALL THE GOOD PRODUCTS GONE?

The quest for cheap products explains why some manufacturers move their supply chains to Asia or cut benefits to their employees working on the domestic front. Discount retailers that sell these products must keep their overhead and expenses low as well; hence, employees are paid less and if benefits are offered, many of the employees cannot afford them anyway.

So where have all the good products gone? The answer of course is that good products abound everywhere, but not necessarily in the discount retail industry. This industry is dominated by Wal-Mart, dollar stores, big box stores, and the like. Because consumers tend to be more price centric, products that sell well in the discount industry may not be the best in quality (Fishman 2005; Shell 2009). This is all fine as long as consumers are aware of the limitations of shopping in this sector. Indeed, there is a sense where consumers must realize that if they pay little now, they will pay more later. Sascha Segan comments in PC Magazine that paying a low price on a laptop up front may mean paying a *price* later in a loss of customer service, durability, and reliability (Segan 2008).

## LINK TO INNOVATION

Even if a company does not operate in the discount retail industry, it still behooves all managers to take note of Gresham's law. The problem is there is always a tremendous force to cut costs in any industry, discount, or otherwise. When the focus becomes on cutting costs, then a loss of innovation may result. Innovation is the force that creates new ideas and products. But when management is preoccupied with cost cutting, innovation, in the form of R&D expenditures, often decreases. If new products and new ideas are not forthcoming from a company, then it is reduced to offering whatever it offers at the cheapest price possible. It focuses on the short-term results of keeping income and stock prices up, while failing to position itself for long-term growth in the future.

For companies that solely operate in the discount industry, writer Ellen Shell (2009, p. 209) offers a bleak legacy of what could happen:

> When competition is mostly about price, innovation too often takes a backseat to cost cutting. Laying off workers and hiring cheaper ones is one sure way to enhance the bottom line. Another is to scour the world for low-wage workers, especially those in countries with lack-luster enforcement of environmental and workers' rights regulations. Neither of these tactics is innovative, and neither is in the long-run contributes to growth.

A preoccupation of scouring the world for low-wage workers in countries with lax environmental and workers' rights does not sound like a legacy to be proud of. Hence, this dilemma remains a hot spot in the supply chain.

## QUESTIONS FOR RESEARCH AND DISCUSSION

1. What categories of consumer products seem to fall into the low-cost, low-quality realm, no matter where these goods are purchased (e.g., The author proposes that it is difficult to buy a durable garden hose at a mainline retailer because hoses often kink and then develop leaks)?
2. In what ways are innovation directly related to supply chain management?

**REFERENCES**

Fishman, C., *The Wal-Mart Effect: How the World's Most Powerful Company Really Works—And How It's Transforming the American Economy*, The Penguin Press, New York, 2005.
Segan, S., Pay now or pay later, *PC Magazine*, 27(3), 68, 2008.
Shell, E., *Cheap: The High Cost of Discount Culture*, Penguin Books, New York, 2009.

## Discussion Questions

1. Describe demand management in your own words. How is it different from strategic planning?

2. Is demand management more closely associated with marketing or operations? Why?

3. What are the demand management strategies described in this chapter? Give an example, for each strategy, of a company that uses that strategy.

4. What factors (characteristics) can be used to describe the differences among service businesses?

5. Which of the *resource* factors shown in Table 5.2 do you think are most important for banks? Hospitals? Accounting or law firms? Retail stores? Wholesale distributors?

6. Which of the *demand* factors shown in Table 5.2 do you think are most important for banks? Hospitals? Accounting or law firms? Retail stores? Wholesale distributors?

7. Which of the *strategic* factors shown in Table 5.2 do you think are most important for banks? Hospitals? Accounting or law firms? Retail stores? Wholesale distributors?

8. Which demand management strategy, or strategies, do you think should be used by banks? Hospitals? Accounting or law firms? Retail stores? Wholesale distributors?

9. Select any industry listed earlier or any other industry of your choice, and list some specific programs that you believe represent their attempt to implement their chosen demand management strategies.

10. Why should marketing and operations work closely together? Under what circumstances would they be most likely to agree? Disagree?

## References

Andrews, C.G., Extending JIT into distribution, *APICS Conference Proceedings*, Las Vegas, NV, pp. 698–701, 1987.
Blackstone, J.H., *APICS Dictionary*, 14th edn., APICS—The Association for Operations Management, Chicago, IL, 2013.
Cecere, L., A changing technology landscape, *Supply Chain Management Review*, 10(1), 15, 2006.
Chase, R.B. and Garvin, D.A., The service factory, *Harvard Business Review*, 67(4), 61, 1989.
Collier, D.A., The service sector revolution: The automation of services, *Long Range Planning*, 16(6), 10, 1983.
Crandall, R.E., Demand management—Today's challenge for service industries. Dissertation, University of South Carolina, Columbia, SC, 1993.

Crandall, R.E. and Crandall, W.R., *Vanishing Boundaries, Integrating Manufacturing and Services Creates Customer Value*, CRC Press, Boca Raton, FL, 2014.

Crum, C. and Palmatier, G.E., Demand collaboration: What's holding us back? *Supply Chain Management Review*, 8(1), 54, 2004.

Heskett, J.L., Sasser, W.E., Jr., and Hart, C.W.L., *Service Breakthroughs: Changing the Rules of the Game*, The Free Press, New York, 1990.

Hill, S., Dream the impossible, *Apparel Industry Magazine*, 61(4), 12, 2000.

Kimes, S.E., Yield management: A tool for capacity-constrained service firms, *Journal of Operations Management*, 8(4), 348, 1989.

Lapide, L., Demand management revisited, *The Journal of Business Forecasting*, 25(3), 17, 2006.

Ling, R.C., Implementing demand management, *APICS Conference Proceedings*, St. Louis, MO, pp. 252–254, 1991.

Ling, R.C., How to implement sales and operations planning, *APICS Conference Proceedings*, Auburn, AL, pp. 307–311, 1992.

Ling, R.C. and Palmatier, G., Demand management: Integrating marketing, manufacturing and management, *APICS Conference Proceedings*, Falls Church, VA, pp. 266–269, 1987.

Littlefield, M., *Demand Driven Manufacturing: Synchronizing Demand with Production*, Aberdeen Group, Boston, MA, 2007.

Lovelock, C.H., Classifying services to gain strategic marketing insights, *Journal of Marketing*, 47(3), 9, 1983.

Lovelock, C.H., *Services Marketing: Text, Cases, & Readings*, Prentice-Hall, Inc., Englewood Cliffs, NJ, 1984.

Mentzer, J.T., A telling fortune, *Industrial Engineer*, 38(4), 42, 2006.

Parker, K., Demand management and beyond, *Manufacturing Systems*, p. 2A, June 1996.

Parker, K., S&OP encompassing broader financial and performance parameters, *Manufacturing Business Technology*, 26(1), 28, 2008.

Sasser, W.E., Match supply and demand in service industries, *Harvard Business Review*, 54(6), 44, 1976.

Showalter, M.J. and White, J.D., An integrated model for demand-output management in service organisations: Implications for future research, *International Journal of Operations and Production Management*, 11(1), 51, 1991.

Taylor, D.H. and Fearne, A., Towards a framework for improvement in the management of demand in agri-food supply chains, *Supply Chain Management*, 11(5), 379, 2006.

Thomas, D.R.E., Strategy is different in service businesses, *Harvard Business Review*, 56(4), 158, 1978.

Vollman, T.E., Berry, W.L., and Whybark, D.C., *Manufacturing Planning and Control Systems*, Irwin, Homewood, IL, 1992.

Zeithaml, V.A., Parasuraman, A., and Berry, L., *Delivering Service Quality: Balancing Customer Perceptions and Expectations*, The Free Press, New York, 1990.

# Section III

# Supply Perspective—Distribution, Production, Procurement, and Logistics

# 6

# *Distribution and Retailing*

## Learning Outcomes

After reading this chapter, you should be able to

- Summarize the history of retailing
- Identify the various types of retail stores
- Identify the various types of nonstore retailers
- Identify the factors to be considered in retail store design
- Identify the factors important to the success of retailers
- Explain the role of inventory in retail operations
- Describe the role of wholesalers and distribution centers
- Explain the role of technology in the distribution function
- Describe how services are a part of the distribution function

## Company Profile: Lowe's

Retail stores are where the consumer buys products and services. It may be a physical location or an online store. Some retailers emphasize a physical presence and others an online presence; however, most major retailers are blending the two means of selling. Both types of retailers need distribution centers (DCs) to deliver goods to the stores or to consumers for online purchases. Good inventory management is essential, to have the right product at the right place at the right time. One of the major problems for retailers is forecasting demand to avoid stockouts or leftover inventory at the end of the selling season. As you read this company profile and the chapter in the book, think about these questions:

- Why is store location important for a retailer? How do you decide where to locate?
- How large should a store be? What should its layout look like?
- Where do you locate DCs? Does the decision differ for online stores?
- Should a retailer buy from foreign suppliers? What are the advantages or disadvantages?

What additional questions do you think relate to retail or distribution functions?

**Company Profile: Lowe's**

Chapter 6 is about the role of retail and distribution functions in a supply chain. Our company profile is about a well-known public company, Lowe's. Lowe's is the second largest, after Home Depot, home improvement retailer in the United States. It is ranked 56th on the Fortune 500 list, just after Google and just ahead of Coca-Cola, with $50.5 billion in revenues and $1959 million in net profits (Fortune 500 2013).

In his 2012 letter to the shareholders, Robert A. Niblock, Chairman of the Board, President, and Chief Executive Officer of Lowe's, outlined a major program, Retail Relevance, which will continue the transformation of the company, which started in 2011. There are two areas within the Retail Relevance program—value improvement and product differentiation. These programs are designed to enrich product mix, increase in-stock levels, and improve product presentation. The *seamless and simple* approach outlines the following objective: "Beyond 2013, we expect to enhance our associates' ability to sell seamlessly across channels, introduce improved project management tools and expand fulfillment capabilities to cultivate personal and simple connections with customers" (Lowe's Companies, Inc. Annual Report 2012, p. 3).

The company recognizes they must change their infrastructure to fit their changing strategies. One change is to align functional heads into two distinct disciplines: customer experience and operations. The customer experience function will "use customer insights to design experiences that serve customers seamlessly across selling channels. The Operations organization will deliver the customer experience in an efficient and effective manner. While the two organizations have distinct responsibilities, they will collaborate intensely" (Lowe's Companies, Inc. Annual Report 2012, p. 3).

This relationship between customer experience and operations falls nicely under the umbrella program of sales and operations planning (S&OP), so widely used by many companies to better match supply with demand. An example of how this works effectively is the way Lowe's works with one of their major appliance suppliers, Whirlpool. Representatives from each company meet weekly through a conference call to discuss what is selling and Whirlpool's capacity to make the better-selling models. If Whirlpool does not have the short-term capacity to meet the higher need, Lowe's can influence, or shift, demand through promotions, in-store displays, and sales incentives to shift demand from the capacity-constrained model to one that Whirlpool has the capacity to deliver (Stank 2011).

Another major transformation for Lowe's is their entry into online selling. Customers can shop either in a store or on a mobile device and have the product shipped directly to their home or to a store for pickup. Another alternative is to have a sales specialist visit the customer's home and generate a quote and close the sale all in one visit. Over 190,000 items are available for customers to consider. Their flexible fulfillment initiative enables the company to deliver parcel orders to customers in over 90% of U.S. markets within 24 h (Lowe's Companies, Inc. Annual Report 2012, p. 7).

**RETAIL**

As of the beginning of 2013, Lowe's had 1754 retail stores, with 1715 in the United States, 34 in Canada, and 5 in Mexico. There was at least one store in all of the 50 states, with Texas having the most (141) and Wyoming the least (1). The average store is 113,000 square feet, with an additional 32,000 square feet in outdoor garden center selling space. The average store carries 40,000 items; however, not all stores carry the same 40,000 items, because of geographical differences in markets. The company employs

160,000 full-time and 85,000 part-time employees and operates 7 days a week (Lowe's Annual Report 2012, Form 10-K, p. 5).

Stores have Wi-Fi capabilities to provide customers with Internet access. Lowe's is rapidly developing its online selling capabilities by providing online product information, customer ratings and reviews, online buying guides, and how-to-videos and information. Lowes.com accounted for approximately 1.5% of total sales in 2012. The company has deployed over 42,000 iPhones for sales associates to check inventory availability, retrieve product rebate information, and help customers access their MyLowe's account information (Lowe's Annual Report 2012, Form 10-K, p. 5).

Lowe's sells a wide selection of national brand name merchandise such as Whirlpool® appliances and water heaters, GE® and Samsung® appliances, Stainmaster® carpets, Valspar® paints and stains, Pella® windows and doors, Sylvania® light bulbs, Dewalt® power tools, Owens Corning® roofing, Johns Manville® insulation, James Hardie® fiber cement siding, Husqvarna® outdoor power equipment, and Werner® ladders. They also have a number of private brands to differentiate themselves from the competition. These include items in tools, seasonal living, home fashions, storage and cleaning, paint, fashion plumbing, flooring, millwork, hardware, fashion electrical, and lumber (Lowe's Annual Report 2012, Form 10-K, p. 3). Store layouts are relatively the same with wide and well-marked aisles for easy access by customers. Seasonal merchandise is located to the right as you enter the stores, since these tend to be impulse items. In contrast, items that are nonimpulse, such as lumber or construction items, are located at the back of the store.

In addition to physical products, Lowe's also offers a line of services that include the following:

- *Installed sales*: Installation services through independent contractors with product categories of flooring, millwork, and cabinets and countertops accounting for the majority of installed sales. Installed sales accounted for approximately 7% of total sales in fiscal 2012.
- *Proservices*: In 2012, the commercial business program was rebranded as proservices to service contractors and other businesses in categories of building materials, plumbing, electrical, hardware, paint and tools, and outdoor equipment.
- *Extended protection plans and repair services*: These provide customers with product protection that enhances or extends the manufacturer's warranty. The service provides repair services for major appliances and outdoor power equipment and tools through the stores or in the home.
- *Credit financing*: This offers a proprietary consumer credit card for retail customers under an agreement with GE Capital Retail Bank. There is also a proprietary credit program for procustomers. Each of these programs offers discounts on purchases and competitive interest rates on credit purchases.
- *MyLowe's*: This is a new online tool that makes managing, maintaining, and improving homes simpler and more intuitive. Customers can create home profiles; save room dimensions and paint colors; organize owners' manuals and product warranties; create shopping, to-do, and wish lists for future projects; set reminders for common maintenance items; and store purchase history from across all Lowe's channels. Since its introduction in 2011, over five million users have registered for this service (Lowe's Annual Report 2012, Form 10-K, p. 4).

As part of their effort to improve customer service, Lowe's will add staffing to better serve customers and increase assistance available in the aisles (Lowe's Annual Report 2012, Form 10-K, p. 15).

## DISTRIBUTION

Lowe's has an extensive distribution network to serve the retail stores and to handle the increasing online sales with direct shipments to customers. Approximately 75% of merchandise at the stores is delivered from regional distribution centers (RDCs). This tightly linked distribution system includes the following types of DCs (Lowe's Website: Distribution Center Information, Effective Date: December 3, 2012):

- *RDC network (15 locations)*: Provides centralized distribution for stock merchandise that is cartoned and under 8 ft in length. A typical RDC covers approximately one million square feet of space.
- *Specialty DC (3 at North Wilkesboro location)*: Provides centralized distribution for merchandise that is noncartoned or irregularly shaped, requiring specialized handling.
- *Coastal holding facilities (2, one on East Coast and one on West Coast)*: Provide storage capacity for import programs above and beyond RDC capacity. These facilities both help Lowe's to absorb peak seasonal storage requirements and appropriately postpone allocation of import programs to better match varying sales rates throughout the country.
- *Transload distribution centers (TDCs) (3 locations)*: Each TDC will transload (100% cross dock) import merchandise from a domestic port to serviced RDCs in Lowe's existing distribution network.
- *Flatbed distribution centers (FDCs) (15 locations)*: The purpose is to serve Lowe's stores with lumber, plywood, boards, and other building materials that can be forklift loaded onto flatbed trailers.

RDCs are strategically located to provide frequent deliveries to the stores. This improves stock availability for customers and improves inventory turnover while minimizing total delivery distances. Each RDC supports approximately 120 stores. Lowe's buys from over 7000 vendors worldwide with no single supplier representing more than 7% of total purchases. Beginning in 2013, Lowe's will further develop their flexible fulfillment capability by deploying the Central Dispatch (CDO). The CDO will employ centralized delivery scheduling and better route planning, resulting in lower fuel cost, greater fleet utilization, and more productive overall delivery (Lowe's Annual Report 2012, Form 10-K, p. 15).

## SUSTAINABILITY AND SOCIAL RESPONSIBILITY

While Lowe's is building their integrated supply chain, they are trying to in a sustainable way. They track carbon emissions and participate in the Carbon Disclosure Project, design new stores with energy-efficient features, select energy-efficient products, and deliver products to stores in an environmentally responsible manner. Among a number of other awards, they were recognized by the U.S. Environmental Protection Agency (EPA) with their third consecutive ENERGY STAR Sustained Excellence Award, honoring their leadership as a retailer of energy-efficient products (Lowe's Annual Report 2012, Form 10-K, p. 6 and Lowes.com/Social Responsibility).

The company also tries to be a good neighbor in their communities through public education and community improvement projects. In 2012, their Lowe's

Charitable and Educational Foundation (LCEF) contributed more than $30 million to schools and community organizations in the United States, Canada, and Mexico. They are also active as a company and a number of individual employees are involved in charitable organizations, such as Habitat for Humanity and the American Red Cross (Lowe's Annual Report 2012, Form 10-K, p. 6 and Lowes.com/SocialResponsibility). They also try to be responsive in times of natural disasters, such as tropical storm Sandy, by having needed materials available, both prior to and after a storm.

Through its tightly integrated network of retail stores and DCs, Lowe's is an example of how well-designed supply chains enhance the consumer's experience.

## REFERENCES

Fortune 500, 2012, http://money.cnn.com/magazines/fortune/fortune500/2013/snapshots/, accessed July 24, 2013.

Lowe's Annual Report, 2012, 10-K Report, http://www.lowes.com/AboutLowes/AnnualReports/annual_report_12/index.html, accessed July 24, 2013.

Stank, T.P., Demand and supply integration: A key to improved supply chain performance, *Supply Chain Management Review*, April 14, 2011, www.scmr.com/article/demand_and_supply_inntegration_a_key_to_improved_supply_chain_performance/, accessed July 10, 2013.

## Introduction

This chapter describes the activities closest to the consumer—the retail and wholesale functions. The traditional flow of goods originated with a manufacturer, progressed to a wholesaler (or a DC), and finally to the retail store. This progression is still the dominant way in which goods and services flow to the consumer. However, in recent years, e-business activities created a flow that is sometimes direct from the manufacturer, or a DC owned by the online retailer, to the customer. An omnichannel presence is becoming essential for those retailers who expect to survive. The emergence of Amazon as a retailer of a wide variety of products puts many retailers in a vulnerable position (Crandall 2014).

The supply chain is concerned with the flow of goods and services from the point of origination to the point of use, usually from manufacturers to consumers. DCs are an intermediate step between manufacturers and retailers, where goods can be stored temporarily and sometimes repackaged from bulk to assortments for use at the retail store. The retailer makes the product available to the consumer by displaying it in convenient, attractive, and functional surroundings. The retailer also assists the consumer in the selection of items and provides suggestions on the care and use of the product.

The role of e-commerce is to provide consumers with a wider selection of items than could be displayed at a retail store and with sufficient product description to enable consumers to make a decision. While the lead times to obtain the goods are longer, the price may be lower and the selections wider than found in a retail store.

Even though the flow of goods is from manufacturer to retailer, the decision about what to make is shifting from the manufacturer to the retailer. For this reason, we begin our discussion of the supply chain participants with the retailers.

## Retail Function

The traditional retail store is the contact point for most consumers. It has a physical presence in a specific location and is referred to as a *brick and mortar* retailer. The primary role of retailers is to offer a selection of merchandise that is readily available to consumers and, in a complementary role, provide service to consumers in helping them to make their choices. They provide presale services that include informative and accurate advertising, inviting and accessible facilities, logical and attractive merchandise displays, knowledgeable and courteous sales people, and a reputation for high-quality products and services. They provide postsale services such as delivery and installation of merchandise, follow-up to assure customer satisfaction, fair and responsive product returns policies, and a functioning reverse logistics program.

## History of Retailing

The following section presents an overview of the growth of retailing in the United States. It does not attempt to describe all of the forms of retailing but highlights the general trends in the development of retail establishments.

The composition of retail stores has changed over the years. *Door-to-door salespeople* were probably the first examples of retailing. They were replaced by the general store, with its wide variety of merchandise, and usually with an owner–manager of a single unit. The general store represented the first formal form of retailing in the United States. This establishment carried a little bit of everything—groceries, animal feed, guns, clothing, furniture, household furnishings, jewelry, and anything else that was available for customers to buy. Customers did not have a lot of choice, either on price or product variety. The level of service varied with the individual storeowner, who was often the store manager. The stores did not have a great deal of choice either, but had to buy their merchandise wherever it was available, especially as the population moved westward from the Atlantic to the Pacific Ocean.

As communities grew to become towns and eventually cities, the general store grew in size to keep up with the evolving needs of its markets. The general store eventually grew to become the department store, a larger facility with larger inventories of merchandise. The early department stores carried a wide assortment of goods, organized by product category into departments. They carried the names of their founders—Wanamaker, Macy's, Gimbel's, and Filene's, to name a few. As cities grew in size, the general store expanded into multistory full-line department stores with a wide variety of items. The department store was the primary retail outlet for a long time, from mid-1800s to mid-1900s.

Department stores flourished in the cities, but in the sparsely populated farm country, retail stores were few and far between. To satisfy this need, companies such as Sears, Roebuck and Co., and Montgomery Ward published catalogs, with thousands of items, from which customers could order goods to be delivered by the U.S. Post Office. This was the forerunner of today's e-business where shipments are made directly to the consumer.

Department stores became increasingly complex; some carried over one million stock keeping units (SKUs) of merchandise. As the population moved from downtown toward

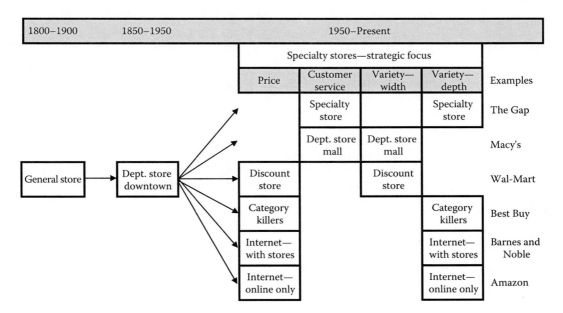

**FIGURE 6.1**
Evolution of retail stores.

the suburbs, large department stores began to lose business to the suburban shopping malls, where a consumer could find almost any type of merchandise at one of the many smaller stores in the mall. Shopping malls were anchored by two or more department stores and offered a social experience as well as a shopping experience. Malls became the *place to go.* Even those in the farm communities began to find that malls were not too far away.

As customers satisfied their pent-up demand for consumer goods, they became more selective. Some wanted lower prices, while others wanted more service. This dichotomy resulted in specialty stores. Some specialty stores competed on the basis of low prices, while others competed on their ability to offer more personalized service. Some had a wide assortment of goods, while others had a deep variety of items within a given product area. Eventually, larger specialty stores began to emerge, such as in electronics and furniture, which further eroded the large department store's business.

These special retail formats came about because of the opportunities to compete successfully with department stores. Figure 6.1 shows how specialty stores entered segments of the mass market once serviced by department stores. It includes examples of a specialty store type for each combination of *price or service* and *breadth or depth* of merchandise. For a more comprehensive coverage of the types of retail specialty stores, see Table 6.1 (Churchill and Peter 1998).

While the department store was the dominant player for over 100 years, other forms emerged rapidly in the latter part of the twentieth century. Today, the department store remains a major player in retailing; however, it is under considerable competitive pressures and will have to modify its role in order to survive.

Consumers have a wide choice of retail establishments. They range from small specialty stores, such as those that offer ski equipment, to large wide assortment stores, such as Wal-Mart. Department stores are struggling because of intense competition and lack of

**TABLE 6.1**

Major Types of Retail Stores

| Type | Definition | Examples |
|------|------------|----------|
| Specialty stores | Stores that handle a deep selection in a limited number of product categories | Foot Locker |
| Limited-line stores | Stores that offer a large assortment of a few related product lines | The Gap |
| Supermarkets | Large, departmentalized retail establishments offering a relatively broad and complete stock of grocery-related items | Kroger, Publix |
| Department stores | Retailers that carry several lines of merchandise, such as women's and men's clothing cosmetics, jewelry, and small appliances | Macy's, J.C. Penney |
| Mass merchandisers | Large retail stores with relatively low prices, a large variety of items, and limited customer service | Wal-Mart, Target |
| Warehouse stores | Retailers that offer groceries, drugs, hardware, and home furnishings in low-cost environments with little customer services | Costco |
| Variety stores | Establishments selling merchandise in the low and popular price range such as stationery, toilet articles, and housewares | Dollar General |
| Off-price retailers | Retailers that offer lower price for goods late in the season or with a limited selection of colors and sizes | T.J. Maxx |
| Convenience stores | Retail institutions whose primary advantages to consumers are locational convenience, high-margin, and high-turnover retailers | 7-Eleven |

*Source:* Adapted from Churchill, G.A., Jr. and Peter, J.P., *Marketing: Creating Value for Customers*, 2nd edn., Irwin McGraw-Hill, Boston, MA, 1998. With permission.

brand identity (most department stores were named after their founder and that name carried only limited geographic recognition). The winners today are national chain stores with brand identity and a specific marketing niche—low cost or a marketing focus in a limited product or age group, sometimes both. However, even specialty stores are coming up against increasing competition from online retailers, with Amazon being one of the notable entrants into the field.

There are several theories to explain why the form of retail establishments changes. The retail accordion theory, or the general–specific–general cycle, suggests that retailers with excessive inventory diversification open opportunities for other outlets to specialize in one segment of inventory. The wheel of retailing model proposes that a firm enters the market as a low-cost, low-margin operation. Over time, however, they acquire better locations, improved merchandise, and wider advertising and promotion and eventually mature as high-cost, high-margin businesses. This opens opportunities for the next low-cost, low-margin entry. A third theory is the polarization principle that suggests that fewer, but larger, retail establishments are eventually counterbalanced by smaller but more convenient stores (Brown 1987). While none of these theories are completely satisfactory, they all agree that the retailing industry is dynamic and new forms of retailing continue to emerge.

Figure 6.1 shows several types of retail stores. The list is not exhaustive, because the category designations change over time, as do the examples of companies that represent each category. However, the table shows the variety of strategies used by companies to appeal to consumers.

If this were not enough variety, there is another category of retailing called nonstore retailing that is becoming a more significant factor as consumers have less time to shop and who want the convenience of shopping from home. Table 6.2 shows the types of nonstore retailers available today (Churchill and Peter 1998).

**TABLE 6.2**

Types of Nonstore Retailers

| Type | Definition | Examples |
|---|---|---|
| Vending machines | Machines that deliver a product when the buyer inserts payment | Photocopies, soft drinks, stamps, live fish bait |
| Direct marketing | Personal selling or advertising media to solicit orders from consumers where they live on work | Telephone sales calls, catalogs |
| Direct selling | Personal explanation and demonstration of a product, with an opportunity to buy | Kitchen products, personalized scrapbooks |
| Direct-mail marketing | Mailing of brochures, letters, and other materials that describe a product and solicit orders | Letter inviting subscription to magazine: catalog to buy seeds |
| Telemarketing | Telephoning prospects, describing the product, and seeking orders | Phone call from a theater inviting the consumer to buy season tickets |
| Direct-action advertising | Advertisements containing ordering information | Television infomercial with a toll-free number for placing an order |
| Online marketing | Computer display of product information providing ordering via the Internet | Website that invites visitors to use a credit card to place orders |
| Integrated direct marketing | Combining several types of direct marketing into a single effort | Direct mail offer coupled with follow-up phone calls |

*Source:* Adapted from Churchill, G.A., Jr. and Peter, J.P., *Marketing: Creating Value for Customers*, 2nd edn., Irwin McGraw-Hill, Boston, MA, 1998. With permission.

Although enclosed shopping malls are still viable, many are losing market share to large specialty stores located in smaller shopping centers without frills but with convenient parking and access. Many malls have closed in the United States, as consumers no longer seem interested in spending hours strolling leisurely around the mall (Brown 1987). They want to get in and out of the retail store; they want to *buy*, not *shop*. Customers want variety and convenience; as a result, many are turning to electronic shopping (Plant 2007). For an interesting account of the malls that have closed in the United States, visit the website Dead Malls at http://www.deadmalls.com/.

E-commerce, in the form of business to consumer (B2C), is increasing rapidly. Consumers can find variety and up-to-date information on products by browsing company web pages. They find the prices are often lower than in the retail stores, and they find it convenient—the goods are delivered directly to their home. Most buyers started with simple, low-cost items, such as books. However, Dell demonstrated that customers will buy more complex products, such as computers. Consumers can even buy automobiles through websites. Although still a small portion of the total retail business, the potential for B2C selling is vast. With the advent of so many different forms of retailing, it appears that the consumer can not only choose the specific merchandise they want to buy but also choose the way they buy it, pay for it, get it delivered, have it serviced, and finally dispose of it. On the other hand, having a wide variety of choices for some products may be a bit of overkill. After all, how many different flavors and sizes of toothpaste do consumers really need?

There is another trend to watch for in the future. There is an excess of retail floor space in the United States. As new retail forms emerge, it is likely that some of the existing forms of businesses will cease to exist. Retailing is a very competitive business and it requires the combined efforts of all of the business functions—marketing, operations, and finance—to be successful.

Although the supply chain for food and grocery items is somewhat different from other retailers, it is not significantly different. The corner grocery store has given way to *big box* supermarkets that are part of a company with hundreds of stores. Eating establishments are similar. While there are still plenty of one-unit, mom and pop restaurants, there are also many more multinational chains with hundreds of stores, especially in the fast-food category.

Regardless of the product category, the retail industry is still the point at which most consumers connect with a supply chain designed to provide goods and services. Consumers want a wide variety of both products and services, and retailers are trying their best to provide it.

We are all familiar with at least some portion of the retail industry, because we are all customers. We buy groceries, clothing, fast food, home improvement items, and a host of other products from small independently owned craft stores to large corporate-owned department or discount stores. Some of the major considerations in retail management include high customer contact and the need for employees with good people skills, the need for effective inventory management, the importance of facility location, the need for variable staffing requirements, the occurrence of rapidly changing product lines, and the need for general managers to oversee relatively small operations.

## Characteristics of the Industry

Before we look at the details of retailing operations management, consider where the retail industry fits with the other major industries. Table 6.3 shows the total estimated U.S. business revenues for 2006–2011 (in billions of dollars) for all industries (BEA).

At the macrolevel, retailing represents a significant portion of the total U.S. economy. Retailing is the link to the ultimate consumer, the individuals who make the purchase of the goods or services for their own use. The inputs for the retail industry come from wholesalers or directly from the manufacturers. Table 6.3 shows the classifications are by type of products sold. While this classification is useful, it does not provide an accurate view of the type of customer who receives these products and services. Such information is becoming more important to the strategic planning of retailers.

### Customers and Markets

Customers have unique buying preferences. For a long time, businesses treated this collection of individuals as a homogenous group, or a mass market. They provided standardized products with a limited number of variations, such as sizes and colors in clothing. Over time, the variety of products increased in an attempt to reach niche markets. Businesses now classify customers by target markets and view them in terms of age, geographic location, income level, occupation, areas of interest, and a host of other ways in order to decide how best to provide them with their desired goods and services. As online retailers gather more information about their individual customers, they can greet them by name when the customer signs on and even recommend items that relate to previous purchases.

As consumers become more selective, they desire different products, because they have different wants or needs—tastes in foods or styles in clothing. They may even

**TABLE 6.3**

Relative Size of Major Industries in the United States

| Industry | Gross Output by Industry (in Billions of Dollars) | | | | | | Total (%) | Annual Increase (%) |
|---|---|---|---|---|---|---|---|---|
| | 2006 | 2007 | 2008 | 2009 | 2010 | 2011 | | |
| Agriculture, forestry, fishing, hunting | 299 | 347 | 382 | 346 | 367 | 420 | 1.5 | 8 |
| Mining | 423 | 467 | 579 | 355 | 440 | 533 | 2.0 | 5 |
| Utilities | 387 | 407 | 458 | 361 | 390 | 377 | 1.4 | −1 |
| Construction | 1,328 | 1,317 | 1,260 | 1,095 | 1,002 | 982 | 3.6 | −5 |
| Manufacturing: durable goods | 2,528 | 2,671 | 2,592 | 2,040 | 2,299 | 2,523 | 9.3 | 0 |
| Manufacturing: nondurable goods | 2,394 | 2,569 | 2,737 | 2,296 | 2,553 | 2,897 | 10.6 | 4 |
| Total, private goods-producing industries | 7,359 | 7,778 | 8,007 | 6,493 | 7,052 | 7,731 | 28.4 | 1 |
| Wholesale trade | 1,138 | 1,182 | 1,224 | 1,021 | 1,177 | 1,296 | 4.8 | 3 |
| Retail trade | 1,286 | 1,318 | 1,271 | 1,189 | 1,297 | 1,355 | 5.0 | 1 |
| Transportation and warehousing | 758 | 793 | 837 | 710 | 773 | 846 | 3.1 | 2 |
| Information | 1,111 | 1,165 | 1,198 | 1,161 | 1,195 | 1,258 | 4.6 | 3 |
| Finance and insurance | 2,136 | 2,274 | 2,238 | 2,294 | 2,298 | 2,274 | 8.3 | 1 |
| Real estate and rental and leasing | 2,513 | 2,594 | 2,633 | 2,559 | 2,589 | 2,628 | 9.6 | 1 |
| Professional and business services | 2,338 | 2,544 | 2,649 | 2,512 | 2,629 | 2,770 | 10.2 | 4 |
| Educational services, health care, and social assistance | 1,646 | 1,747 | 1,853 | 1,924 | 2,011 | 2,121 | 7.8 | 6 |
| Arts, entertainment, recreation, accommodation, food service | 920 | 977 | 999 | 961 | 990 | 1,051 | 3.9 | 3 |
| Other services, except government | 537 | 554 | 569 | 542 | 552 | 572 | 2.1 | 1 |
| Total: service-producing industries | 14,383 | 15,148 | 15,470 | 14,873 | 15,510 | 16,170 | 59.3 | 2 |
| Total: private industries | 21,742 | 22,926 | 23,477 | 21,366 | 22,562 | 23,901 | 87.7 | 2 |
| Government: federal | 916 | 955 | 1,040 | 1,093 | 1,163 | 1,173 | 4.3 | 6 |
| Government: state and local | 1,819 | 1,939 | 2,045 | 2,090 | 2,133 | 2,181 | 8.0 | 4 |
| Total: all industries | 24,477 | 25,820 | 26,562 | 24,549 | 25,858 | 27,255 | 100.0 | 2 |

*Source:* Adapted from the Bureau of Economic Analysis, http://www.bea.gov/iTable/iTable.cfm?ReqID=5&step=1#reqid=5&step=4&isuri=1&402=15&403=1.

want the same product but for different reasons—a TV set to watch movies or to watch the news.

As businesses segment markets in more detail, providing customized products and services to the individual consumer will become more commonplace. This practice fits well with the idea of customer relationship management (CRM) in which businesses think of the individual consumer as a lifelong customer, instead of considering them only as an individual transaction/sale.

## Outputs

The outputs for retailers are both products and services. Usually, the products are in finished form when displayed at the retail store; the store adds the services. Product segments fit the intended market, that is, grocery stores sell food products and department stores sell clothing items. The trend is to segment products even more to fit even smaller market segments. Grocers package milk as whole milk, 2% fat, 1% fat, or skim milk in quart, half gallon, or gallon containers to satisfy an increasing variety of consumers.

Services are becoming more important and are increasingly included as part of the product/service package. The service may be as direct as having the products arranged in the store to be easily accessible to the customer or as subtle as having well-maintained shopping carts (with no squeaks or wobbly wheels) easily available to gather the merchandise at a supermarket. Another service that is becoming more popular is the self-checkout lines for customers with only a few items or an affinity for new technology.

In terms of outputs, retailers must offer an increasing variety of products and services. Carried to the ultimate, the combination product/service package will become diverse enough to have a unique package for each customer. This result goes beyond mass customization of products; it will include the mass customization of the entire product/service package. While it may be impractical to prepare the customized package before the customer actually buys it (Amazon is considering how to do this), it will be necessary to design the transformation process to quickly customize the product/service package at the time of the purchase.

## Inputs

Inputs can either be things/materials or people. In most businesses, both kinds of input are present. For example, a department store has the product in the form of a man's shirt to be unpacked; priced; arranged on the shelf by type, size, and color; and ready for sale. The store provides sales personnel who work to transform a prospective customer into a satisfied customer through a pleasant greeting, providing information about the shirt, and completing the checkout process quickly and courteously. However, this personal service appears to be declining in some stores, presumably because of the pressure to reduce labor costs.

Other than unpacking the item, the tangible products for most retail stores are already in their finished product state when they arrive at the retailer. Sometimes, the retailer makes cosmetic changes in packaging modifications or by adding prices or security tags. The biggest impact on retailers is product proliferation. The number of variations of products is becoming almost overwhelming. Either the retailer must build larger stores to house the greater variety or must limit the number of choices available to the consumer. This increase in variety has caused department stores to reduce product lines in furniture or appliances and has led to the appearance of new specialty stores for those same products.

Viewing people as inputs has led retailers to recognize they must be more aware of and responsive to the differences in consumers. As with product variety, people variety is impossible to leave unmanaged. Consequently, retailers attempt to target segments of the total market instead of trying to meet all of the needs of that market. For example, they may decide a certain age group—young adults—as their target market and design their business offerings around that target market. While they may not turn away business from other groups, their main thrust is only the segment of the market in which they are interested.

## Transformation Process

The transformation process is the step during which the inputs are converted to outputs. For retailers, it is a combination of elements, such as the classic four Ps of marketing, as shown in Figure 6.2. Kotler classifies the first four Ps as tactical—product, price, place, and promotion. He also stresses that businesses must first perform the strategic four Ps of

- *Probing*: Analyzing the marketplace, commonly called market research
- *Partitioning*: Segmentation of markets into clusters
- *Prioritizing*: Ranking the segments on which to concentrate
- *Positioning*: Pinpointing the competitive options in each segment to target (Kotler 1989)

While the marketing function plays a major role in the design of the elements shown in Figure 6.2, the operations management function plays an equally important role in the final design and implementation of these elements, especially in the product and place/distribution elements (Zikmund and d'Amico 1995).

Retailers are trying to put more emphasis on providing customer service; to do this, they want their salespeople to have more time available for selling and less time required for moving and modifying merchandise. They are asking upstream distributors or manufacturers to provide *floor-ready* products by adding packaging, pricing, and sorting features before the merchandise arrives at the retailer.

The activities at most retailers combine marketing and operations management tasks. Selling is a marketing function; inventory management is an operations management function. The checkout process at a point-of-sale (POS) register is a combination of both selling and inventory management, so the employee is performing a combination function.

| | |
|---|---|
| Product—the goods or services that are designed, produced, and delivered to satisfy the needs of the customer | Price—the amount the customer pays for the goods or services; may be based on cost or competitive pressures, or a combination of both |
| Place—the location at which the goods or services can be purchased, such as a physical store or the Internet | Promotion—the communications that a seller may use to attract the customer and describe the product or service |

**FIGURE 6.2**
The four P's of marketing.

The transformation process is an integrated system of functions. Activities related to elements shown in Figure 6.2 cannot be performed without considering the effect on the other elements of the transformation process. Store design must account for product assortment; location of POS registers must be consistent with the desired level of customer service; and advertising or promotions must consider inventory management.

Some functions, such as advertising or commodity purchasing, can be handled from a centralized point in the organization, to assure consistency of message and purchasing efficiency. Other functions, such as employee hiring and end-of-aisle specials, will be more influenced by local management. As mass customization of the product/service package becomes more important, good local management will become more critical.

### Impact on Operations Management

How do all of the changes described earlier affect operations management? Remember that operations management is concerned with managing the transformation process of converting inputs to outputs. While marketing plays a leading role in deciding what the outputs are to be, the operations management function must actually manage the conversion process.

Look back at Figure 6.2. Operational management plays a leading role in handling the place/distribution element by participating in the design and location of retail outlets and in the design and implementation of the supply chain. Operations management is also heavily involved in the product element. They manage inventory, arrange the store layout, train employees in customer services, and assure the store presents the correct image to the consumer. While the promotion element may be the primary responsibility of marketing, operations management must prepare for special promotions by providing adequate inventory and employee staffing to handle the expected increase in sales. The price element also carries implications for operations management in that they must apply correct prices on the merchandise and monitor the movement of sales items. Markdowns reflect on the quality of the inventory management. Promotions bring extra sales; they also help to dispose of excess inventories. Promotions become even more complex when stores e-mail customers with loyalty card notices of unmarked specials, which only show up when the customer checks out.

While operations management works closely with marketing managers in designing and managing the transformation process, they also need to work closely with the accounting function in documenting sales and cost information. Operations management also must consult with the finance function when capital funds for new investments are required.

---

### Designing the Retail Process

This section examines the initial design of a transformation process for a newly conceived retail business or the redesign of an existing retail business. The why and what questions of this process will be considered first; later, the how and who questions will be explored.

One of the first design considerations involves selecting the strategic objectives of the system. Most retail businesses focus on either the price or the service side of customer orientation.

## Strategic Orientation

Usually, a business tries to satisfy either the price or the service orientation. Price represents an orientation toward the minimal allocation of resources to the transformation process, while a service orientation requires a more generous allocation of resources to the transformation process.

An orientation toward price appeals to customers with limited income, while customers with a more comfortable income level generally look for more service. Designing the customer interface considers these differences. A price orientation involves little direct customer contact, perhaps only a *hello* from a salesperson, who is busy stocking or rearranging merchandise as the customer enters the door. A service orientation may involve a warmer and more committed offer of assistance by a salesperson that drops whatever task they were doing to be completely available to the incoming customer for as long as the customer is in the store. A particularly effective action may involve greeting the customer by name. Some women clothing stores offer a service whereby a salesperson will select a number of combination outfits, including accessories, for a prospective customer to consider so that the customer does not have to wander through the store to *mix and match* separate items. Often, this selection process is done before the customer arrives to avoid waiting. Salespersons can cultivate a following of loyal customers who seek out that salesperson when they come to the store.

In a price-oriented business, the product assortment will focus on fast-moving items. This strategy reduces the commitment of resources to slow-moving or excess inventory items and reduces the need for markdowns. In a service-oriented business, the product assortment will be broader or deeper or both.

Product quality will be high in a service-oriented business, while it will be lower in a price-oriented business. Lower prices can be achieved through fewer features in the product, less costly materials, or lower quality acceptance standards.

Price-oriented businesses aim for higher inventory turns. This strategy means the store will have a more limited selection of goods and will probably experience a higher stockout rate than for service-oriented businesses that carry a more liberal assortment of goods in their inventories.

In service-oriented businesses, employees will be more knowledgeable, because of better training and a company culture that emphasizes excellent service, than in price-oriented businesses. They will possess more product knowledge and will interact with the customer more effectively than those employees in a price-oriented business. This is not to imply that price-oriented businesses are inferior in some way, but the trade-off for lower product prices is less customer service, which is a simple economic reality.

The retail facility in a price-oriented business will contain only the necessities. Customers will tend to select their items and buy quickly. On the other hand, the service-oriented business requires a facility with many customer-oriented extras, such as attractive décor, inviting product displays, and convenient checkout counters. Customers will not rush to buy or leave. Indeed, the experience of shopping is part of the goal of the customer-oriented business. Price-oriented businesses will tend to have large stores, often freestanding, in lower-cost areas where customers can get in and out quickly. Service-oriented businesses will tend to have smaller, or departmentalized, stores in malls, where customers can enjoy a shopping experience.

Information systems are critical in the design of retail stores. In a price-oriented business, the emphasis will be on keeping the costs low, while service-oriented businesses will build more customer-friendly features into the system, such as remembering customer birthdays, providing alerts for upcoming specials, catering to shopping preferences, and encouraging frequency of visits.

## Critical Design Points: Keys to Success

In the following section, we describe some of the key design decisions that a retail business needs to make. The discussion is not meant to be exhaustive, but, rather, to provide a feel for the decision area and some of the variables to be considered.

*Customer interface*: One of the most important design considerations is the customer interface, or the point at which the customer first comes in contact with the business. It may be an ad in a newspaper, a catalog that arrived through the mail, a website on the Internet, or more traditionally, a salesperson in the store. First impressions are lasting, and it is important for a retail business to choose the best ways to make contact with the prospective customer. Some businesses use a mass-market approach with low expectations of successful contacts, while other businesses use a very selective approach with high expectations of success. Table 6.4 shows an array of approaches.

All approaches can be successful, if the support infrastructure is properly designed and in place.

*Type and size of facility*: If the retail business operates without the need for the customer to visit a store, such as through Internet or catalog selling, the design of the store is immaterial. It can be whatever is available and functional. However, if the business requires a facility where the customer comes to see and buy the merchandise, the type and size of the facility become more important. Almost all retail stores are designed to be appealing to the customer. Each company has its own ideas about what constitutes appealing. Take eating establishments, for example. Some restaurants thrive on a noisy environment with music playing loudly in the background. Others seek to place their customers in quiet settings such as private booths for a more somber eating experience. While some retail businesses may have unique designs (such as a building shaped like a hot dog to indicate the product sold), most business today elect to use more conventional shapes and designs.

The size of the building is usually a result of the type and variety of merchandise offered. Department stores have a wide variety of product lines; hence, they need a large store. Furniture stores have a more limited product line; however, the product is large and bulky, so they need a larger store. Specialty clothing stores tend to be smaller, because their merchandise is more compact and usually limited in variety. Pipe and tobacco stores are quite small for the obvious reason they do not need much space to display their limited

## TABLE 6.4

Levels of Personal Contact

| Level of Customer Selectivity | Level of Personal Contact | | |
|---|---|---|---|
| | Low | Medium | High |
| Low (no attempt to identify) | Internet or newspaper ad (Amazon.com) | Grocery store with staffed checkout lanes | Salesperson in a store (Macy's) |
| Medium (identify as part of a group) | Magazine ad or catalog mailing (L.L. Bean) | Mail-merge letter from a database of names | Open house for invitation-only customers |
| High (identify as a specific individual) | Written invitation to an advance sale | Availability of a personal shopper at store | Invitation to a private party (Tupperware) |

line of products. At the smallest level are kiosks, usually run by one employee in a more open location, such as the center of a shopping mall.

Selecting a design for a building implies a long-term commitment, whether a lease in a mall or a company-owned facility. Most multistore retail businesses adopt a standard format and design, both for internal simplicity and consistency and for customer recognition. They do not change quickly or easily. When a change is made in store design, it is usually an expensive endeavor that involves all of the existing stores. The restaurant chain Ruby Tuesdays, for example, recently revamped the physical structure of its existing stores to incorporate a more modern look. The move was a high-profile event, which included a webcast of the demolition of one of its older stores (Ruby Tuesday 2008).

*Location*: Retail stores are located near their market. This means they must have more locations to reduce the need for customers to travel to find them (except for those businesses that do business over the Internet or through some other forms of indirect selling). For example, it would not make sense for McDonald's to have one giant restaurant in Iowa and expect to serve the entire United States.

While there are a number of techniques for choosing locations, most businesses select a consistent way to decide on locations. Except in the largest of cities, department stores are usually the anchor stores in a mall. Grocery stores are the anchor stores in strip malls. Specialty stores are part of a mall complex. Building supply stores are *big box* stores on a main highway on the outskirts of town. Drug stores, at least in recent years, tend to be freestanding stores for easy access. Automobile dealerships group together along the bypass around a town. Fast-food restaurants locate along the interstates or main highways. Family dining restaurants locate near motels. While there are always exceptions, most small retail businesses find the need to be near other retail businesses in order to gain the benefit of the traffic generated by the better-known names.

*Layout*: Some retail stores, such as grocery stores or building supply stores, have straight aisles with large signs at the end of each aisle to identify the type of merchandise in the aisle. This helps the customer to find what they are looking for quickly and without much help from the store employees. Customers usually know what they want and desire to get in and out of the store quickly.

Where the customer may be less direct in their buying, as in department stores, the store layout may take the customer through the store in a circuitous route so they can see all of the departments in the store. In this environment, the staple items, such as children's clothing, tend to be in the back of the store; the customer has to walk past the impulse items to get to what they want. Usually, the cosmetics department is at the mall entrance to the department store because it is an attractive area for shoppers to see as they enter the store. The men's clothing department is near the parking lot entrance for easy access. Department stores group like products together, such as women's dresses or shoes.

Table 6.5 identifies the dimensions of atmospherics and factors to consider in designing a retail store layout in order to achieve a desired image (Churchill and Peter 1998).

*Inventory management*: The business of a retailer is to sell merchandise. Consequently, inventory management is a critical part of the product/service package design. At the design level, the questions to answer include the following: What type of merchandise will we carry? How will we display the merchandise? How much will we show in the store? Will it be brand names or private label?

Retail outlets will vary in how much inventory is actually *on the floor* versus *in the back*. Smaller establishments will typically carry most of their in-store inventory out front

**TABLE 6.5**

Considerations in Retail Store Layouts

| | Some Dimensions of Atmospherics | |
|---|---|---|
| **Dimensions** | **Description** | **Examples** |
| Architecture | Imposing room heights and elegant details; period architecture; small or large rooms; modern or old fashioned | Warehouse look of stores emphasizing low price; waterfalls, pillars, and mosaics in upscale department stores |
| Layout | Basic grid of straight aisles; clustering of related goods; main aisle with merchandise along walls and in center | Grid pattern of most supermarkets for efficient shelving; boutique pattern of upscale fit shops to encourage browsing |
| Lighting | Bright or dim; purely functional or attention getting | Bright lights to stimulate sales in a supermarket; special effects to create an exciting atmosphere in a nightclub |
| Color scheme | Warm colors to draw in customers and stimulate quick decisions; cool colors to relax customers | Fast-food restaurant decorated with heavy use of red and yellow; elegant restaurant decorated in blues and grays |
| Sounds | Music (loud or soft, fast or slow); high or low noise level | Pianist playing soft music to create sense of elegance; loud music to generate excitement and fast turnover |
| Merchandise display | Huge volumes of merchandise stacked high on displays; few product items well spaced in store | Large quantities to suggest good sellers and low prices; wide open spaces suggest affluence and exclusivity |
| Odors | Pleasant smells like cologne, chocolate, leather, fresh produce, pizza | Food smells can generate hunger in restaurants; other odors for products such as perfume or leather goods |
| Salespeople's appearance | Well-groomed salespeople for store selling expensive luxuries; costumes for theme restaurant and stores | Athlete's Foot sales personnel wear referee outfits to look athletic and knowledgeable about shoes |

*Source:* Adapted from Churchill, G.A., Jr. and Peter, J.P., *Marketing: Creating Value for Customers*, 2nd edn., Irwin McGraw-Hill, Boston, MA, 1998. With permission.

where the customers are shopping, with little inventory at all in a back room somewhere. Larger stores, such as department stores or grocery stores, may carry some inventory in back rooms for replacement out front when necessary.

*Supply chain*: The retailer is at the end of the supply chain and is dependent on a number of businesses and suppliers to have the correct merchandise available when needed for the customer. Traditionally, manufacturers were larger than retailers and dictated the design and supply of merchandise to the retailer. The manufacturers *pushed* their products out to the retailers. In recent years, with the emergence of such retail giants as Wal-Mart, Home Depot, Safeway, and Target, retailers have gained the power and are more aggressive in linking customer preferences to the design and supply facets of the business. Consumers are now *pulling* what they want from the retailers, who, in turn, are pulling the merchandise from the manufacturers.

*Use of technology*: Technology is sweeping across retail organizations with a swiftness that is almost overwhelming. Even the smallest of retail businesses has a POS register not only to record the sales transaction but also to preserve the sales item data for further use in inventory management and product selection. Every item has a bar code; in the future, they may have radio-frequency identification (RFID) chips. Bar code readers help to track inventory, and electronic data interchange (EDI) communication systems place computer-generated orders with suppliers. Many retail businesses have an Internet site, even when they have a store—the *bricks and click* mode. Large chains have automated DCs. Design departments use computer-aided design (CAD) systems to design new products, from furniture to clothing. Automated fund transfer systems are commonplace. Embedded security tags are attached to clothing to reduce shoplifting. Automated self-checkout systems are becoming commonplace in grocery stores and in other types of retail stores. Homeowners can elect to have their utility payment deducted from their checking account automatically. Computer-generated purchase orders replace some of the judgment decisions formerly required of buyers.

*Technology is here to stay*: The dilemma for retailers is what technology to use. In some cases, the choice is obvious or dictated by their industry, such as the use of bar codes or RFID tags. In other cases, they may have to choose trade-offs between lower operating costs or enhanced customer service, such as the use of automated checkout stations or computer responses to telephone calls. Nonetheless, the trend to use more technology appears to be inevitable. One of the rapidly growing technologies is the use of mobile devices for checkout and payment purposes.

## Additional Factors to Consider in Retail Store Design

In addition to the store-specific considerations described earlier, there are some general factors to consider in retail store design.

### Multidiscipline Project

All the major functions in a business have a role to play in designing the retail store. The marketing staff wants the store to be attractive to the customer and the merchandise to be readily available. Operations managers want the store to be accessible to persons restocking the merchandise and all fixtures and equipment functional. The finance department wants the facilities and equipment to be durable and well maintained. Human resource managers seek an internal environment conducive to good employee performance and retention. Top management wants the store to operate successfully.

### Continuing Need to Readjust or Redesign

No matter how well the store is designed, it will need to be updated and modified over time. Just as there are product life cycles, there are store and location life cycles. Merchandise mix changes and displays need refreshing to attract new customers and keep old customers with renewed interest. Traffic patterns may need changing to make it easier for customers to access new areas. The convenience store chain, 7-Eleven, realized several years ago that it needed to use a different format for its stores located in downtown urban areas compared to stores in the suburbs. Downtown stores utilize an urban strategy, which focuses on serving smaller *neighborhoods* with deli quality food, gourmet coffee offerings, and a variety of fresh foods. The typical urban store is smaller in square feet than its suburban counterpart and is more likely to be located closer to other 7-Eleven stores in the downtown areas (7-Eleven, Inc. 2009).

### E-Commerce Considerations (Direct vs. Indirect Selling)

E-commerce is introducing new decision areas for the traditional *brick-and-mortar* retailers. Should they add kiosks inside the store for customers to access online displays?

Should they adjust their inventory practices to mesh with their online websites? Should they have a *pickup* area for online purchases? For example, like many retailers now, car parts chains have an arrangement where customers can order their car/truck parts online and then pick them up at their local neighborhood store when those parts arrive. This online purchasing is being extended to grocery stores where the customer order is *picked* by a store employee and stored in temperature-controlled coolers, awaiting the arrival of the customer for pickup. The customer can park in a designated area, close to the store, and a store employee will deliver the order and place it in the car. These and other options will force retailers to consider how they can best operate their combination of store and online businesses.

### Strategic or Long-Term Considerations

Investment in a physical store is a major, and long-term, investment for a business. Management should consider all of the factors described earlier in order to make this decision pay off for the company. Failed locations, even if the retail chain itself is successful in other locations, still represent unnecessary negative cash flows to the company.

## Managing a Retail Business

Once the retail system is designed, how do managers operate it on a continuing basis? Deciding on the steps to take is relatively straightforward; achieving the desired results is much harder. Generally, the steps involved in operating a successful retail business include the following:

- Determine the expected demand.
- Plan the capacity needed to meet the demand.
- Implement the plan by applying the resources.
- Measure the results.
- Replan for the next period.

### Determine the Expected Demand

What volume of sales should a store plan for? Retail stores struggle with this question because if they are not able to handle the flow of traffic in their stores, they not only lose the immediate sale, but they may also lose the customer forever.

Forecasting future demand is a combination of art and science. There are a number of techniques that companies use to forecast; usually, they involve statistically extending past history into the future. Past history helps to determine demand patterns, such as the peaks and valleys of demand during the day at a restaurant. Other retail businesses have fluctuations in demand during the week or during the month.

Most retail businesses have seasonal demand variations. Examining past sales history can help identify specific demand trends. Where demand information is available for product segments, it is helpful to see trends in different types of products. However, even the most sophisticated forecasting technique is usually not sufficient. We described the demand forecasting process in Chapter 3.

A company must modify their base forecast to include other factors. First, they must schedule their own *planned events*, such as promotions or advertising campaigns. Failing to have the necessary inventory on hand for a major promotion is a sure way to alienate customers. Next, the company must include any new formal contracts or agreements that did not previously exist in the forecast. Finally, they must consider the effect of external forces on demand. Is the economy better or worse than the last period? Will the new government's tax law increase or decrease spending? Has the customer base moved toward or away from buying their goods? For example, furniture businesses are finding that customers are buying more *high-tech* furnishings, such as computers and high-definition televisions, instead of a traditional line of furniture.

One of the current topics of interest is *big data*, or data analytics. Some companies are finding they can even collect information from social media sites to get a sense of what the public is interested in and use this information as a guide for future demand forecasts.

## Plan Capacity to Meet Demand

In a retail business, the physical capacity of the facility—the size of the building, the number of parking spaces, access, and egress—is relatively fixed for most stores. The more adjustable variables to consider in planning capacity are inventory levels and employee staffing levels. While inventory levels can be changed, most stores want the appearance of having an abundance of inventory for the customer to buy. The trick is to have the right assortment of inventory to fit the season or the market area. Snow shovels do not sell as well in the spring in northern climates and never in areas that do not have snow. In recent years, some industries developed programs to gain more rapid replenishment capabilities. With this, retailers can quickly reorder those items that are selling and avoid having excess inventory that must be discounted at the end of the season. The *quick response* (QR) program in the apparel industry and the *efficient consumer response* (ECR) in the grocery industry are examples of this type of program.

Employee staffing levels must be adjusted to meet fluctuating demand. In a fast-food restaurant, this means scheduling more employees during the rush hours for breakfast, lunch, and dinner. In a specialty dress shop, it means staffing more employees at night or on weekends, when the general buying public can shop. In a toy store, it means adding more employees during the pre-Christmas rush. During these higher-demand periods, many retailers find using part-time or temporary employees to supplement the regularly scheduled employees is preferable to incurring overtime. Scheduling part-timers can be risky however, as customer service may suffer with less experienced employees *on the floor*. In general, retailers cannot build an inventory of serviced customers; they must vary their employee staffing to change capacity levels.

## Implement the Operating Plan

Eventually, the operating plan must be implemented. The true test of retail management comes when the doors open and the customers come pouring in (hopefully). If the right goods are in inventory and the sales employees are courteous and knowledgeable, the customers will buy. The POS registers take over and begin to accumulate the sales information that triggers replenishment orders and builds the financial story of the store. Often, the task of a store manager is to react to variances from the original plan. This may mean rescheduling employees, placing a rush order for goods, or satisfying a disgruntled customer. Plans, actions, and results are all necessary for the success of a retail business.

The key steps in implementing the operating plan include the following:

- Apply resources where needed.
- Assure quality of the product and service.
- Measure customer response.
- Measure store performance.
- Assure employee quality and productivity.
- Utilize resources efficiently.
- Continue to develop improved performance measures.
- Replan for the next period.
- Continue to refine management capability.

While all of these steps are important, store managers have to decide which to focus on most carefully.

### Measure Performance

The purpose of performance measurement is to make improvements for future operations. Results of store operations should be measured on a regular basis and compared with projected operational plans. Some results, such as sales revenue, product costs, inventory levels, and labor costs, can be measured accurately and quickly. Other results, such as number of stockouts, late deliveries, and customer complaints, take longer to determine. Some results, such as lost customers or damaged image, may never be determined.

### Replan for the Next Period

To complete the feedback loop, the performance results should be analyzed and used to plan for the next operating period. Retail managers should try to improve the planning process as well as the execution process. In addition, efforts should be made to allow adjustments as needed for problems and other developments that will have an effect on the company's operation.

---

## Retail and Inventory Management

Distributors and retailers deal primarily with finished goods inventory. In most cases, the product is produced and stocked in the retail stores in anticipation of customer demand. This means the manufacturers use a make-to-stock (MTS) strategy, an approach described more fully in Chapter 7. A number of factors are important to the success of retailers. They include the value of the product, its availability, product variety, merchandise presentation, sales service, and replenishment response time.

### Value

Products must have value to entice the customer to buy. Value is more than price; it is the balance between the perceived benefit to the customer and the price. If the customer thinks the item is worth the price, the item has a value to the customer.

## Availability

Retail store sales depend on availability, or actually having the product in stock when the customer arrives. A picture of the item is usually not enough. The exception is when the retailer is selling through its Internet website. Even in this case, the customer needs some assurance that the item will be available to them in a reasonable amount of time.

## Variety

Customers may need variety, in the form of different sizes of clothing. They may want variety in the form of colors, styles, and materials. Consequently, manufacturers are producing increasing arrays of products from which customers can choose. Ultimately, the retailer will merchandise the item to the customer.

## Presentation

Even when the products are available in a wide variety of choices, they must be displayed in a way that makes it convenient for customers to find and choose them. Sales may be lost if the customer cannot find an item because it is in the wrong place or covered with dissimilar items or if the item is in the back room of the store, instead of out on the floor where the customer can select it. Attractive displays of goods also encourage the customer to buy, whether it is an expensive dress or an end-of-aisle display of soft drinks.

## Service

Sometimes, the product sells itself with its appearance and the customer's knowledge. However, in many cases, the sale may need an assist from an experienced customer service representative. No matter how much a customer may know about the automobile they think they want, they may welcome the assistance of a sales representative if that person is not too aggressive. Even in a grocery store, the customer may ask a stocking clerk if the produce is fresh, or where to locate a certain hard-to-find item, or help in reaching to the top shelf for an item.

## Response Time

Despite the best efforts of retailers, their actual sales often differ from their forecasts, especially in the product mix sold. This means they have some items that do not sell as well as expected and some that sell better than expected. For the former, they must mark them down, move them to another store, return them to the supplier, or donate them to charity.

For those items that sell well, retailers must have a QR system that enables them to reorder and receive replenishment inventory from the manufacturer or distributor, quickly enough to make the sale. This practice is especially important in seasonal items.

## Present Situation in Retailing

At the present time, there is excess capacity in the retail industry. Daniel Gross, business columnist for *Newsweek*, cites sources that estimate there were 159 million square feet of vacant retail space in 2008 (Gross 2008). Howard Davidowitz, retail consultant, estimated there were 19.5 square feet of retail selling space for every American person, and that should be reduced to 12 square feet per person, requiring a massive adjustment in the commercial real estate market (Kangas 2009). This number increased even more by 2009 to 23.06, as a result of expansion in anticipation of a recovery from the recent economic slow-down (Lal and Alvarez 2011). This means some companies, such as Wal-Mart and Lowe's, are still expanding, while other companies, such as Kmart/Sears, are closing stores. Malls are in trouble; customers are demonstrating a preference for stand-alone stores with easier access, such as Bed Bath & Beyond or Kohl's.

Customers are changing. They want greater variety and faster service. Carried to an extreme, this means some customers would prefer buying from an Internet seller, rather than take the time to go to a store and purchase the goods from a store employee. As a result, some believe that the greater variety of choices may actually be creating stress among buyers in their efforts to make choices (Schwartz 2004).

Increased competition and customer demands are causing turnover in the makeup of the retail industry. This means that companies that are to survive will have to adapt as their external environment changes.

## Future in Retailing

The form of the retail business continues to evolve. While some stores get larger, others get smaller. Through it all, retailers try to get closer, at least in convenience, to the customer, whether by adjusting physical locations or using the electronic accessibility of the Internet.

Retailers continue to extend their use of technology. Although starting slowly, RFID tags will eventually find a place in retail operations, though they are unlikely to completely replace bar codes. E-commerce will grow as a means of buying goods and services, as the public becomes more comfortable with it. As consumers become more comfortable with online shopping, the number of product categories increases. Books and electronic items are especially vulnerable, as killer category stores such as Borders and Circuit City have disappeared and stalwarts such as Barnes and Noble and Best Buy are struggling to rein-vent themselves.

Grocery stores are sending e-mails to loyal customers highlighting *specials* available only to those on the store's e-mail list. In addition to the use of credit cards, retailers are seeing a higher usage of debit cards. Retailers are using loyalty cards to track the purchases of their regular customers and to reward those loyal shoppers with special promotions. To provide a wider selection of merchandise, some retailers have added ATM machines and even more extended services (Walker 2009). Although many will object, personal selling will diminish in many retailing initiatives.

With the rise of the Internet, consumers have more choices, not only in product selection but also in seller selection. They can buy most consumer items online, even at 2 am, when most retail stores are closed. As Dave Blanchard (2013a), a long-time observer of changes in supply chains, puts it: "Thanks to the rise in Interest in omni-channel distribution - a retailing strategy that represents a convergence of brick-and-mortar stores, online, catalog and mobile sales - manufacturers are being pushed by their retail customers to adapt their supply chains accordingly. The basic premise is that smartphone-toting consumers can place orders for products at any time, so retailers want to be able to fulfill those orders right away, not just next-day but increasingly same-day."

## Role of Wholesalers and Distribution Centers

Earlier, we described the retailing industry and its role in providing goods and services directly to consumers. Retail stores are supported by wholesalers and DCs. A wholesale company is one that buys products from a manufacturer and then sells and delivers those products to retailers. A DC is usually a part of the retail company. The retailer buys product from a manufacturer and has it delivered to its own RDC, where it performs certain operations to make the merchandise *floor ready* for the retail store. The RDC transports the goods to the retail store, usually on a regular schedule.

Although there may be differences in ownership, both the wholesaler and the RDC perform similar functions in ordering, receiving, storing, picking, loading, transporting the goods to the retail store, unloading, and displaying the goods.

### Ordering

Placing purchase orders is generally considered a procurement function and will be described in depth in Chapter 8. An order for goods originates from the customer—either the wholesaler or the retail organization—to the manufacturer. The customer is no longer the individual consumer; the customer is an organization that conducts business with another organization.

### Receiving

The manufacturer ships the order to the DC by the most appropriate mode of transportation. We will discuss those transportation modes, and how they are selected, later in Chapter 9. When the goods arrive at the DC, they go through a process known as receiving. DC employees check the arriving shipment to make sure the goods are in good condition and are in the correct quantity. Once the goods are received, the quantities are added to the inventory of the DC.

Receiving methods have changed over the years. Instead of unpacking the cases and inspecting each item, receivers often accept cartons or pallets to reduce the time required. Traditionally, they have done this by scanning bar codes on the containers. Today, there is a movement to use RFID tags that contain more information and are easier to read, because they do not require a direct line of sight to be read, as bar codes do. The DC may use a sampling procedure to check a portion of the shipment, but, increasingly, they are relying on their suppliers to send the correct amount and quality of goods. This practice is also a result of more diligence in selecting and working closely with suppliers. DCs are increasingly challenged to adopt reusable pallets and shipping cartons to reduce costs and negative impact on the environment.

### Stocking

After the goods are received, they are moved to designated sections of the DC and stored in open racks or *slots*. This is usually only a temporary storage—hours or days, not weeks or months—because the objective is to keep the inventory moving or turning. Goods may be stored in designated slots—canned peas always go in slot No. 77—or in random locations if the computer system is programmed to retain that information, when it comes time to pick the peas for shipment to the store.

Some received goods may not even be moved into storage but instead go through a process called *cross docking*, to move it more quickly to the outgoing trucks for transport to the stores. Cross docking is a process that transforms the goods from the form in which they arrive at the receiving dock to the form to be loaded onto the shipping dock. For example, the containers from the manufacturer may have a gross of small brown gloves in one container, a gross of large brown gloves in a second container, a gross of small black gloves in a third container, and large black gloves in a fourth container. During the cross docking process, the gloves are reassembled into assortments of one dozen of each type of glove for the first store, one dozen of each type of glove for the second store, and so on until all of the gloves are repackaged in the form that is most useful to the retail stores.

## Picking

Picking is the process of finding an item in the storage area, selecting it, loading it on the picking conveyor such as a forklift, and moving it to the loading dock area. Often, the loading dock has an area designated as a staging area that offers an opportunity to visually inspect the goods for damage and to ensure that the order is complete before it is loaded on the truck. During the picking process, the inventory records are changed to reflect the transfer of the inventory from the DC to the retail store.

## Loading the Trucks

If the truck is destined to go to only one store, the order of loading is not as critical. However, most trucks deliver to more than one store; therefore, it is important to load the truck in reverse order of delivery—the first store to be delivered to is loaded last. The orders also need to be separated to prevent errors in delivery. The loading is also done carefully to avoid damage during transit. Modern container design has helped prevent damage.

## Transporting to Stores

The current trend is to make more frequent deliveries to retail stores. This practice reduces inventory and allows more variety at the store, but it imposes greater demands on the DC to process orders quickly and provide consistent deliveries free of damage and incorrect counts. Accomplishing this means greater dependence on information technology in order processing and delivery scheduling.

This process most commonly involves trucks; however, it could include rail transport, especially in moving goods from the manufacturer, farm, or mine to the DC.

## Unloading and Display at Stores

Trucks deliver goods to stores in what is called a *floor-ready* state, such as for clothing stores. This means that employees at the DC put the dresses on hangers, add price tags, and do whatever else the store may request to reduce the amount of work that has to be done by the store employees.

Trucks must also follow a rigid schedule to arrive at the store at a scheduled time so the store can have the right employees available to help unload, verify the accuracy of the shipment, and put away the goods. For example, many restaurants will only accept deliveries at times when the dining room is not busy with customers. While the schedule is less demanding where the store has only a few suppliers, a schedule becomes imperative where there are multiple suppliers. One of the reasons some companies have set up their own DCs is to reduce the inbound traffic at the stores. Even with fewer deliveries, the truck order will be fuller and more time will be needed by the store manager and receiving employees to check in the order when it does arrive.

## Critical Success Factors for Distribution

Distribution is an increasingly important function within the supply chain. As shown in Table 6.3, the wholesale or distribution function represents approximately 5% of the total output in the United States. It no longer just moves goods from one place to another; it performs a number of customizing tasks previously done by the manufacturer or the retailer. The distribution function must also adhere to a more rigid and compressed time schedule—it must get the goods to the right place at the right time and faster. In addition to the four Ps of marketing—product, pricing, promotion, and placement—some suggest there are three additional Ps—people, process, and physical evidence (Tracy 2004). Distribution also handles a greater variety of goods within the same shipment, as customers place more frequent orders for smaller quantities of an increasing variety of goods. In addition, it must do all of this in a more competitive environment because of government deregulation in the trucking industry!

## Inventory within Distribution Functions

Inventory management is a continuing challenge for retailers and distributors. The bulk of retail sales occurs, because the store has the merchandise available when the customer walks in. The e-commerce activity, where goods are shipped directly to the customer, is an exception. This means that the retail store must anticipate the demand, not only in total but also by individual SKU. This level of detailed forecasting is extremely difficult, but not impossible. Consequently, stores must have a way to quickly replenish their stock of goods when they run out. This is where the DC must be prepared to respond quickly. They must be flexible enough to respond to orders of varying volume and product mix. In some cases, they may have to customize the products to make them more floor ready for the retail store.

The distributor can help the retailer manage inventory in several ways. One of the most important ways is to be the first responder when the retail store needs stock replenishment. The DC can deliver to a store much faster than if the item had to be manufactured and then shipped to the store. With high turn SKUs that have minimal inventory at both the retail store and the DC, a significant portion of the inventory could be in transit at any given time. The DC's inventory provides some degree of risk pooling to inventory kept at the retail store (Walker 2009). Another way the DC can help is to make deliveries at times that are convenient to retailers, like restaurants, which have peak operating periods. Even providing the same truck driver over a period of time is helpful, as the driver becomes knowledgeable of the unique receiving characteristics of each store and its inventory holding areas.

## Inventory Management between Retailer and Distributor

One of the key decision areas in the retail–distribution link is where to put the inventory. Obviously, there needs to be an assortment of products in the retail stores that are adequate to support sales without incurring too many stockouts. Adequate usually means some inventory level that reconciles the trade-off between too much inventory, which can lead to extra holding costs, and running out of the product altogether, a problem that can lead to lost sales.

The distributor will also have some inventory and, in many cases, be able to quickly replenish the inventory at the retail store. However, the distributor is primarily a receiver and mover of goods, not a warehouse for storing backup inventories. For this reason, it is important that the distributor be fully informed about marketing strategies and programs at the retail store level. Otherwise, they will not be able to forecast future demand accurately and place orders with their suppliers, the manufacturers.

Some customizing functions occur at the distributor level. This enables the retail store personnel to concentrate on the presentation and selling of the goods. Customizing may mean making additions or modifications to the base product, or it may mean kitting or repackaging products into quantities and formats suitable for the retail store.

## Technology in Distribution Functions

Technology plays an important role in distribution functions, both in the physical movement of goods and in the movement of information. Information is essential to the sales and marketing function; it is also a vital part of the operations function to increase reliability of deliveries and reduce handling and inventory costs.

### At the Retail Store

The POS terminal is the backbone of the retail information system. It collects sales information by item, for marketing and sales information, and by dollar, for accounting. It also reduces checkout times and increases accuracy for operations.

POS terminals may be staffed with a person to check out the goods or they may be located at self-checkout stations, such as in grocery stores. The use of credit cards is rapidly displacing the use of cash, even in small purchase items, such as groceries. In high-dollar items, credit cards are the accepted form of payment. The computer system that supports the POS terminal maintains perpetual inventory records and initiates stock replenishment orders, based on algorithms built into its logic network. The system also makes it easier to communicate within the organization's multiple locations.

All products are identified through bar codes. Some retailers are moving to RFID tags to make it easier to take physical inventories and to capture increased amounts of information from the products. To help those of us, including at least one of the authors, who have spent time in grocery stores looking for an item that has no logical location, some stores are exploring the addition of a GPS app that could help customers find what they want. Many retail stores have added websites to increase their availability to consumers. The consumer can find a wider assortment of items and considerable information about the products on a website; consequently, they are better prepared to make a purchase when they come to the store.

## Movement of Goods

Improved information flow has greatly enhanced the flow of goods. It is easier for the retail store and the DC to track the movement and location of orders. This reduces the need for expediting and duplicating orders, and it helps the retail manager to plan for the upcoming delivery. Such a feature is important when inclement weather is in the area and the arrival of a delivery may be delayed.

The mode of transportation is more carefully analyzed today; therefore, there is a more efficient matching of products and orders with the appropriate means of transportation. While there is a heavy dependence on trucks, other modes of transportation such as rail, air, and water still have significant roles.

The use of more sophisticated computer software makes it possible to improve delivery schedules. This increases the reliability of deliveries and reduces the delivery costs and the negative environmental effect of fossil fuel emissions from the delivery vehicles.

Redesigned product packaging can increase space utilization and reduce in-transit damage. At the same time, many companies are working to reduce the negative effects of their transportation activities on the environment.

## At the Distribution Center

One of the most revolutionary technologies at the DC is the use of automatic guided vehicles (AGVs) for materials handling activities. The AGV can put goods away, pick them, and move them to staging areas.

Computer guidance makes it possible to use a random slot-selection process to improve space utilization. It selects the slot where the goods are stored and remembers that location when it is time to pick the goods for shipment.

Computer intervention can also generate a picking ticket that arranges the items to be picked in a logical path through the DC, so that the person (or AGV) who picks the order does not have to backtrack to get an item.

The use of cross docking techniques reduces inventories and lead times for delivery to the retail stores. Cross docking requires an information system, specialized materials handling, and layout modifications of the DC.

Independent distributors, who are becoming more marketing focused, are expanding their use of wireless e-mail/Internet access for customers, online ordering, and customer relationship management (CRM). ERP systems, so popular in manufacturing companies, have not reached the same level of use in distributors (Wells 2013).

## Distribution Center Design

In recent years, demand patterns have increased in variability in product mix and item volumes. Consequently, it has become more difficult to decide how to design DCs. In general, the need has been to invest in technology and design configurations that assure greater flexibility in meeting changing demand. One report offers very detailed ideas about how to create a comprehensive design of the modern DC. The authors stress the need to compile source data and point out that one of the most common reasons for suboptimal designs is suboptimal data (Hobkirt and Diebold 2013).

As the distribution function becomes more critical, industrial DCs are getting larger (Andel 2013) and more complex as both manufacturers and retailers strive to add technology and systems that will enable them to increase customer service while reducing costs (Blanchard 2013b).

## Positioning Services within the Distribution Functions

When companies operated with the mindset that they were not part of a supply chain, the disposition of service activities was influenced by the competitive environment. If the competitive emphasis was on costs, companies tended to limit the services they added to their product package. If the competitive emphasis was on customer service, companies tended to expand the services they provided. As companies consider themselves more a part of a supply chain, they still consider costs and customer service, but do it within the purview of the supply chain, not the individual company.

### Presale Services

At the retail store level, presale services include store design, layout, location, and presentation of the goods to be sold. In general, stores try to position their goods so that customers can see and touch them as part of their shopping process. Some of the more expensive goods, such as watches and jewelry, are displayed for customers to see, but not to handle without the aid of a sales person. Some goods, such as appliances, have information about their sizes and features displayed alongside the item. One of the latest innovations is for fast-food restaurants to display the calorie counts for menu items. To prepare the consumer for in-store shopping, company websites also offer a great deal of information about products as well as the company. Some companies maintain lists of customers who are notified of impending sales by mail or by telephone, such as for cosmetics or automobile maintenance.

The retail shopping experience depends heavily on the availability of a salesperson to complete the sale, although this is changing as automatic checkout stations are becoming more popular, especially in grocery and general merchandise stores.

### Postsale Services

Once the sale is made, additional services are commonplace. Grocery stores offer free assistance in taking the items to the customer's car. For bulky and higher-value items such as appliances, many stores offer free delivery and in-house placement. They may include free hookups or installation at a modest price.

Many stores follow the sale with a telephone call or e-mail to ask about the customer's level of satisfaction with their purchase experience. This increases the likelihood that a satisfied customer will respond favorably and return. If the customer is dissatisfied, it offers the store an opportunity to correct a problem. Often, correcting a problem can reestablish and strengthen a good customer relationship.

For more expensive items, retailers provide extended credit terms to make it convenient for the customer. It is not only an added convenience for the customer but also additional income for the company.

### Role of Third-Party Service Providers

Third-party service providers are becoming more evident. Banks make credit purchases convenient; delivery companies like United Parcel Service (UPS) and Federal Express (FedEx) reduce the complexity and cost of a retail store's operations; website managers simplify the in-store talent required; and specialized inventory companies take physical inventories faster, more accurately, and often at a lower cost than in-store staff could do.

Some third-party providers are extending their range of services considerably. For example, UPS offers financing for export and import transactions, operates repair centers for computers, and coordinates the flow of goods globally (UPS 2013). We will discuss third-party service providers more fully in Chapter 9.

---

## Distribution Performance Measurement

Inasmuch as the distribution function is increasing in importance, it follows that companies will be looking more carefully at the function's performance. Performance measures are needed to measure not only the effectiveness but also the efficiency of distribution. Performance measurement is further complicated, because it must consider how well the distribution function services the retail function.

### Financial Performance Measures

Financial measures should be linked to the overall performance of a business. These measures include the traditional performance measures such as profit margins and asset utilization. Other measures needed include gross profit and inventory turns. More advanced measures such as return on assets and economic value added (EVA) may also be desirable.

### Operating Performance Measures

In addition to financial performance measures, most companies use a variety of measures to evaluate operating performance. These include percentage of on-time deliveries, number and percentage of stockouts, and average lead times. In addition to these tangible measures, businesses should also attempt to evaluate supplier responsiveness to change and contribution of innovative ideas. These measures can be included in an overall vendor rating system.

### Collaboration Performance Measures

Finally, there is a need for retailers and distributors to measure how well they collaborate with each other and the value of that collaboration. This is a difficult measure and most companies are still trying to develop meaningful and available measures.

## Retailer–Distributor Relationship

The interface between retailers and distributors is the first link in the supply chain as it moves upstream away from the consumer. The distribution function can be owned by the retail company, or the retail store can buy from an independent external distributor. Whether owned or outsourced, the distribution functions are essentially the same. The functions were described earlier in this chapter.

Since this relationship between the retailer and distributor involves two parties, it is often convenient to build a fully developed relationship. A solid collaborative relationship can reduce delivery and inventory costs, achieve faster and more reliable deliveries, and assure a closer matching of supply with demand. When the retailer owns the distributor, it is easier to develop performance measures that encourage the distribution function and the retail function to work together for the good of the entire company. When the distributor is independent, the retailer is often larger and will usually be in control of the relationship-building process.

Success in the retailer–distributor relationship depends on the correct matching of supply (product mix, quantities, and delivery timing) with demand (product mix, quantities, and availability). The retail store wants to never be out of stock of an item, while maintaining the lowest possible inventories. Communication between the two parties is essential to achieve these desired objectives. Effective communication breeds small successes that, in turn, can lead to cooperation and even greater results. It is difficult to build effective relationships; therefore, both the retailer and the distributor benefit by viewing the relationship as a long-term investment and working hard to make it a success.

The pressure on retailers to have product on hand or quickly available puts increased pressure on distributors to respond. This will require changes in supply chains, both physical and cultural. There will be an increased need for enhanced planning and analytic capabilities, location selection strategies, DC design, material handling technologies, and replenishment processes. The line between manufacturer, distributor, and retailer will blur as they become more tightly linked (Blanchard 2013a).

## Summary

Retailing is one of the oldest businesses. Far from being staid and slow, it is one of the most rapidly changing areas of business. It is also becoming the most powerful member of many supply chains. Up until the latter quarter of the twentieth century, manufacturers were generally dominant and used distributors and retailers to move their products to the consumer. However, by the end of the century, companies such as Wal-Mart (general merchandise), Home Depot (building supplies), and Safeway (groceries) were representative of retailers who were large enough to dictate the terms and conditions of sales to their suppliers, the manufacturers (Davis-Sramek et al. 2008).

The distribution function is the link between manufacturers and retailers. It has the primary responsibility to move and store merchandise to balance supply and demand. It also performs a number of functions to transform the output from the manufacturer to a form usable by the retailer. The transportation process in distribution is important. Matching the right transportation mode to the goods being distributed is a demanding task.

Tompkins International, a leading logistics company, suggests that the distribution function is at a crossroad. Historically, independent distributors tended to be industry oriented—automobile parts, groceries, and the like. With the entrance of Amazon Supply, with its starting offering of 500,000 items, distributors will have to consider becoming more diversified, not only in products carried but in channels by which customers can buy—websites, branches, catalogs, integrated supply, social media, mobile devices, and peer-to-peer integration. Customers are looking for solutions, not just products (Tompkins International 2013).

## Hot Topic: Contaminated Milk

The interim suppliers in the supply chain can have a major impact on the retailer, especially if their product is defective. The hot topic for this chapter is about such a problem in the milk supply chain of a grocery retailer.

### Hot Topics in the Supply Chain: The Problem of the Middleman and a Milk Crisis in China

One of the concerns of supply chain managers is what goes on in the tier two and tier three levels of the supply chain. It is this unseen area of the supply chain where companies can lose control of the quality of the products they are manufacturing and selling. The melamine milk crisis described next is an example of what can go wrong when the unseen tiers of the supply chain operate in an unethical manner.

### BACKGROUND

Melamine is used in the making of plastics and laminates, hardly a product one would expect to find in milk. And yet, in parts of China, it was used in a powdered form to boost the protein levels of milk. Adding the artificial protein supplement was a common practice among dairy middlemen who first watered down the milk to extend its volume. The higher protein content enabled middlemen to command higher prices on the dairy market (Long et al. 2010). However, melamine is toxic when consumed and can cause kidney stones and renal failure (Pickert 2008).

The now defunct Sanlu Group was a dairy company that was both state owned and a joint venture. The New Zealand–based company, Fonterra Cooperative Group Ltd, owned 43% of the Sanlu (DeLaurentis 2009). Sanlu was one of several companies that had been aggressive in meeting the growing demand for infant formula in the Chinese market. However, in May 2008, company general manager Tian Wenhau found out that it was purchasing melamine-laced milk for use in its baby formula (Chinaview 2009). This breach in the supply chain was caused by middlemen who were watering down milk supplies and then adding melamine as an artificial protein booster. See Figure 6.3 for a description of the supply chain.

As the melamine-laced milk moved into the supply chain, illnesses began to be reported. As early as December 2007, the Sanlu Group received complaints about its baby milk formula, but the cause of the problems had not yet been determined (DeLaurentis 2009). The extent of the melamine-laced milk supply crisis eventually caused 300,000 illnesses and six fatalities (*Chemical Week* 2009). Within international

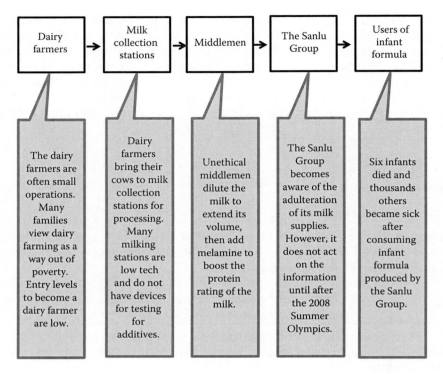

**FIGURE 6.3**
The supply chain for melamine-laced milk.

supply chains, the melamine scandal resulted in a ban on Chinese-made milk products in Japan, Malaysia, Bangladesh, Tanzania, Gabon, and the 27-nation European Union (Long et al. 2010).

## COMPANY RESPONSE

When there is a breach in the supply chain, a company needs to act swiftly to address the problem. In the case of Sanlu however, the response was slow and ambiguous. Even though Sanlu was aware of the melamine presence in its milk supply in May, it continued to produce and distribute the infant formula for several months after the discovery (DeLaurentis 2009).

Several reasons for the slow response have been offered and these should be noted, not only for their significance to this case but also as a reference guide for all companies sourcing products from China. First, saving face is a cultural norm in China that manifests itself by seeking to not put the blame for a problem directly on any one entity (Ye and Pang 2011). Instead, ways are found to not publically communicate the full extent of the crisis (Yu and Wen 2003) so that the reputation of the company and its management can be maintained. The Sanlu Group did communicate it had met all government standards for quality and inspections with its products. In addition, it reminded its stakeholders that it had a good reputation and a strong track record and if anything, perhaps its products, had been mislabeled (Ye and Pang 2011). In other words, it sought to protect its reputation.

A second factor should also be acknowledged. The 2008 Summer Olympics was being held in Beijing from August 8 to 24. Sanlu formally announced to the public on September 11 that melamine was in its milk supply. Some speculate that announcing a food scare before the Olympics would have been damaging to the country and the games itself (Liu 2008). A further irony was that Sanlu was both a joint venture and a state-owned company. Quality inspectors who monitor food safety reside with the government, as does the ownership of Sanlu. Such a situation creates a conflict of interest as inspectors and the dairy work for the same entity (DeLaurentis 2009).

## QUESTIONS FOR RESEARCH AND DISCUSSION

1. The impact of this crisis was felt mainly in the Chinese supply chain. However, what implications does this case have for companies sourcing its products from China?

2. What is the relationship between cultural and ethical considerations? In other words, can a decision be ethical in one culture but unethical in another? Provide your own examples.

## REFERENCES

China court sets death penalty in melamine case, *Chemical Week*, 171(3), 5, January 26, 2009.

Chinaview.com., Tian Wenhau, industry leader to disgraced prisoner, January 22, 2009, http://news.xinhuanet.com/english/2009-01/22/content_10704519.htm, retrieved January 1, 2014.

DeLaurentis, T., Ethical supply chain management, *China Business Review*, 36(3), 38–41, 2009.

Liu, M., Saving face goes sour, *Newsweek*, p. 7, October 6, 2008.

Long, Z., Crandall, W., and Parnell, J., A trilogy of unfortunate events in China: Reflecting on the management of crises, *International Journal of Asian Business and Information Management*, 1(4), 21–30, 2010.

Pickert, K., Brief history of melamine, *Time Online*, September 17, 2008, Retrieved July 25, 2012, from http://www.time.com/time/health/article/0,8599,1841757,00.html.

Ye, L. and Pang, A., Examining the Chinese approach to crisis management: Cover-ups, saving face, and taking the "upper level line". *Journal of Marketing Channels*, 18, 247–278, 2011.

Yu, T. and Wen, W., Crisis communications in Chinese culture, *Asian Journal of Communication*, 13(2), 50–64, 2003.

## Discussion Questions

1. What are the functions of a retailer?
2. How have the types of retail stores changed over the last century?
3. What are the key considerations involved in designing a retail facility?
4. What steps are involved in managing a retail business?
5. How do you think retailing will change in the future?
6. What are the main functions accomplished by DCs?
7. Describe the role of technology in the distribution function.
8. What is cross docking? How does it benefit the supply chain?
9. What is *floor-ready* merchandise? What are the benefits to the retailer?
10. Discuss the role of fulfillment centers in online sellers such as Amazon.

# References

Andel, T., Demand for big box DCs rising, *MHL News*, www.mhlnews.com, October 10, 2013.

Blanchard, D., Retail-centric supply chains are evolving into something else entirely, www.industryweek.com, September 2013a.

Blanchard, D., Supply chain & logistics: Warehouses are evolving into high-tech fulfillment centers, *Industry Week*, www.industryweek.com, November 1, 2013b.

Brown, S., An integrated approach to retail change: The multi-polarisation model, *The Service Industries Journal*, 7(2), 153, 1987.

Bureau of Economic Analysis, http://www.bea.gov/industry/gpotables/gpo_action.cfm?anon=96705&table_id=24547&format_type=0, accessed July 9, 2013.

Churchill, G.A., Jr. and Peter, J.P., *Marketing: Creating Value for Customers*, 2nd edn., Irwin McGraw-Hill, Boston, MA, 1998.

Crandall, R.E., Omnichannel retailing, *APICS Magazine*, 27(2), 26–29, 2014.

Davis-Sramek, B., Mentzer, J.T., and Stank, T.P., Creating consumer durable retailer customer loyalty through order fulfillment service operations, *Journal of Operations Management*, 26(6), 781, 2008.

Dead Malls, http://www.deadmalls.com/.

7-Eleven, Inc. celebrates 25,000 stores worldwide; Landmark opening commemorated in downtown Chicago, *PR Newswire*, New York, p. 1, July 9, 2009.

Gross, D., America has too many stores, *Slate*, www.slate.com/toolbar.aspx?action=print&id=2184492, February 16, 2008.

Hobkirt, I. and Diebold, J., Confidently committing to a distribution center design, when demand is unpredictable, Commonwealth Supply Chain Advisors, LCC, www.commonwealth-sca.com, March 2013.

Kangas, P., Malaise at the mall, Nightly Business Report, www.pbs.org/nbr/site/onair/transcripts/090114b, January 14, 2009.

Kotler, P., From mass marketing to mass customization, *Planning Review*, 17(5), 10, 1989.

Lal, R. and Alvarez, J.B., Retailing revolution: Category killers on the brink, Harvard Business School Working Knowledge, Boston, MA, October 10, 2011.

Plant, A.C., *The Retail Game, Playing to Win*, Douglas & McIntyre, Vancouver, British Columbia, Canada, 2007.

Ruby Tuesday blows up old-style unit to push new look, *National Restaurant News*, 42(31), 8, August 11, 2008.

Schwartz, B., *The Paradox of Choice: Why More Is Less*, Harper Collins, New York, 2004.

Tompkins International, Distribution at a crossroads, *Industrial Distribution*, pp. 12–13, www.inddist.com, May/June 2013.

Tracy, B., The seven Ps of marketing, Entrepreneur.com, 2004, http://www.entrepreneur.com/article/70824, accessed July 9, 2013.

UPS, http://www.ups.com/content/us/en/shipping/index.html?WT.svl=PriNav, accessed July 9, 2013.

Walker, W.T., Supply chain consultant, personal interview, March 20, 2009.

Wells, A., Tech usage & investments, *Industrial Distribution*, www.inddist.com, p. 26, May/June 2013.

Zikmund, W.G. and d'Amico, M., *Effective Marketing: Creating and Keeping Customers*, West Publishing Company, St. Paul, MN, p. 348, 1995.

# 7

## Production and Service Processes

### Learning Outcomes

After reading this chapter, you should be able to

- Describe the evolution of the production function
- Identify the critical success factors (CSFs) needed by manufacturing companies
- Describe the types of manufacturing strategies and explain how they relate to response times and levels of customization
- Explain the differences between batch flow and lean flow production
- Identify the advantages and disadvantages of outsourcing
- Discuss the factors that should be considered when increasing capacity
- Describe how service production strategies affect the supply chain
- Describe the nature of the relationship that manufacturers have with their downstream customers
- Describe the difference between a transaction approach and a process approach to business operations
- Justify why a process approach is preferable to a transaction approach
- Discuss the major trends in production
- Identify the various ways supply chain performance can be measured

### Company Profile: Caterpillar

Manufacturing companies make products; in many cases, they also provide services to maintain and repair their products. Manufacturers sell primarily to other companies—agriculture, mining, other manufacturers, and service providers—or to retailers. They employ skilled workers and use automatic equipment to assure quality products at low cost. They also outsource some of their operations so they have to be able to manage extended supply chains. As you read this company profile and the chapter in the book, think about these questions:

- How would you describe a typical hourly manufacturing worker's job?
- Where do manufacturing companies locate their factories? Why?

- How do manufacturers move their goods to retail stores or consumers?
- What kinds of inventory do manufacturers have to manage?

What additional questions do you think relate to manufacturing?

---

### Company Profile: Caterpillar Inc.

Caterpillar Inc. is a manufacturing company with headquarters in Peoria, Illinois. It traces its beginning to the Holt Manufacturing Company, founded in 1905, which built steam traction engines to help repair the San Francisco cable car system. The Holt Manufacturing Company developed the concept of using treads, instead of wheels, on tractors to move over soggy and uneven terrain. The tractors were used in World War I to move artillery and supplies to the front line. The tractors were also used in a number of major projects to build dams and bridges. They also found application in rescue efforts after earthquakes and hurricanes. Caterpillar Inc. was formed in 1925 with the merger of the Holt Manufacturing Company and the C. L. Best Tractor Company (Caterpillar—Our History 2013).

Caterpillar is ranked No. 42 on the 2013 Fortune 500 list of U.S. companies and No. 155 on the 2012 Fortune International list of companies. They had sales of over $65 billion dollars in 2012, with income of over $8 billion and are the largest industrial equipment manufacturer in the United States. Their closest competitor, John Deere, is about one-half the size, with $34 billion in revenues (Fortune 500 2013).

Caterpillar has grown by expanding their internal product lines. They have also grown through acquisitions that have enabled them to enter new lines of business. They made 37 acquisitions from 1951 to 2012, while at the same time divesting eight businesses. Some of their most recent acquisitions include Electro-Motive Diesel, a major maker of locomotives; MWM, a German maker of industrial engines; and Bucyrus International, a 131-year maker of mining equipment (Caterpillar—Our History 2013).

### LINES OF BUSINESS

They have four manufacturing business segments—construction industries, energy and power systems, resource industries, and customer and dealer support. They also have a Financial Products business segment to provide financing and insurance to customers and dealers. Table 7.1 shows a summary of each business segment and its contribution to the total company.

### PROCESSES

Caterpillar uses a variety of processes to make their products. They use a build-to-stock (BTS) process to make parts that are kept in inventory at the dealers if a replacement is needed by a customer to repair a piece of expensive equipment.

An assemble-to-order (ATO) process is used for complete machines such as excavators and earth-moving machinery. Caterpillar has what they call a lane strategy that enables a customer to order from four levels of options for the piece of equipment. The first lane is a standard model with basic options, and each subsequent lane offers a wider array of customization choices with increasingly longer lead times (Katz 2011). As a comparison with larger equipment, the 450F Backhoe Loader has an engine with 127 horsepower and a gas tank that holds 44 gal.

For even larger machines, such as mining trucks, a make-to-order (MTO) process enables the customer to order a wide variety of options but may have a lead time of several months.

**TABLE 7.1**

Business Segments with Sales and Profits by Segment

| Business Segment | Industries Served | Product Groups | 2012 Revenues (Millions) | 2012 Profits (Millions) | Geographic Markets (Sales) North America | Latin America | EAME | Asia Pacific |
|---|---|---|---|---|---|---|---|---|
| Construction industries | Infrastructure (roads), buildings, mines, and quarries | Loaders, excavators, tractors, graders, pipelayers, select work tools | $19,334 | $1,789 | $7,101 | $2,650 | $4,633 | $4,950 |
| Energy and power systems | Electric power, industrial, petroleum, marine, and rail industries | Reciprocating engine-powered generator turbines, diesel-electric locomotives | $21,122 | $3,434 | $8,720 | $2,191 | $6,043 | $4,168 |
| Resource industries | Mining, quarry, forestry, paving, tunneling, industrial and waste industries | Shovels, mining equipment, borers, scrapers, dozers, off-highway trucks | $21,158 | $4,318 | $6,037 | $3,662 | $4,374 | $7,085 |
| Customer and dealer support and all other | All industries who use Caterpillar equipment throughout the world | Distribution, spare parts, remanufacturing, telematics | $1501 | $1014 | $777 | $65 | $395 | $264 |
| Financial products | Customers and dealers around the world | Finance leases, installment sales, lease purchase, insurance | $3090 | $763 | $1675 | $397 | $408 | $610 |
| Corporate charges and adjustments | | | –$330 | –$2,745 | –$231 | –$29 | –$26 | –$44 |
| Total (consolidated) | | | $65,875 | $8,573 | $24,079 | $8,936 | $15,827 | $17,033 |
| Percent of total sales | | | 100% | | 37% | 14% | 24% | 26% |

*Source:* Caterpillar, Year in review, 2013, http://www.caterpillar.com/companym, accessed August 15, 2013.

To give you some idea of the immensity of these trucks, the 797F Mining Truck is 25 ft high and over 48 ft long and has an engine of 4000 horsepower and a gas tank holding 1000 gal.

The most challenging is the engineer-to-order (ETO) process in which a product, such as a tunnel borer or a mining conveyor belt, is designed for a specific tunneling project, such as digging an underground tunnel underneath a river or removing coal from a seam less than 36 in. high. Pre-engineered systems are designed to be mobile, modular, and flexible. Easily transported for use in different areas of the mine, the pre-engineered belt terminal groups reduce lead times and can be used in multiple bulk material-handling applications.

## SERVICES

Many manufacturing companies have also added a line of services to their product line. Many Caterpillar products have multiyear lives, and customers want to extend that life as much as they can because of the heavy investment in the equipment. Caterpillar makes or supplies spare parts to dealers. Some dealers are large. Gregory Poole equipment, the Cat dealer in Raleigh, NC, stocks $6 million in parts and has repair and rebuilding facilities for most of Caterpillar's equipment. The dealer has 850 employees and prides itself on fast service to their customers. As the CEO Greg Poole puts it, "If you need a part, we probably have it here. If we don't, you can order it as late as 7 p.m., and we'll probably have it here by 6 a.m. the next morning" (Colvin 2011, p. 144). The company also has repair and remanufacturing capabilities to refurbish equipment throughout the world. One of Caterpillar's key selling points to customers is that their equipment may be somewhat more expensive, but it has the longest useful life of any of the company's competitors.

As another service to help customers get the most effective, and efficient, use of their equipment, Caterpillar offers training courses through Caterpillar University, which offers classes, e-learning, and development opportunities to employees, dealers, suppliers, and customers (Caterpillar University 2013).

Caterpillar also offers a variety of financing alternatives to customers to help them purchase equipment. They also can provide accident and life insurance for customers who need it.

They even offer CAT products for sales at dealer stores and online, such as caps and license plates (see www.shopcaterpillar.com).

## LOCATIONS

Caterpillar is a global company, with two-thirds of its 2012 sales coming from outside North America. Many of their products are used to build roads and extract materials from mines and quarries and in commercial building. These are products used in developed countries, and, to an increasing degree, they are essential in emerging countries that are trying to bolster their economies by producing goods for export. As these emerging countries grow, they need to improve their transportation infrastructure in order to better serve their domestic markets.

The company has manufacturing plants and dealers spread throughout the world, as shown in Table 7.2. As of the end of 2012, there were 189 dealers, also scattered throughout the world. Many of their products are large and expensive; therefore, customers need to have support services located nearby. The company builds manufacturing plants and sets up dealers in the markets served. By doing this, they learn to operate in each market, and they shorten their supply chains to reduce the logistics costs of moving product to the consumer. They also are closer to provide services to help keep the equipment operating efficiently.

**TABLE 7.2**

Manufacturing Locations throughout the World

| Business Segment | United States | | Outside the United States | | |
|---|---|---|---|---|---|
| | Number of Manufacturing Plants | Number of States with Facilities | Number of Manufacturing Plants | Number of Countries with Facilities | Total |
| Construction industries | 7 | 4 | 13 | 9 | 33 |
| Energy and power systems | 22 | 8 | 25 | 9 | 64 |
| Resource industries | 19 | 11 | 27 | 16 | 73 |
| Total | 48 | 23 | 65 | 34 | 170 |

*Source:* Adapted from Caterpillar, 10-K report, 2012, p. 21.

**TABLE 7.3**

Manufacturing Locations outside the United States

| | Construction | Energy and Power | Resource |
|---|---|---|---|
| Australia | | | 1 |
| Belgium | 1 | 1 | |
| Brazil | 2 | 4 | 1 |
| Canada | | | 1 |
| China | 3 | 2 | 4 |
| Czech Republic | | 1 | 1 |
| France | 1 | | 2 |
| Germany | | 3 | 2 |
| India | 1 | 2 | 2 |
| Indonesia | 1 | | 1 |
| Italy | | | 2 |
| Japan | 2 | | 1 |
| Mexico | | 2 | 4 |
| Russia | 1 | | 1 |
| Switzerland | | 1 | |
| Thailand | | | 1 |
| The Netherlands | | | 1 |
| United Kingdom | 1 | 9 | 2 |
| Total | 13 | 25 | 27 |

*Source:* Adapted from Caterpillar, 10-K report, 2012, p. 21.

Table 7.3 shows the number of manufacturing locations in 18 different countries throughout the world. Many of the locations are located to be near the market in which the business segment competes.

## EMPLOYEES

At the end of 2012, Caterpillar had 125,341 employees, with 70,783, or a little over 56%, outside the United States. Of the 54,558 employees in the United States, 13,666 belonged to unions, principally the United Auto Workers (UAW), International Association of Machinists (IAM), and the United Steelworkers (USW). Caterpillar has had several long strikes during its history but is now enjoying a period of calm, with some union contracts extending through 2017 (Caterpillar 10-K Report, p. 8).

## RESEARCH AND DEVELOPMENT

Caterpillar continues to look for ways to improve its products so that they ensure that "customers make more money using Cat equipment than using competitors' equipment" (Colvin 2011, p. 142). In the last 3 years, they have increased the dollar amount spent on research and development (R&D) from $1.905 million to $2,446 million. Because of their increase in sales, the percentage of sales spent on R&D declined from 4.5% to 3.7% (Caterpillar 2012, p. 8).

## SUSTAINABILITY

The company is regulated by federal, state, and international environmental laws governing the use, transport, and disposal of materials and substances, especially in the control of emissions. They are engaged in remedial activities at a number of locations. Caterpillar believes they have made progress, and will continue to make progress, in their efforts to be a good corporate citizen.

## STRATEGY DURING THE RECENT ECONOMIC DOWNTURN

Caterpillar is in a cyclical business. When business is good, customers want to buy. Conversely, when economic conditions are bad, as they were the last few years, customers want to make their equipment last as long as possible. Consequently, Caterpillar sales fluctuate during the economic cycle. They survived the recent downturn by developing contingency plans prior to the downturn and then rigidly following the plan during the downturn. As a result, Caterpillar came through the tough period from 2007 to 2009 with a strong cash flow, their credit rating intact, and continued paying a stock dividend (Colvin 2011).

## REFERENCES

Caterpillar, 10-K report, Peoria, IL, p. 8, 2012.
Caterpillar, Our history, http://www.caterpillar.com/company/history, accessed August 15, 2013.
Caterpillar, Shop Caterpillar 2013, http://www.shopcaterpillar.com/, accessed July 15, 2013.
Caterpillar University, 2013, http://www.caterpillar.com/en/company/dealers-customers-suppliers/cat-university.html, accessed July 15, 2013.
Caterpillar, Year in review, 2013, http://www.caterpillar.com/companym, accessed August 15, 2013.
Colvin, G., Caterpillar is absolutely crushing it, *Fortune*, pp. 136–144, May 23, 2011.
Fortune 500, 2013, http://fortune.com/fortune500/2013/, accessed July 15, 2013.
Katz, J., Creating high-value supply chains, *Industry Week*, 260(3), 34, 2011.

## Introduction

In Chapter 6, we described the distribution and retail functions designed to move goods and services from the manufacturer to the consumer. In this chapter, we will describe how these goods or products are produced. As an overview, we will begin by providing some definitions from the *APICS Dictionary* (Blackstone 2013):

- *Manufacturing*: A series of interrelated activities and operations involving the design, material selection, planning, production, quality assurance, management, and marketing of discrete consumer and durable goods.

- *Operation*: (1) A job or task consisting of one or more work elements, usually done essentially in one location; (2) the performance of any planned work or method associated with an individual, machine, process, department, or inspection; and (3) one or more elements that involve one of the following: the intentional changing of an object in any of its physical or chemical characteristics; the assembly or disassembly of parts or objects; the preparation of an object for another operation, transportation, inspection, or storage; and planning, calculating, or giving or receiving information.

- *Operations management*: (1) The planning, scheduling, and control of the activities that transform inputs into finished goods (FGs) and services and (2) a field of study that focuses on the effective planning, scheduling, use, and control of a manufacturing or service organization through the study of concepts from design engineering, industrial engineering, management information systems, quality management, production management, inventory management, accounting, and other functions as they affect the operation.

- *Production*: The conversion of inputs into finished products.

- *Production and operations management* (POM): Managing an organization's production of goods or services and managing the process of taking inputs and creating outputs.

- *Production management*: See operations management.

For simplicity, we will use *production* as the general term to describe the preparation of both goods and services for distribution to the consumer. If a company takes inputs of raw materials such as wood or steel and converts them into finished products—outputs—such as furniture or automobiles, they are manufacturers. If they take inputs such as untreated patients at a doctor's office and convert them into treated patients—outputs—they are in services. In either case, they are performing a production function. Figure 7.1 shows these relationships.

Production functions occur at several points along a typical supply chain. However, in every case, the production function is concerned with transforming some form of input into an output. At the supply chain's origination point, the production function may involve transforming corn seeds into mature ears of corn or transforming seams of coal

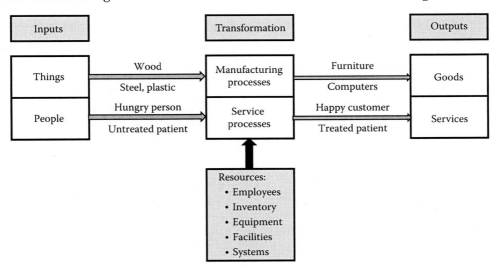

**FIGURE 7.1**
Basic input–transformation–output (ITO) model.

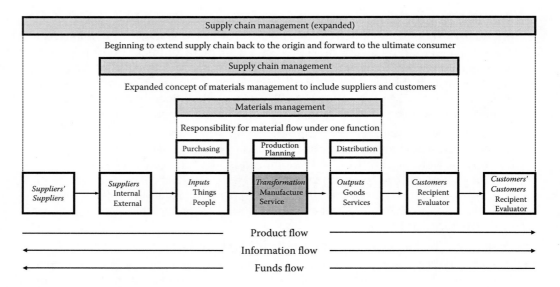

**FIGURE 7.2**

Evolution of the expanded supply chain. As companies extend their supply chain, both forward to the customers and back to the suppliers complexity increases, making communications more critical; variability in demand decreases, making reduced inventories possible; interdependence increases, making delivery reliability more important; supplier dependability increases, making fewer suppliers necessary; and material flow smooths, making lead (response) times shorter.

into processed coal use at a power-generating station. As the product moves through the supply chain, additional production operations are performed at interim points. Figure 7.2 shows the relationships among the multiple customers and suppliers in the supply chain. We described the supply chain as a system in Chapter 4. In this chapter, we will describe a production function that could exist in several places along a supply chain.

## Evolution of the Production Function

From the earliest of times, people produced things by taking raw materials and converting them into usable products and services. For centuries, this production process was largely done by individuals for other individuals. There were no large companies to plan and organize the production; the work was performed by individuals who learned how to perform all the steps necessary to make a finished product. They were called craftsmen, because they tended to specialize in a specific line of work, such as carpentry, stonemasonry, baking, weaving, or tailoring. In order to pass along their knowledge to the succeeding generations, they hired apprentices who learned by working with the master craftsman. Sometimes, it took years to learn all of the skills necessary to perform all of the steps necessary to complete a job. Much of the work was completed in an individual's home. The products were customized to satisfy the needs of individual customers.

In some cases, entire communities organized themselves to develop all of the needed crafts for the community to be self-sufficient. Although individual communities attempted to be independent of each other, they soon found certain products in another community that were better than they could make themselves; this was the beginning of trade among communities. In this arrangement, communities designated work areas that housed the tools and materials needed to do the work, such as in a carpenter's shop or a weaving room.

## From Craft to Mass Production

As populations grew and trade became an accepted practice among communities and countries, there was a need for a system that could produce the higher volumes of products necessary to satisfy increasing demand. This need led to the discovery that job specialization, or breaking the total work necessary to make a product into smaller activities, could lead to increased productivity and shorter training requirements. Frederick Taylor studied how industrial tasks could be performed more efficiently through specialization and came up with the concept of scientific management to describe his philosophy.

Figure 7.3 shows the movement from the craft age to mass production was achieved by standardizing products and processes. This work practice encouraged the idea of job specialization because the use of interchangeable parts reduced the need for finishers who could *fit* the pieces together, a step that sometimes required skillful technicians.

The movement to mass production, especially in the United States, was assisted by the need for high-speed equipment. There were not enough workers to make all of the products needed. In order to meet the demand, workers needed additional equipment to do some of the work. Initially, the equipment took the form of hand tools—saws, awls, planes, and clamps—that aided the worker. Later, as steam and waterpower became available, the equipment could be powered to reduce the physical effort required by the worker. The power-driven equipment was also faster than the worker and, in most cases, more consistently accurate (Hounshell 1985).

Mass production has been the major paradigm for over a century. It helped create an enormous variety of products with relatively low costs and high quality. The major feature that was lost was the ability to customize products that was present during the craft age.

## From Mass Production to Mass Customization

Figure 7.3 shows that mass production is beginning to be replaced by a new paradigm—mass customization, which is defined as "The creation of a high-volume product with

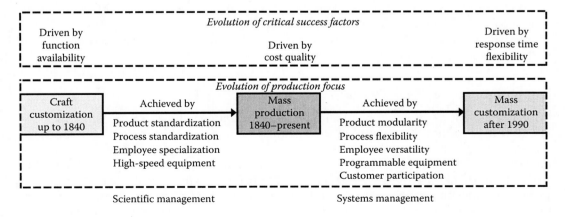

**FIGURE 7.3**
From the craft age to the mass customization age.

large variety so that a customer may specify an exact model out of a large volume of possible end items while manufacturing cost is low due to large volume. An example is a personal computer order in which the customer may specify processor speed, memory size, hard disk size and speed, removable storage device characteristics, and many other options when PCs are assembled on one line and at low cost" (Blackstone 2013).

Mass production is a way to make a large volume of standard products at a low cost and, yet, with high quality. Mass customization is a way to make a large volume of customized products, at near the cost and quality of standard products. We discussed the need for customizable product design in Chapter 3. We will describe the processes necessary for mass customization in more detail in Chapter 11. While mass production is still the dominant mode, a number of companies are looking at ways to move closer to mass customization practices.

## Critical Success Factors for Manufacturers

CSFs are those things a company must do well to be successful. In general, the number and variety of CSFs have been increasing for manufacturing companies.

### Cost and Efficiency

Historically, manufacturing companies have been concerned with making products at a low cost. To accomplish this, the company worked to increase employee and machine efficiency and the overall productivity of their operations. In many cases, they introduced new machines and material-handling equipment to improve the flow of goods through the manufacturing facility and reduce the labor content of the product.

### Quality

The quality movement came into prominence during the 1960s and grew into a major requirement for almost every form of manufactured goods. In recent years, product quality has become even more important. Japanese automobile manufacturers gained market share in the United States and worldwide by making cars of higher quality than those produced by U.S. automobile manufacturers. As a result of global competition, in recent years, the American and European car makers have, in most cases, improved the quality of their cars to a level equal to that of the Japanese manufacturers.

### Responsiveness: Timing of Delivery

With the advent of just-in-time (JIT) and lean production movements, companies are attempting to reduce the amount of inventory they have, not only at their own locations but also along the entire supply chain. One way of accomplishing this is to reduce the amount of safety stock needed. We pointed out in an earlier chapter that reducing the variation in demand patterns would reduce the need for safety stock. The same is true when variations in supply are reduced. If suppliers consistently meet scheduled delivery times, the receiving company can carry less safety stock. Consistent deliveries require coordinated production and delivery schedules—production to make the right quantities at the right time and to be delivered at the right time.

## Responsiveness: Product/Service Mix

Responsiveness also means the manufacturer must have the right product mix to meet customer demand. Because the customer wants a greater variety of products, manufacturers must design and produce a greater variety. This task requires greater coordination among the various functional activities within the manufacturing organization (Crandall 2006).

### *Flexibility*

In order to meet all of the requirements outlined earlier, manufacturing organizations must become more flexible, that is, develop the ability to move from one product to another quickly and effectively without losing their ability to make a high-quality product at a low cost. Flexibility means being able to work seamlessly to satisfy demand by functioning within the existing constraints of product, process, capacity, human resources, and organizational structure.

## Agility

In addition to being flexible, manufacturers must also be agile. This implies that they can quickly develop new products, design new processes, accommodate different customer requirements, and adapt to changing open system requirements. Even more than being flexible, they must be agile in order to adapt to the changing business environment.

## Information Technology

To accomplish all of the things necessary to make products efficiently, manufacturers must have effective and far-reaching information systems. It is no longer sufficient to know what is going on within their own organization; they must be aware of what is happening in their far-reaching network of customers and suppliers. They must also be able to supply information about products to their customers. We will describe the technologies needed later in this chapter.

---

## Manufacturing Strategies

In order to meet the demands of customers, manufacturers can select from a variety of strategies, which include make to stock (MTS), ATO, MTO, and ETO. These strategies were described earlier in Chapter 4.

Figure 7.4 shows the relationship among the strategies graphically. The general relationship can be seen by examining the trade-off between the response time and the degree of customization. The greater the level of customization desired, the longer the response time required. For some products, such as bread or milk, customers will not wait if the product is unavailable; they simply go to another store to get the items they want. However, for other products, with a lower frequency of purchase—computers, furniture, wedding rings—the customer is usually willing to wait longer to get more nearly what they want.

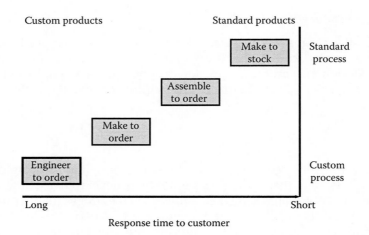

**FIGURE 7.4**
Traditional production.

Inventory positioning is another consideration. The manufacturing concept of postponement focuses on waiting as long as possible, before putting the customizing features on a product. In an MTS situation, the final product is carried as inventory in its finished state, with all of the costs included. In an ATO strategy, the inventory can be carried as modules ready to be assembled. This practice reduces the number of items carried in inventory and also postpones the application of the labor needed to assemble the final customization features until the very end of the process. Inventory for the MTO strategy is even more constrained; it is in a more flexible state and contains even less of the labor required. For the ETO strategy, no inventory is required.

## Make to Stock

Manufacturers have long used the MTS strategy to make standardized products for mass markets. The objective is to manufacture products based on forecasted demand and to stock those products in convenient locations, such as retail stores or automobile dealerships, so that customers can buy them without having to wait. An early example of this strategy was Henry Ford's quest to make the Model T in "any color as long as it's black!" Today, manufacturers face making an increasing variety of products in order to meet the growing wants of customers. This is especially true in everyday items, such as those found in grocery stores. Breakfast cereals, canned vegetables, and breads are just a few examples of products where the number of items within a general category has grown almost beyond reason. One writer expressed the conviction that the increased variety is causing consumers increased stress because of the need to make choices among the large variety of goods offered (Schwartz 2004).

## Locate to Order

Although not formally recognized as a separate manufacturing strategy, an offshoot of the MTS strategy in the service sector is one that we will call locate to order (LTO). When a customer wants to buy a car and the particular model is not available on a dealer's lot, the dealer tries to locate the desired car at a nearby dealer. While the manufacturing strategy is MTS, modern communications capabilities make it possible for dealers to work with each other to gain increased customer satisfaction. Sometimes, clothing stores use this same approach, although the practice seems to be more related to the individual salesperson's willingness to help the customer rather than to company practice.

### Assemble to Order

In an effort to be more responsive to customer demands for increased variety but, at the same time, limit the number of finished products items, some companies use an ATO strategy. Dell made this strategy popular when they took orders for computers and quickly assembled the major modules together with selected software to configure a computer to meet the customer's order. Under the ATO strategy, the customer must be willing to wait a little longer to get a product that is more nearly what they want. A simple example of this is getting your sandwich assembled to your specifications at a delicatessen.

### Make to Order

If a customer is willing to wait longer, they can have products that are even more specifically made to their specifications. In this strategy, the manufacturer can make, or have made, the component parts to be assembled into the final configuration ordered. Examples of this approach could include a piece of furniture, such as a sofa, which is built to fit within the dimensions of a particular room in a house, or a picture window that is made for that same house. Usually, the product requires only slight modifications from the standard product that the manufacturer already makes.

### Engineer to Order

Where the product requires extensive change from a standard product, the manufacturer can use an ETO strategy. In this case, the company designs the product, usually in consultation with the customer, before any raw materials are processed in any way. A wedding cake fits this category, because often they are a collaborative effort between the bride and the baker. Another example is an automatic assembly line where the requirements are specific to the needs of the manufacturing process for which the line is designed.

In Chapter 11, we will describe how companies can move from the traditional mass production strategies to mass customization strategies.

## Batch to Lean Operations

In addition to the movement toward mass customization, many businesses are implementing lean production operations. The lean concept is an outgrowth of the JIT movement that started with Toyota. Lean manufacturing was a result of research by Womack and Jones in the automobile industry but has spread to other manufacturing industries and even to some service operations (Womack and Jones 1996). The primary objective is to create a sustained flow of goods through the process, thereby reducing inventories and response times.

Moving from batch flow to lean production sounds attractive, and many companies have made this transition, at least in part. However, it is not simple or easy. In the following example, we focus on the changes required to make the transition a successful one and point out why it is a difficult change management project.

## Present Batch Flow

Figure 7.5 portrays the order flow in a typical batch flow process. The diagram shows three major steps, or operations, in the process, such as fabricating components, assembling the final product, and packing and shipping. Each operation occupies different physical areas of the plant and is a self-contained unit with its own supervision, work force, and facilities. We call them Operations 1, 2, and 3 in Departments A, B, and C. For clarity in our following discussion, we call the company Ultralean.

Figure 7.6 shows a possible physical layout and the flow of materials from one department to the next and finally to the customer. Accounting can classify each department as a work center and give production credit to the department when it moves its output to the next department. There are plenty of raw materials, an ample supply of work in process (WIP) to keep the machines running, and adequate finished goods (FGs) to fill most orders. It takes each department 1 day to process an order, so that, with a normal number of orders, the lead time is 30 days. Let us look at what happens under several situations.

| Batch flow | | | | |
|---|---|---|---|---|
| | Order no. | Order no. | Order no. | |
| Stores | 30 | 20 | 10 | Fin. gds. |
| Stores | 29 | 19 | 9 | Fin. gds. |
| Stores | 28 | 18 | 8 | Fin. gds. |
| Stores | 27 | 17 | 7 | Fin. gds. |
| Stores | 26 | 16 | 6 | Fin. gds. |
| Stores | 25 | 15 | 5 | Fin. gds. |
| Stores | 24 | 14 | 4 | Fin. gds. |
| Stores | 23 | 13 | 3 | Fin. gds. |
| Stores | 22 | 12 | 2 | Fin. gds. |
| Stores | 21 | 11 | 1 | Fin. gds. |
| | Department A | Department B | Department C | |
| Assume: | 1 order completed each day | | Lead time | 30 days |

| What do we do about: | |
|---|---|
| Machine breakdowns? | Quality problems? |
| Lack of materials? | Operator skills? |
| Customer demand patterns? | Customer service? |

| Advantages of batch flow: | Disadvantages of batch flow: |
|---|---|
| Long runs | Long lead times (marketing) |
| High utilization | High cash investment (finance) |
| Low purchase cost | Physical count harder (accounting) |
| Job specialization | Slow to detect defects (quality) |

**FIGURE 7.5**
Review of batch flow characteristics.

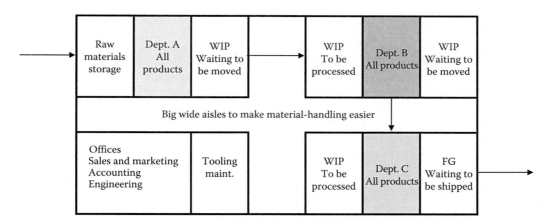

**FIGURE 7.6**
Physical layout—a process layout.

## Machine Breakdowns

Suppose the machines in Department B break down. Department C keeps running because they have a backlog of orders. Department B will surely get their machines running again in time to catch up, even if they have to work overtime. Department A keeps fabricating parts and *pushing* the parts along to the assembly area, so they can get production credit for their work. Having extra inventory just in case a machine breaks down looks like a good idea.

## Supplier Failures

Suppose a supplier has a problem—strike, fire, or other—and fails to deliver on time. It is inconvenient but that is why Ultralean has a supply of raw materials in inventory. They may even have time to find another supplier if their regular supplier is down for an extended period. Again, having extra inventory looks like a good idea.

## Worker Flexibility

Suppose the workers are assigned permanently to their home departments: A, B, and C. They can be transferred to another department to work, but that move requires a temporary transfer. Having this level of job specialization keeps the job grades lower, thus reducing labor costs, but this is offset by a reduced ability to shift resources around.

## Customer Lead Times

Assume in this scenario that 30-day lead times used to be acceptable for most customers. Recently, however, a number of customers, including Ultralean's best ones, indicate they are transitioning to lean production and need more frequent deliveries and shorter lead times. Meanwhile, competitors have found a way to reduce their lead times and have been chipping away at Ultralean's market share. Having excess WIP (inventory) causes orders to wait until moved or worked on. This was evident recently when Ultralean's best customer insisted on a rush order, and the company was able to ship the order in 3 days. Ultralean assigned the order the highest priority and moved it to the front of the line at every operation. Of course, other orders were delayed a little, but it did satisfy their best customer, at least until the next order.

*Customer Order Size*

Recently, several customers insisted on more frequent deliveries of smaller orders. This helped them keep their inventories low, improved their cash flow, and reduced the obsolescence of inventory when they changed the design of their product. Ultralean minimized setups by building FGs inventory but have been stuck with some obsolete inventory when their customers changed their specifications.

Upon review, the batch flow system has some plusses; however, it also has some minuses. Could lean manufacturing work better? What does it look like and what does it achieve? What would Ultralean have to do to implement a lean system?

## Proposed Lean Production Flow

Figure 7.7 shows the ideal lean production flow. To achieve this ideal flow, Ultralean categorized their product line into families of similar products. They rearranged the plant layout to form manufacturing cells that grouped all of the equipment, workers, and support staff needed to make each family of products. This new layout is shown in Figure 7.8. Ultralean managers were more than a little surprised, and pleased, at the amount of empty space available as they eliminated some of the WIP. Consider how the new arrangement would work.

*Customer Lead Time*

The lead time is now 3 days on a regular basis. Ultralean's customers are pleased, even amazed, at how fast they can promise delivery. In fact, the response has been so overwhelming that Ultralean managers now see they are not going to be able to keep up with the surge in demand.

| JIT flow (ideal) | | | | |
|---|---|---|---|---|
| Stores | 3 | 2 | 1 | Shipments |
| | Department A | Department B | Department C | |
| Assume: | 1 order completed each day | | Lead time | 3 days |

| What do we do about: | |
|---|---|
| Machine breakdowns? | Quality problems? |
| Lack of materials? | Operator skills? |
| Customer demand patterns? | Customer service? |

| Advantages of JIT: | Disadvantages of JIT: |
|---|---|
| Short lead times | Needs stable demand |
| Low WIP | Requires excess capacity |
| Multiskilled operators | Lower utilizations |
| Supplier relationships | Time to develop suppliers |

**FIGURE 7.7**
Review of JIT characteristics.

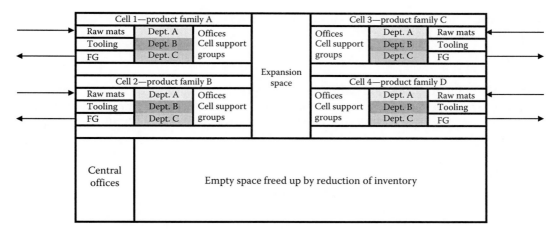

**FIGURE 7.8**
Physical layout—with manufacturing cells.

### Extra Available Space

The availability of so much empty space is a real advantage. Ultralean will not need to build an addition to the existing plant; they now have space to introduce another line of products.

### Faster Detection of Quality Problems

Under this new arrangement, quality problems are detected sooner than under the previous arrangement. This occurs because there is so little WIP that the operator who receives defective materials or product alerts supervision immediately. In fact, the operators are so close in proximity they can tell the operator producing the defective work about the problem. No more formal reports are necessary.

### Ease of Transferring Operators

This new arrangement also makes it easier to move operators from one operation to another without having to write temporary transfers because the operators are in the same group with the same job grade. Of course, the job is now of a higher grade because the operators must be able to perform any of the jobs in the cell; however, the flexibility is worth it.

### Reduced Scheduling Requirements

The product is moving smoothly, and lead times are so short that you can use a first-in, first-out, sequence for processing orders. You have almost eliminated the need to juggle orders caused by delays in your processing or changes in customer requirements. Because your customers know when to expect their orders, they stop calling to expedite them.

### Smaller Lot Sizes

The lot sizes are smaller because your customers want more frequent, smaller shipments. You reduced your setup times and the operators are more versatile, so the smaller lots are not a problem. It has leveled your production load because you can produce about the same mix of orders every day.

### No Buildup of WIP Inventory

With product flowing smoothly through the process, you no longer have a buildup of inventory when one of the operations shuts down. With fewer quality problems, there are fewer work stoppages. In fact, as you reduce the variances in the process, you find that you have no need for excess WIP.

### Empowered Employees

Your employees are more versatile, and they have the training and motivation to partici- pate in teams that tackle process improvement projects. Some of the projects contribute significantly to the reduction of product costs, the improvement of product quality, and the reduction of lead times.

### Reduced Equipment Breakdowns

One of your new challenges is that Ultralean managers have to reduce equipment break- downs because they do not have the buffers of inventory previously present. However, they have empowered production employees who can perform some of the preventive maintenance needed to keep the equipment running.

### Reduced Late Material Deliveries

Ultralean's suppliers had to improve their operations, too. They cannot afford to have late deliveries or back orders, because the flow of materials into Ultralean's production process is so closely integrated that any delays would cause you to lose a substantial amount of output. They are able to operate with fewer, but better, suppliers.

### Reduced Write-Off of Inventory

Because Ultralean has less inventory on hand, they have almost completely eliminated inventory write-offs. The inventory does not deteriorate, it does not become obsolete, it is much easier to account for, and it is less prone to damage.

## Obstacles to Implementing Lean

The expectation for lean production is attractive, and many companies are moving suc- cessfully in this direction. However, it is a change that demands planning and dedication. Some of the most frequently encountered obstacles include the following:

*Investment required to change layouts*: Moving equipment takes time and may take special equipment if the machines are heavy. Floors have to be refinished and the plant loses production time.

*Resistance from employees*: Employees, like most managers, think changes make the work harder, at least temporarily. The change may also lead to layoffs.

*Lack of support from top management*: Top management may not understand why there is a need to change layouts, job descriptions, or any of the other myriad of tweaks needed.

*Temporary reduction in profit because of reduction in inventory*: Reducing inventory means producing less than is being sold. This means what is produced carries a larger amount of overhead expense and causes reduced profits until the equilibrium point is reached.

*Need to negotiate smaller and more frequent deliveries from suppliers*: Suppliers may not understand why they have to change, especially if they are expected to sell their products at the same price.

*Need to retrain employees*: Lean operations usually require more versatile employees; they must be trained in the correct, perhaps vastly different, procedures.

*Need to assure higher quality of product*: Defective products disrupt the flow of goods; there is little margin for disruption in a lean process.

*Need to change process flow*: Changes in layout and materials handling means that products move more frequently and with the need for greater precision. Changes will also uncover process improvements.

*Need to involve marketing and finance*: Marketing needs to capitalize on the reduced lead times; finance needs to understand why the investment is necessary and why profits will decline temporarily.

*Need to convince customers of the benefits*: Customers may not realize the benefits of getting shorter lead times, more flexibility in ordering, higher quality, and other benefits, unless it is pointed out to them.

While lean production has made a great difference in the world's production, it has not come without a major commitment by those companies who have implemented the changes.

---

## Make or Buy Strategies

As we described earlier, manufacturing companies are changing in form and focus. While they perform the same functions of materials processing, fabrication, and assembly, the organizational structures have changed from vertical integration to loosely connected supply chains.

### Vertical Integration

During the first half of the twentieth century, many manufacturers believed that vertical integration was the preferred strategy, at least insofar as making their own component parts. Some even ventured into the retailing of their own products. Vertical integration was considered a good way to control the cost and quality of products. Henry Ford epitomized this approach with the design of the River Rouge plant that would take iron ore brought off Lake Michigan to the Rouge River and convert it into a Model T within the next 48–72 hours (Ford 1988). An interesting bit of trivia is that the River Rouge plant first produced submarine chaser ships during World War I before ever building automobiles (River Rouge 2008). Vertical integration worked well when product lines were limited, and the strategy of mass producing for mass markets was still a viable choice. It also was a satisfactory approach when manufacturers sold primarily to domestic markets and did not have the opportunity or need to consider foreign markets or suppliers.

After World War II, the situation began to change. Manufacturers in the United States could sell practically any products they could produce because the capacity of other major industrialized countries—in Europe and Japan—had been largely destroyed. This opened markets for U.S. manufacturers in other countries and provided an opportunity for establishing their own facilities in new markets or to enter into joint ventures or alliances with companies in those market areas.

In the 1960s and 1970s, competitors from Germany and Japan began to make inroads into consumer goods markets in the United States, such as automobiles and electronics—radios and televisions. Manufacturers in the United States seemed to be more interested in financial manipulation than in improving their basic manufacturing capabilities. Teledyne, IT&T, and others achieved notoriety as progressive in their pursuit of synergistic acquisitions. During that time, the competitive advantage that U.S. manufacturers held began to weaken and, in some industries, faded away completely.

## Outsourcing

The concept of outsourcing, or hiring external companies to perform some part of the total manufacturing process, is not new. It was earlier known under the banner of *make or buy*, where companies could choose between performing the manufacturing or service activity within their own company and buying that good or service from another company. Manufacturers, both large and small, had already begun to outsource service functions, such as custodial and food services, because they were considered extraneous to their primary functions and, in most cases, a distraction from their main business (Crandall 2005).

Another function that soon became a popular area to outsource was payroll administration. In most cases, it was a relatively routine function that could easily be handled by an outside supplier. It also relieved the internal information technology function from having to keep up with all the tax changes that differed among local, state, and federal agencies.

Companies knew how to outsource, and while there were some undercurrents of dissatisfaction among employees, especially those in highly unionized industries over the loss of jobs, the effect was not great enough to be disruptive. This was soon to change as manufacturers began to recognize the effect of foreign competition on their sales volume.

In the 1980s, U.S. manufacturers began to react to the foreign competition. In some cases, such as in steel and consumer electronics industries, they had little opportunity to do more than reduce prices. These companies had ample manufacturing capacity, although it was older and less efficient than some of the technologies available to foreign competitors, who had rebuilt their capacity base after World War II with the latest technologies.

Inasmuch as U.S. manufacturers could not justify investment in new process technology, many elected to close down their obsolete facilities and outsource the manufacturing to low-cost suppliers, many of whom were located in foreign countries. This trend started during the 1990s and has continued at an increasing rate to the present. The outsourcing has been largely directed at obtaining component parts that could be assembled in the United States; however, some manufacturers have turned their entire manufacturing process over to contract manufacturers and have become shell companies that focus primarily on the product development and marketing functions as part of the shift from tactical to strategic outsourcing (Engardio 1998; Gottfredson et al. 2005).

While outsourcing has made it possible for manufacturers to obtain lower-cost goods, it has been disruptive to the smooth flow of goods through the supply chain. Attempts to minimize inventories and reduce response times have been thwarted by increased transportation times and uncertainties from foreign suppliers. The effect on product quality is mixed. While some suppliers maintain acceptable levels of quality, the general perception is that product quality is lower than when manufactured in the United States. Extreme examples of disasters from poor product quality show up as lead paint in toys (Palmeri 2007) or salmonella in food products (Reinberg 2008).

In addition to the problems associated with their supply chains, some companies are now beginning to reconsider the cost aspects of offshore outsourcing. As they begin to assess the total cost of ownership of goods, they are finding there are hidden costs that are both materially and potentially disastrous. One of the most disturbing is the potential loss of intellectual property that could enable a supplier to become a competitor. Table 7.4 shows a hierarchy of total costs to consider in an outsourcing decision. Beginning at the bottom with the basic price of materials, the total cost column builds with incremental cost until it reaches the ultimate customer cost at the top of the column. While the costs at the bottom of the column are relatively tangible, they become less tangible nearer the top of the column (Cavinato 1992).

Although cost must be a consideration in outsourcing decisions, one of the potential benefits of offshore outsourcing is that it enables a manufacturer to establish a presence in a foreign market. If they first learn to do business through an independent supplier, it may be prudent to consider establishing their own manufacturing facility in order to become a more acceptable supplier to the market in that country. A number of U.S. manufacturers have become global manufacturers in this way. GE, Cisco, and Caterpillar are examples of companies that have set up manufacturing facilities in foreign market countries.

Offshore outsourcing has become a political issue in industrialized countries because it causes imports to exceed exports, creating a balance of trade controversy. Many politicians want to limit outsourcing and thereby reduce imports because they claim that this will protect jobs in their own country. However, it appears that jobs will continue to move

**TABLE 7.4**

Total Cost of Ownership Model

| Ultimate Customer Cost/Value | Strategic Business Factors |
| --- | --- |
| Marketability<br>Downstream channel costs | Intermediate customer factors |
| Product improvement<br>Supplier cost commitment<br>Supplier R&D | Tactical input factors |
| Transaction overhead costs<br>Payment terms | Indirect financial costs |
| Logistics chain costs<br>Production costs<br>Lot size costs<br>Receive/make-ready costs | Operational/logistics costs |
| Quality costs<br>Warranty costs | Quality costs/factors |
| Transportation terms<br>Transportation costs | Landed costs |
| Maintaining supply relationships<br>Fob terms | Supply relational costs |
| Cost of transaction method | Direct transaction costs |
| Basic price of materials | Traditional basic input costs |

toward low-cost countries. Perhaps, the better strategy would be to increase exports to those developing countries that need manufactured products—road-building equipment and technically advanced medical equipment—that can still be economically produced in highly developed countries, such as the United States. GE is counting heavily on their efforts in the areas of infrastructure technology and health care as future growth areas (General Electric 2012).

The recent wave of negative publicity about foreign suppliers and the increased awareness of hidden costs of outsourcing raise the question of whether some companies will consider a reversal of their outsourcing efforts and return to a more vertically integrated strategy. While it may appear to be a remote possibility, there are some industries where vertical integration could become the strategy of choice. In traditional vertical integration, as with Ford's River Rouge plant, companies owned all of the functions within its immediate supply chain. While companies may not want to own all of the functions, they can move toward virtual integration by more closely linking participants along the supply chain. This feature of supply chain integration will be covered more fully in Chapters 11 through 13.

As an interim step toward vertical integration, a number of companies are considering moving supply operations closer to their final destination. Instead of sourcing from China, they buy the same goods from Mexico, where the labor costs are slightly higher than China, but the distance goods have to move is reduced. This trend is called near sourcing.

An even more extreme move is to find suppliers in the United States, or any country of origin, to reduce even more the distance between user and supplier. This movement is called reshoring or onshoring. As wage rates increase in China and other low-cost countries, the cost benefits are reduced, making it easier to see that the increased logistics complications (more about this in Chapter 9) may offset the reduced labor costs.

## Capacity Planning

The evolution of supply chains will have an effect on the capacity planning for manufacturing companies. Historically, the easiest way to justify additional capacity was to satisfy increased demand. This will still be true for new and evolving industries, such as in telecommunications and solar power generation. However, today, and in the foreseeable future, the need will be more for replacement capacity that will improve quality and flexibility in established industries, such as automotive and appliances. Justifying replacement capacity is a more difficult concept to sell to the financial community who want to see tangible returns on their investment.

### How Much Capacity? When? What Kind?

Manufacturers will continue to consider how much capacity is needed and when it is needed. A complication is the fact that, as product life cycles shorten, there is less certainty as to what kind of processes will be needed to make the new products. This dilemma suggests that new equipment and related technologies will need to be more flexible and adaptable to new products. This may mean a movement toward more general-purpose equipment and away from the high-speed, special-purpose equipment that has been a favorite of many companies in the past. One noted authority on management, in speaking of the need for flexible manufacturing, said, "General Motors largely wasted $30 billion on automating the traditional process, which only made its plants more expensive, more rigid and less responsive. Toyota (and to some extent, Ford) spent a fraction of what GM spent. But it spent the money restructuring production around market information—on 'flexible manufacturing'" (Drucker 1992).

### Location and Ownership

Perhaps, an even more critical consideration is the location and ownership of the additional capacity a manufacturer needs. Offshore outsourcing generally requires that suppliers provide the capacity needed. While this relieves the base manufacturer of the investment requirements, it also involves a degree of risk concerning the ongoing availability of that capacity. The final manufacturer may decide they need to own the capacity in some form to assure continued flow of goods; the question then is, do they own it and manage the facility or do they own the facility and leave the management to an external supplier? These are critical management questions that can generate a variety of choices. It is unlikely that there will be one universal approach to this complex situation.

The question of location is an integral supply chain consideration. Where along the supply chain is the capacity needed? In the past, there has been an excess of capacity, just as there was an excess of inventory, because of the lack of coordination, cooperation, and collaboration along the supply chain. Just as increased collaboration among supply chain members can reduce the total inventory along the supply chain, increased collaboration can also reduce the amount of capacity needed.

## Service Production Strategies

Supply chains for service operations differ from supply chains for tangible products. As described earlier, services involve the transformation of inputs into outputs. The difference is that the inputs tend to be human beings, not things. As a result, it becomes more difficult to see the flow of services along the supply chain. It also means that service supply chains are usually complementary to product supply chains because product companies need services to accompany their product (a computer manufacturer needs to establish a helpline) and service companies need products to accompany their services (a restaurant needs food to sell).

In Chapter 4, we showed how supply chains are a combination of goods-producing and service-producing entities. In general, a supply chain moves from producing goods (mining, agriculture, and manufacturing) to producing services (distribution and retailing), as it moves toward the ultimate consumer. In addition, there are support services (finance and insurance) needed to facilitate the supply chain in their management of the flow of goods and services to the consumer.

Schmenner developed a classification of service operations. They include

- The service factory (low degree of customer interaction and service customization, low labor intensity—airlines, trucking, hotels)
- The service shop (high degree of customer interaction and service customization, low labor intensity—hospitals, auto repair)

- Mass service (low degree of customer interaction and service customization, high labor intensity—retailing, wholesaling, schools)
- Professional services (high degree of customer interaction and service customization, high labor intensity—physicians, lawyers)

Despite their differences, they all include some combination of goods and services. Even professional services include some goods, such as medications or written reports (Schmenner 1995).

Although lean production techniques originated in manufacturing, some believe that it can be even more important in services where long, complex, variable processes with multiple decision points can be simplified and improved through the application of lean techniques (Ross 2013).

Applying the supply chain concept to some less obvious service areas could be interesting. Consider the following examples:

- *Formal education*: It is apparent in most societies that there is a generally accepted flow of students within the formal education system. The process begins at kindergarten, and students progress to elementary school, to junior high school (sometimes called middle school in certain areas), to high school, to college, and perhaps on to postgraduate school. After school is completed, the student moves to an employer and hopefully becomes an educated citizen in society. In each stage, the output from the previous stage becomes the input to the next stage, where it is transformed into another output to be passed on to the next stage in the supply chain. However, most of us do not think of the need to closely connect the stages of this supply chain to make it more effective. If we did, there would be greater collaboration among the different entities.
- *Professional athletics*: Professional sports teams, such as in the National Football League (NFL) or the National Basketball Association (NBA), are very much aware of the need to develop a flow of product (athletes) for their consumption. In some cases, they build their own organizations—minor league baseball. In other cases, they organize ways to monitor (scouting) and select (drafts) the outputs from high school or college teams. They are finding ways to improve the flow of product, just as much as Wal-Mart works to improve the flow of goods to their stores.

Figure 7.9 shows how processes must become more flexible to handle the increasing demand for variety and customization of products and services. This diagram was described more fully in Chapter 4.

## Relationships with Downstream Customers

Manufacturers are changing their relationships with downstream customers. In part, this is because the bargaining power has shifted from manufacturers to retailers. At one time, manufacturers had the power to make a product and then push it through the retailers to sell. Retailers bought only what was available to them and had little room to negotiate

Product and service variety and volume

| High | Variety—number of items | Low |
|------|-------------------------|-----|
| Low | Volumes of individual units | High |
| MTO | Product and service orientation | MTS |

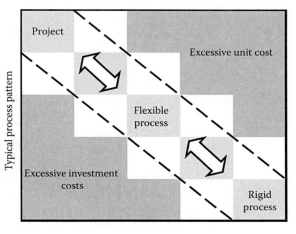

| Short | Length of product or service life | Long |
|-------|-----------------------------------|------|
| Short | New product development time | Long |
| Short | New service development time | Long |

Product and service expected life and development time

**FIGURE 7.9**
Product–process matrix.

with their manufacturers on prices or changes in model styles. Today, retailers have the information about what the customer wants and have used that to become increasingly influential in dictating what they are willing to buy from the manufacturer.

Consequently, manufacturers must provide greater service to their downstream customers. One of the most obvious ways is to provide more options in packaging their products to meet their customers' needs. Some examples of this include

- Blister packs of shoe laces, razors, cookies, and a myriad of other products that come in consumer-ready form, so retailers only need to position them on display racks
- Adding price tags on merchandise before shipping to the retail store
- Imprinting customer brand names on merchandise
- Stocking the shelves of the retail store (health and beauty aids, CDs) (Dutta 2004)

In addition to providing more *floor-ready* merchandise, manufacturers are becoming more responsive to the stock replenishment needs of their customers, usually retail stores. In an increasing number of product areas, retailers no longer have to order a full season's volume of goods; they can order just enough to get the season started and then reorder those items that are selling. This practice reduces lost sales because of stockouts and also reduces the end-of-season merchandise that must be discounted or transferred. To provide this quick response capability, manufacturers must receive more current sales information from the retailers. This is another example of the value of supply chain collaboration.

## From Transactions to Processes

You have no doubt heard a lot about the need to change—to improve, to innovate, to collaborate, and to do a lot of things that sound good but also sound like a lot of work. Are these changes necessary or are companies all right the way they are? The answer, as always, is "It all depends." Generally, there are two drivers of change. One is that global competition requires it. The other is that companies have the resources necessary to bring about that change, especially with the expansion of information technology capabilities. Both conditions—must and can—are necessary for a successful change program. While it may not make sense to change just because companies know how, it makes even less sense to recognize that they should change but do not know how.

One of the reforms advocated by many business writers is that companies should change from a transaction orientation to a process orientation. What does that mean? Is it easy or hard? Is it worthwhile or an exercise in busyness? Can we do it ourselves or do we need outside help?

### Transactions versus Processes

First, what is a transaction? The *APICS Dictionary* defines transactions as "individual events reported to the system, e.g., issues, receipts, transfers, adjustments" (Blackstone 2013). If we build on that definition, we can say that transactions are individual activities or operations that go on throughout a business every day as part of accomplishing the total work of the business. They are distinct and probably separated from one another by organizational structure, location, functional specialty, and historical practice. They are internally focused and operated to be efficient. Are transactions bad? No, because they are necessary; however, they may cause delays and increased costs unless properly managed.

The next question is what are processes? The *APICS Dictionary* defines a process as "A planned series of actions or operations (e.g., mechanical, electrical, chemical, inspection, test) that advances a material or procedure from one stage of completion to another" (Blackstone 2013). As a side area of interest, the dictionary also has definitions for 37 other terms beginning with *process* such as process batch, process capability, and process train. Operations management practitioners are certainly getting their money's worth from the word *process*. This observation suggests that a process is organized (planned) and connects individual activities/transactions into an integrated system to accomplish an objective. Simply stated, a process connects individual transactions into a sequence of activities that accomplish a desired end result. If you accept this idea, it is easy to see that a process orientation leads nicely into the supply chain concept. Processes tend to be both internally focused toward operations and externally focused toward the customer or supplier. While we want our processes to be efficient (doing things right) and of high quality, we want them especially to be customer focused and effective (doing the right thing) (Drucker 1974).

In describing lean management, Womack also describes a process when he states that "All value created in any organization is the end result of a lengthy sequence of steps—a value stream. These steps must be conducted properly in the proper sequence at the proper time. The flow of value toward the customer is horizontal, across the organization" (Womack 2006). In their book about implementing *Six Sigma* in service operations, Snee and Hoerl insist that a process view is necessary if improvement programs are to be successful. They believe that all works—whether done in manufacturing, financial services, health care, e-commerce, or anywhere else—occur through a system of interconnected processes (Snee and Hoerl 2005).

Consider this simple example—customer order processing. Think about all of the functional areas that an order could pass through before the customer receives the goods—order entry, credit, materials management, production, accounting, distribution, and shipping. Each function performs a series of transactions. When combined in a flow of activities, the customer experience is enhanced. Another example comes from a hospital setting—a surgery patient. The patient moves through preadmitting (administrative), preadmitting preparation (physical), admitting, final preparation, surgery, recovery, release, and therapy. Again, each function performs a series of transactions. When linked together in a smooth process, the patient experiences faster and higher-quality service.

## Basic Processes of a Business

Some examples of processes that exist in almost every kind of business, whether manufacturing or service, include

- Customer sales order
- Supplier purchase order
- New product development
- Capital acquisition
- Stock replenishment
- Strategic planning
- Sales and operations planning (S&OP)
- Continuous improvement program
- Crisis management
- Employee hiring and training
- Outsourcing

We could go on, but you get the idea. These processes transform inputs into outputs. They have an objective and a purpose. Think about the functional areas involved in each of these processes. None of the processes would work without cross-functional participation.

It gets even more interesting to think about linking processes together to form what could be considered as master processes. If we link the customer sales order process involving accounts receivable, with the stock replenishment process involving inventory, with the supplier purchase order process involving accounts payable, we can form a cash-to-cash cycle process that measures how well a business is using its cash (Farris and Hutchison 2002).

## Benefits of a Process Orientation

What are the benefits of a process orientation? As mentioned earlier, processes tend to be customer oriented and, as a result, should improve customer service. One of the ways this is accomplished is through reducing response times, whether it is in product development or normal order processing. A transaction orientation tends to be batch oriented where in-process inventories build up to assure efficient processing, whether it is in manufacturing washing machines or approving supplier invoices. A process orientation is designed to achieve a smooth and regular flow of work through a series of workstations. This is desirable within an entity; it is even more desirable if the flow can be extended to external entities such as customers and suppliers to form supply chains. A process perspective will not only enable a business to provide better customer service; it will also enable it to reduce processing costs and errors (Snee and Hoerl 2005).

Those are some of the tangible benefits. One very important, but less tangible, benefit is in knowledge transfer. Have you ever learned anything from another functional area that was helpful in your own work? Of course you have, whether your transaction processing accidentally collided with a transaction in another department or if you were part of a smooth-functioning process team.

## Effect of Process Orientation

How does the concept of process orientation fit in with the management of a business? While it has an effect on the day-to-day operations, it also has an effect on longer-range considerations. For instance, what is the effect on organizational structure, strategies, and knowledge management?

### Organizational Structure

As mentioned earlier, a process orientation tries to achieve shorter response time. To do this requires that product and services flow smoothly in a horizontal direction. To facilitate the flow of goods and services, information must also flow horizontally. In the traditional organizational structure, information flowed primarily in a vertical direction (within a company) and ever so slowly in a horizontal direction (between departments or companies). To accommodate horizontal information flow, organizations must become flatter without the need for excessive layers of management to summarize and analyze the information. Quinn was one of the first management writers to describe ways in which alternative organizational forms facilitate the flow of information (Quinn 1992). To date, many businesses have used teams to provide a cross-functional approach. For example, Wang reports companies often implement business process reengineering projects by forming cooperative task teams, instead of changing their formal organization structures (Wang 1997). However, teams usually are temporary in nature and do not constitute a permanent change in organizational structure. Womack believes all organizations, including Toyota, are organized vertically by department—engineering, purchasing, production, sales—because it is the best way to create and store knowledge and the most practical way to channel careers (Womack 2006). Earlier, Drucker proposed organizations would retain their traditional functions while integrating teams and network organizations as required (Drucker 1974).

In speaking of learning organizations, Senge subscribes to the need for organizational redesign. However, he also believes the changes required will be not only within an organization but also within individuals. It is only through changing how people think that will lead to changes in deeply embedded policies and practices. When people interact, they can share visions, understandings, and new capacities for coordinated action (Senge 1994).

Another study looked at how readily employees *switch* between standardized production work and continuous improvement work within the context of a total quality management (TQM) program. They found employees perceive a difference in organizational forms. When doing standardized work, they perceived the structure to be hierarchical and rigid. When doing continuous improvement work, their perceptions changed so that the structure seemed more flexible, horizontal, and more open to empowered employees (Victor 2000).

At this point, not enough businesses have moved from the traditional pyramidal organization shape to a contemporary structure that incorporates both functions and processes in an effective manner. Consequently, the benefits from a process orientation are still not completely realized.

## Strategies

What are the effects of a process orientation on business strategies? We can only mention a few of the most current strategic considerations.

Mass customization involves a transition from the traditional MTS to MTO—from standard to customized products and services. While desirable from the customer viewpoint, it offers problems to the provider. Customization breeds complexity, and, while good complexity is necessary and beneficial to the customer, a company must weed out bad complexity. Taking a process perspective will help to blend customization and complexity (Anderson et al. 2006).

Another current movement is outsourcing both the production of goods and the provision of services. The introduction of outsourcing of any magnitude almost forces a company to take a process perspective. It must decide what to outsource—a transaction or a process—and how best to restructure the organization to bridge over the area outsourced. As suggested earlier, outsourcing itself is a process.

One of the current initiatives that many companies are pursuing is lean production, Lean Sigma, or some other adaptation of lean in both manufacturing and services. The lean approach stresses the need to improve flow by eliminating barriers such as excess inventories, equipment malfunctions, employee limitations, and variation of any kind.

## Knowledge Management

The process orientation stimulates knowledge transfer. This can be desirable or undesirable. It is desirable if knowledge transfer enhances the performance of the company. It is undesirable if the transfer strengthens the competition and weakens the donor company's market position. This might occur when there is a transfer of knowledge to suppliers who share it with their customers or to customers who use it to vertically integrate that knowledge into their own operations. When it is not possible or practical to have the teams meet face-to-face—the desired state—it may be useful to use virtual teams that communicate knowledge among the members with electronic communication technology (Mohamed et al. 2004).

## Change Management

How do companies manage the changes required to move from a transaction orientation to a process orientation? While technology is a driver of change, successful implementation requires changes in the culture of the work force and modification of the infrastructure (Crandall and Crandall, 2014). Major change, such as the move to a process orientation, takes time—usually longer than most managers would like—and skipping steps only creates the illusion of speed and never produces a satisfactory result (Kotter 1995). Aberdeen reports that 85% of the companies cited that significant culture change was required and that 35% of the companies reported that top management commitment was the biggest challenge (Jutras 2006).

A collection of articles entitled Managing Change and Transition provide several perspectives on the subject with topics such as examining the different types and approaches to change, preparing for organization change, using a systematic approach, putting your plan in motion, considering social and human factors, helping people adapt, and staying competitive through change (Harvard 2003). Having an action plan that addresses the critical factors for dealing with significant change enables an organization to increase its chances of a successful transformation, and they may also become a more flexible organization better able to deal with future change (Chrusciel and Field 2006).

Changing from a transaction orientation to a process orientation is one of the major challenges facing businesses today. It involves a blend of the old with the new. Like moving to a new location, it is not easy deciding what to throw away and what to keep.

## Trends in Production

In addition to the movement from a transaction orientation to a process orientation, there are several distinct trends in production that relate to the evolution of supply chains. These include

- Continued evolution from manual to automated operations
- Extension from domestic to global operations
- Movement from standard products to customized products
- Use of postponement as a strategy to achieve mass customization
- Acceptance of S&OP as a means of improving customer fulfillment
- Use of additive manufacturing as an alternative manufacturing method

### From Manual to Automated

There are two main drivers in the movement toward higher levels of automation in both manufacturing and service processes. The first is the push to reduce costs and improve quality. The present thinking is that increased automation will achieve both objectives; therefore, there is pressure to innovate with automated equipment. The second driver is the mirror of the first. Sometimes, companies develop technologies before there is a recognized need. The latter is especially true in the rapid development of electronic capabilities that create the need to figure out how and where to apply them. Technologies such as videoconferencing and global positioning systems (GPS) began as novelties but are rapidly evolving into important applications of automated systems.

One of the major effects of automation is the effect on the employees required. Automation reduces the need for the number of employees; however, it increases the skill levels required of employees. There is a shortage of skilled workers to fill the openings, and workers displaced by outsourcing do not have the skills required. Who should do the retraining and who should pay for it are just two of the many questions confronting the manufacturing industry in industrialized countries. To confound the problem, today's millennials are not attracted to manufacturing because they are not sure it offers what they are looking for (Panchak 2013).

### From Domestic to Global

The movement from domestic to global operations is obvious to everyone. Even before the offshore outsourcing movement came to the forefront, companies were expanding into other countries to move closer to their potential markets. As more countries moved from state-controlled to market-driven economies, the attractiveness of these new markets increased. Whether buying or selling, the effect on supply chains is to increase their scope and complexity, thereby increasing the challenge of managing them effectively.

### From Standard Products to Customized

Earlier in this chapter, we described the progression from mass production to mass customization. This trend will likely continue as production managers find ways to accomplish this transition. While we do not expect that standard products will vanish altogether, we do expect that customized products will become increasingly popular.

### Postponement

As a corollary to mass customization, businesses will move to postponement as a strategy to make customization a practical approach. Postponement means building a standard product composed of standard modules and then adding customizing features at a later time in the production process. The signal to end the postponement is when the customer places their order for the product. The postponement may be done at the factory, at the distribution center, or even at the retail store, depending on the complexity and skill requirements of the customizing to be done.

Some of the challenges for continued success of the postponement approach include the following:

- Assess application of postponement relative to performance.
- Select appropriate postponement points.
- Decide how to apply postponement in the service industry.
- Further develop postponement as a response to uncertainty.
- Investigate postponement adoption in companies (Boone et al. 2007).

There are several forms of postponement companies can consider. These include product design postponement, purchasing postponement, production postponement, logistics postponement, price postponement, and product postponement (Boone et al. 2007). These strategies roughly correspond to the manufacturing strategies of MTS, ATO, MTO, and ETO described earlier in this chapter.

## Sales and Operations Planning

S&OP was a technique that was first promoted over three decades ago and found little acceptance. Today, S&OP is becoming an integral program in many manufacturing companies as they work to improve their customer service. S&OP involves a combination of marketing and production collaboration to try to build exactly what the customer wants when they want it. "In order to be successful, S&OP approaches must evolve to adopt the attributes of 'integrated business planning'—incorporating truly cross-functional, multidimensional processes that include all elements of demand, supply and financial analysis in relation to the business goals and strategy" (Ball 2008).

## Additive Manufacturing

Additive manufacturing, or 3D printing, is one of the more promising major changes in manufacturing. Parts are produced by printing successive layers of material, much like spreading successive layers of icing on a cake. Using CAD software, the 3D design specifies where the material will be collected for each layer. Complex parts can be quickly produced. The process is called additive, as contrasted with the subtractive process where material is cut away from a base stock, such as cutting a steel fitting from bar stock or a piece of furniture from a log. See Crandall (2013) for a fuller discussion of this innovation in manufacturing.

Additive manufacturing could be revolutionary. It will enable companies to design and bring products to market faster, it will reduce the need to outsource, and it will make mass customization more realizable. Companies are already producing usable products, such as dental braces, aircraft fuel nozzles, and hearing aids (Dzieza 2013). GE Aviation is planning to use additive manufacturing to build fuel nozzles for its jet engines. A single part can be produced that will replace nearly 20 component parts needed previously. The company expects to be building 30,000–40,000 of these fuel nozzles by 2020 (Hessman 2014).

---

## Performance Measurement

Performance measurement is difficult in individual companies. It is even more difficult in supply chains, where independent companies are obligated to work together for a common purpose, even when they often have different, or even conflicting, company objectives. Nonetheless, we can attempt to measure performance through accounting means, through production measures, or along certain points of the supply chain.

## As Measured by Accounting

Accounting measurement usually works in dollars or the equivalent monetary unit in the respective country. Even within a specific business, it is sometimes difficult to accurately determine revenues, costs, and income. That is why public companies are required to have audited financial statements, in the hopes of reporting reliable financial results. A limitation of accounting measures is their timeliness. By the time the final results are available, it is difficult to relate them back to the actual operations of the company; therefore, financial measures are not always useful to operations managers in determining how well they are doing.

If it is difficult to quantify the financial results of a single business, it becomes even more difficult to determine the financial results of an entire supply chain, especially when there are multiple supply chains serving a number of independent companies.

## As Measured by Production

Production managers do not use dollars as often to measure performance or progress against goals. They use physical units, such as units produced, machine or labor hours used, days of inventory on hand, and average number of days of lead time. Because they have different information needs, they often require separate information reports. They need these reports daily or weekly to be able to quickly identify problem areas. In some cases, they need real-time feedback to assess product quality or machine performance.

## Measures along the Supply Chain

It follows that a realistic approach to measuring the performance of the entire supply chain involves some measures that do not require confidential financial information from the participants. At best, these measures can be approximated from information gathered from each member of the supply chain. Some measures that have been tried include the number of days of lead time from origin to consumer or the number of days of inventory throughout the supply chain.

Performance measurement of supply chains will be one of the challenges for both practitioners and academics in the future. We will discuss this area in more detail in Chapter 16.

---

## Summary

The production function evolved from a craft orientation to a mass production orientation during the nineteenth and twentieth centuries. During the latter quarter of the twentieth century, a new transition began, from mass production to mass customization. While mass production is still the dominant mode, many companies are moving toward mass customization because of consumer demand for greater product variety.

Manufacturers use several different production strategies in mass production—MTS, ATO, MTO, and ETO. All of these have the potential to be adapted to fit mass customization requirements. Lean production is another strategy that is moving some manufacturers along the path to mass customization by removing obstacles to the smooth flow of product.

Another factor that affects supply chain performance includes the trend toward offshore outsourcing. This extends the supply chain geographically, sometimes completely around the world. While costs may decrease, lead times increase, and integration of supply chains becomes more difficult.

The orientation toward supply chain understanding is moving from a transaction approach to a process approach. A transaction is micro in nature, and does not tend to account for the immediate impacts of the action on other departments. A process, on the other hand, is a collection of transactions that have been purposely linked to each other by the understanding that a good or service needs to transition through various stages of completion. These processes can be then linked with other processes in the supply chain.

Finally, the trends in production management must be noted. There is a movement from manual to automated operations, from domestic to global processes, and from making only standardized products to creating customized products. These three immediate trends have created the need for postponement in the production process—the delaying of adding the customized features to a standard product module, to the last possible moment in the production process, usually when the customer places the order. To accomplish these goals, a rebirth of S&OP concept has taken place.

---

## Hot Topic: Clothing Manufacturing

Many clothing manufacturers have outsourced large portions of their supply chain to offshore contract manufacturers. This is the subject of this chapter's Hot Topic.

---

### Hot Topics in the Supply Chain: Clothing, the Achilles Heel of the Supply Chain

Clothing is an interesting product because of its uniqueness from other products that are massed produced. Much clothing is not really made with high technology; it is still stitched the old-fashioned way, by hand and on a sewing machine.

The making of clothing has a checkered past. Images of female immigrant workers laboring in substandard sweatshops abound. One of the main stories we remember is the disastrous 1911 Triangle Shirtwaist Factory fire in New York City that killed 146 workers, many of whom were burned in the fire or jumped to their deaths from the ninth floor of the Asch Building, a structure that still exists today (Miller, 2003).

Unfortunately, fires and other accidents in clothing factories have continued to be a problem to this day. Indeed, the making of clothing represents the Achilles Heel, or weak spot, in modern-day supply chains. But how did we get to this point? Some of the reasons are fairly easy to understand. First, much of the manufacturing of clothing is done in developing countries in Asia. Because of the concern of sweatshops in these countries, large multinational corporations (MNCs) and their associated brands are starting to require codes of conduct from their suppliers. However, such codes can be counterproductive for the host government that desperately seeks the business of MNCs. A second reason for this Achilles Heel in the supply chain is the role host governments play in attracting business to their countries. Such governments find themselves in the precarious position of balancing the MNC's interest with those of the local worker. Both of these reasons are discussed next.

#### CODES OF CONDUCT

In response to the problem of sweatshops, many companies that contract to suppliers in Asia have started to implement codes of conduct for those suppliers. Such codes stipulate the working conditions and plant standards that need to be abided by in order for that supplier to be used by the company. Codes of conduct sound like a good idea, but they can be difficult to enforce. Furthermore, the motivation to abide by a code of conduct by a supplier may be low. The irony is that such codes may actually hurt the very stakeholder they are supposed to help, the worker (Lund-Thomsen 2008). In fact, it may be the MNC that stands to gain the most from codes of conduct because of the enhanced publicity it generates (Hoang and Jones 2012). After all, a positive message is sent when a company claims on its website that it requires all suppliers to abide by a code of conduct.

To the supplier in Asia however, abiding by a code of conduct creates problems. First, it ultimately results in higher costs to the supplier. When operating expenses increase, the price advantage that particular supplier holds is decreased relative to other suppliers that could perform the same job. Hence, for codes to be effective, they need to be enforced across the entire industry in the host country.

A second problem is that the suppliers have other customers besides the large MNC that wants to generate a positive image in the public's eye. Many suppliers in the clothing industry work through third-party middlemen, who then move the goods to the MNC. For example, some garment factories in Vietnam supply as many as 30 different clients (Hoang and Jones 2012). This realization is important as most of us in the West are conditioned to think that each clothing retailer works directly with one supplier. The reality is that these dyad relationships are not the norm, instead clothing supply chains are more like networks (Hoang and Jones 2012; Nassimbeni 2004) with clothing suppliers contracting with multiple partners via third-party intermediaries. What this means is that a supplier may be held accountable to meet a code of conduct by perhaps one major MNC retailer, while other clients do not require a code. However, codes are enforced for a whole factory, not just for one client. The codes, if enforced, would need to apply to every company that supplier services. This drives up operating costs, which normally would be passed on to the customer. However, MNCs put great pressure on their suppliers in developing countries to keep costs low. This scenario puts that supplier in an almost impossible situation—abide by a code of conduct, and yet keep costs as low as possible. At this point, it would seem necessary that the local host government step in and provide some assistance. However, this too is a difficult task.

## ROLE OF THE HOST GOVERNMENT

Governments in developing countries are desperately seeking MNCs to contract with local suppliers. These countries desire to be part of global supply chains in order to prosper economically (Gereffi and Lee 2012). The competition to attract business has been compared to the race to the bottom, a bleak economic scenario in which countries have to restrict welfare initiatives for their workers (for cost reasons) in order to attract business (Chan and Peng 2011). A debate exists though as to how much raising worker standards really affects the marketplace. Some hold the viewpoint that raising worker standards will only result in minimal price increases to the customer (Pollin et al. 2001). However, MNCs and host governments tend to react in the other direction. MNCs pressure suppliers to keep prices low and governments restrict improving worker welfare, lest they lose the business to another country.

One country, Bangladesh, has been in the middle of this economic drama in regard to the clothing industry. Issues involving building safety and fires have resulted in the deaths of approximately 1800 garment workers since 2005 (MacDonald 2013). The government seems to be unable to cope with the vast amount of clothing factories that need inspections. The situation has resulted in the creation of an Accord on Fire and Building Safety in Bangladesh. The agreement, while voluntary to sign on to, is a legal contract that requires companies sourcing clothing in Bangladesh to (1) make their suppliers submit to independent inspections, (2) make the results of those inspections public, and (3) help fund any needed improvements to the factory buildings (Cline 2013). It is hoped that the Accord will accomplish what the government has not been able to, the assurance of safe buildings for garment workers.

## QUESTIONS FOR RESEARCH AND DISCUSSION

1. In regard to the situation in Bangladesh, some companies have chosen to sever their ties with suppliers in that country while others are agreeing to provide resources to make the factories safer. What are the advantages and disadvantages of each of these strategies?

2. What role can a consumer (final end user) play in improving the working conditions for those workers in the garment making industry?

## REFERENCES

Chang, C. and Peng, Z., From iron rice bowl to the world's biggest sweatshop: Globalization, institutional constraints and the rights of Chinese workers, *Social Service Review*, 85, 421–445, September 2011.

Cline, E., No more Rana Plazas, *The Nation*, 296(24/25), 6–8, 2013.

Gereffi, G. and Lee, J., Why the world suddenly cares about global supply chains, *Journal of Supply Chain Management*, 48(3), 24–32, 2012.

Hoang, D. and Jones, B., Why do corporate codes of conduct fail? Women workings and clothing supply chains in Vietnam, *Global Social Policy*, 12(1), 67–85, 2012.

Lund-Thomsen, P., The global sourcing and codes of conduct debate: Five myths and five recommendations, *Development and Change*, 39(6), 1005–1018, 2008.

MacDonald, G., Is it ethical to keep buying clothes from Bangladesh? *Christian Science Monitor*, May 27, 2013, Retrieved January 20, 2014, from http://www.csmonitor.com/Business/2013/0527/Is-it-ethical-to-keep-buying-clothes-from-Bangladesh.

Miller, J., Why economists are wrong about sweatshops and the antisweatshop movement, *Challenge*, 46(1), 93–122, 2003.

Nassimbeni, G., Supply chains: A network perspective, in S. New and R. Westbrook (eds.), *Understanding Supply Chains: Concepts, Critiques, and Futures*, Oxford University Press, Oxford, U.K., pp. 43–68, 2004.

Pollin, R., Burns, J., and Heinz, J., Global apparel production and sweatshop labor: Can raising retail prices finance living wages? Political Economy Research Institute, Amherst, MA, Working Paper Series, No. 19, 2001.

## Discussion Questions

1. Discuss the role of manufacturing in the supply chain. How does it link to downstream parts of the supply chain—distribution and retail?

2. Explain the relationship between production lead times and level of customization provided.

3. What changes are needed to move from mass production to mass customization?

4. Compare mass production and mass customization on the basis of cost, quality, lead time, and flexibility.

5. What is the difference between vertical integration and outsourcing? What are the advantages and disadvantages of each?

6. What is the difference between a transaction and a process? Why is the process approach to business operations preferred over the transaction approach?

7. Describe the major trends emerging in the area of production management.

8. Why is measuring performance along the supply chain difficult? What remedies to this problem would you suggest?
9. Discuss the differences between MTS, ATO, MTO, and ETO.
10. What is the role of a manufacturing company in a supply chain?

---

# References

Anderson, B., Hagen, C., Reifel, J., and Stettler, E., Complexity: Customizations' evil twin, *Strategy & Leadership*, 34(5), 19, 2006.

Ball, T., Transformative supply chain strategies, *Industry Week*, p. 1, June 2, 2008. www.industryweek.com/PrintArticle.aspx?ArticleID=16437.

Blackstone, J.H., *APICS Dictionary*, 14th edn., APICS—The Association for Operations Management, Chicago, IL, 2013.

Boone, C.A., Craighead, C.W., and Hanna, J.B., Postponement: An evolving supply chain concept, *International Journal of Physical Distribution & Logistics Management*, 37(8), 594, 2007.

Cavinato, J.L., A total cost/value model for supply chain competitiveness, *Journal of Business Logistics*, 13(2), 285, 1992.

Chrusciel, D. and Field, D.W., Success factors in dealing with significant change in an organization, *Business Process Management Journal*, 12(4), 503, 2006.

Crandall, R.E., Device or strategy: Exploring which role outsourcing should plan, *APICS Magazine*, 15(7), 21, 2005.

Crandall, R.E., Beating impossible deadlines, *APICS Magazine*, 16(6), 20, 2006.

Crandall, R.E., Where will additive manufacturing take us? *APICS Magazine*, 23(1), 20–23, 2013.

Crandall, R.E. and Crandall, W.R., *Vanishing Boundaries, How Integrating Manufacturing and Services Creates Customer Value*, CRC Press, Boca Raton, FL, 2014.

Drucker, P.F., *Management: Tasks, Responsibilities, Practices*, Harper & Row, New York, 1974.

Drucker, P.F., Drucker on management: The economy's power shift, *Wall Street Journal*, Eastern edition, A16, 3, September 24, 1992.

Dutta, D., Floor-ready merchandise (FRM)—Management briefing: Key differences in non-FRM and FRM-ready supply chains, *Just-Style*, p. 13, August 2004.

Dzieza, J., HP's 3-D printer is a marvel that you probably won't own, *MIT Technology Review*, www.technologyreview.com/news/52071, October 30, 2013.

Engardio, P., Souping up the supply chain: Today's supercontractors are turning manufacturers into models of efficiency, *Business Week*, Issue 3593, p. 110, August 31, 1998.

Farris, M.T., II and Hutchison, P.D., Cash-to-cash: The new supply chain management metric, *International Journal of Physical Distribution & Logistics Management*, 32(3/4), 288, 2002.

Ford, H. (In collaboration with Crowther, S.) (Reprint edition), *Today and Tomorrow*, Productivity Press, Cambridge, MA, 1988.

General Electric (GE) Annual Report, 2012, http://www.ge.com/ar2012/, accessed July 1, 2013.

Gottfredson, M., Puryear, R., and Phillips, S., Strategic sourcing: From periphery to the core, *Harvard Business Review*, 83(2), 132, 2005.

Harvard, *Managing Change and Transition*, Harvard Business School Press, Boston, MA, 2003.

Hessman, T., Will 3-D supply meet GE's sky-high demand? *Material Handling & Logistics*, www.mhlnews.com, accessed July 30, 2014.

Hounshell, D.A., *From the American System to Mass Production, 1800–1932: The Development of Manufacturing Technology in the United States*, The John Hopkins University Press, Baltimore, MD, 1985.

Jutras, C., Enhancing lean practices: Lean for industrial machinery and components manufacturers, *Industry Week Webcast*, December 6, 2006.

Kotter, J.P., Lifetime learning: The new educational imperative, *The Futurist*, 29(6), 27, 1995.

Mohamed, M., Stankosky, M., and Murray, A., Applying knowledge management principles to enhance cross-functional team performance, *Journal of Knowledge Management*, 8(3), 127, 2004.

Palmeri, C., Mattel takes the blame for toy results, *Business Week*, p. 1, September 24, 2007. http://www.businessweek.com/bwdaily/dnflash/content/sep2007/db2007    0921_569200. htm?chan=top+news_top+news+index_businessweek+exclusives.

Panchak, P., The manufacturer's agenda: How to attract millennials to solve the skilled workforce shortage, *Industry Week*, December 5, 2013, www.industryweek.com/print/corporate-culture.

Quinn, J.B., *Intelligent Enterprise: A Knowledge and Service Based Paradigm for Industry*, The Free Press, New York, 1992.

Reinberg, S., Salmonella outbreak tied to dry dog food continues, *Business Week*, November 6, 2008. http://www.businessweek.com/print/lifestyle/content/healthday/621133.html.

Ross, K., Lean is even more important in services than manufacturing, *Industry Week*, December 13, 2013, www.industryweek.com.

Schmenner, R.W., *Service Operations Management*, Pearson Custom Publishing, Boston, MA, 1995.

Schwartz, B., *The Paradox of Choice: Why More Is Less*, Harper Collins, New York, 2004.

Senge, P.M., *The Fifth Discipline: The Art and Practice of the Learning Organization*, Currency Doubleday, New York, 1994.

Snee, R.D. and Hoerl, R.W., *Six Sigma beyond the Factory Floor*, Pearson Prentice Hall, Upper Saddle River, NJ, 2005.

The Henry Ford, River Rouge factory tour, www.hfmgv.org/rouge/index.aspx, accessed July 16, 2013.

Victor, B., Boynton, A., and Stephens-Jahng, T., The effective design of work under total quality management, *Organization Science*, 11(1), 102, 2000.

Wang, S., Impact of information technology on organizations, *Human Systems Management*, 16(2), 83, 1997.

Womack, J.P., From lean tools to lean management, *Lean Institute Memo*, November 21, 2006.

Womack, J.P. and Jones, D.T., *Lean Thinking: Banish Waste and Create Wealth in Your Corporation*, Simon & Schuster, New York, 1996.

# 8

# *Procurement/Purchasing*

## Learning Outcomes

After reading this chapter, you should be able to

- Describe the role of procurement in the supply chain
- Identify the differences between traditional purchasing and contemporary purchasing
- Identify and define the four stages of purchasing sophistication
- Describe the critical success factors (CSFs) needed in the purchasing function
- Describe how purchasing is a key participant in integrating the members of a supply chain
- Identify the key functions of purchasing
- Construct a vendor rating system when making an important purchase
- Describe differences between the make-or-buy decision and the outsourcing decision
- Evaluate how offshore outsourcing has affected purchasing
- Identify how the performance of the purchasing function can be measured
- Describe how the purchasing function will change in the future

## Company Profile: Nestlé

All organizations purchase something. Purchasing is one of the most important functions in an organization, because purchased goods represent a high percentage of total costs in a company. It may be 80%–90% in some retail or distribution operations and is probably at least 50% in most manufacturing companies. Even hospitals have high and critical purchasing requirements. Only high-value service operations, such as law firms, public accountants, and consulting firms, have low purchase costs. As you read this company profile and the chapter in the book, think about these questions:

- How has the purchasing function changed over the last few decades?
- Does purchasing have a role in product design? Explain.

- Should purchasing be centralized or local in a multilocation company? Why?
- How should a purchasing group's performance be measured?

What additional questions do you think relate to purchasing?

### Company Profile: Nestlé

Chapter 8 is about purchasing. We will highlight a company that has an awesome purchasing challenge. Their supply chains are long and complex, with some sources in remote parts of the world. In addition, the volume of materials purchased is huge and varied. The company also buys materials that go into food, a product that is highly regulated and carefully watched by both governments and special interest groups.

Nestlé S.A. (the company) is a Swiss multinational food and beverage company headquartered in Vevey, Switzerland. It is the largest food company in the world measured by revenues. Nestlé's products include baby food, bottled water, breakfast cereals, coffee, chocolate and confectionery, dairy products, ice cream, pet foods, and snacks. Twenty-nine of Nestlé's brands have annual sales of over 1 billion Swiss francs (about $1.1 billion), including Nespresso, Nescafé, Kit Kat, Smarties, Nesquik, Stouffer's, Vittel, and Maggi. Nestlé has around 450 factories, operates in 86 countries, and employs around 328,000 people. It is one of the main shareholders of L'Oréal, the world's largest cosmetics company (Nestlé 2014).

Nestlé was formed in 1905 by the merger of the Anglo-Swiss Milk Company, established in 1866 by brothers George Page and Charles Page, and Farine Lactée Henri Nestlé, founded in 1866 by Henri Nestlé. The company grew significantly during World War I and again following World War II, expanding its offerings beyond its early condensed milk and infant formula products. The company has made a number of corporate acquisitions, including Crosse & Blackwell in 1950, Findus in 1963, Libby's in 1971, Rowntree Mackintosh in 1988, and Gerber in 2007 (Nestlé 2014).

### PRODUCTS AND MARKETS

Table 8.1 shows the distribution of sales for 2012 by geographic area and product group. They are a true global company. They have grown over the years by expanding their product groups, often through acquisition, as described earlier, and by moving into new geographic areas.

Their sales are from countries throughout the world; therefore, it is consistent that their employees and factories are also spread over a wide geographic area. Table 8.2 shows this distribution.

As Table 8.2 shows, the company is invested heavily in both developed and emerging markets. They report that "we invested responsibly and sustainably, expanding our manufacturing footprint, while continuing to reduce the environmental impact of our business" (Nestlé Annual Report 2012, p. 40).

### VALUE STATEMENTS

Nestlé puts great emphasis on its efforts to provide nutritional foods, especially to emerging markets. The company has not been without its share of controversies, as will be discussed later. Because they are in the food business, much of which is directed toward infants and growing children, they must be very careful to provide products that are nutritional but also tasty. While it is important in developed countries, where consumers have a wider selection of foods, it is essential in emerging countries,

**TABLE 8.1**

Nestlé Sales for 2012

| | CHP (Swiss Francs) in Millions | U.S. Dollars in Millions | Sales (%) |
|---|---|---|---|
| *Sales by geographic zone* | | | |
| Europe | 26,529 | 28,651 | 28.8 |
| Americas | 25,894 | 27,966 | 28.1 |
| Asia | 18,604 | 20,092 | 20.2 |
| Latin America + Caribbean | 15,218 | 16,435 | 16.5 |
| Africa | 3,332 | 3,599 | 3.6 |
| Oceania | 2,609 | 2,818 | 2.8 |
| Total | 92,186 | 99,561 | 100.0 |
| *By-product group* | | | |
| Powdered and liquid beverages | 20,038 | 21,641 | 21.7 |
| Water | 7,178 | 7,752 | 7.8 |
| Milk products and ice cream | 18,564 | 20,049 | 20.1 |
| Nutrition and health care | 10,726 | 11,584 | 11.6 |
| Prepared dishes and cooking aids | 14,432 | 15,587 | 15.7 |
| Confectionary | 10,438 | 11,273 | 11.3 |
| Pet Care | 10,810 | 11,675 | 11.7 |
| Total | 92,186 | 99,561 | 100.0 |

*Source:* Adapted from Nestlé Annual Report, 2012, p. 43, http://www.nestle.com/asset-library/documents/ library/documents/annual_reports/2012-annual-report-en.pdf, accessed August 28, 2013.

**TABLE 8.2**

Nestlé Employees and Factories

| | Number of Employees | Total (%) |
|---|---|---|
| *Employees by geographic region* | | |
| Europe | 96,276 | 28.4 |
| Americas | 112,548 | 33.2 |
| Asia, Oceania, and Africa | 130,176 | 38.4 |
| Total | 339,000 | 100.0 |
| *Factories by geographic region* | | |
| Europe | 153 | 32.7 |
| Americas | 171 | 36.5 |
| Asia, Oceania, and Africa | 144 | 30.8 |
| Total | 468 | 100.0 |

*Source:* Adapted from Nestlé Annual Report, 2012, p. 40, http://www.nestle.com/asset-library/documents/ library/documents/annual_reports/2012-annual-report-en.pdf, accessed August 28, 2013.

where the food supply is not as abundant. The following statement from their 2012 Annual Report reflects their recognition of this obligation:

> We took further steps to enhance our position as the trusted leader in Nutrition, Health and Wellness. We continued to reformulate products to make them healthier and tastier. We leveraged our research and development capabilities to deliver good nutrition and develop solutions to help people manage diet-related illnesses. We continued to build partnerships with organisations active in the fight against

non-communicable diseases. We acquired Wyeth Nutrition and a number of new capabilities for Nestlé Health Science. We inaugurated the Nestlé Institute of Health Sciences, added two new R&D units in China, a new R&D centre in India and opened a global centre for clinical trials in Switzerland. (Nestlé Annual Report 2012, p. 42)

They also list the following accomplishments in their 2012 Annual Report:

- 75.7% of Nestlé products meet Nutritional Foundation criteria.
- 5.4 million children reached by the Nestlé Healthy Kids Global Programme, in 64 countries.
- 6692 renovated products for nutrition or health considerations.
- 11,700 equivalent tons of salt removed by *Maggi* from its portfolio over the last 8 years.
- 100 billion servings of iodine-enriched *Maggi* products sold worldwide.
- 217 clean drinking water projects in the South Asia region, helping to improve access and sanitation for more than 100,000 school children.

## PURCHASING

The purchasing group plays a major role in the company's success. To give you some idea of the magnitude of the purchasing task, they deal with 165,000 direct suppliers for all types of raw materials, packaging products, and other supplies. They also deal with over 690,000 farmers, primarily in purchasing cocoa, milk, and coffee.

As a result of this extensive supplier network, they have been accused of buying from sources using child labor and contributing to deforestation through their purchases of palm oil, where hardwood forests were replaced by palm oil plantations.

In August 2010, Nestlé published a supplier code to establish the guidelines for suppliers. The preamble to the code ends with this statement: "By acceptance of the Code, the Supplier commits that all existing and future agreements and business relationships with Nestlé will be subject to the provisions contained herein" (The Nestlé Supplier Code 2010). The code contains the following sections, which have been condensed from the full document:

1. *Business integrity*: Suppliers must comply with applicable laws and regulations and must not promise any personal or improper advantage in order to obtain or retain a business or other advantage from a third party.
2. *Sustainability*: Nestlé expects the supplier to continuously strive toward improving the efficiency and sustainability of its operations, which will include water conservation programs.
3. *Labor standard*: Suppliers will not use or benefit from forced or compulsory labor, will not use child labor, will comply with labor standards pertaining to hours worked, will receive compensation covered by laws or collective bargaining agreements, will not discriminate, and will grant employees the right of collective bargaining.
4. *Safety and health*: The supplier will provide employees with safe and healthy working and, where provided, housing conditions; all products and services delivered by the supplier must meet quality and safety standards required by applicable law.
5. *Environment*: The supplier must operate with care for the environment and comply with applicable laws and regulations.

6. *Supply farmers*: The supplier shall ensure that supplying farmers become fully aware of the code and will provide education and training sessions as necessary.

7. *Audit and termination of the supply agreement*: Nestlé reserves the right to demand corrective measures and to terminate the agreement with any supplier who does not comply with the code.

## PALM OIL

In 2009, Nestlé joined the Roundtable on Sustainable Palm Oil (RSPO). The company's guidelines ask their suppliers to source palm oil from plantations that are legally compliant, respect local and indigenous communities, protect peatlands and high-carbon-stock forests, and respect all other RSPO principles and criteria. By 2103, the company expects to be sourcing 100% RSPO-certified palm oil (Responsible Sourcing Highlights 2013).

## SOYA

Soya comes primarily from Brazil and Argentina. The traditional supply chain has a high degree of complexity. In 2011, Nestlé partnered with Conservation International, a global environmental nonprofit organization, to develop Responsible Sourcing Guidelines (RSGs). These guidelines allow for the assessment of soya producers with a view to eliminating any occurrence of deforestation or loss of high conservation values (Responsible Sourcing Highlights 2013).

## PAPER AND PAPERBOARD

In 2010, the company partnered with The Forest Trust to develop RSGs for paper and paperboard to ensure that the products come from fiber sources that have not led to deforestation or loss of high conservation values. In 2011, Nestlé started mapping and assessing the supply chains of more than 260 suppliers in Brazil, China, Europe, India, Indonesia, Malaysia, and the United States (Responsible Sourcing Highlights 2013).

## WATER

Water is a critical ingredient in Nestlé products and processes. The company is a founding signatory of the CEO Water Mandate, an initiative led by the United Nations Global Compact, an organization committed to improving and reporting water usage. Since 2000, Nestlé has increased food and beverage production volume by 63% while reducing water withdrawal by 33% (Creating Shared Value Report 2009).

Nestlé reports the following results in their sourcing efforts:

- 89.5% of suppliers complied with the Nestlé Supplier Code.
- Sourced 11% of their cocoa through the Nestlé Cocoa Plan, trained more than 27,000 farmers, and distributed more than 1,000,000 high-yield, disease-resistant cocoa plantlets.
- Helped 14 cocoa cooperatives achieve UTZ or Fair Trade certification.
- Purchased 133,000 tons of green coffee through Farmer Connect, trained more than 48,000 farmers, and distributed 12 million coffee plantlets in 2012.
- 80% of the palm oil purchased was RSPO compliant.
- More than 8000 farmers joined the Nespresso AAA Sustainable Quality™ Program in 2012 and provided 68% of the Nespresso coffee purchased by Nestlé (Responsible Sourcing 2013).

A study for the Thunderbird School of Global Management examined Nestlé's purchasing practices and found the company has been active in supporting cocoa growers in Ecuador, Venezuela, and Cote d'Ivoire, the world's largest cocoa producers. The study also determined Nestlé has implemented new initiatives that give them more control over and information about their supply chains. The company teamed with The Forest Trust to develop sourcing guidelines for palm oil to assure the products would not contribute to deforestation. The study concluded: "What is clear in their operations is that Nestlé is taking corporate responsibility increasingly seriously, both for the economic and image benefits they are reaping" (Bright et al. 2011).

## SUSTAINABILITY

Nestlé views itself as a sustainable company. The following achievements were reported in the 2012 Annual Report:

- A 24% reduction in direct greenhouse gas (GHG) emissions since 2002
- 489 water-saving projects in their factories, saving 6.5 million cubic meters of water
- 39 factories generating zero waste for disposal

In addition, they have a major effort in reducing deforestation, as described earlier.

## CONCLUSION

Nestlé is a gigantic company with customers and suppliers throughout the world. Their supply chains are long and complex. They have many opportunities for minor disruptions in the flow of goods. Even more critical, their vast array of supply sources requires constant attention to prevent contaminated products from entering the world's food supply.

## REFERENCES

Bright, A., Buhrman, J., Ellingson, J., Harrop, J., Longo, M., and Narayan, A., Can Nestlé be a good global citizen? Thunderbird School of Global Management, Glendale, AZ, April 18, 2011, www.thunderbird.edu/blog/faculty/washburn/2011/04/18/nestle-global-citizen, accessed August 23, 2013.
Nestlé Annual Report, 2012, http://www.nestle.com/asset-library/documents/library/documents/annual_reports/2012-annual-report-en.pdf, accessed August 28, 2013.
Nestlé History, http://www.nestle.com/aboutus/history, accessed January 15, 2014.
Nestlé Responsible Sourcing, 2013, www.nestle.com/csv/responsible-sourcing/highlights-challenges, accessed August 20, 2013.
Nestlé Shared Value Report, 2009, www.nestle.com/CSV, accessed August 25, 2013.
The Nestlé Supplier Code, August 2010, www.nestle.com/suppliers, accessed August 15, 2013.

## Introduction

This chapter deals with the purchasing of goods and services by various customers in the supply chain—manufacturers, wholesalers, and retail companies. This is an important function, because purchased materials and services represent over half of the product costs in most manufacturing industries and are even higher for wholesale and retail

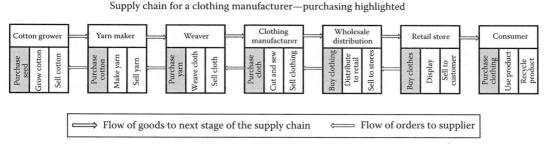

**FIGURE 8.1**
Purchasing along a clothing manufacturer supply chain.

companies, sometimes exceeding 90% (Joyce 2006). As more manufacturing companies move to a *core competency* strategy, they will outsource more of their component manufacturing to suppliers and become primarily an assembly operation. In some cases, they have moved even further by outsourcing the entire manufacturing function to concentrate on product and brand development and marketing that product.

Purchasing is also important in the service sectors of supply chains, including wholesale and retail operations. In both areas, the value of purchased goods and services is even higher, as a percentage of the sales dollar, than in manufacturing operations. Figure 8.1 illustrates the idea that every type of business performs a purchasing function. In this chapter, we will first describe the general functions and responsibilities. Then, we will identify some of the differences in purchasing, depending on the type of business in which they are involved.

## Role of Procurement in the Supply Chain

The title of this chapter includes both purchasing and procurement. While these terms have somewhat different meanings, they are often used interchangeably. We include the following definitions of related terms:

*Procurement*: The business functions of procurement planning, purchasing, inventory control, traffic, receiving, incoming inspection, and salvage operations.

*Purchasing*: The term used in industry and management to denote the function of and the responsibility for procuring materials, supplies, and services.

*Buyer*: An individual whose functions may include supplier selection, negotiation, order placement, supplier follow-up, measurement and control of supplier performance, value analysis, and evaluation of new materials and processes. In some companies, the supplier scheduler handles the functions of order placement and supplier follow-up.

*Product*: Any good or service produced for sale, barter, or internal use (Blackstone 2013).

These definitions from the *APICS Dictionary* indicate that purchasing is an activity within the broader function of procurement. In practice, there may be some overlap in the use of these terms and even in the job titles. Purchasing seems to be the term of choice among industrial businesses, while procurement seems to be more generally associated with government agencies. A buyer is an individual who performs a purchasing or procurement function. We will use the term product to indicate both goods and services. Services have become an increasing share of the purchasing responsibility in recent years.

Another reputable source reports, "Terms such as purchasing, procurement, materiel, materials management, logistics, supply management, and supply chain management, are used almost interchangeably. No agreement exists on the definition of each of these terms, and managers in public and private institutions may have identical responsibilities but substantially different titles" (Leenders et al. 2002).

The Institute of Supply Management (ISM) Career Center lists dozens of job openings with different job titles, showing variations on assistant buyer, buyer, purchasing agent, director of purchasing, director of procurement, and sourcing specialist (ISM 2014). Other jobs within the purchasing function may include expediter, buyer–planner, or value analyst, depending on the scope and scale of a particular purchasing department.

In keeping with what appears to be a general practice in the business community, we will use purchasing as the most widely used term and interchangeable with procurement (Gundlach et al. 2006). We will consider a buyer as someone who performs at least a part of the purchasing function. There may be other jobs within the purchasing function, such as expediter or value analyst, depending on the scope and scale of a particular purchasing department.

## Traditional Purchasing

The traditional or *old school supplier management techniques* (Vonada 2008) used by most buyers were to send a request for quotation (RFQ) to two or three potential suppliers—some of which they may have known and some that perhaps they did not know. They tried to exclude poor vendors at the beginning; however, sometimes, they did not have enough information and it was not considered critical because they had multiple choices and were not committing to long-term relationships. When they received the RFQ back, they decided to whom they would award the order, usually based on the lowest quoted cost, assuming the delivery time and product quality were acceptable. Once the order was placed, the buyers often had to expedite deliveries. Placing repeat orders for the same product offered the buyer the opportunity to pressure the supplier for a lower price.

The traditional purchasing function was concerned primarily with the buying process in which they placed the orders and followed up to assure delivery of those orders. The purchasing agent was concerned with obtaining products to meet the requirements of functionality—does it do what we want it to? Availability—can we get it when we need it? Cost—does it meet our target costs? Quality—does the quality meet our requirements?

In response to having to evaluate vendors in a more systematic manner, many companies have resorted to using vendor rating systems that cover a number of both tangible and intangible factors. Table 8.3 shows a typical vendor rating system. First, it is necessary to select the factors that will be used to rate the suppliers. In this example, there are

**TABLE 8.3**

Vendor Rating System

| Factor | Factor Weight | Supplier A | Supplier B | Supplier C |
|---|---|---|---|---|
| Cost | 2 | 10 | 8 | 7 |
| Quality | 2 | 7 | 8 | 8 |
| Delivery | 4 | 3 | 9 | 9 |
| Technical | 1 | 5 | 5 | 8 |
| Financial | 1 | 4 | 6 | 7 |
| Total | 10 | 55 | 79 | 81 |

five factors, but a company should decide which factors best fit their circumstances. It is best to identify at least four to eight factors to provide sufficient scope without excessive detail. Product cost, quality, and delivery are almost always factors to consider. In addition, we show another factor called technical. This is the supplier's ability to not only make the desired product but also offer suggestions on how to offer new approaches to making the product benefit the customer better. The last factor, financial, is a measure of the supplier's financial stability. The purchasing department should be hesitant about selecting financially weak suppliers. They may do it, but with caution.

Once the factors have been determined, the vendor selection team must assign weights to each factor. Often this requires considerable discussion and compromise before a consensus weighting is achieved. We show all of the weights adding up to 10; they could be assigned weights that add up to 1, or any number, just to make it convenient. In this example, delivery is assigned the highest weight, with technical and financial being relatively low weights and cost and quality in the middle.

The final decision requires assigning scores to each supplier. This example shows scores that range from 1 to 10, with 10 being the highest possible score. In this example, Supplier A received a 10 for its ability to keep costs low. Assigning a score can sometimes result from specific measures, such as relative costs, number of defects, or on-time and correct quantity deliveries. However, judgment will usually be required to arrive at a consensus score.

Once the numbers are assigned, an overall score can be determined by multiplying the weight by the supplier's score for each factor. In the example, Supplier A had an overall score of 55, while Supplier B had a score of 79, and Supplier C had a score of 81. Supplier A had the lowest price but fell short on some of the other factors. Based on this rating system, they are not an attractive supplier. Both B and C are close. While C has the highest cost, they have higher scores in both technical and financial factors. If only one supplier is to be selected, the vendor selection team still has a difficult choice to make.

This type of vendor rating system helps, but never eliminates, the judgment required in making the final supplier selection. Often, when two suppliers are close, the decision is to hedge and award a portion of the purchase to both suppliers.

Today, some tactical purchasing functions are routine, such as stock order replenishment; indeed, the actual order placement has been automated through enterprise resource planning (ERP) systems, at least in larger companies (Austin 2008). Another study forecast that "tactical purchasing activities such as ordering, quoting, expediting, and so forth will be automated and/or outsourced, and headcounts will be reduced" (Carter et al. 2000).

## Contemporary Purchasing

Purchasing functions and responsibilities have expanded, especially with the advent of supply chain management and outsourcing. Today, companies recognize that purchasing should be involved in both tactical and strategic decisions. The movement of purchasing from a tactical implementer to a strategic participant is making a big change in how members of an organization view the purchasing function.

In business practices, increased globalization, technological advances, and increased demands by management require that purchasing change from an administrative function to a strategic activity (Giunipero and Pearcy 2000). Minahan warns, "Being efficient in the day-to-day details of procurement is fine. Today, it is not enough. Leading companies are transforming their transaction-based procurement operations into a strategic resource that can deliver real value to the enterprise" (Minahan 2005, p. 47). As recently as 2003, however, others pointed out that while the literature has claimed purchasing is a strategic function, other research has found the level of strategic participation is low. They suggest purchasing functions must demonstrate a history of cooperation to secure their place as a strategic contributor (Goebel et al. 2003).

The contemporary approach requires the buying organization to become partners with its suppliers. In the broadest sense, the relationship is designed to work for the long-term well-being of all supply chain participants. That is a broad objective and requires an extensive change in the cultural orientation of the purchasing department members. Many buyers who were trained to exercise a *do it or else* approach to suppliers find it difficult to reorient their thinking to a *let's do it together* approach. They view it as a loss of power and many are unable or unwilling to change. Even today, some old-timers in the field are having difficulty in making this transition.

Training for a purchasing career embodies a number of new skill requirements. One of the most important is the human relations skills of negotiation. *Negotiation* is the process by which a buyer and a vendor agree upon the conditions surrounding the purchase of an item (Blackstone 2013). Even when the buying organization has considerable power over the negotiations because of its size or uniqueness, it is important for them to exercise considerable skill in developing a win–win contract. A *supplier partnership* is the establishment of a working relationship with a supplier organization, whereby the two organizations act as one (Blackstone 2013). Human relations skills are also necessary in addressing the other functions within the buying organization that influence the purchasing process. Professionals in marketing, engineering, accounting, and operations are generally both knowledgeable and insistent on getting what they need; it takes a skillful facilitator to lead this sometimes disparate group to an acceptable consensus solution. As one supply chain executive expressed when discussing requirements for purchasing professionals, "If all my wishes came true tomorrow, I would hire professionals from this day forward who have multi-disciplined experiences and expertise in program management, project management, engineering, operations and supply chain" (Bernstein 2005).

Purchasing organizations are expected to participate, often as part of a cross-functional team, in decisions concerning strategic sourcing.

These decisions include

- Supplier evaluation
- Supplier selection for long-term relationships
- Contract structuring and negotiation

- Supplier relationship management (SRM)
- Supply chain coordination and collaboration (Kocabasoglu and Suresh 2006; Strategic Direction 2005)

Contemporary purchasing is more oriented to building long-term relationships with fewer, but more dependable suppliers. While there is continued, and sometimes fierce, insistence on low prices, there is a need to include other criteria in describing the good or service being purchased. In the broadest sense, the relationship is designed to work for the long-term well-being of the entire supply chain. This is a broad assignment and requires an extensive change in the orientation of the purchasing philosophy and of the members of the purchasing department.

In addition to their expanded role in strategic sourcing, some advocate that purchasing should take a more inward-looking perspective by participating in activities such as

- Total cost of ownership (TCO) and value analysis (Joyce 2006)
- Product design and specifications (Nelson et al. 2005)
- Target costing (Ellram et al. 2007)
- Inventory management (Flanagan 2005; Monczka and Morgan 2000)
- Monitoring strategic materials cost drivers, commodity indices, and availability (Austin 2008)

All of the preceding activities should lead to an effective level of supply chain coordination and eventually collaboration among supply chain participants. The broadened role for purchasing involves more than obtaining goods and services at the lowest cost; it also requires that purchasing "fits the need of the business and strives for consistency between its capabilities and the competitive advantage being sought throughout the supply chain" (Gundlach et al. 2006). The purchasing function also has the responsibility in many organizations to act as an interface between suppliers and functional areas within the buying organization. As companies move to cross-functional teams with responsibilities to the customer, the teams must also have access to the corresponding suppliers to assure that they meet the needs of the customers.

There is no doubt that purchasing responsibilities are expanding and their value to a company is increasing. Some reasons for this include increased levels of outsourcing, increased use of the Internet, greater emphasis on supply chain management, globalization, and continuing efforts to reduce costs and increase quality (Joyce 2006). This statement captures the essence of purchasing importance in managing inventory: "As inventory levels are being drastically cut and shipments are arriving daily rather that weekly or monthly, purchasing is in the greatest position to either achieve stunning success or be the source of failure" (McCollum 2001, p. 57).

## Changing Role of Purchasing

How does purchasing move from the traditional approach to the contemporary? Over three decades ago, Kraljic (1983) envisioned the need for the change. Table 8.4 shows a progression as complexity of the supply market increases and the importance of purchasing increases.

**TABLE 8.4**

Stages of Purchasing Sophistication

| Key Performance Criteria | Purchasing Management — Low Complexity of Supply Market, Low Importance of Purchasing | Materials Management — Low Complexity of Supply Market, High Importance of Purchasing | Sourcing Management — High Complexity of Supply Market, Low Importance of Purchasing | Supply Management — High Complexity of Supply Market, High Importance of Purchasing |
|---|---|---|---|---|
| Procurement focus | Noncritical items | Leverage items | Bottleneck items | Strategic items |
| Key performance criteria | Functional efficiency | Cost/price and materials flow | Cost and reliable sourcing | Long-term availability |
| Typical sources | Established local suppliers | Multiple suppliers, chiefly local | Global, new suppliers | Established global suppliers |
| Time horizon | Limited: normally 12 months or less | Varied, typically 12–24 months | Variable, depending on availability | Up to 10 years |
| Items purchased | Commodities | Commodities, specified materials | Mainly specified materials | Scarce and high-value materials |
| Supply | Abundant | Abundant | Production-based scarcity | Natural scarcity |
| Decision authority | Decentralized | Mainly decentralized | Centrally coordinated | Centralized |

*Source:* Adapted from Kraljic, P., *Harv. Bus. Rev.*, 61(9), 109, September 1983.

*Notes:* Complexity of supply market criteria: availability of supply, pace of technological advance, entry barriers, logistics cost, and complexity. Importance of purchasing criteria: cost of materials/total costs, value-added profile, profitability profile, risk management, strategic importance.

The top half of the table shows how changing requirements from routine or commodity items to more specific items increase the importance of purchasing. This moves purchasing from a purchasing management status to a materials management status.

The bottom half of the table shows the further advancement when the supply market becomes more complex. Purchasing now has to acquire more critical, even strategic, items from global sources where there may be scarcity of capacity constraints. These additional challenges move purchasing into a sourcing management status and eventually into a supply management status, considered by Kraljic as the ultimate goal for purchasing in a company (Kraljic 1983).

## Critical Success Factors for Purchasing

Before describing the purchasing activities in more depth, let us look at the CSFs for the purchasing function. The first four belong to the traditional form of purchasing and include obtaining products that function as intended, are available in sufficient quantity, sell at an acceptable price, and have satisfactory quality. The last six have become more important with the increased interest in supply chains and include the following: (1) match inflow of goods with outflow, (2) reduce variations in deliveries, (3) increase supplier dependability, (4) reduce the bullwhip effect, (5) become an intercompany facilitator, and (6) find sustainable suppliers.

### Functionality

The first CSF for purchasing was to obtain goods or services that satisfied the needs of the users. While purchasing managers had some flexibility in selecting the supplier, the expectation was that the product purchased would work as expected. In specifying more technical purchases, user departments often specified the model number and what the vendor wanted, and were not willing to accept substitutes. In this case, the buyer became a routine order placer.

### Availability

Purchasing has always had the primary responsibility to make goods or services available when needed by the production line, by the retail store when having a sale, or for a surgeon about to perform an open-heart surgery. The presence of the expediter job in many companies still attests to the importance of availability. Late deliveries were bad; early deliveries were generally acceptable, perhaps even considered a positive indication of supplier performance. However, in recent years, neither early nor late deliveries, or overamounts or underamounts, are acceptable.

### Cost

Close behind availability was the requirement to buy goods and services at the lowest possible cost. The use of purchase price variances (difference between the actual purchase price and planned, or standard, price) has been a bedrock measure of purchasing performance from the beginning. Often, the best way to obtain reduced unit prices was to buy larger volumes. While this helps the buyer's purchase price variance, it often resulted in excess inventory or the wrong mix of inventory, or sometimes both.

## Quality

Obtaining a good quality of purchased goods and services has always been an implicit responsibility of purchasing. Some companies used quality as a measure of purchasing performance; however, many did not. Production managers may have grumbled about inspecting many incoming parts to screen out the defective products, but if the price was right, the purchasing department usually did not receive much criticism.

Functionality, availability, cost, and quality were all important and remain so today. The better the purchasing department, the better they were able to satisfy these basic requirements. However, purchasing was not often considered a key member of the top management organization. As the transition to contemporary purchasing evolved, a number of CSFs have been added.

## Match Inflow with Outflow

One of the best ways to obtain low purchase prices was to buy in large quantities or to make *early buys*—buy out of season to help suppliers balance their production. Buying in large quantities resulted in excess inventories that were subject to holding costs, damage, obsolescence, and a premature commitment of funds. One of the authors remembers an overzealous restaurant manager who purchased ice tea bags at a discount, quite a few cases, to say the least. There were cases of ice tea bags in almost every storage area of the facility. The problem though was that ice tea bags cannot last for a long period of time. Several years of tea bags were eventually discarded due to old age!

Buying in large quantities also meant that, sometimes, too much of some items meant too little of other items. Under this system, the flow of goods into a facility did not match the flow of goods out of the facility. Just as matching the product amounts and mix produced with the amounts sold required collaboration between the production and marketing departments, matching inflows with outflows required collaboration between production and purchasing as well. Some companies addressed this requirement by establishing the materials management function. *Materials management* is the grouping of management functions supporting the complete cycle of material flow, from the purchase and internal control of production materials to the planning and control of work in process to the warehousing, shipping, and distribution of the finished product (Blackstone 2013).

## Reduce Variances in Delivery

In their attempt to more closely match the arrival of goods with the production department's needs, purchasing departments found, in addition to having to moderate their large volume buys, they had to assess the effectiveness of the delivery practices of their suppliers. Some suppliers consistently made on-time deliveries; however, many were inconsistent—sometimes early, but more often late. Some suppliers shipped complete orders; some shipped multiple partial orders. It became necessary for purchasing departments to monitor the performance of their suppliers and encourage them to be more consistent in their performance. If they did not improve, suppliers faced the prospect of reduced business, or even elimination as a supplier.

## Increase Supplier Dependability

The attention to supplier delivery performance opened the opportunity and need to evaluate suppliers on other tangible criteria as well, such as lead times and quality of merchandise, and even to explore such intangible services such as the supplier's help in suggesting ways to improve the buyer–seller relationship. Some suppliers could provide input about new product developments or market conditions that were of value to the buying organization. As buyers learned more about their suppliers, it was a natural development to group the suppliers into categories representing their value to the buying organization. This closer look also made it easier for purchasing departments to reduce the number of suppliers they needed and enabled them to place increased business with their more dependable suppliers.

## Reduce the Bullwhip Effect

As purchasing departments learned more about their suppliers and how to match the flow of incoming goods to outgoing goods, they demonstrated their increased knowledge to other parts of the business, and even to other members of their supply chains. One of the more troublesome problems for supply chains was the bullwhip effect. The *bullwhip effect* is an extreme change in the supply position upstream in a supply chain, generated by a small change in demand downstream in the supply chain. The impact is that inventory can quickly move from being backordered to being excess. This problem is caused by the ripple effect nature of communicating orders up the chain with the inherent transportation delays of moving product down the chain. The bullwhip effect can be markedly improved by synchronizing the supply chain. If purchasing departments could reduce the variances with suppliers, then each stage of the supply chain could work together to reduce variances and thereby reduce the bullwhip effect.

## Become an Intercompany Facilitator

As customers demanded more from fewer suppliers, it became necessary for other functions in an organization to have access to their counterparts in supplier organizations— engineers with engineers, marketers with marketers, accountants with accountants, and the like. Someone had to initiate and help sustain these relationships; this facilitating function was a natural function for most purchasing groups. They generally had the most information about suppliers and could expand their information base to include information desired by other functions in the buying organization.

## Find Sustainable Suppliers

The reputation of a firm is closely linked to the social, environmental, and ethical profiles of an organization's spending (Reeve and Steinhausen 2007). Purchasing departments must add finding responsible suppliers to their list of operating objectives.

These added CSFs for purchasing heralded the changing role of purchasing.

## Purchasing Functions: Participating

Purchasing functions and responsibilities have expanded, especially with the advent of supply chain management and outsourcing. Companies recognize purchasing personnel should be involved in a number of strategic and tactical areas from which they had previously been excluded. The movement of purchasing from a tactical implementer to a strategic participant is a transition making a big change in how organizations view the purchasing function. "Changes in business processes, increased globalization, technological advances, and increased demands by upper management are forcing concomitant changes in the purchasing/supply management function. These changes are coupled with the evolution of purchasing/supply management from an administrative function to a strategic activity" (Giunipero and Pearcy 2000, p. 4).

In its quest for reducing costs, the financial function expects that purchasing can contribute best by having the senior purchasing officer (the chief purchasing officer [CPO]) sit on the highest decision-making body in the company. However, "the CPO and the purchasing team need to have the right knowledge and set of skills that will earn the respect of people throughout the organization" (Gerardo and Spanyi 2008, p. 28). The following functions are some in which contemporary purchasing is involved as a participant in the decision-making process.

### Product Design

Product design had been largely the responsibility of marketing and engineering. Recently, however, it is becoming more obvious that other functions should be involved in the product design process. One of the leading concepts in this area is concurrent engineering or participative design/engineering—a concept that refers to the simultaneous participation of all the functional areas of the firm in the product design activity. Suppliers and customers are often included in the product design process. The intent is to enhance the design with the inputs of all the key stakeholders. Such a process should ensure the final design meets all the needs of the stakeholders and should ensure a product that can be quickly brought to the marketplace while maximizing customer value and minimizing costs (Blackstone 2013).

### Product Specifications

Once the product has been designed, there is a need to further define it by developing the product specifications. A specification is "a clear, complete, and accurate statement of the technical requirements of a material, an item, or a service, and of the procedure to determine if the requirements are met" (Blackstone 2013). Product engineering is a key player at this point; however, as they specify the materials and services required, purchasing personnel must also participate in decisions involving the availability and costs of those materials and services.

### New Product Introduction

In considering new product introductions, engineering and marketing are the key players, and they are essential to successful new product launches. However, purchasing plays an equally important role. Managerial Comment 8.1 describes an approach that worked successfully in a major manufacturing company (Walker 1992).

## MANAGERIAL COMMENT 8.1   ROLE OF PURCHASING IN NEW PRODUCT INTRODUCTION

One element of reducing time to market for new product design is concurrent project materials management from the investigation phase through the pilot run phase. By moving material's involvement with the new product right up front, long lead time procurement delays have been eliminated during production ramp-up. The objective of this discussion is to describe the operation and systems support of an organization called *SPRINT* (Systems and Product Introduction), which is dedicated to new product materials management.

SPRINT people report to purchasing but reside within the engineering community. As the new design evolves, the SPRINT team first adds value by structuring the bill of materials (BOM) for engineering. They scrutinize new material for preferred part status and quickly order in limited quantities. Several strategies are used to procure material inside of normal lead times and to remain flexible during a steady stream of design changes. When product performance must be tied to components from a new supplier, SPRINT adds additional value by acting as the focal point for new supplier qualification. Outputs from SPRINT activities include inventory control and kiting of prototype parts along with periodic material cost roll-ups for project management. Prior to a pilot run, SPRINT transfers the final BOM and unique parts inventory to production's MRP system.

### MATERIALS MANAGEMENT

The SPRINT organization combines elements of purchasing, scheduling, materials engineering, information systems, and manufacturing specs into one integrated whole. The SPRINT organization is centrally located within engineering and is a part of the engineering expense budget but reports to purchasing. An ideal team consists of a buyer, a materials engineer, a planner, and a BOM expert. Three or four people, each with the right skill mix, can achieve the critical mass necessary to make SPRINT work.

It is essential that SPRINT becomes involved with each new product design very early in its development. By overlapping the materials effort with product design, tasks are accomplished concurrently. Whenever work can be done in parallel rather than sequentially in time, the total project time to market is reduced. This usually results in a more competitive introduction and in an enhanced return on project investment.

Mature project design teams operate within the context of a product design life cycle. While the names of each project phase differ from company to company, the sequence flows along the lines of investigation phase to design phase to prototype phase to pilot run phase. Table 8.5 provides a list of tasks.

### WHERE SPRINT CAN ADD VALUE

The introduction of each new product generation represents a unique opportunity to improve manufacturing's competitiveness. The front end of a new design is the place to reduce the number of suppliers required to support production and to eliminate usage on purchased parts that have long since reached their maturity.

**TABLE 8.5**

Steps in the SPRINT Project

| Project Phase | SPRINT Activity |
|---|---|
| 1. Investigation | Assess the business risk of strategic technology sole-sourced suppliers. |
| 2. Design | Capture an accurate parts list. |
| | Develop a complete BOM. |
| | Consolidate the number of suppliers. |
| | Provide a direct material cost roll-up. |
| 3. Prototype | Quick procurement of all materials. |
| | Provide inventory control function. |
| | Review component life cycle on old parts. |
| | Group all suppliers by business risk. |
| | Consolidate the number of part numbers. |
| | Continuous updating of the BOM. |
| | Regular updates of material cost roll-up. |
| 4. Pilot run | Transfer final BOM to manufacturing. |
| | Transfer start-up materials to inventory. |

*Source:* Walker, W.T., *APICS—The Performance Advantage*, March 1992, pp. 48–51. With permission.

Current suppliers with very high business risk should be eliminated from the list of preferred suppliers for the new design. This is also the time to reduce the cycle time required to produce the product by carefully engineering its product structure.

As the design team selects the technology to be used to implement the new design and specific components are specified, the important decision must be made about who will be the suppliers. The purchased parts on the BOM can be grouped into two lists. One list should include the vast majority of parts; this is a list of standard, commonly available materials with multiple sources of supply. The company may decide for business reasons to work with a single supplier, but each of the parts on the first list is produced by multiple sources. The second will be a very short list of those parts that determine the price/performance competitive advantage of the product. The parts on the second list are often sole sourced because of the technology involved. The buyer reviews the suggested suppliers from both lists and adds value in the process by evaluating the business risk associated with each supplier. In addition, the SPRINT buyer sees all the purchased material requirements from all the engineers and is in a position to recommend consolidation and/or elimination of some sources.

*Source:* Walker, W.T., *APICS—The Performance Advantage*, March 1992, pp. 48–51. With permission.

## Target Costing

Traditionally, companies designed and specified the product. Then, based on the projected cost, they could add the markup and calculate a selling price for the product. In recent years, however, it has become apparent that the market strongly influences prices of many products. Consequently, many companies are now working with a concept called *target costing*. This approach first determines what an acceptable selling price will be to compete in the marketplace and subtracts a desired contribution margin to arrive at an acceptable, or target, cost. The equation for this process is

Projected selling price – the desired profit contribution = the product target cost

Ellram (2000) provides a comprehensive description of this process. She explains that target costing is an emerging process, whereby organizations calculate the allowable cost (i.e., target cost) for buying/producing the product or service they offer for sale by first determining the acceptable selling price in the marketplace and the organization's required internal margin on the product. She highlights the need for purchasing and supply management (PSM) to maintain a close working relationship between buyer and design/R&D personnel. This goal involves identifying capable suppliers, facilitating early supplier involvement, qualifying suppliers to ensure only qualified suppliers are selected, negotiating with suppliers, developing long-term contracts/agreements with suppliers, obtaining supplier cost breakdowns where possible, working with suppliers on cost reduction ideas, and suggesting changes to target cost.

## Strategic Sourcing

Strategic sourcing, or supplier selection, is critical in the contemporary purchasing approach. In the past, vendor turnover was more common when orders were awarded based on the most recent low bid. There were no guarantees that getting today's order gave a supplier an advantage on getting the next order. Buyers pursued low prices with a vengeance and did not place a high value on continuing relationships. In describing the experiences of Motorola, Honda, and Toyota, one author said, "Closeness to suppliers has become the not-so-new mantra as the rough and tumble of negotiation hardball has been replaced by deeper, more constructive supplier relationships" (Strategic Direction 2005, p. 29).

Today, suppliers are selected for their potential to sustain a long-term relationship. Companies consider factors other than price in the supplier selection process. These factors include product quality, on-time delivery performance, facility proximity, industry position, financial stability, product development assistance, continuous improvement capability, operational flexibility, and ethical operating standards. To accomplish this selection process, companies must elevate the purchasing function from a transactional position to a strategic position. This transition involves coordinating purchasing with other functions in the organization and sharing information with key suppliers (Kocabasoglu and Suresh 2006). While the transition is far from complete, it is gaining momentum in the purchasing practices of many companies.

The purchasing organization is a key member of the team that identifies, evaluates, and selects new suppliers. They participate in multifunctional teams that perform due diligence and risk assessment visits to potential suppliers. Purchasing, more than any of the other functions, wants to select vendors that have not only the capability to perform as a dependable supplier but also the willingness to work through the problems and changes that are sure to occur during a long-term relationship.

This has become such an important area that many companies, especially larger ones, are using third-party application suites to assure an objective and integrated approach. The applications suites are a set of related solutions that support upstream procurement activities, including identifying suitable suppliers, finding savings opportunities, quantifying and reducing supply risk, negotiating and managing contracts, tracking ongoing supplier and category-level performance, and monitoring the overall success of the procurement team (Wilson et al. 2013).

Strategic sourcing implies building relationships with suppliers, which takes time to develop and a higher level of transparency on both sides. When done successfully, there are a number of benefits, including

- A more cost-efficient supply chain
- Better supply chain partners
- Improved service levels and coordination
- Improved partner network capabilities
- Enhanced flexibility and responsiveness
- Increased revenue

While cost reductions may be the primary driver, other gains may be in gaining innovative ideas from suppliers and new products that generate added sales (Zubko 2008).

Strategic sourcing usually involves strategic negotiating of contracts. Negotiating is a core requirement, especially as purchasing evolves into a high-level strategic function. Negotiations often center on complex issues that include risk management, intellectual, innovation opportunities, and long-term strategic partnerships. Consequently, they must become a win–win agreement, not an adversarial win–lose proposition (Siegfried 2013).

## Supplier Location

Supplier location is one of the criteria used in selecting vendors. It is especially important because of the emphasis on reducing supply chain response time. For this reason, it is desirable to have suppliers located close by. Some companies are even experimenting with having their suppliers on-site to reduce delivery time; however, these facilities are largely experimental at this point. Ironically, at the opposite end of the spectrum, some buying organizations are using suppliers that are on the other side of the globe because of the attractiveness of lower costs.

Resolving the conflict between a desire for the close proximity of suppliers and a desire for low purchase costs requires careful analysis. At the present time, the attractiveness of low costs is fueling the offshore outsourcing movement, a development discussed later in this chapter. At the same time, there is growing interest in the reshoring, or onshoring, movement in which companies move offshore operations closer to home.

## Inventory Management

Inventory management is an important component of supply chain management. Inventory is defined as "those stocks or items used to support production (raw materials and work-in-process items), supporting activities (maintenance, repair, and operating supplies), and customer service (finished goods and spare parts). Demand for inventory may be dependent or independent. Inventory categories include anticipation, hedge, cycle (lot size), fluctuation (safety, buffer, or reserve), transportation (pipeline), and service parts" (Blackstone 2013).

We discussed the role of finished goods inventory in Chapter 6 and the place of work-in-process inventory in Chapter 7. While purchasing has a primary responsibility for obtaining raw materials inventory for manufacturing companies, it also has a responsibility for finished goods inventory when the purchasing function is in distribution or retail companies. Purchasing also becomes a key participant in supply chain inventory management.

At the strategic level, inventory management involves deciding where to position the inventory along the supply chain. Will it be stored in the retail store at the end of the supply chain, where it will need to be in the largest number of configurations, but will be readily available to the customer? Or should it be stored upstream in the supply chain, where it can be in a semifinished state to reduce the variety required, but less accessible to the customer? This decision involves a trade-off between cost and availability to the customer.

At the tactical level, there are numerous opportunities to determine how best to control the level and flow of inventory through the supply chain. Determining inventory levels, order quantities, and order timing is a necessary decision for thousands of items in even the simplest of supply chains. While purchasing aims to buy at a rate that corresponds with the consumption rate of their business, minimum buy quantities required by suppliers may sometimes upset the desired flow of materials. Buyers need a way to monitor the turnover of inventory items for which they have a responsibility.

Effective supply chain inventory programs require collaboration among all supply chain participants. While one party may be more influential than others in a supply chain, there is little room for nonparticipants. The purchasing function is usually able to act as an enabler to provide effective interfaces along the supply chain.

## Supplier Risk Management

As supply chains get longer and more complex and as companies strive to become leaner, the risk of disruptions in the flow of goods becomes potentially more serious. As part of supply chain design, managers must prepare for these seemingly inevitable, but often unpredictable, events. Whether it is a tsunami in Japan, a building collapse in Bangladesh, a flood in Thailand, or a fire in Texas, some preplanning is necessary to mitigate the potential negative effects on the supply chain. Often, purchasing must find alternate sources in case of an emergency; however, other functional areas may also be called upon to help human resources in case of loss of life, marketing to reassure customers, finance to make special funds available, engineering to consider substitutes, operations to rearrange schedules, and top management to coordinate the team and represent the company in public. Risk management will be discussed more in later chapters because of its increasing importance in supply chain management.

## Purchasing Functions: Directing

There are tactical and operational functions that purchasing manages directly. These include the purchasing process, supplier evaluation, and building supplier relationships. Other functional areas may participate on occasion, but the primary responsibility rests with the purchasing organization.

## Purchasing Process

Product design and specification, target costing, supplier selection and location, and inventory management are all strategic decision areas. Once these phases have been completed,

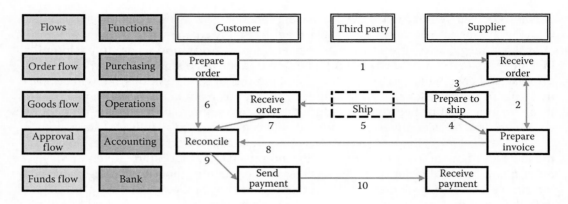

**FIGURE 8.2**
Purchase order process flow. 1, Prepare order and send to supplier; 2, supplier checks credit of customer before shipping; 3, supplier prepares order for shipment; 4, supplier prepares invoice; 5, supplier ships to customer, directly or through third party; 6, purchasing sends purchase order to accounting; 7, operations sends receiving report to accounting; 8, supplier sends invoice to customer (accounts payable); 9, accounts payable reconciles purchase order, receiving report and invoice; approves payment; 10, customer bank sends electronic funds transfer to supplier bank.

it is time to begin the tactical phase of the purchasing process. These activities include determining the order quantity, timing the order placement, monitoring the orders for on-time delivery, assessing the accuracy of the order, checking the quality of the product, approving the supplier invoice for payment, and reconciling problems that occur during the process.

Joyce describes the purchasing cycle as follows: "The purchasing cycle begins with a request from within the organization to purchase material, equipment, supplies, or other items from outside the organization, and the cycle ends when the purchasing department is notified that a shipment has been received in satisfactory condition" (Joyce 2006, p. 203). He lists the main steps in the cycle as follows: (1) purchasing receives the request; (2) purchasing selects a supplier; (3) purchasing places the order with a vendor; (4) orders are monitored; and (5) orders are received. Figure 8.2 shows the general model of the ordering activities. The diagram shows payments, step 10, are made electronically. The traditional ordering process included considerable paperwork; today's ordering systems are largely electronic.

## Supplier Evaluation

An important part of the purchasing function should be to evaluate each supplier's performance. The evaluation should include tangible measures, such as product quality and on-time delivery, but it should also include less tangible measures such as supplier cooperativeness and trust. We describe these performance measures more fully in the Performance Measurement section.

As part of the evaluation process, suppliers should be encouraged to present an evaluation of their own performance. Doing this should help to obtain supplier *buy-in* to the evaluation process and often results in constructive suggestions about how to improve their value to the buying organization. A variation of the rating system shown in Table 8.3 could also be used for ongoing supplier evaluations.

## Supplier Relationship Management

SRM is a relatively new concept and serves as a mirror image of the customer relationship management (CRM) program designed to establish the relationship with customers. SRM is defined as "a comprehensive approach to managing an enterprise's interactions with the organizations that supply the goods and services the enterprise uses. The goal of SRM is to streamline and make more effective the processes between an enterprise and its suppliers. SRM is often associated with automating procure-to-pay business processes, evaluating supplier performance, and exchanging information with suppliers. An e-procurement system often comes under the umbrella of a SRM family of applications" (Blackstone 2013).

Although purchasing has a lead role in establishing and maintaining effective supplier relationships, they often benefit from the involvement of other functions. A major electronics manufacturing company initiated international procurement organizations (IPOs) and involved the collaboration of purchasing, logistics, and materials engineering to assist in domestic procurement (Walker 2009).

In Chapter 4, we described CRM and pointed out the need to group like customers into distinct segments to better serve them. The same approach can be used with suppliers—to group them by the type of material or commodity purchased. Commodity buying is the grouping of like parts or materials under one buyer's control for the procurement of all requirements to support production. A commodity procurement strategy is the purchasing plan for a family of items. This would include the plan to manage the supplier base and solve problems (Blackstone 2013).

## Supply Chain Coordination/Collaboration

All of the aforementioned activities should lead to an effective level of supply chain coordination and eventually to collaboration among supply chain participants. The purchasing functions in each organization will not only be participants in the process, they should be drivers of the change.

---

## Purchasing along the Supply Chain

As we outlined in Figure 8.1, the purchasing function is present at each stage of the supply chain, beginning at the ultimate consumer and working upstream through all of the supplier organizations.

### Consumer

Individual consumers trigger the purchasing activity along the supply chain, even though they are located at the last stage of the process. While individual consumers do not usually have training or exercise a formal purchasing process, they are becoming much more knowledgeable buyers. Through access to the Internet and company websites, they can learn a great deal about products—both the one they may be interested in buying and the competing products. They can read about the attributes of products, and through informal blogs or formal reports through magazines like *Consumer Reports*, they can often have an understanding of the product before they buy. One of the areas in which this increased knowledge can be useful is in buying automobiles. Consumers have a good idea of what a fair price should be and can even get competitive quotes over the Internet. While consumers may be primarily interested in price, they are also aware of other factors, such as customer complaint levels and repair histories of products.

Consumers are also willing, even eager, to build lasting relationships with suppliers. Because they have limited time available, many consumers do not want to *shop* for products; they want to *buy*. The stereotypical shopper who spends leisurely hours strolling along the enclosed malls before deciding what to buy is becoming a relic of the past. Today's consumer is more apt to want to shop online or, if they must, travel to their store of choice, park close to the entrance (often in an open strip mall or stand-alone location), go in, quickly make their purchase, and then go on to other activities. Consumers are learning to be buyers that are more discriminating and to shop more expediently.

## Retail

Retail stores are interested in finished products. Their primary objective is to have products ready to sell to the consumer. Retailers often use the job function of merchandiser to invoke the responsibilities of both buying the goods required and developing the selling programs that will move the product through the stores. The merchandiser must train salespeople in the critical features and benefits of the product. At one time, it was sufficient for salespeople to make the consumer feel welcome and be helpful in locating the merchandise in the store for the consumer to look at. They also completed the sales at the point-of-sale (POS) register. However, today, most retail stores are finding the need to train their salespeople in the technical aspects of the product; otherwise, the consumer may know more about the product than the salesperson. This knowledge is especially important in the appliance and electronics industries, where customers may need help in discerning the different choices available to them. In some cases, building supply stores, such as Lowe's and Home Depot, have technically trained employees throughout the store to answer consumer questions, with the checkout registers staffed by another set of employees, trained for a more limited scope of duties.

Inasmuch as the success of any retail store depends on its merchandise selection and the quality of its salespeople, it is imperative that the purchasing and merchandising functions be thoroughly involved in the selection of the product line and in developing the marketing approach that will be successful. Some of the current concerns for retailers include reducing inventory levels, managing exclusive brands, managing commitment and production cycles, and overseeing direct sourcing from manufacturers versus indirect sourcing third parties (Flanagan 2005).

## Wholesale

Historically, wholesalers, or distributors, have been concerned only with handling finished goods. That is changing as more distributors begin to collaborate with manufacturers in the postponement concept. Postponement "is a product design strategy that shifts product differentiation closer to the consumer by postponing identity changes, such as assembly or packaging, to the last possible supply chain location" (Blackstone 2013). This practice may involve such procedures as putting price tags on clothing, combining items into assortments of the correct quantities for individual stores, or arranging merchandise in unique packages for preventing damage and loss. Carbone points out, "Electronics distributors have seen the future of value-added services and it is supply chain management" (Carbone 2000).

As described in Chapter 6, the distribution function is becoming more critical as customers want faster response times. However, one problem in this regard is that offshore outsourcing is causing supply chains to be longer. These conflicting objectives put more pressure on distributors to improve their methods and effectiveness. It also means the purchasing function has a wider scope of products to buy—both finished goods and raw materials—with a wide range of volumes and suppliers.

## Manufacturer

Manufacturers are concerned with producing and distributing finished goods downstream to wholesalers and retailers. On the supply side, they are involved with obtaining raw materials and complementary services to be able to produce the finished products. As we described in Chapter 7, the scope of most manufacturing businesses is changing to a more fragmented mix of internal and external operations. As outsourcing increases, the remaining internal operations are apt to be less continuous and synchronized. This means that maintaining continuity and smooth flows depends on the effective coordination of inputs (purchases) of goods and services from external suppliers. As a result, the purchasing function becomes a prime player in helping to provide that coordinated flow.

Some manufacturing companies are outsourcing more of their internal operations, even to the point of outsourcing all of their manufacturing. This is a drastic change in operating philosophy and requires careful planning and execution of the transition plan. Contract manufacturing has become commonplace in the electronics industry with the emergence of major players such as Flextronics and Solectron in the United States and Foxconn in China and Taiwan.

## Mining and Agriculture

Even in the most basic industries, purchasing has an expanded role. Mining requires extensive investment in capital equipment to mine and transport the mined materials from the mine to further processing destinations. Because mining is performed in harsh environments, the equipment required needs extensive maintenance and repair services often performed by outside suppliers who must be responsive to the mining company needs. When a major piece of equipment costs thousands of dollars an hour to operate, it must not be out of service for extended periods of time. It is the purchasing function's responsibility to be sure that suppliers with replacement parts and repair services are available when needed.

In agriculture, the emphasis shifts from fast response to technical knowledge requirements. Suppliers provide information about the latest developments in seeds, fertilizers, effects of climate changes, and growing methods. As in mining, there are large amounts of expensive equipment that can perform an increasing array of tasks on farms, reducing, although not eliminating, the labor requirements.

## Services

Buying services can often be more challenging than buying materials, because it is more difficult to measure the quality of services. In addition, services are often packaged in extended-term contracts, requiring less formal ways of evaluating the performance of services suppliers.

Services cover a wide scope of activities from custodial and employee cafeteria food services that are supplemental to the main mission of a firm, to services that are in the mainstream of operations, such as testing and materials handling, to strategic considerations, such as product design, and facility location. The purchasing function also involves buying services less visible to day-to-day operations but essential to the well-being of the workforce, such as supplying health benefit and retirement programs.

While purchasing services is growing in importance and magnitude, the resources to manage them are not. Consequently, organizations can improve their services purchasing in terms of cost and value, by effectively dedicating more resources to services purchasing. Developing an outstanding capability to purchase services, and to manage that purchase, could truly be the next frontier for improved supply chain and organizational performance (Ellram et al. 2007). Because of the diversity and complexity of the purchased services, it is apparent the purchasing function cannot be the sole group involved, as most purchases of services require a cross-functional team.

Purchasing logistics services, including transportation services, may be one of the most important areas for purchasing. Transportation costs represent a substantial part of the total landed costs, especially on offshore purchases. Purchasing must arrange for incoming delivery of goods that are on time, with minimal damage to goods, all at a competitive price. If the company does not have its own delivery fleet, purchasing must arrange for outgoing deliveries that also satisfy time, quality, and cost objectives (Walker 2009).

---

## Offshore Outsourcing

Offshore outsourcing has become a major topic in the business literature. While outsourcing is not confined to the procurement function, purchasing plays a major role in the design and implementation of outsourcing programs. In this section, we will look at outsourcing briefly. For a detailed coverage of this topic, see the articles by Crandall (2005) or Walker (2000). This topic will be covered again in Chapter 11 in discussing the need for tighter integration of supply chains.

### As a Strategic Concept

The three concepts that define offshore outsourcing as a strategic concept are the make-or-buy decision, outsourcing, and subcontracting.

*Make-or-buy decision*: The act of deciding whether to produce an item internally or to buy it from an outside supplier. Factors to consider in the decision include product costs, capacity availability, proprietary and/or specialized knowledge needed, quality considerations, skill requirements, volume expectations, and timing.

*Outsourcing*: The process of having suppliers provide goods and services that were previously provided internally. Outsourcing involves substitution—the replacement of internal capacity and production by that of the supplier (see subcontracting).

*Subcontracting*: Sending production work outside to another manufacturer (see outsourcing).

The preceding definitions from the APICS Dictionary (Blackstone 2013) do not make a major distinction between the make-or-buy decision and the outsourcing decision. We view the make-or-buy decision as belonging to a simpler period when the decision was primarily concerned with manufactured parts and a decision was easily reversible, or even temporary, because of seasonal, or other peak demand, requirements. This was a tactical decision that purchasing could make or, at the most, purchasing in conjunction with some input from operations.

Conversely, outsourcing is a more contemporary concept. It involves both products and services, is usually of a permanent nature, and is often associated with using a supplier in a country other than where the buying company is located. Outsourcing is a strategic decision requiring cross-functional analysis and consensus. Table 8.6 highlights the key distinctions between the make-or-buy decision and outsourcing.

Outsourcing where the supplier is located in another country has come under the more formal term of low-cost country sourcing (LCCS). Some of the challenges in LCCS include

- Low supplier maturity levels for best-in-class global trade practices
- Extremely complex supplier assessment and development efforts
- Inefficient logistics infrastructure
- Low level of IT infrastructure for business-to-business (B2B) transactions and collaboration
- Problems in the protection of an organization's intellectual property
- Understanding country-specific business processes and methodologies
- Differences in culture, lifestyle, and languages between outsourcer and supplier countries
- Poor information visibility for control and improving supply chain responsiveness
- Massive requirements for bureaucracy management and government regulations
- Cycle time variances due to understanding *tricks of the trade* or underlying business dynamics (Saha and De 2008)

As a result, there can be benefits in utilizing an alternate arrangement, a procurement service provider (PSP), as a means of entering a new area. While the purchasing department may be the focal point in the outsourcing movement, it is only a part of the outsourcing program.

## TCO Considerations

Offshore outsourcing intensified the search for a way to determine the total cost of outsourcing (TCO) of the purchased good or service. The TCO concept is not new, with origins as far back as pre–World War II. However, there is not a universal model that is common to all companies that use the TCO approach. In a comprehensive survey of the ISM members, researchers found the types of product purchased under TCO valuation methods included capital goods (79% of respondents), manufactured parts (46%), raw materials (38%), services (29%), maintenance, repair, and operating supplies—MRO (26%), subassemblies (25%), and packaging (22%). They also found that companies used as many as 127 different factors. Consequently, they concluded that while the use of TCO analysis is worthwhile, a standard TCO model does not exist, and probably never will. In fact, even within a firm, about 85% of the respondents reported they found moderate-to-high variation among the factors they used to make their TCO evaluation. The researchers found the following functions had moderate-to-heavy involvement

**TABLE 8.6**

Stages of Purchasing Sophistication

| Key Performance Criteria | Purchasing Management<br>Low Complexity of Supply Market, Low Importance of Purchasing | Materials Management<br>Low Complexity of Supply Market, High Importance of Purchasing | Sourcing Management<br>High Complexity of Supply Market, Low Importance of Purchasing | Supply Management<br>High Complexity of Supply Market, High Importance of Purchasing |
|---|---|---|---|---|
| Procurement focus | Noncritical items | Leverage items | Bottleneck items | Strategic items |
| Key performance criteria | Functional efficiency | Cost/price and materials flow | Cost and reliable sourcing | Long-term availability |
| Typical sources | Established local suppliers | Multiple suppliers, chiefly local | Global, new suppliers | Established global suppliers |
| Time horizon | Limited: normally 12 months or less | Varied, typically 12–24 months | Variable, depending on availability | Up to 10 years |
| Items purchased | Commodities | Commodities, specified materials | Mainly specified materials | Scarce and high-value materials |
| Supply | Abundant | Abundant | Production-based scarcity | Natural scarcity |
| Decision authority | Decentralized | Mainly decentralized | Centrally coordinated | Centralized |

*Source:* Adapted from Kraljic, P., *Harv. Bus. Rev.*, 61(9), 109, September 1983.

*Notes:* Complexity of supply market criteria: availability of supply, pace of technological advance, entry barriers, logistics cost, and complexity.

Importance of purchasing criteria: cost of materials/total costs, value-added profile, profitability profile, risk management, strategic importance.

in the TCO valuation process: purchasing (88% of respondents), manufacturing (69%), design engineering (67%), logistics (58%), information technology (52%), accounting (41%), and marketing (37%). This heavy, and widespread, cross-functional involvement illustrates the importance companies associate with the outsourcing decision (Ferrin and Plank 2002).

Most companies have multiple supply chains. Consequently, it would be useful to develop a cost model to consistently evaluate the outsourcing decision. Cost modeling can produce information to help supply chain managers make sound business choices and reduce procurement total costs.

A new focus study examines how optimization techniques can help in sourcing activities. "Analyzing multiple scenarios gets the buyer closer to the 'sourcing sweet spot'—that is, those sourcing decisions that balance the needs of internal users with supply market realities at competitive prices" (Donnalley 2009, p. 32).

## As a Project, with Project Management Needs

The make-or-buy decision had limited effect on the total operations of a company and rarely was of sufficient importance to affect the supply chain. On the other hand, outsourcing, especially offshore outsourcing, is definitely of sufficient magnitude and complexity to require a project management approach (Murray and Crandall 2006).

## Other Considerations: Intangible Costs and Public Acceptance

TCO analysis attempts to identify tangible costs for outsourcing decisions. Even with an in-depth analysis, there are other factors to consider that require judgment on the part of decision makers. One report cautioned executives to carefully evaluate the importance of speed, the availability of skilled talent, the potential for further productivity gains in Asia, one-time transition costs, the local import and tax implications, and organizational interfaces (Ajay et al. 2008).

One of the most intriguing considerations is the extent to which the decision affects the public perception of the outsourcing firm. A number of companies have been criticized or boycotted because of their offshore outsourcing activities, particularly in those geographic areas hard hit with job losses.

Intangible costs also need to be examined with respect to globalization trends that are occurring in the business world. The last two decades in India and China illustrate the fact that increased globalization and global outsourcing have resulted in a more complex business environment and can create crisis events in organizations (Crandall et al. 2013). As Western firms rely more heavily on their Asian counterparts, they will be affected by political or cultural unrest in those countries. In addition, a number of American firms, including the retail giant, Wal-Mart, have encountered crisis events of their own such as negative publicity and boycotts resulting from their ties to countries where *cheap labor* is more prevalent. However, not all press reporting has been negative. Jason Furman of New York University notes Wal-Mart's economic benefits cannot be ignored, as the retailer saves its customers an estimated $200 billion or more on food and other items every year (Mallaby 2005).

Intangible costs ultimately result from offshoring or offshore outsourcing. Evidence suggests globalization may contribute to organizational and even domestic industry crises in the long term. The growing Chinese automobile market will likely lead to upheaval in the U.S. auto industry, and perhaps to those in Britain, Germany, and Japan as well.

Virtually every leading global automaker has ventured to China and entered into partnerships with one or more Chinese automakers. The Chinese companies benefit from the experienced automakers' knowledge. In addition, the visiting automakers can take advantage of the lowered labor costs. In the United States, however, labor unrest, particularly from unions, questions the long-term benefits these relationships can bring (Crandall et al. 2013). In recent months, the major participants in the automobile industry have indicated a greater desire to locate their supplier network closer to the final assembly locations. They have also indicated a greater interest in locating their final assembly plants nearer the market to better match supply with demand.

### Supplier Location as a Strategy for Entering an Offshore Market

Most companies view offshore outsourcing as a way to reduce their product and operating costs. A secondary reason is to gain experience in a foreign country for establishing a manufacturing or retail presence, either by direct investment or through some sort of alliance with an already existing company. Aircraft manufacturers such as Boeing, and virtually every major automaker in the United States, Japan, and Europe, are using this approach in gaining an entry into China. While companies may first enter a foreign country to buy, they often stay to sell. For this reason, the market potential is an important consideration, even in the initial offshore outsourcing decision. A survey by the Aberdeen Group found cost savings on procured goods was the primary driver for LCCS, with penetration of new markets as the second most cited driver (Fitzgerald 2005).

---

## Performance Measurement

Just as the scope of purchasing has changed over recent decades, so has the way in which purchasing performance is measured.

### Traditional: Positive Purchase Price Variance

Under the traditional view of purchasing, with its emphasis on the cost of purchased materials, one of the most popular measures was purchase price variance, or the difference between the budgeted, or standard cost of an item, and its actual cost. Often, the buyer purchased large volumes to receive the price discounts. Consequently, buyers might have been less diligent about the quality of the materials in order to buy from a low bidder.

### Contemporary: Enhanced Value for the Consumer

In the contemporary purchasing environment, performance measures are more global in nature, with the ultimate measure being the level of enhanced value from the supply chain for the consumer. Inasmuch as determining this value for an entire supply chain is difficult, if not impossible, companies are still looking to develop measures that move from cost to value criteria.

In speaking of measuring performance improvement, one study suggests the use of such traditional measures as inventory turns, on-time delivery, supplier responsiveness, product quality, purchase prices, and total cost (Petersen et al. 2005). Another measure that is gaining favor is the cash-to-cash cycle time.

One forecast for purchasing and supply predicted "there will be an intellectual fight over designing metrics that are very specific for particular chains" (Carter et al. 2000). They doubt there will be a standard set of metrics that can be used throughout the supply chain. Others advocate the use of economic value added (EVA) as a measure, because of its recognition of the contribution of revenue, operating costs, and capital investment (Monczka and Morgan 2000). A system designed for the health-care industry suggested a system built on the balanced scorecard concept that included the specific measures in each of the following areas: customer, supplier, process, IT system, learning and growth, as well as a category labeled *overall* (Kumar et al. 2005).

## Future of Purchasing

What does the future hold? Purchasing is becoming increasingly important, because of increased levels of outsourcing, increased use of the Internet, greater emphasis on supply chain management, globalization, and continuing efforts to reduce costs and increase quality (Joyce 2006).

The introduction of new information technologies affects the way purchasing does its job. Although most companies want to build long-term relationships with suppliers for their critical and mainstream materials and services, they may want to take advantage of electronic auctions for spot buys or some other nonrecurring purchases. A spot buy is a purchase made for standard off-the-shelf material or equipment, on a one-time basis (Blackstone 2013).

There is an increasing need for connective links among supply chain members and the role of purchasing in achieving these links. This increasing contribution by purchasing is a profound indicator of the major changes still to be seen (Andersen and Christensen 2005).

Purchasing has coming challenges in the following areas: electronic commerce, strategic cost management, strategic sourcing, supply chain partner selection and contribution, tactical purchasing, purchasing strategy development, demand-pull purchasing, relationship management, performance measurement, process uncoupling, global supplier development, third-party purchasing, virtual supply chain source development, competitive bidding and negotiations, strategic supplier alliances, negotiation strategy, and complexity management. PSM professionals will require greater *general management* (i.e., interdisciplinary) training than they have had in the past (Carter et al. 2000).

A Delphi study of key procurement and supply management executives identified procurement and supply management strategies that could lead to significant improvements over the coming years or decade. The study concluded, "Increased integration, information sharing and collaboration among supply chain members are most likely to be implemented and will have the largest impact on organizations" (Ogden et al. 2005).

Another study looked at the issue of ethical behavior, especially as it relates to conflict of interest in supply chains. The study concluded, "Companies no longer have the luxury of ignoring the importance of integrating ethical considerations into decision making, with particular focus on supply chain management" (Handfield and Baumer 2006, p. 48).

A study by the Aberdeen Group found industry leaders are working to transform procurement's role from one of cost containment to one of value generation (Minahan 2005). They recommend the following:

- Improve supplier development and collaboration.
- Enhance and integrate procurement automation infrastructure.
- Adopt low-cost country supply initiatives.
- Transition to a center-led procurement organization.
- Increase the amount of expenditure under procurement management while improving compliance.

Organizations are encouraged to recognize the need to use *sustainability purchasing* to enhance their position in the marketplace. Companies should "carry a wide range of product and services that encourage health and well-being, support healthy and productive jobs, and reward the operations of responsible businesses" (Reeve and Steinhausen 2007). Although desirable, developing a reliable measure of sustainability is still a challenge for most companies. Without widespread acceptance of sustainability measures, most purchasing organizations are having difficulty incorporating these requirements in their buying procedures.

While performance measures across the entire supply chain are elusive, work is being done to make them available. The CAPS Research organization reviews the results of 20 cross industry benchmarks reported by 153 different organizations representing 12 different industry sectors. Some of these measures were in monetary units (dollars) such as average purchase order process cost; others were physical measures, such as average cycle time (days) from requisition approval to PO placement, while many were in percentages, such as supply management operating expense as a percent of spend (CAPS Research 2012). No doubt, these measures will be refined in the future.

Current B2B e-procurement practices are the focus of recent research efforts. After a study of success factors and challenges, researchers concluded that E-procurement is a very important initiative with significant cost savings potential for firms (Angeles and Nath 2007).

Research provides solid evidence that in both manufacturing and services, today's CPOs have greater responsibilities, report higher in the organization, and carry more significant titles than their predecessors. The conclusion is that, at least in large North American companies, supply has grown substantially in corporate status and influence since 1987, a particularly welcome discovery (Johnson et al. 2006).

The future for purchasing professionals is an exciting challenge for those who want it. More than ever before, supply chain management and purchasing professionals will have a breadth of experience in a company unmatched by any other functional area. This realization should prepare them for positions in general management and even consideration for selection as a chief executive officer (CEO).

---

## Summary

The traditional role of purchasing was to obtain goods and services for the organization by using a bid system. Typically, potential suppliers were given *a list* of items to be purchased. The suppliers would provide their price for that item to the purchasing agent, who would select the lowest bid for each item and purchase from the suppliers accordingly.

What is the present status of purchasing? Although there are certainly differences in purchasing practices among companies, some general trends seem to be appearing:

- Companies are making progress in building stronger, and more lasting, relationships with their suppliers. With the active involvement of purchasing, organizations are reducing the number of suppliers as the result of more carefully selecting and working with those suppliers that remain.

- Purchasing is accepted as a more important member of management, not just because of its direct contribution in the purchasing process but also as a member of cross-functional decision-making teams, at least at the tactical level.

- With the emergence of supply chains and offshore outsourcing, the purchasing function is becoming a core competency for successful companies. It has a vastly expanded and more critical role, both internally within a company and as a key integrator with suppliers.

- The trend toward global sourcing shows little indication of reversing itself. While there is resistance from some influential stakeholders such as labor unions, there does not seem to be sufficient opposition to prevent companies from establishing a presence in many foreign countries.

- The transition from traditional purchasing to contemporary purchasing is slow. While procurement excellence offers competitive advantage, one of the biggest roadblocks is the built-in resistance to change (Quinn 2005).

## Hot Topic: Apple Juice

One of the requirements of a good purchasing organization is that their suppliers will not withhold information about the products that could be detrimental to the purchasing organization. The hot topic for this chapter describes such a surprise.

### Hot Topics in the Supply Chain: Why Apple Juice Should Be Made from ... Apples

The procurement of supplies and raw materials is a fundamental part of supply chain management. However, it is fraught with a number of challenges, one of which is assessing the integrity of the supplier. For example, imagine you are a manufacturer and distributor of apple juice and one day you find out that your main supplier of apple juice concentrate has been providing your company with a product that does not contain apples. The astute reader might think that this could hardly happen, as apple juice concentrate must be made from primarily, apples, right? Precisely. But the Beech-Nut Nutrition Corporation found itself in this exact situation in 1977. In fact, the Beech-Nut case has remained a classic for those who study business ethics as well as supply chain issues.

#### BACKGROUND

The story begins rather innocently during the late 1970s, when Beech-Nut discovered it was the victim of an apple juice concentrate scam. Its supplier at the time, Universal Juice, Inc., had been selected for cost reasons. However, there were two warning signs that signaled this supplier was suspicious. First, Universal was offering its product at 20% below the regular market price (Bad Apples 1989). This fact alone should have been cause for further research on the part of Beech-Nut since component products that are far below the market price are practically nonexistent. The second warning sign was even more surreal: Beech-Nut was forbidden to tour the production facilities at Universal (Hartley 1993). Nonetheless, Beech-Nut used its supplier because the cost advantages were enticing.

The warning signs concerning Universal Juice, Inc. were valid. Beech-Nut's then director of Research and Development, Jerome LiCari, conducted his own research on the apple juice concentrate and confirmed that it was nothing more than corn and cane sugar syrup with apple flavoring added to it. In short, there was no apple juice in the apple juice concentrate. Unfortunately, Beech-Nut advertised its apple juice as "100% fruit juice, no sugar added," a claim that was false.

## THE AFTERMATH

Faced with this ethical quagmire, Beech-Nut could have revealed the truth about its supplier, pleaded ignorance, and most likely escaped any prosecution because it was an innocent victim (Bad Apples 1989; Hartley 1993). However, rather than ceasing to conduct business with Universal, Beech-Nut chose to continue its relationship with its supplier. From 1977 to 1983, Beech-Nut sold its juice as 100% pure when, in fact, it was nothing more than a "100% fraudulent chemical cocktail," according to an investigator close to the case (Welles 1988, p. 124).

The decision to not change suppliers became an ethical sourcing crisis. Beech-Nut's president, Neils Hoyvald, and vice president for operations, John Lavery, were the main executives who instigated the cover-up. When the case went to court, both men were found guilty of violating federal food and drug laws (Crandall et al. 2014). Management researcher, Robert Hartley (1993), estimated this supply chain cover-up cost Beech-Nut $25 million in fines, legal costs, and lost sales.

So why revisit a case that occurred over three decades ago? Because the integrity of suppliers is still at stake, only today, those suppliers may be far off in a different culture on another continent. Today more than ever, suppliers must be assessed and monitored closely, with the hope that another bad apple will not be detected in the supply chain.

## QUESTIONS FOR RESEARCH AND DISCUSSION

1. For the industry in which you work:
   a. What are the protocols for visiting a supplier?
   b. At what percentage below market price does a component part in the supply chain appear to be "suspicious"?
   c. What examples of component fraud have you seen?
2. Jerome LiCari, the research and development director discussed earlier, tried unsuccessfully to convince upper management at Beech-Nut to change suppliers. He was unsuccessful in his attempts and he eventually resigned, not wanting to work for a company that was breaking the law.

   At your place of business, would you be able to successfully convince upper management to change its supplier if it were discovered a supplier was selling a bogus product, but at a desirable low price? If so, what would your strategy be in convincing management to change suppliers?

## REFERENCES

Bad apples: In the executive suite, *Consumer Reports*, p. 294, May 1989.
Crandall, W.R., Parnell, J., and Spillan, J., *Crisis Management: Leading in the New Strategy Landscape*, 2nd edn., Sage Publishing, Thousand Oaks, CA, 2014.
Hartley, R., *Business Ethics: Violations of the Public Trust*, Wiley, New York, 1993.
Welles, C., What led Beech-Nut down the road to disgrace? *Business Week*, 124–128, February 22, 1988.

## Discussion Questions

1. Compare the traditional and contemporary roles of purchasing. What are the major differences between the two approaches?

2. Identify and describe the CSFs needed in the procurement process.

3. Describe the differences between the make-or-buy decision and the outsourcing decision.

4. Construct a vendor rating system (see Table 8.3) to help you in an important purchase that you need to make in the near future. Most students may find this useful in purchasing an item, such as automobile or a laptop computer.

5. What are the major functions that must be accomplished in the purchasing process?

6. Draw out the ordering cycle that takes place when your school's academic department needs to order paper for its copy machines. Note that you may need to consult with the school's departmental secretary or administrative assistance to help you in your answer.

7. How does the purchasing function change at different levels of the supply chain?

8. What potential problems can be encountered when a company chooses to outsource?

9. What are the intangible costs associated with offshore outsourcing?

10. What items would you consider in evaluating the TCO of a product that is completely outsourced to a company in China?

## References

Ajay, K.G., Nazgol, M., and Vats N.S., Time to rethink offshoring? *The McKinsey Quarterly*, 4, 108, 2008.
Andersen, P.H. and Christensen, P.R., Bridges over troubled water: Suppliers as connective nodes in global supply networks, *Journal of Business Research*, 58(9), 1261, 2005.
Angeles, R. and Nath, R., Business-to-business e-procurement: Success factors and challenges to implementation, *Supply Chain Management*, 12(2), 104, 2007.
Anonymous, Strategic sourcing, *Strategic Direction*, 21(11), 29, 2005.
Austin, L., CPFIM, C.P.M. Supply chain consultant, personal interview, June 2008.
Bernstein, M., Raytheon goes from traditional purchasing to an integrated supply chain, *World Trade*, 18(11), 36, 2005.

Blackstone, J.H., *APICS Dictionary*, 14th edn., APICS—The Association for Operations Management, Chicago, IL, 2013.

CAPS Research, Summary review: Cross-industry report of standard benchmarks 2012, https://knowledge.capsresearch.org/publications/pdfs-protected/CI201207Metric.pdf.

Carbone, J., Buyers look for supply chain management services, *Purchasing*, 129, 7, 83, 2000.

Carter, P.L., Carter, J.R., Monczka, R.M., Slaight, T.H., and Swan, A., The future of purchasing and supply: A ten-year forecast, *Journal of Supply Chain Management*, 36(1), 14, 2000.

Crandall, R.E., Device or strategy? Exploring which role outsourcing should play, *APICS Magazine*, 15(7), 21, 2005.

Crandall, W.R., Parnell, J.A., and Spillan, J.E., *Crisis Management in the New Strategy Landscape*, 2nd edn., SAGE, Thousand Oaks, CA, 2013.

Donnalley, C., Optimization in strategic sourcing, *Inside Supply Management*, 20(7), 32–33, September 2009.

Ellram, L.M., Purchasing and supply management's participation in the target costing process, *Journal of Supply Chain Management*, 36(2), 39, 2000.

Ellram, L.M., Tate, W.L., and Billington, C., Services supply management: The next frontier for improved organizational performance, *California Management Review*, 49(4), 44, 2007.

Ferrin, B.G. and Plank, R.E., Total cost of ownership models: An exploratory study, *Journal of Supply Chain Management*, 38(3), 18, 2002.

Fitzgerald, K., A step-by-step approach to low-cost sourcing, *EDN*, p. 131, November 2005.

Flanagan, M., How retailers source apparel—Why what concerns buyers isn't always what concerns sellers—Management briefing: Key strategic sourcing topics, *Just-Style*, p. 17, January 2005.

Gerardo, R.A. and Spanyi, A., The CFO's best friend, *Strategic Finance*, pp. 25–30, December 2008.

Giunipero, L.C. and Pearcy, D.H., World-class purchasing skills: An empirical investigation, *Journal of Supply Chain Management*, 36(4), 4, 2000.

Goebel, D.J., Marshall, G.W., and Locander, W.B., Enhancing purchasing's strategic reputation: Evidence and recommendations for future research, *Journal of Supply Chain Management*, 39(2), 4, 2003.

Gundlach, G.T., Bolumole, Y.A., Eltantawy, R.A., and Frankel, R., The changing landscape of supply chain management, marketing channels of distribution, logistics and purchasing, *The Journal of Business & Industrial Marketing*, 21(7), 428, 2006.

Handfield, R.B. and Baumer, D.L., Managing conflict of interest issues in purchasing, *Journal of Supply Chain Management*, 42(3), 41, 2006.

Institute of Supply Management (ISM), www.ism.ws/CareerCenter/JobDescriptions, 2014, accessed January 15, 2014.

Johnson, P.F., Leenders, M.R., and Fearon, H.E., Supply's growing status and influence: A sixteen-year perspective, *Journal of Supply Chain Management*, 42(2), 33, 2006.

Joyce, W.B., Accounting, purchasing and supply chain management, *Supply Chain Management*, 11(3), 202, 2006.

Kocabasoglu, C. and Suresh, N.C., Strategic sourcing: An empirical investigation of the concept and its practices in U.S. manufacturing firms, *Journal of Supply Chain Management*, 42(2), 4, 2006.

Kraljic, P., Purchasing must become supply management, *Harvard Business Review*, 61(9), 109, September 1983.

Kumar, A., Ozdamar, L., and Ng, C.P., Procurement performance measurement system in the health care industry, *International Journal of Health Care Quality Assurance*, 18(2/3), 152, 2005.

Leenders, M.R., Fearon, H.E., Flynn, A.E., and Johnson, P.F., *Purchasing and Supply Management*, 12th edn., McGraw-Hill Irwin, Boston, MA, 2002.

Mallaby, S., Wal-Mart: A progressive dream company, really, *Fayetteville Observer*, p. 11A, November 29, 2005.

Minahan, T.A., 5 strategies for high-performance procurement, *Supply Chain Management Review*, 9(6), 46, 2005.

McCollum, B.D., How changing purchasing can change your business, *Production and Inventory Management Journal*, 42(2), 57, 2001.

Monczka, R.M. and Morgan, J.P., Why economic value add needs to be measured, *Purchasing*, 129(2), 77, 2000.

Murray, M.J. and Crandall, R.E., IT offshore outsourcing requires a project management approach, *SAM Advanced Management Journal*, 71(1), 4, 2006.

Nelson, D., Moody, P.E., and Stegner, J.R., The 10 procurement pitfalls, *Supply Chain Management Review*, 9(3), 38, 2005.

Nestlé History, http://www.nestle.com/aboutus/history, accessed January 15, 2014.

Ogden, J.A., Petersen, K.J., Carter, J.R., and Monczka, R.M., Supply management strategies for the future: A Delphi study, *Journal of Supply Chain Management*, 41(3), 29, 2005.

Petersen, K.J., Ragatz, G.L., and Monczka, R.M., An examination of collaborative planning effectiveness and supply chain performance, *Journal of Supply Chain Management*, 41(2), 14, 2005.

Quinn, F.J., The power of procurement, *Supply Chain Management Review*, 9(9), 6, 2005.

Reeve, T. and Steinhausen, J., Sustainable suppliers, sustainable markets, *CMA Management*, 81(2), 30, 2007.

Saha, R. and De, S., A blueprint for LCCS start-ups, *Inside Supply Management*, 19(5), 14, 2008.

Siegfried, M., A negotiating transformation, *Inside Supply Management*, 24(7), 26–29, 2013.

Vonada, K., Update old-school SRM, *Inside Supply Management*, 19(5), 48, 2008.

Walker, W.T., SPRINT your new products to market, *APICS—The Performance Advantage*, pp. 48–51, March 1992.

Walker, W.T., Rethinking the reverse supply chain, *Supply Chain Management Review*, 4(2), 52–59, May/June 2000.

Walker, W.T., Supply chain consultant, personal interview, March 23, 2009.

Wilson, D.R., Bergfors, M., and Adams, R., Magic quadrant for strategic sourcing application suites, Gartner, July 1, 2013, www.gartner.com/technology/reprints, accessed July 11, 2013.

Zubko, N., Sources of strength, *Industry Week*, 257(4), 48, 2008, www.industryweek.com/ReadArticle.aspx?ArticleID=15927.

# 9

# Logistics: The Glue That Holds the Supply Chain Together

## Learning Outcomes

After reading this chapter, you should be able to

- Describe the logistics function and its role in supply chains
- Describe the evolution of integrated logistics
- Identify the various modes of transportation used in supply chains
- Explain how offshore outsourcing has changed the role of logistics providers
- Explain what a third-party logistic (3PL) provider is
- Describe the services provided by 3PLs
- Explain what the Reshoring Initiative is
- Clarify the difference between offshore outsourcing and reshoring
- Identify and describe the various steps involved in the decision to outsource
- Discuss how logistics helps link supply chain participants together
- Identify the various high-tech industry issues present

## Company Profile: Transportation Insight

Logistics is the term used to describe the moving of goods from point to point along the supply chain. Logistics providers not only move the product, they are often the ones who plan how and when to move the product. Moving goods from foreign countries to distribution centers or retail stores usually involves different modes of transportation, different regulations and fees for each country, and potential delays from overloaded ports, weather disruptions, and other causes. As supply chains are extended throughout the world and become more complex, the logistics function is emerging as a critical part of supply chain management (SCM). As you read this company profile and the chapter in the book, think about these questions:

- How would you move bars of soap from the factory to the grocery store?
- How would you move a locomotive from the factory to the railroad?

- What are the major costs of moving products? How do you estimate them?
- Should a company outsource the logistics function? Why or why not?

What additional questions do you think relate to logistics?

<hr>

**Company Profile: Transportation Insight**

Chapter 9 is about logistics. In some cases, products are moved from one remote location to an even more remote location. As supply chains become global in scope and complexity, more companies are turning to third-party logistics providers (3PLs) to help them find better ways to move products from one location to another. We will highlight a company—Transportation Insight (TI)—that has grown rapidly as a 3PL.

## HISTORY

TI was started in 1999 by Paul Thompson, an industrial engineering graduate from the University of Arkansas, who had almost two decades of experience with one of the largest motor carriers in the United States. His timing in entering the 3PL business was impeccable as the company became one of the most innovative and fastest growing in its category of a non-asset-based 3PL. At the end of 2013, the company had revenues of over $600 million and employed over 200 employees with a wide range of professional education and industry experience in logistics and related fields (Transportation Insight 2013b). In 2013, it was ranked No. 17, moving up from 23rd in 2012, in the list of top private companies in North Carolina by *Business North Carolina Magazine* (Lacour 2013).

The company is headquartered in Hickory, North Carolina, with secondary offices in Bentonville, Arkansas (home of Walmart), and Charlotte, North Carolina. They have more than 40 client support offices across North America. TI serves over 300 clients who have over 400,000 shipping and receiving locations (Transportation Insight 2013b). In 2013, they received national recognition as the first fourth-party logistics provider (4PL) asked to join the International Warehousing and Logistics Association (IWLA) and for their continuous improvement methodology, Extended LEAN™ (Transportation Insight 2013d).

TI has achieved recognition in the 3PL field. Global Trade magazine selected the company as one of 10 3PLs with the best range of transportation services. Other companies in the top 10 included UPS Supply Chain Solutions, FedEx Supply Chain, and BNSF Logistics (Market Watch 2014).

In 2013, *Food Logistics* magazine named TI as a top 100 software and technology solution provider to the food and beverage industry. This award is in recognition of the company's cloud-based and scalable systems that offer blended technology. "Transportations Insight's customizable logistics technology solutions are designed to meet supply chain challenges from the loading dock to the C-suite, adding value throughout the enterprise" (kctv5 2013).

The company has been repeatedly honored by Inc. Magazine for 6 years in a row as a member of the 500|5000 group of companies achieving significant and continuous growth. After winning the recognition 5 years in a row, TI earned a spot on the Inc. honor roll. As a part of the list, the company joined a group of major companies such as Microsoft, Facebook, Oracle, Levi Strauss, and Zappos (Globe Newswire 2012).

TI has also been honored with a green supply chain award by *Supply & Demand Chain Executive Magazine*. This award recognizes companies making sustainability a core part of their supply chain strategy. In receiving the award, Paul Thompson cited the use of

Insight TMS® as a key factor in optimizing transportation modes and carrier selection so that energy costs, mileage, and fuel usage are reduced (Hawaii News Now 2013).

## ABOUT 3PLs

As we have described in earlier chapters of this book, supply chains are systems that move products from the point of origin, mine or farm, to the ultimate consumer. In Chapter 9, we describe in more detail the logistics activities that make it possible to move products around the world in an effective and efficient manner. While retailers, distributors, manufacturers, growers, and miners are participants in supply chains, often they are not the businesses that actually move the product. That is left up to the logistics companies, such as railroads, trucking companies, ocean-going shippers, airlines, and pipelines. These are the asset-based logistics companies who are clearly visible and essential to the logistics activities.

However, there is another group of professional companies who are less visible but equally essential to logistics—the non-asset-based 3PLs. These are the companies that figure out the best way to move the goods and perform many of the administrative functions required to move products between carriers, between countries, between companies, and over long distances. While the consumer may read the label and see that a dress was made in a foreign country, they may never think about all of the steps required to move that dress from its point of manufacture to the local clothing store.

Many 3PLs grew out of the asset-based side of their business—trucking, railroad, or package carrier. Other 3PLs were started by professionals who understood how to manage the exchange points—the steps when a product had to change from one mode of transportation to another, go through a customs check point, or be detoured around a bottleneck in the supply chain, to name a few.

3PLs can be classified as brokers, freight forwarders, warehouse 3PL, and enterprise 3PL. The first three groups provide specialized services, while the enterprise 3PL provides a full range of logistics services. TI is an enterprise 3PL and expands on that classification by considering itself as a Co-managed Enterprise 3PL. Through their Co-managed Logistics™ program, they become a consultative strategic partner to the client and in a position to provide whatever 3PL services the client may need and want.

## TRANSPORTATION INSIGHT SERVICES

TI is a full-service 3PL—it can perform all of the services normally expected of a 3PL. Using their Co-managed Logistics approach, it can provide core services in the logistics area by investigating to spot problems, analyzing to identify opportunities, and recommending a plan that fits their client's needs and capabilities. Some of the specific services they offer in this area include service needs assessment, carrier sourcing, contract and rate administration, claims processing, outsourced execution options, warehouse sourcing, international transportation, and continuous improvement of processes.

Another core service is in the area of financial settlement where the company can recommend some of the following: freight bill audit, consolidated electronic billing, general ledger (GL) coding, carrier payment, freight refunds for service provider failures, and freight accruals. Where the client so desires, the company can take over the actual administration of some of these functions.

TI also has the capability to perform more advanced services through the use of proprietary software. Their Insight TMS enables them to optimize transportation routes, monitor shipment execution, and provide visibility and collaboration capability to their clients. Their system can be integrated with an ERP system such as SAP or Oracle or used as a stand-alone web-based application. This system offers shipment notification and tracking and automatic transportation price quotes to customer service and sales, supplier shipment routing and supplier shipment notification to purchasing, and enterprise and divisional reporting along to accounting and finance (Transportation Insight 2012b).

One of the strengths of their information systems is the capability to provide beneficial reporting for all areas of a company. Some of the reports include

- *Executive leadership*: Transportation savings reports, transportation margin reports, transportation profitability reports, revenue per month, and plan versus actual
- *Finance*: Freight payment metrics, accrual reports, and costs by GL code
- *Logistics*: Aggregation opportunities, least cost report, and carrier scorecards
- *Information technology*: Invoice detail report and analyst detail
- *Customer service, sales, marketing*: Freight cost by customer or customer location, freight margin by customer
- *Purchasing*: Freight cost by vendor or vendor location and vendor compliance

TI can maintain a centrally managed repository of a client's logistics data and related company information. With this, the reporting system uses web-based technology to provide comprehensive supply chain reporting, interactive and dynamic analytics, and customized logistics engineering studies (Transportation Insight 2013c).

TI possesses global trade expertise to help companies move products around the world. They can provide freight forwarding services of international shipments. Available services include support and recommendations regarding most cost-effective mode options, finding specific ports that best fit the company's needs, documentation, inland transit options, and compliance management. Other international services include providing total supply chain visibility through their transportation management system (TMS), import/export compliance, security compliance, and custom house brokerage.

The Extended LEAN initiative is one of the company's most unique and valuable programs. It is designed to help companies incorporate the lean manufacturing concept along their supply chains, not just in individual organizations. This approach aligns all of the companies along the supply chain, from raw material production, to manufacturing, to distribution, to point of sale and all logistics activities between these facilities. At TI, the program is led by Eric Lail, Vice President of Total Insight LEAN Performance Solutions, who is one of only 200 Shingo Prize examiners in the world.

One of the newest services is in supply chain analytics, which involves taking traditional reporting and combining it with contemporary data analytics to provide not only insight about the past and current performances but also a look at where a company should go in the future. Figure 9.1 shows the progression from static reporting to strategic modeling.

With all of these services available, TI can offer services ranging from the most basic to the most sophisticated, depending on the client's needs.

## SUCCESS STORIES

The following are examples of the type of companies TI has worked with and how they have helped them achieve positive results.

A national manufacturer of air filtration products was incurring frequent product damages across their transportation network and did not have an effective way to

Customized, insightful reporting to help clients measure progress and manage performance

**FIGURE 9.1**
Supply chain analytics. (From Gentry, C., The inner workings of a 3PL, a presentation to the Foothills Chapter of APICS, Hickory, NC, September 2013. With permission.)

process freight claims. TI analyzed the claims filings and claims-handling process and recommended preferred carriers and a carrier bid process. They worked with carriers to fully define the product characteristics, handling requirements, and existing damage problems. As a result, damages declined and the claims were processed faster and with less cost (Transportation Insight 2012a).

A health and beauty aid supplier was experiencing high costs associated with shipping from their distribution centers to retail customers. With the implementation of Insight TMS, TI developed optimized routes, using a combination of less-than-truckload (LTL) and parcel delivery, which reduced rates and overcharges. The new systems also provided custom reports that enhanced their supply chain visibility and control (Transportation Insight 2012c).

A heavy equipment rental company had a complex shipping network that included thousands of moves between more than 1500 branch locations and hundreds of original equipment manufacturers. Through their Co-managed Logistics approach, TI developed a plan in which the client had visibility of all inbound shipments, but TI made all carrier decisions on those shipments. The result was a multimillion dollar reduction in freight costs (Transportation Insight 2013a).

## BENEFITS

The obvious benefit for clients of TI is a supply chain that moves products faster and at a lower cost. However, the president of the company, Chris Baltz, offers a loftier goal: "To meet the rapidly changing needs of 400 clients across North America, our team continues to raise the bar in the logistics technology space. Our primary objective is to develop solutions that help our clients maintain end-to-end supply chain visibility and continuously improve their supply chain value streams. The innovative tools we deliver help our clients grow faster than the competition" (kctv5 2013).

## CONCLUSIONS

TI is a successful company. Their line of services fits well with the needs of many businesses that have extended and complex supply chains. The company has built a strong organization of knowledgeable people. Reducing logistics costs is one of the prime objectives of most supply chains. The top three executives in TI all have industrial engineering degrees; they are naturals to build and manage a company to seek out cost reduction opportunities.

Paul Thompson views the rapid growth of the company he founded, even during the economic slowdown of recent years, this way: "As we saw the economy collapsing in 2008, we decided not to change our course. We made all the previously planned investments in people and technology. Our ability to offer innovative solutions and process improvements has helped our clients to progress and grow faster than the economy. What really makes TI different is our people. Business models can be replicated, but our strength and core advantage is that our people execute better than the competition" (Globe Newswire 2012).

TI has been successful financially. They have been recognized for their efforts in sustainability. They have also been a socially responsible company. They are a sponsor of Samaritan's Feet, 10,000 Homes, and the National Foundation on Fitness, Sports and Nutrition (Transportation Insight 2013b). In addition, the company contributed one million dollars in support of the Transportation Insight Center for Entrepreneurship at Appalachian State University in Boone, North Carolina, one of the 16 universities in the University of North Carolina system.

It is somewhat ironic that the more successful a 3PL company is, the less the public hears about them. If products are moving smoothly through the supply chain, consumers take it for granted and rarely know who is making that smooth flow possible. It is only when there are problems that 3PL companies may share the spotlight with the manufacturer or retailer. Although the individual consumer may not know about TI, supply chain professionals do.

## REFERENCES

Appalachian State University, Transportation Insight Center for Entrepreneurship, Boone, NC, www.entrepreneurship.appstate.edu, accessed February 5, 2014.

Gentry, C., The inner workings of a 3PL, a presentation to the Foothills Chapter of APICS, Hickory, NC, September 2013.

Globe Newswire, Transportation Insight recognized five years straight on the Inc. 500|5000, 2012, http://globenewswire.com/news-release/2012/09/04/488785/10003980/en/Transportation-Insight-Recognized-Five-Years-Straight-on-the-Inc-500-5000.html, accessed February 5, 2014.

Hawaii News Now, Transportation Insight honored with Green Supply Chain Award, 2013, http://www.hawaiinewsnow.com/story/23998112/transportation-insight-honored-with-green-supply-chain-award, accessed February 5, 2014.

Kctv5, Transportation Insight Ranked as top 100 logistics technology provider by Food Logistics Magazine, 2013, www.kctv5.com/story/24243299/, accessed February 5, 2014.

Lacour, G., Laying it online, How the top private companies in the state are opening up commerce on the Internet, *Business North Carolina Magazine*, pp. 39–45, October 2013.

Market Watch, Transportation Insight among America's top 10 3PLs with best range of transportation services, http://www.marketwatch.com/story/transportation-insight-among-americas-top-10-3pls-with-best-range-of-transportation-services-2013-11-12, accessed February 5, 2014.

Transportation Insight, Increased control improves freight flow, air filter manufacturer reduces damage claims, 2012a.

Transportation Insight, Insight TMS®, The most versatile transportation management system on the market today, 2012b.

Transportation Insight, Supply chain makeover, health and beauty aid supplier reduces costs and improves operational efficiency, 2012c.

Transportation Insight, Customized solution drives continuous improvement of complex transportation network, world's largest equipment rental company saves millions, 2013a.

Transportation Insight, Fact sheet, 2013b.

Transportation Insight, If you can measure it, you can manage it, game-changing insight for complex business decisions, 2013c.

Transportation Insight, Transportation Insight honored as top 100 3PL by Inbound Logistics Magazine, 2013d.

## Scope of Logistics

The logistics function links multiple participants of a supply chain together. Its geographical sphere of influence has grown from being locally oriented to one of global orientation. While its primary objective is to facilitate the flow of goods and information along the supply chain, it is becoming an integrating force—the glue that holds the supply chain together.

Logistics is defined as follows: (1) "In an industrial context, the art and science of obtaining, producing, and distributing material and product in the proper place and in proper quantities; (2) In a military sense (where it has greater usage), its meaning and also include the movement of personnel" (Blackstone 2013). A definition sponsored by the Council Supply Chain Management Professionals (CSCMP) is the following: "Logistics: The process of planning, implementing, and controlling procedures for the efficient and effective transportation and storage of goods including services, and related information from the point of origin to the point of consumption for the purpose of conforming to customer requirements. This definition includes inbound, outbound, internal, and external movements" (Vitasek 2006).

Logistics management is a broader term and is defined by CSCMP: "Logistics management is that part of supply chain management that plans, implements, and controls the efficient, effective forward and reverse flow and storage of goods, services, and related information between the point of origin and the point of consumption in order to meet customers' requirements. Logistics management activities typically include inbound and outbound transportation management, fleet management, warehousing, materials handling, order fulfillment, logistics network design, inventory management, supply/demand planning, and management of third party logistics services providers. To varying degrees, the logistics function also includes sourcing and procurement, production planning and scheduling, packaging and assembly, and customer service. It is involved in all levels of planning and execution—strategic, operational, and tactical. Logistics management is an integrating function which coordinates and optimizes all logistics activities, as well as integrates logistics activities with other functions, including marketing, sales, manufacturing, finance, and information technology" (Vitasek 2006).

The military is generally credited with coining the term logistics, to describe efforts to move equipment, supplies, and troops to the area needed. Coordinating the flow of goods to remote locations was the challenge for many armies, and the failure to move support goods for the troops often led to defeat. Even today, the movement of assets (a general term to include whatever is needed) is a key element of military strategy and tactics.

The earlier definitions provide a broad scope for the logistics functions. Its scope spans the entire supply chain and there is no doubt it is an essential component of any supply chain. This importance is described more fully in comments from some of the leading writers about the role of logistics as a vital component of a successful supply chain.

> Logistics involves the management of order processing, inventory, transportation, and the combination of warehousing, materials handling, and packaging, all integrated throughout a network of facilities. The goal of logistics is to support procurement, manufacturing, and customer accommodation supply chain operational requirements. Within a firm the challenge is to coordinate functional competency into an integrated supply chain focused on servicing customers. In the broader supply chain context, operational synchronization is essential with customers as well as material and service suppliers to link internal and external operations as one integrated process. (Bowersox et al. 2013, p. 29)

> The advent of collaboration, extended enterprise visioning, and the increased availability of integrated service providers combined to drive radically new supply chain solutions. The notion of shared and synergistic benefits served to solidify the importance of relationships between firms collaborating in a supply chain. The extended enterprise logic stimulated visions of increased efficiency, effectiveness, and relevance as a result of sharing information, planning, and operational specialization between supply chain participants. The deregulation of transportation served as a catalyst for the development of integrated service providers. This development served to redefine and expand the scope of specialized services available to facilitate supply chain operations. In combination, these three drivers helped create integrated supply chain management. They serve to identify and solidify the strategic benefits of integrated management. They combined to reinforce the value of core-competence specialization and cast the challenges and opportunity of creating virtual supply chain. (Bowersox et al. 2013, p. 19)

Logistics has also evolved from being thought of as merely a materials handling or a transportation function to one that has greater responsibilities in providing support services to companies. "The logistics of internationalization involves four significant differences in comparison to national or even regional operations. First, the distance of typical order-to-delivery operations is significantly longer in international as contrasted to domestic business. Second, to accommodate the laws and regulations of all governing bodies, the required documentation of business transactions is significantly more complex. Third, international logistics operations must be designed to deal with significant diversity in work practices and local operating environment. Fourth, accommodation of cultural variations in how consumers demand products and services is essential for successful logistical operations" (Bowersox et al. 2013, p. 26).

Historically, companies have tried to anticipate what customer demand would be and plan their production to meet that anticipated demand. The lack of good information made this an approach that could result in lost sales because of the failure to have the right product available when needed. It could also result in higher costs because of excess inventory that had to be marked down or scrapped because of lack of demand. A more desirable approach is to discover what the customer demand will be and plan to meet that demand consistently. Change from an anticipatory business model to a responsive business model is occurring because it is easier to get reliable information about what the ultimate consumer wants or needs (Bowersox et al. 2013, p. 11).

Some of the forces driving the rapid change in SCM include globalization, technology, organizational consolidation, empowered customers, and government policy and regulation (Coyle et al. 2013, p. 7).

## Evolution of Integrated Logistics

As supply chains grew in size and complexity, they morphed from a collection of individual and fragmented functions, such as demand forecasting, purchasing, production planning, materials handling, warehousing, transportation, and receiving, to combinations of the individual functions. These could be classified as inbound and outbound logistics. Finally, these two groups of activities were combined into an integrated logistics concept. Jim Tompkins describes integrated logistics as having three elements: materials management, material flow systems, and physical distribution. Integrated logistics starts at the origin—the farm or mine—and follows the product through every step of the process until it reaches the ultimate consumer (Tompkins 1996).

Figure 9.2 shows the components of a supply chain for an automobile manufacturer. The arrow at the bottom shows the logistics functions present from the origin of the supply chain to the ultimate consumer.

Figure 9.3 shows the development of the supply chain, from a series of independent functions within and between companies to an integrated system that moves products and services from origin to consumer.

Purchasing, production planning, and logistics were often separate functions within an organization, operating independently from each other. Purchasing was concerned with getting raw materials needed by the production people to make the products. The purchasing department was often, but not always, part of the total operations group that included other functions like production engineering, production, and quality. Production planning, part of the operations group, determined the schedule for producing the products, based on a forecast from marketing. The logistics group was concerned with getting the product to the customer in a timely manner and was closely aligned with marketing. Although these different departments were loosely connected, they sometimes had differing objectives that prevented a smooth flow of goods from input of raw materials to output of finished goods.

In order to smooth the flow of goods through the organization, the materials management concept was adopted by many manufacturing companies. The materials manager was charged with making sure the production schedule closely aligned with what the marketing forecast specified and that the purchasing organization was buying what the production group needed to make the product the marketing people wanted. Materials management was primarily concerned with the flow of goods within an organization and was the forerunner of SCM.

As companies became more concerned with the need to improve response times to customers, they built closer relationships with their customers. In order to improve

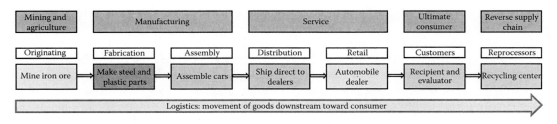

**FIGURE 9.2**
Integrated logistics in a supply chain.

Evolution of the supply chain
Progression is from the bottom of the chart to the top

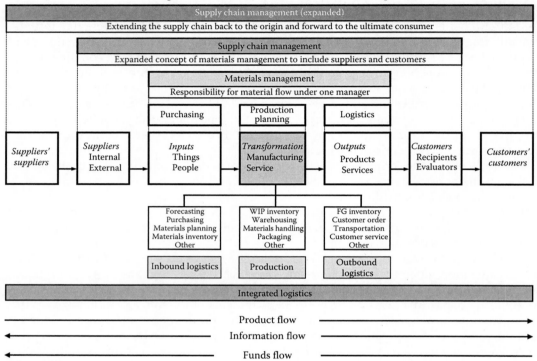

**FIGURE 9.3**
Evolution of the supply chain.

responsiveness and reduce variability in their deliveries, companies needed to build closer relationships with their suppliers. By linking with customers and suppliers, they created the supply chain. As time passed, companies found that extending their supply chains forward to their customers' customers and backward to their suppliers' suppliers, they were able to achieve even greater benefits.

As the supply chain model expanded and became an integrated system, so also did logistics move from independent functional areas to an integrated system. Figure 9.4 shows this evolution, using the manufacturer as the focal company. A number of logistics-type functions are performed in the input, transformation, and output areas of the company. These become grouped into inbound logistics, production logistics, and outbound logistics sets. As stronger links are forged between supply chain participants, the integrated logistics system emerges.

The logistics function was at one time closely associated with the marketing function. Early definitions of marketing included activities involved in the flow of goods from production to consumption. In the 1950s, the marketing discipline was divided between two research streams: business logistics and marketing channels. These two areas were integrated when the supply chain perspective appeared in the 1980s (Gripsrud et al. 2006). Figure 9.3 shows the outputs from an organization as aligned with the marketing function. This responsibility is included within the four Ps of marketing—product, price, promotion, and place. Figure 9.4 extends this association of marketing and logistics by

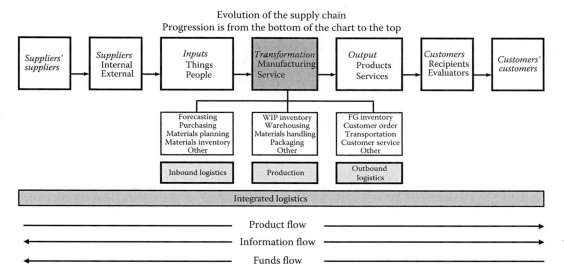

**FIGURE 9.4**
Evolution of integrated logistics.

connecting such activities as customer order and customer service within the outbound logistics sphere. Logistics becomes a means of encouraging collaboration among major participants in the supply chain to assure a smooth and economical flow of goods and services to the ultimate customer.

Integrated logistics includes the physical movement of goods (transportation), determining what and how much is to be moved (materials management), interim storage and processing, facilitating the flow of goods through exchange points (loading, unloading, and paperwork), providing product traceability, and integrating all activities along the supply chain.

## Transportation

Initially, the logistics function was concerned primarily with movement of goods, or transportation. It remains the major component of total logistics costs. Moving goods within a company is generally termed materials handling. A variety of methods are used, such as forklift trucks, conveyors, automatic guided vehicles (AGVs), and manual methods. Materials handling considerations were covered in Chapters 6 and 7. This chapter will be concerned with the movement of goods between companies. It is important to recognize that the information systems supporting the physical transportation systems are essential as supply chains have grown in distance spanned and intermodal complexity has increased.

There are five basic methods of moving conventional goods—truck, rail, water, air, and pipeline. Each has its own niche in the transportation arena, with cost and lead time being the most likely trade-off considerations. In some cases, the mode is seemingly more naturally aligned with the product, such as oil and pipelines; however, even here, there is a trade-off between cost and convenience.

In this increasingly electronic age, there are three other means of transportation that should be noted. Small package deliveries have become a unique transportation method to get fast, dependable delivery of small quantities, replacement parts, special deliveries, and a host of other items. Another unique delivery is the shipment of electrical energy over high-voltage transmission lines. Finally, the shipment of data over fiber-optic networks or wireless networks is an essential part of sending information to accompany products.

## Truck: Privately Owned or Third-Party Carriers

Deliveries by truck have become the primary mode of shipping goods domestically. The most obvious advantage is door-to-door convenience. Trucking represents a fast mode of transportation with good geographic coverage. These advantages are offset by the high costs; however, with shorter deliveries, costs do go down due to the direct costs of fuel. The interstate highway system and good side roads make it possible for trucks to pick up and deliver to almost any business or consumer. As the amount of business increases, it becomes more difficult for companies to find qualified and reliable drivers. The increase in fuel costs is causing some businesses to consider other modes of transportation, mainly rail.

Although trucking is by far the largest mode of transporting goods, it is not without its problems. The American Transportation Research Institute (ATRI), part of the American Trucking Associations Federation, is a nonprofit research trust. They list the following *Top 10* challenges for the trucking industry:

- Fuel costs (current and long term)
- Economy (pressed by increasing regulation, slumping demand, and excess capacity)
- Driver shortage and retention
- Government regulation (safety plus other local regulations)
- Hours of service
- Traffic congestion
- Tolls/highway funding
- Environmental issues
- Tort reform (civil judgments against trucking firms)
- Onboard truck technology (Outsourced Logistics 2008)

## Rail: For Selected Goods

The use of rail as a transportation mode is a medium-cost, medium-speed option, with limited geographic coverage. Since the latter third of the nineteenth century, railroads have been a major shipper, largely in heavier freight categories. The primary disadvantage is they lack the convenience of trucks. Only a few suppliers have a rail spur at the side of their facility; almost no modern retail stores do. Consequently, rail shipments usually require multiple loadings.

One of the alternatives is the use of intermodal freight transport, often called *piggybacking*, where truck trailers are loaded onto flat railcars for movement. This arrangement combines the load-carrying capability of the railroad with the convenience of the truck. Railroads may see an increase in their traffic as gasoline prices increase.

Trains are being used more extensively to move goods from Western China to Europe. As Chinese manufacturers moved west to access more affordable labor, they moved farther from the east coast of China, making the time to load onto ships longer. Companies are now moving the goods by train (21 days over almost 7,000 miles) instead of by container ship (40 days and over 12,000 miles) (Roberts et al. 2013).

## Waterways

Shipments by barge on inland waterways and cargo ships on the oceans play an important role. Using waterways as a mode of transportation represents a slow speed, low-cost option, but with obvious geographic limitations. Some cargos, such as crude oil and metallic ores, must be transported from foreign countries. Because of their heavy weight, the only practical way to transport them is by ship or barge. The other driver of increased shipping business is the increase in offshore outsourcing. As companies extend their supply chains into other countries, there will be an increased demand for shipping by sea.

## Air

Air freight for small and expensive goods is becoming more popular. Air is a high-cost, high-speed transportation mode and is offset with limited geographic coverage. Using air transport usually means coupling deliveries with trucks as well. However, it is unlikely heavier cargos will be shipped by air unless there is an emergency. At the present time, companies such as UPS and FedEx are the major cargo air carriers.

## Pipeline

Oil is the principal commodity shipped by pipeline. Where appropriate, pipelines are an economical, fast, and safe way to ship goods. However, these advantages are offset by the limited geographical coverage of using pipelines. At present, pipelines are used primarily for oil, natural gas, and chemicals. Although there have been proposals for carrying alternative goods, none have yet reached a significant stage of development.

## Parcel

In addition to the U.S. Postal Service, private carriers like UPS and FedEx have grown as carriers of lightweight parcels. All of them stress reliability and speed of their deliveries, along with good geographical coverage. Costs are similar to trucking. This mode will probably see continued growth as providers become increasingly agile in their capabilities. Increased online buying with direct-to-home deliveries will also increase the importance of parcel deliveries. These companies are also expanding their roles as third-party providers in such areas as handling returned goods for repair. One of the more innovative approaches is the proposed use of drones to deliver packages.

## Transmission Lines

Electricity is an essential resource produced in large generating stations and delivered to the point of use. Some businesses have their own power-generating plants, but they are exceptions. Portable generators can be used for limited power availability in emergency situations. None of the conventional transportation modes described earlier can deliver electricity; this task requires specially constructed transmission lines, substations for reducing voltages, and customized power lines to businesses and houses. As such, transmission lines are a fast way to move electricity, with good geographical coverage. The cost of electricity is more a function of how it is generated, from coal, water, or nuclear sources, as opposed to how it is transported. Indeed, transmission lines are the only known way to move large amounts of electricity.

Electricity is an essential part of the supply chain, because it is required to drive equipment at manufacturing plants and a variety of office equipment for service operations. It is also required to move data around the world. There are indications it could become a replacement for fossil fuels, such as gasoline, as a power source for automobiles.

### Fiber-Optic Cable Networks

The world depends on the flow of information. Information was transported, in the form of bits of electrical energy, through copper lines since the development of the telegraph in the mid-nineteenth century. In the past couple of decades, fiber-optic cables, with their larger carrying capacity and lighter weight, are rapidly replacing the copper lines. These specialized networks represent a fast, low-cost transportation mode with good geographic coverage. In a number of cases, electronic information is replacing physical goods. For example, books can be sold as either a hard copy that requires movement by truck to the retail store or as an electronic download that requires movement by telephone lines to a consumer's computer.

### Materials Management

Logistics is concerned with moving goods, but it is even more concerned with moving the right goods at the right time to the right place. Consequently, it is important that the logistics network has some way to determine what is to be shipped in order to determine the correct packaging to prevent damage of the goods in transit and the best mode of transportation to meet *need by* dates at the most economical cost. A good information system is necessary to link transportation companies together.

Inventory management is an important part of any logistics system. When the integrated logistics system is functioning smoothly, less inventory is needed. However, inventory management is a complex process extending beyond logistics management, and the logistics system needs to be linked with the inventory management systems of all participants in the supply chain, a most challenging requirement.

### Interim Storage

As goods move from origin to destination, there will likely be a need for interim storage. This may be at docks where goods move from ship to train or truck. It could be at distribution centers where goods undergo some transformation, such as repackaging, to make them ready for the next destination. It could be storage tanks for oil or gas before movement to the factory or home. An integrated logistics system must accommodate these transfer points because they often represent a change in ownership of goods and a need for inspections to assure the quality of the goods being moved. As with other parts of the logistics systems, a good information flow is necessary.

## Exchange Points

Methods for loading and unloading goods from one mode to another have evolved over time. In general, the evolution has been from manual to equipment assisted. Shipping containers have revolutionized the movement of goods on ships. Goods are loaded into the container at the factory and are loaded onto trucks for delivery to the port where the containers are loaded intact onto ships. The process is reversed at the destination. This intermodal system has been refined over the past few decades and has been a key factor in enabling companies to move to offshore outsourcing.

Moving goods through exchange points requires, as minimum, documentation of quantities and product condition, at the time of exchange. It often requires compliance with governmental agencies and the accompanying paperwork. When the exchange is between countries, the paperwork is even more involved and requires specially trained personnel, either at the shipping or receiving company or, more often, at a 3PL.

## Traceability

As supply chains get longer and companies get leaner, knowing where a shipment is becomes much more important. Tracking technology is enabling companies to track their shipments, almost on an hourly basis. This is most critical for food and drug shipments where customers not only want to know where their shipment is but also want to know that it is being handled correctly. Food products have to be stored in refrigerated conditions to preserve their quality and medicines stored in secure conditions to protect them from theft or damage.

## Integration

Linking all of the logistics requirements into a total system is sometimes more than any single company can handle, especially small companies with limited personnel. Consequently, in order to concentrate on their core competencies, an increasing number of retail and manufacturing companies are turning to 3PLs to design and manage the logistics part of their supply chains. We will describe 3PLs more fully later in this chapter. Before that, we want to cover the topic of outsourcing, a management action that is making the logistics function more complex and difficult to manage.

## Outsourcing

One of the major trends in supply chains in recent years is the movement toward offshore outsourcing. This strategy has extended supply chains around the world. While offshore outsourcing has made it possible to reduce costs, at least for direct product costs, it has increased lead times and complexity of supply chains. In addition, it has increased the risk of supply chain disruption.

Outsourcing is not new. It is "the process of having suppliers provide goods and services that were previously provided internally. Outsourcing involves substitution— the replacement of internal capacity and production by that of the supplier" (Blackstone 2013).

On the manufacturing side, the *make or buy* decision has long been a basic consideration. On the services side, all types of companies have outsourced such business support functions as food and custodial services. As information technology facilitated global communications, it became possible to outsource call centers and help desks, hire well-trained persons to write programs, and have medical technicians read x-rays, accountants prepare your taxes, and even business journalists interpret companies' financial statements (Thottan et al. 2004).

However, there are growing pains in outsourcing. A few years ago, the Aberdeen Group found the biggest challenge for companies going global was how to keep the supply chain moving without damaging the sourcing savings that outsourcing offered. This required synchronizing logistics, compliance, and finance processes. Over 90% of the companies felt pressure to improve their global trade process because (1) longer lead times were inhibiting their ability to respond to market demands and (2) expected product cost savings were being eroded by unanticipated global supply chain costs (Enslow 2005).

Today, there is an apparent rethinking of the offshore outsourcing decision by some companies, especially in view of the recent earthquake, tsunami, and nuclear disaster in Japan. Risk management articles abound about the need to not only consider risks but also to develop strategies to mitigate or prevent major disruptions to the supply chain. See the following for just a sampling (Arnseth 2011; Crandall 2010a; Goldberg et al. 2011; Katz 2011; Marks 2011).

In addition to risk management concerns, there are also signs of companies beginning to modify their offshore outsourcing decisions by considering nearshoring (Mexico instead of China) (Smith 2011) or even reshoring or inshoring by bringing offshored work back to the United States (Isidore 2011).

The pressure is not to forsake outsourcing but to establish an integrated system of outsourcing that more carefully selects and manages the outsourced projects, especially after considering the total costs and benefits. We will view outsourcing from the U.S. perspective; however, it is a global issue. It involves both strategic and tactical decision areas; consequently, it has major implications for a business.

As the business environment becomes more complex, it is increasingly difficult for a company to be sufficiently competent in all facets of a business. Therefore, they must seek help from more qualified sources. Yuva and Trent (2005) offers the following reasons for going global: (1) to gain a global perspective, (2) the cost/value benefits, (3) greater access to product and process technology, and (4) to facilitate the transition from selling to buying in a region.

There is a caveat, however. It is a complex decision that requires both a good decision-making process that is systematic and comprehensive to supplement good judgment.

Outsourcing is far from a *no-brainer* decision. The decision to outsource should follow a comprehensive analysis, rather than reacting to short-term considerations (Gottfredson et al. 2005; King and Malhotra 2000; Yuva and Trent 2005).

### Drivers of the Outsourcing Movement

Short-term drivers are often tangible economic considerations, such as the following:

- Reduce product costs.
- Improve product quality.
- Reduce response time.
- Increase flexibility.

Of these, only the first—reduce product costs—is a likely result from offshore outsourcing. Offshore outsourcing may have a negative effect on the remaining three. Consequently, the trade-offs should be explored carefully in the financial analysis.

The following questions can help to focus on some of the issues in this initial decision to outsource:

- *Why should companies outsource?* Companies increasingly view outsourcing as a strategic decision, not a tactical decision. As a result, they consider the impact of the decision over a longer time horizon and include a greater number of factors in making the decision.
- *What should they outsource?* Companies continue to reduce the scope of support functions performed internally. At what point does the concept to preserve the core competencies fall apart?
- *When do they outsource?* As part of the longer-range perspective, more companies will consider what their costs could be if they implemented continuous improvement programs, such as just in time (JIT), lean production, Six Sigma, and SCM. With internal improvements, outsourcing may not be as attractive.
- *Where do they outsource? To whom should they outsource?* Should a company distribute their outsourced products or services among a number of suppliers to lessen the risk or concentrate its outsourcing in a few or even one company? Some major consulting companies are gearing up to provide integrated services. Accenture offers a *bundled outsourcing* consulting service (Accenture 2011). The International Business Machines Corporation (IBM) has also announced a variety of outsourcing-related services (IBM 2011). This decision boils down to a choice between spreading and concentrating risk.
- *What will happen if we don't outsource?* Sometimes, it is important to consider the potential outcomes of taking no action. One issue that is emerging is the lack of available skilled workers in the United States. Manpower Inc. (2011) reports that the immediate problem is a talent mismatch. Employers are seeking not just technical skills but also critical thinking skills, and there are not enough sufficiently skilled people available.

The longer-term drivers are often less tangible, but can be of equal, or greater, significance. Some of them are described in the following.

### Sustainability

The sustainability movement is known by different names—triple bottom line, cradle to grave, cradle to cradle, and green supply chains (Crandall 2010b). This movement is gaining traction and will become an accepted goal for most companies.

### Social Responsibility

Just as with the sustainability movement, there is an increasing need for companies to carefully weigh their strategies with respect to their social responsibilities. Does offshore outsourcing enhance this position or degrade it?

### Ethical Responsibility

The ethical decision is even more difficult. In general, it involves several stakeholder groups—employees, managers, shareholders, consumers, and the public. The overriding question is "Will customers buy more if a company's outsourcing results in lower prices, or will customers buy less because they are more negative toward the company because of its outsourcing practices?"

### Future Company Well-Being

In making the decision, companies balance the tangible benefits of offshore outsourcing, such as lower costs and increased capabilities, with the increased risks and uncertainties of remote suppliers. What will the situation be in 5–10 years? How easy, or difficult, will it be to transition from one country to another or back to the United States, without sacrificing customer service?

The political implications are also beyond normal analysis methods. What will governments at all levels and in all countries do with respect to outsourcing practices? Will they try to regulate the practice? If so, will they encourage outsourcing as a step toward expanding global markets, or will they restrict outsourcing because it causes disruptions in jobs, in the form of layoffs and the need for retraining and reeducation?

In considering the future, top management must also consider their long-term marketing goals. If they want to sell more in foreign countries, will it help to buy from suppliers in those countries? Some major multinational U.S. companies are offshoring (building and operating facilities in other countries) as a means of entering or expanding their foreign sales.

While these areas have different issues, at some point, all of the considerations must be wrapped up into the decision to outsource or not. In this paper, we deal primarily with the economic analysis, although both ethical and political issues have economic implications.

## Steps in the Outsourcing Decision

In the following, we outline a general approach in making the outsourcing decision, which corresponds to Steps 1–8 in Figure 9.5. This approach emphasizes the need to consider both strategic and tactical factors in the decision-making process.

### Step 1. Determine Project Feasibility

Before spending time in comprehensive analyses, top management should decide if outsourcing is a sound strategic direction for the company. What are their objective(s) in

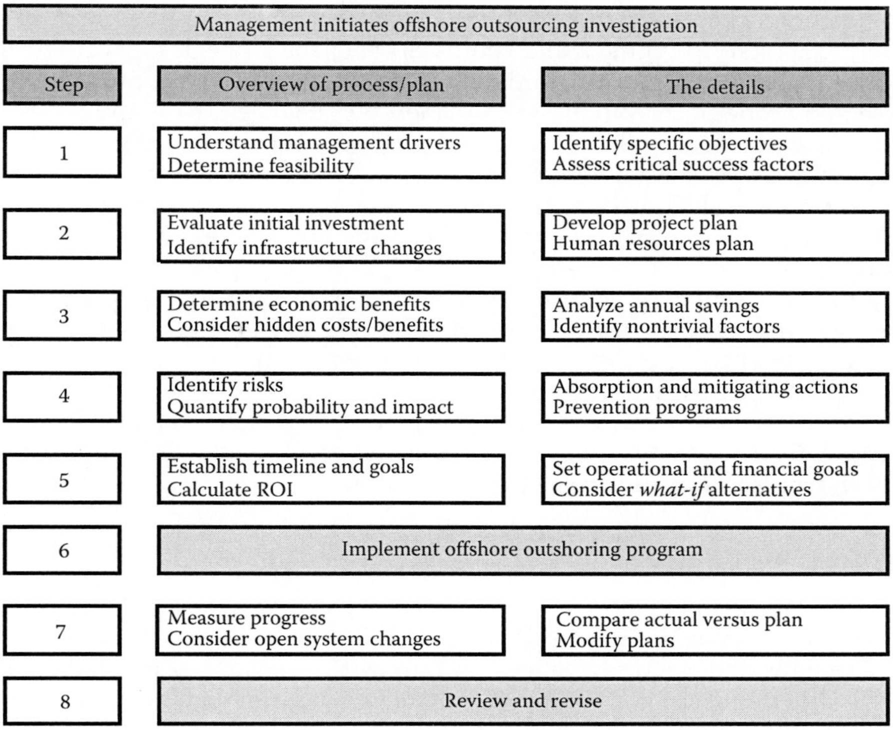
Management initiates offshore outsourcing investigation
Step
Overview of process/plan
The details
1
Understand management drivers
Determine feasibility
Identify specific objectives
Assess critical success factors
2
Evaluate initial investment
Identify infrastructure changes
Develop project plan
Human resources plan
3
Determine economic benefits
Consider hidden costs/benefits
Analyze annual savings
Identify nontrivial factors
4
Identify risks
Quantify probability and impact
Absorption and mitigating actions
Prevention programs
5
Establish timeline and goals
Calculate ROI
Set operational and financial goals
Consider *what-if* alternatives
6
Implement offshore outshoring program
7
Measure progress
Consider open system changes
Compare actual versus plan
Modify plans
8
Review and revise

**FIGURE 9.5**
An approach to outsourcing decision making. (Based on input from Robert Crandall, former Vice-President, Lowe's, Inc. in a personal interview, January 15, 2013.)

making the change? Is it to reduce costs, increase revenues, a combination of both, or for some other reasons? While the initial objective is often to reduce costs, there is also the possibility of added revenues, especially in emerging markets. As Figure 9.5 indicates, it is important to understand the management drivers to consider.

### *Step 2. Prepare a Project Plan*

As part of the project plan, select those decision variables most significant to the company's decision. Preparing a project identifies the critical success factors for the outsourcing initiative and verifies its feasibility. A wide cross section of the organization will usually be required to develop a meaningful plan. However, each organization is different and the project plan must be adapted to each outsourcing project.

These costs include the costs of searching for a supplier and all of the activities necessary to establish a working relationship. These costs could include site visits, updating specifications, changing the existing organization structure, and investing in new tooling. It could also include revenues from selling equipment deemed expendable when the outsourcing change is complete.

As businesses outsource services, the remaining organization becomes more streamlined. While streamlining may suggest it is simpler, it is not. The relationships, both within the business and between internal and external entities, become more complex and require attention to assure satisfactory results. Davidson (1990) emphasizes that the human resource function is necessary to help in the outsourcing program by providing help in internal organization realignment and coordination of the strategic alliances formed by outsourcing. This requires greater coordination among the involved entities.

The role and scope of the purchasing function will change significantly. It will have greatly increased responsibilities for internal and external coordination of the various outsourced projects. As the amount of outsourcing increases, some businesses will probably create an *outsourcing* function within the organization with greater cross functional responsibilities, perhaps initially a part of purchasing but eventually a more senior level function responsible to top management.

The information technology function is necessary in the outsourcing program to provide help in setting up the electronic communication and interorganizational interfaces necessary for outsourcing. It also means that some elements of the information technology function should be retained, not outsourced.

In addition to the time invested by various members of the organization, other costs may include travel to visit potential outsourcing locations, consultation with third-party firms specializing in outsourcing, and related implementation costs.

### Step 3. Estimate Annual Savings

Outsourcing costs include both the direct product costs and the indirect support costs. Some recommend a regular (annual or at the end of a contract) review to reevaluate the effect of economic and political changes, as well as the vendor performance (Yuva and Trent 2005). Others suggest that the reviews should be often enough to prevent the loss of in-house critical resources and competencies (Gottfredson et al. 2005). Companies should view outsourcing as dynamic, not a *one and done* kind of decision. They need to make a careful decision to outsource and should periodically review that decision.

In addition, there are costs necessary to manage the outsourcing project on a day-to-day basis. These costs would not be incurred if the production remained in-house. Each organization must come up with its own way to estimate these costs, which are over and above the costs for setting up the project.

In addition to the tangible costs, there may be intangible costs, such as the loss of technical knowledge, establishment of a business relationship with foreign contacts that may spawn a competitor, and loss of public acceptance in the marketplace. Assigning dollar amounts will be difficult and will require a *best estimate* approach. However, some of these factors are significant and could be the basis for rejecting the decision to offshore outsource.

If the estimated savings are still positive, after an analysis of both direct and indirect costs, the next step will be to more carefully assess the cost of making the transition. While a cursory estimate of these costs is considered in preparing the project plan, it is important to accurately verify the transition costs before going ahead.

### Step 4. Identify Risks or Disruption Costs

Up to this point, the cost analyses are based on ongoing *normal* operations. When outsourcing offshore, the likelihood of disruption increases, because of longer and multimode transportation requirements, as well as problems inherent in intercountry movement of goods.

These disruptions can include risks ranging from normal everyday variations to an enterprise-threatening crisis (Crandall 2010a). The probability of occurrence also varies. Some variations may occur often, but with minor effect, and can be absorbed as part of the normal, tangible costs. On the other extreme, a crisis may occur infrequently but could be very costly, even disastrous. An expected value (EV) analysis can be helpful in projecting the possible economic impact of such disruptions.

### Step 5. Estimate the Return on Invested Capital

Use a time value of money approach to assess the total return on invested capital for the offshore outsourcing project. Selecting a reasonable time horizon is critical because it is unlikely that the initial arrangement will last for very long. A discounted cash flow analysis can show the return on invested capital. It should include the initial costs of dropping the present system and recovering these costs over a relatively short period. Subsequent costs could include the cost of changing supplier, perhaps in another low-cost country. A final cost could involve making the transition back to in-house or shutting down the total operation. At this point, management must consider whether or not to go ahead.

The cost at the end of the expected life is the cost of reversing the decision or otherwise terminating the project. However, there is a good probability that some disruptive force will cause the outsourcing decision to be reevaluated before the end of the proposed project duration.

After making the individual analyses described earlier, there should be a final review by senior management before implementing the outsourcing program. This final summary can take a variety of forms, depending on the preferences by the decision makers. At such a review, it is likely that senior management will add their own judgment to the analysis, especially in the areas of risks and benefits. They may have insight or inclinations that are not obvious, or available, at the lower analyst level.

### Step 6. Implement the Offshore Outsourcing Program

If the decision to offshore outsource is supported, the project should be implemented. Implementation requires a project management approach. No matter how well the implementation plan was thought out, the actual implementation will probably not go according to plan; consequently, changes will have to be made. Some of the changes will be minor and easily accommodated; however, some changes will be substantial and require the implementation team to reach an agreement on the corrective action needed. Most offshore outsourcing programs will require time for the changes to be accepted, especially at the company doing the outsourcing. Employees affected by the program will have to adapt to new job content and learn to work with new individuals, often with language and culture differences. Perseverance and dedication are necessary traits for those involved in offshore outsourcing implementations. While technology has made many forms of communication possible, many companies are finding there is no substitute for face-to-face meetings to resolve issues in the implementation process.

### Step 7. Measure Progress

Progress should be monitored periodically to be sure the results are as expected. Many of the metrics used will be nonfinancial measures. While financial results are important, nonfinancial measures are likely to give an indication of how the program is going well before financial results can detect the amount of progress with the program. It should be easy to measure direct purchase costs and product quality to see if they are as planned. It should also be easy to measure the increase in delivery times, although they will probably vary more than before. It will not be as easy to measure the effect on employee morale or customer perception of the change, both of which can affect the financial results of a company.

### *Step 8. Review and Revise*

Although informal reviews may identify major deviations, it is appropriate to prepare a formal analysis of the results, probably 6 months or a year after implementation. If called for, changes should be made to assure ongoing success or, if necessary, effect a reversal, or major modification to the outsourcing project.

### *Summary*

Who will do the analyses needed to arrive at a good decision? While the process requires extensive participation of different functional areas of the business, such as purchasing and operations, finance and accounting will bear the primary responsibility of bringing the analysis to an effective conclusion. The approach described in this chapter will allow all functions to become partners in the strategic outsourcing process of a company.

The outsourcing trend will likely continue in the United States; however, many companies are being more cautious and taking a more exhaustive look at the initial decision to outsource. Although there is increased interest in the reshoring movement (return manufacturing to the United States), there is insufficient evidence that it has stopped the flow of outsourcing work to offshore countries.

Offshore outsourcing has been one of the reasons that improved logistics has become a more visible challenge for management. It has increased indirect costs and required more comprehensive decision making. Offshore outsourcing has increased risk, a topic we will discuss more fully later in this and other chapters, because it is becoming one of the most critical concerns of many organizations. Risk includes not only economic considerations but also societal and environmental issues. Offshore outsourcing has also moved the transportation functions from a simple modal choice to a multimodal coordination challenge.

As a result of its expanded scope, logistics activities are becoming a greater portion of total product cost and, consequently, warrant greater management attention. In many cases, companies that are primarily retailers or manufacturers are outsourcing the logistics functions to specialized logistics providers (3PL or 4PL).

---

## Reshoring Initiative

In recent years, there has been some movement toward reversing offshore outsourcing with what is being known as a reshoring initiative. This movement is being championed by Harry Moser, who believes that many companies outsourced work without conducting a careful, comprehensive analysis of the total costs associated with outsourcing. Consequently, some companies are not recognizing that the economic benefits of outsourcing are less than they originally anticipated (Gephart 2012).

How widespread is the reshoring movement? Kent (2013) reports there is some evidence that companies are exercising a greater level of due diligence when reviewing their supply chain design; as a result, some are revising their decisions to offshore outsource. This is especially relevant for high-tech firms with rapid innovation cycles. Kent goes on to report the results from several surveys:

- In a 2012 survey, Boston Consulting Group (BCG) reports 37% of companies with annual sales above $1 billion were actively considering shifting production facilities from China to America.
- The Hackett Group reports that offshoring is moving toward equilibrium (zero net offshoring) (Kent 2013).

Although the reshoring initiative sounds desirable for American interests, Gartner warns that companies should look as carefully at reshoring as they did in making the decision to outsource. Just because some companies are finding reshoring to be a favorable decision does not apply to all (Kent 2013). Even if companies move to reshore, it does not necessarily follow that there will be a big increase in American jobs. A new report by the NAIOP Research Foundation concludes that the reshoring trend of manufacturing industries to the United States yields stabilization of jobs, but not new growth. Some industries, such as chemicals and technology, will expand, while more labor-intensive industries, such as textiles, will continue to contract (Selko 2013). It appears the gap is narrowing between outsourcing and reshoring. Companies will have to be even more diligent in their analysis of this decision area.

There is no doubt that outsourcing has increased the complexity of supply chains, often to the point where more companies are realizing they need external help in making and implementing this phase of their supply chain design and implementation. As a result, they are turning to third parties to gain the needed services.

## Rise of 3PLs

Global competition and outsourcing are the primary reasons why a number of companies have moved, and continue to move, to the use of third-party logistics providers (3PLs). Global competition requires companies to concentrate on their core competencies and outsource more of their noncore activities to outside suppliers. Outsourcing, especially offshore outsourcing, has extended supply chains into countries with different business environments, and many companies do not have the internal capability to move through the labyrinth of laws, paperwork, business practices, and cultures. Consequently, they are turning to 3PLs who have specialized knowledge and capabilities to move product smoothly along the global supply chain.

The use of 3PLs is growing. More than 85% of the Fortune 500 companies use 3PLs, with companies in the technology industry being the largest users. A report by Armstrong & Associates found that many of the top companies use multiple 3PL companies, with Wal-Mart, Procter & Gamble, and General Motors each using more than 50 3PLs (American Shipper 2013).

3PLs can perform a variety of services. Some of the most often used services include (1) inventory management; (2) order acceptance and processing; (3) pick-and-pack operation; (4) order fulfillment; (5) assembly/packaging/value-added activities; (6) credit card verification; (7) invoicing, credit, and collection; (8) presort capabilities; and (9) returns handling (Ackerman 2000).

The electronics industry has seen a growth in how third-party providers contribute to improvements in SCM. Electronics buyers use distributors for inventory management, logistics management, bonded and consignment inventory, kitting, engineering assistance, and supplier research to make their supply chains operate more efficiently (Carbone 2000).

According to Berglund et al. (1999), the rise of 3PLs has come in three phases:

- The first wave dates back to the 1980s, when organizations with a strong position in either transportation or warehousing began to provide these services to retail and manufacturing companies. These were traditional logistics service providers who added these services to other companies to their mainstream business.

- The second wave began in the early 1990s, when parcel and express companies, such as DHL, TNT, and UPS, added 3PL services to capitalize on their air express networks and experience in moving in countries throughout the world.

- The third wave, beginning in the late 1990s, brought in companies with extensive knowledge in information technology, consultancy, and financial services. These companies included IBM, Accenture, and GE Capital Services.

Most companies added the 3PL services as an addition to their main business. However, as the 3PL market grew, many spun the 3PL business off as a separate entity, to provide growth opportunities and to reduce any possible conflict with customers of their primary business. Deregulation in the transportation industry made it possible for 3PLs to expand their services, and increasing global competition made it necessary for many retail and manufacturing organizations to outsource their logistics functions (Berglund et al. 1999).

Table 9.1 shows the estimated size of the 3PL market, as provided by Armstrong Associates, Inc. (2013), one of the leading consulting and information providers for the 3PL industry. As expected, the Asia Pacific (China and Japan), North America (United States), and Europe (Germany, United Kingdom, and France) are the leading areas. In the industrialized countries (North America and Europe), logistics costs average 8.9% of the GDP, while the percentage of logistics costs to GDP run higher in the rest of the world. The use of 3PLs is also larger in North America and Europe (about 10% of logistics costs) and lower in the remaining global countries.

Hertz and Alfredsson (2003) developed a two-by-two matrix to show the different types of 3PL providers. First, Figure 9.6 shows the origins of 3PLs. The upper left quadrant, the lower

**TABLE 9.1**

Global 3PL Market Size Estimates

| Region | 2011 GDP | Logistics (as % of GDP) | 2011 Logistics Costs | Revenue % to 3PL | 2011 3PL Revenue | Total 3PL Revenue (%) |
|---|---|---|---|---|---|---|
| North America | $18,004 | 8.9 | $1,598 | 10.0 | $160 | 26.0 |
| Europe | 17,690 | 8.9 | 1,567 | 10.2 | 160 | 26.0 |
| Asia Pacific | 19,208 | 12.8 | 2,458 | 7.8 | 191 | 31.0 |
| South America | 4,178 | 12.3 | 514 | 7.7 | 40 | 6.4 |
| Remaining | 11,081 | 15.9 | 1,762 | 3.7 | 65 | 10.6 |
| Total | $70,160 | 11.3 | $7,899 | 7.8 | $616 | 100.0 |

*Source:* Adapted from Armstrong Associates, Inc., ABA's top 50 global third-party logistics provider (3PL) list, largest 3PLs ranked by 2011 logistics gross revenue, 2013, www.3plogistics.com/Top_50_Global_3Pls.htm, accessed May 30, 2013. With permission.

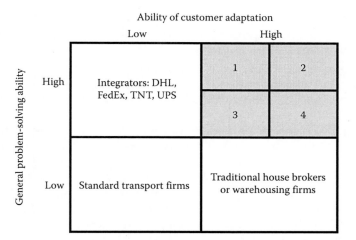

**FIGURE 9.6**
Origins of 3PLs. (Adapted from Hertz, S. and Alfredsson, M., *Ind. Market. Manag.*, 32(2), 139, 2003. With permission.)

3PL providers
Ability of customer adaptation

|  | Relatively high | High |
|---|---|---|
| **High** | *Service developer* Example: An advanced modular system of a large variety of services and a common IT system used for all customers | *Customer developer* Example: The 3PL firms develops advanced customer solutions for each customer. The role is of a consultant or 4PL |
| **Relatively high** | *Standard 3PL provider* Example: Standardized modular system offering customers relatively simple combination of standardized services | *Customer adapter* Example: Totally dedicated solutions involving the basic services for each customer. 3PL firm is seen as a part of the customer organization |

General problem-solving ability

Customer adaptation ⟶

**FIGURE 9.7**
Classification of 3PLs. (Adapted from Hertz, S. and Alfredsson, M., *Ind. Market. Manag.*, 32(2), 139, 2003. With permission.)

left quadrant, and the lower right quadrant all show the basic business from which many 3PL activities started. The upper right quadrant shows four types of 3PL providers.

These classifications are expanded in Figure 9.7. The authors explain their classification as follows:

- The *standard TPL provider* (3 in Figure 9.6) supplies standard TPL services such as warehousing, distribution, and pick and pack.
- The *service developer* (1 in Figure 9.6) offers advanced value-added services such as forming specific packaging, cross docking, track and trace, and special security systems.

- The *customer adapter* (4 in Figure 9.6) is a TPL firm that takes over a customer's existing activities and improves the efficiency in the handling, such as a warehouse operation.

- The *customer developer* (2 in Figure 9.6) is the most advanced form and involves a high integration with the customer often by taking over its whole logistics operations. The customer developer would be similar to what is now known as a 4PL.

These classifications represent a hierarchy of services, beginning with the *standard TPL provider* and progressing ultimately to the *customer developer*. This does not mean that all 3PLs aspire to the highest level or that 4PLs want to invest in transportation or warehousing systems. It does represent a progressive outsourcing of logistics services by the buying company. Consequently, as a company increases its outsourcing to a 3PL or a 4PL, they must build increasingly strong relationships with their provider and have an increasing level of trust with that provider. As a TPL company increases the scope of its portfolio of services, it tends to provide more services to fewer customers.

Perhaps there is a need for another rung on the hierarchy of TPLs. Should we add a 5PL or supply chain integrator? We will discuss that need later in Chapter 16.

## Benefits of 3PL Services

A number of benefits can be achieved through the use of 3PLs. They include lower total cost, shorter total time from origin to destination, and more consistent processing with less variability in elapsed time; individual participants can concentrate on core functions and can lead to further outsourcing through 3PL contacts, and 3PLs can lead to specialized knowledge not normally internal to company.

### *Reduces Total Cost*

A 3PL should be able to reduce the logistics costs along a supply chain. It may come from improved route planning, lower negotiated transportation rates, optimized selection of mode of transportation, or any other means of gaining economies in the movement of goods.

### *Decreases Total Time from Origin to Destination*

Lead time to replenish inventory is important to companies and a 3PL should be able to help in this area through close coordination of movement of goods, especially when changing from one mode to another (e.g., truck to ship or ship to train), accurate paperwork to smooth flow through customs or other transfer points, and tighter scheduling of routes to reduce delays in transit.

### *Results in More Consistent Processing: Less Variability in Elapsed Time*

A 3PL with a global perspective should provide a holistic view of the total supply chain so that each movement of goods can be planned to minimize interruptions and delays. Reduced variability in elapsed time allows inventory managers to avoid stockouts and excess inventory.

### Allows Individual Participants to Concentrate on Core Functions

Logistics management in extended supply chains is a complex process. No matter how carefully planned, changes are commonplace. Not having to worry about the day-to-day logistics activities can allow managers in manufacturing and retailing companies to concentrate on their primary, or core, business.

### Increases Flexibility of the Supply Chain

If a manufacturer or a retailer has to respond to new developments in the marketplace, they may need changes in the logistics system to find a new supplier or movement of goods to a new location. A 3PL can provide this flexibility better than a company with other concerns, such as new product development or new market penetration.

### Leads to Market Development through 3PL Contacts

A 3PL can help on the demand side also, by suggesting new market opportunities or by assisting the sourcing company to learn how to improve their marketing program in new territories. A global 3PL understands how to do business in many countries, and sharing this with their customer is a win–win situation for both sides. In speaking of trends in logistics outsourcing, Robinson (2013) reported that "while companies move to cut spending and reduce both headcount and infrastructure, they are looking for new ways to improve service levels to existing customers and efficiently reach new markets."

### Taps into Specialized Knowledge Not Normally Internal to Company

A 3PL provider accumulates knowledge about activities along the supply chain beyond just the logistics functions. As they build closer relationships with their customers, they can begin to share some of this knowledge, whether it is in new products, new processes, trends in the business climate in foreign countries, emerging trends in transportation modes, or any number of related items.

## Obstacles to Successful Implementation

Just as there are benefits to the use of 3PLs, there are also some obstacles or negatives associated with the change. They include lack of qualified personnel within the company to manage the 3PL, lack of full-service providers, perceived loss of control by outsourcing company, difficulty in coordinating disparate participants, and lack of a comprehensive financial analysis before implementing.

### Lack of Qualified Personnel within the Company to Manage the 3PLs

Managing an outsourced activity, such as a 3PL, requires knowledgeable and dependable managers. Managing a remote activity is usually more difficult than managing a local activity, even if the 3PL is fully qualified to do the job. There are always questions relating to company policy or handling of difficult, or incompetent, suppliers.

### Lack of Full-Service Providers

Just as there may be a lack of qualified employees within a company to perform all of the logistics functions, it may be difficult to find a 3PL who can perform the full range of services needed. Outsourcing logistics activities on a piecemeal basis is probably not a desirable approach; therefore, it is necessary to find a 3PL who provides all of the services.

### Perceived Loss of Control by Outsourcing Company

Outsourcing carries with it the potential for loss of managerial control of the function outsourced. Loss of control can be disastrous when experienced with a critical function such as logistics. Even minor disruptions in the supply chain can be serious and major disruptions could lead to severe loss of market position or even company failure.

### Difficulty in Coordinating Disparate Participants

Most companies have multiple supply chains to handle multiple products or markets. This requires multiple suppliers, often in multiple foreign countries. Managing all of these suppliers is more difficult through a third party. Therefore, outsourcing companies should expect to assign qualified persons to handle these challenging and time-consuming relationships.

### Lack of a Comprehensive Financial Analysis before Implementing

The decision to outsource to a 3PL requires careful analysis, just as outsourcing manufacturing or other services. The analysis can be difficult because there are a number of less tangible considerations, such as the effect of improved or deterioration of delivery times, supplier relationships, transportation suppliers, or government officials in countries of origin, to identify a few.

### Difficulty in Selecting and Using Appropriate Performance Measures

Although measuring the performance of a 3PL may be an obvious objective, actually doing it may be more difficult. One research study by Cooper et al. (2012) developed a general model that looked at the activities to be measured and developed possible measures within each activity:

- Incoming order management, with measures of order entry accuracy, no touch orders, document invoice accuracy, response to order inquiry, released same day, and orders received
- Transportation to the regional distribution center (RDC), with measures of on-time delivery, percent defect-free delivery, inbound cost per unit, and transit damage frequency
- Inventory management, with measures of weeks forward coverage, inventory accuracy, inventory turnover, inventory obsolescence, inventory carrying cost, days sales outstanding, days payable outstanding, and warehouse efficiency

- Transportation RDC to customer DC, with measures of packing shipping accuracy, on-time delivery, outbound cost per unit, and transit damage frequency
- Delivery management, with measures of customer service level, order cycle time, fill rate, response to customer inquiry, # of returns, and back orders

While collecting the data for all of these measures would be a major task for a company, it is likely that a subset of the measures, along with input from persons dealing directly with the 3PL, would be adequate to gauge their performance.

### Trend toward Outsourcing the Distribution Function

Retailers recognize the importance of the distribution function. Some large retailers— Wal-Mart, Lowe's, and Publix—have developed their own network of RDCs and believe it provides them a competitive advantage. Other retailers do not consider distribution a core competency and are moving toward outsourcing the distribution function to third-party providers. Manufacturers are also either putting more emphasis on developing their own distribution function internally or outsourcing it to a third-party provider.

RDCs are becoming a key part of most supply chains. An RDC is not the traditional warehouse where inventory is stored for indefinite periods; it is a center of activity in which goods are received from suppliers, stored temporarily or cross docked, picked for loading onto delivery vehicles, and shipped to retail stores. An RDC is also becoming a center where product customization is being done to make the goods *floor ready* when received at the retail outlet.

RDCs are strategically located to reduce the delivery time to retailers. A de facto standard is that the RDC will be within a 1-day delivery distance to all retail stores. This makes it possible to keep the retail inventories low because they can be quickly replenished. It also means that delivery costs can be planned and controlled by scheduling optimum routes and planned arrival times at the store.

### Major Companies

A number of companies have aggressively moved into providing an array of distribution services. Armstrong Associates, Inc. (2013) published a list of the largest global 3PLs. Their list for 2011 include DHL Supply Chain and Global Forwarding, Kuehne + Nagel, DB Schenker Logistics, Nippon Express, C.H. Robinson Worldwide, CEVA Logistics, UPS Supply Chain Solutions, Hyundai Glovis, DSV, and Panalpina. It is obvious that major 3PLs are located throughout the world. It is difficult to measure the size of 3PL operations because some of these activities are located within the larger corporate structure.

UPS has been one of the leaders. In addition to their traditional package delivery, they provide repair services for Toshiba computers in their Louisville, Kentucky, center (Hesseldohl 2004). They also coordinate global delivery services, provide kiting services, and offer short-term financing and a variety of other supply chain solutions (Morton 2007; van Hoek and Chong 2001).

### Role of 4PL in Building Supply Chain Relationships

As supply chains, and the corresponding logistics problems, have become more complex, many 3PLs have begun to add consulting services to their services. This has moved them into the fourth-party logistics provider, or 4PL, category (Norall 2013). Innovative solutions are being explored in logistics to provide additional benefits to users. Some of them require increased collaboration among supply chain partners. Most supply chain collaboration is vertical (along the supply chain). One venture involved horizontal collaboration (across the supply chains). Collaboration in this type of venture could provide benefits of reduction in excess inventory and duplicate activities and increased coordination. This involved using an external company to jointly manage the supply chains for two different companies. The study envisioned this as moving from a traditional 3PL to a collaboration specialist as the complexity of collaborative distribution and intensity of collaboration increased. The study concluded that suppliers would be more willing to participate in such a venture than retailers, who feared loss of competitive advantage (Hingley et al. 2011).

## High-Tech Industry Issues

In 2012, IDC Manufacturing Insights did a survey of 125 U.S.-based high-tech companies (Ellis 2012). The survey found three key areas have emerged as catalysts for gaining market share and increasing profitability. These areas are the following:

- *Targeted innovation*: This means designing products and services for narrowing niche markets with product variations.
- *Emerging markets*: The transition of China, India, South America, and Eastern Europe from low-cost suppliers to more prosperous consumers, especially of electronic products, makes this an attractive target market.
- *Complementary services*: High-tech companies are adding services, content management and service support, to their products. They seek added revenue after the sale as well as from the sale itself.

The companies were asked to rank the top three issues driving changes in their approach to import/export in their supply chains over the next 3–5 years. The results are shown in Table 9.2.

**TABLE 9.2**

Priorities of High-Tech Supply Chains

| Issues | Top Priority | Second Priority | Third Priority | Total Top 3 |
|---|---|---|---|---|
| Cost | 48 | 24 | 10 | 82 |
| Lead times | 14 | 26 | 19 | 59 |
| Responsiveness | 14 | 4 | 17 | 35 |
| New free trade agreements | 10 | 17 | 7 | 34 |
| Intellectual property | 5 | 8 | 14 | 27 |
| Security risks | 3 | 4 | 10 | 17 |
| Resilience | 2 | 6 | 9 | 17 |
| Complexity in import and export processes | 2 | 5 | 2 | 9 |
| Increased international regulation | 2 | 2 | 4 | 8 |
| In-market competence | 0 | 9 | 8 | 17 |

*Source:* Adapted from Ellis, S., Change in the chain: Import/export in the U.S. high-tech supply chain, *IDC Manufacturing Insights*, September 2012.

The three top issues for high-tech supply chains are cost, lead times, and responsiveness. However, these priorities are at odds with one another. To achieve lower costs, many companies are outsourcing to Far East countries, notably China, but more recently to other Pacific Rim countries. Outsourcing extends the supply chain, thereby increasing the lead times and reducing the ability to be responsive.

The IDC survey also examined where companies plan to source their product supply now and in the future. They found a number of companies plan to reduce their sourcing in the United States and expand it significantly in Asian, notably India, and South American countries (Ellis 2012). No doubt some of this increase stems from their expectation of increased market opportunities in those countries.

## Risk Management

The logistics part of the supply chain can be a source of disruptions in the flow of goods. It follows that the longer the distance moved and the more times a product has to be transferred from one mode of transportation to another, the greater the opportunity for delay or damage. It may not be the result of mishandling on the part of the carriers; it could be a combination of circumstances, such as the delay in docking because of a queue of ships arriving at the same time. Regardless of the cause, a delay is a delay in the minds of the customer waiting for the goods to arrive at the manufacturing plant, the RDC, or the retailer.

There are a number of actions an organization can take to prevent, or at least mitigate, these negative occurrences. They generally fall under the umbrella of risk management, one of the latest management topics to be added to the list of concerns for management of an organization. We will describe risk management more fully in Chapter 17. At this point, there are two key points to consider. Outsourcing increases the risk in a supply chain and the use of 3PLs reduces the amount of direct control an organization may have in managing their supply chains.

## Status Report

The ninth annual 3PL market research report by *Inbound Logistics* (2013) highlights the fact that connectivity between a company and its 3PL(s) is an important consideration. While transportation and warehousing are the core services, more 3PLs are providing a wider array of services and are shifting from transactional relationships to more integrated partnerships. Integrated logistics services are becoming the norm, with 80% of 3PLs reporting this capability. The survey (O'Reilly 2013) highlights several trends in 3PL logistics:

- Over the past decade, non-asset-based providers have emerged to compete with traditional warehousing and trucking-based solutions providers. These 3PLs offer functional areas of expertise in freight bill audit and payment, freight brokerage, import/export, and customs brokerage, all essential in global trade.

- E-commerce is also generating demand for third-party logistics services, even among big box stores. This requires an extension of services into home deliveries, which almost half of the surveyed 3PLs now provide.
- The emergence of the cloud is also recognized by 3PLs. Over half of them are exploring investment in cloud-based solutions such as TMS, optimization technologies, and warehouse management systems (WMS). An increasing number are developing capabilities in sustainability, predictive analytics, and global trade management.
- The geographic area covered is increasing. Over 90% of the surveyed 3PLs operate in North America, but only 13% operate exclusively in the United States. Domestic shippers are operating across borders, especially in Mexico, a result of the NAFTA agreement and resolution of the cross border truck movement controversy. Europe and Southeast Asia are areas targeted for growth, especially among global carriers UPS and FedEx.

Shippers and 3PLs face similar challenges, although they do not rank them all the same in order of importance. Table 9.3 shows the results of the survey. In general, shippers

**TABLE 9.3**

Shipper and 3PL Challenges

| Challenge | Shippers (%) |
|---|---|
| Cutting transport costs | 63 |
| Business process improvement | 33 |
| Improving customer service | 30 |
| Supply chain visibility | 22 |
| Managing inventory | 20 |
| Expanding/selling to new markets | 19 |
| Reducing labor costs | 19 |
| Finding, retaining, training qualified labor | 18 |
| Regulations/security/compliance issues | 17 |
| Expanding/sourcing to new markets | 15 |
| Vendor management | 14 |
| Technology strategy and implementation | 13 |
| | **3PLs (%)** |
| Technology investment | 57 |
| Capacity | 55 |
| Rising operational costs | 52 |
| Finding, retaining, training qualified labor | 47 |
| Regulations | 47 |
| Finding/retaining customers | 40 |
| Making a profit | 30 |
| Meeting customer service requirements | 27 |
| Contingency planning/risk management | 23 |
| Global coverage | 15 |
| Corporate social responsibility (incl. sustainability) | 13 |

*Source:* Adapted from O'Reilly, J., 3PL perspectives 2012: Shippers and 3PLs—Perfect together, *Inbound Logistics*, July 2012, www.inboundlogistics.com/cms/article/3pl-perspectives-2012, accessed June 17, 2013.

appeared most concerned about cutting transport costs and improving processes. 3PLs are focused more on investment in technology and capacity, perhaps with an eye on future expansion of services (O'Reilly 2013).

Intermodal capabilities are also becoming more important for 3PLs, especially rail-specific capabilities, as there is evidence that movement by rail is less costly and more environmentally sustainable than by truck. An additional concern is the U.S. highway infrastructure and increasing congestion, without any long-term solution in sight.

There are two developments that may have significance for the logistics field—larger ships and the expansion of the Panama Canal. Maersk, the world's largest shipper, has ordered 20 new ships, 1,312 ft long, 194 ft wide, and weighing 55,000 tons empty. Each ship will be capable of carrying 18,000 TEU (twenty-foot equivalent unit), or the equivalent of 9,000 40 ft containers. These ships offer the possibility of reduced shipping costs by being more fuel efficient; however, they require major investment capital, costing $185 million each (Bennett 2013).

The Panama Canal is being expanded to handle container ships of up to 13,200 TEUs. While perhaps not able to handle the larger ships described, the expanded canal offers new decisions for shippers. In anticipation of increased business, the ports on the East Coast of the United States are investing billions of dollars to expand their capacities. This introduces new considerations for companies. Will it be better to move products to the East Coast primarily by ship, through the Panama Canal, or ship to the West Coast and move the goods by train to the East Coast? Variables to be considered include direct costs, time and cargo value, inventory, and risk management (Siegfried 2013).

As previously noted, there is increasing recognition that no single 3PL is likely to meet all of the needs of a customer, especially large companies. The other trend is in the growing strategy of 4PLs, where the service provider "orchestrates operations among the others" (O'Reilly 2012). Logistics is a critical part of SCM, and its importance will increase in the future.

---

## Hot Topic

Obtaining products from foreign countries often involves shipping containers on ocean-going ships. This introduces additional risks, as described in this chapter's hot topic.

### Hot Topics in the Supply Chain: Container Shipping and its Risk Points

There is a silent industry out there we rarely think about. It hauls 90% of just about everything in the global supply chain (George 2013). When it runs smoothly, nobody gets noticed or praised for that matter. However, when things go bad, supply chains are interrupted and managers at all levels are suddenly in crisis mode. The shipping by container industry operates in these critical realms.

Shipping goods in containers makes sound economic sense. Economies of scale are at work as ships capable of holding thousands of cargo containers can efficiently sail from continent to continent. As a result, clothing and toys from Asia can be transported to other countries for just pennies an item. But there are some considerations supply chain managers need to be aware of. Like any mode of transportation, shipping has its weak points.

## CONTAINER SHIPPING AND ITS RISKS

*Pirates*: Who would have thought that pirates would be a factor in modern-day supply chains? Most of us probably have outdated views of pirates, thinking of images of Captain Kidd and other ne'er-do-wells who once roamed the sea, looking for cargo and bounty to steal. Today however, pirates are not looking for cargo, but large ransoms from the insurance companies of the shippers. A special type of insurance even exists for such unfortunate events, kidnap and ransom insurance (K&R). It is the insurance companies that get involved when a ship is hijacked and held for ransom. Negotiations can last up to −100 days with ransoms reaching $10 million for a small family-owned ship (George 2013).

Today's pirates are typically from the failed country of Somalia. Desperate for cash flow, groups of Somali pirates work together to apprehend an unsuspecting ship at sea and then use it as collateral. Once they capture a ship, it may also become a mother ship that can be used as a base for launching other attacks. Hence, even pirates have to think about their supply chains. The Indian Ocean, from Africa all the way to India, is a breeding area for Somali pirates using captured mother ships to extend their geographical range (Leach 2011).

Shipping companies must contend with this risk. Sailing away from dangerous waters is an option, but it increases costs and slows down the delivery of goods. Equipping the ship with high-pressure water hoses to ward off pirates has been done by some shippers. Blinding searchlights and earsplitting loudspeakers have also been used. Building a safe room where the ship's crew can retreat to for safety is also a best practice (Leach 2011). A growing number of companies have hired armed security teams to sail with their ships, but the largest shipping company, Maersk, currently prohibits this practice (George 2013).

*Terrorism*: The placement of a weapon of mass destruction (WMD) inside of a cargo container is a security risk for the shipping industry. The detonation of a bomb could close down a port, sink a ship, and disrupt supply chains. Even with a tip that a WMD may be within a port, looking for the suspected weapon could take days and cost in the billions of dollars. Port security is a crucial issue as the detonation of a WMD could seriously hamper commerce and trade and devastate the economy (Ortiz 2007).

With so much cargo being shipped in just one vessel, trying to identify hazardous cargo can be overwhelming. One initiative in place to combat the threat of a WMD is the Cargo Security Initiative (CSI). Operating under the U.S. Customs and Border Protection, the CSI has a threefold approach:

1. Identifying containers that may pose a risk for terrorism
2. Prescreening suspected containers at the port of departure before it arrives at its port of destination
3. Using the latest technology to scan containers that pose a risk

The CSI operates in 58 ports throughout the world and is capable of prescreening up to 80% of all maritime cargo destined for the United States (Container Security Initiative in Summary May 2011).

*War*: A war or other conflict near a shipping region can interrupt the shipping of cargo in close-by areas. The Middle East is an obvious hot area because of the potential for conflicts and its proximity to the Suez Canal, a key passageway that links the European Union and Asia commerce routes. The closing of the Suez Canal would necessitate routing ships around the Cape of Good Hope, an expensive endeavor that would also require up to 100 more ships to handle the industry traffic (Bonney 2011).

## QUESTIONS FOR RESEARCH AND DISCUSSION

1. Using a map or Google Earth, look up the geographical area of east Africa, particularly around the country of Somalia. What areas of vulnerability do you see for potential pirate attacks?

2. Using Google Earth, look up the geographical area around the Suez Canal in Egypt. Notice the various ports, the length of the canal, and the pictures that are posted on Google Earth adjacent to the canal. What observations can you make in terms of risk potential?

## REFERENCES

Bonney, J., Shipping's chokepoint risks, *The Journal of Commerce*, 12–15, February 21, 2011.

Container security initiative in summary, May, 2011. Retrieved February 11, 2014, from http://www.cbp.gov/linkhandler/cgov/trade/cargo_security/csi/csi_brochure_2011.ctt/csi_brochure_2011.pdf.

George, R., *Ninety Percent of Everything: Inside Shipping, the Invisible Industry That Puts Clothes on Your Back, Gas in Your Car, and Food on Your Plate*, Metropolitan Books, New York, 2013.

Kelman, J., Hazards in the maritime transport of bulk materials and containerized products, *Loss Prevention Bulletin*, pp. 28–36, October 2008.

Leach, P., The rising costs of piracy, *The Journal of Commerce*, 24, 25, May 2, 2011.

Ortiz, P., Container Security Initiative helping to ward off terrorist attacks. *Caribbean Business*, p. S2, August 9, 2007.

## Discussion Questions

1. How has the logistics function changed over the past two decades?
2. What mode of transportation has become more important in global supply chains?
3. What is meant by the term *exchange points*?
4. What are container ships? How have they changed movement of goods?
5. What are the sustainability issues related to the movement of goods?
6. What is the difference between offshore outsourcing and reshoring?
7. What are the main drivers of the offshore outsourcing movement?
8. What are the main drivers of the reshoring movement?
9. Why have 3PLs become more important in recent years?
10. Why is risk management becoming more important in today's supply chains?

## References

Accenture, An argument for bundled outsourcing, http://www.accenture.com/us-en/landing-pages/dynamic/Pages/fy11-ps-outsourcing-bundled.aspx?group=bundled&c=tec_usbunpsgs_1210&n=g_Integrated_Outsourcing/a_0_k/integrated_outsourcing&KW_ID=7ee4cfd3-df0d-aba9-f3ff-000022fd48b7, accessed May 27, 2011.

Ackerman, K., How to choose a third-party logistics provider, *Material Handling Management*, 55(3), 95, 2000.

American Shipper, 3PL spend growing at Fortune 500 firms, *AS Daily News*, 2013, www.american-shipper.com/main/ASD/54442.aspx, accessed July 15, 2013.

Armstrong Associates, Inc., ABA's top 50 global third-party logistics provider (3pl) list, largest 3PLs ranked by 2011 logistics gross revenue, 2013, www.3plogistics.com/Top_50_Global_3Pls.htm, accessed May 30, 2013.

Arnseth, L., Japan's disasters highlight force majeure, *Inside Supply Management* (Supplemental Article), 22(4), 28–31, May 2011.

Bennett, D., Risk ahoy: Maersk, Daewoo build the world's biggest boat, *Business Week*, p. 44, September 2013.

Berglund, M., van Laarhoven, P., Sharman, G., and Wandel, S., Third-party logistics: Is there a future? *International Journal of Logistics Management*, 10(1), 59–70, 1999.

Blackstone, J.H., Jr., *APICS Dictionary*, 14th edn., APICS—The Association for Operations Management, Chicago, IL, 2013.

Bowersox, D.J., Closs, D.J., Cooper, M.B., and Bowersox, J.C., *Supply Chain Logistics Management*, 4th edn., McGraw-Hill Irwin, Boston, MA, 2013.

Carbone, J., Buyers look to distributors for supply chain services, *Purchasing*, 128(2), 50, 2000.

Cooper, O., Tadikamalla, P., and Shang, J., Selection of a third-party logistics provider: Capturing the interaction and influence of performance metrics with the analytical network process, *Journal of Multicriteria Decision Analysis*, 19(3–4), 115, 2012.

Coyle, J.J., Langley, C.J., Jr., Novack, R.A., and Gibson, B.J., *South-Western Cengage Learning*, Mason, OH, 2013.

Crandall, R.E., Risk Management in Supply Chains, Minimizing disruptions to streamline flow regardless of complexity, *APICS Magazine*, 20(1), 30–33, 2010a.

Crandall, R.E., Putting together a global sustainability movement, *APICS Magazine*, 20(6), 26–29, 2010b.

Crandall, W.R., Personal interview, January 15, 2013.

Davidson, W.H. and Davis, S.M., Management and organization principles for the information economy, *Human Resource Management*, 29(4), 365–383, 1990.

Ellis, S., Change in the chain: Import/export in the U.S. high-tech supply chain, *IDC Manufacturing Insights*, September 2012.

Enslow, B., Global trade management: Globalization growing pains, Aberdeen Group Research Brief, Boston, MA, April 6, 2005.

Gephart, F., Reshoring to boost the bottom line, *Target*, pp. 12–14, Fall 2012. www.ame.org/Target.

Goldberg, S.N., King, J.M., and Singleton, K., Insure against disruptions, *Inside Supply Management* (Supplemental Article), 22(4), May 2011.

Gottfredson, M., Puryear, R., and Phillips, S., Strategic sourcing: From periphery to the core, *Harvard Business Review*, 83(2), 132–139, 2005.

Gripsrud, G., Jahre, M., and Persson, G., Supply chain management—Back to the future? *International Journal of Physical Distribution & Logistics Management*, 36(8), 643–659, 2006.

Hertz, S. and Alfredsson, M., Strategic development of third party logistics providers, *Industrial Marketing Management*, 32(2), 139–149, 2003.

Hesseldahl, A., Toshiba will have UPS fix its laptops, *Forbes*, http://www.forbes.com/2004/04/27/cx_ah_0427ups_print.html, April 27, 2004.

Hingley, M., Lindgreen, A., Grant, D.B., and Kane, C., Using fourth-party logistics management to improve horizontal collaboration among grocery retailers, *Supply Chain Management*, 16(5), 316–327, 2011.

IBM, Outsourcing services, http://www-935.ibm.com/services/us/en/it-services/outsourcing.html, accessed May 27, 2011.

Isidore, C., Made in USA: Overseas jobs come home, CNNMoney, June 17, 2011, http://money.cnn.com/201106/17/news/economy/made_in_usa/index.htm?iid=HP_LN.

Katz, D.M., All in the timing, CFO.com, June 1, 2011, http://www.cfo.com/printable/article. cfm/14577188.

Kent, D., Do you know how "reshoring" impacts your supply chain? May 28, 2013, www.sdecex. com/article/10950803/do-you-know-how-reshoring-impacts-your-supply-chain, accessed June 10, 2013.

King, W.R. and Malhotra, Y., Developing a framework for analyzing IS sourcing, *Information and Management*, 37(6), 323–334, 2000.

Manpower Inc., Supply/demand, 2010 Talent Shortage Survey results, 2011.

Marks, N., How to manage risk management, CFO.com., 2011, http://www.cfo.com/printable/ article.cfm/14582289, accessed December 15, 2013.

Morton, R., Keeping the supply chain in focus, *Logistics Today*, 48(7), 12, 2007.

Norall, S., 3PL vs. 4PL: What are these PLs, anyway? Layers of logistics explained, http://cerasis. com/2013/08/08/3pl-vs-4pl/, accessed August 6, 2014.

O'Reilly, J., 3PL Perspectives 2012: Shippers and 3PLs—Perfect together, *Inbound Logistics*, July 2012, www.inboundlogistics.com/cms/article/3pl-perspectives-2012, accessed June 17, 2013.

Robinson, C.H., Collaborative outsourcing, preparing for a successful logistics outsource, C.H. Robinson Worldwide, Inc., Eden Prairie, MN, 2013, www.chrobinson.com.

Siegfried, M., Navigating impacts of an expanded Panama Canal, *Inside Supply Management*, 24(6), 18–21, August 2013.

Smith, L., Will on-shoring be the trend for 2011/2012? *Industry Week*, June 16, 2011, http://forums. industryweek.com/showthread.phy?t=26551&cid=NLIWMB.

Thottan, J., Tumulty, K., and Rajan, S., Is your job going abroad? *Time*, New York, 163(9), 26–35, Mar 1, 2004.

Tompkins, J., Integrated logistics: Journey to the ultimate customer, *Material Handling Engineering*, 51(3), 63–72, March 1996.

van Hoek, R. and Chong, I., Epilogue: UPS logistics—Practical approaches to the e-supply chain, *International Journal of Physical Distribution & Logistics Management*, 31(6), 463, 2001.

Vitasek, K., Logistics References, Logistics Service Locator, 2006, http://www.logisticsservicelocator. com/resources/glossary03.pdf, accessed January 15, 2014.

Yuva, J. and Trent, R.J., Harnessing the potential of global sourcing, *Inside Supply Management*, 16(4), 33–40, 2005.

# 10

## Reverse Supply Chains

**Learning Outcomes**

After reading this chapter, you should be able to

- Explain the difference between a supply chain and a reverse supply chain (RSC)
- Briefly describe the history of reverse logistics (RL)
- Describe the main activities of an RSC program
- Identify the principal drivers of the current movement for RSCs
- Explain the role of private industry, the various levels of government, and consumers in promoting RSCs
- Discuss how product design can affect the RSC
- Discuss why companies often outsource their RSC
- Discuss how IT can augment the functioning of RSCs
- Discuss the future of RSCs

**Company Profile: GENCO**

Chapter 9 was about moving goods from the mine or farm to the consumer, along each step in the supply chain. Chapter 10 is about moving goods in the reverse direction, from the consumer back toward the point of origin. Reverse logistics (RL) is about reusing, repairing, and recycling products to satisfy both economic and sustainability objectives. Often, the RSC is not as well organized as the forward supply chain. However, it is becoming an important consideration for all companies who make products, especially those that contain potentially harmful ingredients if deposited in landfills, such as automobiles and computers. As you read this company profile and the chapter in the book, think about these questions:

- Does your school or organization actively recycle waste products?
- What will happen to your smartphone when you want to dispose of it?
- Should the government require companies to do more in RL?
- How can you contribute to the RL movement?

What additional questions do you think relate to RL?

---

### Company Profile: Reverse Logistics GENCO

---

This chapter is about RL, or the RSC. While the forward supply chain moves products and services from the point of origin to the point of use, the RSC is concerned with what happens to the product after its useful life ends or the product is not sold to a consumer. The company featured in this company profile is a leader in RL even though it is still privately owned.

## COMPANY HISTORY

GENCO was started in 1898 as the H. Shear Trucking Company, by Hyman Shear, to deliver commodities with a horse and wagon through the streets of Pittsburgh, Pennsylvania. Mr. Shear purchased a gas-powered truck in 1917 and, over the next few years, expanded its fleet of trucks and service area. In the early 1940s, Hyman's son, Sam, took over the leadership of the company and expanded the delivery fleet and added warehousing and distribution services. The company name changed to General Commodities Warehouse and Distribution Company (GENCO, for short).

In 1971, Herb, son of Sam, joined the company. Through the 1970s and 1980s, they added third-party logistics (3PLs) services to include public warehousing, distribution, and product postponement services. The company entered a major new service area in the late 1980s when Herb Shear developed a centralized returns process known today as RL. The company developed proprietary RL software and expanded its services to help companies manage and maximize the value of their returned products.

In the 1990s and through the first decade of the twenty-first century, the company continued to grow through both organic growth and acquisitions. They added forward logistics and warehouse management software companies to become one of the leading providers of 3PL logistics and supply chain solutions. They expanded their service offerings to include transportation logistics, parcel negotiation and audits, unsaleables management, health-care logistics, and government and defense logistics.

In 2010, GENCO merged with ATC Technology Corp. (ATC) to create one of the largest and most innovative product life cycle logistics companies. ATC had been active in the high-tech industry, and the merger positioned the company as a supply chain management leader in the consumer electronics industry (GENCO History 2013).

"Today, GENCO is a recognized Global Top 25 third party logistics (/3pl-third-party-logistics.php) provider, the industry leader in reverse logistics (/reverse-logistics/reverse-logistics.php) and the leader in Product Lifecycle Logistics solutions" (GENCO History 2013).

GENCO has more than 10,000 employees with over 130 operations in the United States, Canada, and Mexico. It manages more than 38 million square feet of warehouse space, handles more than $1.5 billion in freight annually, and liquidates more than $5 million worth of returns and excess inventory daily (GENCO Information at a Glance 2013).

With the acquisition of ATC Technology Corp. in 2010, the combined company was renamed GENCO ATC with annual revenues of $1.5 billion. The combined company became one of the top 3PL companies in the United States and the top company in RL and product life cycle management (PLM) (GENCO Supply Chain Solutions 2010). In an organization announcement, Herb Shear, former chairman and CEO of GENCO, was appointed executive chairman, and Todd R. Peters, former CEO of ATC Technology, was appointed CEO of the combined company (PR Newswire 2013).

As a further indication of their prominence in the 3PL field, GENCO ATC was rated 25 among global 3PLs and 9th among U.S. companies by Armstrong & Associates, a leading consultancy for the 3PLs sector. In that listing, GENCO ATC was listed as having gross logistics revenues of $3,372 million in 2011 (Burnson 2013).

## DISCUSSION OF REVERSE LOGISTICS

RL is the process of receiving returned components or products, or products at the end of their useful life, for the purpose of recapturing value or properly disposing of them. RL includes the handling of product recalls, end-of-life returns, and seasonal returns and parts in a way to recover value from items that might otherwise be dumped into landfills. By processing these items through a series of evaluation and refurbishing steps, the products can eventually be returned to stock, returned to the original equipment manufacturer (OEM), liquidated in a secondary market, or repaired and reshipped to the original customer. If the complete product cannot be used, spare parts can be recovered, or the base materials can be recycled. If none of these recovery alternatives are available, the materials can be destroyed (Greve and Davis 2012).

Returns have become an increasingly important consideration for retailers, especially online retailers, where the return rate can be between 20% and 30% of purchases, up from the return rate of 8%–10% at traditional retailers. At GENCO, they do a significant amount of analysis prior to implementing RL operations by looking at each client's needs, that is, where should the RL operation be: close to their distribution centers, their stores, or their suppliers? (McCue 2013).

## DESCRIPTION OF REVERSE LOGISTICS AT GENCO

GENCO is a full-service 3PL with services in distribution and warehousing, transportation logistics, contract packaging services, and specialized services for specific industries, such as high tech, retail, and health care. However, it is probably best known for its pioneering work in RL. As mentioned earlier, GENCO entered the RL field in the late 1980s. This came about as a result of their work with Phar-Mor, a discount drug chain that went bankrupt in 1992. GENCO had been working with Phar-Mor to help the drug company reduce its glut of excess inventory. They built a software-based system to process Phar-Mor's returns, a move that propelled GENCO into the RL business.

In 2002, Target bought the RL service from GENCO, and they were soon followed by Sears, Kmart, and Wal-Mart. As a result, GENCO moved from a regional company to one covering the United States and Canada. It has more than 150 customers, including many Fortune 500 companies. Other customers include PPG, Heinz, Dick's Sporting Goods, Levi's, Nike, Dell, CVS, and Harley-Davidson (Cato 2012).

Table 10.1 shows a summary of the returns management services offered by GENCO.

In the retail industry, GENCO developed their proprietary R-Log RL software "to manage the reverse flow of products, information and cash from the point of receipt through final disposition" (GENCO Retail 2013). The services for the retail industry include

- Automation of the returns authorization process
- Validation and crediting of online customer returns
- Improved reconciliation of customer, store, and vendor compliance
- Virtual elimination of store labor to process returns
- Reduced inventory and improved cash flow through fast, automated processing

**TABLE 10.1**

Returns Management Services at GENCO

| | |
|---|---|
| Return center operations | Management of centralized or regional return centers |
| Return to vendor management | Management of the flow of returned product from retailer to vendor |
| Product recall management | Management of recalls because of product defects and other circumstances |
| Test, repair, and refurbishment | Inspection, testing, and repair of products for resale |
| Product recycling | End-of-life recycling to maximize the reuse of all component parts and raw materials |

*Source:* Adapted from GENCO, Reverse logistics, 2013, http://www.genco.com/reverse-logistics/reverse-logistics.php, accessed July 19, 2013.

- Immediate application of vendor credits to the accounts payable system
- Complete visibility and control through a web-based tool that provides real-time information on returns inventory (GENCO Retail 2013).

GENCO is the leading provider of RL services to the technology industry. Some of the services include the following:

- Focus and help implement industry best practices in returns management.
- Integration of harvesting, testing, and repacking good parts.
- Serialized cradle-to-grave tracking to monitor final disposition of end-of-life components.
- Provide detailed reports on product disposition measure compliance.
- Rapid location and recall of relevant models (GENCO Technology 2013).

GENCO also provides RL in liquidating products that do not have sufficient recovery value to warrant returning to the original vendor (GENCO Liquidation 2013). They also have an unsaleables management practice that help companies identify the cause of unsaleable products (damaged or expired) and jointly, with the retailers and vendors, determine responsibility and corrective actions to reduce the amount of future unsaleable products (GENCO Unsaleables 2013).

## DESCRIPTION OF PLM AT GENCO

With the acquisition of ATC Technology in 2010, GENCO became a leader in PLM, with emphasis on the high technology industry. They explain their approach in their video at http://www.genco.com/about/product-lifecycle-logistics-video.php. The PLM approach provides a cradle-to-cradle approach to a product, tracking it from origin to consumer and then from consumer back to the disposition through the RL process.

RL has become an integral part of supply chain management. It is no longer possible for companies to complete the design of a new product without considering both the requirement for its use as well as its eventual disposition.

## REFERENCES

Burnson, P. The top 50 global & top 30 domestic 3PL providers. SupplyChain247, 2013, www.supplychain247.com/article/the_top_50_global_top_30_domestic_third_party_logistics_providers/ihs, accessed September 3, 2013.

Cato, J. GENCO goes forward in reverse logistics, *McClatchy—Tribune Business News*, April 8, 2012.

GENCO History, It began in 1898, www.genco.com/about/history.php, accessed September 2, 2013.

GENCO Information at a glance, www.genco.com/about/at-a-glance.php, accessed September 2, 2013.

GENCO Liquidation, Inventory liquidation, 2013, www.genco.com/liquidation/inventory-liquidatin.php, accessed September 3, 2013.

GENCO product lifecycle logistics, http://www.genco.com/Videos/product-lifecycle-logistics-video.php, accessed September 4, 2013.

GENCO Retail, Reverse logistics in the retail industry, 2013, www.genco.com/retail/retail-reverse-logistics.php, accessed September 2, 2013.

GENCO supply chain solutions announces completion of merger with ATC Technology Corporation, 2010, www.genco.com/press-releases/2010/10125_GENCO-ATC-Complete-Merger.php, accessed September 4, 2013.

GENCO Technology, Reverse logistics for the technology industry, 2013, www.genco.com/technology-industry/ce-reverse-logistics/php, accessed September 2, 2013.

GENCO Unsaleables, Unsaleables management, 2013, www.genco.com/unsaleables/unsale-ables-management.php.

Greve, C. and Davis, J., Recovering lost profits by improving reverse logistics, A report commissioned by UPS, Atlanta, GA, 2012.

McCue, D., Dealing with zombie inventory, *World Trade, WT 100*, 26(3), 39–42, 2013.

PR Newswire, GENCO announces leadership change, Pittsburgh, PA, January 9, 2013.

## Introduction

Earlier chapters of this book described the activities involved in the supply chain that moves goods and services toward the ultimate consumer. This chapter describes the activities that move those goods, and their accompanying services, in the reverse direction or back to the supplier or third-party processor. This process is commonly termed RL or RSC. We use the terms interchangeably in this book.

## Description of Reverse Supply Chain Networks

Reverse logistics (RL) or reverse supply chains (RSCs) are an integral part of the supply chain process. Here are definitions offered by two well-known sources:

> Reverse logistics—A complete supply chain dedicated to the reverse flow of products and materials for the purpose of returns, repair, remanufacturing, and/or recycling. (Blackstone 2013)

Blumberg offers a more exhaustive description:

> There are a number of reasons why attention needs to be placed on the processes, dynamics, and structure involved in the return of goods, material, and parts from the field at the end of the direct supply chain. In addition to the normal situation, where the products and goods at the end of the supply chain are no longer wanted or needed or have little value because they are obsolete or not operating, there is, of course, a requirement for disposal, which deals with specific questions involving solid waste, liquid waste, and hazardous materials. Furthermore, many other products and materials in the field have value that could be recovered through repair, disposition, and recycling,

such that it may not be efficient to simply throw them away. These could include the following types of goods and products:

- Products in the field that have failed and need to be repaired or properly disposed
- Parts and subassemblies of products that can be reused, because they are perfectly good (no trouble found), or that can be repaired or reworked
- Products that are perfectly good but have, nevertheless, been returned by the purchaser as well as products sitting on retailer's shelves that have not been sold
- Products and materials that have been recalled or are obsolete, but still have a useful life
- Products, materials, and goods that have been thrown away, but can be recycled and reused
- Products at the end of lease, but not at the end of life

In these situations, and others, reverse logistics is important from an economic recovery standpoint, and because of the reduction or elimination of trash and junk. In addition, it has been increasingly linked to the general *direct* supply chain since in many cases the original supplier is in the best position to control the return process as well. This full process of shipment out and back by the organization, found particularly in high tech service and support roles, is defined as a *closed loop supply chain* (CLSC). (Blumberg 2005)

We make a distinction between *reverse logistics* and *green supply chains*. RL refers to the process for handling returns and how best to reintroduce the returns, as whole units or component parts, back into the forward supply chain, through repackaging, remanufacturing, or reprocessing. The intent is to avoid costs or, in more forward-looking companies, generate added revenue. A green supply chain implies a concern with the eventual disposition of the materials in the finished products and their effect on the environment. While both elements may be part of the same system, they have different drivers. We will discuss green supply chains as part of the broader sustainability movement in Chapter 11.

## Benefits of Reverse Logistics*

The most direct benefit from RL is cost avoidance. Ironically, upper management is often the biggest obstruction to achieving that benefit, because they do not understand what returns are costing them (Andel 2004b). Although most of the attention is on product recovery, even entire buildings can be part of the RL process. The Pillowtex complex of buildings in Kannapolis, North Carolina, was demolished to make way for developer David Murdock's North Carolina Research Campus. The D.H. Griffin Wrecking Co. will recycle 10 million bricks (enough for 625 brick houses), 10,000 tons of steel (enough for 6,667 sedans), and 5,000 pine beams (enough for flooring 500 living rooms) (Bell 2006).

There are also environmental benefits. The private sector will be driven to recycling, both for profit and for public relations purposes. "I have no doubt that every industrial product will be subject to recycling and remanufacturing. At some level, state or federal, guidelines will be established to influence people to do it. The environmental political capital achieved by pressing on recycling issues can't be ignored. Politically it's hot to support recycling" (Andel 2004a, p. 35).

---

* This section has been adapted from Crandall (2006).

## Barriers to Reverse Logistics

As with all programs, there are barriers to their successful implementation. An earlier survey of over 300 logistics managers by the Reverse Logistics Executive Council reported the following major barriers—relative unimportance, company policies, lack of systems, competitive issues, management inattention, lack of personnel resources, lack of financial resources, and legal issues (Caldwell 1999).

More recently, similar barriers were discovered:

- No executive overseer or champion to take responsibility for RL and drive it.
- RL is seen as *just a cost.*
- Management does not see customer satisfaction and service differentiation opportunities that RL can bring.
- There is a perceived inability to quantify the cost of returns.
- There is a lack of systems/information visibility.
- A silo versus a cross-functional mentality exists.
- Poor manual processes for conducting RL exist (Andel 2004b).

Others studied the variables of timing RL programs (early or late) and the commitment of resources. They found that early entrants have an advantage over late entrants, because it is more difficult for the late entrants to catch up. However, late entrants can offset their delay somewhat by committing additional resources to the effort. They recommend the following:

- Companies that have not started a formal RL program should begin so now.
- Companies already involved in RL should make sure they make enough of a resource commitment to be able to gain the potential benefits.
- Companies not willing or able to commit sufficient resources to RL should consider outsourcing this function (Richey et al. 2004).

Perhaps, the greatest barrier is thinking returns can be handled on an ad hoc basis when most authorities insist it requires a systematic approach.

## Continuation of Forward Supply Chains

RL represents a continuation of the supply chains described in earlier chapters. The typical supply chain carries goods and services forward from the raw materials stage through to finished goods to be distributed to the consumer. RL, or the RSC, takes the finished products, after they have been used or rejected by the consumer, and returns the products back along the RSC until the materials have been refitted for reuse or returned to landfills or hazardous waste disposal sites.

While the forward supply chain is generally portrayed as a continuous flow of products, the RSC is still in a fragmented state. Consequently, while all of the stages of the RSC are being performed to some extent, they are somewhat isolated from each other and have seldom been linked together in a continuous reverse flow.

Figure 10.1 shows the RSC is a continuation of the forward supply chain, and the two are linked together at each stage. In Figure 10.1, the upper level shows the forward supply chain as a flow from the source of most materials, the mines, or agriculture. The lower

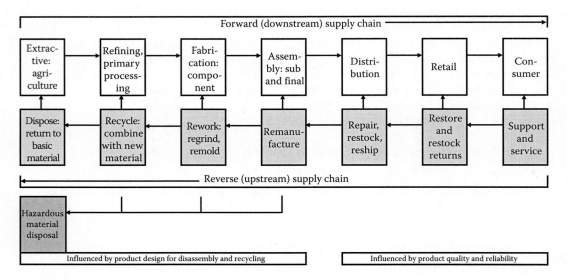

**FIGURE 10.1**
Forward (downstream) supply chain.

level shows the reverse flow and indicates the multiple links between the forward supply chain and the RSC. The diagram also shows the need to extract hazardous materials from the RSC and dispose of it in an acceptable way.

Another way of viewing the RSC is as an extension of the product life cycle. All products go through the stages of design and development, introduction, growth, maturity, and decline. The RSC has its own life cycle that can be similar in shape to the product life cycle, although lagging in time. Figure 10.2 shows this relationship. In the early stages of a

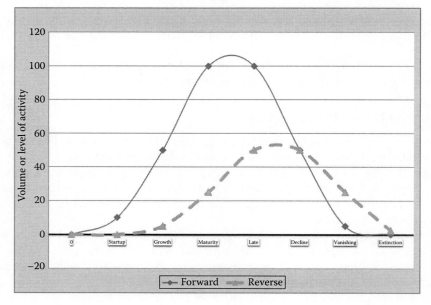

**FIGURE 10.2**
Product life cycles.

product's life cycle, the major activity in the RSC will be in processing returns for transfer or repair before reentering it into the supply chain. As the product enters its mature stage, remanufacturing increases and recycling begins. The RSC continues to operate beyond the end of a product's life cycle as parts continue to be recycled. The last stage, if it occurs at all, will be returning the raw materials back to some holding area, either for future use or for hazardous materials disposal.

The reverse life cycle is a combination of returns, restoration, remanufacturing, recycling, and disposal. Some writers characterize the combination of forward supply chain and RSC as a cradle-to-grave or, even more completely, cradle-to-cradle cycle (Dyckhoff et al. 2004).

## History of Reverse Logistics

Managerial Comment 10.1 describes the modern evolution of the RL movement and its emergence as an auxiliary business for a company. It also points out that RL can be the beginning of a whole new industry.

Even before the formal recognition of RSCs, society has been finding ways to informally process goods along RSCs. Examples include

- Patching clothes to extend their useful life
- Using worn-out clothing for cleaning rags
- Using stale bread for toast or croutons
- Placing kitchen scraps and such in a compost bin
- Recycling aluminum cans, glass, and plastic containers
- Using old tires and a rope to make a swing
- Using an empty wine bottle to make a candle holder

The list could go on, but the idea is there; an informal RSC has always been in operation. It has been formalized through the emergence of consignment shops for clothing and recycling excess food supplies from restaurants or retail grocery stores through food outlets for the hungry.

## Principal Drivers of the Movement

What are the drivers of the RSC? There are at least six—individual consumers, individual businesses, society as a group, business sectors, educational institutions, and governments at all levels.

### Individual Consumers

Individual consumers were among the first to recognize the benefits of the RSC. Perhaps, the first application was to use animals for both food and clothing. Another early driver was convenience—when clothes were hand sewn at home, it was easier to patch a pair of

**MANAGERIAL COMMENT 10.1    EVOLUTION OF REVERSE LOGISTICS**

The idea of RL businesses has been around at least since the early industrial age when merchants recognized that old clothing and rags, linens, and so forth could be reproduced and used to produce new textile products. The business gained significant momentum, gaining visibility in the 1980s when environmental issues became emotionally charged sensitive topics. This led to the emergence of waste (general and hazardous) processing and recycling business.

The closed-loop supply chain (CLSC) concept, embedding RL principles, developed in the early 1980s as a direct result of microminiaturization, large-scale integration, and modularization design; it was first implemented in the electronics industry (computing, office automation, telecommunications process, control, and plant automation, etc.).

Until this time, the typical process for repairing a product or piece of equipment was to make the corrections in the field (i.e., fix in place). However, the use of large-scale integrated digital circuits and circuit boards, with built-in diagnostics and defined test points, led to two changes:

1. Increased reliability of the subsystems and components
2. Reduced repair time through pull and replacement of modules rather than field repair

The increased sophistication and complexity of circuit boards and pull-and-replace modules increased the value of these components, which, in turn, increased the economic value of returning the material for repair and reallocation. This trend has now been extended from electronics into other technologies, including mechanical and electromechanical equipment. In addition, sensors and control mechanisms originally provided in analog mechanical form are increasingly converted to digital functions to increase reliability and reduce the size, weight, and power requirements. This in turn created the business need to apply RL and CLSC processes in the high-tech market.

The key is to understand the concept of life cycle management and support—viewing a product or technology over its entire life cycle of use.

*Source*: Adapted from Blumberg, D.F., *Introduction to Management of Reverse Logistics and Closed Loop Supply Chain Processes*, CRC Press, Boca Raton, FL, 2005.

pants than to make a new pair. It was also common practice in large families to pass down clothing from the older child to a younger child.

Another driver was scarcity or unavailability of needed items. Early settlers used wooden pegs to connect timbers before iron nails were available. Animal skins provided warm coats before woven cloth became available.

The desire to save money was a driver in many communities. Even if the product was available, some households could not afford to buy new items; consequently, they recycled or reworked many products, especially food and clothing to avoid the need to buy new. For example, hash and soup recipes extended the life of a hambone or chicken.

## Individual Businesses

Businesses are also beginning to recognize the potential problem of resource scarcity. Is there a danger of running out of critical resources, such as oil or precious metals? Will population growth increase to a point that exceeds the capacity of the earth's food production? Friedman points out this is a real possibility if today's trends are not changed (Friedman 2008).

In an increasing number of cases, new businesses are emerging to manage RSC activities. Rather than original product manufacturers setting up to handle the reverse activities, new companies are emerging to perform these activities. Some examples include

- Recycling plants to process paper and plastics
- Remanufacturing plants to rebuild capital equipment—water meters, locomotives
- Used books sold through Amazon.com
- Plants to recover precious metals from computers

Many individuals are interested in becoming entrepreneurs, and the RL venue offers many opportunities.

## Society as a Group

Society, as a group, is beginning to recognize, and accept, the need for action in RSCs. Adverse public reaction is also a strong driver at the business level. In today's society, individual consumers are encouraging the movement toward RSC. Communities are at work recycling, in fact, as this book was being written, the campus where one of the author's works just implemented an aggressive new recycling program. Recycling and promoting green efforts are seen by many as a way to identify with a social cause that is helpful in the stewardship of natural resources.

### Environmental Concerns

One of the most obvious reasons for RL is a concern about the adverse effects of past practices on the environment. Contemporary examples of practices that reduce landfill requirements include

- Using recycled oil to heat a manufacturing plant (not only does this reduce the amount of oil disposal, but it also minimizes heating costs)
- Using broken wood pallets to make biodegradable woven blankets that hold grass and fertilizer for restoring excavated areas of new roads or patching bare spots in lawns
- Putting filters and other cleaning devices on exhaust systems to reduce the amount of smoke and other contaminants emitted into the atmosphere
- Grinding up worn-out automobile tires to use in the fill for road beds and for building new running tracks
- Recycling floor coverings such as carpet and tile and using them to make new floor coverings

Reducing the adverse effects on the environment is a reason why RSC is an important concept, not only for society but also for business. While businesses often were forced to begin some of these practices through either societal or government pressure, benefits such as less waste and lowered costs can also be realized.

### Resource Scarcity Concerns

Conserving natural resources is another reason the RSC has become important. Natural resources such as precious metals, rain forests, endangered animals, air, water, and land for growing food are just a few examples of the need to conserve.

Societal opposition stems first from environmental concerns. Smoke, smog, polluted rivers, blighted forests, receding shorelines, and the uncertain effects of global warming have stirred the formation of numerous special-interest groups to focus their attention on the need to reduce greenhouse gases and liquid runoffs into streams and rivers.

The concern about resource scarcity is closely allied with environmental concerns. Pollution of rivers and lakes raises the question of the availability of pure drinking water. The solution of bottled drinking water answers one concern only to introduce another concern—how to dispose of the plastic water bottles.

The proliferation and enormity of environmental problems are of concern to business and government leaders. In terms of promoting a sustainability program, it is difficult to decide where and how to start. In an effort to assign priorities, Edwards identifies what he considers the most pressing problems—climate temperature and uncontrolled population growth. He believes that if we could solve these two problems, the remaining environmental problems would become more manageable. He goes on to suggest that solutions exist for these problems. Global warming could be largely controlled by shifting from fossil fuels to alternative energy sources such as hydrogen, solar, wind, and geothermal. Population growth could be controlled through educational programs, although such efforts would have to overcome political and sociocultural resistances (Edwards 2005).

## Business Sector

In the past, adverse public reaction has also been a strong driver at the business level. Steel companies in the Pittsburgh, Pennsylvania, area found increasing animosity among local residents with the smoke-filled atmosphere. Citizens in Los Angeles, California, were vocal in their resentment of smoggy conditions in their area. Paper companies found adverse opinions surrounding the discovery of dead fish in the rivers below their discharge of deadly chemicals. In response to public pressure and eventually government regulation, businesses found they had to clean up their practices in order to reduce their pollution output (Hartley 1993).

However, today, the principal driver of business interests is the opportunity for increased profits from going *green*. Economic benefits at the company level and the spawning of new business and even industries is now possible.

### Economic Benefits Possible

Many businesses find that, contrary to their initial beliefs, the cost of transitioning to new technologies that enable them to recycle materials is more than offset by the savings they receive from reusing those materials. The initial skepticism some managers display toward sustainability efforts may be a combination of resistance to change coupled with the fact that recycling is not a part of their basic business of making products to sell. In these cases, management probably views the need to recycle as a distraction from their corporate mission.

Today, a primary driver is cost savings. Eastman Kodak found that they could recover silver they were washing down their drains (Poduska et al. 1993). Many businesses have eased into sustainability practices by using hot air from exhaust gases to heat buildings or drive processes.

### New Businesses, Even New Industries, Possible

In an increasing number of cases, new businesses are springing up to meet the needs of RSC activities. Rather than original product manufacturers tooling up to handle these reverse activities, new companies are emerging to perform these activities on an outsourcing basis.

### Educational Institutions

Educational institutions have two basic responsibilities in the RL movement—to teach about it and to practice what they teach. They must teach students about the topic and encourage them to become responsible citizens as they assume their adult role as members of society. At the higher level, universities and technical schools have the added responsibility of educating students for careers in the field of supply chain management or, more specifically, in the growing field of RL. Many universities have introduced programs of study about sustainability, of which the RSC is an integral part.

In addition, educational institutions have a responsibility to organize and maintain recycling programs and use recycled products. They also have a responsibility to practice in the broader realm of sustainability. Universities, many with numerous buildings, have an opportunity to put ideas into action, such as wind turbines, solar thermal systems, energy-efficient appliances, and LEED©-certified buildings. One of the authors teaches at a university recognized by the Sierra as one of the *coolest* schools in the United States (ASU News 2011).

### Governments: At All Levels

Governments face the difficult task of deciding the following: first, should they act? And, second, what should they do? They face a variety of opinions from special interest groups that interpret tentative scientific evidence to suit their own purposes. Often, the dilemma is to determine if the problem is serious enough to warrant action.

They also have the problem of trying to decide which level of government has the jurisdiction to act. Should local governments decide on the allowable content of the discharge from a local paper company? Should state governments decide on allowable automobile emissions? Should the federal government impose mandatory requirements for fuel-efficient cars? Although many special interest groups push for more government intervention, government agencies are finding it difficult to agree on the most appropriate level of government control.

## Activities in Reverse Logistics

The RSC consists of several phases: support services, returns, restoration, remanufacturing, recycling, and disposal.

### Service: Assure Proper Use of Product

Although it may not be thought of as a part of the RSC, it is important that the company selling the original product provides the support and service to meet customer needs. Support activities include informing the customer how to remedy common problems that may occur with the installation or life of the product. Many such items can be handled with a simple phone call or email. Otherwise, the product may be discarded as useless and prematurely enter the RSC. The producer may incur extra costs to replace a product that was good but not properly maintained by the customer. It also means that the RSC has an unnecessary addition.

### Returns: Repackaging or Relocation

Customer returns are a way of life for most businesses, especially those in the retail end of the supply chain. It is especially true for the e-commerce companies who incur a higher level of returns than bricks and mortar retailers.

Handling returns vary according to product type. It is unlikely that the product can be returned to the selling area without some work being done. It may need a new price tag or repackaging to make it salable. Usually, these procedures can be performed at the store or distribution center that accepts the return. However, some returns may need to be shipped to other locations if the selling season is over or if there is an excess amount of inventory at the location.

Analyzing the reasons for returns is also an important function. As companies determine the causes of returned items, they can take actions to reduce the level of returns (Rogers et al. 2002). This type of analysis becomes especially important as online purchases increase. The rate of returns from online purchases can be upward of 30% of purchases, as compared to 10% for purchases from physical retailers.

### Restoration: Minor Modification or Repair

If the returned product remains the property of the customer and is being returned for repairs, it probably will not be handled at the retail store level but sent to a third-party site for repair. For example, UPS is handling the repair function for Toshiba laptops in their Louisville, Kentucky, facility (UPS 2004).

Another cause of returns is the issuing of product recalls, where the manufacturer has to modify the original product to assure its full and safe functioning. Product recalls have become a regular occurrence, especially in the automobile industry.

In the preceding examples, the customer retains ownership and use of the product. In some cases, products returned to the selling company may need minor repairs before they can be resold into the forward supply chain. Examples include replacing batteries in a cell phone or repainting a scratch on a clothes dryer before restocking it in a store's inventory.

## Remanufacturing: Overhaul and Major Rebuilding

Remanufacturing is a term that has come into common usage during the past half century. It is defined as "An industrial process in which worn-out products are restored to like-new condition. In contrast, a repaired product normally retains its identity, and only those parts that have failed or are badly worn are replaced or serviced" (Blackstone 2013).

Remanufacturing implies a substantial renovation of the product before its reuse. It is more than just a simple *remove and replace* operation. Train locomotives and airplanes are examples of products that incur substantial overhaul or component replacement to make them ready for continued use. Automobiles that have been in accidents can sometimes be rebuilt to a satisfactory level. Truck fleets such as delivery step vans are also common products that are remanufactured.

At one time, the idea of being rebuilt carried with it a somewhat inferior connotation—using a rebuilt carburetor may not have been considered the equivalent of a new one. Today, reconditioned, or rebuilt, items may be superior to new ones, because they will have the latest, and improved, versions of components in them. Just as replacement gauges in recalled automobiles are presumed to be an improvement, reconditioned machines may also be improvements over the originals (Andel 2004b).

## Recycling: Reconstitution as Part of Another Product

If a product is no longer usable in its original form, it can often be recycled by changing its form to blend it with a newer version of a similar product. Scrap paper can be recycled and combined with new paper to make a composite that can be used for newsprint, greeting cards, or other applications.

In some cases, the recycled product reemerges in a form completely different from its original state. Aluminum house siding is melted down to make kitchen pans. Corn stalks are ground to become a component for animal feed.

There is another form of recycling and that is the movement of equipment or special furnishings from one location to another. This could occur when companies close down one facility and move to another or sell the equipment from the closed facility to other users. Paul Clemens (2011) tells of watching, over a period of months, the dismantling of the Budd Detroit plant, built in 1919, in 2007. Some of the equipment was huge; one press was 26 ft tall and weighed a million pounds. One purchase required 120 truckloads to get a press line to Mexico. Over 1200 lots of equipment—mechanical presses, power shears, mills, lathes, grinders, band saws, boring machines, radial arm drills, blast cabinets, threading machines, robotic welders, planers, air compressors, cooling towers, boom lifts, and forklifts—were sold at public auction. This was a gigantic recycling project.

## Disposal: Return to Natural State

In some cases, the product may not be considered suitable for recycling into other products. In this case, it is transported to landfills or other disposal sites where it is expected to eventually return to its original state. Obviously, this does not always occur within a reasonable time period. While paper and wood products may be compatible with the disposal objectives, plastics and metals may not.

At the present time, a substantial portion of products that enter landfills can be recycled. The challenge for RSCs is to improve the collection and recycling systems to reenter a much higher percentage of products into extended useful lives. We describe the need for an integrated system later in the chapter. One of the most active disposal programs is the disposal of close to 300 million automobile tires each year, just in the United States (Crandall 2012).

## Hazardous Waste Disposal

The intent is that eventually only benign materials are put into normal landfills and toxic or hazardous materials become a part of specially designated hazardous waste disposal sites. Unfortunately, that ideal situation is not always achieved today. Many hazardous items—electronic appliances or computers containing toxic, even radioactive, materials—find their way into the community landfill, which can eventually pose a risk to the local water supply.

One of the most aggressive efforts in waste disposal is taking place in Europe, where the Waste Electrical and Electronic Equipment (WEEE) and the Restriction on Hazardous Supplies (RoHS) directives were put into effect in 2008. The WEEE and the RoHS directives are a step toward requiring more formal requirements on the correct identification and disposition of materials. WEEE aims to minimize the impacts of electrical and electronic equipment on the environment during their product life times and when they eventually are discarded and become waste. It applies to a huge spectrum of products. It encourages and sets criteria for the collection, treatment, recycling, and recovery of waste electrical and electronic equipment and makes producers responsible for financing most of these activities. Private householders are able to return to the WEEE without charge.

The RoHS Directive, which became effective on July 1, 2006, bans placing on the European Union (EU) market new electrical and electronic equipment containing more-than-agreed-on levels of lead, cadmium, mercury, hexavalent chromium, polybrominated biphenyl (PBB), and polybrominated diphenyl ether (PBDE) flame retardants. There are a number of exempted applications for these substances. RoHS takes its scope broadly from the WEEE Directive. Manufacturers will need to ensure that their products—and their components—comply in order to stay on the single market. If they do not, they will need to redesign products (Yang 2008). U.S. companies that plan to sell their products in the EU will also have to comply with these directives.

This is an area that requires the combined efforts of business and government to come up with an acceptable and workable solution. In the preceding section, we described the major steps in the RSC process. In the next section, we will outline actions needed to achieve these desired objectives.

## Role of Private Industry

Private industry will be a key participant in developing RSCs. While consumers and governments will support these ongoing efforts, private industry will be the driver needed to actually perform the work to make it happen. Some of the specific things they need to achieve are described as follows.

## New Paradigms in Product Design (Design for Sustainability)

One of the most important areas will be the design of new products for greater sustainability (Guide and Wassenhove 2002). This fits with the other *design for* efforts such as design for manufacturability (DFM) or design for assembly (DFA). It also aligns with the PLM concept (Crandall 2008).

If products are designed with sustainability objectives, it will prevent the need for corrective efforts later in the supply chain. This move will be a positive step, but it places additional demands on product designers to plan further into the future. For example, an increased number of decisions will be needed concerning both direct materials that go into the product and indirect materials, such as cleaning solutions that are used in the manufacturing processes. One of the more interesting examples of a product designed with sustainability in mind is Procter & Gamble's cool water detergent. The company set out to develop a clothes washing detergent that was as good as their present product but that would not require using electric energy to heat the energy (Crandall 2010).

One of the trade-offs that always arises in product design is the effect of the improved sustainability design on product costs. Will the consumer understand the reason for the increased price and accept it? Will the consumer willingly endorse the product as being more desirable? How will the originating company consider the salvage value of the product? Will the salvage value be increased because of the efforts along the RSC to maximize the utilization of the product? If so, how do these factors affect the product design?

## Design and Operate Green Supply Chains

Regardless of how well the product is designed, management must pay careful attention to the processes used in making the product to avoid contamination and modification that reduces the compatibility of the product in a green supply chain. It is important not only to design green processes within a company but also to extend the processes throughout the supply chain (Hervani et al. 2005).

One of the more interesting challenges is the need to develop accounting systems that can identify and assess not only the costs at each stage of the green supply chain but also the value of the product. How will the costs and values be assigned to each participant so that profits can be equitably distributed? Companies are finding this task is difficult for their forward supply chains; it is certain to be even more difficult for RSCs.

## Develop Systems to Manage Reverse Logistics

Original manufacturers may eventually be responsible for managing the RSC for their products. These responsibilities will most likely be spelled out in government legislation, such as in the WEEE and RoHS directives mentioned earlier. However, it is possible that in the future, original manufacturers will have a cradle-to-cradle responsibility for their products. In this scenario, automobile manufacturers will be responsible for the eventual disassembly and recycling of their cars. Consequently, SCM must be expanded to closed loop management (CLM) (Dyckhoff et al. 2004).

If the original manufacturer decides to outsource the activities along the RSC, they have a challenge even greater than designing and managing the forward supply chain—a challenge many companies are still trying to handle effectively. In the forward supply chain, the ultimate consumer is identifiable, and marketing efforts have been adapted to focus on their target markets. In the RSC, there are different customers, and most of these markets are still to be defined. Will the present marketing organizations be the ones to handle the RSC markets, or will that fall to other functional areas, such as engineering or operations? (Crandall and Crandall 2003).

## Participate in Joint Ventures to Seek Social Objectives

In managing forward supply chains, private industry generally resists governmental participation, unless it is in the form of subsidies without too many strings attached. In the RSC, private industry may be more receptive to government participation. It may take government initiatives to mandate that individual consumers comply with the directive to recycle. In practice though, it is uncertain whether consumers will fully comply unless they face economic penalties (Technology International Group).

Another way in which industry and government may collaborate is in locating hazardous waste disposal sites. For example, operators of nuclear power plants have no way to require a state to allow a disposal site to be built in their area; only the government can require that. Getting a major effort to recycle materials started may require governmental subsidies. While private enterprises collect the roadside materials from homeowners, this effort may require government promotion and financial encouragement to make it a success.

Green supply chains and waste disposal programs are a prime target for many socially active organizations. It will be difficult to develop a workable consensus. It may be necessary for private businesses to form alliances with outside stakeholder groups to share and validate information in order to design and implement action programs that satisfy mutually acceptable objectives.

---

## Role of Government

Private industry cannot achieve operational green supply chains or fully functioning RSCs alone. They will need help from government entities at all levels—local, state, and federal. Some actions that should be expected from governmental agencies include research, legislation, regulation, and participation.

## Research: To Identify Threats and Opportunities

One of the most important functions of government, especially at the federal level, is to have agencies that conduct and support research to determine cause-and-effect relationships in a number of areas related to RSCs. A few of the existing agencies and their assigned areas of responsibility are the following:

1. *U.S. Environmental Protection Agency (EPA)*: The mission of the EPA is to protect human health and the environment. Since 1970, EPA has been working for a cleaner, healthier environment for the American people (http://www.epa.gov/epahome/aboutepa.htm).
2. *U.S. Bureau of Land Management (BLM)*: The BLM is responsible for carrying out a variety of programs for the management and conservation of resources on

258 million surface acres, as well as 700 million acres of subsurface mineral estates. These public lands comprise approximately 13% of the total land surface of the United States and more than 40% of all land managed by the federal government (http://www.blm.gov/wo/st/en/info/About_BLM.html).

3. *U.S. Department of Energy (DOE)*: The DOE's overarching mission is to advance the national, economic, and energy security of the United States, to promote scientific and technological innovation in support of that mission, and to ensure the environmental cleanup of the national nuclear weapons complex (http://www.energy.gov/about/index.htm).

4. *U.S. Department of Interior (DOI)*: This department has several agencies for which it is responsible, including the Bureau of Reclamation, U.S. Geographical Survey, BLM, U.S. Fish and Wildlife Service, National Park Service, Minerals Management Service, Bureau of Indian Affairs, and Office of Surface Mining. Each agency has its own mission statement (http://www.doi.gov/).

This is not a complete list, but it should give the reader an idea of the scope of the U.S. government's interest in environmental issues. It is likely that any of these agencies could have a need to support research in the area of RSCs that affect environmental issues.

State governments also participate in environmental matters. The Environmental Council of the States (ECOS) is the national nonprofit, nonpartisan association of state and territorial environmental agency leaders. The purpose of ECOS is to improve the capability of state environmental agencies and their leaders to protect and improve the human health and the environment of the United States.

The rationale behind the ECOS is that state government agencies are the keys to delivering environmental protection afforded by both federal and state laws. Further, the ECOS (http://www.ecos.org/) can play a critical role in facilitating quality relationships between federal and state agencies in the fulfillment of that mission. That role is to

- Articulate, advocate, preserve, and champion the role of the states in environmental management
- Provide for the exchange of ideas, views, and experiences among states and with others
- Foster cooperation and coordination in environmental management
- Articulate state positions to congress, federal agencies, and the public on environmental issues

At the international level, the United Nations (UN) has sponsored three agencies that have responsibilities throughout the world—the Intergovernmental Panel on Climate Change (IPCC), the United Nations Environment Programme (UNEP), and the Food and Agriculture Organization of the United Nations (FAO).

The IPCC was established to provide decision makers and others with an objective source of information about climate change. The IPCC does not conduct any research nor does it monitor climate-related data or parameters. Its role is to assess on a comprehensive, objective, and transparent basis of the latest scientific, technical, and socioeconomic literature concerning climate change. This includes literature relevant to the understanding of the risks of human-induced climate change, its observed and projected impacts, and options for adaptation and mitigation. IPCC reports should be neutral with respect to policy, although they need to deal objectively with policy-relevant scientific, technical, and socioeconomic factors. They should be of high scientific and technical standards and aim to reflect a range of views, expertise, and wide geographical coverage.

The IPCC is a scientific intergovernmental body set up by the World Meteorological Organization (WMO) and by the UNEP. Its constituency includes the following:

- *Various governments*: The IPCC is open to all member countries of the WMO and UNEP. Government representatives participate in plenary sessions of the IPCC where decisions concerning the IPCC work program are taken and reports are accepted, adopted, and approved. They also participate in the review of IPCC reports.
- *The scientists*: Hundreds of scientists all over the world contribute to the work of the IPCC as authors, contributors, and reviewers.
- *The people*: As a UN body, the IPCC work aims at the promotion of the UN human development goals (http://www.ipcc.ch/about/index.htm).

A second agency is the UNEP. Its mission is to provide leadership and encourage partnership in caring for the environment by inspiring, informing, and enabling nations and peoples to improve their quality of life without compromising that of future generations. For more information, see http://www.unep.org/Documents.Multilingual/Default. asp?DocumentID=43.

A third agency under the guidance of the UN is the FAO. It leads international efforts to defeat hunger. Serving both developed and developing countries, the FAO acts as a neutral forum where all nations meet as equals to negotiate agreements and debate policy. The FAO is also a source of knowledge and information. It helps developing countries and countries in transition modernize and improve agriculture, forestry, and fishery practices and ensure good nutrition for all. Since its founding in 1945, it has specially focused on developing rural areas, home to 70% of the world's poor and hungry people. For more information, see http://www.fao.org/about/about-fao/en/.

The existence of multiple agencies indicates there is concern over the importance of environmental and related issues. It is also evidence of the diversity of interests and potentially conflicting approaches to identifying and solving these problems. As mentioned earlier, the list is not meant to be complete but, rather, representative of international programs.

## Legislation: To Standardize Business Requirements

Once problems are identified, the role of governments is to enact legislation that will provide a focused effort on solving the problem. This is easier said than done. Some problems, such as global warming, or climate change, as some prefer, have remained controversial for years since there is not universal agreement that it is even a problem. Even among those who do agree that warming is taking place, there is not complete agreement as to the consequences and even less agreement about the actions to take. The questions are complex and invariably involve short- or intermediate-term economic considerations that almost always seem to conflict with the long-term environmental solutions.

However, that is the job of governments—to make laws, or issue directives, which will direct and control the efforts of private industry and consumers. In the case of RSCs, it is difficult to evaluate how well the governments are doing during the short term. Traditionally, government objectives are usually sincere in motive, but often inefficient in implementation of enacting and enforcing policy changes. Sincere individuals agree on the objectives; they do not agree on the means of achieving those objectives. In his book *The Sustainability Revolution*, Andres Edwards overviews a wide assortment of legislation throughout the world at local, regional, national, and international sustainability efforts (Edwards 2005).

### Regulation: To Monitor Performance

Once governments enact legislation, the designated agencies need to monitor the performance of affected organizations and individuals to assure compliance. The discharge from paper mills into rivers has to be measured; the emissions from a steel mill into the atmosphere have to be monitored; and the disposition of hazardous materials by computer repair shops into community landfills has to be controlled.

Regulation works reasonably well when organizations cooperate and are supportive of easy-to-decipher rules with which the organization can comply; however, vague directives and uncooperative organizations make enforcement difficult for governmental agencies and frustrating for those organizations who do try to comply.

A somewhat trivial example is the ban on smoking within 50 ft of a university's buildings. Smoking is an obvious act, and 50 ft is a tangible measure, so enforcement should be easy. Unfortunately, the smokers do not comply. As a result, entrances to otherwise attractive buildings are littered with cigarette butts. There are no containers in which to place the discarded butts, because the university administration maintains that would be an endorsement to violate the smoking ban and, even if there were containers, some smokers would ignore them. So what is the solution? Should the campus security patrol the building entrances to fine those who violate the ban? Should the custodial crew clean up the butts every day to preserve a semblance of attractiveness to the entrance? Should the university initiate a program to convince smokers that they should voluntarily obey the rules? Should harsher penalties be invoked? Or should everyone just ignore the situation and hope for the best? Unfortunately, that cannot be the case when environmental laws are broken. Enforcement is needed and, when done effectively, serves to encourage other businesses to comply as well.

### Participation: To Encourage and Support Ongoing Programs

Ideally, governments should provide the results of well-grounded research on the critical topics of interest; enact thoughtful and well-written legislation; establish agencies with clear missions and personnel who want to collaborate with industry to achieve the objectives of the legislation; support the efforts of private industry and individuals to comply with the legislation; provide feedback about the results of the combined efforts; and outline future steps that will enhance existing and proposed programs. While this may sound utopian, it is not impossible.

## Role of Consumer

Governments and industry are the prime movers in the efforts to make RSCs a reality; however, the individual consumer also has a vital role to play.

## Participant in Reverse Supply Chain Programs

There are opportunities for consumers to participate in RSC programs in a number of ways:

- Recycle paper and plastic products through their city's recycling collection program.
- Return hazardous materials—used oil, paint, and fertilizers—to proper disposal sites.
- Return purchased goods promptly and in good condition to the store of purchase.
- Recycle used clothing to nonprofit agencies for distribution or resale.
- Return electronic products containing hazardous materials—computers—to designated sites.
- Recycle printer cartridges and paper through their employer's recycling program.

Most individuals mean well and are inclined to comply; however, there is still a need for processes to make it easier for them to cultivate the recycling habit.

One of the more interesting decision areas is in the sorting of different materials to be recycled. Again, we will use an example from one of the author's schools. Previously, participants were asked to presort materials—paper, plastic, metal—before placing the items into designated containers. This met with indifferent responses. Today, there are two containers: the first with the label *It all goes in* and the other *Landfill*. The sorting is done later by knowledgeable persons. The same transition is also present in the town's recycling program. The recycling containers are sorted by the collection person, relieving the homeowner of that responsibility.

## Educated Consumer

In addition to being a participant, the individual can become educated about RSC topics. They should learn about key environmental issues—climate change, hazardous materials disposal, green purchasing, sustainable home building, and energy- and resource-efficient appliances. Then, they can better decide how they can best support the efforts of governments and industry to promote RSC initiatives.

There are both formal and informal opportunities to become educated. Colleges and universities are adding courses about sustainability, green supply chains, environmental conservation, and topics in an effort to provide both traditional and nontraditional students an opportunity to learn. There are numerous websites and television programs that present information, both objective and biased. Information about RSCs abounds, but assimilating and understanding it is a challenge.

## Supporter of Green Supply Chain Efforts

Finally, individuals can support the RSC efforts. They can contribute to their own organization's ideas about what to do. When in a position of managerial responsibility, they can comply with the directives of their governmental agencies. They can vote for political candidates who are supportive of appropriate sustainability programs. They can buy products from companies that comply and even go beyond the minimum requirements to develop and manage RSCs that work. Through some dedicated effort, they can modify their lifestyles to become more responsive to the need for environmental responsibility.

One area that consumers may find interesting to learn about is the role of eco-industrial parks. These parks are typically an arrangement of businesses that operate in a closed-production system with each other. Wastes from one business become input to another business. For example, a paper mill, serving as the hub of an eco-industrial park, could supply pulp to makers of fiberboard, cellulose insulation, and furniture. The businesses in the park are located in close proximity to each other so that the transportation of outputs and inputs is kept to a minimum. Enthusiasm for eco-industrial parks began in Europe, with one of the more famous in Kalundborg, Sweden. This park started with a single power plant in 1972 but has evolved over time to include an oil refinery, a pharmaceuticals plant, a plasterboard manufacturer, and the municipality of Kalundborg. Treated waste-water from the refinery is used as cooling water by the power station. Surplus heat from the power station is used for warming homes in the community and a local fish farm. The fly ash produced by the power station is used in cement manufacturing and road build-ing. Surplus gas from the refinery is delivered to the power station as an energy source. By-products from the pharmaceutical plant are used by local farmers as fertilizers. While the initial motivation for the clustering of industries was cost savings, industry managers and local residents realize they are generating environmental benefits (IISD 2013). The Green Enterprise Initiative (2013) has prepared a list of a number of U.S. sites.

## Reverse Logistics Network

The opportunities, and difficulties, described earlier were intended to show that design-ing and developing RSCs is a difficult, but worthwhile, endeavor for the combined efforts of industry, government, and individuals. The question is not should organizations build RSCs, but how to do it quickly and effectively. The RSC should become an integral part of every organization's strategic and operating plans.

### Continuation of the Forward Supply Chain

We pointed out earlier that the RSC should be viewed as an extension of the forward sup-ply chain. When companies take this perspective, they are more likely to consider the consequences of their product design decisions. Some of the areas in which product design may affect the RSC include the following:

- *Materials*: Designers should attempt to avoid hazardous materials. While there are limited prohibitions on materials today, there will probably be more in the future. Conversely, designers should attempt to use materials that are easier to rework, recycle, or dispose of at the end of its life cycle.

- *Processes*: Designers, and manufacturing engineers, should select processes that do not require the use of undesirable procedures or indirect materials, such as cleaning solvents or soldering fluxes that require special treatment before release to the water treatment systems.

- *Method of disassembly*: Some companies are already beginning to consider methods of disassembly of their products. BMW is designing automobiles that are easier to disassemble in anticipation of easier compliance with the WEEE directive (Routroy 2009).

- *Method of transportation*: Transportation and related packaging costs have become a substantial part of the total costs of products. Products have not normally been designed with the transportation and storage requirements in mind; however, that will change in the future to optimize the total costs of products as they move through both the forward supply chain and RSC. The first manufacturers who use environmentally conscious shipping choices will not only reduce the costs of transportation but will also differentiate themselves to customers, strategic partners, and investors and gain a significant competitive advantage (Mallett 2008).

- *Design for recyclability*: Designers will work to reduce the flow of goods back through the RSC by reducing the need for returns, repairs, and remanufacturing. They will make products that need fewer adjustments and repairs; therefore, the products of the future will tend to work better and last longer. In other words, there will be a movement in quality back to the features of reliability and durability. This return to quality and pride of craftsmanship, particularly in the small appliances industries, will be a welcome relief for many consumers, who have endured years of paying cheap prices, for cheap products that do not last.

- *Organize for the RSC*: While product designers can handle the more technical parts of product design, top management must begin to decide how best to organize the RSC. Should the company establish in-house the capability to manage the activities in the RSC, or should they outsource those activities?

## Open System Environment

Just as for the forward supply chain, the RSC operates in an open system environment. This means a number of external forces, including competitors, technology, economy, governments, environment, and social mores affect the RSC design and operation. At any point in time, the RSC could be affected by a different external force. At the present time, social pressure seems to be one of the drivers advocating for more effective RSCs. Within the next few years, government legislation may become a more influential driver. Social activism usually precedes government regulation if businesses and industries do not choose to self-regulate. At present, companies vying for a market niche in the RSC industry are minimal and unorganized. However, in the future, the RSC industry will move from its current state of being small and fragmented into a growth industry.

## Heavily Outsourced by Major Businesses

The RSC has grown piecemeal and not in a planned way. Each step of the RSC has experienced growth, but not always in an orderly fashion or directly within the operation of the company that produced the original product or service.

Support and service activities in the RSC have generally been handled by third parties. Independent dealers serviced appliances and answered questions about how best to operate them. More recently, computer manufacturers have set up call centers, often in a foreign country, to answer questions that consumers may have about their computer. Too often, businesses have viewed post-sale service as a necessary inconvenience and not as an opportunity to further reinforce their relationship with the customer.

Businesses have always had to deal with product returns and transfers. They have often done this in an ad hoc manner and have not been organized to be effective and efficient. That is changing, particularly in the e-commerce arena, where returns are heavy and must be handled as a regular part of doing business.

Most OEMs do not make repairs to their products. Automobile owners may go back to the original dealer for repairs, but many others prefer taking their vehicles to their favorite repair shop. In the case of household appliances, third-party companies usually service these items as opposed to the original manufacturer. Consumers who purchase cheap appliances at a discount retailer may rarely have that appliance repaired, opting instead, to simply throw it away and purchase another one instead.

When more extensive rebuilding is required, remanufacturing is also an activity that most OEMs outsource. When early industries were heavily integrated, such as the railroad industry in the latter part of the nineteenth century and the early part of the twentieth century, they had their own repair shops and performed all of the needed remanufacturing. Remanufacturing generally implies that in addition to repairs, more extensive rebuilding of the product is required. Today, this is an activity that is usually not handled by most original manufacturers. For example, the airline industry outsources much of its heavy maintenance and overhaul work (Smith and Bachman 2008).

If the equipment cannot be preserved for its original purpose, its parts may be salvaged by regrinding or otherwise reworking them as part of a new product. Steel scrap yards separate the various materials in an automobile and rework them for use in subsequent product uses. As with most other RSC activities, this work is usually performed by independent companies and not connected in any way with the original manufacturer.

Recycling can occur in-house or on a contracted basis. If the reworked materials go back into the original product, such as candy coating or plastic powder that has been reground from defective molded toys, the work is probably done in-house.

However, there may be independent companies that recycle the materials into completely different products, like the earlier example of grinding rubber tires into particles that become part of the road-paving materials. This type of processing is usually performed by companies other than the tire manufacturer.

---

## Need for a Life Cycle Systems Approach*

The OEM does not usually use a strategy of vertical integration in designing their forward supply chain. It is even more unlikely they will vertically integrate into their RSC. However, in the future, they will probably have increased responsibility for the RSC of their products. To reduce their risk and increase the acceptability of their products, they will need to develop at least a virtual integrated network for both their forward supply chain and RSC.

The need for an RSC should be viewed from several perspectives. First, there should be a systems orientation on three fronts—as an integrated supply chain system; as a management system involving strategic, tactical, and operating plans; and as a system with global geographic implications. Second, external and internal stakeholders need to be recognized. External stakeholders include customers and suppliers. Internal stakeholders, the employees, include all functional departments of a company that contribute directly or indirectly to the creation of excess inventories. Third, excess inventories evolve through several stages, or a life cycle, from the time they are created until they have been eliminated. Fourth, their resolution requires a cross-functional effort from internal groups in cooperation with external entities. Finally, all of the aforementioned elements need to be

---

* Some of the text in this section was adapted from Crandall and Crandall (2003).

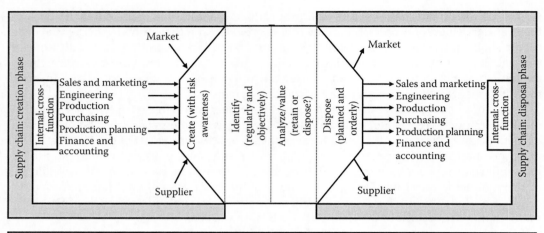

**FIGURE 10.3**
Excess inventory life cycle model.

integrated so that managing excess inventories is an ongoing part of the management process and not a piecemeal attempt at a quick fix.

A life cycle model implies excess inventory occurs as a regular part of doing business and must be managed in a systematic manner. Managers are aware of their potential, recognize quickly when they do occur, and have a preplanned way to dispose of them without causing a major disruption in the business. The following discussion looks at this life cycle in more detail.

Figure 10.3 is a diagram of the life cycle model. The left side shows the creation of excess inventories, the result of decisions by supervisors and managers within the internal functional areas. These decisions are influenced by the market and suppliers. However, we have added the prevent stage, because the decisions that cause excess inventories also offer the opportunity to prevent or at least reduce the amount and severity of excess inventories. Decisions to increase inventory are made with some awareness of the risk and knowledge that unacceptable amounts of excess inventory will not be allowed to remain in the system very long. The merging of the lines indicates there is cross-functional coordination in the decision making that leads to excess inventories. One of the current topics in supply chain discussions is the use of demand management (Chapter 5), an approach that avoids the static forecasting of yesterday and replaces it with a more fluid, ongoing view of determining demand that involves soliciting input from all demand-chain constituents.

Figure 10.3 also illustrates excess inventories have their own life cycle that is definable and consistent. Excess inventory is created, identified, analyzed in terms of value, and then disposed. This sequence of events occurs almost with the regularity of sunrises, rather than the randomness of hurricanes. The Identify/Classify phase includes a process whereby excess inventory is more quickly recognized through a measurement and control system. The Analyze/Value phase requires excess inventory be revalued and a decision made as to its course of disposal. The horizontal lines at the top and bottom of the diagram signify that the creation phase and the disposal phase are an integral part of the total model. The final phase—Dispose—is handled, as before, by internal functions that, in this model, work together among functions and with customers and suppliers to create

| Create | Identify, classify | Analyze/value | Dispose |
|---|---|---|---|
| Demand variation<br>  Inaccurate forecast | Damaged | Initial valuation<br>  Market value | Return to supplier |
| Economic cycle<br>  Customer vacillation | Defective | Cost | To other locations |
| | Obsolete | | Modify for spare parts |
| Supply variation | | Revaluation | |
| Lack of capacity | Replaced | Adjust over time | Sell at discount |
| Quality problems | | Adjust to market | |
| Delivery time | Aged | | Barter for services |
| Internal variation | Deteriorated | Value | Donate |
| Process variation | | Book (accounting) | |
| Capacity limitations | Abnormal amounts | Physical (salability) | Scrap |
| Output variation | | | |

**FIGURE 10.4**
Factors to consider during RSC life cycles.

disposal outlets. Contrary to current practice where disposal is often viewed as an expediency instead of a rational activity, many of the methods of disposal have been prearranged and do not require a discovery process to be activated.

Figure 10.4 expands on the life cycle model by showing, in more detail, some of the considerations in each phase of the life cycle.

## Need for IT

To receive the most benefit from the RSC, excellent information management systems are needed. The emerging importance of RL requires the execution and management of new inter- and intraorganizational processes. Given the complexity and uncertainty of these processes, information and communications technology (ICT) support is necessary, even more than in traditional (forward) supply chain management. Moreover, the handling of reusable materials requires the involvement of new chain partners, such as material recovery organizations and e-marketplace intermediaries.

ICT vendors have not yet given high priority to incorporate RL functionality in their systems. Recent research efforts recognize the importance of managing reversed flows in the supply chain, and Internet-based technologies have created new opportunities. Besides the development of new design, planning, and control components in existing enterprise resource planning (ERP) and advanced planning and scheduling (APS) systems, electronic marketplaces can be developed to effectively manage return flows in the supply chain (Hillegersberg et al. 2001). GENCO (2013), a leading 3PL provider, offers a system of RL services that includes

- *Return center operations*: Management of centralized or regional return centers
- *Return to vendor management*: Management of the flow of returned product from retailer to vendor
- *Product recall management*: Management of recalls because of product defects and other circumstances

- *Test, repair, and refurbishment*: Inspection, testing, and repair of products for resale
- *Product recycling*: End-of-life recycling to maximize the reuse of all component parts and raw materials as well as regulatory-compliant disposal

Information systems are equally important in RL system as in forward supply chain systems.

## Other Considerations in Designing Reverse Supply Chains

William T. Walker, noted supply chain practitioner and consultant, offers some additional points to consider.

- The RSC design should include a typical cost model to quantify the expected revenues and costs of operation. Some of the costs that would not be incurred in the forward supply chain include warranty expenses, returns credits, and restocking charges.
- During a new product introduction, it is essential to recover information about early field failures, analyze the causes of the failures, and redesign the product to prevent future failures. An RSC design is necessary to actuate this redesign process.
- Many IT systems do not run backward. Businesses usually need a separate system to adequately manage the RSC.
- There is a growing need for serialized product, product tracking, and source traceability capability. The manufacturer no longer has the luxury of making a product, shipping it to a customer, and forgetting about it. Future legislation will likely require extended participation in the RSC.
- A bill of material (BOM) for the disassembly should be prepared for each product. It can be used to help in forecasting the rate of finished product returns and the usability of each part in the returned product.
- In designing the RSC, just as in designing the forward supply chain, businesses should consider flows of material, information, and cash (Walker 2009).

## Future

What does the future hold for RSCs? Is this an unstructured collection of worthwhile efforts that will grow in stature and focus into a prominent management program? The likelihood is it will continue to evolve into a prominent part of supply chain management because of its rapidly growing acceptance and practicality.

### Growth in Amount of Materials Recycled

It is almost certain the amount of recycled materials will increase. Programs initiated at the local level where the city collects roadside materials have been endorsed by the public, even though not all individual consumers are participating to the fullest. While there have been some very successful programs, the future potential is far greater. It will take the combined efforts of consumers, producers, and government to increase the effectiveness of these recycling programs. While today's programs are well intentioned, they do not have the focus, infrastructure, and acceptance to make them an unqualified success story.

### Increase in Number of Companies Performing Reverse Logistics Activities

As the volume of activity grows in the area of RL, more companies will begin to participate. There is profit potential as well as good public relations for those companies that demonstrate they can perform needed functions along the RSC. While the potential for public acceptance is an attractor, the real payoff must be an economic benefit; otherwise, companies will not be able to sustain a position in RL activities in the long run.

### Joint Ventures between Private Business and Government

There will be more collaboration between industry and government in the design of RL systems and in developing acceptable standards to guide the decisions along both the forward supply chain and the RSC. In some cases, the government will have the responsibility for setting standards and mandating compliance. They will be more successful if they involve the business community in developing these guidelines. In other cases, the government will provide economic incentives, in the form of tax relief, or financial support, such as in grants to develop new technologies for use in the RSC.

### Increased Emphasis on Prevention, Not Just Reusing

Although there will be increased efforts in making RSCs more effective, there will also be increased efforts in making better products so that the RSC does not become overburdened. This strategy requires prevention, or at least reduction, of product problems that cause them to enter the RSC in the first place. Products will fit consumer needs better, product life will increase, and improved transportation methods will reduce damage. *Reduce, reuse, and recycle* are the new watchwords for manufacturing managers looking to set a new standard for manufacturing excellence (Kenney 2008).

### More Companies Will Design Integrated Reverse Logistics Systems

As outlined earlier, more companies will design integrated RSC systems. They will probably outsource most of the RSC activities, but they will monitor the effectiveness of those activities more closely. The RSC will be an extension of the forward supply chain, not an adjunct or auxiliary support function.

## Summary

Mollenkopf and Closs present an excellent summary of the RL environment:

> Reverse logistics has often been viewed as the unwanted step-child of supply chain management. It has been seen as a necessary cost of business, a regulatory compliance issue, or a "green" initiative. But more companies are now seeing reverse logistics as a strategic activity—one that can enhance supply chain competitiveness over the long

term. To understand how reverse logistics can create value, it is necessary to understand both the marketing and logistics components of this process. Companies can use the returns process to enhance marketing efforts by analyzing reasons for returns and conducting ongoing defect analysis. In many cases, the returns management process had been ad hoc, meaning that the strategies, operations, and guidelines were not well defined or thought out. However, all the organizations surveyed have quickly seen compelling benefits, to the point where they have all opted to build reverse logistics capabilities into their overall logistics and supply chain strategies and processes. (Mollenkopf and Closs 2005)

As returns management and RL processes become more refined, the volume of returns will increase. This may cause some businesses to rethink their position on handling returns. Rather than considering outsourcing as a nuisance activity, it may become significant enough to warrant bringing it in-house as a regular part of the company's business.

In this chapter, we described the growing importance of RSCs as extensions of the forward supply chains being developed by most companies. We explained the activities in an RSC and how they differ from the activities in the forward supply chain.

There are several stakeholders in RSCs, in addition to the business itself. The stakeholders include the consumer, the government, and society in general. The consumer needs ways to prolong the life of their purchase or dispose of it in an enlightened manner. The government has a responsibility to ensure the disposal process, especially of hazardous materials, is accomplished without endangering the lives of the general population. Society is interested because of the potential threat to the environment and their safety.

We also described the need for the RSC to be designed as an integrated system and offered suggestion on how this can be done. Finally, we outlined some of the most likely changes in store for RSCs in the future.

### Hot Topics in the Supply Chain, Reshoring: Revisiting the Make or Buy Decision

There is a movement afoot whereby companies that have offshored to suppliers in Asia are returning production back to the home country. In the United States, this movement has been called reshoring. Apple has announced that it will manufacture a line of Macs in the United States. In addition, GE has announced it will invest $1 billion to reshore appliance manufacturing back to factories in the United States (Gray et al. 2013). There are a host of other well-known companies that are reshoring, including NCR and the Ford Motor Company. Ford is moving the making of hybrid transmission components and battery packs because of concerns over quality. NCR is reshoring the making of ATMs because of slow response times from foreign contract suppliers (Moser 2013).

Some analysts may be taken back by this trend that was preceded by the offshoring movement, which sought companies in Asia to make products, primarily for low labor cost reasons. Why now, the sudden urge to move back home again? Those reasons are explored next.

### REASONS FOR RESHORING

There is no single reason that drives the reshoring process. Instead, a number of factors are at work.

*Quality*: Some companies have discovered that their products are just not made as well as they were when they were manufactured in the home country (Moser 2013).

*Transportation costs*: Moving the product made in Asia back to the home country such as the United States adds to the total cost of making the product. Rising fuel costs have made transportation costs rise (Ellram et al. 2013).

*Theft of intellectual property*: Companies operating in China must have a majority partner from the Chinese side. These partnerships involve the exchange of technologies and processes, some of which have been pirated (Navarro 2013).

*Labor wages increasing in Asia*: Although low wage rates attracted the first major wave of outsourcing to Asia in the 1990s, wages are now rising in China (Moser 2013).

*Public relations problems*: Companies that have suppliers in Asia have often been the target of public criticism because of their association with sweatshops. Efforts to monitor factories for code compliance can be difficult (Frank 2008).

*The problem of "too many eggs in one basket"*: Some companies find themselves in the situation of having most of their production taking place in China. If a great deal of their revenue enhancement is also in China, then the risk of having too much presence in one country is possible. With regard to China, a trade war with the United States or a diplomatic dispute could sour the prospects for that company (Navarro 2013).

*Total cost of ownership*: When calculating all of the tangible and intangible costs of offshoring, companies are discovering that the price differential between the United States and Asia is not that far apart. In some cases, it swings in favor of manufacturing in the United States (Moser 2013).

## THE BIG PICTURE

Lisa Ellram noted in her editorial that a time of rethinking the offshoring movement is necessary:

> Most of the direction of the movement of manufacturing locations in the past several decades has been to low-cost countries. With all of the attention that is being given to reconsidering optimal supply chain configurations, it is time to reconsider the manufacturing location decision from some different perspectives. (Ellram et al. 2013, p. 3)

Certainly, one perspective suggests that searching for the lowest-cost supplier, even if it is on the opposite side of the world, does not make intuitive sense. The lowest-cost supplier approach may indeed be a myth to be grappled with in the future, particularly since there are so many other factors that affect the smooth running of the supply chain.

We would be amiss if we did not acknowledge at least one infamous force that led to the offshoring movement to begin with, the movement from a stakeholder perspective to a shareholder perspective. The stakeholder view maintained that the corporation needed to balance its responsibilities to a number of stakeholders, including to employees, customers, owners, the local community, special interest groups, and the government (Carrol and Buchholtz 2012). While this perspective was prevalent for decades, the shareholder perspective is now firmly entrenched. This viewpoint maintains that the only consideration of the corporation should be to its stockholders. The catalyst to this perspective was the movement to tie CEO compensation with shareholder value (Navarro 2013). What has resulted in the corporate world and Wall Street is an emphasis on short-term firm performance, which boosts stock prices and compensates the CEO and the shareholders. This perspective does not consider the long-term viability of the company but, instead, seeks short-term strategies to keep production costs low.

One man who has spearheaded the reshoring movement is Harry Moser, head of the Reshoring Initiative, an organization that is encouraging bringing manufacturing back to the United States. The premise of the Reshoring Initiative is to consider the total cost of making the product, including shipping it back to the United States. Considering the total cost of making the product, not just the labor cost, reveals a different picture of what that product is costing the company. Moser (2013) cautions though that if reshoring is to occur, the United States will need to make significant investments in education so that manufacturing employees are ready for the demands of highly automated factories. Such factories will combine the demands of mass customization under the guise of lean management techniques.

## QUESTIONS FOR RESEARCH AND DISCUSSION

1. What costs are parts of making a product in a developing nation? Which of those costs would not be a factor if the product was made locally?

2. It is generally agreed that reshoring all industries is not a good idea. Instead, selective reshoring is more viable. What industries do you think are good candidates for reshoring? Which industries should continue to be remaining offshore?

## REFERENCES

Carroll, A. and Buchholtz, A., *Business and Society: Ethics Sustainability, and Stakeholder Management*, 8th edn., South-Western/Cengage Learning, Mason, OH, 2012.

Ellram, L., Tate, W., and Petersen, K., Offshoring and reshoring: An update on the manufacturing location decision, *Journal of Supply Chain Management*, 49(2), 14–22, 2013.

Frank, T., Confessions of a sweatshop inspector, *The Washington Monthly*, 40(4), 34–37, April 2008.

Gray, J., Skowronski, K., Esenduran, G., and Rungtusanathan, M., The reshoring phenomenon: What supply chain academics ought to know and should do, *Journal of Supply Chain Management*, 49(2), 27–33, 2013.

Moser, H., Manufacturing, *Economic Development Journal*, 12(1), 5–11, 2013.

Navarro, P., China 2013: The year of reshoring to America? *Financial Executive*, 29(1), 33–35, January/February, 2013.

## Discussion Questions

1. How do RSCs differ from forward supply chains?
2. How do RSCs differ from green supply chains?
3. What are the major activities needed in an RSC?
4. What are the drivers of the RSC movement?
5. What is the role of the consumer in promoting RSCs?
6. What is the role of special interest groups in encouraging RSCs?
7. What is the role of business in supporting RSCs?
8. What is the role of governments in legislating RSCs?
9. How will the RSC movement look in the future?
10. How can you participate in RSCs?

# References

Andel, T., Rethinking recycling, *Material Handling Management*, 59(5), 35, 2004a.

Andel, T., Putting returns to work, *Material Handling Management*, 59(9), 35, 2004b.

ASU News, Appalachian named one of America's Top 20 "Coolest" Schools by Sierra Magazine, *University News*, Appalachian State University, Boone, NC, www.nes.appstate.edu/2011/08/17/top-20-%e%80%9ccoolest%e2%/80%/9d-schools/, August 17, 2011.

Bell, A., Rubble full of riches, *The Charlotte Observer*, pp. B1–3, February 11, 2006.

Blackstone, J.H., *APICS Dictionary*, 14th edn., APICS—The Association for Operations Management, Chicago, IL, 2013.

Blumberg, D.F., *Introduction to Management of Reverse Logistics and Closed Loop Supply Chain Processes*, CRC Press, Boca Raton, FL, 2005.

Caldwell, B., Reverse logistics, *Information Week*, Manhasset, NY, 729, 48–52, April 12, 1999.

Clemens, P., Disassembly required, tearing down an American factory, ton by ton, *Bloomberg Business Week*, pp. 98–103, January 24, 2011.

Crandall, R.E., Opportunity or threat? Uncovering the reality of reverse logistics, *APICS Magazine*, 16(4), 24, 2006.

Crandall, R.E., The evolution of product lifecycle management, *APICS Magazine*, 18(5), 21, 2008.

Crandall, R.E., Putting together a global sustainability movement, how many strands does it take to make a cable? *APICS Magazine*, 20(6), 26–29, November–December 2010.

Crandall, R.E., Out of the landfill, examining alternative uses for throwaway materials, *APICS Magazine*, 22(6), 18–20, November/December 2012.

Crandall, R.E. and Crandall, W.R., Managing excess inventories: A life-cycle approach, *The Academy of Management Executive*, 17(3), 99, 2003.

Dyckhoff, H., Souren, R., and Keilen, J., The expansion of supply chains to closed loop systems: A conceptual framework and the automotive industry's point of view, *Supply Chain Management and Reverse Logistics*, Springer, Berlin, Germany, 2004.

Edwards, A.R., *The Sustainability Revolution: Portrait of a Paradigm Shift*, New Society Publishers, Gabriola Island, British Columbia, Canada, 2005.

Friedman, T.L., *Hot, Flat and Crowded: Why We Need a Green Revolution—And How It Can Renew America*, Farrar, Staus and Giroux, New York, 2008.

GENCO, Reverse logistics, 2013, http://www.genco.com/reverse-logistics/reverse-logistics.php, accessed July 19, 2013.

Green Enterprise Initiative (GEI), Existing and developing eco-industrial park sites in the U.S., 2013, http://gei.ucsc.edu/eco-industrial_parks.html, accessed July 19, 2013.

Guide, V.D.R., Jr. and Wassenhove, L.N., The reverse supply chain, *Harvard Business Review*, 80(2), 25, 2002.

Hartley, R., *Business Ethics: Violations of the Public Trust*, John Wiley & Sons, New York, 1993.

Hervani, A.A., Helms, M.M., and Sarkis, J., Performance measurement for green supply chain management, *Benchmarking*, 12(4), 330, 2005.

Hillegersberg, J., van Zuidwijk, R., van Nunen, J., and van Eijk, D., Supporting return flows in the supply chain, *Communications of the ACM*, 44(6), 74, 2001.

International Institute for Sustainable Development (IISD), Kalundborg, Denmark, 2013, www.iisd.org/business/viewcasestudy.aspx?id=77, accessed July 19, 2013.

Kenney, B., The zero effect: How to green your facility, *Industry Week*, 257, 36–43, July 1, 2008, www.industryweek.com/PrintArticle.aspx?ArticleID=16515.

Mallett, T., Shipping: Six steps to achieving manufacturing's holy green grail, *Industry Week*, June 20, 2008, www.industryweek.com/PrintArticle.aspx?ArticleID-16624.

Mollenkopf, D.A. and Closs, D.J., The hidden value in reverse logistics, *Supply Chain Management Review*, 9(5), 34, 2005.

Poduska, R., Forbes, R., and Bober, M., The challenge of sustainable development: Kodak's response, *Columbia Journal of World Business*, 27(3/4), 286, 1993.

Richey, R.G., Daugherty, P.J., and Genchev, S.E., Reverse logistics: The impact of timing and resources, *Journal of Business Logistics*, 25(2), 229, 2004.

Rogers, D.S., Lambert, D.M., Croxton, K.L., and Garcia-Dastugue, S.J., The returns management process, *International Journal of Logistics Management*, 13(2), 1, 2002.

Routroy, S., Antecedents and drivers for green supply chain management implementation in manufacturing environment, *ICFAI Journal of Supply Chain Management*, 6(1), 20, 2009.

Smith, G. and Bachman, J., The offshoring of airplane care, *Business Week Online*, p. 3, April 10, 2008.

Technology International Group, WEEE and RoHS directives, www.techintl.com/weee.cfm?OVRAW=RoHS%2FWEEE&OVKEY=rohs%20weee&OVMT, accessed August 12, 2013.

UPS, Toshiba and UPS join to set new standard for laptop repair, 2004, http://www.pressroom.ups.com/pressreleases/printer/0,1052,4421,00.html?ct=press_releases&at=domain_mainpressroom&tn=pressreleases_archives_archive&id=4421&srch_page=1&srch_phr=Toshiba, accessed July 7, 2013.

Walker, W.T., Rethinking the reverse supply chain, *Supply Chain Management Review*, 4(3), 52, 2000.

Walker, W.T., Supply chain consultant, personal interview, March 23, 2009.

Yang, W., Regulating electrical and electronic wastes in China, *Review of European Community and International Environmental Law*, 17(3), 337, 2008.

# Section IV

# Need for Integration

# 11

## The Need to Integrate

### Expected Outcomes

After reading this chapter, you should be able to

- Identify the reasons to integrate supply chains
- Describe the differences between mass production and mass customization
- Explain why there is a need to move from vertical integration to virtual integration
- Describe how organizations are moving toward more diverse cultures
- Discuss the use of the triple bottom line to facilitate supply chain integration
- Identify the drivers of change for integrating supply chains
- Describe the steps needed to integrate supply chains
- Discuss why developing relationships and trust is needed to integrate the supply chain
- Justify the use of strategic planning when integrating the supply chain
- Explain why integrating supply chains requires a project management approach

### Company Profile: Cisco

In the earlier chapters of this book, we have described the separate functions along the supply chain. In the remaining chapters, we will describe the need to combine the separate functions into an integrated supply chain. Chapter 11 outlines the need to change because most organizations will have to change their internal operations as well as their relationships with customers and suppliers. Change is often difficult. In addition to changes in individual organizations, changes will also be required in communities and entire countries. Short-term fixes will have to give way to long-term program planning. As you read this company profile and the chapter in the book, think about these questions:

- Why is change difficult? Do you like change? Why or why not?
- Why is mass customization important to consumers? To suppliers?
- Why is it difficult for state or federal governments in the United States to change?
- How should companies organize to facilitate supply chain integration?

What additional questions do you think relate to the need for supply chain integration?

## Company Profile: Cisco

Chapter 10 is about the need to tightly integrate the participants in a supply chain. As outsourcing has moved to disconnect the supply chain, many companies have had to work harder to maintain a steady flow of products, especially those companies with rapidly changing products and whose customers and suppliers are scattered throughout the world. Cisco is such a company.

### HISTORY OF COMPANY

Cisco Systems was started in 1984 by two members of the Stanford University computer support staff: Leonard Bosack and Sandy Lerner. They were joined by Kirk Lougheed, also at Stanford University; Greg Satz, a programmer; and Richard Troiano, who handled sales, as the early Cisco team. The company went public in 1990 on the NASDAQ Stock Exchange and, through internal growth and acquisitions, continued to grow through the early 1990s. The rapid growth of the Internet during the late 1990s fueled the phenomenal growth of Cisco until the economic downturn at the beginning of the new millennium (Cisco 25th Anniversary Time Line). Additional information about the history of the company can be found at http://www.youtube.com/watch?v=W9SWUYHgBaU.

In the mid-1990s, Cisco was primarily a router and switch vendor. Over time, Cisco moved from making gear for data networks to providing equipment for voice communications and video systems and has become more focused on providing software to support its hardware (Waltner 2009).

Figure 11.1 shows the net sales since the company went public in 1987. Sales come primarily from products in the early years; however, in recent years, services have become

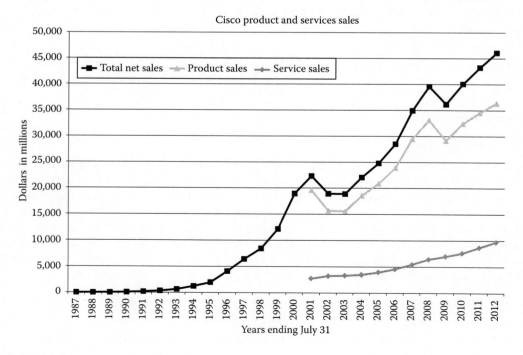

**FIGURE 11.1**
Cisco sales. (Graph prepared with data from Mergent Online, http://www.mergentonline.com, accessed September 12, 2013.)

a larger part of their sales. Even in the down years of 2002, 2003, and 2009 for products, services sales continued to increase.

They also recognized that to sustain their rapid growth, they would have to outsource much of their manufacturing. This gave them more flexibility and reduced the investment required to set up their own manufacturing facilities. They turned to contract manufacturers, as did other large electronic companies, such as Sony and Apple. While it has proven to be a viable strategy for rapid growth, it has not been without some disruptions along the way.

In the third quarter of 2001, after a dramatic decline in sales resulting from an economic downturn, Cisco was forced to incur a $2.25 billion write-down in excess inventory. The company had been overly optimistic about continued increases in sales and, in order to assure the supply of critical components, had guaranteed their purchases to suppliers. When sales declined, they were committed to accepting components from suppliers; unfortunately, they were not able to realize full value of this inventory in subsequent products because of the rapid change in product designs (Barrett 2001). The only other downturn in sales was in 2009, when the economic slowdown in the United States was a major factor in the reversal of continuing sales increases. Although net income declined from 2008, the company remained profitable.

## MARKETS

Cisco sells products throughout the world, as shown in Table 11.1. While over half their sales are in the Americas, they are aggressively extending their scope, especially in emerging markets.

In his letter to shareholders, Cisco CEO John Chambers said, "We believe that our long-term strategy, which is focused on delivering intelligent networks and technology and business architectures built on integrated products, services, and software platforms, is the right one to help our customers achieve their top priorities and fuel our success over the long run" (Cisco Annual Report 2012, Letter to Shareholders).

## PRODUCTS

The Company designs, manufactures, and sells Internet Protocol ("IP")–based networking and other products related to the communications and IT industry and provides services associated with these products and their use. These products, primarily integrated by Cisco IOS Software, link geographically dispersed local-area

**TABLE 11.1**

Cisco Markets

| | Sales and Gross Margin for 2012 (Millions of Dollars) | | | |
|---|---|---|---|---|
| | **Americas** | **EMEA** | **APJC** | **Total** |
| Net sales ($) | 26,501 | 12,075 | 7,485 | 46,061 |
| Sales (total %) | 58 | 26 | 16 | 100 |
| Gross margin ($) | 16,639 | 7,605 | 4,519 | 28,763 |
| Gross margin (%) | 63 | 63 | 60 | 62 |

*Source:* Cisco Annual Report, 2012, p. 129, http://www.cisco.com/web/about/ac49/ac20/about_cisco_annual_reports.html, accessed September 2012.

*Note:* Americas—United States, Canada, Latin America; EMEA—Europe, United Kingdom, Ireland, Africa, Russia; APJC—Asia Pacific, Japan, China.

**TABLE 11.2**

2012 Net Sales by Product Groups

| | Millions of Dollars | Total (%) |
|---|---|---|
| Switching | $14,531 | 32 |
| NGN routing | 8,425 | 18 |
| Collaboration | 4,139 | 9 |
| Service provider video | 3,858 | 8 |
| Wireless | 1,699 | 4 |
| Security | 1,349 | 3 |
| Data center | 1,298 | 3 |
| Others | 1,027 | 2 |
| Total product | 36,326 | 79 |
| Service | 9,735 | 21 |
| Total sales | $46,061 | 100 |

*Source:* Cisco Annual Report, 2012, p. 130, http://www.cisco.com/web/about/ac49/ac20/about_cisco_annual_reports.html, accessed September 2012.

networks ("LANs"), metropolitan-area networks ("MANs"), and wide-area networks ("WANs") (Cisco Annual Report 2012, p. 130).

One recent example of Cisco's integrated products is the network infrastructure provided for the London 2012 Olympic Games. Cisco partnered with the British Government, the London Organizing Committee of the Olympic and Paralympic Games, British Telecom, and broadcasters such as NBC. Cisco's system enabled 1,800 Wi-Fi hot spots and 80,000 data connections with a capacity of four times larger than any previous Olympic Games. Cisco technology helped connect 10 million spectators, 76,000 volunteers, and 22,000 athletics and coaches.

Cisco also believes they are uniquely positioned in the market transition currently underway toward more programmable, flexible, and virtual networks. Their focus on five core businesses is as follows: routing, switching, and services; data center (virtualization/cloud); collaboration; and video/and architectures for business transformation (Cisco Annual Report 2012, Letter to Shareholders).

Table 11.2 shows the breakdown of net sales by product category.

Although switching and routing continue to be the prime sources of revenue, other product groups are growing, especially in the security and data center areas.

## SUPPLY CHAIN STRUCTURE

Cisco has extended supply chains because it outsources much of the manufacturing of their products. One report indicates Cisco has more than 1,000 suppliers, 4 contract manufacturers, and 50,000 purchased parts. It also outsources assembly of its finished products (Preparing for Pandemics 2011). In their 2012 Annual Report, Cisco states they "rely on contract manufacturers for all of our manufacturing needs." "Our arrangements with contract manufacturers generally provide for quality, cost, and delivery requirements, as well as manufacturing process terms, such as continuity of supply; inventory management; flexibility regarding capacity, quality, and cost management; oversight of manufacturing; and conditions for use of our intellectual property. We have not entered into any significant long-term contracts with any manufacturing service provider" (p. 13). This understanding shows Cisco has very tightly linked themselves with their contract manufacturers.

Cisco has a continuing need to integrate acquired companies into the supply chain network. Many acquired companies have supply chains that may, or may not, readily mesh with those established within Cisco. To illustrate their activity in this area, Cisco made seven acquisitions during 2012, most in related lines of business.

## RISK MANAGEMENT

As a result of its extensive and complex supply chains, the company started a formal risk management program in 2007. The risk management team identifies current and future potential risks in the supply chain and develops plans and strategies if potential risks develop into full-fledged crises.

Despite the efforts to improve their risk management program, Chad Berndtson (2010) reported that several prime customers of Cisco reported that delays and partial shipments of products had adversely affected their business. This was during the period following the business downturn in 2009. Even so, most customers recognized the difficult circumstances Cisco was operating under and acknowledged that Cisco representatives worked closely with them during this period.

One of their most challenging tests was the 2011 earthquake and tsunami in Japan. James Steele, program director for Cisco, reports the following sequence of events following the earthquake:

- The earthquake occurred on a Thursday (West Coast time), and by Friday morning, "we had a war room set up, and identified all the people that needed to be involved. By noon we had an accounting of all our major suppliers within Japan, at least all the tier ones. Within 15 hours of the quake, we had a good understanding as to which suppliers were impacted. There were about 160" (Carbone 2011).
- The risk management team determined that suppliers within a 100-mile radius of the Fukushima plant would be impacted by the disaster, as a result of direct damage, power outages, evacuations, and employees not able to get to work.
- Within 2 days, the team knew which suppliers were up and running and which had been damaged or shut down. They knew which parts required mitigation to avoid shutdowns of the Cisco supply chain.
- In some cases, Cisco had second sources (previously determined prior to the earthquake) who could supply parts. In other cases, Cisco made spot buys from authorized distributors, independent distributors, and parts brokers.
- As a result, Cisco did not suffer any serious disruption of their operations. A spokesperson of Cisco credits the minimal disruption to operations to the planning that had been done in anticipation of a catastrophic event such as the earthquake (Carbone 2011).

Steele stated there are more risks in the supply chain than there were in the past. "Generally, the world has become more of a risky place. I think we're hitting an inflection point right now with world economics, political unrest, the recent debt crisis and the U.S.-China relationship. Things are much more intertwined from a financial standpoint within the world financial system. There's a lot more interdependence now just in general. Such interdependence means more risk for everyone in the supply chain" (Carbone 2011).

Cisco's submission for the Institute of Supply Management (ISM) 2012 Award for Excellence in Supply Management described how the company's approach to risk management is centered in its customer value chain management (CVCM) organization, which is responsible for the planning, design, manufacture, delivery, and quality of the company's products and solutions. The program includes the following four key processes:

- *Product resiliency*: Process for engaging with Cisco Engineering and product operations to identify risk trade-offs in the early design and development phases of Cisco products
- *Supply chain resiliency*: Process for working with CVCM's global supplier management (GSM) and global manufacturing operations (GMO) functions to assess and improve resiliency across Cisco's supply base, manufacturing, and test equipment partners
- *Business continuity planning (BCP)*: Semiannual process to engage all critical supply chain partners in providing Cisco with over 36 resiliency data points such as emergency contact information, availability of alternate power supplies, and estimated time to recover (TTR)
- *Supply chain incident management*: Process for monitoring worldwide events on a 24/7 basis, identifying and escalating any incident of concern, assessing impact, and organizing a cross-functional response team to mitigate the risk to resolution (O'Connor et al. 2012)

After their experiences in 2001, they were better prepared to deal with the economic downturn in 2009 and the earthquake in 2011.

## CONCLUSION

Cisco is a company that has shown the ability to adapt to changing conditions. It has grown rapidly in less than 30 years to become one of the premier companies in the world and a recognized leader in their industry. Although they have experienced setbacks, they have learned from those experiences and taken steps to be able to deal with future negative events, whether from economic downturns or from physical catastrophes. Ken Pesti, a technology consultant, thinks that Cisco will continue to face complex technological, logistical, and competitive issues. However, he has confidence in the company. As he puts it, "People have bet against Cisco before, and they have been wrong" (Walther 2009).

## REFERENCES

Barrett, L., Cisco's $2.25 billion mea culpa, *CNET News*, 2001, http://news.cnet.com/2100-1033-257278.html, accessed February 1, 2010.
Berndtson, C., Cisco response to supply chain woes lacking, say partners, *CRN*, May 11, 2010, www.crn.com/224701527/printablearticle.htm.
Carbone, J., Cisco strives to identify and mitigate risk in its supply chain, Digi-Key Corporation, Thief River Falls, MN, October 8, 2011, www.digikey.com/supply-chain-hq/us/en/articles/buying-conditions/cisco-strives-to-identify-and-mitigate-risk-in-its-supply-chain, accessed September 5, 2013.
Cisco Annual Report, 2012, http://www.cisco.com/web/about/ac49/ac20/about_cisco_annual_reports.html, accessed September 2012.
Cisco 25th anniversary time line, http://www.youtube.com/watch?v=W9SWUYHgBaU, accessed August 7, 2013.
Mergent Online, Cisco Systems, http://www.mergentonline.com, accessed September 12, 2013.
O'Connor, J., Steele, J.B., and Scott, K., Supply chain risk management at Cisco: Embedding end-to-end resiliency into the supply chain, 2012, http://www.ism.ws/files/RichterAwards/CiscoSubmissionSupportDoc2012.pdf.

Preparing for Pandemics: How Cisco immunized its operations, The educated executive, October 27, 2011, www.educated-exec.com/news/2011/10/preparing-for-pandemics-how-cisco-immunized.

Waltner, C., Cisco's silver anniversary: A history of reinvention, December 10, 2009, http://newsroom.cisco.com/dlls/2009/ts_121009.html, accessed September 9, 2013.

## Introduction*

Increased competition is forcing businesses to consider how best to gain a competitive advantage or at least to hold their own against global competitors. As a result, effective supply chain management (SCM) is becoming increasingly important. Today's supply chains are comprised of selected customers and suppliers who build lean and agile supply networks.

The benefits of successful supply chains are direct and measurable—lower product costs, reduced inventories, higher-quality products, faster response times, fewer stockouts, and higher on-time deliveries. Other benefits that are more difficult to measure and verify include increased market share, improved customer satisfaction, and increased customer retention. In speaking of global strategies, Edward Davis and Robert Spekman suggest, "the extended enterprise is really about -creating a defensible long-term competitive position through strong supply chain integration, collaborative behaviors, and the deployment of enabling information technology" (Davis and Spekman 2004).

Before we begin our discussion, we would like to draw your attention to one of the most comprehensive descriptions of how a company can build a supply chain. In his book, *Supply Chain Architecture: A Blueprint for Networking the Flow of Material, Information, and Cash*, William T. Walker calls on his 30+ years of experience with Hewlett-Packard and its spin-off, Agilent Technologies, to provide the reader with a detailed description of the individual components and how they fit together to form a world-class supply chain (Walker 2005).

Manufacturing, or the production of tangible goods, has changed substantially over the past 50 years and continues to change. Some of the most notable changes include the following:

- Product design has moved from a *what can we make?* attitude to a *what does the customer want?* attitude. As a result, there is increased collaboration between marketing and manufacturing in the product design and development process.

- Products are moving from a standard *one size fits all* focus to one that can be customized to fit individual customer wants and needs. Manufacturing is moving from a mass production mode to a mass customization mode. In some cases, customizing is completed downstream from the primary manufacturing company at the wholesaler or retailer level.

- Manufacturers have moved from being vertically integrated, where they produce the component parts and then assemble them, into being assemblers with independent suppliers providing the component parts.

- Some manufacturers have moved even further to outsource more of their former internal functions to outside suppliers. Increasingly, these outside suppliers are in foreign countries where lower labor rates have become a major source of cost savings.

---

* A major portion of this section is adapted from Crandall and Crandall (2014).

- Some manufacturers have gone to the ultimate in outsourcing by having contract manufacturers do the entire manufacturing function. In this case, the company that was once a manufacturer is now a *shell* manufacturer, that is, a manufacturer in name only.

- While certain manufacturing activities have been removed from the traditional manufacturer's role, a number of service activities have been added. It is essential that today's manufacturers have services to accompany their product if they are to be successful.

- The focus that was once primarily on low cost and efficiency now includes flexibility and responsiveness as manufacturers try to be more attuned to their customers.

- The net effect of these changes is manufacturers have become more concerned with the management of widespread operations that have become more complex with the addition of product variety, customization, faster response times, higher-quality demands, and larger and more demanding customers, especially at the retail level.

As the future unfolds, more changes can be expected.

## Setting the Stage*

Many managers are finding their supply chain is not operating as they expected—the reality is far different from the dream. What did they expect? Apparently, SCM is more than finding a few good vendors who will deliver the right product at the right time in the right quantity and with good quality.

What about demand? Is the customer part of the supply chain? In case there is any doubt, another definition clears up that point. The lean supply chain is "a set of organizations directly linked by upstream and downstream flows of products, services, finances and information that collaboratively work to reduce cost and waste by efficiently and effectively pulling what is needed to meet the needs of the individual customer" (Manrodt et al. 2005). The bottom line is that the customer is an integral part of the supply chain.

Does the supply chain have to be lean, too? Ideally, it should and according to some authorities, the supply chain should be agile, adaptable, and aligned (Lee 2004). It should also be integrated, innovative, and globally optimized, involve interorganizational collaborative relationships, and share the benefits equitably among the supply chain members (Heckman et al. 2003). SCM is a strategic, cooperation-oriented, business process management concept, cutting across organizational boundaries (i.e., integration oriented), which leads to increased results for all supply chain members. SCM demands integration and cooperation beyond the logistics dimension. SCM processes are customer and/or end user driven (Kotzab and Otto 2004).

Should the company have an integrated supply chain organization with a C-level executive in charge, much as materials managers presided over the flow of materials? Some believe they should, although one study found that only about 25% of the companies studied had a single executive in charge of the total global supply chain, while a third of the

---

* A major portion of this section is adapted from Crandall (2005).

companies did not even list a supply chain officer on their website. While the absence of an executive in charge does not indicate the company did not have an integrated supply chain, it poses doubts (Gilmore 2008).

## Reasons to Integrate

Do all companies have world-class supply chains? Not unless they are among the handful of companies that are continuously cited in the literature as having the best supply chains (Apple, Procter and Gamble, and Wal-Mart, to name a few). However, for many companies, effective supply chain implementation has been difficult to achieve. Research on supply chain implementation has revealed the following:

- An extensive review of the literature spawned the conclusion that while integration is beneficial, it is also difficult. Adopting a supply chain perspective requires trading partners to think and act strategically. This is difficult within a single organization; it is even more difficult across a diverse and dispersed group of trading partners (Power 2005).

- A working paper from Michigan State University, a leading SCM school, made a distinction between today's supply chain and tomorrow's supply chain. They considered today's supply chain to be relatively simple, unidimensional, and focused on price reduction and on-time delivery. On the other hand, tomorrow's supply chain will be strategically coupled and value driven. It will have to deal with more complexity and issues and be able to adapt to rapidly changing conditions and situations (Melnyk et al. 2011).

- Individual supply chains are becoming more complex. In addition, companies must build multiple supply chains because of differences among products, customers, suppliers, geographic locations, economic cycles, political actions or inactions, and other confounding occurrences. In speaking of this complexity, one paper poses the question, "Should we give up now?" The authors respond, "None of this is to say that the supply chain world is too complex to effectively plan and execute. Our professional challenge is to do just that, and it is a large part of what makes our professional lives more rewarding than most" (Van Bodegraven and Ackerman 2013).

- Another study sponsored by CAPS Research, the research arm of the ISM, also identified competitive pressures as having a strong and direct effect on supply chain strategy and integration. The study focused on two key issues—alignment and linkage—both inside an organization and across organizations. The study concluded that well-integrated supply chains are not ubiquitous at this time. While there are success stories of excellent supply chain integration, there are also many cases of failures. The authors identified 14 challenges that organizations must overcome to achieve true supply chain integration. They conclude that good strategies and practices originating with leading companies will provide guidance for companies to follow in pursuing supply chain integration (Carter et al. 2009).

- Another study found that leading companies in SCM consider three capabilities that are essential to supply chain success today—visibility, analytics, and flexibility. Supply chain leaders are building systems and developing partnerships that give them greater visibility into a wider set of factors. To take advantage of this visibility, they have to be able to analyze the vast amount of data collected.

Finally, they must possess the flexibility to take advantage of their increased knowledge. The study found that firms with these capabilities are outperforming their rivals in many dimensions, notably in growth and profit (CSC 2012).

• Some writers are pointing out that it may be too simplistic to think of supply chains as linear; they should be thought of as a network or web of interconnected entities with built-in redundancy and resiliency to confront today's rapidly changing business environment. This type of network could also mediate risks and encourage longer lasting relationships (Siegfried 2013).

The gap between reality and goals does not mean supply chains are not a good idea, but it does indicate that building an effective supply chain is not easy and takes time. What factors make it so difficult?

Supply chains are complex. Consider that each company in the supply chain has its own set of customers that are unique from other companies in the supply chain. Now consider that each company offers its own set of products, different from other companies in the supply chain. The result is a maze of supplier/product/customer combinations. Even in a small company, the supply chain is complex; in a large company, it is overwhelming. A study by Deloitte Touche Tohmatsu describes the complexity paradoxes of optimization, customer collaboration, innovation, flexibility, and risk (Deloitte 2003).

If the inherent complexity described earlier were not enough, most companies are making the relationships even more complex by outsourcing both the manufacturing of goods and the providing of services. Outsourcing is here to stay and assimilating it into everyday practice is a challenge for even the most experienced global participant. Competition is increasing; this means every business has to continually improve its performance. The supply chain that was innovative just a few years ago is barely adequate today. Variability is rampant along the supply chain. Not only are the relationships complex, they are constantly changing, and change breeds variability. It is difficult enough to build interpersonal relationships among individuals and even more difficult for companies to build interorganizational relationships with their supply chain partners. It is difficult and takes time, longer than many managers have the patience to devote.

Another major element of change is the introduction of e-commerce in the equation. That is another whole subject, but the impact on supply chains is indisputable. Companies cannot use identical supply chains for their e-commerce efforts as they do for their regular products.

Supply chains, as a management concept, are becoming more comprehensive. What started as a transactional process is rapidly becoming a strategic process. What started in the purchasing department is becoming a cross-functional, horizontal flow process that includes all of the functions of a business.

### Research in Support of Integration Efforts

While leading companies have developed effective supply chains, some businesses are still in the early stages of supply chain development. However, the gap is getting wider between those companies that are successful at implementation versus those that are not (Beth et al. 2003). Whether or not to have an effective and efficient supply chain is no longer an option, it is an imperative!

Several sources provide direction and encouragement. The consulting group, PRTM, describes the transformation required to move from a functionally focused supply chain to the more desired cross-enterprise collaboration supply chain. The stages begin with a

(1) functional focus and proceed to an (2) internal integration, then to an (3) external integration, and finally to (4) cross-enterprise collaboration (PRTM 2005). Cigolini and colleagues offer a similar progression from (1) traditional logistics, with a focus on reducing inventory; to (2) modern logistics, with a shift from cost reduction to improving customer service; to (3) integrated process redesign, which provides a systemic vision of the supply chain; to (4) an industrial organization, which focuses on the strategic alliances among the various members of the same supply chain (Cigolini et al. 2004). Others describe the need for logistics synchronization, information sharing, incentive alignment, and collective learning necessary for an integrated supply chain. There are plenty of answers; it is just that none of them is easy.

Some see a brighter future for effective implementation, if managers are willing to work for it. Kempainen and Vepsalainen emphasized, "Information sharing and coordination are often considered the preconditions for successful supply chains. Our analysis of supply chain practices in industrial supply chains shows that visibility is still limited" (Kempainen and Vepsalainen 2003).

A survey of APICS members found that information sharing was not among the top criteria used by companies in selecting suppliers; yet, information sharing was one of the major positive influences on firm performance, in such areas as market share, return on assets (ROA), product quality, and competitive position. Not surprisingly, they suggest increased attention to effective information sharing (Kannan and Tan 2003).

In the interest of promoting improved coordination, there are advantages in loosely coupled supply chains. Tightly coupled supply chains exist when partners closely connect in a one-to-one relationship, such as through traditional electronic data interchange (EDI). In contrast, "a loosely coupled supply chain is an extensively integrated but loosely connected supply chain network. These concepts have emerged from the Web services, Web portal and Extended Mark-up Language (XML) technologies" (Wong et al. 2004). A loosely coupled supply chain has a specific niche for short- and medium-term buyer–seller relationships, SMEs, and secondary partners and for a large number of partners.

Charles Poirier and Frank Quinn also offer cautious optimism. In reporting the results of a survey, they comment, "The findings further suggest that those running the business today continue to think of SCM as a short-term cost-reduction effort, not worthy of strategic importance. Only 37% of the companies indicated their supply chain initiatives were 'mostly' or 'fully' aligned with corporate strategy. The survey results indicate that collaboration across an extended enterprise is still more theory than practice. Collaboration was the single most pressing need—both internal collaboration and external collaboration with suppliers and customers. Perhaps the most important insight from our survey is that the real business benefit of advanced SCM remains largely untapped" (Poirier and Quinn 2006).

Consider this scenario. Companies report that developing reliable demand forecasts is one of their major challenges. In order to make better forecasts, a company needs more information from downstream members of the supply chain, and the company needs to collaborate with downstream customers and upstream suppliers, in order to develop a mutually beneficial demand forecast. To obtain better information, partners need effective interorganizational communications. Improved communications involves more than technology. Not only do they need to have a better way to communicate, such as through electronic interorganizational systems (IOS), but they also need to agree on what information to communicate. Here, it gets tricky.

It is at this point that the element of trust becomes important. Unfortunately, trust has been largely lacking in many supply chains. Most of those who have studied trust, and the list goes well beyond business researchers to include those in organizational science, industrial psychology, economics, and operations research, have found it a difficult ingredient to integrate into working relationships (Handfield and Bechtel 2004). Much of the research concludes that in order to trust someone or some entity, the parties involved have to work with each other to verify that the benefits of trust truly outweigh the risks. At some point, companies have to make a judgment decision to trigger the cycle from trust to collaboration, to better demand forecasts, to improved execution, to increased profits along the supply chain.

Business success involves recognition of the need to move to an integrated supply chain. It is a major decision for a company and requires top management support and participation because it requires major expenditures and commitment of resources, especially of key management employees. At present, few companies have moved to a completely integrated supply chain. To accomplish the change, management will need to cope with four separate, yet related, paradigm shifts:

- From mass production to mass customization
- From vertical integration to virtual integration
- From homogeneous cultures to diverse cultures
- From bottom line to triple bottom line (TBL)

Each of these transitions takes time and dedication on the part of the companies involved.

In Chapter 7, we described how companies were moving from a batch flow process to a lean flow process. In making this transition, they are able to reduce the amount of inventory they carry, especially the work-in-process inventory. When they reduce inventory, they reduce the response time to customers. However, most companies cannot make the transition to a lean supply chain unless other members of the supply chain also incorporate lean production techniques in their operations.

## From Mass Production to Mass Customization

For over a century, mass production has been the dominant theme in manufacturing, not only in the United States but also throughout the industrialized world. Mass production aimed to produce high volumes of relatively standard products and make them available to consumers within a reasonable lead time. The emphasis was on cost and quality, not variety, so it was possible to make a limited number of products using a make-to-stock (MTS) approach.

### From Craft to Mass Production

Up until about 1840, most products were produced by small, family-owned businesses. Products were customized to meet the needs of individual consumers. The critical success factors for producers were function and availability. Several factors limited the growth and scope of businesses. The lack of distribution capability made it difficult for companies to sell their products very far from their home location—they just could not move products over great distances because the predominant way to ship goods was horse-drawn wagons or small canal boats.

The invention of the steam engine at the beginning of the industrial revolution to power train locomotives heralded the beginning of a new age in manufacturing. Companies could now expand their market area. This meant increased sales and the need for increased manufacturing capability. In order to produce in greater volumes, manufacturing companies had to develop new methods of making their products. The major driver of this transition was standardization, which began with the concept of interchangeable parts. One of the more publicized efforts in this direction was the work by Eli Whitney to make muskets for the government that featured using interchangeable parts. Although Whitney's work was only partially successful, it did represent the beginnings of work to make interchangeable parts a standard component of the manufacturing process (Hounshell 1984).

Companies found that to make interchangeable parts, they had to carefully control the process of making those parts. This led to the standardization of manufacturing processes. Tooling, equipment setups, and raw materials were items that needed to be standardized, so the manufacturing processes could produce consistent parts with acceptable quality.

While equipment was an important part of the movement to mass production, workers played an equally important role. Just as with products and processes, workers had to be standardized, in order to mass produce standard products. Usually, this meant dividing the job into small tasks for each worker, a concept known as job specialization. Workers could learn to perform small increments of work quickly and achieve high levels of productivity.

As companies made progress in standardizing materials, processes, and workers, they also began to develop more consistent and higher-speed equipment. In the United States, there was a labor shortage and equipment was developed to replace the need for workers. The movement to higher speed was aided by advances in methods of transmitting power to these machines. Electric power eventually replaced steam power, a revolutionary improvement.

Beginning in the later part of the nineteenth century, Frederick W. Taylor combined all of these standardizing elements into a concept that became known as scientific management. The principles of scientific management were easily understood, although not always popular among traditionalists. The age of mass production had fully arrived (Stevenson 2007).

In Chapter 7, we described various strategies by which companies provide products to customers. These stages include MTS, assemble-to-order (ATO), make-to-order (MTO), and engineer-to-order (ETO). Standard products are made in anticipation of a sale and stocked in a location easily accessible to the customer, such as a retail store, to reduce the response time to the customer. As the amount of customization required increases, it is being achieved by using ATO, MTO, or ETO methods. This satisfies the desire for customization but increases the response time to the customer.

Mass production has been a successful way to produce goods for mass markets, where the emphasis has been on low costs and high quality. However, it sacrifices the capability to produce customized products for small markets.

## Prelude to Mass Customization

Beginning in the last quarter of the twentieth century, the manufacturing capability of the global industrial world caught up with the demand of mass markets. As companies looked for new ways to compete, it was given that everyone had low costs and high quality. In order to differentiate themselves from their competitors, manufacturers began to offer a greater variety of products. While this movement was most apparent in consumer products, it also emerged, to some extent, in industrial products. Appliance manufacturers offered greater variety to consumers, and integrated circuit manufacturers also offered greater variety to assemblers of electronic products.

Changing customer preferences also contributed to the need for greater variety. In order to capture their own personality and tastes, affluent individuals began to search for *something different* in the way of products and services.

In order to more effectively compete, manufacturers began to *push* their greater variety of products out to the consumer. On the other side, consumers wanting greater variety began to *pull* those products from the manufacturers. The age of mass customization emerged as businesses began to decide how best to operate in the changing environment. Joseph Pine ushered in the age of mass customization with his book by the same name (Pine 1993).

One approach to providing increased choices to consumers is to manufacture a greater variety of products using the traditional MTS approach. However, increased variety can quickly add complexity and costs to production planning and inventory management. Companies found there were limits to the number of products they can profitably support in an MTS environment. Responding to the demands for faster response times and greater variety necessitated a different approach.

The transition from mass production to mass customization requires radical changes in products, processes, employees, and equipment. One of the key changes is in the design of products. Building on the interchangeable parts idea from mass production, mass customization requires that products include combining interchangeable parts into modular subassemblies that can be quickly customized to meet a customer's requirements. This approach is an adaptation of the ATO process.

The need for product flexibility triggered the movement to make processes more flexible. Manufacturers did not want to lose the high productivity and quality the mass production approach offered. However, changes were needed to find ways to make their processes more flexible. One obvious improvement was to reduce setup times when manufacturing changes from one product to another. Under mass production, the emphasis was on long runs and setup times were insignificant. If a line could run 80 hours before a change to another product, then 4 hours to set up the equipment was not a problem. However, as product variety increased, run times decreased and a 4-hour setup for an 8-hour run became a problem. With careful analysis, companies found that they could reduce setup times from hours to minutes. Toyota's work in reducing setup times became known as the single minute exchange of dies (SMED) system to convey the idea that setup times had to be drastically reduced (Shingo 1985).

Employees also had to change. Where job specialization was the accepted approach under mass production, job enlargement became the preferred approach for mass customization. Job enlargement was emerging in progressive companies even before mass customization because of the need to reduce job boredom and monotony that led to high employee turnover. Because employees can think (unlike machines), they can adapt to changing conditions and, hence, are more flexible than machines. Under mass production, many companies tried to automate their processes to eliminate the need for workers. The *lights out* factory was the ultimate goal, where the plant would be completely automatic and few workers would be required. As companies moved toward mass customization, they realized production-level employees who could think and make decisions would be a natural complement to add versatility to high-speed equipment.

Even with the changes in machine setups and employee capabilities, equipment designers also tried to add more flexibility and versatility to the equipment through computer programming. Computer-numeric-controlled (CNC) equipment became commonplace.

**TABLE 11.3**

Examples of Movement toward Mass Customization

| Industry | Approaches to Mass Customization |
| --- | --- |
| Automobiles | Deliver a customized car in 3 days |
| Information software | PC software that can be customized by the user |
| Telecommunications | Cell phone apps to fit customer needs |
| Personal care products | Match makeup to individual customer |
| Beverage industries | Provide a wide variety of beverages |
| Breakfast cereals | Provide a wide variety of breakfast cereals |
| Clothes | Custom fit clothes |
| Insurance | Variety of group insurance plans |
| Banking | Variety of bill payment plans |

The scientific management approach facilitated improvements in mass production. The systems management approach is driving the movement to mass customization. By definition, a systems approach implies that different activities, often involving different entities, have to be linked together to create a smooth functioning result.

Numerous industries illustrate movement toward mass customization. Table 11.3 lists some of these ventures. While mass customization is far from replacing mass production, it is making inroads.

The manufacturing strategies of MTS, ATO, MTO, and ETO can be adapted from a mass production perspective to a mass customization perspective. Businesses continue to search for ways to produce high volumes of customized products with significantly decreased lead times.

## From Vertical Integration to Virtual Integration

The eminent business historian, Alfred Chandler, Jr., provided a detailed explanation of business organizations from the early eighteenth century until late in the twentieth century (Chandler 1977). The progression was from small owner-managed local businesses, to large professionally managed global businesses that were often tightly vertically integrated, to today's large loosely connected supply chains.

The owner usually managed these early businesses. An individual would start a company and continue to run it over his or her lifetime and then pass the ownership and management to the children. This was the dominant form of organization in the United States until about midway in the nineteenth century.

During the Industrial Revolution, companies grew larger and new forms of organizations began to appear. The railroads and oil companies grew rapidly and needed to move toward a group ownership format, as opposed to an individual ownership arrangement. Not surprisingly, these businesses were too large to continue under the old owner-managed practices. While owners were often active in the business, they moved to top management decision making, such as overseeing the allocation of resources, while day-to-day operations became the responsibility of professional managers. Even so, both managers and owners were more concerned with short-term decisions than long-term, or strategic, decisions. The line-and-staff form of organization began to appear.

From the 1880s until World War I, a number of industries blossomed and saw the growth of large companies—U.S. Rubber in rubber products, General Electric in manufactured goods, and DuPont in explosives and chemicals. Companies found there were two areas necessary in their efforts to expand their sales: (1) a production system that connected the flow of materials and the processes into a continuous stream and (2) an in-house marketing organization that could sell the product into increasingly larger and more remote markets (Chandler 1977). Ford's focus on the assembly line for automobiles represented the most progressive movement toward product flow. Whereas companies had used jobbers and agents to sell their products earlier, they found that having their own sales and marketing organizations increased their ability to direct and control those efforts. The line-and-staff form of organization became popular to the extent that almost all larger companies, such as General Electric and DuPont, used it in some form. Companies also decided that vertical integration was the most effective way to organize in terms of securing resources and retailing their products.

From World War I to World War II, organizations became more formal and professional managers almost completely took over the day-to-day management of the companies. Owners may have still managed the smaller organizations, but the large public corporations used professional management. Long-range planning—the forerunner of strategic planning—appeared, but was still not a well-entrenched practice in most companies. Companies began to refine their accounting practices, and DuPont introduced a way of combining profitability with asset utilization in their return on investment model. They used the following equation to look at both elements:

Return on Investment (ROI) = (Income/Investment) = (Income/Sales) × (Sales/Investment)

The first part of the equation measures profitability, or percent income on sales, and the second part measures asset turnover. For internal performance measurement, the ROI can be changed to ROA.

Following World War II, the United States was the only major industrial country whose manufacturing capability remained intact. Much of the manufacturing capacity in Europe and Japan was destroyed by bombing during the war. At the time, the rest of Asia did not possess much manufacturing capability. However, Europe and Japan rebuilt quickly, with the latest technologies, and, by the 1960s, were looking to enter the U.S. market. While they had some initial difficulties with automobiles, Japan quickly became a major player in the consumer electronics industries. Meanwhile, U.S. companies had plenty of capacity and operations managers found it difficult to convince their company financial executives on the need to invest in newer technology. Consequently, American manufacturers in basic industries, such as steel, found they were beginning to lose market share.

During the 1960s and 1970s, U.S. companies became fascinated with growth through acquisitions. Teledyne and IT&T became the darlings of the investment community with their ability to add seemingly disparate companies through stock acquisitions and increase the stock value of the acquiring company more than the absolute value of the acquired company. *Synergy* became the CEO's mantra. As a result of combining dissimilar businesses, different organizational forms appeared. One of the more popular forms was the matrix organization, an attempt to assemble a variety of functional representatives into a functioning project organization.

By the 1980s, it became apparent that foreign competition was taking large segments of the U.S. market. Companies were also finding the conglomerates they had created through acquisitions were not necessarily that profitable. As a result, many companies began to divest those businesses that did not fit well with the mission of the organization.

Improvements in production as well as services were required, and the 1980s saw the rise of a number of management programs designed to provide those improvements, including just in time (JIT) followed by lean production, total quality management (TQM) followed by Six Sigma, and EDI followed by Internet EDI. In addition, the organizational structure was changing as new flatter forms were used to reduce the slow decision-making process inherent in the tall, line-and-staff hierarchical organizations.

Another organizational form that developed was an extension of the project, or matrix, structure. This structure, the virtual organization, involves "short-term alliances between independent organizations in a potentially long-term relationship to design, produce, and distribute a product. Organizations cooperate based on mutual values and act as a single entity to third parties" (Blackstone 2013). Boeing used this type of organization for their latest model airplane, the 787 Dreamliner (Crown and Epstein 2008).

In the last century, businesses have moved from a tightly coupled vertically integrated form of organization to a loosely coupled virtual organization form. Vertical integration provided direct control over the processes and afforded businesses the opportunity to grow to large sizes with relatively standard products and services. The virtual integration approach enables businesses to change their product lines as market demands rapidly change. Consequently, as businesses move from mass production to mass customization, they must change their organizational structures to fit with their new strategies.

## From Homogeneous Cultures to Diverse Cultures

When businesses were smaller and confined to local or regional market areas, they prided themselves on having stable work forces where it was not unusual for employees to spend their entire career with one company. During his tenure as a division manager of a manufacturing operation in a small Midwestern community, one of the authors recalls that the work force included three generations from some families. Low labor turnover was a desirable objective. The culture of this business was well entrenched and somewhat predictable.

Today, it is rare that multiple generations work in the same company. Mobility among employees is accepted and necessary during this period of product proliferation, offshore outsourcing, and accelerated technology development. Younger members of the work force must be prepared to consider movement a way of life, in jobs, careers, companies, and locations. Consequently, organizational cultures are more diverse and less predictable.

The diversity of cultures makes it imperative that businesses learn to become members of integrated supply chains. Integration, in this case, implies that diverse entities work together effectively and that these entities will themselves be diverse.

## From Bottom Line to Triple Bottom Line

There is another form of supply chain integration that is receiving an increasing amount of attention—the integration of environmental and social concerns. Just as outsourcing is exposing the risks of disruptions in supply chains, it is also focusing attention on poor working conditions among supplier organizations (a building collapse in Bangladesh that kills hundreds of workers) and environmental disasters arising from oil spills or train or truck wrecks as goods are transported to their destinations.

What can an organization do to achieve this integration? Over a decade ago, John Elkington (1998) suggested an approach he called the triple bottom line (TBL), a term that has gathered support over the years. Elkington proposed that businesses should aim for success in three areas: economic, societal, and environmental. He recognized it takes all three to achieve sustainability.

Sustainability is another term that deserves some explanation. Sustainability, for a business, is the ability to keep operating successfully. The Brundtland Commission report, *Our Common Future*, defined sustainable development as "development that meets the need of the present world without compromising the ability of the future generations to meet their own needs" (Anderson 2006).

As the TBL concept implies, the mission of business is to make a profit. That mandate has long been a *given* in the management literature and in the management of companies. Efforts to introduce social and environmental concerns were often viewed as obstacles, or at least distractions, by executives, especially at the highest levels. Just as victories count in measuring a coach's performance in football, profits were the measure of executive performance. Economic performance is a tangible, short-term game in which only the strong can survive. This is the first bottom line.

Social issues, such as providing employees with suitable working conditions, fair wages, and a host of other matters, were topics to be debated with unions during contract negotiations. Other matters, such as increased traffic flows, noise, and tax relief, were to be considered as the necessary evils of bringing jobs into a community. Lead-laden toys, polluted milk, and side effects of drugs were too often considered the unfortunate consequences of the drive for business continuity. Companies often used human resource or public relations departments as buffers against the barrage of new social agendas. Achieving social responsibility is difficult to measure, report, and set goals for, although a number of companies are beginning to come to grips with the need to report their activities in this area. However, it has been difficult for many businesses to resolve their obligations in the societal arena, the second bottom line.

Environmental issues have moved through a transition. Protecting the environment was originally viewed by most businesses as simply a need to comply with government regulations. Saving endangered species and trying to assess the effects of global warming was for special interest groups and the government, not business. As businesses began to reluctantly comply with mandates to reduce polluted discharge water or exhaust gases, or to substitute benign for hazardous materials, or to reduce the waste going into landfills, they made an astounding discovery—the changes could actually save them money! Companies began to associate doing good—environmental improvement projects—with doing well or making a profit (Laszlo 2008). A separate, and important, topic is the role of reverse logistics and its contribution to improved resource utilization as we described in Chapter 10. Environmental projects can be planned and measured; they are often multi-year, and they represent the third bottom line.

In a follow-up to his 2008 book, Chris Laszlo, along with coauthor Nadya Zhexembayeva, emphasizes the need for an integrated approach in business to the economic, social, and environmental issues. In speaking of the widespread coverage of social and ecological topics, the authors conclude, "With topics ranging from $CO_2$ emissions, water rights, and deforestation to child labor, peace, and social equity, the needs of society and the environment present a perfect storm for the average manager: complex, disorienting, and maddeningly inscrutable... How is one to understand the vast landscape of seemingly disconnected concerns? Looking deeper into the economic, social, health and ecological pressures that fall under the sustainability umbrella, one finds three distinct but interconnected trends: declining resources, radical transparency, and increasing expectations. Together, these trends are becoming a major market force that is redefining the way companies compete. It has now reached a critical point, changing the rules for profit and growth in almost every sector of the economy" (Laszlo and Zhexembayeva 2011, p. 6). The sustainability movement is gaining momentum. It is much like a bridge supported by cables, with each cable requiring thousands of strands of wire woven together. The sustainability movement has thousands of individual strands of effort from special interest groups, the public, business, and government being woven together into an integrated global effort. While it has not yet reached the tipping point, it clearly portends the need for all businesses to decide how to participate.

## Drivers of Change

What is driving the changes required to make supply chain integration a desirable, even necessary, move for most businesses? Four of the principal drivers are global competition, global markets, economic advantage, and building relationships and trust.

### Global Competition

Several authors have written books recently that address the rapidly changing global business environment. In his book *The World Is Flat* and later in *Hot, Flat and Crowded*, Thomas Friedman vividly describes the growth of competition from countries that, until very recently, were considered to be undeveloped or, at least, underdeveloped. Friedman points out that the United States must improve its capabilities to compete with these countries or lose its position as world leader (Friedman 2005, 2008).

In the *J Curve*, Ian Bremmer makes a similar argument. He describes the struggles of many countries in trying to move from a state-controlled economy to a market-driven economy. When they achieve this transition, the economic benefits are huge. When they try but fail to make the transition, their economies and political state are often precarious. When they do not try but remain in a state-controlled state, the results are often disastrous for the general population, and the political situation is unstable to the point of crisis (Bremmer 2007).

Yergin and Stanislaw make a similar case in their book *The Commanding Heights*. This book also looks at the changes over the past few decades in the political revolutions that have transpired in many countries. As the countries become more open and the citizens have the opportunity to engage in relatively free business enterprises, they can achieve phenomenal results (Yergin and Stanislaw 2002).

Fareed Zakaria echoes this theme in his book, *The Post-American World*, by pointing out that the rise of other countries will increase the level of competition among a greater number of countries throughout the world. He also suggests that as economic conditions improve, political environments change, which usually involves moving toward a more democratic society (Zakaria 2008).

Peter Navarro also echoes Bremmer's comments in his book, *The Coming China Wars.* Navarro maintains that China is the world's *factory floor,* due to its low-cost labor availability. As a result, China has benefited economically, and this has resulted in its presence in many worldwide supply chains. The movement toward a market economy has also helped in this respect. However, wealth distribution has been unequal, with many Chinese citizens not being able to benefit from the growing economy (Navarro 2007).

The aforementioned presents the rise of global competition from a political perspective—the movement from state-controlled countries to market-driven economies. Emerging countries often have lower labor costs; this gives them an advantage because they can sell products at lower prices, assuming they have acceptable quality. The increased competition is forcing U.S. businesses to look for ways to improve their competitive position, and supply chain integration is one of those ways.

## Global Markets

U.S. firms are starting to recognize the need to defend their position in the domestic marketplace. However, most are not content to be on the defensive. They want to aggressively pursue growing markets in developing countries.

However, it is not always possible to build an integrated facility in all market areas. Consequently, it may be more appropriate to divide the supply chain into modules that can be located separately but with close coordination among them.

This decision must consider financial, structural (organization), political, and risk management considerations. Even recognizing these factors that go into the decision-making process is a challenge; trying to design a system that adequately reflects the relative significance of each factor is even more daunting.

## Economic Advantage

Changing from a batch, or intermittent, flow of products and services to a lean and integrated supply chain will help to meet competition and to exploit new markets. It will also provide lower operating costs and increased revenues for already-existing products and markets.

### Lower Costs

Integrated supply chains reduce costs in several ways. One of the major benefits is that they reduce inventories along the supply chain. This is accomplished by sharing information about the actual demand at the end point of the supply chain, the retailer. This reduces the bullwhip effect, described in an earlier chapter. Another benefit is the reduction of unnecessary information within the supply chain; it becomes easier for management to know what information is needed and what is irrelevant.

Integrated supply chains also enable management to find ways to improve the quality of the product. If quality is what the customer wants, integration makes it easier to provide only what the customer truly wants and not what an engineer or marketer thinks the customer wants.

Integrated supply chains increase resource utilization. By eliminating duplication, fewer resources are required to meet demand. It also makes it possible to schedule resources more effectively and reduce unnecessary capital investment.

### Higher Revenues

Lower costs can mean lower prices for the customer, which can result in increased revenues from higher unit volume sales. For high-end items, higher product quality can also lead to increased selling prices, which can lead to increased sales dollars. One member of a supply chain can adversely affect the quality of a product or service; therefore, it is important that all members work to achieve the target quality level.

## Relationships and Trust among Supply Chain Participants

In order to integrate supply chains, it is necessary to build relationships between companies, one at a time. Building relationships requires trust between entities, and trust is difficult to achieve. There are several ways to develop trust between organizations. We will describe it briefly in this section and explore this topic more fully in Chapter 13.

### Trust between Individuals

Often, trust between businesses comes about because individuals develop trust for one another through regular contact in their normal work. If enough key employees in one company trust enough key employees in another company, then a critical mass of trust builds up, and finally, it follows that one company begins to trust another company. While the relationship may become a strong one, it can sometimes be easily disrupted if key individuals change jobs or companies.

### Formal Contracts or Agreements

Sometimes, companies build trusting relationships by preparing formal agreements and then operating in accordance with those agreements. After a reasonable time, if both sides are satisfied with the performance of the other party, they begin to trust one another. Conversely, if either party does not live up to the formal agreement, a trusting relationship is never realized. While formal contracts may not have the emotional ties that bind individuals, it is a more permanent relationship and often does not depend on individual relationships.

### Common Interests or Projects (Enforced Trust)

Often, companies have common interests that require them to work together. As a result of working together, individuals realize that it is beneficial to both if they refine their relationship into one of mutual collaboration and trust. For example, two companies are selected to participate as members of a virtual organization, such as when Boeing starts on a new airplane project. Suppose one is designing the electrical system and the other the hydraulic system. They must work together because the hydraulic system includes the need for electric power to operate some of the hydraulic valves. They also may need to decide how best to allocate space in order to accommodate both systems in the plane. In this kind of enforced collaboration mode, they must quickly learn to trust each other.

In a different example, trust took a while to develop between the Tennessee Bun Company (TBC) and McDonalds. The CEO and president of the Dickson, Tennessee bakery, said it took 4 years and 30 interviews before McDonalds was ready to trust her bakery to deliver the goods. Today, TBC, a highly automated factory, provides English muffins and hamburger rolls to 600 McDonald restaurants in the southeast United States (Sarma and Harrington 2005).

## Involves Change Management

The transitions described earlier require changes. While the amount of change required varies by company, all companies have to change to remain viable. In some cases, the change extends throughout an entire industry, such as automobile manufacturing, from gas guzzlers to fuel-efficient retailing; shopping centers, from enclosed malls to stand-alone convenience; and health care, from cure to prevention. The change, while necessary, will not be easy.

## Change Is Difficult within a Company

"Change that I suggest is good; change that others suggest is suspect." That seems to be the basic attitude of many managers. We all embrace some changes and, just as enthusiastically, resist other changes. Effecting change within a company begins with figuring how to get each person to accept and support the proposed change. Universal acceptance is unlikely; there must be a sufficient level of acceptance to enable the company to go ahead with the proposed change without undue resistance or interference from the employees.

### *Embedded Culture*

Changing the culture of a company is usually a difficult chore. It may be most difficult in those firms that have been successful. When Alan Mulally, formerly with Boeing in the commercial aircraft industry, became CEO of Ford Motor, Bill Ford, great-grandson of Henry Ford, warned Mulally that one of his major challenges would be to change the culture within the organization. "Mulally's biggest challenge, Ford said, would be breaking down silos, specifically the operating regions around the world—Europe, Asia, South America, and Australia—that were more interested in defending their turf than working together. It was a culture, Ford explained, where one's career had come to mean more than the company" (Kiley 2009).

Sometimes, careful planning is the key to change. If management wants to implement a Six Sigma program, they must prepare the employees, at both the supervisor and the worker levels, to accept and support the program. Too often companies fail to recognize that company cultures are difficult to change.

Sometimes, necessity is the key to change. If a company is facing reduced earnings from a competitor's new product, it may be sufficiently obvious to management and employees that a fast reaction is necessary. In this case, the employees may be willing not only to change but also to contribute their ideas to help a new product development program succeed. Andrew Grove, former CEO of Intel, describes this as a *strategic inflection point* and goes on to relate how he faced this situation when the company was facing the need to move from memories to microprocessors (Grove 1999).

External factors and timing may even cause employees to suggest and welcome change. High gasoline prices and the resultant high costs of commuting encouraged a number of organizations to move to four 10 hour days a week from the more traditional five 8 hour days, a change employees actively sought.

Establishing closer relationships with customers and suppliers may not appear to be an obvious need or beneficial action to many within a company. However, many companies are finding it necessary and beneficial.

### Policies and Procedures

While changing the embedded culture of a company is difficult, it is also necessary to change the written culture—the policies and procedures that guide the day-to-day operation of the organization. This is time consuming and tedious, at least in those organizations that have written procedures. In integrating supply chains, the major changes will be in redefining policies and procedures dealing with closer relationships with customers and suppliers. While procedures within the company may not change appreciably, key points may need to be revised to reflect that some information will be shared with other companies. Even something as simple as spelling out an abbreviation or explaining a unique term may become part of this type of change.

### Organization Structure

If companies are to integrate their supply chains, their organizations will need to become flatter to enable more direct and faster communication between a functional area of a company and its counterpart in another company. Engineers will have to talk with engineers, accountants with accountants, quality assurance persons with quality assurance persons, and so on. This expanded exposure to interorganizational communications will change the dynamics of communications within the company. Individuals will have to be able to act more independently of their supervisors and, in some cases, may even infringe on the responsibilities of other functional areas, if their counterpart in another company has somewhat different areas of responsibility.

### Customer Relationships

In Chapters 3 and 4, we described customers and customer relationship management (CRM) in some depth. In this section, we reiterate that establishing effective and collaborative relationships with customers is a continuing challenge. It is a dynamic environment filled with tension and edge-of-precipice decisions, where cordiality and smiles may obscure the real feelings of the participants.

### Supplier Relationships

We described supplier relationships in Chapter 8. Just as with customer relationships, the environment is dynamic and even the best of relationships may be fragile. Considering that most companies of any size will have dozens, perhaps hundreds, of suppliers, it is easy to see that maintaining solid relationships with all of them requires continuous attention.

### Union Relationships

The influence of unions in manufacturing has declined in recent decades. Their strength was in the basic industries—steel, automobiles, and consumer products. They have not been as successful in new industries such as electronics and telecommunications and have shifted their emphasis to services industries, such as retailing and local governments.

Some of the elements of integrating supply chains can pose threats to a union's influence with its members. Job enlargement and employee empowerment expand direct communication between managers and individual employees and reduce the hierarchies of line-and-staff organizations. To some extent, this conflicts with the traditional union structure of shop stewards who represent the individual employees in communications with managers.

The broader question of outsourcing and redistribution of functions within supply chain participants makes it difficult to maintain formal relationships. Employees must be flexible and have broader knowledge, beyond their formal job description. This amorphous situation means that both management and unions have to find a more flexible relationship.

At one time, employees could look forward to spending an entire career with one company; today, that prospect is unlikely. Are individuals on their own in making job, or even career, moves? Or should there be some more formal process established to make these transitions easier for the individual and improve the outcome for society? Should that responsibility belong to the company, the government, or to some other entity?

## Change Is More Difficult for a Community

As difficult as change is for a company, it is even more difficult for a community, where there are multiple businesses and organizations and where there are diverse interest groups.

### Company–Community Relationship

The relationship between a business and a community is complex. While a business is an integral part of a community—it sells to the population, it hires its employees, it pays taxes, and uses community resources—its interests and influence usually go beyond the immediate community. When it begins to integrate its supply chain, its interests span multiple communities.

If a business adds jobs by selling more products to a customer, the community views that as desirable. If a business reduces jobs by transferring an activity to a customer or supplier, the community views that as undesirable. If a company communicates its intentions ahead of its action, it can expect reactions from the diverse interest groups. If the company acts without communicating its intentions, it can expect reactions from diverse interest groups. The potential for inadequate, or excessive, communication of a company's intent is one of those nagging decisions that always seem to be difficult.

The decision to integrate supply chains may not cause a ripple in the normal operations of a community; therefore, not much change is required. On the other hand, if a company closes a plant or otherwise disrupts the business landscape of a community, the reaction may reach well beyond the local community and have adverse effects on the company. In 2008, two of the *Big Three* automakers had to appeal to the federal government for emergency funds to keep operating. As part of the agreement, they had to close a number of their manufacturing plants, some of which had operated for decades.

### Difficulty in Changing Laws and Regulations

Aside from causing a disruption in the informal activities and attitudes of a community, some changes required in integrating supply chains may require formal action by a community, in the form of changing laws and regulations. Adding a process that can introduce carbon emissions into the atmosphere may require communities to impose legislation limiting the amount of emissions. Adding a second shift to accommodate increased business may require new limitations of noise. Either regulation, along with a host of others, will likely trigger objections from some of the special interest groups. If the special interest groups are strong enough, they may effect a change in the elected officials that enacted the new legislation.

### Difficulty in Changing Infrastructure

Changing laws and regulations may be controversial, but they usually do not require large expenditures of funds or resources. However, some changes may require significant changes in the infrastructure of a community. There may be a need for new roads, new signs, added security, diverted traffic flow, temporary disruptions from construction, or a myriad of other physical changes. These changes cost money and are disruptive, even when welcomed by citizens in the community.

### Change Is Most Difficult for an Entire Country

While the pace of change in individual businesses and local communities may often seem slow, it is usually at warp speed when compared with the rate of change in an entire country. The decision by one company to integrate supply chains may be insignificant; if most companies do it, it can even require change within an entire country.

### Political Implications

Just as in local communities, change at the national level has political implications. No matter what the change, some will support it and some will oppose it, while some will be indifferent, at least temporarily. Major issues such as offshore outsourcing, green supply chains, and computer privacy are all issues connected with supply chain integration. Change at the national level may take decades to crystallize and gain acceptance. Even legislation, when finally enacted, may not resolve the issue.

### Regional Differences

As far-reaching as national legislation is, it may not resolve the issues at the state, or regional, level. Offshore outsourcing does not affect all sections of the United States equally. It may depress sections of the midwest, in states like Michigan and Ohio, while enhancing the potential of the southeast, in states like Tennessee and Alabama. The center of the automobile industry has moved south in the United States, driven more by economic considerations than legislative. Companies made the move because they wanted to, not because they were required by the government.

## Steps in the Integration Process

This section will be an overview of the integration process, to be covered in more detail in Chapter 13. The general approach is to establish a tentative relationship and then to expand and solidify that relationship as events and results dictate.

### Build Interfaces with Customers and Suppliers

An interface is a beginning point of contact. It may begin with an individual relationship between a buyer at the customer's location and a customer service person at the supplier's location. If this works, it can extend to other individual relationships. An interface can also be established between computer systems, where information is transferred effectively, although not always efficiently.

### Change Interfaces to Interlaces to Make the Relationships Closer

Individual interfaces can be expanded into more mutually dependent relationships that we can call interlaces, to indicate a more lasting and important connection. Individual relationships become departmental relationships where individuals can be interchanged without detrimental effects. Computer systems are made sufficiently compatible to exchange all types of information both effectively and efficiently.

### Change Interlaces to Integrated Relationships

The ultimate objective is to develop integrated relationships where participants from different companies not only communicate well but also have a sufficiently trusting relationship to collaborate. This may mean sharing resources or making joint decisions. The relationship is solid enough to weather the swirl of problems and issues that are sure to appear over time.

## Need for Strategic Planning

Most businesses recognize that they should use strategic planning as a foundation for their future sustainability. How does an integrated supply chain relate to strategic planning? Building an integrated supply chain should become one of the strategies employed by a successful company.

   At the risk of oversimplification, the root cause of strategic planning failures appears to be that the plans do not adequately reflect reality. Mintzberg distinguished between the intended strategy (plan) and the realized strategy (actual). He proposed that realized strategy resulted from a combination of intended strategy and emergent strategy—"patterns or consistencies realized despite, or in the absence of, intentions" (Mintzberg and Waters 1985). Nonoka and Toyama present strategic management as distributed practical wisdom. They claim, "Strategy is not created from the logical analysis of environment and a firm's resources... Strategy emerges from practice" (Nonoka and Toyama 2007). Lengnick-Hall and Wolff describe three types of strategies that require facing reality—capability logic, guerilla logic, and complexity logic. They defined these strategies in terms of the dominant logic supporting each strategy. Capability logic is the premise that firms seek to develop and implement strategies that will create a sustainable competitive advantage. Guerrilla logic is shaped by the emerging research into high-velocity firms and industries embroiled in not only extremely competitive but intentionally disruptive interactions. Complexity logic links strategic success with the natural consequences of understanding, shaping, and moving with the paradoxical forces that shape organizational systems. They compared the three approaches on a number of factors that reflected the need to be realistic (Lengnick-Hall and Wolff 1998).

Before we can show how strategic planning should reflect reality, we must first describe what we mean by reality. In the next section, we outline the major activities of an organization.

## Categories of Operations

There are four major categories of activities in a business. The first is *normal ongoing operations*—the bulk of the business activities. This type of operation involves transforming inputs, such as steel and wood, into products. It also involves transforming inputs into service outputs; often, the input is a person to whom a business provides the service. In most organizations, the ongoing operations are what the company considers its primary business and around which they build their strategic plans.

The second category is *improving operations*, in which the company uses planned programs to make improvements in its ongoing operations. These improving programs can be either incremental or radical. Incremental improvements represent a continuing flow of improvements that involve a large number of employees. Each improvement may not be major; however, the cumulative effect can be significant. Programs such as TQM and JIT represent incremental improvement efforts. Radical improvements are major in scope and often pose a disruptive effect on the ongoing operations. Lean manufacturing is an example of a program that may provide considerable benefit but may be disruptive to the normal operations of a company.

The third category is *problem-solving activities*. This involves the normal kind of actions to correct a reoccurring problem, such as late deliveries or invoice errors. Employees within the company can usually resolve these problems within the framework of their normal workday.

The fourth category is *crisis management* and this is quite different from normal problem-solving activities. It is a different story. A crisis occurs because of a major unexpected event, such as a toy recall because of lead paint, a fire that destroys a manufacturing plant, or a building collapse that kills hundreds of workers. Often, this type of event requires outside, expert help.

As a result, strategic plans can never be completely oriented toward desired goals; they must consider the actual situation as it exists. Organizations must start from where they are, not just from where they would like to be.

### Normal Operations

As we pointed out earlier in this chapter, most businesses are members of a supply chain as part of their normal operations. If the supply chain is composed of well-integrated entities, each member of the supply chain should benefit. The strategic plan should recognize the need to keep the supply chain integrated and functioning smoothly.

## Improvement Programs

In planning improvement programs, companies should consider the effect on their supply chain. While some improvement programs have only a local effect, some may affect the supply chain, either positively or negatively. Changing an internal wage payment plan may affect the company's employees but will not have an effect on other participants in the supply chain. On the other hand, changing a component material from steel to plastic will affect suppliers, customers, and members of the reverse supply chain. Improvement programs are necessary but must be carefully coordinated with other members of the supply chain.

## Problem-Solving Programs

While some problems arise within the confines of an individual business, many occur because of transactions between businesses. Deliveries may miss their scheduled arrival time, fabric colors may be different from ordered, or help desks may not resolve the software problem. Closer relationships between members of a supply chain will help resolve the problems faster and more effectively.

Strategic plans should include some contingency planning by anticipating some potential problems that may not have a routine solution, that is, the loss of electric power during a hurricane or snowstorm or the bankruptcy of a key supplier.

## Crisis Management

A crisis is a situation that can have life-threatening consequences for a business—contaminated toys or food or a fire in a sole supplier's manufacturing plant. In a tightly integrated supply chain, it will be easier to find a satisfactory solution by working with other supply chain members. We will spend more time on discussing crisis management, as part of a total risk management program in most organizations, in later chapters.

---

# Need for a Multiyear Project Plan

Building an integrated supply chain requires a project management approach. A project involves different entities working on a variety of activities over an extended period of time. This approach requires coordination and communication and a commitment of significant amounts of resources, both human and financial.

In Chapter 12, we describe some of the obstacles preventing supply chain integration. Overcoming these obstacles takes time. In addition, companies should not expect that supply chain integration is a *once and done* effort. Building an integrated supply chain is only the first step; it must then be maintained, often over an extended period. We describe this more fully in Chapter 13.

In addition to taking time, supply chain integration requires the coordinated participation of multiple and diverse members of one company working with multiple and diverse members of other companies. There are a number of *touch points* to be resolved requiring patient, enlightened, and adaptable participants.

## Performance Measurement across the Supply Chain

How do you know that an integrated supply chain is an improvement? Even among the most cordial of relationships, it is probably a good idea to have some form of measurement system to keep track during sharing situations. A company needs some way to measure performance for their own operation and for the supply chain in total. It is not enough to make local improvements; there must be improvement throughout the supply chain. We will discuss performance measurement in supply chains in more depth in Chapter 16.

At the same time, care must be taken that investments in supply chains are monitored closely and overspending does not occur. Suvankar Ghosh and colleagues warn that some companies have invested two to three times the level that would be desirable from a stakeholder perspective (Ghosh et al. 2007).

## Integration Requires Sharing

An integrated supply chain offers both benefits and responsibilities. In earlier sections, we outlined many of the more tangible benefits. Sharing also requires assuming responsibilities.

Sharing responsibilities means that companies begin to treat one another with confidence and respect. Sometimes, it is not possible to spell out all of the specifics in a relationship; the participants must just understand them. When a party is responsible, they do all of the obvious things and are prepared to do more if the situation requires.

One of the benefits of an integrated supply chain is the sharing of resources—facilities, equipment, and employees. Resource sharing can be both efficient and effective. It can also lead to disputes over control and usage but that can be resolved among collaborating participants.

## Summary

In this chapter, we described some of the conditions that make an integrated supply chain desirable. We looked at major transitions in manufacturing—from mass production to mass customization, from vertical integration to virtual integration, and from homogeneous cultures to diverse cultures. Drivers of these changes include global competition, global markets, economic benefits, and a growing awareness of the benefits of building trust among supply chain participants. Although change is difficult within a company, it is even more difficult in communities and governments, both local and national.

Moving toward integrated supply chains requires strategic planning, for normal operations, for improving operations, and for problem-solving activities. In many cases, project planning is required to coordinate diverse entities. Supply chain integration involves sharing resources and responsibilities among the supply chain participants. Finally, there is a need to measure progress along the journey to an integrated supply chain.

In Chapter 12, we examine more deeply the obstacles that prevent a company from participating in an integrated supply chain. In Chapter 13, we describe some ideas about how to go about developing the integrated supply chain.

---

## Hot Topic: AECL, Part 1

The need for integrating information flow along the supply chain is demonstrated vividly in this chapter's hot topic about Atomic Energy Canada Limited (AECL) and its cancer-fighting machine.

### Hot Topics in the Supply Chain: Atomic Energy Canada Limited Encounters Problems with Its Cancer-Fighting Machine, Part 1

This chapter discussed the need for integration along the supply chain. One of the goals of integration is to achieve a state of information symmetry, whereby the primary companies in the supply chain have uniform information so that business can transpire in a mutually beneficial manner. However, information asymmetry can still be a problem and can manifest itself in various forms. In one form, information asymmetry occurs when similar incidents involving the same technology transpire over a wide geographic area (Boin et al. 2003). Put a bit differently, a company may have knowledge of a product defect that is not widely known by its customers. In terms of information exchange, such a situation puts the customers in the forward supply chain at a disadvantage because of the obvious information asymmetry. The example that follows concerns a company that made a cancer-fighting machine, the Therac-25 that used radiation beams aimed at tumors to fight the disease. The machine was ahead of its time because of its ability to direct high-energy radiation beams at the tumor without destroying surrounding tissues. In addition, the Therac-25 had the ability to fight skin cancers or direct its waves toward areas of deeper tissues (Leveson 1995). The machine had been used successfully for 2 years in over a thousand treatments; however, six events occurred from June 1985 to January 1987 that resulted in three fatalities and three major injuries (Fauchart 2006). Although the company that made the Therac-25 was aware of the incidents, none of the four medical centers using the machine were aware of the events of overradiation at the other facilities.

### BACKGROUND

The Therac-25 machine was introduced in 1982 by AECL. It was a computer-controlled radiation device classified as a medical linear accelerator, also known as a linac (Leveson 1995). The Therac-25 was unique because of its ability to position the patient, set the amount and type of radiation, and to turn itself off in case of a malfunction, all by software control (Fauchart 2006). Despite the fact that the machine had worked well for 2 years since its introduction in 1982, six separate malfunctions occurred over a 20-month period that resulted in massive doses of harmful and, in three cases, lethal radiation.

Figure 11.2 displays the six incidents and the information flows that existed between the company and the four medical centers. In all six incidents, an overdose of radiation was administered by the Therac-25 machine.

The incident that started the Therac-25 crisis occurred at Kennestone Regional Oncology Center in June 1985. The overdose of radiation resulted in a burn to the

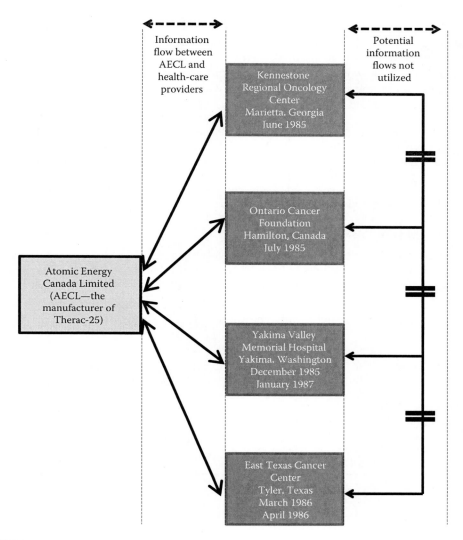

**FIGURE 11.2**
Information asymmetry along the supply chain. (Adapted from Fauchert, E., *Journal of Contingencies and Crisis Management*, 14(2), 97–106, 2006; Crandall, W.R. et al., *Crisis Management: Leading in the New Strategy Landscape*, 2nd edn., Sage Publishing, Thousand Oaks, CA.)

patient, who later filed a lawsuit against AECL. The second incident occurred at the Ontario Cancer Foundation in July 1985. Again, the patient received a burn and an overdose of radiation. Yakima Valley Memorial Hospital experienced two similar incidents, one in December 1985 and another in January 1987. East Texas Cancer Center also experienced two overdose incidents in both March and April 1986 (Fauchart 2006).

## PROBLEM OF INFORMATION ASYMMETRY

The Therac-25 was a very specialized and sophisticated medical device. In fact, at the time of the accidental overdoses, only 11 had been delivered to medical centers throughout North America. With such low numbers, one would think that the information of the accidents would have been readily available to all 11 customers. However, this was not the case.

Fauchart (2006) noted in his analysis of the case that communication took place between each of the four medical centers and AECL. However, the reports of the accidents were not made available by AECL to other users of the Therac-25; hence, each medical center was not aware that overdoses had occurred at other facilities. Information asymmetry was part of the forward supply chain, and this, at a time, when patient safety had been compromised.

> The manufacturer should have informed all the users that a number of accidents had occurred, but he did not do so. Instead, he told every user who asked for information about other possible incidents that he was not aware of any. He thus used the information asymmetry to pretend that each accident was a one-off fluke. This clearly delayed the instauration of a learning process aimed at fixing the problem and preventing other accidents from occurring (p. 101).

## QUESTIONS FOR DISCUSSION

1. It is interesting to note that the incidents involving the Therac-25 occurred from 1985 to 1987. This time period was before the use of e-mail and the Internet. What disadvantages does this time period present to this case?

2. If the same situation existed today, how could AECL managed it differently so that all partners in the forward supply chain had adequate information?

## REFERENCES

Boin, A., Lagadec, P., Michel-Kerjan, E., and Overdijk, W., Critical infrastructures under threat: Learning from the anthrax scare, *Journal of Contingencies and Crisis Management*, 11(3), 99–104, 2003.

Crandall, W.R., Parnell, J.A., and Spillan, J.E., *Crisis Management: Leading in the New Strategy Landscape*, 2nd edn., Sage Publishing, Thousand Oaks, CA.

Fauchart, E., Moral hazard and the role of users in learning from accidents, *Journal of Contingencies and Crisis Management*, 14(2), 97–106, 2006.

Leveson, N., *Safeware: System Safety and Computers*, Addison-Wesley, Reading, MA, 1995.

## Discussion Questions

1. Describe an integrated supply chain.
2. Discuss why membership in an integrated supply chain is beneficial for a company.
3. What are some of the major drivers to integrate supply chains?
4. What are some of the most important benefits of an integrated supply chain?
5. Discuss some of the major obstacles to supply chain integration.
6. Why does building an integrated supply chain require a project management approach?
7. Why is change difficult for supply chain members?
8. What is the role of strategic planning in supply chain integration?
9. What are problem-solving programs? How do they help or hinder supply chain integration?
10. What is the role of crisis management in supply chain integration?

# References

Anderson, D.R., The critical importance of sustainability risk management, *Risk Management*, 53(4), 66, 6, 2006.

Anonymous, The challenge of complexity in global manufacturing, Special report, Deloitte Touche Tohmatsu, New York, 2003.

Beth, S., Burt, D.N., Copacino, W., and Gobal, C., Supply chain challenges: Building relationships, *Harvard Business Review*, 81, 7, 64, 2003.

Blackstone, J.H., *APICS Dictionary*, 14th edn., APICS—The Association for Operations Management, Chicago, IL, 2013.

Bremmer, I., *The J Curve: A New Way to Understand Why Nations Rise and Fall*, Simon & Schuster, New York, 2007.

Carter, P.L., Monszka, R.M., Ragatz, G.L., and Jennings. P.L., Supply chain integration: Challenges and good practices, CAPS Research, Institute of Supply Management, Tempe, AZ, 2009.

Chandler, A.D., Jr., *The Visible Hand, The Managerial Revolution in American Business*, Harvard University Press, Cambridge, MA, 1977.

Cigolini, R., Cozzi, M., and Perona, M., A new framework for supply chain management: Conceptual model and empirical test, *International Journal of Operations & Production Management*, 24(1/2), 7, 2004.

Computer Sciences Corporation (CSC), The ninth annual global survey of supply chain progress, CSC, Neeley Business School at TCU, and *Supply Chain Management Review* (SCMR), 2012.

Crandall, R.E., Dream or reality? Achieving lean and agile integrated supply chains, *APICS Magazine*, 15(10), 20, 2005.

Crandall, R.E. and Crandall, W.R., *Vanishing Boundaries, How Integrating Manufacturing and Services Creates Customer Value*, CRC Press, Boca Raton, FL, 2014.

Crandall, W.R., Parnell, J., and Spillan, J., *Crisis Management: Leading in the New Strategy Landscape*, 2nd edn., Sage Publishing, Thousand Oaks, CA.

Crown, J. and Epstein, K., Boeing's audacious allies, *Business Week (Online)*, March 10, 2008.

Davis, E.W. and Spekman, R.E., *The Extended Enterprise: Gaining Competitive Advantage through Collaborative Supply Chains*, Prentice Hall, Upper Saddle River, NJ, 2004.

Elkington, J., *Cannibals with Forks: The Triple Bottom Line of 21st Century Business*, New Society Publishers, Gabriola Island, British Columbia, Canada, 1998.

Friedman, T.L., *The World Is Flat: A Brief History of the Twenty-First Century*, Farrar, Straus & Giroux, New York, 2005.

Friedman, T.L., *Hot, Flat and Crowded, Why We Need a Green Revolution—And How It Can Renew America*, Farrar, Straus and Giroux, New York, 2008.

Ghosh, S., Thorton, J., DeHondt, G., and Faley, R.H., The paradox of overinvestment in enterprise integration, *Align Journal*, 1(2), 40, 2007.

Gilmore, D., First thoughts, the integrated supply chain organization, *Supply Chain Digest*, June 5, 2008, www.scdigest.com/assets/FirstGhoughts/08-06-05.php.

Grove, A.W., *Only the Paranoid Survive: How to Exploit the Crisis Points That Challenge Every Company*, Currency Doubleday, New York, 1999.

Handfield, R.B. and Bechtel, C., Trust, power, dependence, and economics: Can SCM research borrow paradigms? *International Journal of Integrated Supply Management*, 1(1), 3, 2004.

Heckmann, P., Shorten, D., and Engel, H., *Supply Chain Management at 21: The Hard Road to Adulthood*, Booz Allen Hamilton Inc., New York, 2003.

Hounshell, D.A., *From the American System to Mass Production, 1800–1932: The Development of Manufacturing Technology in the United States*, The John Hopkins University Press, Baltimore, MD, 1984.

Kannan, V.R. and Tan, K.C., Attitudes of US and European managers to supplier selection and assessment and implications for business performance, *Benchmarking*, 10(5), 472, 2003.

Kempainen, K. and Vepsalainen, A.P.J., Trends in industrial supply chains and networks, *International Journal of Physical Distribution & Logistics Management*, 33(8), 701, 2003.

Kiley, D., Ford's savior? *Business Week*, p. 31, March 16, 2009.

Kotzab, H. and Otto, A., General process-oriented management principles to manage supply chains: Theoretical identification and discussion, *Business Process Management Journal*, 10(3), 336, 2004.

Laszlo, C., *Sustainable Value, How the World's Leading Companies Are Doing Well by Doing Good*, Stanford Business Books, Stanford, CA, 2008.

Laszlo, C. and Zhexembayeva, N., *Embedded Sustainability, the Next Big Competitive Advantage*, Stanford Business Books, Stanford, CA, 2011.

Lee, H.L., The triple-A supply chain, *Harvard Business Review*, 82(10), 102, 2004.

Lengnick-Hall, C.A. and Wolff, J., Achieving consistency of purpose, *Strategy and Leadership*, 26(2), 32, 1998.

Manrodt, K.B., Abott, J., and Vitasek, K., Understanding the lean supply chain: Beginning the journey, 2005 Report on Lean Practices in the Supply Chain, APICS, Chicago, IL, 2005.

Melnyk, S.A., Sandor, J., and Burns, L.A., Crossing the supply chain chasm: Obstacles and issues, Working Paper, Department of Supply Chain Management, Michigan State University, East Lansing, MI, 2011.

Mintzberg, H. and Waters, J.A., Of strategies, deliberate and emergent, *Strategic Management Journal*, 6(3), 257, 1985.

Navarro, P., *The Coming China Wars: Where They Will Be Fought and How They Can Be Won*, Financial Times Press, Upper Saddle River, NJ, 2007.

Nonoka, I. and Toyama, R., Strategic management as distributed practical wisdom (phronesis), *Industrial and Corporate Change*, 16(3), 371, 2007.

Pine, B.J., II, *Mass Customization, The New Frontier in Business Competition*, Harvard University Press, Cambridge, MA, 1993.

Poirier, C.C. and Quinn, F.J., Survey of supply chain progress: Still waiting for the breakthrough, *Supply Chain Management Review*, 10(8), 18, 2006.

Power, D., Supply chain management integration and implementation: A literature review, *Supply Chain Management*, 10(3/4), 252–263, 2005.

PRTM, Supply chain maturity model, www.prtm.com/services/supply_chain_-maturity_model.asp, 2005, accessed January 15, 2014.

Sarma, A. and Harrington, C., The bun lady, *Fast Company*, p. 76, May 2005.

Shingo, S., *A Revolution in Manufacturing: The SMED System*, Productivity Press, Portland, OR, p. 113, 1985.

Siegfried, M., Creating a resilient supply web, *Inside Supply Management*, 24(1), January 2013.

Stevenson, W.J., *Operations Management*, McGraw-Hill, New York, 2007.

Van Bodegraven, A. and Ackerman, K.B., The myth of THE supply chain, *DC Velocity*, March 18, 2013, www.dcvelocity.com/articles?20130318-the-myth-of-the-supply-chain, accessed March 19, 2013.

Walker, W.T., *Supply Chain Architecture: A Blueprint for Networking the Flow of Material, Information, and Cash*, CRC Press, Boca Raton, FL, 2005.

Wong, C.Y., Hvolby, H.H., and Johanses, J., Why use loosely coupled supply chains? Report from the Center of Industrial Production, Aalborg University, Aalborg, Denmark, 2004.

Yergin, D. and Stanislaw, J., *The Commanding Heights: The Battle for the World Economy*, Free Press, New York, 2002.

Zakaria, F., *The Post-American World*, W. W. Norton & Company, New York, 2008.

# 12

## *Why Integration Is Difficult*

### Learning Outcomes

After reading this chapter, you should be able to

- Discuss the importance of setting strategic objectives and ROI expectations when designing the supply chain
- Recognize the obstacles to building integrated supply chains
- Explain why there is a need for a company to have multiple supply chains
- Discuss the role of technology when integrating the supply chain
- Discuss the role of infrastructure when integrating the supply chain
- Explain how company culture plays a role in building an integrated supply chain
- Identify ways of building collaborative relationships
- Identify potential ways to measure the effectiveness of an integrated supply chain
- Identify the obstacles to international supply chain management (SCM)

### Company Profile: Boeing

In Chapter 11, we explained that supply chain integration is needed. In this chapter, we point out that supply chain integration is difficult. Organizations have to change their strategic focus. This may mean changing their technologies (online communication systems), their infrastructure (flatter organizations), and culture (empowered employees). They must also build relationships with customers and suppliers that depend on trust and shared benefits and are designed to be long lasting, not temporary expedients. As you read this company profile and the chapter in the book, think about these questions:

- Why does an organization need multiple supply chains? How do they decide how many?
- How do you build trust with a customer? With a supplier?
- Why are communication systems important to supply chain integration?
- How do you justify spending resources—time and money—to integrate supply chains?

What additional questions do you think relate to the difficulties in integrating supply chains?

## Company Profile: Boeing

### INTRODUCTION

In Chapter 11, we described the need to integrate participants along the supply chain. In Chapter 12, we show supply chain integration can be difficult, especially as supply chains are extended throughout the world and become more complex because of differing cultures, technical knowledge, and business practices. Boeing has long been a premier aircraft manufacturer with a string of successful new products. However, they have had major difficulties with their most recent product, the Dreamliner 787. While revolutionary in design concept, the company has had difficulty in building a supply chain to support the manufacture of the airplane.

### HISTORY

Boeing was started by William E. Boeing in 1916 in Seattle, Washington. Mr. Boeing became wealthy in the timber business and used his knowledge of wood in some of early airplanes built by the company. This knowledge of wood was especially helpful immediately after World War I when demand dropped and the company built dressers, counters, furniture, and flat-bottom boats to survive. The company began supplying airplanes to the U.S. Navy in World War I and has continued to be a supplier of planes to the government in subsequent years. Over the years, Boeing has become the leading airplane manufacturer in the United States and vies with Airbus as the leader in the world. Table 12.1 lists the evolution of models over the years. Their emphasis

**TABLE 12.1**

Boeing Aircraft over the Years

| Model | Year Introduced | Attributes |
| --- | --- | --- |
| C | 1917 | Seaplane for U.S. Navy |
| B-1 | 1919 | Flying boat with one pilot and two passengers |
| P-12 | 1923 | Fighter |
| 40 | 1927 | Mailplane |
| 80 | 1928 | Biplane—3 engines, 12 passengers |
| Monomail | 1930 | Low-wing mailplane |
| 247 | 1933 | First truly modern airliner |
| 314 Clipper | 1938 | 90 passengers on day flights, 40 at night |
| 307 Stratoliner | 1938 | First pressurized-cabin transport |
| B-17 and B-29 | 1938 | Bombers used during World War II |
| B-47 and B-52 | 1950 | Advanced bombers |
| 707 | 1958 | First commercial jet airliner |
| 727 | 1960 | First commercial jetliner to reach 1000 sales |
| 737 | 1967 | Has become the best selling commercial jet in history |
| 747 | 1968 | Required world's biggest factory |
| 757 | 1983 | Single-aisle narrow body; discontinued in 2004 |
| 767 | 1983 | Twin aisles |
| 777 | 1994 | In between 767 and 747 in size |
| 737 | 1995 | Next-generation version |
| 787 | Early 2000s | Highly popular with record prelaunch sales |

*Source:* Adapted from Boeing, Milestones in innovation timeline, http://www.boeing.com/stories/timeline.html, accessed October 10, 2013.

**TABLE 12.2**

Revenues and Earnings by Business Segment

| Business Segment | Revenues (Millions of $) | Earnings (Millions of $) | Earnings as % of Revenue |
|---|---|---|---|
| Commercial airplanes | 49,127 | 4,711 | 10 |
| Defense, space, and security | 32,607 | 3,068 | 9 |
| Boeing capital | 441 | 82 | 19 |
| Other segments | 133 | −159 | −120 |
| Unallocated items and eliminations | −610 | −1,391 | |
| Total | 81,698 | 6,311 | 8 |

*Source:* Boeing, Annual report, 2012, https://materials.proxyvote.com/Approved/097023/20130301/CMBO_157699/.

**TABLE 12.3**

Number of Airplanes Delivered by Model

| Deliveries | 737 | 747 | 767 | 777 | 787 | Total |
|---|---|---|---|---|---|---|
| 2012 | 415 | 31 | 26 | 83 | 46 | 601 |
| Cumulative | 4293 | 1458 | 1040 | 1066 | 49 | 7906 |

*Source:* Boeing, Annual report, 2012, https://materials.proxyvote.com/Approved/097023/20130301/CMBO_157699/.

shifted from military and mail to commercial over the years, although they continue to be a major supplier of military aircraft, including being active in missiles.

Table 12.2 shows the 2012 results for each of the major segments of the company. Commercial airplanes dominate; however, defense and space operations are also a major portion of the company's business.

One of the strengths of the company's product line is its ability to upgrade and modernize the basic models to maintain their presence as a viable airplane. Each model is designed to fit a different niche in the industry, and they offer a full line of aircraft to meet every carrier's needs. Table 12.3 shows the number of airplanes delivered in 2012 and cumulative during the life of the model.

Boeing has shown perseverance and adaptability over the years by introducing new airplanes to meet changing customer needs, both commercial and military.

## DESCRIPTION OF DREAMLINER

In response to the competition from Airbus, Boeing developed the concept of the 787, or Dreamliner, as it came to be known. The new design was to be revolutionary with significant reductions in both initial costs and ongoing operating costs. It will be more fuel efficient and, as a result, fly longer distances between stops, thereby eliminating the need for some connecting flights. The design required new materials and electrical systems.

The Dreamliner is a radical and innovative design. One of the main features is the use of composite materials (carbon fiber, aluminum, and titanium) that would be lighter and stronger, thereby reducing fuel costs per mile. The composite materials also make it possible to increase the humidity and pressure to be maintained in the passenger cabin, improving the comfort of passengers. Increased space inside the plane would provide larger windows for passenger viewing and more room for passenger seating.

The plane's new electrical system involves lithium-ion batteries. The new design will require 20% less fuel for comparable flights and reduce the cost-per-seat mil by 10% than for any other aircraft (Denning 2013b).

The corporate strategy also changed. In previous models such as the 737 and 747, Boeing outsourced approximately 40% of the work. For the 787, the company planned to outsource 70% of the total work. The increased outsourcing was expected to reduce the lead time to develop the plane and also the cost of the plane. The idea of increased outsourcing was patterned after Toyota who regularly outsources about 70% of their car. However, Toyota outsources fully designed parts to suppliers with whom they have a long history. This enables them to maintain close control over the manufacture of the components and subassemblies that go into the final car.

## DESCRIPTION OF 787 SUPPLY CHAIN

The supply chain was also revolutionary. The 787 involved major technological innovations unproven in an airplane. This risk would suggest Boeing should have taken greater involvement in the development and manufacture of the aircraft. However, they did the opposite and delegated much of the detailed engineering and procurement to subcontractors (the Tier 1 companies) (Denning 2013b).

The 787 supply chain included manufacturers from all over the world. In the past, Boeing had maintained contact with hundreds of suppliers directly in sourcing parts for their airplanes. For the 787, they elected to select Tier 1 suppliers who would, in turn, select Tier 2 and Tier 3 suppliers as needed. The Tier 1 suppliers would be responsible for the complete assemblies and, in some cases, even be responsible for some of the design features. Some writers also reported that the team charged with the 787 project consisted primarily of marketing and finance personnel, and did not include any supply chain representatives.

## DIFFICULTIES

The concepts for the 787 design and the supply chain to support it were not without merit. They wanted to build a new and innovative airplane that would help them regain their position as the premier airplane designer and manufacturer in the world.

The decision to outsource was not new to Boeing, although the extent of outsourcing was greater than for any previous airplane. The difference was that Boeing decided to organize the supply chain network differently. They chose their primary (Tier 1) suppliers and assigned them responsibility for design details as well as manufacture of subassemblies. Allworth (2013) suggests this was at the heart of subsequent problems. As he points out: "In the creation of any truly new product or product category, it is almost invariably a big advantage to start out as integrated as possible." Being integrated gives the design engineers more degrees of freedom when the product is still being designed. The parent company does not have to understand what all the interdependencies are going to be for a product that has not yet been created. As a result, you do not have to ask suppliers to figure out how to manage relationships with other suppliers that have not been created yet.

If Boeing had completed the design and outsourced finished modules to the subcontractors, which is what Toyota and Apple do, the process would probably have worked better. "But with the 787, it appears that Boeing tried a very different approach: rather than having the puzzle solved and asking the suppliers to provide a defined puzzle piece, they asked suppliers to create their own blueprints for parts" (Allworth 2013).

The problems stemmed from the inadequacies of some of the Tier 1 suppliers and the lack of communication between those Tier 1 suppliers and Boeing. In turn, Boeing depended too much on the Tier 1 suppliers and failed to adequately monitor their progress. In one of the most extreme cases, Boeing had to buy the Vought Company to bring that portion of the supply chain back under the direct supervision of Boeing engineers.

A Boeing aerospace engineer, Dr. L.H. Hart-Smith, had warned of this potential neglect of close monitoring of subcontractors, in a paper at a 2001 conference, in which he warned that it is necessary for the prime contractor (Boeing) to provide on-site quality, supplier management, and sometimes technical support. "If this is not done, the performance of the prime manufacturer can never exceed the capabilities of the least proficient of the suppliers. These costs do not vanish merely because the work itself is out-of-sight" (Denning 2013b).

As part of their efforts to identify and correct problems, Boeing management recalled some of their retired engineers to brainstorm with current Boeing engineers and project managers to come up with some ideas that could lead to corrective actions. One member of the group, Joe Sutter, joined Boeing after World War II and had a guiding hand in every Boeing jetliner through the 1980s (Sanders 2010).

## OUTLOOK

The management at Boeing continues to be optimistic about the future of the 787. The latest version, the 787-9, has a redesigned system so battery cells, the most frequent source of problems, are separated and insulated. The new system was also designed to ventilate outside the plane any smoke that would come from overheating batteries. As of the end of April, the 787 grounding was lifted. The 787-9 is 20 ft longer than the 787-8 and holds 40 more passengers (210–250). Boeing is not finished with the design of the 787. The 787-10 is scheduled for the first delivery scheduled in 2018. It will fly up to 7000 nautical miles and have seating for 300–330 passengers. Early responses from passengers have been favorable (Cha and Hetter 2013).

## CONCLUSIONS

Stephen Denning (2013a) has written several articles on the 787 history. In one, he outlines seven lessons that he believes all CEOs of companies involved in outsourcing should heed:

1. *Use the right metrics to evaluate offshoring and include total costs*: Firms must go beyond basing decisions just on direct cost of the product and include the total cost and risk involved in global supply chains.
2. *Review whether earlier outsourcing decisions made sense*: There is a big difference between outsourcing coffeemakers and airplanes.
3. *The engineering is mission critical*: Boeing outsourced some of the engineering along with the manufacturing. As a result, the components did not always fit together.
4. *Bring some manufacturing back*: The low costs in offshore countries are not not as attractive as previously. GE and Whirlpool are companies that are finding it makes sense to reshore some of their manufacturing to assure they still have the capability.
5. *Adequately assess the risk factors of offshoring*: Offshore outsourcing, and the resultant extended supply chains, increases the risk of product flow disruption, unexpected quality problems, and extended response times to solving problems.

6. *In outsourcing, explicitly evaluate the risk of intellectual property*: There is little risk of losing intellectual property in outsourcing pencils; there is considerable risk in outsourcing airplanes. Manufacturing is not a generic; it requires continuous innovation.

7. *Replace the goal of maximizing shareholder value with the goal of continuously creating value for customers*: The new world of manufacturing will require a radically different management from the hierarchical bureaucracy now prevalent. "It will require a different goal (adding value for customers), a different role for managers (enabling self-organizing teams), a different way of coordinating work (dynamic linking), different values (continuous improvement and radical transparency) and different communications (horizontal conversations). Merely shifting the locus of production is not enough. Companies need systemic change—a new management paradigm" (Denning 2013a, p. 34).

The total backlog for Boeing of $390.2 billion is the largest in the history of the company and is almost five times the 2012 revenues. "With an expanded backlog of $390 billion and an enduring focus on productivity, we are poised to deliver sustained growth and strong business performance in the years ahead" (Boeing Annual Report 2012, Letter to Shareholders, p. 4).

## REFERENCES

Allworth, J., The 787's problems run deeper than outsourcing, HBR Blog Network, January 30, 2013, http://blogs.hbr.org/2013/01/the-787s-problems-run-deeper, accessed September 18, 2013.

Boeing, Annual report, 2012, https://materials.proxyvote.com/Approved/097023/20130301/CMBO_157699/.

Boeing, Milestones in innovation timeline, http://www.boeing.com/stories/timeline.html, accessed October 10, 2013.

Cha, F. and Hetter, K., Latest Boeing dreamliner completes test flight, CNN travel, September 18, 2013, www.cnn.com/2013/09/17/travel/boeing-787-9-dreamliner-takes-off/indes/html?hpt=hp_bn10, accessed September 19, 2013.

Denning, S., Boeing's offshoring woes: Seven lessons every CEO must learn, *Strategy & Leadership*, 41(3), 29–35, 2013a.

Denning, S., What went wrong at Boeing? *Forbes*, January 21, 2013b, http://www.forbes.com/sites/stevedenning/2013/01/21/what-went-wrong-at-boeing/.

Sanders, P., Boeing brings in old hands, gets an earful, *Wall Street Journal (Online)*, July 19, 2010, http://online.wsj.com/news/articles/SB10001424052748704746804575367350030750626, accessed October 5, 2013.

## Introduction

In Chapter 11, we described the need to develop integrated supply chains, or closer relationships with customers and suppliers. In this chapter, we describe some of the obstacles to achieving integrated supply chains in more detail. The obstacles fall into several categories: determining strategic objectives, evaluating the potential return on investment (ROI), designing for participant differences, recognizing the need for multiple supply chains, selecting and implementing technology, realigning infrastructure, transforming cultures, building relationships, measuring performance, and maintaining the system. Change management skills are required in abundance.

## Determining Strategic Objectives

It is difficult to implement an integrated supply chain. Aside from the complexity of the task, it requires more than making positive moves with technology; it requires overcoming resistance within the organization and with potential partners in the project. It also requires that the actions taken by an individual supply chain company fit within the framework of that company's strategic and operating plans.

Some of the challenges for supply chain management (SCM) and future prospects include as follows:

- SCM can be seen as part of a wider set of trends involving outsourcing, cross boundary working, and new organizational forms characterized by flattened hierarchies, teams, employee empowerment, and so on rather than rigid command and control practices (Ruigrok et al. 1999).
- The trend toward outsourcing and the increasing importance of intangibles heightens the need for, and the potential of, SCM.
- The trend toward fragmentation and variety in product and service offerings necessitates greater thought and skill in managing decoupling points and postponement of final product composition.
- Globalization necessitates greater attention to logistics and to other component elements of SCM (Storey et al. 2006).

Some of the more specific objectives of an integrated supply chain include seamless flow from initial source(s) to final customer, demand-led supply chain (only produce what is pulled through), shared information across the whole chain (end-to-end pipeline visibility), collaboration and partnership (mutual gains and added value for all), information technology (IT)–enabled procedures, all products direct to shelf, batch/pack size configured to rate of sale, customer responsive, agile and lean, mass customization, and market segmentation (Storey et al. 2006).

With these objectives in mind, we describe some of the major barriers to supply chain integration being encountered by companies today.

## Evaluating the Potential Return on Investment

Management must assure itself that building an integrated supply chain will provide a satisfactory ROI. While there may be compelling intangible benefits, managers always face the need to demonstrate that they are using company resources wisely. Consequently, they usually try to establish a satisfactory ROI, or more specifically, return on assets employed (ROA), in the program to integrate the supply chain. To accomplish this, they must evaluate the incremental benefits, costs, and assets employed in the program. While some of this information is readily available, much of it is not.

### Uncertainty of Benefits

Benefits can be either tangible or intangible. They should be attributable to the program to integrate the supply chain. Even for tangible benefits, assigning them to a specific program is difficult. On the tangible side, how much of the increase in sales was the result of the integration efforts? On the intangible side, how much of the improved customer trust was the result of the integration efforts? Deciding on incremental benefits usually requires a combination of facts and judgment.

### Tangible or Direct

Tangible benefits include increased sales, either from existing or from new customers. Businesses must work harder to retain their existing customers by building closer relationships. There is general agreement that it is better to retain existing customers than to continually work to develop new customers. However, the difficulty arises when trying to assign credit for the increased sales. Did they result from closer customer relationships, or were they the result of the improved quality of the product or the addition of the added technical features on the product?

Another benefit could be increased selling prices. Just as with increased sales volume, it is difficult to assign credit to one cause. Integrating the supply chain may reduce the cycle times for customers and, for that, they may be willing to pay more for the product. However, just as for the increased sales volume, the price increase could also have been the result of improved quality of the product or the addition of the GPS technology on the product.

### Intangible or Indirect

Intangible benefits may come in the form of reduced inspection requirements at the customer's receiving dock. While the customer benefits from reduced inspection costs, the supplier will eventually benefit from reduced time required to resolve their customer's receiving questions. However, does this come from the closer working relationship or improved quality of the product?

Another intangible benefit that is even more elusive is an award by a customer for delivery service excellence. It may come in the form of a *Supplier of the Year* award. It is difficult to determine the benefit of this type of recognition, even in total, much less to divide it among a number of action programs by the supplier.

## Uncertainty of Costs

There are two main categories of costs—product and overhead. Product costs are those costs that can be assigned directly to the product. For a manufacturing company, they include direct materials, direct labor, and direct overhead expenses, such as repairs, maintenance, and materials handling. For wholesale and retail businesses, product costs are largely their cost of purchases.

All businesses have overhead costs that cannot be charged directly to their products. These costs have to be assigned, or allocated, through some sort of rational method. In recent years, activity-based costing (ABC) has grown in popularity. More recently, lean accounting systems have been promoted as appropriate for companies using lean production methods.

### Product Costs

While the increase, or decrease, of direct product costs can be determined with precision, it is considerably more difficult to assign a specific cause to their increase or decrease.

This is especially true in trying to decide how much of the cost is the result of supply chain integration efforts.

Let us look at one example. A company finds that they can reduce the cost of their product by replacing a machined part with a stamped part. They consult with their customer's engineering department, a result of closer customer relationships, and reach an agreement to make the change. How much of the product cost reduction should be attributed to the supplier's engineering department and how much to the closer relationships built through supply chain integration?

### *Support (Overhead) Costs*

Overhead costs are difficult to classify and evaluate, even within a single company. It is almost impossible, or impractical, to accomplish when assessing their response to supply chain integration efforts. Suppose the accounts payable department of the supplier works with their customer to effect a change in their billing process to the customer. The result is a reduction in overhead costs to both the supplier and the customer. But how do you assign a specific number to this action?

### Uncertainty of Assets Employed

The ROA equation shown in Chapter 11 requires the value of the assets employed in the denominator. Each company can select the assets they employ in this calculation; however, most businesses use some combination of working capital, equipment, and facilities.

While there may be ways to isolate the assets employed in the supply chain integration process, it is intuitively obvious that such an exercise would be time consuming, especially in trying to separate the effect of the supply chain integration from other local activities that are ongoing within most businesses.

The examples provided earlier illustrate the difficulty in identifying an ROA that results from supply chain integration. While most companies would like to be able to see a tangible result, the obstacles are formidable.

## Designing for Participant Differences

There are differences in participating companies that slow, or even prevent, the transition from mutually dependent entities to a smooth-functioning integrated team.

### Participants Are Not Equal

One of the most obvious difficulties arises from the fact that the participants are not equal. They do not have the same level of commitment; they do not all have the same level of contribution; and some must be drivers of change, while others must be followers.

### *Commitment*

Some companies have a higher level of commitment to building the integrated supply chain, often because they see greater returns to their own company. In recent years, the emergence of large retailers, with their emphasis on customer service, has increased their dependence on effective supply chains. Wal-Mart, Target, Home Depot, and Lowe's, to name a few, have repeatedly displayed their emphasis on establishing integrated supply chains. They are committed to the concept. Conversely, their suppliers have not always shown the same level of commitment.

In some cases, manufacturing companies have been the ones showing the commitment. This can be seen in the automobile industry where the major players have recognized their dependence on dealer networks to sustain and improve market acceptance throughout the world.

Usually, the level of commitment relates to the level of return expected. Consequently, the larger participants expect a larger return; therefore, they are more committed. William T. Walker makes the distinction between a trading partner and a nominal trading partner. A trading partner participates to a significant level and depends on the supply chain for financial success. The nominal trading partner participates only to a small extent, is easily replaced, and does not depend on the supply chain for its financial success (Walker 2005).

### Contribution

Building integrated supply chains also requires contribution by the participating members. While it may seem desirable that all participants contribute equally, this is not usually the case. It is not even true that those who expect the greatest return will demonstrate the greatest contribution.

When Wal-Mart mandated its top suppliers would have to begin using RFID technology at the case and pallet level, they expected a level of contribution from their suppliers to make the transition work. Some suppliers believed they would incur increased costs, and Wal-Mart would receive most of the benefits. This perception did not invoke a wide acceptance of the program by suppliers, especially small ones, who did not see great benefits to themselves, other than the hope they would retain their business with Wal-Mart. This program has since progressed as suppliers and Wal-Mart have made accommodations to assure a more equitable arrangement.

### Different Roles: Drivers and Followers

The net effect is that some companies will be drivers and push the efforts toward building integrated supply chains. They see the benefits to themselves and, to some extent, to the supply chain in total. Other companies will be followers, because either they do not see the same level of benefits or they just do not have the resources to evaluate the situation sufficiently. Regardless of the motives, some companies will be the leaders in the effort.

### Technical Differences

In addition to the motivational differences among supply chain participants, there will be technical differences among them.

### Size

There is always the issue surrounding company size. Large companies have greater influence in deciding to pursue or resist. Small companies may be progressive; however, the likelihood is they will be followers of large company actions.

### Proximity

There is also the issue of proximity. Some companies try to reduce the physical separation from customers or suppliers. This is especially true in the global marketplace. Suppliers find it beneficial to be close to their customers. This means automobile parts suppliers tend to congregate around their large customers' assembly plants. It also means many companies establish offshore operations to be closer to their existing, or planned, markets. Examples of this include the experiments by automobile manufacturers to locate their suppliers within the same facility as the assembly operations. This approach aims to reduce both the response time from suppliers and the variability in response times. When achieved, the customer can reduce their inventory levels without sacrificing confidence in deliveries to their customers. While this approach has created interest, it is still more experimental than regular practice.

On the other hand, some companies are finding that improved communications systems compensate for their remote locations. A physical presence is not required. Cisco is developing virtual meeting capabilities to provide the *feel* of human presence that is lacking in electronic text communication (Thibodeau 2008). As a result, each supplier can seek low-cost locations and still provide exceptional customer service.

### Type of Operation

The type of operation, or process used, varies from company to company. As described in Chapter 7, manufacturers may use a job shop, batch flow, line flow, or continuous process approaches. Retail companies may be specialty shops with a limited product line or a full-line retailer with a wide variety of products. Some retailers focus on the high end of a product line, while others focus on low-cost items. These differences require that adjustments are necessary to enable all of the members of a supply chain to fit with each other.

In addition to physical differences, there are differences in attitudes. These differences arise from two sources—the implicit culture of a company and the explicit statement of their mission and strategies.

Company cultures are different from company to company. In some cases, the differences may be slight and easily accommodated; however, they may also be significant and difficult to reconcile. Although implementing new technology is often difficult, changing corporate cultures is always difficult.

Most companies develop a mission statement as part of their strategic planning process. Their stated mission and strategies may be compatible with the culture of the company; however, it may be different. When there is inconsistency within a company, it will be difficult to accommodate differences with other members of the supply chain.

## Need for Multiple Supply Chains

Most businesses have multiple supply chains. A single supply chain is difficult to integrate; integrating multiple supply chains is even more difficult. While it may be theoretically possible to design and build a universal supply chain that accommodates all aspects of a company's business, it is not likely to be the most practical approach.

There are a number of variations that could motivate the need for multiple supply chains, primarily involving customer differences and supplier differences. As a starting point, it is a good idea for a company to map the supply chains in which they are involved. Walker suggests the following approaches to supply chain mapping:

1. Start midstream and imagine finished goods setting in the warehouse.
2. Now, use the bill of materials and work upstream to reach each raw material supplier.
3. Then, identify each different fulfillment channel used to reach the end customer.
4. Determine which organizations are trading partners versus nominal trading partners.
5. Logistics service providers, information service providers, and financial service providers are not a part of the network map (Walker 2005).

## Different Customer Segments

Different customer segments require differing supply chains. A clothing manufacturer who supplies to a high-end retailer may be expected to provide high quality and extra services, while a clothing manufacturer who supplies to a low-price retailer may be expected to eliminate all aspects of their process that do not add value to the product. Such differences are difficult to accommodate in a single supply chain.

An approach called category management can be helpful in assessing the differences among customer segments. Category management involves classifying products into compatible categories with an objective of fitting the product categories with the appropriate customer segments. When correctly matched, a consistent approach can be used for each product category—customer segment combination.

As the markets of the twenty-first century evolve, suppliers have to provide higher levels of customer service. Companies need to consider market characteristics such as duration of life cycle, time window for delivery, volume, variety, and variability. To be successful in these markets, companies will need not just one supply chain solution but many (Aitken et al. 2005). The authors go on to describe a lighting company that developed four distinctly different supply chains to satisfy different segments of their customer base.

In Chapter 7, we described the basic manufacturing processes as make to stock (MTS), assemble to order (ATO), make to order (MTO), and engineer to order (ETO). Figure 12.1 shows how inventories can be shifted upstream to increase flexibility, although increasing response time to customers.

Naylor, Naim, and Berry explain how creating multiple supply chains can increase flexibility in serving customers. They use the term *leagile* to describe a supply chain that is lean in the upstream section and agile in the downstream section (Naylor et al. 1999). The primary area of inventory concentration differs for each strategy. MTS requires finished goods inventory to be at the retail store for immediate availability to the consumer. ATO maintains the primary inventory in the form of modular subassemblies that can be quickly assembled into customized finished goods. MTO inventory is primarily in the form of raw materials, while ETO does not require inventory at all.

## Different Supplier Segments

The customer usually dictates the supply chain structure, because a supply chain is designed to serve a customer segment. Once the customer segment is established, the

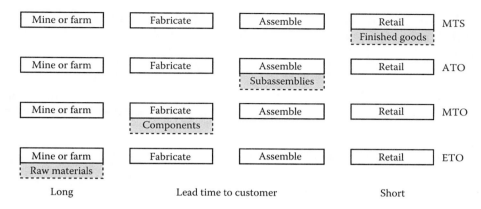

**FIGURE 12.1**
Examples of inventory locations in different supply chain strategies.

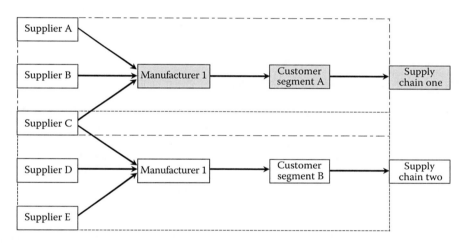

**FIGURE 12.2**
Multiple supply chains for Manufacturer 1.

supply side is made up of the suppliers needed. While the suppliers may represent a diverse group, they become a part of a single group for that particular supply chain. This could result in a supplier becoming a part of more than one supply chain for the same customer. This occurs when the supplier provides an assortment of products or services that are needed for different customer segments. Figure 12.2 shows how this situation could exist.

Manufacturer 1 separates their customers into two segments. Suppliers A, B, and C supply Manufacturer 1 for Customer Segment A. Suppliers C, D, and E supply Manufacturer 1 for Customer Segment B. Supplier C is a supplier in both supply chains. Even within this simple illustration, there could be additional supply chain combinations.

## Different Logistics Networks

The way in which products are moved along the supply chain could have a major impact on how supply chains are integrated. A retailer that has its own distribution centers may focus on low costs and outsource to a low-cost country. On the other hand, a grocery store that sells fresh produce may buy locally and have products delivered directly to the store.

## Separating Interwoven Networks

Companies often have multiple networks and have to separate them to facilitate further analysis and action. Walker describes four types of networks that must be separated from one another:

*Serial*: Self-contained networks arranged in a series.

*Tangential*: The customer end of one network is tangential to the midstream zone of another network.

*Crisscrossed*: One or more organizations buy and sell as part of another network.

*Competing*: Other supply chain networks that compete (Walker 2005).

Careful analysis during the initial design of a supply chain will prevent problems later on.

## Selecting and Implementing Technology

Technology is one of the most visible obstacles to achieving integrated supply chains. Supply chains are designed to facilitate the flow of goods and services from suppliers to customers. The supply chain must also accommodate the flow of information and the flow of funds. All three flows require the use of technology to achieve the desired effectiveness and efficiency.

### Product and Service Processes

The primary purpose of supply chains is to achieve a smooth and dependable flow of goods and services between members of the supply chain. This requires four important steps. Companies must remove any barriers to the flow; eliminate the duplication, or redundancy, in the flow; develop compatible processes; and strive for effectiveness and efficiency.

#### Remove Barriers

The first step is to remove any barriers to the flow. Barriers can be thought of as delays in processing goods or services. Some barriers that can exist include temporary storage to accumulate a truck load batch size, delays in receiving incoming inspections, and stock-outs that require additional partial deliveries. While it may not be possible to completely eliminate all of these barriers, they should be identified and reduced to an acceptable level.

#### Eliminate Redundancy

In addition to eliminating barriers to flow, supply chain participants may find they are performing duplicate operations or, even worse, doing operations that have to be reversed by the next member of the supply chain.

Inspections represent one activity that may be duplicated by both the supplier and the customer. As the supply chain becomes more closely integrated, it is usually possible for a supplier to demonstrate a history of acceptable quality, such that the customer can eliminate their inspections of the incoming goods.

Suppose a supplier of women's dresses wants to assure that the dresses arrive at their customer's store wrinkle free. They decide to hang the dresses on temporary cardboard hangers and ship them. The customer receives the dresses, removes them from the temporary hanger, and puts them on a permanent hanger for display in the store. Of course, a more desirable process would have the customer send the permanent hangers to the supplier, who would use them to hang the dresses for shipment. Often, these duplications only begin to show up once the participants begin to work together more closely.

### Develop Compatible Processes

Supply chain participants should try to make their processes more compatible. Often, this involves the transfer packaging of goods. It could be as simple as shipping resistors in reels of 1000 for immediate loading onto the automatic insertion machine, instead of in specially designed hermetically sealed packages of 25.

It could also involve packing dress gloves in assortments of one dozen each of four colors and three sizes, instead of sending each of the 12 combinations in cases of 100 each. If sorted at the manufacturer, it eliminates the need for a cross-docking process at the distribution center.

Developing processes that are more compatible often involve small, incremental changes, not breakthrough innovations. It is a tedious process along the path to an integrated supply chain.

### Strive for Effectiveness and Efficiency

Once the supply chain is smoothed, companies should work to make them most effective, especially from a timing perspective. The correct goods and services should arrive when needed and depart when scheduled. This task represents the effectiveness of responsiveness and flexibility. Each activity should be performed efficiently to minimize cost and assure the expected quality.

## Incomplete Interorganizational Systems

We describe interorganizational systems (IOS) in more detail in Chapter 14. An IOS enables companies to share relevant information with one another quickly and confidentially.

### Lack of Systems Compatibility

One of the primary barriers to effective information sharing is the lack of systems compatibility. The customer's computer does not talk to their suppliers' computers easily. In the days of paperwork systems, people could interpret and compensate for the incompatibility of computer information systems. Today, with the need for faster and more accurate information transmission, computer systems must be compatible. Sometimes, this is easier to accomplish by sending information to a third party, a value-added network (VAN), which will accept information from the customer in its format, adapt the information to the supplier's format, and send it on to the supplier. The VAN can also achieve this same compatibility in the reverse direction.

### Lack of Information

One of the biggest barriers to supply chain integration is the lack of information, or worse, inaccurate information. The lack of information is an obvious problem; however, it may be difficult to identify inaccurate information until it becomes so blatant that it causes major problems.

Sometimes, information sharing is inhibited by individuals or company cultures, a topic we discuss later in this chapter. However, there may be potential technology barriers to information, or knowledge, sharing, such as those in the following list:

- Lack of IT systems and process integration
- Lack of technical support
- Unrealistic expectations of technology users
- Lack of compatibility between diverse IT systems and processes
- Mismatch between individual needs and total system needs
- Lack of understanding or experience
- Lack of training
- Lack of demonstrated advantages (Riege 2005)

Deciding what information should be available and how best to provide it is a continuing challenge.

### Lack of a System

Even with the existence of compatibility, standards, and accurate information, supply chain integration depends on having systems that achieve the desired results. Some of the major systems required in an integrated supply chain include the following:

- *Demand forecasting*: A demand forecast is the starting point for a number of other necessary activities.
- *Production planning*: With a demand forecast, manufacturers can plan for capacity, inventory, and employee requirements.
- *Order processing*: Until there is a system for order processing, goods cannot be ordered, delivered, and paid for.
- *Returns and reprocessing*: Some provision must be made for product returns and reprocessing, to return the product either to the regular supply chain or to the reverse logistics process.
- *Information preservation*: Once entered into the supply chain IOS, there needs to be a system to preserve that information for continuing use.
- *Performance measurement*: Performance measurement systems must be flexible and relevant. This means that they are often customized to a number of measures and dyadic relationships within the same supply chain.

## Financial Funds Flow

We will describe flow of funds in more detail in Chapter 15. Here, it is sufficient to say that, if the goods do not flow and the information does not flow, it is likely that the funds will not flow very smoothly either.

## Realigning Infrastructure

Technology is often the driver; however, supply chain integration requires more than technology. It also requires that the infrastructure of the participants be restructured to facilitate communications and decision making. This often requires changes in the organization structure and almost always requires changes in the formal policies and procedures of the firms.

Although companies may recognize the need to change, altering organizational structures remains a work in progress. A round table convened by the *Harvard Business Review* concluded "functional silos hinder communication and efficiency, but many companies still struggle to tear down the walls" (Beth et al. 2003).

Another study suggests an even more dramatic change in supply chain organizations. It proposes that disintegration of supply chains via outsourcing and virtual organizations will eventually require a third-party manager. "As companies rely increasingly on external suppliers, there is an emerging and compelling need for 'maestros'—neutral third parties that can coordinate the entire network and align the incentives for all the participating players" (Bitran et al. 2007).

### Internal Organization

Many organizations have used a hierarchical organization structure with a vertical chain of command that sends decisions down through the lower levels and requires information developed at the lower levels to be sent up toward the decision makers. While effective, it was not always efficient and usually caused delays in decision making and implementation.

In the rapidly changing business environment, it is necessary to make decisions faster. This means decisions must be made nearer the point of an activity or transaction. In order for this to happen, employees at lower levels are empowered to make the decisions. This requires a change in the organization structure, from vertical to horizontal. Decisions can then flow along with the goods and services.

This movement toward horizontal organization structures has been evolving for several decades. One of the reasons for the vertical hierarchical structure was to facilitate the information flow in large companies. Intermediate levels, the middle managers, were set up to analyze and summarize data collected at the lowest levels of the organization. Middle managers were responsible to identify and solve some problems, but, above all, they were to be interpreters of the information and extractors of the most important information for consumption by top managers. Today, information systems can be designed to do most of the analyzing and summarizing activities; therefore, middle managers become less essential in this role. They may even be barriers to the smooth flow of information along the supply chain.

At one time, the primary contact with customers was through the sales and marketing organization, and the main contact with the suppliers was through the purchasing organization. Today, it is likely that almost every function within a company has a counterpart with customers and suppliers. Instead of one contact point, there are several to be managed. This arrangement may be expedient, but it is a more complex relationship to be built and managed.

The formal organization structure must change to reflect the changes in the methods of operation. The formal organization is built around job or position descriptions. In addition, every company has an informal organization that is built around individual employees, many of whom work largely within their position description, but may extend beyond their formal description to perform duties for which they are uniquely qualified.

## Effect of Outsourcing Movement

We described outsourcing in previous chapters. Outsourcing, especially offshore outsourcing, introduces a major obstacle in integrating supply chains. Offshore outsourcing has several desirable objectives, such as cost savings, capital investment reduction, acceleration of business process reengineering, increased focus on core competence, flexibility enhancement, and increased globalization (Lau and Zhang 2006).

However, offshore outsourcing also presents some significant obstacles or potential problems, including

- Loss of control
- Loss of critical skills
- Inadequate capabilities of the service provider
- Loss of flexibility
- Failure to realize hidden costs of the contract
- Difficulty in obtaining organizational support
- Indecisiveness on which activities to outsource
- Inadequate cost and benefit analysis systems
- Fear of job loss (Lau and Zhang 2006)

The benefits of outsourcing are often tangible, while the obstacles are often intangible. It takes a thorough analysis to arrive at a carefully reasoned decision.

## External Organization

It is almost too obvious that, as difficult as it is to change the internal organization structure, it will be considerably more difficult to adapt the structure in one company to make it compatible with other members of the supply chain. Job titles, reporting relationships, and organization levels are just the beginnings of the variables that exist and must be understood before meaningful supply chain integration is possible. As globalization becomes standard practice, the need for reconciling organizational differences will become more important.

## Policies and Procedures

As a result of the changes in organization structure, company policies and procedures must also be changed. This is another time-consuming process, sometimes easy to postpone. Written policies and procedures take a long time to approve, often because they affect a number of functions, and anything written is always subject to reviews, disputes, and delays. As a result, it is difficult to have a completely up-to-date procedure manual in a rapidly changing environment. Today, as electronic formats become the standard, it is easier to make changes and communicate them to those affected by the change.

## Physical Infrastructure

The physical infrastructures involving ports, highways, loading docks, and transportation equipment all become important considerations in considering how best to integrate supply chains. In a study focusing primarily on China, but applicable to any emerging country, one firm specializing in outsourcing services reported that the road to high performance in China is still problematic. They listed underdeveloped infrastructure, fragmented distribution systems, insufficient technology, and onerous regulations as some of the challenges in doing business (Byrne 2006).

## Transforming Company Cultures

With all of the changes required to achieve supply chain integration, it is almost a certainty that some of the changes will be significantly different from past practices. In a company with long established ways of doing things, these changes may run counter to the company culture, or the embedded informal practices. When this happens, it may be difficult to get employees to accept the changes. "But this is the way we have always done it" becomes the battle cry of the most senior and heretofore loyal employee. Most individuals resist change, not to be belligerent, but because they truly believe the old way to be best. In addition, change often involves a loss of control to the employee involved.

### Internal

Within an organization, cultures can be changed by transforming the individual employees from a position of resistance, to acceptance, to participation. If this strategy fails, it may be necessary to transfer or replace the existing employees with new ones who do not have the embedded resistance.

Sales and operations planning (S&OP) is one of the most active programs designed to link the marketing and operations functions together to arrive at the best production plan to meet the targeted sales. S&OP transitioned from an approach where the finance department and operations operated independently to forecast demand and production plans to an integrated process. Making this transition required companies to adopt a business culture of planning, a culture change that sometimes took years to accomplish (Kosk 2013).

One of the more interesting examples of this latter approach occurred with the Ford organization when they moved from the Model T to the Model A in the latter part of the 1920s. During this period, the president fired most of the managers and supervisors who had so successfully built the Model T. He did not want any of the Model T thinking to be used in building the Model A. This was an extreme measure to change the culture (Hounshell 1985).

### External

Developing integrated supply chains will involve a culture shift on the part of the external entities as well. Suppliers will need to change their mindset toward each other from distrust to collaborative.

## Building Relationships

It takes time to build collaborative relationships. Most companies go through several steps before they achieve that objective.

### Communicate

The first step is to communicate, or find a way to have an interchange of information or ideas—one that is available to both parties. Companies use many technologies, including overland mail, faxes, telephone, and a variety of electronic means. While technology is required to communicate, it does not guarantee effective communication. Communication requires both a sender and a receiver. Effective communication requires both parties to participate in the process.

### Coordinate

Coordination is to "harmonize in a common action or effort" (*American Heritage Dictionary* 1982). Once companies begin to communicate, the next step is to coordinate the activities in which they are both involved. This could include setting the price and terms of the purchase order and selecting a place and time for its delivery. The customer may specify conditions acceptable to the supplier; it does not necessarily imply a give-and-take arrangement. Coordination requires a blending of the technology with participation by employees.

### Cooperate

To cooperate means "to act together toward a common end or purpose" (*American Heritage Dictionary* 1982). At this level, the two parties may adjust their desired positions to accommodate their mutual interests. The customer may be willing to accept partial shipments of their order to get some of the product by the date they need it, if the supplier is willing to make an additional setup in order to make the partial shipment. It could involve the customer sharing its sales information with suppliers so the suppliers will have a better idea of their customers' future demand. Participants must be willing to compromise their positions in the interest of getting an acceptable solution for the entire supply chain. This often requires employees to be empowered to make decisions at lower levels of the organization.

### Collaborate*

To collaborate means "to work together, especially in a joint intellectual effort" (*American Heritage Dictionary* 1982). This could involve representatives from the customer and the supplier analyzing the sales information together, in order to develop a demand forecast with which they both agree. Collaboration goes beyond the technology used; it requires a level of mutual trust that must be present if the resultant forecast is to be meaningful.

One of the major programs developed for this purpose is the collaborative planning, forecasting, and replenishment (CPFR) process. This is a process whereby supply chain trading partners can jointly plan key supply chain activities from production and delivery of raw materials to production and delivery of final products to end customers.

---

* Some of the following section has been adapted from Crandall (2008).

Collaboration encompasses business planning, sales forecasting, and all operations required to replenish raw materials and finished goods (Blackstone 2010).

Ron Ireland and Colleen Crum have written an excellent book entitled *Supply Chain Collaboration: How to Implement CPFR and Other Best Collaborative Practices*. They describe in detail what is required for collaboration among supply chain participants and how Wal-Mart and Warner-Lambert demonstrated the value of collaboration in their pilot project with Listerine (Ireland and Crum 2005). There is a growing consensus that the *best practice* is to reject the historic focus on adversarial buyer relationships with suppliers in favor of a more long-term collaborative approach based on trust and partnerships/ alliances (Cox 2001).

CPFR evolved from earlier programs that were developed to reduce response time from supplier to customer. These programs included quick response (QR), efficient consumer response (ECR), and vendor-managed inventory (VMI). The earlier programs involved coordination and cooperation, but CPFR was the program designed specifically to incorporate collaboration among participants (Barrett 2004). Crandall (2006) provides a fuller description of these programs.

A key element in collaboration is preparing the demand and order forecasts. Historically, members of a supply chain developed their demand forecasts independently, usually by statistically extending past sales (order) history into the future (Crandall 2005). These forecasts often failed for at least three reasons: (1) they incorporated the bullwhip effect of order batching; (2) they were based on orders (sales), not demand, a distinction that sometimes caused timing errors; and (3) they did not accurately reflect the timing or magnitude of events—sales promotions, new product introductions, product phaseouts, and other possibilities—for customers.

Collaboration is an attempt to share information so that upstream suppliers have a better understanding of not only what happens downstream but also why it happens. Furthermore, they can have information of what their customers are planning in the future. However, information sharing is not enough; there is evidence that shared decision making is also beneficial. For example, a supplier and a customer work together to develop mutually agreed demand and order forecasts.

## Lack of Trust Inhibits Collaboration

Collaboration sounds simple enough, but not everyone is doing it. Barratt points out a number of issues such as inadequate internal integration, multiple plans, uncoordinated events, poor communication, lack of understanding of processes, delegation conflicts, inconsistent performance measures, and information overload. He also indicates there has been an overreliance on technology, a failure to understand when and with whom to collaborate, and a lack of trust between partners (Barrett 2004).

Crum and Palmatier list the most common reasons why demand collaboration has not realized its potential:

- The pace of adopting new ways of doing business is slow.
- Demand information supplied by customers is not put to use in their trading partners' own demand, supply, logistics, and corporate planning in an integrated manner.

- Demand management and supply management processes are not integrated, and S&OP is not utilized to synchronize demand and supply.
- There is a lack of trust among trading partners to share pertinent information and collaborate on decision making.
- There is a desire to partner but not commit to executing the communicated plans.
- A common view exists that demand collaboration is a technology solution and that the current technology is too complex (Crum and Palmatier 2004).

Developing the technical capability to share information along the supply chain is a difficult and time-consuming process. However, it can all be for naught if the parties involved are not willing to trust one another sufficiently to allow the information to flow. Trust evolves from a combination of demonstrated performance and the intangibility of judgment. It can take a long time to establish and only an instant to destroy. At best, it is a temporary relationship, often dependent on individuals as much as facts.

The underlying reason companies do not collaborate more seems to be that companies do not trust each other (Drickhammer 2007). They can share information because the technology exists to make it possible. They do coordinate because it makes sense and does not require that they give up their *inalienable rights* to manage. But they just cannot get to the point of collaborating, even when it makes sense to do so.

A practitioner also echoes the need for trust in building effective supply chain relationships. "The more trust there is reduces the need for controls between individuals and organizations. However, do you actually trust your suppliers? And do they trust you? Unfortunately the answer to both is probably 'no'" (Hawkins 2007). Another writer adds, "The weakest link in the most advanced supply chains is not technology, not software or hardware, but people—or rather the level of trust between people who must cooperate and collaborate to get results" (Mariotti 1999).

However, if collaboration provides tangible benefits, how do companies learn to overcome this barrier of lack of trust? Is it something that top management can decree to be done by the 15th of the month or is it a more complex transformation? While a management decree may get the transformation started, the full achievement requires changing some well-entrenched practices and beliefs. As it turns out, a number of scholars have studied the trust issue from many different perspectives. Indeed, trust has been examined within the context of sociology, international politics, law, marriage, and business organizations (Grossman 2004).

How do companies build mutual trust? Does it start with one buyer trusting one salesperson and then grow from that beginning until companies reach a critical mass of individual trusts to become a corporate trust? That sounds slow and subject to disruption when individual relationships change through promotions, transfers, or retirements. Can companies build mutual trust between companies?

Before coming to a conclusion, it may be worthwhile to look at some of the ways that trust grows between individuals or companies. Trust implies confidence that something will happen as expected. Other synonyms for trust include faith, belief, hope, conviction, expectation, reliance, and dependence. Trust implies that good things will happen; otherwise, companies would not enter into the relationships.

There are a number of ways to build confidence and, subsequently, trust. Repetition provides increasingly large samples of what suppliers or customers will do. Just as large sample sizes in quality control reduce the statistical variance, repeated experiences do the same for supply chain partners.

The performance of partners goes along with repetition. It is easier to increase trust in those whose performance closely matches their promises. Consistently, poor performance could also build a level of expectation but would tend to defeat the purpose of collaboration.

Sharing information also helps to build confidence in a partner. If a customer or supplier is willing to share, it becomes easier to believe they have at least some interest in your welfare, and the concept of fair play requires that you have more confidence in them. One company developed a collaborative online system that lets the company integrate data from supply chains, customer inventories, design tools, and project management software. They warn that trusting your supply chain partners and believing they will behave responsibly "is absolutely critical to the success of collaborative business" (Wilder and Soat 2001).

Sharing information and decision making lead to common goals; this means businesses have a mutual interest in meeting objectives. Like it or not, partners have to trust each other to meet their obligations. This can lead to standards and compliance agreements such as RosettaNet and ISO (Strausl 2001).

Handfield and Bechtel found that contractual agreements increase the level of trust between organizations. They believe such contractual safeguards are important in determining the level of trust the buyer has in the supplier, in that they help define the nature of the relationship between the partners from the outset (Handfield and Bechtel 2002).

Knowledge is a key element in building trust. If lack of confidence, and trust, results from lack of knowledge about certain facets of a supply partner, then filling in the gaps with knowledge should go a long way in eliminating uncertainties about each other. An organization must include perpetual learners (Schein 2004). This may mean changing the culture of the organization, a difficult, but not impossible, transition.

The potential benefits of collaboration are huge, be it integrating *supply chains*, product development, or even business processes. Companies and their *supply chains* can cut out waste, speed time to market, and be more responsive to customers' needs by sharing information. The age of specialization demands that companies lay down their age-old adversarial relationships with trading partners. Six steps to developing trust include the following:

1. Begin by collaborating on a small scale, such as synchronizing one type of sales data, and with a partner you already trust (hopefully, the feeling is mutual).
2. Look inward. The necessary precondition for establishing trust with outside partners is establishing trust with one's internal peers.
3. Gather around. This is no time to use e-mail or videoconferencing. There is no more sophisticated method to build up trust than to meet with people the old-fashioned way, around a table.
4. Go for the win–win. It is a cliche, but be prepared to hear, and say, *win–win* (or even *win–win–win*) over and over again.
5. Do not give away the store. Declare that no one has to, or should, share all information.
6. Just do it. One of the best ways to build trust is simply to start (Paul 2003).

The conclusions and suggestions outlined suggest that most scholars and practitioners consider trust an important element in building integrated supply chains. It is also one of the most elusive. As James Heskett, noted professor at Harvard Business School, sums it up, "Trust is a big issue these days judging from the volume of responses to this month's column. Its importance in management is agreed on... Yet many managers are experiencing a trust deficit" (Heskett 2012, p. 1).

All companies build a reputation or image. Their culture, which is often a result of their founder and their years of practice, is often a precursor of how they will perform in an interdependent relationship. The names of Penney, Walton, Kroc, Hewlett and Packard, Thomas, and Gates may mean different things to different people, but they often carry well beyond microlevel transactions.

Despite the progress over the years, the lack of trust remains one of the most commonly cited reasons for the lack of complete supply chain integration.

## Measuring Performance

What is the value of an integrated supply chain? Most managers want some objective way to measure the benefits of this change. While desirable, to date, very few companies have been able to develop a reliable measure. There are at least two major obstacles to developing effective measures.

The first obstacle is in deciding on the measures to use. Some performance measures used in supply chain activities include

- Lead times, measured in days
- Inventory levels, measured in days of supply or inventory turns
- On-time deliveries, measured as a percentage of total deliveries
- Quality of products and services, measured in number of defects per million

Most of these measures are based on physical measures, such as the number of orders or number of units. This information is not found easily in most accounting systems; therefore, special data collection and reporting systems have to be developed to make this information available to the users. Such systems may not be feasible in many companies.

Morgan traces the broad history of performance measurement from both local and global managerial perspectives. He identified four outstanding challenges: (1) the response to the increasing environmental challenges, (2) the effect of the lean/agile debate on the development of supply networks and their measurement systems, (3) the need to develop performance measurement systems to break through the *dyadic relationship barrier* and make management of the supply network a more realistic aspiration, and (4) the challenges faced in making performance measurement a multicultural phenomenon and through this make management of international *supply chains* more effective.

Morgan's review of performance measurement history enabled him to identify five phases in time:

- *Step 1, 1450+:* Performance measurement dominated by transaction costs and profit determination
- *Step 2, 1900+:* Performance measurement enhanced to include operations and value-adding perspectives
- *Step 3, 1970+:* Performance measurement enhanced to include process, quality, and customer focus

- *Step 4, 1990+*: Performance measurement enhanced to give a balanced view of the organization
- *Step 5, 2000+*: Performance measurement enhanced to include the supply chain and interprocess activities (Morgan 2007)

His description of the five phases does not mean that all companies have successful performance measurement systems. Many are still working to develop their internal reporting systems.

Another obstacle is the difficulty in getting similar information for the entire supply chain. If it is difficult to get useful information in one company, it is considerably more difficult to get it from all participants. Even if the information were available, to whom would it be sent for consolidation and distribution? As we will describe later, most supply chains exist through collaborative agreements among participants and do not have a supply chain manager who oversees the total supply chain operation.

Supply chain members should expect they will share equitably in the benefits achieved by the supply chain. This is an area requiring additional attention. Chapter 16 covers this topic in more detail.

---

## Maintaining the System

Supply chains are like chameleons—they change to fit the situation. Conditions, objectives, and participants change over time. Supply chains would be difficult to build under static conditions. Under continuously changing circumstances, they seem to be in an eternally amorphous state. Companies can usually prepare for incremental change and manage it in an orderly fashion. However, changes caused by revolutionary shifts—need for compliance with a new green supply chain government directive or under crisis conditions—and global product recall often cause major disruptions in supply chains.

### During the Implementation Process

A new supply chain is not built with a rigid procedure or exact time schedule. It involves more than getting the correct contact information for the key players at each participating company and providing each with a list of other contacts in the designated supply chain. As the supply chain begins to form, some customers may be added, some suppliers may be replaced, and some practices may be changed. It is a dynamic period and those responsible must be prepared to mold the supply chain to fit their needs.

### During the Operation of the Supply Chain

Even after the supply chain is formed, there will be continuing disruptions and adjustments. Some examples of the changes that supply chain managers must routinely handle include new customers added to a market segment, new suppliers added, existing suppliers replaced, changes in customer requirements, and changes in personnel at customers or suppliers or internally. These changes could be considered as adjustments to the supply chain.

There will also be dramatic or disruptive changes requiring changes or redesign of the supply chain. These could include natural disasters that destroy supplier capability, such as hurricanes or fires; rapid expansion, or contraction, of customers; and new product introduction requiring new processes with new suppliers.

These changes require immediate, and extraordinary, action on the part of the supply chain manager.

### Extension into Reverse Logistics

One of the major changes taking place in most supply chains is the need to consider end-of-life consequences for products. Companies will have to build supply chains not only to get their products to the customer but also to return the product back through the reverse supply chain to its final destination. The increase in returns resulting from online buying is presenting major problems to companies who are trying to integrate the reverse flow of good with the outflow of goods (Banjo 2013). We described the reverse supply chain in Chapter 10. It is sufficient to point out that this requirement will add greatly to the difficulties in integrating supply chains.

## Obstacles to International SCM

Most of this chapter has been about the obstacles that prevent integration of all supply chains. We have only touched on the added obstacles that companies confront when trying to integrate global supply chains. A study of international supply chain management (ISCM) found many of the same roadblocks apply, including a history of local optimization, insufficient communication between supply chain partners, firefighting that reinforces the short-term orientation in the supply chain, functional silos as root causes for a lack of common goals, inadequate participation of top management, and attitudinal resistance. They concluded the following:

- A first striking and rather surprising observation is that the international dimension was not identified as a major obstacle on the road to ISCM.
- The mechanisms that enable or block effective ISCM management appear to be fairly generic across industries.
- Finally, many of the observed roadblocks for effective ISCM appear to be deeply embedded in the organizational structures and cultures of companies (Akkermans et al. 1999).

While the barriers to global supply chain integration are similar to those encountered in domestic supply chains, they are likely to be more persistent and difficult to resolve.

## Summary

In this chapter, we identified a number of obstacles to supply chain integration. We listed some of the objectives and benefits that integrated supply chains can provide. However, the first decision most companies try to deal with is the question of return on their assets;

consequently, we pointed out that it is difficult to calculate the expected return because both the benefits and costs are somewhat nebulous. We then discussed the fact that companies have different goals and expectations, and these differences are sometimes hard to reconcile. It is also difficult to measure performance, especially in trying to evaluate the financial gains.

While companies are trying to integrate supply chains, they are also pursuing two major strategies that are tearing supply chains apart. They are diversifying their product lines, thereby requiring multiple supply chains. In addition, they are outsourcing parts of their processes, and this makes it even more difficult to keep products, information, and funds flowing freely.

We then turned to some of the difficulties in making the necessary changes in technology, infrastructure, and cultures, both internally and among supply chain partners. Technology is often viewed as the driver of change, but without appropriate modifications in organization structure, processes, systems, and attitudes, technology will not be enough to carry the project to a successful conclusion.

The difficulty in integrating supply chains is highlighted in this chapter's "Hot Topic" section, a continuation of the discussion started in Chapter 11, about Atomic Energy Canada Limited (AECL) and its cancer fighting machine.

---

**Hot Topics in the Supply Chain: Atomic Energy Canada Limited Encounters Problems with Its Cancer Fighting Machine (Part 2)**

---

The AECL case was introduced in Chapter 11. Recall that their machine, the Therac-25, had been involved in six cases of overdose accidents involving patients. Three of those patients died as a result of the radiation overexposure, while the other three had serious injuries. While AECL had full knowledge of the accidents, each of the four medical centers involved was not aware that an accident had occurred at other facilities (Fauchart 2006). As mentioned in part 1 of this case, such a situation constitutes a state of information asymmetry, a state whereby not all of the important stakeholders have complete information on a given matter.

Integration of the supply chain is facilitated when the partners have adequate information to conduct their affairs. Cases such as AECL illustrate how situations can arise when one partner is privy to information that should be available to another partner in the supply chain. Trust is a key factor in making this information available. However, there will be times when the unexpected occurs and the manager is left with the question "How much information should be shared, especially if the situation is somewhat murky?" The AECL case illustrates one of those scenarios.

### BACKGROUND

As a supply chain partner, the AECL was in a situation not unlike many companies today. The items in the following show how AECL was similar to many contemporary manufacturers:

- The Therac-25 was based on a previous model, the Therac-20. Hence, the new model was an upgrade, which also consisted of more software features from its predecessor.
- The Therac-25 had been installed in 11 locations in North America, 5 in the United States, and 6 in Canada (Leveson 1995). Because of the multiple locations, the six accidents occurred not only at different times but different locations, hence, the space–time dilemma.

- Prior to the first accidental overdose in June 1985, the Therac-25 had been used thousands of times without an incident (Fauchart 2006). Such a track record can create a false feeling of confidence for the manufacturer. This would be true for any manufacturer, not just AECL.
- The Therac-25 was an example of a complex technology. Because of this feature, it also contained hidden software bugs that were not apparent, even after several years of successful use.
- AECL had introduced the Therac-25 in order to gain market share, especially since many medical facilities were replacing their Cobalt-60 machines (Fauchart 2006).

AECL, like many companies today, was trying to enlarge its market by introducing a better machine. What took the company by surprise was the series of six accidents, this, after many successful hours of use at 11 different medical centers in two countries.

## POTENTIAL OBSTACLES TO INTEGRATION

For supply chain managers, the AECL case is noteworthy because the company did not accurately inform its customers of problems that were occurring with the Therac-25. After all, a case involving overexposure to radiation, by a machine that is supposed to accurately direct radiation, is of interest to all stakeholders. Had the forward supply chain been perfectly integrated, each of the four medical centers affected, plus the other seven centers not affected, would have had complete information of the overdose accidents. In actuality, this was not the case.

But why would AECL not inform its customers that potentially hazardous problems existed with its product? Two reasons can be offered, each of which applies to managers today: (1) AECL really did not think the machine could administer an overdose of radiation and (2) it did not want to hurt market share.

The fact that the Therac-25 had operated successfully for many operating hours might have given its management and engineers a false feeling of invulnerability. Table 12.4 summarizes the accidents and the preliminary responses by AECL. For the first four accidents, AECL engineers conducted their own investigations and concluded that an overdose of radiation was just not possible. However, the research by Leveson (1995) suggested that the engineers were focused more on hardware and user operator problems as opposed to potential software problems. Fauchart (2006, p. 101) notes that "AECL may have expected the incidents to be rare events, induced by the entrance of the machine into very low probability states." Hence, the company may have assumed that the accidents were flukes.

Fauchart (2006) also suggests that AECL might have been more concerned with losing market share. However, in fairness to the company, the exact causes of the overdoses had not been determined as of the first four accidents. It was not until the fifth accident at Yakima Valley that the software bug problem was considered as the source of the problem. With uncertainty about what was even causing the accidents to begin with, AECL might have been reluctant to raise too much publicity about its machine, especially since it was seeking to build market share. As Fauchart (2006, p. 101) states, "AECL had a new and unproven technology and it was crucial, during this phase of its market penetration, that is reputation should not be damaged by the publicity of accidents to patients cause by radiation overdoses."

**TABLE 12.4**

Overview of the Accidents

| Accident Date and Location | Details | Outcomes |
|---|---|---|
| Kennestone Regional Oncology Center, June 1985 | Female patient received an overdose of radiation that felt like a heat burn. Patient files a lawsuit against AECL and the hospital in November 1985. | AECL is not officially notified of the accident until the lawsuit is filed in November. Information about the accident and lawsuit is not made available to other Therac-25 users. |
| Ontario Cancer Foundation, July 1985 | Female patient received an overdose of radiation that felt like a heat and electrical burn. | AECL is informed of the accident and sends a service representative to investigate. The FDA and Canadian Radiation Bureau and all Therac-25 users are informed of the accident. The Therac-25 users claim they are not told that a patient injury occurred. |
| Yakima Valley Memorial Hospital, December 1985 | Female patient received an overdose of radiation that resulted in a parallel striped pattern on her hip. | AECL investigates the accident and concludes it could not have resulted from a radiation overdose. No official cause of the accident is determined at the time. |
| East Texas Cancer Center, March 1986 | Male patient received an overdose of radiation on his back. Patient dies from the overdose 5 months later. | Two AECL engineers investigate the accident at the site. AECL tells the East Texas Cancer Center physicist that it is not aware of any cases of the Therac-25 causing overdoses of radiation. AECL engineers conclude the Therac-25 cannot administer an overdose of radiation. They conclude the problem is electrical in nature. |
| East Texas Cancer Center, April 1986 | Male patient received an overdose of radiation on his face. Patient dies from the overdose 3 weeks later. | A *Malfunction 54* problem emerges in the Therac-25. Engineers at AECL and the physicist at East Texas Cancer Center trace the problem to the data entry speed of the operator. This indicates a potential software problem with the machine. |
| Yakima Valley Memorial Hospital, January 1987 | Male patient received an overdose of radiation to his chest. Patient dies from the overdose in April, 1987. | A software problem is discovered by AECL engineers. This same problem is thought to have caused the first accident at Yakima Valley Memorial Hospital in December 1985. AECL communicates to all Therac-25 users that new software and hardware changes will be forthcoming. |

## QUESTIONS FOR DISCUSSION

1. Supply chain integration includes the component of providing a state of information transparency to all affected parties. Yet, this case illustrates that information uncertainty as to the cause of an accident can make the task of communication dissemination difficult. What should be the role of a manufacturer when a product defect has been found but the exact cause of that defect has not been determined?

2. Software integration into tangible products today is common. Discuss examples of products that have had software problems. How did manufacturers assure a state of information transparency?

## REFERENCES

Fauchart, E., Moral hazard and the role of users in learning from accidents, *Journal of Contingencies and Crisis Management*, 14(2), 97–106, 2006.

Leveson, N., *Safeware: System Safety and Computers*, Addison-Wesley, Reading, MA, 1995.

Weick, K., Sense and reliability: A conversation with celebrated psychologist Karl E. Weick, *Harvard Business Review*, 81(4), 84–90, 2003.

## Discussion Questions

1. What are the benefits from integrating supply chains?
2. What are some of the obstacles to achieving integrated supply chains?
3. Why does a company need multiple supply chains?
4. What are some of the obstacles in implementing IT systems?
5. How does increased outsourcing act as a barrier to supply chain integration?
6. How does organizational culture play a role in the integration of supply chains?
7. In what ways is organizational structure changing to enable better integration of the supply chain?
8. How can trust be built between a purchasing organization and a supplier?
9. Why is it difficult to measure the effectiveness of an integrated supply chain?
10. What are the added difficulties in integrating a supply chain that spans different continents?

## References

Aitken, J., Childerhouse, P., Christopher, M., and Towill, D., Designing and managing multiple pipelines, *Journal of Business Logistics*, 26(2), 73, 2005.

Akkermans, H., Bogerd, P., and Vos, B., Virtuous and vicious cycles on the road towards international supply chain management, *International Journal of Operations & Production Management*, 19(5/6), 565, 1999.

*American Heritage Dictionary*, 2nd college edn., Houghton Mifflin Company, Boston, MA, 1982.

Banjo, S., Rampant returns plague E-retailers—Sellers get proactive, suggesting sizes and redirecting discounts, to break shoppers of bad habits, *Wall Street Journal*, B.1, December 23, 2013.

Barrett, M., Understanding the meaning of collaboration in the supply chain, *Supply Chain Management*, 9(1), 30, 2004.

Beth, S., Burt, D.N., Copacino, W., Gopal, C., and Lee, H.L., Supply chain challenges: Building relationships, *Harvard Business Review*, 81(7), 64, 2003.

Bitran, G.R., Gurumurthi, S., and Sam, S.L., The need for third-party coordination in supply chain governance, *MIT Sloan Management Review*, 48(3), 30, 2007.

Blackstone, J.H., *APICS Dictionary*, 13th edn., APICS—The Association for Operations Management, Chicago, IL, 2010.

Byrne, P.M., Five trends support logistics success in China, *Logistics Management (2002)*, 45(6), 22, 2006.

Cox, A., The power perspective in procurement and supply management, *Journal of Supply Chain Management*, 37(2), 4, 2001.

Crandall, R.E., Just trust me, lack of confidence inhibits effective supply chain collaboration, *APICS Magazine*, 18(2), 14, 2008.

Crandall, R.E., Beating impossible deadlines, a variety of methods can help, *APICS Magazine*, 16(6), 20, 2006.

Crandall, R.E., Demanding discipline, even with greatly refined forecasting methodologies, the challenges remain, *APICS Magazine*, 15(6), 16, 2005.

Crum, C. and Palmatier, G.E., Demand collaboration: What's holding us back? *Supply Chain Management Review*, 8(1), 54, 2004.

Drickhammer, D., Beware the big words, *Logistics Today*, 48(4), 5, 2007.

Grossman, M., The role of trust and collaboration in the Internet-enabled supply chain, *Journal of American Academy of Business, Cambridge*, 5(1/2), 391, 2004.

Handfield, R.B. and Bechtel, C., The role of trust and relationship structure in improving supply chain responsiveness, *Industrial Marketing Management*, 31(4), 367, 2002.

Hawkins, D., Firstperson: A question of trust, *Supply Management*, 12(5), 16, 2007.

Heskett, J., Why is trust so hard to achieve in management? Harvard Business School Working Knowledge, Boston, MA, July 5, 2012, http://hbswk.hbs.edu/cgi-bin/print/7034.html, accessed January 2, 2013.

Hounshell, D.A., *From the American System to Mass Production, 1800–1932: The Development of Manufacturing Technology in the United States*, The John Hopkins University Press, Baltimore, MD, 1985.

Ireland, R.K. and Crum, C., *Supply Chain Collaboration: How to Implement CPFR and Other Best Collaborative Practices*, J. Ross Publishing, Inc., Boca Raton, FL, 2005.

Kosk, N., Is supply chain ready for the next generation of S&OP? *Supply & Demand Chain Executive*, February 21, 2013, www.sdcexec.com/article/10882462/is-supply-chain-ready-for-the-next-generation-of-sop?, accessed April 22, 2013.

Lau, K.H. and Zhang, J., Drivers and obstacles of outsourcing practices in China, *International Journal of Physical Distribution & Logistics Management*, 36(10), 776, 2006.

Mariotti, J., Plenty of technology, but a shortage of trust, *Industry Week*, 248(11), 128, 1999.

Morgan, C., Supply network performance measurement: Future challenges? *International Journal of Logistics Management*, 18(2), 255, 2007.

Naylor, J.B., Naim, M.M., and Berry, D., Leagility: Integrating the lean and agile manufacturing paradigms in the total supply chain, *International Journal of Production Economics*, 62(1–2), 107, 1999.

Paul, L.G., Suspicious minds; Collaboration among trading partners can unlock great value. Mistrust is the barrier. Here are six ways to build confidence, *CIO*, 16(7), 1, 2003.

Riege, A., Three-dozen knowledge-sharing barriers managers must consider, *Journal of Knowledge Management*, 9(3), 18, 2005.

Ruigrok, W., Pettigrew, A., Peck, S., and Whittington, R., Corporate restructuring and new forms of organizing: Evidence from Europe, *Management International Review*, 39(Special Issue), 41, 1999.

Schein, E.H., *Organizational Culture and Leadership*, 3rd edn., Jossey-Bass (John Wiley & Sons, Inc.), San Francisco, CA, 2004.

Storey, J., Emberson, C., Godsell, J., and Harrison, A., Supply chain management: Theory, practice and future challenges, *International Journal of Operations & Production Management*, 26(7), 754, 2006.

Strausl, D., Four stages to building an effective supply chain network, *EBN*, 1251, 43, February 26, 2001.

Thibodeau, P., Rising fuel prices prime pump for more telecommuting, virtual meetings, *Computerworld*, 42(22), 6, 2008.

Walker, W.T., *Supply Chain Architecture: A Blueprint for Networking the Flow of Material, Information, and Cash*, CRC Press, Boca Raton, FL, 2005.

Wilder, C. and Soat, J., The trust imperative, *Information Week*, 848, 34, July 30, 2001.

# 13

## How to Build an Integrated Supply Chain

### Learning Outcomes

After studying this chapter, you should be able to

- Contrast and discuss past, present, and future methods of supply chain management (SCM)
- Discuss the implications of lean management in regard to integrated supply chains
- Describe the transition from a functionally focused supply chain to one that involves cross enterprise collaboration
- Describe a comprehensive model of supply chain integration (SCI)
- Discuss how a lean and agile supply chain can be achieved
- Identify the steps involved in changing to an integrated supply chain
- Explain why complexity and clairvoyance are important to understanding future SCI

In this chapter, we outline the steps required to integrate a supply chain. The steps involve all functions in an organization. Companies must build internal relationships (within the company) before they can build external relationships (other companies). Integration is a demanding and continuous process; it is ongoing because conditions and relationships are never permanent. Simplicity sounds desirable; however, supply chains are complex. Cooperation is good, but collaboration is better. Integration is a worthy goal, but integrated supply chains require innovation and added change. As you read this company profile and the chapter in the book, think about these questions:

- How long should it take to integrate a supply chain? How can the time be shortened?
- Why is project management important in integrating supply chains?
- How do you measure your progress along the SCI timeline?
- How do you decide that your present level of SCI is *good enough*?

What additional questions do you think relate to building an integrated supply chain?

---
**Company Profile: Interface, Inc.**
---

## INTRODUCTION

Interface®, Inc. is a company with a mission—"To be the first company that, by its deeds, show the entire industrial world what sustainability is in all its dimensions: People, process, product, place and profits—by 2020—and in doing so we will become restorative through the power of influence." Their primary mission is called "Mission Zero: our promise to eliminate any negative impact our company may have on the environment by the year 2020" (Interface Inc. 2013).

If Interface were a legal firm or a consulting firm whose final outputs were paper documents, the objective stated earlier might seem reasonable. However, Interface is a carpet manufacturer and the material traditionally used to make carpets includes oil. In addition, carpet has long been a major component of many landfills. Is Interface serious? How do they expect to achieve this lofty goal? To better understand how far the company has already progressed, it is necessary to understand how they started their journey toward ultimate sustainability.

## HISTORY

Ray Anderson started the company in 1973 as a joint venture with Carpets International Pic (CI), a British company, and a group of American investors to produce and market modular soft-surfaced floor coverings. The company participated in the office building boom during the 1970s and sales reached $11 million by 1978. The company went public in 1983 and through acquisitions, gained entry into the European and Middle Eastern markets. With the acquisition of Heuga Holdings B.V., Interface became the world leader in carpet tiles (Interface's History 2013).

In 1994, Anderson received a note from an associate in the research division who said that customers were asking what the company was doing for the environment. The note asked, "How should we answer?" Anderson decided the inquiries came from responsible customers—architects and interior designers—who wanted to do the right thing and deserved an answer. In looking for an answer, he read Paul Hawken's book *The Ecology of Commerce* that espoused the idea that living systems are in decline and the industrial system is responsible. Anderson experienced what he described as a *spear-in-the-chest* moment. He decided to do something about it—"to put back more than we take from the Earth, and to do good for the Earth" (Posner 2009, p. 48).

Needless to say, this was a major turning point in the life of the company, especially for a company whose products and production methods are petroleum intensive. Once employees and suppliers realized Anderson was serious, they began to move toward the objective. Anderson coined the term Mount Sustainability to not only represent the task ahead but to use the mountain to symbolize the Earth and its resources to be preserved (Posner 2009, p. 48).

## PRODUCTS

Interface has been a pioneer in modular carpet tile, or carpet that is composed of 50 cm × 50 cm pieces (tiles). This is in contrast to broadloom carpet that is cut to fit an entire room. Carpet tile has many advantages, not the least of which is the flexibility to combine different patterns within the same area. Different colors and patterns can be used to satisfy the creative instincts of homeowners or interior decorators. Carpet tile is also least wasteful in that a portion of a room can be renewed by replacing a few tiles instead of carpeting the entire room. This means less used carpet goes to the landfill. Interface prides itself on its efforts in sustainability. See the following excerpt from their website's product description.

> Our products are designed to meet your needs. When we introduced carpet tile to North America in the '70s, it was a square idea in a broadloom world. But it was also more flexible and created less waste. It was a breakthrough design platform that set the stage for our sustainable future.
>
> Using the principles of biomimicry, we created i2™, our second design platform. Biomimicry seeks sustainable solutions by using nature as a model. Like leaves on a forest floor, i2 tiles vary from one to the other – yet they come together to form a beautiful floor.
>
> Whereas i2 exemplifies how nature would design a floor, Convert™, our third design platform, considers how nature would manufacture a floor. Convert platform styles are created using the first responsible resource of post-consumer nylon – including type 6 and 6,6. Convert products contain 64–75% total recycled content, including 32–35% post-consumer content depending on style and color. Convert reduces our use of virgin resources, moving us closer every day to a closed-loop system and our Mission Zero™ goal.
>
> And we successfully combine responsibility with performance and beauty. In addition to a wide range of colors and patterns, our products also have several high performance backing options, durable face construction, less installation waste, mold and mildew resistance with our Intersept® antimicrobial preservative, stain resistance and more.

**Interface about our Products (2013)**

As noted earlier, the Intersept® product has antimicrobial preservative to reduce mold and mildew, a feature valued by hospitals. The Florida Hospital Memorial Medical Center (FHMMC) in Daytona Beach is proud of their new facility located along the I-95 interstate. It is designed to be efficient while at the same time has features, such as more natural lighting in patient and operating rooms, to make it a more pleasant environment for both patients and staff. The first name on the spec sheet is carpet and carpet tile from Interface, Inc. (Eagle 2010).

Interface is increasing their capability to provide custom samples quickly and on-time delivery of customized final products. They can generate realistic digital samples from which they can create an almost unlimited number of new designs. Approximately 75%–80% of their modular products are now made to order. This not only is an attractive feature for customers, but it always helps the company improve their inventory turns.

## GLOBAL MARKETS

Interface's primary emphasis has been the corporate office market. In recent years, they have focused more attention on the noncorporate office markets such as government, education, health care, hospitality, retail, and residential space. In the Americas, the noncorporate office market represented 36% of total sales in 2001; by 2012, it had grown to 53%.

The FLOR products were designed to bring high-style modular carpet to the North American residential market, estimated at $11 billion. Interface offers FLOR directly and over the Internet, in a FLOR catalog, and in 18 FLOR retail stores. Additional retail stores are planned to further extend the reach of the FLOR products.

**TABLE 13.1**

2012 Sales and Assets Deployed (Millions of Dollars)

| Country | Sales | Total Sales (%) | Long-Lived Assets | Total Assets (%) |
|---|---|---|---|---|
| United States | $421 | 45 | $79 | 47 |
| United Kingdom | 83 | 9 | 10 | 6 |
| Australia | 87 | 9 | 12 | 7 |
| Netherlands | | 0 | 33 | 20 |
| China | | 0 | 16 | 9 |
| Other foreign countries | 340 | 36 | 16 | 10 |
| Total | $932 | 100 | $166 | 100 |

*Source:*   Interface 2012 Annual Report, p. 79.

The company has a large and diverse customer base, with no single unaffiliated customer representing more than 10% of total sales. Sales, outside the United States, were primarily in Canada, the United Kingdom, Australia, and Latin America. Interface is also making progress in Asian countries, especially China with local customers in regional mainland cities. Table 13.1 shows how sales and assets are distributed among the key market areas (CEO 2012 Letter to Shareholders).

As part of their strategy to concentrate on carpet tile, the company sold their Bentley Prince Street business, which was primarily a broadloom carpet operation. Despite a fire that destroyed a part of their Australia operation, they remain committed to the area and expect to open a new plant by the end of 2013.

To support their marketing efforts, they maintain marketing offices in over 70 locations in over 30 countries and distribution facilities in approximately 40 locations in 6 countries. They also have manufacturing operations in Bangkok, Thailand; Craigavon, Northern Ireland; Minto, Australia; Scherpenzeel, the Netherlands; and Taicang, China, in addition to their U.S. locations in LaGrange and West Point, Georgia. These manufacturing facilities are located close to the markets they serve.

## SUPPLY CHAIN

The supply chain involves the purchase of raw materials from multiple sources—both regionally and globally—except for synthetic fiber (nylon yarn). For yarn, the company has two major global suppliers with significant relationships with at least two other suppliers. They also have some flexibility in manufacturing carpet, using face fiber produced from two separate feedstocks, if needed. They are heavily involved in using reclaimed and waste carpet as part of their *ReEntry*®*2.0* carpet reclamation program (Interface 2012 Annual Report, p. 9).

The company has manufacturing operations located throughout the world, as listed earlier. They believe that locating their manufacturing near their markets enables them to offer the most advantageous delivery times and costs to their customers and enhances their ability to develop a strong local presence in foreign markets. They also try to standardize their manufacturing procedures throughout their system by establishing global standards. For example, they have settled on a carpet size of 50 cm × 50 cm, because the metric system is universally used. Their manufacturing facilities are certified under International Standards Organization (ISO) Standard No. 14001, an environmental certification.

The company distributes their product through two primary channels: (1) direct sales to end users and (2) indirect sales through independent contractors or distributors.

They work with architects, engineers, interior designers, contracting firms, and other specifiers who can often make or substantially influence purchasing decisions (Interface 2012 Annual Report, p. 8).

The company also offers installation services. One of their recent innovations is called TacTiles, which uses small squares of adhesive plastic film to connect intersecting carpet tiles. This eliminates the need to apply adhesive to the floor and makes replacement of tiles easier.

## REVERSE SUPPLY CHAIN

Modular tile carpeting reduces the amount of carpet recycles because only a portion of a carpet may need replacing where heavy traffic or stains may have created unattractive areas. Interface is also heavily committed to using recycled carpet materials in making new carpet.

One of the more innovative efforts is the reclamation of fishing nets. The U.N. Food and Agriculture Organization estimates that 640,000 tons of fishing gear are discarded annually, creating a serious hazard for marine life. Interface set up a test case in the Philippines to have villagers take the nets to a central spot and exchange them for cash. The nets are then sent to a processing plant in Ljubljana, Slovenia, where it is made ready for recycling into carpet. Interface uses the recycled nets in the Net Effect carpet tiles, which contain 81% total recycled content and 100% recycled nylon face fiber. While still in the early stages, the company has great hopes for the future of this program (Minter 2013).

## SUSTAINABILITY

Interface is committed to their sustainability efforts and has an initiative called EcoSense. EcoSense, which includes the QUEST waste reduction initiative, aims to eliminate energy and raw material waste and ultimately to reclaim and restore shared environmental resources. The program has two objectives:

- To learn to meet our raw material and energy needs through recycling of carpet and other petrochemical products and harnessing benign energy sources
- To pursue the creation of new processes to help sustain the earth's nonrenewable natural resources

To help in their quest, Interface has engaged some of the world's leading authorities on global ecology as environmental advisors. The advisors include Paul Hawken, author of *The Ecology of Commerce: A Declaration of Sustainability and the Next Economy* and coauthor of *Natural Capitalism: Creating the Next Industrial Revolution*; Amory Lovins, energy consultant and cofounder of the Rocky Mountain Institute; John Picard, President of E2 Environmental Enterprises; Bill Browning, founder and former director of the Rocky Mountain Institute's Green Development Services; Janine M Benyus, author of *Biomimicry*; and Bob Fox, renowned architect. In a further indication of their goal of becoming sustainable, the company launched the Mission Zero global branding initiative to represent their mission to eliminate any negative impact their organization may have on the environment by the year 2020 (Interface 2012 Annual Report, p. 11).

As an example of their research and development efforts, they have developed a flexible-input carpet backing line called Cool Blue™. This process uses next-generation thermoplastic technology to improve the use of reclaimed and waste carpet in the production loop and opens up the possibility of using other plastics and polymers as inputs.

## COOL CARPET™

Our climate neutral Cool Carpet* zeros out all greenhouse gas (GHG) emissions associated with the entire life cycle of our carpet. Offsetting GHG emissions helps keep the planet cool and reduces our contribution to global warming.

- Cool Carpet comes standard on all Interface products sold in North America.
- Earn an LEED Innovation Credit with Cool Carpet.
- Help build a market for greener materials.
- Get the same product and performance you expect from Interface with an extra benefit to the planet.

In other words, we are not just blowing hot air (Interface Products 2013).

## CONCLUSIONS

Interface, Inc. is a company that is sustainable in all three areas outlined in the triple bottom line concept of sustainability. They are financially sound, they are socially aware, and, above all, they are committed to being an environmentally responsible company. Ray Anderson's vision has propelled the company to its unique position among global businesses. The present management has promised to carry on, after Mr. Anderson's death in September 2011.

## REFERENCES

Eagle, A., Traffic stopper, *Health Facilities Management*, 23, 12–17, May 2010.
Interface about our Products, 2013. https://www.interfaceflor.com/Default.aspx?Section=2&Sub=3.
Interface, 2012. Annual Report, CEO Letter to Shareholders, p. 2.
Interface Inc., 2013. Mission/Vision Values, http://www.interfaceglobal.com/Company/Mission-Vision.aspx, accessed September 21, 2013.
Interface's History, 2013. www.interfaceglobal.com/Company/History.aspx, accessed September 21, 2013.
Interface Products, Cool Blue, 2013. https://www.interfaceflor.com/Default.aspx?Section=3&Sub=4&Ter=11.
Minter, S., The global manufacturer: A net gain for sustainable manufacturing, it takes a village to help make a new carpet tile. *Industry Week*, August 1, 2013. www.industryweek.com, accessed September 5, 2013.
Posner, B.G., One CEO's trip from dismissive to convinced, *MIT Sloan Management Review*, 51(1), 47–51.

# Introduction

This chapter outlines an approach to building an integrated supply chain. We first briefly describe the major steps in the integration process and explain them in more detail later in the chapter. The steps are as follows:

- Investigate to determine what needs to be done and why.
- Involve top management to support the program.

---

* Not just any carpet can be cool. Our Cool Carpet is verified by a third party. Environmental experts have reviewed our data and projects to make sure we are investing in credible, high-quality products.

- Include necessary cross-functional members of the project team.
- Initiate the action—inform and gain support within the organizations.
- Identify and analyze areas of resistance and resolve.
- Implement in stages and solidify the gains at each stage.
- Integrate to form the seamless supply chain.
- Institutionalize to assimilate the supply chain as a normal part of doing business.
- Innovate—change the supply chains as events and opportunities present themselves.

In the past, vertical integration was necessary, and successful, in industries where administrative coordination provided competitive advantages. Where technology did not lend itself to mass production, and where volume distribution did not benefit from specialized scheduling or services, vertical integration failed to bring concentration (Chandler 1977).

## Who Manages the Supply Chain?*

Most businesses have supply chains. But how do they manage them? Put another way, when we speak of supply chain management (SCM), are we really talking about management of the supply chains, or are we just describing the concepts and techniques of how supply chains operate?

To give you a feel for the magnitude of the problem, one study categorized the supply chain as having three strategic dimensions: *synthesis*, with insights from the industrial organization, institutional economics, and network theory literature; *synergy*, drawing primarily from the interorganizational relationships and strategic management literature; and *synchronization*, founded on research in operations management, logistics, operational research, and systems engineering. They offer a beginning, not a complete answer (Giannakis 2008).

Another study looked at theoretical explanations of how to structure and manage supply chains from three different perspectives—an economic perspective, a socioeconomic perspective, and a strategic perspective. They concluded there is no such thing as *a unified theory of SCM* at the present time (Halldorsson et al. 2007).

## Past and Future of Supply Chain Management

We will first consider the extremes of SCM (the past versus the future) and then see if we can find some places in between that fairly represent the present. First, take a look at the past. A suitable starting point is Ford's River Rouge plant. This highly integrated factory brought iron ore off the Great Lakes and processed it into steel for use in assembling the Model T (Ford 1988).

---

* This section has been adapted from Crandall (2009).

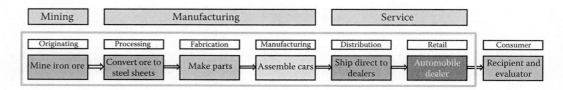

**FIGURE 13.1**
Tightly linked or controlled supply chain under one major participant.

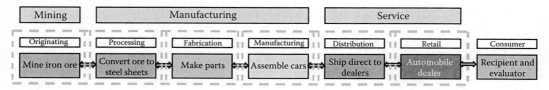

**FIGURE 13.2**
Loosely coupled supply chain with aligned direction through collaborative links throughout.

This is about as tightly controlled as companies got—a period when vertical integration was a commonly accepted way to manage. Their supply chain started at the ore mines and carried forward to their sale of cars, but it was largely under the control of one organization. If problems arose, they were resolved on-site. Figure 13.1 shows a tightly integrated supply chain under the control of the owning or focal company.

Now for the future. Many writers portray the supply chain of the future as a series of tightly connected links (chain) in which each link is a separate entity that collaborates with its customers downstream and suppliers upstream so that products and services flow smoothly to the ultimate consumers. In this arrangement, the supply chain participants have a unified view of their mission, and all are motivated to do what is best for the ultimate consumer, believing that objective will create prosperity for all members of the supply chain. The participants trust each other completely and collaborate to plan and execute their mutually agreed responsibilities. Should problems arise, they work together to resolve them quickly and equitably. In a manner similar to self-directed teams within a company, the supply chain becomes a self-directed supply chain with members that use their understanding of the big picture to manage their own operations. Incidentally, the participants in the supply chain may be any place in the world. Figure 13.2 shows a supply chain that achieves the singularity of purpose through collaborative links between participants.

While the future supply chain model has many advocates, they also point out that it is not easily achieved. "All managers recognize technology, information, and measurement systems as major barriers to successful supply chain collaboration. However, the people issues—such as culture, trust, aversion to change, and willingness to collaborate—are more intractable" (Fawcett et al. 2008).

## Present Supply Chains

Most supply chains are somewhere in between the end points described earlier. But where are they? How are companies dealing with the problem of moving from the security of vertical integration to the uncertainty of loosely coupled globally dispersed independent

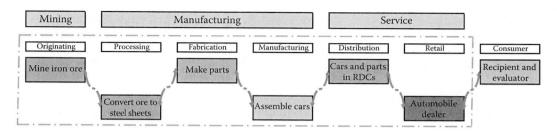

**FIGURE 13.3**
Loosely linked supply chains without direct route to ultimate consumer.

operations? Figure 13.3 shows a supply chain that moves toward the ultimate consumer through loosely coupled links that work at varying levels of effectiveness and efficiency.

In reviewing the literature on supply chains, it is difficult to find articles that deal directly with this issue of SCM or governance. Much of the forward-looking research describes efforts in moving toward trusting and collaborative relationships. However, there is not a lot about what to do in the interim, although there are some suggestions. Crandall (2009) describes five major approaches that fit into the present state of supply chain linkages:

## Virtual Supply Chains

The APICS Dictionary defines a virtual corporation as "the logical extension of outpartnering. With the virtual corporation, the capabilities and systems of the firm merge with those of their suppliers, resulting in a new type of corporation where the boundaries between the suppliers' systems and those of the firm seem to disappear. The virtual corporation is dynamic in that the relationships and structures formed change according to the changing needs of the customer" (Blackstone 2013).

A further definition of a virtual organization is "short-term alliances between independent organizations in a potentially long-term relationship to design, produce, and distribute a product. Organizations cooperate based on mutual values and act as a single entity to third parties" (Blackstone 2013). This latter definition has more relevance to supply chains. An example of this type of virtual supply chain is the one established by Boeing to create their Dreamliner 787. They created a tightly linked supply chain for the entire life cycle of that specific airplane. Participants are clearly identified and bound together with contractual agreements involving mutual commitments.

## Contractual Alliances

A step removed in formal arrangement from virtual supply chains is a supply chain linked together by contractual agreements. These agreements can clarify many issues, but are not guarantees that disputes won't arise or adjustments won't be required as the supply chain evolves over time and as conditions change.

Williamson (2008) suggests that, as bilateral dependency increases between participants, the relationship moves from a simple market exchange to a hierarchical form involving contractual safeguards.

## Dominant Party Management

Supply chains are composed of companies ranging in size from small to large. Although it varies by industry, in recent years, retail companies have tended to become the largest entities in the supply chain. Companies such as Wal-Mart, Target, Home Depot, Lowe's, Best Buy, and Macy's are huge in comparison with most of their suppliers. As a result, the retail company is often the dominant player in a supply chain and can exercise significant direction over other members of the supply chain. While they may not have direct control, they have considerable influence on what their suppliers do.

Some manufacturing companies are equally dominant in their supply chains. Automobile manufacturers, computer manufacturers, oil drilling and refining companies, and pharmaceutical companies can also effectively dictate what other members of their supply chains do.

### Third-Party Direct Management

Another possible alternative is to have a third party, not a member of the supply chain, assume a measure of management responsibility for the supply chain. This is occurring when companies use third-party logistics providers (3PLs) described in Chapter 9. UPS (United Parcel Service) provides a variety of services, including the operation of warehouses that receive orders and ship products to customers for other companies. Amazon provides website management and order processing services for a vast number of companies. Finally, contract manufacturers provide manufacturing services for a number of companies that design and market their products. The outsourcing movement offers numerous opportunities for transferring not only production work but also the management of a portion of the supply chain.

One recent study looked at the role of logistics service providers (LSPs) in supporting SCI. In their literature review, they were surprised to find that very few articles consider LSPs in discussing SCM. LSPs are divided about whether they should be pure *resource providers* or the riskier role of *supply chain designers*. Apparently, there is no complete agreement about the role of third-party providers in SCM (Fabbe-Costes et al. 2009).

### Third-Party Indirect (Third-Party) Management

There is a growing interest in the use of third parties to assist in the indirect management of supply chains. We will highlight several approaches that may be considered.

*Systems integrator*: As supply chains expand in size and complexity, the interfaces between partners become more difficult to maintain. One study's broad hypothesis is that the process of disintegration in many industries is not sustainable from a coordination and control viewpoint and therefore will be followed by eventual reintegration, although it may take different forms in different industries. This will be an expanded role for a systems integrator, which, in many cases, goes beyond critical coordination services and extends into issues related to control and governance of portions of the supply network (Bitran et al. 2007).

*Advisory board*: A variation of this would be to select an advisory board, composed of representatives from companies participating in the supply chain. The advisory board would probably be more concerned with strategic and jurisdictional issues; however, they could be empowered to become more directly involved should the partners agree on the increased responsibilities.

*Auditor*: Still another variation would be to use third parties to audit the functioning of a supply chain and identify problems or opportunities for improvement. This would be similar to public accounting firms who audit public companies, but would not carry the same regulatory requirements. This approach could lead to the development of standards for use in designing effective supply chain partnerships.

*Program manager*: Designing and building an integrated supply chain have characteristics of a project. Companies could use a third party (consultant) to help them design their supply chain and assist them in its implementation. Once the supply chain is working satisfactorily, its management could revert to the companies involved.

*Arbitrator*: Another type of third-party involvement could include an arbitration function to resolve disputes and, in so doing, establish policies that could serve the supply chain members in resolving future disputes.

There are variations for each of these approaches, depending on industry and individual company practices. As we pointed out in Chapter 9, 3PL companies are moving rapidly to fill this need for overall SCM.

## World of Lean Production*

In Chapter 7, we used lean production as an example of one of the major improvement programs that has become popular in the past few years. While it arose in the manufacturing sector, many now advocate its use in service operations. The terms lean production, lean manufacturing, and lean thinking have become almost synonymous in the management literature and have largely replaced just-in-time (JIT) as the preferred designation of programs in this area. It is necessary to consider lean production when considering a movement to integrated supply chains. The key values for lean production are as follows:

- *Create value for the customer*: Lean production creates value for the customer by reducing the amount of waste within companies and along the supply chain.
- *Create a smooth flow of goods*: One of lean production's objectives is to smooth the flow of goods along the supply chain by eliminating problems and other delays that interrupt flow.
- *Provide a fast response*: By eliminating waste and creating a smooth flow, lean production reduces the response times in product and service deliveries.
- *Detect quality problems quickly*: Lean production continues an emphasis on maintaining the quality of products and services. The need for fast, even flow makes it possible to detect quality problems sooner.
- *Develop cross-trained employees*: Cross-trained employees are necessary to maintain the even flows and high quality required by lean production.
- *Illustrate a contemporary outlook*: Lean production reflects some of the latest thinking in operations management and fits nicely with supply chain concepts.

---

* Much of the remainder of this chapter has been adapted from Chapter 10 of Crandall and Crandall (2014).

While lean production provides a number of benefits, it is not without some limitations or constraints in its implementation:

- *It takes time to implement lean production*: Companies that implement lean production should not view it as a quick fix. It takes time to implement it effectively. We will describe some of these steps in the following section.
- *Lean production requires significant change*: To implement a complete version of lean production, a company will probably have to make substantial changes in the way it operates. Changing the physical layout of a plant and moving from job specialization to job enlargement and enrichment are two areas that involve the greatest adjustment.
- *It is difficult to achieve and maintain*: The road to lean production is long and difficult. For those companies that achieve a level of competency, consistent effort is needed to maintain that level of competency. Just as it is difficult for world-class athletes to maintain their physical conditioning, it is difficult for companies to stay lean. To continue with that metaphor, they can become *lax and fat*.
- *It is difficult to extend throughout the supply chain*: Attaining a lean condition in one company is difficult; achieving that condition throughout the supply chain is even more difficult, because there may be trade-offs that benefit one member of the supply chain while adversely affecting another. For example, reducing inventory at the retailer level may impose greater inventory levels upstream in the supply chain.

The potential benefits of a lean manufacturing world sound wonderful. So, how does a company get there? To oversimplify, a company needs to reduce variation in its products or services, its purchasing process, its production process, and its delivery process. These steps relate to the internal management of a business.

### Product

Many companies are attempting to increase revenues by expanding the variety of products or services they offer. While this approach may generate sales in the short term, or until a competitor offers the same products or services, it also increases complexity and variation in the product lines. This complexity inhibits the desire to go lean. To resolve this dilemma, a company needs to develop product or service modules that can be configured to suit the variety required. An example is kitchen cabinets that can be customized by assembling modules to fit the available space. A service example is a will prepared by a lawyer who adapts modules of features to prepare a customized will for specific clients.

### Purchasing Process

The traditional purchasing process has involved placing a variety of potential suppliers in a competitive bidding situation to obtain the lowest cost. This introduced variation in the suppliers used, quality levels, response times, and confidence levels in the suppliers. To reduce these variances, companies need to develop long-term relationships with carefully selected suppliers. Contemporary lean purchasing practices dwell on the total cost of ownership and consider both tangible and intangible factors in their decision making.

## Production Process

In production, variations can take the form of uneven workflow, smaller lot sizes, extra set-ups and changeovers, employee work assignments, more rush orders, and a host of other problems that makes the production of goods and services a demanding task. To reduce the adverse effects of variation, the production process must become more flexible with shorter setup times, more versatile employees, and general-purpose equipment—just to name a few. In addition, the planning horizon must be extended, with increased feedback from downstream customers, so that there is an increased opportunity to cope with the variation.

## Delivery Process

Increased variation means more frequent, but smaller, deliveries. The increased frequency of deliveries has increased the scheduling difficulties, especially in large distribution centers. It is no longer possible for suppliers to deliver at their convenience; they must deliver at the time specified by the customer. If the supplier is early, they wait. If they are late, they pay both financial penalties and perhaps face the loss of future business. The answer is a focused approach to their delivery process that is dedicated to eliminating potential problems before they occur and in designing an integrated delivery system that considers customer needs, delivery equipment, employees, external factors, and potential acts of nature such as hurricanes or snowstorms.

## Demand Variation

Companies also need to reduce the variation in demand. There may be greater variation in demand than in any of the internal areas described earlier. Reducing demand variation may sound counterintuitive at first. Who wants to tell the customer when to place an order or request a service? However, there may be situations where it is to the benefit of the customer if they change their ordering habits to obtain faster service or avoid long waits. Collaboration with customers can sometimes pay dividends in reducing demand variation. Demand management was described more fully in Chapter 5.

Why is it so difficult to achieve the lean status? The literature suggests there is a gap in understanding the transition and change process. Managers understand the need, but they do not always understand how to make the necessary changes. Even if they understand, it is difficult to anticipate the problems and challenges in making the implementation. They need an integrative model of the common drivers (barriers and facilitators) of the change process. In the following section, we will describe why achieving a lean and integrated supply chain is difficult.

## Moving from Functional Focus to Cross-Enterprise Collaboration

The PRTM Group describes the transformation from a functionally focused supply chain to a cross-enterprise collaboration arrangement. The stages are the following:

*Stage 1. Functional focus*: Functional departments within an organization focus on improving their own process steps and use of resources. Managers typically focus on their individual department's costs and functional performance. Processes that span across multiple functions or divisions are not well understood, resulting in limited effectiveness of complex supply chain processes.

*Stage 2. Internal integration*: Division or company-wide processes are now defined, allowing individual functions to understand their roles in complex supply chain processes. Cross-functional performance measures are clearly defined, and individual functions are held accountable for their contributions to overall operational performance. A well-defined demand/supply balancing process that combines forecasting and planning with sourcing and manufacturing is evident at this stage.

*Stage 3. External integration*: Stage 2 practices are now extended to the points of interface with customers and suppliers. The company has identified strategic customers and suppliers as well as the key information it needs from them in order to support its business processes. Joint service agreements and scorecard practices are used, and corrective actions are taken when performance falls below expectations.

*Stage 4. Cross-enterprise collaboration*: Customers and suppliers work to define a mutually beneficial strategy and set real-time performance targets. Information technology now automates the integration of the business processes across these enterprises in support of an explicit supply chain strategy (PRTM 2005).

An alternative approach is to describe the evolution of a supply chain along the following steps:

1. Enterprise integration (functional/process)
2. Corporate excellence (intraenterprise)
3. Partner collaboration (interenterprise)
4. Value chain collaboration (external)
5. Full network connectivity (total business system) (Poirier and Quinn 2004)

Another study proposed the following stages in the progression toward an integrated supply chain:

- *Ad hoc*: The supply chain and its practices are unstructured and ill defined.
- *Defined*: Basic SCM processes are defined and documented.
- *Linked*: This is the breakthrough level. Managers employ SCM with strategic intent and results.
- *Integrated*: The company, its vendors, and suppliers take cooperation to the process level.
- *Extended*: Collaboration between entities is routine with advanced SCM practices in place (Lockamy and McCormack 2004).

All of these models suggest that the movement to an integrated supply chain is an extended journey.

Another model based on research from Gartner shows the following progression:

1. *Market focused*: Organized by business unit, disconnected functional areas (silos), ad hoc coordination, and with an urgency culture.

2. *Cost focused*: Build and extend global logistics capabilities, leverage scale for cost reduction, and emphasize projects that reduce cost, not customer satisfaction or total cost to serve.

3. *Demand driven*: Focus shifts from costs and compliance to risk and outcomes. Cross-functional collaboration aligned to customer value.

4. *Value driven*: Creating value with trading partners through strategic alignment and collaboration. Outcome focused and integrated across value chain (Amber Road 2013).

## Comprehensive Supply Chain Model

Figure 13.4 is a diagrammatic representation of a model to describe the journey from a batch flow state to an integrated, agile supply chain. The general structure of the model is as follows:

The horizontal axis, reading from left to right, shows the progression over time, from batch flow to an integrated and agile supply chain. Although the progression occurs over time, the axis does not represent a specific timescale. The journey varies for each supply chain, but often, the time is in years, not months or weeks.

The vertical axis is composed of three major levels. The top level shows the stages that a company goes through: (1) present batch flow, (2) make the strategic decision to change, (3) develop a strategic direction for their program, (4) effect the internal changes needed, (5) establish linkages with key external entities, (6) attain an effective level of SCI, (7) enhance the supply chain with agility, and (8) enter the virtual environment state. The change agents needed to move from stage to stage are interspersed between the stages.

The middle area of the model shows the key change components, or *Cs*, needed to make the change successful. In addition to the Cs, the diagram shows the initial starting conditions of functional departments, legacy systems, and intermittent process flow. Further to the right, it shows the changes to a responsive organization, integrated information flow, and synchronous product flow. Even further to the right, it shows the beginnings of a supply chain and subsequent progression to integrated and agile supply chains.

At the bottom of the diagram, we show the action steps necessary, or the *Is* of the program. These are the activities necessary to bring about the Cs described earlier.

As an overview, the left third may be viewed as *starting the journey*, the middle portion may be seen as *achieving internal synchronization*, and the right third as *integrating the supply chain*. We believe the model shown in Figure 13.4 illustrates several salient points about the development of effective supply chains:

- Developing a supply chain is an evolutionary process. The movement from batch to a lean enterprise precedes the progression to a supply chain, then to an integrated supply chain, and finally, to a lean and agile supply chain.

- The progression along the model shows movement from discrete, even fragmented, functions of an entity to the integration of multiple entities into a seamless vibrant entity.

- The movement to integration requires the blending of interdisciplinary functions within a company and between companies.

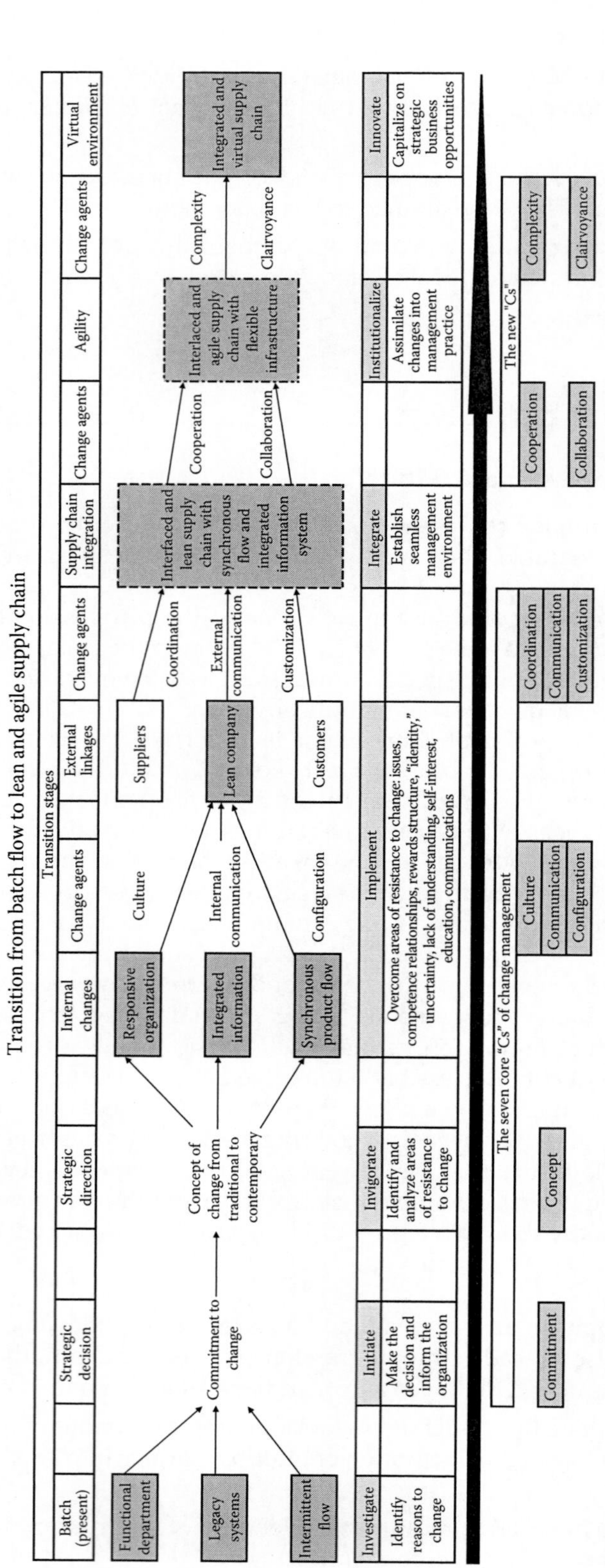

Transition from batch flow to lean and agile supply chain

| | | | | | Transition stages | | | | | | | |
|---|---|---|---|---|---|---|---|---|---|---|---|---|
| Batch (present) | Strategic decision | Strategic direction | Internal changes | Change agents | External linkages | Change agents | Supply chain integration | Change agents | Agility | Change agents | Virtual environment | |

Functional department →

Legacy systems → Commitment to change

Intermittent flow →

Commitment to change → Concept of change from traditional to contemporary

Responsive organization (Culture)

Integrated information (Internal communication)

Synchronous product flow (Configuration)

Suppliers

Lean company

Customers

Interfaced and lean supply chain with synchronous flow and integrated information system

Coordination — External communication — Customization

Interfaced and agile supply chain with flexible infrastructure

Cooperation — Complexity

Collaboration — Clairvoyance

Integrated and virtual supply chain

| Investigate | Initiate | Invigorate | Implement | | Integrate | Institutionalize | Innovate |
|---|---|---|---|---|---|---|---|
| Identify reasons to change | Make the decision and inform the organization | Identify and analyze areas of resistance to change | Overcome areas of resistance to change: issues, competence relationships, rewards structure, "identity," uncertainty, lack of understanding, self-interest, education, communications | | Establish seamless management environment | Assimilate changes into management practice | Capitalize on strategic business opportunities |

The seven core "Cs" of change management | The new "Cs"

Concept

Commitment

Culture
Communication
Configuration

Coordination
Communication
Customization

Cooperation
Collaboration

Complexity
Clairvoyance

**FIGURE 13.4**

Transition to an integrated supply chain.

- As the evolution progresses, there is a shift in emphasis from transactions to strategies. While attention to detail persists throughout, there is also a need to develop a holistic perspective of the program.
- Just as there is a need for interdisciplinary activity, there is also a requirement for interactive movement from connectivity to coordination to cooperation to collaboration.

## Decisions Needed to Achieve a Lean and Agile Supply Chain

It is easy to be enthusiastic about the possibilities of change, but it is difficult to reap the benefits of those changes. There are several important phases a company must go through to achieve success.

### Commitment

The first requirement is a commitment to change. A decision to change should come only after a careful investigation reveals the necessity to change. A commitment to change requires top management support. While this support does not necessarily require day-to-day participation in the details of the program, it does mean that top executives support the program and understand its importance in the company's future. Management should provide adequate funding for the project and commit to future resource allocations as needed. Top management should remember its role as the drivers of change in the structure of the organization and its practices. They monitor the progress of the program and recognize the achievement of milestones in the program's progress.

As part of the commitment to change, the entire organization should recognize and accept a role in the program. This realization means that employees must understand why the program is important and what their role is in its implementation. They should have a role in the design of the program to ensure their *buy-in* as a stakeholder in the process.

The commitment phase should involve enough of a plan to make sure the initial presentations about the objectives, scope, and potential consequences of the program are accurately and objectively portrayed to employees at all levels of the organization. The intentions of the company should also be shared with other key stakeholders, whether it is customers, suppliers, financial institutions, or major investors. Planning should mean that the company is thinking several steps ahead and not just responding to random events.

Once a company has made the commitment to change, they have to act. However, action without a plan is premature. At this point, planning should focus on developing the concept that will carry the program through to a successful conclusion.

### Concept

The next step is to develop a concept of the program—an understanding of what to do, what to accomplish, who will be responsible, and when it will be completed. The concept should be a *big picture* view of the program that provides a direction and scope for those who will participate in the design and implementation process.

Now is the time to involve the entire organization. This type of program needs the participation of the management team and nonmanagerial employees. Program managers should anticipate where they will find enthusiastic acceptance of the change plan and where they may encounter some resistance. While it is desirable to gain acceptance throughout the organization, it is essential to have support during the initial change stages.

It is also important to adapt the program to fit the company. While there are many elements of the program that may be generic to most companies, there are unique characteristics in most companies that will be positive factors if correctly incorporated into the improvement program. On the other hand, these same characteristics may become negative factors if ignored or disturbed.

As part of the program plan, project managers should decide what changes will be required in the structure of the company—the organization, policies, processes, and especially the effect on job contents and employee responsibilities and relationships. They can then decide how best to present these required changes to the rest of the organization.

As the planning progresses, it will become obvious that the program to adopt a lean and agile supply chain consists of a variety of activities and events. In other words, it is a project and needs a project planning approach. As with most improvement programs, the implementation efforts run parallel with the regular sustaining operations of the business. Unless there is a way of monitoring progress, the program will likely falter or even lose its identity as an improvement effort.

### Configuration

This phase involves the physical changes necessary in the marketing, purchasing, manufacturing, distribution, finance, and top management processes. While these changes may be relatively straightforward, they can cost significant amounts of money and often take substantial amounts of time. Many of the changes involve physical or process changes; however, in all cases, the employees involved have to change their operating paradigms. We will discuss this aspect later in the "Culture" section. In the following sections, we list a number of changes for each of the major functions of a business. The changes listed are representative; however, each company will be able to supplement them with their own requirements.

#### *Top Management*

Top management must become an active participant in building the supply chain. There are many decisions, both strategic and operational, that require their support. The supply chain basics for top management appear in the accompanying table. In many companies, the transition to agile and integrated supply chains is a new and complex transition; top management must lead the way.

The configuration change is one of the most visible changes that a company makes in moving toward supply chains. There are many articles written about it because it is often one of the most analyzed and debated areas of change. While it is a necessary step, it is not sufficient in itself. More is required. Many businesses work to make their chains faster or more cost effective, assuming that those steps are the keys to competitive advantage; however, companies must also build supply chains that are agile, adaptable, and aligned. Great companies create supply chains that adapt when markets or strategies change (Lee 2004). The basic responsibilities for top management are shown in Table 13.2.

#### *Marketing*

Marketing will assume greater responsibility for retaining customers. The current thinking, represented by customer relationship management (CRM) programs, is that it is more

**TABLE 13.2**

Supply Chain Basics for Top Management

Propose the structural organizational changes needed along with accompanying responsibilities
Consider the benefits and problems of moving into offshore outsourcing
Initiate contacts with other top managers to facilitate developing supply chain partners
Propose agreements on changes in performance measures and compensation incentives
Serve in the role as the visible motivator of change

**TABLE 13.3**

Supply Chain Basics for Marketing

Build databases to capture more data about existing customers
Analyze existing customers to determine why they buy the products and services they do
Develop programs designed to retain the customer
Obtain feedback from customers for use in design of products and services (sense and respond)
Participate in cross-functional teams to develop customer service programs

beneficial to a company to cultivate lifelong relationships with customers than it is to keep trying to find new customers. They will have to change from a *make and sell* paradigm to a *sense and respond* paradigm (Haeckel 1999). Supply chain basics for the marketing function are included in Table 13.3.

To accomplish these and other changes, the marketing department will need to develop a structure that is flexible and operates with employees who endorse and participate in this new way of thinking.

### *Purchasing*

Purchased goods and services represent a major portion of a company's expenditures. This cost category can range from 50% to 70% of sales in manufacturing industries and even higher in distribution and retailing industries. Consequently, the purchasing function will assume greater responsibilities in the future. The role of purchasing is changing from processing quotations from a wide variety of vendors, to find the lowest price, to building relationships with suppliers. The following are the supply chain basics for the purchasing function.

This transition will require a change in buyer attitudes and an upgrade in the skills required of the purchasing employees. Purchasing will become a professional-level job. Table 13.4 lists some basics for purchasing.

**TABLE 13.4**

Supply Chain Basics for Purchasing

Reduce the number of suppliers and expand the expectations for the performance of every supplier
Expand the areas of responsibility to include purchasing services as well as materials
Become a facilitator to help other functional areas to develop relationships with suppliers
Participate in a cross-functional team to develop effective outsourcing programs
Become involved in the strategic planning for the company

## *Manufacturing*

Manufacturing will need to continue to make the transition from a progressive bundle (batch) system to a modular (flow) system (Abernathy et al. 1999). A shift in focus from equipment and labor efficiencies to improving customer service will be needed. Table 13.5 shows the supply chain basics for the manufacturing function.

To facilitate the transition from internal efficiencies to improving customer service, the manufacturing function will need to work more closely with purchasing and marketing to coordinate customer orders in reducing response times.

## *Distribution*

The distribution function, which includes the movement of goods from the manufacturer to the distributor, then on to the retailer, and from there to the consumer, will also be increasingly concerned with the speed and timing of deliveries. Businesses will be especially concerned with having the right goods and services available to the customer. At the same time, they want to reduce inventories, especially excess and slow-moving inventories. The supply chain basics for the distribution function are shown in Table 13.6.

Distribution costs have become a significant portion of the total costs of products and services. The distribution function is also becoming a major factor in customer service considerations. It will become an even more important part of the supply chain in the future as the timing and traceability of deliveries become more critical.

## *Finance and Accounting*

The financial function, including accounting, has important roles in improving flow in the supply chain. These activities provide the funds to make the physical and structural changes and monitor the results to assure effective utilization of the organization's resources. Table 13.7 shows the basics for finance and accounting.

Finance and accounting must become participants in the company's efforts to build their supply chains, not just perform the functions of interested bystanders.

**TABLE 13.5**

Supply Chain Basics for Manufacturing

| |
|---|
| Rearrange plant layouts to improve the flow of materials |
| Increase worker versatility and empowerment |
| Increase equipment flexibility, such as reducing setup times |
| Reduce process variation in output and quality |
| Change production performance measures to reflect the new environment |

**TABLE 13.6**

Supply Chain Basics for the Distribution Function

| |
|---|
| Retailers and e-commerce sellers will share POS information with their suppliers |
| Increase its capabilities to customize products and services |
| Use a greater variety of transportation methods—omnichannel movement of goods |
| Continue to optimize costs, especially concerning offshore outsourcing |
| Will be increasingly outsourced to private carriers such as UPS, FedEx, and DHL |

**TABLE 13.7**

Supply Chain Basics for the Finance and Accounting Functions

Help develop a total cost of ownership analysis for use in evaluating offshore outsourcing

Modify reporting systems to provide more meaningful information to operating functions

Change the performance measures to evaluate supply chain outcomes, not just local outcomes

Work with supply chain partners to achieve an equitable distribution of costs and profits

Balance their efforts between external reporting and internal reporting requirements

## *Communication*

Informal and formal networks make it possible to communicate between entities, both within the company and with supply chain partners. It is technology driven, but people dependent; is fraught with lack of incompatible or local standards; involves various modes, telephone, fax, e-mail, intranet, electronic data interchange (EDI), and the Internet, with all of the possible combinations, and includes social media sites; and, above all, is continually evolving.

This phase involves designing and implementing the data exchange systems needed to achieve the instantaneous transfer of data, both within the company and among the entities involved in the supply chain. This network can be expensive and time consuming; in addition, management must be willing to exchange confidential information.

It is insufficient to make the physical changes in the supply chain without upgrades in the communications systems, both within the company and between the company and its suppliers and customers. This transition may require new technology in the form of hardware and software to enable rapid and reliable communication between employees in a department, between departments within a company, and between companies.

The implementation of new systems can be both expensive and time consuming. In addition to the technology, the employees who will use these new systems require time to accept and learn how to effectively utilize the new technology. A failure to get the employees successfully involved can often lead to poor results or even failure of the entire effort.

As difficult as it may be to implement the communications system within a company, it is even more difficult to implement intercompany communications. While communications technology is readily available in the form of EDI or the Internet, the reluctance of companies to adapt between companies often prevents a high level of success. Interorganizational systems (IOSs) are not a dream. They exist, although not all company relationships have reached the collaboration stage. That will come as the technology, infrastructures, and relationships of companies come into alignment. Information sharing among supply chain participants is essential to effective SCI.

## *Culture*

Companies often ignore or deemphasize the need to match the changes necessary with the existing culture of the organization. However, it is often the key to successful implementation and, if ignored, may limit success. It may not cost a lot of money to achieve; however, it requires an understanding of human relations and a willingness, and ability, to effect major changes in the thinking of individuals and the actions of teams.

Making the configuration and communication changes requires a high level of acceptance on the part of employees. In fact, the change from batch to lean requires a major cultural shift in the company. For many in the organization, it may signal the end of an era. No more mass production or local optimization thinking; it is time to shift paradigms from the traditional to the contemporary.

The program to change or adapt the culture of the organization to the program of building supply chains often lacks a systematic approach that can be performed in a series of sequential steps. How best to blend or adapt the program to the culture or modify the culture to the program takes insight that involves an understanding of human nature. Failing to account for the impact changes have on the employees can stall or even derail the most desirable of changes.

When the *Harvard Business Review* convened a panel of leading thinkers in the field of SCM, technology was not the top concern. People and relationships were the dominant issues of the day. The opportunities and problems created by globalization, for example, are requiring companies to establish relationships with new types of suppliers. The ever-present pressure for speed and cost containment is making it even more important to break down stubbornly high internal barriers and establish more effective cross-functional relationships. The costs of failure have never been higher. The leading supply chain performers are applying new technology, new innovations, and process thinking to far greater advantage than the companies that are lagging. This round table gathered many of the leading thinkers and doers in the field of SCM. Together, they took a wide-ranging view of such topics as developing talent, the role of the chief executive, and the latest technologies, exploring both the tactical and the strategic in the current state of SCM (Beth et al. 2003). Cultural differences make a difference in the *fit* among supply chain members. Although trade legislation may influence supply chain combinations, they must consider cultural differences as well (Kidd et al. 2003).

When successfully implemented, the changes described earlier will help a company approach a lean production status. However, is a lean company enough? Or will it be necessary to move to a lean and integrated supply chain?

## Customization

Mass customization is an extension of agile manufacturing. Lean is being able to produce customized goods or services at optimal efficiency. Agile is being able to move from one product or service to another within the known product lines without noticeable interruption. Mass customization is the ability to make any of the known products at a high volume and to customize that product to a customer's individual specification, at a cost comparable to that of making a standard product.

Is the market-of-one imminent? Can businesses provide a unique product or service to each customer? While it appears unlikely, a number of companies are working toward that objective. Wherever possible, companies are using modular components to facilitate postponement of the customizing process. They are moving from the make-to-stock mode to the assemble-to-order mode. In time, they may be able to move back even further to the make-to-stock mode if they can reach an acceptable compromise between the level of customization and response time.

Another interim step is to provide greater variety when it is not possible to use the postponement strategy, such as in grocery or small tool items. Greater variety can be used to appeal to smaller market niches, but it is unlikely to approach the market of one. One of the negatives about greater variety is that it forces consumers to make more decisions about which item they want, a decision-making process that may add stress to their lives (Schwartz 2004). The answer lies in combining product modularity, process flexibility, customer relationships, and employee versatility into a dynamic system. All of these features of a business are necessary.

## Integrated Supply Chain

Everyone talks about an integrated supply chain but how do you get there? In most cases, the transition passes through three distinct and often overlapping phases: interfacing, interlacing, and integrating. During the interfacing phase, two separate entities begin to pass information back and forth through dissimilar information processing systems. The movement of goods improves but everyone realizes it could be better. The next move is interlacing, where changes connect more closely and direct the flow of information between the two entities. In the integrating phase, a level of connection is achieved where information flow is efficient and effective for both entities.

Few companies have reached the final phase of complete integration. To achieve even the beginning levels of SCI requires extending communications into cooperation and collaboration. Increasing competition stimulates independent firms to collaborate in a supply chain that allows them to gain mutual benefits. This collaboration requires the collective knowledge of the coordination mode, including the ability to synchronize interdependent processes, to integrate information systems, and to cope with distributed learning.

A major study by Fawcett et al. (2008) discovered a number of benefits, barriers, and bridges to effective (integrated) SCM. The primary benefits were *customer focused*, including increased customer responsiveness, more consistent on-time delivery, customer satisfaction, and shorter order fulfillment lead times, and *company focused*, including reduced purchasing costs, better asset utilization, ability to handle unexpected events, reduced inventory costs, increased firm productivity, and reduced overall product cost.

The primary barriers resulted from *interfirm* rivalry, which includes inadequate information sharing, inconsistent operating goals, lack of willingness to share risks and rewards, and lack of willingness to share information, and *managerial complexity*, which includes lack of alliance guidelines, processes poorly appraised in terms of costs, nonaligned measures, organizational boundaries, inability to measure supply chain contribution, and inability to measure customer demand.

The bridges, or actions to overcome the barriers, involved *operations, process, and supply management*, including accurate comprehensive performance measures, supplier alignment, effective use of pilot projects, and process documentation and ownership, and *people management*, including managerial and employee support, open information sharing, trust-based alliances, cross-trained experienced managers, supply chain education and training, and using chain advisory councils. The study stressed the need for going beyond technology and considering the importance of people.

The study concluded with the following key points:

- Although cost reduction is a prime motivator to strategic supply chain collaboration, customer satisfaction and service are perceived as more enduring by managers.

- All managers recognize technology, information, and measurement systems as major barriers to successful SC collaboration. However, the people issues—such as culture, trust, aversion to change, and willingness to collaborate—are more intractable.

- The people are the key to successful collaborative innovation. Companies continue to invest in technology, information, and measurement systems. However, managers must not overlook the training, educating, and bringing together of right people to use those systems and to interact with one another (Fawcett et al. 2008, p. 45).

## Coordination

Coordination means working together to achieve a mutual objective. It suggests that companies begin to understand what their suppliers or customers want and need and what each participant should do to help achieve these objectives. It can be as simple as sending an order acknowledgment to the customer or as involved as meeting a 1-hour delivery window imposed by the customer.

Coordination provides the infrastructure and opportunities to make it possible to work together. It includes working in an *open* system environment and linking entities together, both inside and outside the company. It involves policies, procedures, contracts, and other means necessary to work together and is among the early attempts to align the supply chain.

## Cooperation

To cooperate means to work or act together toward a common goal. This suggests that companies do more than coordinate actions—they do it with an outcome in mind that will benefit both parties.

A number of programs have been used to encourage intercompany cooperation. They include quick response (QR), efficient consumer response (ECR), vendor-managed inventory (VMI), and collaborative planning, forecasting, and replenishment (CPFR). These programs are described in Managerial Comment 13.1 that describes a series of programs that illustrate the value of cooperation. These were designed to enable manufacturers and retailers to better match the supply from manufacturers with the demand from retailers.

## Collaboration

Collaboration implies a greater level of interconnectivity between customer and supplier. Collaboration involves freely sharing information and working together to develop a more meaningful plan for the flow of both information and goods through the supply chain. Several programs enhance collaboration, most recently CPFR, a program that attempts to integrate the entire stock replenishment process.

Collaboration includes the motivation to work together effectively for the welfare of the entire supply chain. It builds on the coordination infrastructure and requires less tangible qualities such as trust. See Managerial Comment 13.2. It maximizes the benefits to the supply chain, not the individual entity, and provides the ultimate service to the customer.

While coordination, cooperation, and collaboration are necessary, they are difficult to achieve. Internal company objectives may inhibit the attainment of supply chain objectives. The barriers include economic, technology, and cultural matters. While collaboration is seemingly obvious, it requires work, including overcoming barriers such as big egos, apathy, and turf mentality (Siegfried 2013). Some things change slowly, especially those that may have an unknown effect on the jobs of employees at all levels of the organization.

Once the framework of a lean supply chain has been established, companies must move from a program mode to an institutionalized version, in which the key features of the program become part of the normal management practice. Specific programs can be sustained for a limited period; however, at some point, the main elements of the program must be assimilated into the management practice of the company or become lost as the program declines into ineffectiveness or is terminated by management.

## MANAGERIAL COMMENT 13.1  BEATING IMPOSSIBLE DEADLINES

Global competition is forcing companies to improve—or go out of business! Successful companies continue to search for some competitive advantage, whether it is in price, quality, flexibility, or, more recently, in response times. The drive to reduce the time it takes to get an order from the supplier to the customer and to get new products to the market is being pursued just as relentlessly as costs and quality have been. It is no longer good enough to be low cost and high quality; you must be responsive and flexible, too.

## PROGRAMS

The following programs have a primary goal of reducing the lead times from the supplier to the customer. The primary thrust is to clean the excess inventories out of the supply chain so the needed inventory can flow smoothly and quickly. In so doing, they provide other benefits such as reduced costs, improved quality, and greater flexibility in responding to changes in volume and mix.

### Quick Response Systems

In the early 1980s, a group of manufacturers and retailers in the textile and apparel industries hired Kurt Salmon Associates to recommend a program to reduce the number of stockouts at the retailers by providing a closer matching of demand and supply. The program was to provide a way to reduce lead times for stock replenishment orders from manufacturer to retailer and to reduce the lead times for introducing new products. There were a number of components to the program, but the essential ingredients were a willingness, and a capability, to more quickly communicate the actual demand data from the retailer to the supplier so that the supplier could have more time to prepare for orders from their downstream customers. Another goal was to reduce the bullwhip effect by placing orders on a more regular and predictable schedule.

The potential was high; how was it received? A participant in the development of the program reported "in the 10 years since its formulation, quick response has made only limited progress despite its well-demonstrated benefits to the apparel industry" (Hunter and Valentino 1995). Hunter listed the expected benefits as reductions in pipeline inventories; greater probability of garment designs and colors being acceptable to the consumer; ability to reestimate stock keeping unit (SKU) demand at retail, thus reducing stockouts and markdowns; and greater competitiveness for domestic producers facing increased levels of imports.

Some of the problems that delayed the widespread adoption of QRS included the following:

- Naivety: Participants did not realize the magnitude of the task.
- Difficulty in creating *partnerships*: The retailers received the benefits, while the suppliers incurred the costs.
- Structural issues: These include the staggering number of unique SKUs (1.2–1.4 million at a department store every 4 months), decreasing shelf life of fashion items, and retailers dominating the apparel manufacturers.

- Technical problems: These include inadequate accuracy of bar codes, inadequate storage and manipulation of inventory and sales data, and lack of standards in information transmission (EDI).

The factors necessary for the growth of QR are UPC/EAN compliance and standardization, clarification and acceptance of the role of VANs, and the need to find a way to extend EDI to smaller manufacturers.

Despite its slow beginning, QRS programs provided a model for other industries to follow and for the subsequent programs to build upon.

### Continuous Replenishment Programs

Closely allied with QRS, CRP programs were designed to encourage automatic replenishment ordering so customers could automatically place an order when their inventory management system indicated a need for a reorder. Some companies were able to add this feature to their inventory management programs (Lummus and Vokurka 1999).

### Efficient Consumer Response

Encouraged by the positive results from QRS programs, in 1992, several grocery executives formed a voluntary group, known as the Efficient Consumer Response Working Group, and commissioned a study by Kurt Salmon Associates to identify opportunities for more efficient, improved practices in the grocery industry. The consultants returned in early 1993, claiming that the industry could reduce inventory costs by 10% or $30 billion (Frankel et al. 2002). In addition to efficient replenishment, this group added the requirement of category management, consisting of efficient new product introduction, efficient store assortment, and efficient promotion. The program included collection of demand (sales) data with point-of-sale (POS) terminals and the feedback of these data upstream to suppliers with EDI or some other means of electronic communication. Suppliers could then avail themselves of a variety of techniques, such as cross docking, to move the product more quickly to the customer.

### Vendor-Managed Inventory

Some retailers decided that, inasmuch as the suppliers had the consumer demand data, they (suppliers) could assume the responsibility of managing their (retailer) inventory. While the idea of suppliers managing a retailer's inventory was not new—rack jobbers and service merchandisers did it in the health and beauty aid categories years before—it did have the added element of rapid feedback of demand information. One study reviewed the information technology challenges, especially the effects of information delay and accuracy (Angulo et al. 2004). Another study examined the effect of increasing demand visibility and concluded even products with stable demand, a partial improvement of demand visibility can improve production and inventory control efficiency. However, the value of visibility depends on the target products' replenishment frequencies and the production planning cycle employed by the manufacturer (Smaros et al. 2003).

### Collaborative Planning, Forecasting, and Replenishment

In 1997, the Voluntary Interindustry Commerce Standards (VICS) created a sub-committee to develop CPFR as an industry standard. One year later, in 1998, VICS issued the first document on CPFR: *VICS CPFR Guidelines*, which has been constantly updated since then (VICS 2004). While QRS and ECR provided the flow of demand information from the retailer upstream to suppliers, it was the responsibility of the supplier to anticipate demand and the retailer (except in VMI) to actually carry out the ordering. CPFR attempted to eliminate this disconnect by advocating that both the customer and the supplier would collaborate to plan a joint demand forecast and replenishment schedule.

Barratt and Oliveira provided a comprehensive summary of various CPFR initiatives and listed several issues that they believe were addressed for the first time with CPFR, including

- The influence of promotions in the creation of the sales forecast (and its influence on inventory management policy)
- The influence of changing demand patterns in the creation of the sales forecast (and its influence on inventory management policy)
- The common practice of holding high inventory levels to guarantee product availability on the shelves
- The lack of coordination between the store, the purchasing process, and logistics planning for retailers
- The lack of general synchronization (or coordination) in the manufacturer's functional departments (sales/commercial, distribution, and production planning)
- The multiple forecasts developed within the same company (marketing, financing, purchasing, and logistics)

They also provided an excellent summary of benefits, barriers to implementation, and enablers of the CPFR process and concluded that trust and good information technology must be present for success (Barratt and Oliverira 2001).

### PRESENT STATUS

How are companies doing in their quest to reduce response times? A simplistic answer is that the programs work but that few companies are getting the full benefit of these programs because of incomplete implementations. There are three key components to any program of this type: technology, infrastructure, and change management. The technology, primarily in the form of electronic data collection, with POS terminals, and data transmission, with EDI or the Internet, is adequate and getting better. The infrastructure—organization, systems, and functional relationships—is inconsistent among companies and even industries, but getting better. Managing change is a continuing problem, particularly in the area of company culture and trust. How much do you trust? Which customers/suppliers do you trust? Almost every study finds trust is not only essential for success but often lacking in many less-than-satisfactory programs.

There are several reasons why demand collaboration programs have not realized their potential:

- The pace of adopting new ways of doing business is slow.
- Demand information supplied by customers is not put to use in an integrated manner by these customers.
- Demand management and supply management processes are not integrated, and sales and operations planning is not utilized to synchronize demand and supply.
- There is a lack of trust among trading partners to share pertinent information and collaborate on decision making.
- There is a desire to partner but not commit to executing the communicated plans.
- A common view exists that demand collaboration is a technology solution and that the current technology is too complex (Crum and Palmatier 2004).

Another study (Harrington 2003) reports that, while most companies think of CPFR as primarily a technology, it is the business process supported by the internal culture that makes CPFR successful. Computer Sciences Corp., in conjunction with *Supply Chain Management Review,* recently concluded its second annual Global Survey of Supply Chain Progress. The study's aim was to gauge how successfully practitioners are advancing their SCM capabilities. The findings suggest that some companies understand the advantage of leveraging the buy across a more strategic supply base, while others are content to pursue more limited, tactical improvements. Despite the progress needed in some areas, most companies pursuing the various supply chain initiatives are generally happy with the results. What was surprising (and a bit disappointing) was the relatively low ranking of customer satisfaction as a factor in an initiative's success. Only 28% said that an awareness of the need to increase customer satisfaction ratings was a main factor in the success of their initiatives (Poirier and Quinn 2004).

One seemingly simple problem is the need for consistency in product data and its transmission. UCCnet, a nonprofit unit of the Uniform Code Council standards organization, established a global online registry that requires product data with as many as 151 attributes or descriptors, about 40 of which are mandatory. One well-known company found that it was transmitting information by phone, e-mail, fax, CDs, EDI, PDFs, spreadsheets, websites, and printed price pages. The same survey found that the percentage of companies that felt they have the business processes in place to take full advantage of real-time information varied from 20% in the construction and engineering industry to slightly over 70% in the logistics and transportation industry, with the overall average of all industries around 45% (Sullivan and Bacheldor 2004).

Another nagging question is the relative benefits between retailers and suppliers. Do collaborative relationships with large retailers benefit the supplier? Researchers found that, while suppliers benefit in the economic sense and in capability learning, their perception is that there is an inequity in the sharing of benefits and burdens (suppliers bear more of the burden and receive less of the benefits they deserve) (Corsten and Kumar 2005).

Some activities are increasing the response times, most notably the offshore outsourcing movement. An *Industry Week* study observed that the route to cheaper supplies from overseas sources may look like a clear path, but supplier lead times are elongating, a step back from the improvements wrought from years of lean implementations. "Some companies may find the payoff worth it. Others may find themselves with cheaper raw materials and components, but fewer customers" (Vinas 2005). It just takes longer to get products from another country. The costs may be lower but the response time is higher and the uncertainty of supply is higher. Balancing the trade-offs, such as between supplier responsiveness and order stability, requires a great deal of skill (Hutchison 2006).

## FUTURE

What does the future hold? No doubt, the pressure to clean out the excess inventories in supply chains and to gain a better matching of supply with demand will continue. Individual programs, such as those described earlier, are losing their focus, as they become a part of more general programs such as SCM and integrated demand-driven collaboration systems.

Whatever the program is called, companies must integrate their communication systems and develop sufficient trust with one another to collaborate effectively and to gain the benefits of their efforts. If they do, they can hope to succeed. If they do not, they face an uncertain future.

*Source:* Adapted from Crandall, R.E., *APICS Mag.*, 16(6), 20, 2006. For the latest version of the CPFR model, see Andraski 2014.

---

### MANAGERIAL COMMENT 13.2   TRUST

Robert Handfield and Christian Bechtel report that one of the most misunderstood and ripe areas for research in SCM is trust. Trust (and its cousin, collaboration) seems to be the single most-discussed element in making supply chains function effectively and efficiently. They found that four bodies of theory (transaction cost economics, organizational design, relational theory, and network theory) support the premise that effective communication of requirements with appropriate safeguards and approaches is critical to effective customer and supplier relationship management.

They conclude that, while trust is generally considered critical to successful SCM, it is one of the most difficult attributes to measure. There are eight different conceptual paradigms of trust: reliability, competence, goodwill (openness), goodwill (benevolence), vulnerability, loyalty, multiple forms of trust, combining trust with vulnerability, and nonpartisan proactive-based trust. When it is all said and done, integrity in action and thinking seems to be the foundations of trust between companies. Although much of the research to date has focused on analytical approaches to managing supply chains, the area that requires the greatest work is managing supply chain relationships.

*Source:*   Adapted from Handfield, R.B. and Bechtel, C., *Int. J. Integr. Supply Manage.*, 1(1), 3, 2004.

## Steps in the Change Process

While change is often necessary and desirable, it is rarely easy, especially when viewed in relation to implementing integrated supply chains. Some of the action steps to be considered in integrating supply chains will be described in the following section. In Figure 13.4, along the horizontal axis, we listed a number of action steps necessary to achieve an integrated supply chain.

### Investigate

While it is easy to be carried away with enthusiasm about the possibilities of integrating the supply chain, the first need is to investigate why the change would be beneficial. A company should go through the process of deciding what they want to accomplish from the change. If they can satisfy themselves that the transition to an integrated supply chain will provide the results they want, then they can truly make a commitment to go through with the change.

### Involve

Top management support and involvement is necessary for a successful implementation of an integrated supply chain. While this may appear obvious, it does not always occur. Top management must not only support the effort with financial backing but also demonstrate their visible endorsement and encouragement.

### Include

A program of this magnitude requires the commitment of key employees. It is truly a cross-functional effort and should be recognized as such by appointing representatives from the appropriate functions of the business to participate on the implementation team. While not all members of the team will be required full time, some will. All members of the team should be available when needed to keep the implementation moving according to plan.

### Initiate

Once management has made the commitment to change, they have to initiate action. But action without a plan is premature. At this point, the team needs to develop a step-by-step plan that will carry the program through to a successful conclusion.

### Invigorate

This is the phase in which to get the rest of the organization involved. Participation of both the management team and nonmanagerial employees is required. At this stage, management should anticipate where it will find enthusiastic acceptance of the change plan and where it may encounter resistance. While acceptance throughout the organization is necessary, it is critical to have support when the program is in the beginning stages.

### Implement

The actual implementation stage should follow three main areas. First, the configuration of the physical elements of the production or service processes needs to be addressed.

Second, communication within the organization and with suppliers and customers needs to be established in which the details of the program are outlined. Finally, the organizational culture of the firm must be changed. Of course, it is this third area that may take the longest time to complete, as many employees may resist changes in the workplace.

### Integrate

Integration comes slowly. It begins with communications and gradually progresses through coordination, cooperation, and collaboration. One of the primary requirements is trust among supply chain partners, and trust comes from a series of successful experiences in working together. It begins at the individual level and gradually builds to a collective trust relationship between entities. Integration comes as much from human relationships as from technology and structure changes.

### Institutionalize

Specific programs can only be sustained for a limited period; at some point, the main elements of the program must be assimilated into the management practice of the company, or they are lost as the program declines into ineffectiveness or are terminated by the management. The goal is to make the program part of the normal operating life of the organization. In other words, as the novelty of the new supply chain configuration gives way to familiarity and effectiveness, it gradually becomes institutionalized into the business.

### Innovate

Once a company builds a facsimile of an integrated supply chain, it is time to change it. The business world is constantly changing and there is a need to adjust the supply chains to keep pace.

---

## A Look Ahead

Suppose lean production is achieved and then extended to a lean supply chain. Is that the final answer, or should the company move on to agile manufacturing?

Agile manufacturing is first lean and then agile. It is an extension from lean manufacturing and viewed by some as a competitive asset for the future. The three interrelated support areas for agile manufacturing are technology, organization, and people. These areas were described more fully in earlier chapters. The following quotes point out the need for collaboration among supply chain members:

- "You can be lean by yourself, but you cannot be agile by yourself" (Sheridan 1993).
- Virtual companies are "the epitome of the agile enterprise"—a vehicle to tap the knowledge of participating companies (Sheridan 1993).
- "In agile manufacturing, the aim is to combine the organization, people, and technology into an integrated and coordinated whole" (Kidd 1994).

Is agile manufacturing the final answer? No, there may be a need to move even further, toward mass customization and complexity management.

Up to now, we have described actions that are difficult, but reasonably straightforward. In the next two phases, we believe that companies will have to demonstrate significant innovation skills.

## Complexity

Managing the complexity of a business in today's global competitive environment is in itself complex. It is a test for management even greater than achieving mass customization. It goes beyond customizing a known product for an individual customer. It involves confronting the internal and external issues that may exist while, at the same time, making the product at a competitive cost and lead time. Operating in a complex, or chaotic, environment tests the capabilities of any company.

In Chapter 3, we described open systems and the need for organizations to deal with the complexities of the forces that are external to the organization's control. These forces, especially those resulting from global competition, require that businesses develop global strategies affecting product design (products must be designed to fit the market they serve), processes (to reflect the best blend of resource costs, work rules, and local preferences), and locations (how to blend a myriad of decision criteria in the face of potentially dramatic change).

Complexity is difficult and, even worse, introduces some paradoxes that increase the confusion. Some major paradoxes of complexity include the following:

- *The optimization paradox*: Despite the potentially huge economies from designing supply chains from a global view, most manufacturers optimize locally.
- *The customer collaboration paradox*: Despite the need to be much more responsive to customers, few manufacturers are collaborating closely with them.
- *The innovation paradox*: Product innovation is continuing to accelerate, yet few manufacturers are preparing their supply chains for faster new product introduction.
- *The flexibility paradox*: Flexibility is a key priority, but it is being sacrificed in the drive to cut unit cost.
- *The risk paradox*: Keeping supply chain quality high is critical, yet manufacturers' risk of supply chain failures keeps growing (Deloitte 2003).

Traditional decision making relied on linear thinking—the future will be largely like the past. Today, however, many businesses are finding that using the past as a basis for planning the future may not be very useful; in fact, it may be downright misleading. But how can organizations apply nonlinear thinking? It takes a completely new theory, a theory appropriately called complexity theory. Frederick describes complexity theory as a variant of chaos theory that explains the organizational and evolutionary dynamics that occur as complex living systems interact with each other and their environments. He suggests that both the corporation and community are natural systems interacting and coevolving in response to environmental factors. The corporation is hypothesized as a complex adaptive system that must adapt itself to the environment in which it operates (Frederick 1998).

Dealing with complexity will be a necessary core competency for any successful organization. Frederick suggests that a complexity theorist trying to understand corporate behavior might ask the following question: "Is the corporation a self-organized CAS housing an autocatalytic component, operating on a fitness landscape, and exposed to the risk of chaotic change while being held in its niche by a strange attractor?" (Frederick 1998)

Many companies may decide complexity, or its variant, chaos theory, as too *far-out* to be seriously considered. However, a few companies are already trying to decide how to incorporate this theory in their thinking. A full explanation is beyond the scope of this book; we can only suggest that complexity theory will be an important consideration for the future.

## Clairvoyance

Clairvoyance is the power to perceive things that are out of the natural range of human senses. Few articles or books combine clairvoyance and business, presumably because businesses align with rationality and logic rather than with a discipline that has been associated with frauds and horoscopes. Up until now, businesses have been dealing with the known world, even though it may be complex and unexpected. In the future, they will have to deal with greater uncertainty. The present forecasting techniques will be increasingly inadequate. Present methods rely heavily on extension, usually linear, from present known positions. At some point, organizations will have to build bridges between disconnected islands of products or services. In other words, they will have to be clairvoyant. Organizations will need people who have the *gift* to foretell the future, and that is a skill not taught in business schools.

Perhaps, it would be more palatable to say that businesses need people who possess the wisdom to assimilate fragments of leading indicator information into a holistic and reasonably accurate view of the future. Advocates of *big data* and data analytics claim that emerging techniques can help in detecting trends that serve as a predictor of future events.

Wisdom is the distillation and organization of knowledge in such a way that it has relevance to the decision at hand. Some individuals seem to possess an uncanny ability to make correct decisions, while others struggle in a hopeless maze of contradictory information. For the moment, let us call that ability clairvoyance. It raises several questions. What does it take to see into the future? Is it a skill that is inherited—a sixth sense—or can it be learned? If so, how? Is clairvoyance something like *thinking outside the box*? Does it result from empowered cross-functional teams? Can it be created through the use of biotechnology? Can it be directed and controlled or is it the result of random insights? We touched on knowledge management in earlier chapters and will address it again in Chapter 18.

Just as with chaos and complexity theory, a detailed discussion on clairvoyance is beyond the scope of this book. We can suggest books that we believe make what we have called clairvoyance a bit more tangible and realistic. In *Blink*, Gladwell (2005) describes how some people have the knack of making decisions without the need for exhaustive analyses; they see the big picture and just intuitively know! In *A Whole New Mind*, Pink (2005) describes the emergence of *right-brain* thinking as coming of age in the world. Earlier, Mintzberg (1989) also discussed left and right brain implications (see Managerial Comment 13.3). These ideas represent significantly different ways of making decisions from those approaches that have been practiced for a long time in the business world. While practitioners of these techniques may not be clairvoyants, they will be advanced beyond most of today's managers.

**MANAGERIAL COMMENT 13.3 IMPLICATIONS FOR THE LEFT HEMISPHERE**

**LEFT-BRAIN THINKING**

First, I would not like to suggest that planners and management scientists pack up their bags of techniques and leave organizations, or that they take up basket weaving or meditation in their spare time. (I haven't—at least not yet!) It seems to me that the left hemisphere is alive and well; the analytic community is firmly established, and indispensable, at the operating and middle levels of most organizations. Its real problems occur at the senior levels. Here, analysis must coexist with—perhaps even take its lead form—intuition, a fact that many analysts and planners have been slow to accept. To my mind, organizational effectiveness does not lie in that narrow-minded concept called *rationality*; it lies in a blend of clearheaded logic and powerful intuition.

A major thrust of development in our organizations, ever since Frederick Taylor began experimenting in factories late in the last century, has been to shift activities out of the realm of intuition toward conscious analysis. That trend will continue. But managers, and those who work with them, need to be careful to distinguish that which is best handled analytically from that which must remain in the realm of intuition. That is where we shall have to continue looking for the lost keys to management.

*Source:* Adapted from Mintzberg, H., *Mintzberg on Management, Inside Our Strange World of Organizations*, The Free Press, New York, 1989.

## Summary

In Chapter 11, we described the reasons for considering the achievement of integration in supply chains as important. In Chapter 12, we identified obstacles that make SCI difficult. In this chapter, we outlined steps to take to achieve integration in supply chains.

In the "Introduction," we outlined the need for integration and listed some of the steps necessary to achieve a successful integration. We then described the differences between the traditional vertically integrated model that some companies used in the past and the integrated supply chains that are the objectives of many well-managed companies today.

Lean production is currently one of the most widely discussed ways in which companies hope to achieve improved internal performance and external customer service. We described changes an organization must take to achieve lean status.

We next outlined steps leading to SCI and proposed a model that shows the progression from a *batch and queue* mode to integration. Following that, we expanded on the specific decision sequence necessary to achieve integrated supply chains—the Cs of the process.

The following section described the *I* part of the process—the actions needed to integrate. Finally, we concluded with a look ahead into the exciting and challenging, but somewhat uncertain, future for the further extension and evolution of supply chains.

The difficulty of successfully integrating supply chains is described in the "Hot Topic" section for this chapter. As in Chapters 11 and 12, it is about AECL's cancer-fighting machine.

---

**Hot Topics in the Supply Chain: Atomic Energy Canada Limited
Encounters Problems with its Cancer-Fighting Machine (Part 3)**

---

The Atomic Energy Canada Limited (AECL) case was introduced in Chapter 11, discussed in detail in Chapter 12, and concludes in this chapter. Recall that their machine, the Therac-25, had been involved in six cases of overdose accidents involving patients.

## SUPPLY CHAIN DURING A CRISIS

The Therac-25 incidents are an example of an organizational crisis. Pearson and Clair (1998, p. 60) offer the most widely referenced definition of a crisis: "An organizational crisis is a low-probability, high-impact event that threatens the viability of the organization ..." Organizational crises are an important reality for executives and supply chain managers. Although such events are statistically rare, their occurrence can jeopardize a business. In addition, Coombs (2007) reminds us that organizational crises can affect the company's stakeholders as well. With the Therac-25, those stakeholders included the medical centers using the device, as well as the patients and their families.

## INVESTIGATING THE CAUSE OF THE ACCIDENTS

Finding the root cause of the radiation overdoses took some time as the operators of the Therac-25 and AECL were not approaching the problem from the same perspective. AECL's position was that an accidental overdose was simply not possible due to a fault with the machine; instead, the problem may have been electrical in nature (Leveson 1995). However, hospital physicists at several different medical centers began communications with each other and discovered a computer software glitch might be the problem. Software errors can occur in complicated programs and after extensive investigation, AECL, working with physicists at the hospitals, confirmed that the overdose problem was indeed caused by a software glitch.

The software for the Therac-25 contained code from its previous models, the Therac-20 and the Therac-6. However, the safety of the patient from an accidental overdose was in the control of the technician in these earlier models. The Therac-25, on the other hand, was designed to have software controlling the safety of the patient, and hence, it was this software that was supposed to prevent the occurrence of an accidental overdose. Faulty code from the Therac-20 and Therac-6 models was carried over into the code for the Therac-25 (Leveson 1995). The software code, and not an electrical problem, was at the heart of the accidental overdoses.

## REALITY OF SOFTWARE BUGS

Software is embedded in many products, including motor vehicles, aircraft, and, as this case illustrates, certain types of medical treatment devices. The reality of complicated software is that it can contain bugs that will disrupt the operation of the product when certain conditions are reached. Software bugs were recognized even in the 1980s, as *Wall Street Journal* writer Bob Davis noted, "The tiniest software bug can fell the mightiest machine – often with disastrous consequences. During the past five years, software defects have killed sailors, maimed patients, wounded corporations and threatened to cause government-securities markets to collapse. Such problems are likely to grow as industry and the military increasingly rely on software to run systems of phenomenal complexity" (Davis 1987, p. 1).

This early warning by Davis carries over to the present era. Software bugs are still prevalent and embedded in the products we use and, hence, are spread throughout the supply chain. In the automotive industry, software bugs have been attributed to brake problems in the Toyota Prius (Jenkins 2010) and entertainment system glitches in Ford and Lincoln brand vehicles (Muller 2012). As automobiles become more sophisticated and embedded with complicated software, the number of software code problems is expected to increase. "Today's cars are like rolling computers. Some vehicles have nearly 100 million lines of software code inside, affecting 70 to 100 electronic control units that run everything from the power steering and anti-lock brake systems to the heated seats and sliding moonroof. When something goes wrong with your car, increasingly it's because of a software glitch rather than a mechanical malfunction or an electrical short" (Muller 2012).

## CONCLUSION

Twenty-five years after the Therac-25 accidents, software glitches continue to affect products in the forward supply chain. Automobiles are one example, but software systems are embedded in all forms of transportation, including commercial aircraft. It may be that the future of a product is only as good as its source code.

## QUESTIONS FOR RESEARCH AND DISCUSSION

1. Software code problems are difficult to identify. What factors make identifying source code errors a challenging task?
2. With more products being embedded with computer software, what avenues are available to update these software years after the consumer has bought their product?

## REFERENCES

Coombs, W., *Ongoing Crisis Communication: Planning, Managing, and Responding*, 2nd edn., Sage, Thousand Oaks, CA, 2007.
Davis, B., Costly bugs, *Wall Street Journal*, 1, January 28, 1987.
Jenkins, H., Toyota and the curse of the software, *Wall Street Journal*, A.15, February 5, 2010.
Leveson, N., *Safeware: System Safety and Computers*, Addison-Wesley, Reading, MA, 1995.
Muller, J., You can reboot your computer; why not your car? *Forbes*, 9, April 15, 2012.
Pearson, C. and Clair, J., Reframing crisis management, *Academy of Management Review*, 23(1), 59–76, 1998.

## Discussion Questions

1. What is the difference between tightly coupled and loosely coupled supply chains?
2. Discuss how supply chain structure has changed over the past few decades.
3. Discuss the role of technology in supply chain information. What are its benefits? Its obstacles?
4. Discuss the differences between coordination, cooperation, and collaboration.
5. What are some of the benefits of collaboration among supply chain members?
6. What are some of the obstacles to achieving collaboration in supply chains?

7. Why is information sharing important in achieving SCI?
8. Discuss some of the steps in the change process leading to SCI.
9. Why is complexity management important to understanding SCI?
10. What is clairvoyance? Are you aware of any leaders that appear to have this trait?

---

## References

Abernathy, F.H., Dunlop, J.T., Hammond, J.H., and Weil, D., *A Stitch in Time: Lean Retailing and the Transformation of Manufacturing—Lessons from the Apparel and Textile Industries*, Oxford University Press, New York, 1999.

Amber Road, A model for value chain transformation: Achieving global trade management maturity with Amber Road, 2013. www.amberroad.com/.

Andraski, J., The new model of CPFR, https://www.bisg.org/docs/VICS-CPFR.pdf, accessed August 7, 2014.

Angulo, A., Nachtmann, H., and Waller, M.A., Supply chain information sharing in a vendor managed inventory partnership, *Journal of Business Logistics*, 25(1), 101, 2004.

Barratt, M. and Oliverira, A., Exploring the experiences of collaborative planning initiatives, *International Journal of Physical Distribution & Logistics Management*, 31(4), 266, 2001.

Beth, S., Burt, D.N., Copacino, W., and Gobal, C., Supply chain challenges: Building relationships, *Harvard Business Review*, 81(7), 64, 2003.

Bitran, G.R., Gurumurthi, S., and Sam, S.L., The need for third-party coordination in supply chain governance, *MIT Sloan Management Review*, 48(3), 30, 2007.

Blackstone, J.H., *APICS Dictionary*, 14th edn., APICS—The Association for Operations Management, Chicago, IL, 2013.

Chandler, A.D., Jr., *The Visible Hand: The Managerial Revolution in American Business*, Belknap Press of Harvard University Press, Cambridge, MA, 1977.

Corsten, D. and Kumar, N., Do suppliers benefit from collaborative relationships with large retailers? An empirical investigation of efficient consumer response adoption, *Journal of Marketing*, 69(3), 80, 2005.

Crandall, R.E., Beating impossible deadlines, a variety of methods can help, *APICS Magazine*, 16(6), 20, 2006.

Crandall, R.E., What's really going on here? Exploring approaches to supply chain management, *APICS Magazine*, 19(3), 26, 2009.

Crandall, R.E. and Crandall, W.R., *Vanishing Boundaries, How Integrating Manufacturing and Services Creates Customer Value*, CRC Press, Boca Raton, FL, 2014.

Crum, C. and Palmatier, G.F., Demand collaboration: What's holding us back? *Supply Chain Management Review*, 8(1), 54, 2004.

Deloitte, The challenge of complexity in global manufacturing, Deloitte Touche Tohmatsu, 2003, Special report.

Fabbe-Costes, N., Jahre, M., and Roussat, C., Supply chain integration: the role of logistics service providers, *International Journal of Productivity and Performance Management*, 58(1), 71, 2009.

Fawcett, S.E., Magnan, G.M., and McCarter, M.W., Benefits, barriers, and bridges to effective supply chain management, *Supply Chain Management*, 13(1), 35, 2008.

Ford, H., *Ford Today and Tomorrow* (Special Edition of 1926 Classic), Productivity Press, New York, 1988.

Frankel, R., Goldsby, T.J., and Whipple, J.M., Grocery industry collaboration in the wake of ECR, *International Journal of Logistics Management*, 13(1), 57, 2002.

Frederick, W.C., Creatures, corporations, communities, chaos, complexity, *Business and Society*, 37(3), 358, 1998.

Giannakis, M., Facilitating learning and knowledge transfer through supplier development, *Supply Chain Management*, 13(1), 62, 2008.

Gladwell, M., *Blink: The Power of Thinking without Thinking*, Little, Brown and Company, New York, 2005.

Haeckel, S.H., *Adaptive Enterprise: Creating and Leading Sense-and-Respond Organizations*, Harvard Business School, Boston, MA, 1999.

Halldorsson, A., Kotzab, H., and Mikkola, J.H., Complementary theories to supply chain management, *Supply Chain Management*, 12(4), 284, 2007.

Handfield, R.B. and Bechtel, C., Trust, power, dependence, and economics: Can SCM research borrow paradigms? *International Journal of Integrated Supply Management*, 1(1), 3, 2004.

Harrington, L.H., 9 steps to success with CPFR, *Transportation & Distribution*, 44(4), 50–52, April 2003.

Hunter, N.A. and Valentino, P., Quick response—Ten years later, *International Journal of Clothing Science and Technology*, 7(4), 30, 1995.

Hutchison, C.J., Balancing act, *APICS Extra* (on-line), 1, February 1, 2006.

Kidd, J., Richter, F.J., and Li, X., Learning and trust in supply chain management, *Management Decision*, 41(7), 603, 2003.

Kidd, P.T., *Agile Manufacturing: Forging New Frontiers*, Addison-Wesley, Reading, MA, 1994.

Lee, H.L., The triple-A supply chain, *Harvard Business Review*, 82(10), 102, 2004.

Lockamy, A., III and McCormack, K., The development of a supply chain management process maturity model using the concepts of business process orientation, *Supply Chain Management*, 9(3/4), 272, 2004.

Lummus, R.R. and Vokurka, R.J., Defining supply chain management: A historical perspective and practical guidelines, *Industrial Management and Data Systems*, 99(1), 11, 1999.

Mintzberg, H., *Mintzberg on Management: Inside Our Strange World of Organizations*, The Free Press, New York, 1989.

Pink, D.H., *A Whole New Mind, Moving from the Information Age to the Conceptual Age*, Riverhead Books, Penguin Group, New York, 2005.

Poirier, C.C. and Quinn, F.J., How are we doing? A survey of supply chain progress, *Supply Chain Management Review*, 8(8), 24, 2004.

PRTM, Supply chain maturity model, www.prtm.com/services/supply_chain_maturity_model.asp, 2005, accessed April 15, 2013.

Schwartz, B., *The Paradox of Choice: Why More Is Less*, Harper Collins, New York, 2004.

Sheridan, J.H., Agile manufacturing: Stepping beyond lean production, *Industry Week*, 242(8), 30, 1993.

Siegfried, M., The secrets to collaborative success, *Inside Supply Management*, 24(7), 27–29, September 2013.

Smaros, J., Lehtonen, J.-M., Appelqvist, P., and Holstrom, J., The impact of increasing demand visibility on production and inventory control efficiency, *International Journal of Physical Distribution & Logistics Management*, 33(4), 336, 2003.

Sullivan, L. and Bacheldor, B., Slow to sync, *Information Week*, 992, 55, June 7, 2004.

Voluntary Interindustry Commerce Standards (VICS), Collaborative planning, forecasting and replenishment, CPFR: An overview, May 18, 2004, http://www.gs1us.org/DesktopModules/Bring2mind/DMX/Download.aspx?Command=Core_Download&EntryId=631&PortalId=0&TabId=785, accessed August 7, 2014.

Vinas, T., IW value-chain survey: A map of the world, www.IndustryWeek.com, September 1, 2005.

Williamson, O.E., Outsourcing: Transaction cost economics and supply chain management, *Journal of Supply Chain Management*, 44(2), 5, 2008.

# Section V

# Financial and Information Technology Perspectives

# 14

## Information Flow along the Supply Chain

### Learning Outcomes

After this chapter, you should be able to

- Explain why information flow is important in supply chains
- Describe the types of information transmitted among supply chain participants
- Describe intracompany technology (IT) and its use in supply chain management (SCM)
- Identify and describe three types of linking technologies
- Identify benefits of supply chain information flows
- Identify major obstacles to implementing ITs
- Explain the relationship between ERP and supply chain information systems

### Company Profile

It is obvious that supply chains are designed to move products from the point of origin to the ultimate consumer. What may not be so obvious is that information must flow both downstream and upstream in supply chains. In fact, information must flow before goods flow—someone must place an order. Information systems must initiate the flow of goods, track them during their movement, confirm their arrival, facilitate payments, and report on the results, just to name a few things they do. Information systems must surmount the hurdles of inconsistent or incompatible systems between supply chain participants. Information transparency is desirable, but the greater the transparency, the greater the risk of loss of privacy. As you read this company profile and the chapter in the book, think about these questions:

- Name some kinds of information needed in supply chains. Which is most important?
- What are some ways information is communicated among supply chain participants?

- Who designs information systems? Are the designers employees or consultants?
- Why is it difficult to have one information system for the entire supply chain?

What additional questions do you think relate to information flow in an integrated supply chain?

---
**Company Profile: SAP**

---

## INTRODUCTION

Chapter 14 is about information flow. Until the information flows, the products do not flow and until products flow, funds do not flow. When viewed from this perspective, information flow is the driver of supply chains. SAP is a company that produces software and other information technology (IT) applications that help companies send information to other members of their supply chains. We will describe the wide array of products and services provided by SAP to illustrate the vital role that information flow plays in supply chains.

## HISTORY

SAP was started in 1972 by a group of IBM employees when IBM decided to abandon the project on which the employees were working. Forty years later, SAP AG became the third largest software company in the world, trailing only Microsoft and Oracle in revenues. They have been listed on the New York Stock Exchange (NYSE) as SAP since 1998.

Table 14.1 lists a number of key statistics and events in the company's history. In 2012, they had over 65,000 employees working in 130 countries, with almost a quarter-million customers located in 188 countries. With revenues of €16,223, over $21 billion, SAP is a major provider of IT products and services.

As Table 14.1 shows, they began developing software modules for individual applications: financial accounting (1973), purchasing and inventory management (1975), asset accounting (1978), sales and distribution (1980), production management (1981), production planning and control (1983), and human resource (HR) management (1986). However, early on, they also began to develop integrated systems. SAP R/1 (1973) could handle multiple tasks and R/2 (1979) expanded into nonaccounting areas of production management. In 1992, they introduced SAP R/3, the beginning of their enterprise planning management (ERP) systems.

Following the introduction of SAP R/3, the company has continued to add to their product line. Their objective is to have software solutions for every part of a business. Their integration efforts include the following:

- Enterprise resource planning (ERP) systems were designed to link together all of the functions within a business.
- Supply chain management (SCM) systems were designed to link together all of the participants in a supply chain.

In addition to their software products and services, corporate strategic management is linking together the financial, social, and environmental concerns of a business, as described in their 2012 Integrated Annual Report. Integration is a key theme both in their product portfolio and in their management practices. As they explain:

> This year, for the first time, we are publishing an integrated report – meaning that we are combining the SAP Annual Report and the SAP Sustainability Report, which enables us to highlight the connections between our financial and non-financial

**TABLE 14.1**

Key Milestones in the History of SAP AG

| Year | Key Products | Number of Employees | Revenues (Millions) | Number of Customers | Key Happenings |
|------|-------------|--------------------|--------------------|--------------------|---------------|
| 1972 | | 9 | DM 0.6 | | |
| 1973 | RF—financial accounting (beginning of R/1) | | | 40 | Uses IBM servers |
| 1974 | | | | | Converts from DOS to OS |
| 1975 | RM—purchasing, inventory management | | | | Beginning of integration |
| 1976 | | 25 | DM 3.8 | | |
| 1977 | Installs systems outside Germany | | | | Moves HQ to Walldorf |
| 1978 | Asset accounting | | | | |
| 1979 | Beginning of SAP R/2 | | | | Operates own server |
| 1980 | RV—sales and distribution | | | | |
| 1981 | Production management | | | 200 | Participates in first event trade show |
| 1982 | | 100 | DM 24 | 250 | 10th year anniversary |
| 1983 | RM-PPS—production planning and control | 125 | DM 41 | | |
| 1984 | | 163 | DM 48 | | Starts SAP AG in Biel, Switzerland |
| 1985 | | 250 | DM 61 | | Establishes quality assurance committee |
| 1986 | HR management | | DM 100 | | First international subsidiary in Austria |
| 1987 | Begins developing SAP R/3 | 500 | DM 152 | | Expands throughout Europe |
| 1988 | | 940 | DM 245 | 1,000 | From private to public company |
| 1989 | Introduces SAP R/2; progressing on SAP R/3 | | | | |
| 1990 | SAP R/3 financial accounting and matls mgmt | 1,700 | DM 500 | | Reunification of West and East Germany |
| 1991 | Presents first applications of SAP R/3 | 2,700 | DM 707 | | Becomes active in East Germany |
| 1992 | Introduces SAP R/3 to general public | 3,157 | DM 831 | | 50% of sales outside Germany |
| 1993 | Ports SAP R.3 to Microsoft Windows NT | 3,600 | DM 1,100 | | Begins working with Microsoft |
| 1994 | IBM that begins to use SAP R/3 | 5,229 | DM 1,800 | | U.S. sales 34% of total |
| 1995 | More focus on midsize companies | 7,000 | DM 2,700 | | Burger King becomes 1,000th customer |
| 1996 | Joint Internet strategy with Microsoft | 9,202 | DM 3,700 | | Adds Coca-Cola as customer |
| 1997 | Releases 4.0 of SAP R/3 | 13,000 | DM 6,000 | | Enters NYSE |
| 1998 | Enjoy SAP | 19,000 | €4,300 | 15,000 | 10th SAPPHIRE event in Los Angeles |

*(Continued)*

**TABLE 14.1 (*Continued*)**

Key Milestones in the History of SAP AG

| Year | Key Products | Number of Employees | Revenues (Millions) | Number of Customers | Key Happenings |
|------|--------------|---------------------|---------------------|---------------------|----------------|
| 1999 | mySAP.com | 20,000 | €5,100 | | |
| 2000 | | 24,000 | €6,300 | | Adds Nestlé |
| 2001 | Expands mySAP.com | | | | *New Economy* bubble bursts |
| 2002 | | 29,000 | | | 30th anniversary |
| 2003 | mySAP to be succeeded by SAP NetWeaver | 30,000 | | | 17,000 employees outside Germany |
| 2004 | Introduces SAP NetWeaver; plans enterprise SOA | | | 24,000 | 120 countries |
| 2005 | Opens R&D lab in Budapest | 35,800 | €8,500 | | Number of acquisitions |
| 2006 | SAP and Microsoft introduce Duet | | | | |
| 2007 | SAM business ByDesign for small companies | | | | Acquires Pilot Software |
| 2008 | | | €11,575 | | Fourth *Germany's Best Employer* award |
| 2009 | Unveils SAP Business Suite 7 | 47,800 | €10,672 | | Difficult times; employee cutbacks |
| 2010 | | | €12,464 | 50,000 | Acquires Sybase (solutions for wireless) |
| 2011 | Introduces SAP HANA (in-memory) | | €14,233 | | Acquires SuccessFactors (cloud) |
| 2012 | | 65,500 | €16,223 | 248,500 | Customers in 188 countries |

*Source:* Adapted from SAP, A 42-year history of innovation, www.sap.com/corporate-en/our-company/history.

performance. We believe this shift holds both practical and symbolic importance. First, we are signaling that the business landscape has changed, and the information needed to evaluate performance must change with it. Considering our past financial results and our financial outlook alone does not adequately capture our ability to respond to today's challenges or how we create value. Instead, our future success hinges on how well we holistically navigate the social, environmental, and economic contexts in which we operate (SAP Integrated Report 2012a).

## PRODUCTS AND SERVICES

Table 14.2 shows the total revenue by geographic area.

SAP's products span five major market categories: applications, analytics, cloud, mobile, and database and technology. All of these are powered by SAP High-Performance Analytics Appliance (HANA), their current in-memory appliance for business intelligence (BI) allowing real-time analytics. The following descriptions are from the "Portfolio of Products, Solutions, and Services" section of the SAP Integrated Report (2012c, pp. 60–63).

SAP Business Suite is the next generation that captures and analyzes data in real time on a single in-memory platform. It empowers customers to run their business in real time within the window of opportunity to transact, analyze, and predict instantly and proactively in an unpredictable world. This gives companies the ability to translate real-time insights into action immediately, while removing the complexity of redundant system data. Customers can now manage all mission-critical

**TABLE 14.2**

Total Revenue by Region (in Millions of Euros)

| Region | 2008 | 2009 | 2010 | 2011 | 2012 |
|---|---|---|---|---|---|
| Europe, Middle East and Africa (EMEA) | 6,206 | 5,643 | 6,263 | 6,991 | 7,486 |
| The Americas | 3,880 | 3,620 | 4,435 | 5,091 | 6,100 |
| Asia Pacific Japan (APJ) | 1,489 | 1,409 | 1,766 | 2,151 | 2,637 |
| Total | 11,575 | 10,672 | 12,464 | 14,233 | 16,223 |
| *Percent of total revenue by region* | | | | | |
| Europe, the Middle East and Africa (EMEA) (%) | 54 | 53 | 50 | 49 | 46 |
| The Americas (%) | 34 | 34 | 36 | 36 | 38 |
| Asia Pacific Japan (APJ) (%) | 13 | 13 | 14 | 15 | 16 |
| Total (%) | 100 | 100 | 100 | 100 | 100 |

*Source:* Adapted from SAP, 2012 Combined management report, p. 96.

business processes, such as planning, execution, reporting, and analysis, in real time using the same relevant live data.

APPLICATIONS

SAP's leadership in enterprise applications has been the core competence of the company and continues to fuel growth for the future. SAP Business Suite is a business process platform that helps companies run better every day. The core software applications of SAP Business Suite are as follows:

- SAP ERP application supports critical business processes, such as finance and human capital management.
- SAP customer relationship management (SAP CRM) application improves streamlined interaction with customers with integrated social media and mobile device support.
- SAP product life cycle management (SAP PLM) application manages the product and asset life cycle across the extended supply chain, freeing the product innovation process from organizational constraints.
- SAP supplier relationship management (SAP SRM) application supports key procurement activities.
- SAP supply chain management (SAP SCM) application helps adapt company-specific supply chain processes to the rapidly changing competitive environment.

ANALYTICS

Analytics solutions from SAP enable decision makers at all levels of the business to have a more profound impact on their organizations and include the following categories:

- SAP BusinessObjects™ BI solutions enable users to interact with business information and obtain answers to ad hoc questions without advanced knowledge of the underlying data sources.
- SAP solutions for enterprise performance management (EPM) help companies improve performance, organizational agility, and decision making.
- SAP solutions for governance, risk, and compliance (GRC) provide organizations with a real-time approach to managing GRC across heterogeneous environments.

- Applied analytics solutions address challenges in specific industries and lines of business.
- Edge solutions for small and midsize enterprises are editions of BI and EPM solutions for growing midsize companies.
- SAP Crystal solutions are BI solutions designed for small businesses, addressing essential BI requirements.

### Cloud

SAP's 2012 acquisitions of SuccessFactors and Ariba have allowed them to combine powerful assets from all three companies—including innovative solutions, content and analytics, process expertise, access to a robust business network, and enterprise mobility—to build a comprehensive cloud computing portfolio. Cloud applications and suites are delivered as software as a service (SaaS), in which customers pay a subscription fee to use SAP's software. Cloud offerings are designed to optimize a company's most critical assets:

- *People*: The SuccessFactors Business Execution (BizX) suite enables companies to align employee performance with overall corporate objectives.
- *Money*: SAP Financials OnDemand and SAP Travel OnDemand solutions manage key financial processes.
- *Customers*: A suite of applications that manages all aspects of customer interaction—sales, service, and marketing—while employing next-generation social capabilities.
- *Suppliers*: The offering includes solutions for end-to-end strategic sourcing and procurement processes to enable efficient purchasing decisions.
- *Suites*: SAP Business ByDesign and the SAP Business One OnDemand solution provide a full cloud suite for subsidiaries and small business, respectively.

### Mobile

With SAP Mobile, customers can deliver secure, real-time, business-critical information to their ecosystems of employees, partners, and customers—on mobile devices. The mobile development platform creates many opportunities for SAP's partners to develop their own applications for their employees and customers. Mobile solutions include the following:

- *SAP Mobile Platform (includes Sybase® Unwired Platform)*: It provides the tools needed to support mobile initiatives across an enterprise. It provides a development platform (SDK) consistent, but adaptable, enabling customers to develop apps for various mobile devices deployed.
- *SAP Afaria*: It enables companies to better manage and secure all critical data on, and transmitted by, mobile devices.
- *Sybase 365*: Interoperability services that simplify the deployment and delivery of interoperator messaging over incompatible networks, protocol stacks, and handsets among mobile operators worldwide.
- Additionally, SAP is innovating consumer-facing applications that help improve people's lives. These innovations include the Care Circles mobile app to improve how patients, health-care providers, and family members optimize

treatment strategies; the Recalls Plus mobile app to help parents monitor recalls on children's items via social media; the Charitable Transformation (ChariTra) online network to match volunteers with people and organizations in need for their time, skills, or resources; and the TwoGo by SAP service that connects people so they can share rides and carpool together.

## DATABASE AND TECHNOLOGY

SAP provides a comprehensive approach to the orchestration of business applications, no matter how the applications are deployed. Furthermore, SAP harnesses the power of in-memory databases with SAP HANA, which is the data foundation for the next generation of high-performance in-memory computing solutions. The portfolio includes the following:

- *SAP HANA*: Deployable on-premise or in the cloud, SAP HANA combines an in-memory database with an in-memory application server running on in-memory optimized hardware appliances. At the foundation of this product portfolio is SAP HANA, an in-memory computing technology that simplifies and streamlines complex and expensive IT architectures. It helps customers process massive amounts of data and delivers information at unprecedented speeds. SAP HANA is an open platform, adaptable and extensible, enabling customers to create previously unimaginable applications and to rethink and envision new ways to run their businesses. Furthermore, it is a platform for applications developed by SAP ecosystem and partners.
- *SAP NetWeaver®*: Technology platform that integrates information and business processes across technologies and organizations. SAP NetWeaver facilitates the easy integration of SAP software with heterogeneous system environments, third-party solutions, and external business partners.
- *SAP NetWeaver Business Warehouse (BW)*: Data warehouse that provides a complete view of a company and the tools needed to make the right decisions, optimize processes, and measure strategic success.
- *SAP Sybase IQ*: Analytics server designed specifically for advanced analytics, data warehousing, and BI environments.
- *SAP Sybase Event Stream Processor (SAP Sybase ESP)*: High-performance, complex event processing engine designed to analyze streams of business event information in real time and used to create strategic advantage in low-latency applications for financial trading, smart grids, and telecommunications.
- *SAP Sybase Adaptive Server® Enterprise (SAP Sybase ASE)*: High-performance relational database management system (DBMS) for mission-critical, data-intensive transactional environments. It is optimized for use with SAP Business Suite.
- *SAP Sybase SQL Anywhere®*: Mobile, embedded, and cloud-enabled fully relational database that is embedded in more than 10 million installations worldwide, from laptops to tablets to smartphones.

In addition, the SAP virtualization and cloud management offerings help SAP customers automate SAP systems and landscapes operation, improve business agility, and reduce the total cost of ownership (TCO).

## SOLUTIONS FOR LINES OF BUSINESS

SAP has line-of-business solutions that are relevant across all industries and include the following lines of business:

1. Marketing
2. Sales
3. Customer service
4. Procurement
5. SCM
6. Manufacturing
7. Research and development (R&D)
8. Engineering
9. IT
10. Finance accounting
11. HR
12. Corporate strategy and sustainability

## SOLUTIONS FOR INDUSTRIES

In 2012, SAP supported enterprises in 24 industries with solution portfolios that enable industry best-practice processes. In 2013, they will add sports and entertainment as another industry to their portfolio and are currently building up that portfolio of offerings, which includes the SAP Sports and Entertainment Management solution, as well as mobile apps used by athletes, coaches, and spectators for sailing, tennis, and professional team sports.

Discrete manufacturing
  • SAP for aerospace and defense
  • SAP for automotive
  • SAP for high tech
  • SAP for industrial machinery and components
Process manufacturing
  • SAP for chemicals
  • SAP for mill products
Consumer products SAP for consumer products
Energy and natural resources
  • SAP for oil and gas
  • SAP for mining
  • SAP for utilities
Retail and wholesale distribution
  • SAP for retail
  • SAP for wholesale distribution
Public services
  • SAP for defense and security
  • SAP for higher education and research
  • SAP for public sector
Financial services
  • SAP for banking
  • SAP for insurance

**TABLE 14.3**

Revenue by Industry Group

| Industry | € Millions | Total (%) | Growth Rate (%) |
|---|---|---|---|
| Discrete manufacturing | 3,110 | 19.2 | 19 |
| Services | 2,484 | 15.3 | 13 |
| Energy and natural resources | 2,241 | 13.8 | 12 |
| Process manufacturing | 1,684 | 10.4 | 15 |
| Retail and wholesale | 1,541 | 9.5 | 19 |
| Consumer products | 1,516 | 9.3 | 6 |
| Public services | 1,437 | 8.9 | 3 |
| Financial services | 1,444 | 8.9 | 21 |
| Health care and life sciences | 766 | 4.7 | 20 |
| Total | 16,223 | 100.0 | 14 |

*Source:* Adapted from SAP, 2012 Combined management report, p. 98.

Services
- SAP for engineering, construction, and operations
- SAP for media
- SAP for professional services
- SAP for telecommunications
- SAP for transportation and logistics

Health sciences
- SAP for health care
- SAP for life sciences

Table 14.3 shows the total revenues for 2012 by industry group. Although their initial area of concentration—discrete manufacturing—remains the largest sector, it is closely followed by other sectors. Table 14.3 also lists the growth rates and the smallest sectors—financial services and health care and life sciences—are the fastest growing. The overall growth rate of 14% indicates a healthy market for the company.

In addition, SAP provides services throughout the product's life cycle, including SAP Custom Development (building individualized solutions for users), Maintenance and Support, and Professional Services, including consulting and education. For more information about SAP services, visit www.sap.com/services-and-support.

SAP's product line is broad and deep, reflecting their growth over the past four decades. They continue to develop IT applications, either internally or through acquisition, that enable them to offer whatever information handling an organization may need.

## SUPPLY CHAIN INFORMATION

As listed earlier under applications, SAP has developed software for SCM. These applications include the following:

- *Integrated sales and operations planning (S&OP)*: Drive faster, more accurate S&OP by integrating enterprise master data, planning, and reporting processes.
- *Manufacturing and supply planning*: Boost supply chain profitability with greater control over supply and demand, stronger supplier relationships, and real-time visibility into factory production.

- *Demand management*: Better forecast demand, achieve the optimal demand to supply ratio, and accurately predict customer reactions to pricing, for maximum revenue.
- *Service parts management*: Ensure an organization has the right parts in the right place at the right time, for more efficient management of an extended service parts network.
- *Collaborative response management*: Improve communication with suppliers, contract manufacturers, and customers to streamline collaboration, reduce errors, and cut processing costs.
- *Warehouse management*: Improve warehouse efficiency, optimize operations, and sharpen a company's competitive edge, with end-to-end extended warehouse management tools.
- *Transportation management*: Keep information about real-world constraints, costs, and penalties, while planning and consolidating inbound and outbound shipments.
- *Track and trace*: Improve the visibility and traceability of raw materials and products across your supply network, to ensure finished goods are of the highest quality.
- *Supply chain analytics*: Get the real-time analytics needed to improve supply chain performance, responsiveness, and accountability, for a significant competitive advantage.
- *Mobile solutions for the supply chain*: Mobilize key supply chain processes and information. Mobile solutions for SCM can drive productivity, boost agility, and sharpen an organization's competitive edge.
- *Cloud solutions for the supply chain*: Respond quickly to changing market conditions with a flexible supply chain that improves collaboration and proactively matches supply to demand.
- *Supply chain sustainability*: Ensure product safety and regulatory compliance, and reduce the environmental impact of products, with sustainability solutions for SCM.
- *Trade promotion management*: Empower account and trade managers with end-to-end support and increased visibility into the entire trade promotion process.
- *Manufacturing integration and intelligence*: Integrate shop-floor systems with business operations and gain real-time visibility into manufacturing data, from orders to materials to quality.

All of these applications can be supported with integrated, best-in-class applications that run on a database of choice—including SAP HANA (Supply Chain Management SCM Solutions 2013).

## EXTERNAL APPLICATIONS

SAP has organized user groups to provide feedback about the problems and requirements of their users in technical and functional areas of interest. These groups encourage users to exchange information of mutual interest where the users have the opportunity to share their experiences, knowledge, and ideas (SAP User Groups 2013).

Some companies actively promote or support the use of SAP products. Flexera Software developed software that enables users to optimize their use of SAP-named user licenses. Users can detect idle users, identify duplicate users, and assign the most appropriate SAP license, thereby reducing the cost of the licenses (Flexera 2013).

Accenture, a well-known global consulting company, "applies deep industry knowledge, superior expertise and unsurpassed delivery capabilities to drive break-through results. Developing innovative solutions that deliver real value in areas such as analytics on the SAP HANA platform, cloud and mobility is our passion" (Accenture 2013).

Amazon Web Services (AWS) partners with SAP to allow joint customers to use SAP software along with the pay-as-you-go model that comes with AWS (Amazon Web Services, Inc. 2013).

Another well-known company Cisco also supports the SAP NetWeaver BW Accelerator product. Cisco offers a family of appliances equipped with intelligent auto-mation capabilities that are designed to help companies execute SAP best practices (Stone 2012).

## CONCLUSIONS

SAP has received several *Best Company to Work For* awards, including *Germany's Best Employer* four times. They have also received awards in other countries, including China, Bulgaria, Denmark, India, Japan, and Mexico (SAP Company History 2013).

The company places emphasis on their relationships with their stakeholders—customers, employees, governments, industry analysts, financial analysts and inves-tors, nongovernmental organizations (NGOs) and academia, partners and suppliers, and a sustainability advisory panel. They believe that stakeholder engagement and collaboration helps them in the process of innovation and development of their soft-ware solutions (SAP Integrated Report 2012).

In the "History" section earlier, we described SAP's commitment to sustainability. This commitment is further documented in their reporting of some related nonfinan-cial key performance indicators, as shown in Table 14.4.

As the numbers indicate, they have made progress in reducing carbon emissions and energy used per employee. The total energy consumption has stayed about the same, even as the number of full-time employees increased by 25%.

SAP is a company that is committed to improving information flows within compa-nies and along supply chains. If their financial success is an indication of their contri-bution to other companies, they are doing their job well.

**TABLE 14.4**

SAP Sustainability Measures for the Past 5 Years

| Nonfinancial Key Performance Indicators | 2008 | 2009 | 2010 | 2011 | 2012 | Inc (Dec) in 5 Years (%) |
|---|---|---|---|---|---|---|
| Carbon emissions in kilotons | 560 | 480 | 455 | 490 | 485 | −13 |
| Total energy consumption in GWh | 870 | 860 | 845 | 860 | 860 | −1 |
| Data center energy in kWh per FTE | 3146 | 3001 | 2746 | 2824 | 2598 | −17 |
| Number of FTE employees | 51,544 | 47,584 | 53,513 | 55,765 | 64,422 | 25 |

*Source:* Adapted from the SAP 2012 integrated annual report.

## REFERENCES

Accenture, Application services for SAP, 2013. http://www.accenture.com/us-en/technology/systems-integration/sap-solutions/Pages/sap-solutions-index.aspx, accessed October 17, 2013.

Amazon Web Services, Inc., AWS case study: SAP, 2013. http://aws.amazon.com/solutions/case-studies/sap/, accessed October 15, 2013.

Flexera, FlexNet manager for SAP applications, leverage software license usage data to maximize the return on your SAP investment, 2013. http://www.flexerasoftware.com/products/software-license-management/flexnet-sap-license, accessed October 15, 2013.

SAP, A 42-year history of innovation, Company History, 2013. http://www.sap.com/corporate-en/our-company/history/2002-present.epx, accessed October 15, 2013.

SAP, Stakeholder engagement, Integrated Report, 2012a.

SAP, Why integrated reporting?, Integrated Report, 2012b.

SAP, Portfolio of products, solutions, and services, Integrated Report, 2012c, pp. 60–63.

SAP, User Groups, 2013. http://www54.sap.com/communities/user-groups.html, accessed October 16, 2013.

Supply Chain Management SCM Solutions, 2013. http://www54.sap.com/solution/lob/scm/software/overview/highlights.html, accessed October 16, 2013.

Stone, J.A., Keep your BI tools running smoothly with intelligent automation, *SAPinsider*, Jan. * Feb. * Mar. issue, 2012.

## Introduction

In earlier chapters, we described the flow of physical goods and services along the supply chain. We also indicated that supply chains have two additional flows with which to be concerned—information flow and funds flow. This chapter describes information flows along the supply chain, while Chapter 15 describes the flow of funds. While some aspects of information flow were covered in earlier chapters, this chapter describes in more depth the applications and technologies used in creating the information flow. This chapter addresses the following topics: the need for information flow, types of information transmitted, technologies used, benefits expected, obstacles encountered, and major steps in the design and implementation.

## Need for Information Flow

The primary function of supply chains is to move materials, products, and services from raw material suppliers, through factories, warehouses, and distributors, to the end customers. To facilitate the movement of goods and funds, information exchange among suppliers, manufacturers, distributors, and retailers needs to take place on a regular basis. Timely delivery of accurate information to business partners across the supply chain is indispensable to moving product and funds in an efficient and effective manner. Figure 14.1 shows that information flows both downstream and upstream in a supply chain.

Information must first flow within a company to be sure all functional areas are in agreement with the activities being performed within the organization and those activities being planned or in motion for the supply chain. Once the internal information flow is working, the next step is to design and implement a program to create the flow of information along the supply chain.

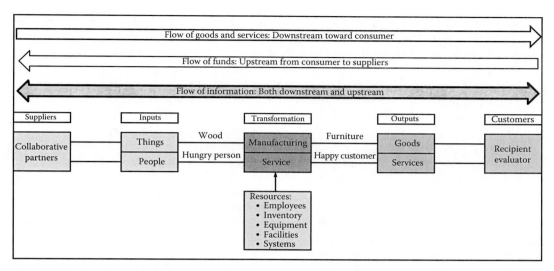

**FIGURE 14.1**
Input–transformation–output–customer (ITOC) model.

In an ideal supply chain, all business partners seamlessly share information on a real-time basis and act together in the best interests of all parties involved. However, both internal and external factors can reduce the motivation of business partners to participate. Internal factors, (within a company) such as ordering policies, opportunity costs, degree of trust with business partners, organizational culture, and customer ordering patterns, vary greatly between business partners. External factors, beyond the control of the company, such as consumers' demands, competition, advances in IT, and even weather, are often uncontrollable and cause uncertainties in demand and supply. In order to cope with these factors and reduce operational risks, business partners across the supply chain often take inappropriate actions— both strategic and operational—because they receive inadequate, inaccurate, or untimely information. Information inaccuracy, information asynchrony, and missing information can disrupt supply chain operations and result in lost sales revenues, overstock or out of stock, long cycle times, unreliable delivery of goods and services, and other problems. When the communication, coordination, and collaboration among business partners are ineffective, information flows are not in synchronization with product and fund flows. The prevalence of the bullwhip effect—erratic changes in orders along the supply chain—demonstrates the importance of accurate and timely information flows along the supply chain.

Information needs vary among managers at different organizational levels. Information for top management is structured differently from information for operational levels. Using a localized, or silo, approach in designing systems to meet functional needs often results in incompatible systems. Business partners customize information systems to meet their own needs. As time passes and the business environment changes, business requirements evolve and new information systems must be developed to cope with these changes. Sharing information requires all information flows to be adequately integrated to meet the varying needs of business partners. Properly used and shared by business partners, information can become an asset to help a company, and its supply chain, achieve competitive advantages. Make to order and mass customization are two business models built upon the ability of selectively leveraging information to better serve customers. Management needs to understand the complexity of information flows and manage them to their company's advantages. The ability to do so can help the company achieve competitive advantages in the area of SCM.

A vast array of ITs are available to facilitate information transmission. These technologies support upstream, internal, and downstream supply chain activities. No matter where a company fits along the supply chain, it must have good internal information flow and external information exchange with its suppliers and customers. All partners who are interested in sharing information with each other should participate in the design and implementation of information integration efforts. Wal-Mart's proactive initiative in involving its top suppliers in the design and implementation of radio-frequency identification (RFID) projects illustrates the importance of participant involvement.

One study looked at the need for supply chain visibility, the sharing of needed information among supply chain participants. The study found gaps between the importance of visibility and the availability of information in such areas as logistics and third-party manufacturing. Many companies had invested heavily in ERP systems, expecting those systems would connect with others in the supply chain. They found this was not working as expected, or as the author of the study puts it, "not as easy as Lego building blocks" (Cecere 2014).

Attainment of the ideal model ensures a company's information systems and information are seamlessly integrated. It takes time, money, and skilled employees to implement an integrated supply chain; therefore, companies interested in integrating their information with business partners need a carefully constructed project implementation plan.

Management must also be aware that not all IT is good or worth the investment. In some cases, the added technology is essential and generates an attractive return on investment (ROI). In other cases, it is difficult to determine ahead of time what the ROI will be, if any. Sometimes, the technology will be for the benefit of customers or suppliers without direct benefit to the company; however, it will be considered necessary if the company is to retain its position in the supply chain. The investment decisions for IT are complex, but necessary, if a company is to remain competitive in this electronic age. Trade-offs, risks, and intangible considerations abound; we will describe some of them in the following sections.

## Types of Information Transmitted

A supply chain includes three interrelated, sequential segments: (1) upstream, (2) internal, and (3) downstream supply chains. Different value chain activities are performed within each supply chain segment. Upstream supply chains deal with supplier relationship management (SRM). Internal activities consist of procurement, transformation operations (manufacturing) management, and outbound logistics activities. Downstream value chains focus on customer-facing activities, including marketing and sales.

### Upstream Suppliers

SRM is part of inbound logistics that includes receiving, storing, inventory control, and transportation scheduling. SRM engages suppliers in the sourcing process, which comprises procuring, spend analysis, requisitioning, sourcing, contracting, invoicing, and supplier management. Information needed to address these issues range from manufacturing and procurement plans, capabilities of sourcing partners, performance standards, product categories, market demand, supplier qualification, negotiation terms, and product categories (Turban et al. 2008). The ability to share this information between business partners provides immediate supply chain benefits, such as cost savings in managing multiple suppliers, improved visibility into spending activities, improved contract compliance, and increased business agility.

## Midstream (Internal)

Internal supply chains convert raw material inputs received from suppliers into complete goods that are ready for distribution. To achieve this transition, all of the key functions of a business must share information. Marketing must provide a demand forecast in sufficient detail to enable operations to plan production. In return, marketing needs to be aware of any delays in shipments to customers or changes in the planned production. Accounting must compile and disseminate cost information to help marketing in pricing and operations in controlling costs. HR needs to know what changes are needed in the number and skills of employees. Finance must be aware of impending cash needs for inventory buildup or capital investment requirements.

Internal supply chains consist of operations management and outbound logistics activities. Operations management includes purchasing, assembly, equipment maintenance, testing, and packaging. Outbound logistics activities include warehousing, order fulfillment, transportation, and distribution management. Operations management involves tactical and execution-level activities, such as master scheduling, material requirements planning, productivity activity control, and vendor order management (Bozarth and Handfield 2005). Information used in the production planning process includes the master schedule, demand and supply forecasts, customer orders, inventory status, detailed material requirements, and vendor information.

Inventory is prevalent throughout the supply chain. A number of factors contribute to the increase in inventory levels. Uncertainty in supply and/or demand can result in the stockpiling of safety stock. Mismatches between the production capacity of upstream suppliers and the demand of downstream customers can contribute to increased inventory. Extended, or variable, lead times to transport goods from suppliers to buyers can also cause inventory to increase. Companies keep inventory in order to mitigate and buffer against uncertainty. However, too much inventory in stock increases operating costs. *Markdown policy* is one of the operational costs evident in industries where goods are perishable and services are constrained with fixed capacity (Gallego et al. 2008). While it is necessary for companies to hold some inventories, balancing the buffer size of inventory against smooth operations is crucial to the sustainability of a company's operations.

## Downstream Customers

CRM systems prevail in downstream activities. Operational CRM is the component of CRM that supports the front-office business processes, those that directly interact with customers (Rainer and Cegielski 2011). Downstream supply chains interface with customers and include self-services, customer support, campaign management, contact center, sales force automation, field service automation, and customer divestment. To support these activities, customer representatives must communicate with customers via multiple channels, such as face-to-face meetings, instant messaging, or phone. With today's technology, organizations are using the Internet, the Web, and other electronic touch points, such as e-mail or point-of-sale terminals, to manage customer relationships. Customers interact directly with these technologies and applications rather than interact with a company representative as is the case with customer-facing applications (Rainer and Cegielski 2011). All functional activities could be involved if customers have questions regarding the billing, maintenance, promotion, financing, and product quality. Business partners can also interact with each other via associations, conferencing, and expos for collaboration opportunities. Companies should collect information about customers to analyze their behavior and strategies. These analyses help a company conceive an effective marketing and sales strategy.

## Reverse Supply Chain

Information flows will also be needed as part of the reverse supply chain. As described in Chapter 10, reverse supply chains are not normally as well organized and routine as forward supply chains. Consequently, the information flows may also be less rigidly structured. However, they are critically important. The type of information transferred in reverse supply chains is significantly different from downstream, midstream, and upstream flows (Walker 2009). Most of the technologies described in this chapter can also be used in reverse supply chains.

## Supply Chain Connectivity

Residing at different segments of a supply chain, business partners need to collect, analyze, and share information with each other to formulate an effective supply chain strategy. The types of information transmitted across the supply chain should be adapted to the needs of the different segments of a supply chain. The process view of information needs can portray a clearer picture of the types of information transmitted across the supply chain. This process-centric perspective is particularly useful in understanding the information needs of supply chains.

In Figure 14.2, we show a framework of ITs used in connecting a company with its supply chain partners. We use specific technologies to illustrate the type of information activity; however, the diagram is not inclusive of all technologies. There are variations of those shown and additional technologies not shown. Our intent is to illustrate the need for connectivity among companies.

At the bottom of Figure 14.2, the ERP system is the information focal point for a company. The ERP system provides integration of information within a company. Support technologies for the ERP system include advanced planning and scheduling (APS), manufacturing execution systems (MESs), and warehouse management systems (WMSs). Most companies make connections with their customers through some variation of a CRM system. They connect with their suppliers through an SRM system. We show these as direct connections because most companies would likely be directly involved in these technologies.

Moving up in the diagram, we show the linking technologies that make connectivity between companies possible. Automatic identification systems (AISs) have value within a company and also carry information between companies. Electronic data interchange (EDI) is an electronic form of communication developed several decades ago to make direct links between companies (1 to 1) and has morphed into Internet EDI with a many-to-many capability. Interorganizational system (IOS) is a generic term to describe a variety of systems to facilitate communication between companies. These three and many similar systems involve heavy use of computers with automatic sensing features that minimize the need for human involvement.

The next section—Linking Applications—shows the ways in which technology plays a supporting role to people in linking companies together. Videoconferencing

Supply chain ITs

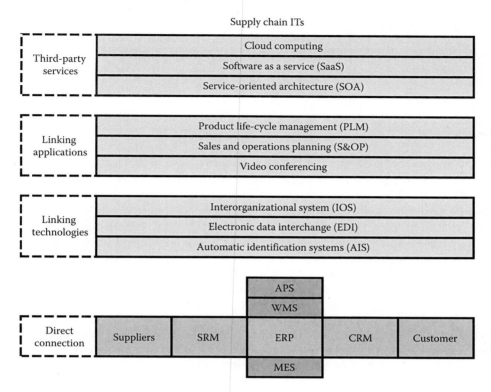

**FIGURE 14.2**
Supply chain ITs.

is new but gaining in popularity, especially as companies face higher travel costs and increased concerns about terrorism or health risks. S&OP is not new but is just coming into more widespread use as a means of enabling collaboration between companies. Product life cycle (PLM) is in the early stages of its life cycle as companies begin to struggle with the concept of preserving information in a usable state throughout the life of a product.

The top section is labeled third-party services because they are being driven by service providers and consultants. Service-oriented architecture (SOA) is an approach that promotes the use of building interfaces between software applications so that a company does not need to have completely integrated systems, such as with ERP. It is an architecture that makes it possible to construct business applications using web services, which can be reused across an organization in other applications (Rainer and Cegielski 2011). The expectation is that SOA provides a more economical, and faster, way of connecting disparate systems. SaaS builds on SOA by making software modules available on the Internet for selection and use by a company. It is a method of delivering software in which a vendor hosts the applications and provides them as a service to customers over a network, typically the Internet (Rainer and Cegielski 2011). Cloud computing extends SaaS by making both hardware and software available on a pay-as-used basis, thereby eliminating the need for capital investment by a company. Tasks are performed by computers physically removed from the user and accessed over a network, in particular the Internet. The cloud is composed of the computers, the software on those computers, and the network connections among those computers (Rainer and Cegielski 2011). One study estimates that about three-fourths of the present SaaS applications could be considered as cloud services, and that could exceed 90% by 2015 (Farb 2011).

## Intracompany Technologies Used

ITs are essential to address the growing information needs of supply chains. They are a means of informing, communicating, collaborating, controlling, and storing information. Companies that effectively manage IT can enhance the competitiveness of their supply chains. Time to market for new products can be shortened. A WMS increases information transparency and enables business partners to manage inventory and control costs. Business activity management (BAM) software can automate the process of tracking business activities and alert management to abnormal business activities, such as a decline in sales or a delay in shipment (Malone 2007). A supply chain professional knows the availability of key supply chain technologies and applies them in supporting business activities at each supply chain segment. IT performs three essential functions within an organization: (1) data capture and communication, (2) data storage and retrieval, and (3) data manipulation and reporting (Hugos 2006). In addition, supply chains require the communication of information between participants.

### Data Capture and Communication

Data are created in product design and process operations. Too often, data are lost and must be recreated because of the lack of connectivity in information flows. Sometimes, the recreated data are not consistent with the original data, creating wasteful or inaccurate processing.

#### *Computer-Aided Design*

Computer-aided design (CAD) capitalizes on the power of computer technology to aid the process of designing, drafting, and engineering new products. This technology speeds up today's engineering and reverse-engineering process. This makes it possible to shorten product development life cycles and reduce development costs.

CAD was a costly IT investment in the 1960s, and only large enterprises in the automotive, aerospace, and electronics industries were able to afford it. As software development methods, for example, object-oriented programming and computer-aided software engineering (CASE) tools, and desktop technologies advanced, commercial applications of CAD for individual use became affordable. Engineers today can develop, modify, share, and test designs via CAD at their desktop without mainframe support. CAD technologies are transforming product development process from the sequential to concurrent engineering process where activities in different stages can overlap with each other, thereby reducing the cycle time of product development (Bozarth and Handfield 2005). This technology will continue to evolve in order to cope with an increasing demand for shorter PLM and lower development costs.

### Point-of-Sale Terminals

Industries, such as restaurants, grocery stores, convenience stores, video rental stores, banks, fast-food restaurant, public transportations, and museums, depend on point-of-sale (POS) terminals for customer interaction and transaction. A typical POS system stores information on products, pricing, promotion, and inventory. A POS system assists in processing customer transactions by processing payments by cash, check, credit and debit cards as well as recording and tracking customer orders. These terminals are available in face-to-face transactions where customers or merchants need to swipe the payment tool (e.g., credit cards, debit cards, checks, and smart cards). Some POS systems have a touch screen interface that eases the transactional process. An increasing number of retail stores are adopting POS terminals to empower customers to self-checkout their purchases. POS systems can automate the shopping experience and speed up the checkout process. Because of the automation, stores can increase traffic, compile timely inventory and pricing information, and increase customer satisfaction. Mobile devices are another convenient way to facilitate POS transactions.

### Automatic Identification Systems

The two primary AIS technologies in use today are bar codes and radio-frequency identification (RFID). Bar codes have been a key to rapid and accurate data collection. They help identify product identities and location along the supply chain. While many see RFID technology replacing bar codes, it is not likely to happen in the short term, nor is it likely that RFID will replace bar codes in all applications. Bar codes can only read one item at a time and need a clear line of sight when scanning an object. Effective use of bar codes requires controlled conditions. In spite of the controlled conditions, bar codes can only capture limited content because of their inability to store information. The scope of applications is limited or restricted primarily to inventory control. Bar codes follow a universally accepted standard. Therefore, information integration is easily accomplished. The cost of manufacturing a bar code is relatively inexpensive. Since bar codes were introduced in commercial use, they have achieved economies of scale and reduced manufacturing and distribution costs.

RFID is superior to bar codes with respect to its technical capability of reading multiple items at a time. The object identification process via RFID is automatic because objects can communicate with each other without human intervention. RFID technologies attach tags or transponders to objects of all sizes and forms, such as books, pallets, containers, bottles of drinks, and medical suppliers. Antenna transmits and receives radio-frequency signal to and from objects at a distance. No line of sight and direct contact are necessary for communication. As such, multiple items can be read quickly. In addition, RFID tags can store and process more information than bar codes. When integrated with the back-end systems (e.g., decision-support systems, BI, and ERP systems), RFID becomes a powerful technology to increase the overall end-to-end supply chain visibility.

Active tags have a battery attached that can extend the reading range. RFID tags can withstand harsh conditions. Therefore, management can creatively apply RFID technology in a much wider range of applications (e.g., hospital, pharmaceutical, fishery, transportation, inventory, and retailer). However, the new technology is still evolving and many companies are reluctant to rapidly adopt this technology. While RFID can improve security control for businesses, it poses significant privacy concerns to some individual consumers.

Strategic business values of RFID applications also have BI and security perspectives. The major focus of RFID applications is on the automation of existing bar code–based tracking features. An organization can deliver greater business value by creating a borderless supply chain. Pharmaceutical companies use RFID in combating counterfeit goods. When a borderless supply chain is built across countries, cargos equipped with RFID tags at the departure country can show the sealed contents to the destination's customs agency. Not only can the customs clearance process be automated on a real-time basis, but security can also be improved in RFID applications. RFID can track and trace tampering activities through the entire delivery process. The initiative of California's state board of pharmacy requires all drugs dispensed within its state borders to have RFID attached and provide detailed information about all activities that take place during the transportation at the pedigree level (Frederick 2007). Item-level tracking has been slow in evolving because of the cost of the RFID tags. However, Johnson (2011) reports that collaboration among some of the major apparel retailers could result in item-level RFID tagging. If so, it could result in a major transformation in the retail industry because of the increased data available for analysis about customer preferences and buying patterns. For example, Zara, a leading retail chain, has announced plans to deploy RFID capabilities in all of its 6,000 stores by 2016 (RFID 24-7 2014).

## Groupware

Collaborative software, or groupware, facilitates computer-supported cooperative work and the idea generation process. Collaborative technologies also have a strong influence on the innovation process of a product over its life span (Merono-Cerdan et al. 2008). Groupware removes the limitations of geographic locations and involves more people into the idea generation process. The chance of having better and newer ideas can increase with the support of groupware.

Many technologies fall into the groupware category. For instance, workflow management applications have been developed to automate the process of moving documents, information, and knowledge across supply chains according to organizational protocols and in a coordinated fashion. Electronic meeting systems allow product design teams to exchange ideas on a real-time basis and to see each other in the virtual sphere. Teleconferencing, web conferencing, screen sharing, and interactive whiteboards allow cross-functional teams to examine prototypes on the same screen and reconcile different opinions. Social technologies (e.g., wikis and blogs) engage business partners into formal and informal conversations on the perfection of prototypes. Best Buy Co. deployed a social networking technology to analyze marketing ideas from its employees. IBM created a web-based corporate directory to help internal employees quickly locate the right person for problem solving. Web technology empowers people to have quick access to knowledge (Brandel 2008). Collaboration systems reduce the number of reworks and shorten product development time. The wide acceptance of groupware applications enables more people to participate in the product development process.

## Data Storage and Retrieval

Database management is a generic function to supply chain technology and other information systems. Data must be collected and stored in a DBMS that provides users with tools to add, delete, access, and analyze data stored in one location. A wide variety of database technologies are available, and the relational database model is one of the most popular and easy to use. Examples of companies who provide relational database software

are Microsoft Access and Oracle (Rainer and Cegielski 2011). Accessibility to a database is an important feature because many of today's businesses are operating extensively with electronic data; paper copies are rapidly disappearing.

## Data Manipulation and Reporting

ERP is a suite of software modules designed to automate business processes by integrating information flows across functional departments within an organization. The sales department executes customer acquisition and retention activities designed by the marketing department. The credit department verifies the credit line of customers before allowing the sales department to sell products to target customers. The accounting department accumulates product and service costs. The warehouse receives, stores, and prepares goods for shipment. The purchasing department prepares orders to ensure that out-of-stock problems do not occur. The production department controls the production process in order to improve quality and meet customer demand. The distribution department arranges delivery of goods with logistics providers. The HR department tracks the performance of employees and motivates them with an effective compensation plan. ERP integrates information generated from all business activities by functional departments and helps deliver products and services to customers.

Vendors design ERP software based on best industry practices; however, not all companies opt to use the best practice. If they do, business process reengineering is a prerequisite to the adoption of ERP software. ERP systems can be implemented in life cycle stages—adaptation, acceptance, routinization, and infusion (Kwon and Zmud 1987). The post implementation stages include routinization and infusion. The extent of customization is a strategic decision that can affect the costs and risks of implementation and the maintenance and upgrade of the adopted ERP system. Rapid changes in the business environment create a need for constant updates to the adopted system. Customization requires the modification of the existing software or the addition of bolt-on modules to the basic ERP package. Too much customization can increase the costs and risks of ERP implementation (Davenport 1998). Excessive customization can cause difficulties for software upgrade and migration to future releases (Botta-Genoulaz et al. 2005). These potential risks of customization discourage many companies from configuring their ERP system and from reengineering their business processes to align with the native features of the ERP packages. Unless customization is adequately justified at the strategic level, this strategy is often the last choice for those companies inexperienced with ERP systems.

SAP is a leading ERP vendor. Its ERP solutions include many modules, such as ERP Financials, Operations, Human Capital Management, Corporate Services, and Performance Management. ERP Financials can help tighten the financial control in order to comply with governmental policies, such as the Sarbanes–Oxley Act. A Human Capital Management module can automate HR processes (e.g., recruiting, personnel cost planning, training, payroll, time management, compensation, promotion, schedule management, and employee benefits administration). An Operations module helps automate the manufacturing process, which converts the raw materials or parts into works in progress or finished goods. Improvement of operational efficiency can minimize the inventory and back order costs and maximize customer satisfaction via the delivery of quality products. Corporate Services modules can streamline the administrative process to maximize transparency of corporate governance, minimize environmental risks, and enhance workplace safety. A Performance Management module can support the integrated measurement strategies, such as balanced scorecard and benchmarking (SAP 2013). As outsourcing practices become prevalent, the functional boundary is blurred. ERP continues to expand its features to service the varying functional requirements inside and outside the organizational boundary. Recent enhancements to ERP are APS, MES, and WMS. All manufacturers have a need to improve decisions in process planning and scheduling. APS enables an enterprise to optimize production planning, fulfillment, and distribution processes in delivering products on time and in full. Major functions of APS range from capacity planning at the plant floor to advanced logistics control for supply chain planning and collaboration (Turbide 1998). Business benefits include reduced lead time, lowered inventories, accuracy in available-to-promise decisions, and increased throughput (Eck 2003). An MES system focuses on management of the shop floor. It takes production plans from the ERP system, converts them into production schedules, and reports results back to the ERP system. A typical warehouse consists of the shipping and receiving areas, storage area, and order picking area. The objectives of WMS are to reduce inventory, increase labor productivity, improve shipping accuracy, and reduce the costs of operating warehouses (Botta-Genoulaz et al. 2005). WMS software solutions can improve inventory accuracy and visibility via optimizing the decision in forecasting variable demand, scheduling of job assignments and labors, cross-docking movement, as well as receiving and order fulfillment process. ERP can add WMS as a stand-alone or a decentralized warehouse system. Variations address the needs of warehouse management, including storage bin management, inventory management at storage bin level, storage bin search, picking, packing, and creating the required shipping documents.

## Supply Chain Direct Links

The wealth of information generated within a company can become more valuable when shared with other members of the supply chain. A company needs to develop closer ties to customers and suppliers. The link with customers can be covered with the umbrella term of CRM, a topic we covered in Chapter 3. The link with suppliers falls under the general heading of SRM, covered in Chapter 8. In Figure 14.2, these are shown as extensions, or add-ons, to the ERP system. Some of the major ERP software companies, such as SAP and Oracle, have integrated CRM and SRM into their ERP offerings.

### Customer Relationship Management

CRM applications were developed to cope with the challenge of integrating information that has been scattered throughout the entire organization or across the supply chain. Customer interactions take place anytime and anywhere. Information collected at each customer interaction point has limited value unless integrated into the total system. CRM applications are added to enterprise systems to address marketing opportunities in downstream supply chains.

A comprehensive CRM system includes three major components: (1) operational CRM, (2) analytical CRM, and (3) collaborative CRM. Operational CRM can resolve problems related to sales force automation, customer service and support, and enterprise marketing automation. Analytical CRM systems provide intelligence based on the data collected and accumulated from all customer interaction points over time. Collaborative CRM systems assist the communication process between customers and organizational functions.

Customers can choose any channel to communicate with service representatives via their preference channels, such as face-to-face, the Internet, telephone, fax, mobile phone, and instant messaging.

## Supplier Relationship Management

Effective procurement performance in the supply chain relies on the creation of a solid relationship with suppliers. SRM applications help an enterprise manage its interactions with suppliers in an effective manner, thereby fostering robust supplier relationships. In the process of building relationships, purchasers need to balance the trade-offs between risks and business benefits. SRM can optimize the balance by streamlining the process of managing supplier relationships, controlling inventory costs, shortening time to market, automating the negotiation process, and simplifying the bidding process. The improvement of relationships with suppliers can realize such benefits as the increased flexibility of managing suppliers, improved visibility of procurement process, and contract compliance. Many ERP providers include SRM in their extended ERP functions.

## Linking Technologies

Three technologies illustrate the way in which companies establish links with one another in the supply chain—IOSs, EDI, and AISs.

### Interorganizational Systems

An IOS is an information system that transcends organizational boundaries. Organizations adopting an IOS share information systems resources, such as database, communication network, and applications (Robey et al. 2008). These commonalities enable business partners across the supply chain to collaborate with each other via exchanging both structured (e.g., spreadsheet, database, invoice, and CAD) and unstructured data (e.g., discussion threads and newsletter). Electronic integration via an IOS enables a firm to expand its business scope and reengineer business processes across the supply chain. The commonality among these IOSs is the ability to connect buyers and sellers and provide their information needs for transaction purposes. IOSs are capable of generating, storing, and disseminating transaction-related (e.g., order, invoice, and fund) and collaboration-related information (e.g., new product design and meeting minutes). An effective IOS can reduce response times and information transparency. In so doing, it improves buyer–seller relationships. IOS can also lower transaction costs, minimize the transaction errors, and reduce cycle time in order fulfillment. Every IOS needs a way to transfer data and information between entities. One of the most common technologies is EDI.

### EDI and Internet EDI

EDI is a conduit to innovative ways of conducting e-business processes as well as facilitating e-business applications and services. EDI is the electronic exchange of trading documents, such as purchase orders, shipment authorizations, advanced shipment notices, and invoices, using standardized document formats (Blackstone 2013).

Traditional EDI, based on proprietary value-added networks (VANs), went through the early adoption stages in the 1990s (Premkumar et al. 1997). Many factors caused the slow diffusion of the technology, including, but not limited to, high investment costs, proprietary standards, poor integration capability with existing corporate systems, rigidity, poor scalability, poor performance in auditing trails, document certification needs, and the perceived need for hardcopies of the documents (Banerjee and Golhar 1994, Peters 2000). In the face of these adoption obstacles, many large and small organizations have been leveraging the open architecture of the Internet to improve their agility and competitiveness. Unlike traditional EDI, Internet EDI adopts the open standard Extensible Markup Language (XML) and entails higher business agility by integrating the information systems of the business partners. Internet EDI is replacing traditional EDI.

Internet EDI is an alternative for small- and medium-sized enterprises (SMEs) from the perspectives of economics, operation (e.g., faster establishment of customer–supplier relationships), and scheduling. Internet EDI allows business partners to conduct business over the Internet and saves money in purchasing and learning expensive software. Traditional and Internet EDIs have natural differences in closed versus open standards, high versus low cost, and business rigidity versus flexibility.

A dedicated virtual private network (VPN) to connect buyers and sellers was an expensive infrastructural requirement of traditional EDI. The ubiquity of Internet infrastructure enabled EDI to be available to more trading partners at a more affordable price. EDI has been evolving in its capability of servicing the needs of trading partners for the electronic movement of business documents.

While social media sites are still in the early stages of development for business purposes, some companies are beginning to evaluate their potential as an information sharing technology. While they may not be practical for everyday applications, they appear to have great potential in time of emergency, such as fires, floods, and other disasters. They do not require the formal structures and standards of most business systems and have the flexibility for rapid text and picture transmissions from on-site observers, who do not have to be employees of the organizations.

## Linking Applications

This section describes how some ITs are applied. Videoconferencing is a way to bring groups together in a simulated face-to-face conference. While still new in many companies, the technology is well developed and effective. S&OP is a concept that has been around for several decades but is finding renewed interest, as supply chain partners find value in developing collaborative relationships. PLM is a relatively new concept that has great potential but is highly dependent on tightly coupled supply chains.

### Videoconferencing

Virtual collaboration is a new tool in the implementations of interorganizational projects. Videoconferences enable all participants of a supply chain to interact with each other on a real-time basis. Videoconferencing involves digital connections that link firms across the supply chain. Although physical proximity has long been seen as important in facilitating the operations of firms inside a supply chain, digital proximity transcends physical boundaries and can enable cooperation among firms. The presence of the videoconferencing channel for participant firms to conduct business transactions, as well as formal and informal communication, is an important addition to achieving operational success in the supply chain.

Businesses service customers by performing marketing, sales, and after-sales service activities. Marketing activities include sales data analysis and BI. Sales force automation, cross selling, up-selling, and campaign management are part of sales activities. Real-time customer support and managing partner relationships are after-sales service activities. The ability to provide these activities can help the firm gain insights about existing and potential customers to better service them each time. In the process of performing these activities, a unified effort among a company and its suppliers, vendors, and industry networks is indispensable. Videoconferencing is an effective communication technology to unite the collaborative efforts across the supply chain. Leading aerospace companies in Europe have been using videoconferencing to collaborate virtually with each other and their business partners across the supply chain (Kenward 2003).

Videoconferencing can be used to conduct daily, weekly, and monthly meetings with local sales representatives and managers to clarify sales and marketing issues. The real-time, two-way mode of communications can also bring in experts to assist the analysis process. For instance, a sales and marketing expert could join the videoconference remotely and share findings on customer trend, customer profitability and sales forecasting, financial performance, and new product development opportunities. BI can also be discovered via virtual meetings between market analysts and sales managers. Videoconferencing removes the constraint of physical distance and promotes many-to-many interactions. Salespersons, managers, and experts deployed locally can save travel time via videoconferences. Prerecorded videos can be shared before meetings begin. For example, in the real estate market, customers can preview houses available in the market in 3D form without physically visiting those houses. A company can also demonstrate product features to business partners via videoconferencing technology.

Videoconferencing technology allows campaign managers and store managers to discuss how to direct sales and marketing efforts for the greatest return. Videoconferencing technology can complement other communication media, including phone, short message service (SMS) or texting, e-mail, fax, and web. Effective campaign management can be supported via videoconferencing-based meetings. Sales teams can collaborate with each other in order to provide after-sales service activities. Staff in different functional departments can work together to provide one-stop solutions to customers. Real-time customer support can be offered as additional services to improve customer satisfaction.

Despite the many advantages of videoconferencing, there are some drawbacks. Aside from the out-of-pocket costs for the physical facility, there are also costs of participants in adjusting to the new technology. It also takes collaboration between sender and receiver and assuring the compatibility of systems. One of the authors recently had the opportunity to have his MBA class in North Carolina engage with his class in Austria. The experience was clearly beneficial to both groups, but the technicians who arranged the videoconference spent several hours working out the arrangements.

At the present time, there are also security issues. Information Management (2012) reports on a demonstration of how companies are vulnerable to hackers accessing their videoconferences. As in many projects involving the Internet, there are some concerns about privacy issues. While these may be minimal in business or educational environments, the possibility exists.

## Sales and Operations Planning

We described S&OP in Chapter 5 as part of the demand management process. It involves the sales function working together with the operations function to determine how best to allocate the production resources to meet demand. It often requires a compromise solution; however, the resultant solution is one in which both functions concur.

When S&OP is extended along the supply chain, it could involve a retailer working with a manufacturer to decide how best to match demand and supply. In the short term, better allocation of resources can optimize the results. In the long term, additional resources of the best type and amount can be added or adapted to meet demand needs.

## Product Life Cycle Management

PLM envisions capturing product information during the product design, using CAD systems to classify and store the data. Those data/information would then be preserved throughout the life of the product and accessible to all participants in both the forward supply chain and the reverse supply chain. We described PLM in Chapter 4. Its implementation throughout the supply chain will be challenging because of the complexity of supply chains and their dispersion around the world. Siemens is one of the major suppliers of PLM software and has announced it will expand its commitment to manufacturers by developing PLM for eight specific industry sectors—aerospace and defense, automotive and transportation, consumer products and retail, electronics and semiconductor, energy and utilities, industrial machinery and heavy machinery, medical devices and pharmaceuticals, and marine (Rowell 2013).

---

## Third-Party Services

IT complexity abounds even within small companies. Despite the unifying results from ERP systems, few companies have completely integrated information flows. That makes it difficult to establish smooth information flows with other companies who often have different hardware/software configurations. Consequently, third-party service providers have opportunities to offer their services. We describe three of the service packages currently offered: SOA, SaaS, and cloud computing.

## Service-Oriented Architecture

SOA is a business-centric IT architectural approach that supports integrating business as linked, repeatable business services (IBM 2009). SOA is an architectural style that allows a community of providers and consumers of services to interact and collaborate with each other regardless of their adopted technologies. SOA enables an enterprise to build reliable distributed systems that deliver business functionality as services and loosely couple interacting services. Because the interface is platform independent, each interaction becomes independent of every other interaction. It means any devices can communicate with each other freely, regardless of the operating systems and software used.

SOA uses standardized architecture that enables various applications to connect with each other and share different data formats. Software services enable parties to communicate with each other in performing business activities. A software service is a self-contained unit comprising functionalities to perform a specific business activity, such as opening a customer account and ordering a new shipment. In order for software services to communicate with each other, disparate information systems need to go through defined protocols to remove system differences and standardize data formats. After removing system incompatibility issues, services are aggregated and customized for business requirements. Identifying shared business services while standardizing business processes is an important task in SOA. The benefits can support not only business agility but also the integration needs for complex systems.

SOA is a web service technology that enables the development of interoperable systems (Casola et al. 2008). SOA removes barriers to systems integration because of its independence from disparate business applications to support vendors and products. The ability of integrating systems can increase agility in systems implementations to cope with dynamic changes in today's business environment. As a result, companies can reduce lead times required to introduce new applications and integrate them with existing applications. This strategy can reduce risks and uncertainties associated with novel systems projects. The life of an existing IT investment can be extended and reused in creative manners despite IT infrastructure complexity. System development costs can be reduced. SOA enables the industrialization of software development and implementation process (Parker 2008). Enterprise software vendors are capitalizing on SOA to consolidate software markets.

ERP systems are large-scale systems integrating applications of different functional departments and business entities. SOA is a natural choice to resolve ERP integration issues.

Although ERP systems are an effective information system, they lack flexibility in supporting agile business operations. An adaptive organization can modify its business process to cope with dynamic environment changes. However, companies often find their systems are incapable of meeting the rapid changes in business processes. SOA empowers companies to quickly and flexibly respond to changing business needs.

## Software as a Service

SaaS is the delivery of software over the Internet as a service on demand (Fan et al. 2009). Google Calendar and Microsoft's OneCare service are two examples of SaaS that users can use their applications to update information into calendar and cleanup virus and spyware without hosting those applications (Richmond 2005). Other applications include accounting, CRM, payroll/HR, e-mail, and business analytics (Lin 2010). As in purchasing electricity from a utility, customers request an application whenever they have a need for it. The application terminates after customers stop using it. SaaS vendors host software applications for customers. Customers have no capital investment in purchasing and maintaining applications. Rather, customers pay for the service based on their frequency of usage. On-demand licensing and use enables a firm to convert the traditional fixed cost model of purchasing software into a variable expense model of leasing.

The adopting firm also does not need to worry about postsales issues, such as software upgrade, patches, and distribution. A reliable SaaS supplier becomes a one-stop solution to address these issues on behalf of clients. Through the bundling of software with delivery and maintenance service, SaaS providers differentiate themselves from traditional shrink–wrap software providers (Fan et al. 2009). SaaS benefits also include lower infrastructure costs, greater functional capabilities, improved application reliability and performance, systematic software updates and upgrades, higher productivity, and lower staff-support requirements (Fontana 2009).

Despite these advantages, SaaS cannot completely replace existing IT applications. One primary reason is hardware/software incompatibility. SaaS providers cannot offer a SaaS application to all clients with a myriad of hardware, software, and legacy systems. The complexity of information systems infrastructure within many large enterprises prevents them from successfully adopting SaaS. In contrast, SMEs can enjoy the benefits of getting enterprise-class functionality with small investment because systems compatibility is less an issue. Therefore, the less complicated their information systems infrastructure, the easier it is for a company to adopt SaaS.

Unique business processes are barriers to SaaS adoption. Too much customization conflicts with the business model of SaaS providers. The SaaS business model looks to standardize data structures and system architecture. A high degree of standardization can save operational costs because SaaS vendors will spend less effort resolving hardware/software incompatibility issues. The top five applications delivered via the SaaS model include CRM, HR, collaboration, travel expense management, and sales incentive management (Hall 2009).

The primary benefit for most companies using SaaS is a lower cost than if performed in-house; however, the primary concern is that SaaS introduces greater security risk. While practically all companies have a disaster recovery plan for on-premise systems, only about half have comparable plans for those applications involving SaaS (Lin 2010).

## Cloud Computing

SaaS applications need to operate on the cloud computing infrastructure (Mitra and Layak 2009). Cloud computing is a unified Internet-based infrastructure to interoperate diverse computing technologies for internal and external customers. Only specialists with years of training, skills, and knowledge are capable of managing complex information system infrastructure and applications. Cloud computing is an abstraction form of information systems and conceals complexity in the configuration of IS components, including hardware, software, data, and telecommunication network. Although not understanding all IS components, general users can access any applications as on-demand services. How applications interact with supporting infrastructure is irrelevant to users because a web browser is the only interface needed to know. Customers can use low-cost PCs, laptops, tablets, and mobile devices to access computing technologies hosted and networked by several companies. Customers only pay for the time they use the computing resources of their providers.

The subscription-based business model is made possible via the cloud computing platform. Customers can enjoy the benefits of reliability, failure resistance, capacity and cost-effectiveness in using diverse computing resources.

Rainer and Cegielski (2011) describe an application of cloud computing in which the New York Times used Amazon's Simple Storage Service to convert the digitized content of all its issues from 1851 to 1922, approximately four terabytes of data, into a web-friendly format. This allows users to view scanned images by hovering their mouse over an article, photograph, or advertisement.

Businesses can derive tangible and intangible benefits from services in the cloud. Tangible benefits include cost reduction in IT investment and time saving in systems implementations. Nearly all costs of using cloud computing services are variable; therefore, the fixed costs are minimal. Users do not own the infrastructure. They pay for the resources only when they are used. Computing resources and costs can be shared among users. Cost savings and ROI are two major reasons driving the eventual adoption of cloud computing (Finnie 2008).

The degree of interoperability increases the difficulty of decoupling customized applications from each other. As infrastructure and services become diverse, challenges are in managing bandwidth, privacy, security, regulatory compliance, and traffic intensify. The emergence of new regulatory requirements may create privacy challenges in hosting data in companies where auditing is not a possibility. However, cloud computing is applicable in even highly regulated sectors such as financial and health-care services (Harris 2014).

Cloud computing can be applied in all areas of SCM. SCM manages eight key business processes: CRM, customer service management, demand management, order fulfillment, manufacturing flow management, SRM, product development and commercialization, and returns management (Lambert 2008). Cloud computing infrastructure can host applications to support these business processes. Cloud computing enables infrastructure and applications to increase quickly to meet the sudden increases for computing resources. After the peak season, computing resources can be scaled down to meet routine needs. This scalability feature is an important benefit to companies, especially to small companies with limited IT resources (Harris 2014). The flexibility of allocating computing resources can increase the agility of an organization to cope with the dynamics of business environment (MacVittie 2008). The major concerns for many companies are recovery, security, and privacy issues.

## Benefits of Information Technologies

IT investment is a costly decision. Management needs to measure tangible and intangible benefits to justify each IT investment decision. Financial measures, such as operating profits and ROI, have been used to measure the overall performance of a company. However, these measures are too general to attribute directly improved supply chain performance to a particular IT investment. For example, companies have utilized CAD to reduce time to market, improve product quality, lower prototyping costs, and optimize product specifications. Groupware has been a primary technology to facilitate communication and collaboration among business partners across the supply chain.

### Tangible Benefits

There are several tangible benefits to the use of ITs. While some are directed at reducing costs, they can also aid in increasing revenues.

#### *Increase Revenues*

CRM software enables 24/7/365 operation and can provide individualized service at the mass level. Consumer behaviors become transparent to marketing managers, who can formulate effective campaign management programs to increase cross selling and up-selling opportunities based on a clear understanding of consumer behaviors. The use of CRM can ultimately increase revenues.

### Reduce Product Costs

Customers are pressuring the supply chain to be more efficient and innovative, while producing quality products. Transformation of a supply chain begins with the product design and development process. CAD can lower prototyping costs, reduce waste, save the cost of reinventing the wheel, and improve the success rate in developing new products. The use of CAD can minimize product development costs and optimize product specifications by integrating the workflow process of product engineering. CAD data are vital to the success of PLM. Other IT applications that can reduce operating costs include robots and automated warehouses, for example.

### Reduce Transaction Costs

EDI automates the process of exchanging business documents between companies. Paper-based transaction costs are much higher than EDI-based transaction costs. EDI transforms the transactional process in the electronic form that can shorten order life cycle, increase inventory availability, lower purchase price, and reduce transaction costs (Leonard and Davis 2006).

### Reduce Product Development Lead Time

Groupware can help research and developers brainstorm new product ideas. This technology can save communication and collaboration costs and promote creativity. Chubu Electric Power Co., Inc. is a utility firm supplying electricity to five prefectures in central Japan, home to more than 15 million people. Chubu's employees have been primarily relying on e-mail, mailing lists, and file service to communicate and share files with each other. However, an inability to manage the information exchange has resulted in outdated or incorrect information, lowered employee productivity, and dissatisfied customers. After adopting the appropriate groupware, information sharing and decision-making processes between employees have become more effective and efficient. Chubu has also enjoyed the additional benefit of reducing product development lead time and increasing the number of product and service inventions that can resolve emerging energy and environmental issues (IBM 2013).

### Reduce Capital Investment Costs

Computer-aided manufacturing (CAM) and CAD can reduce manufacturing and design cycle times. Capital costs are low for the investment of CAM and CAD, in comparison with the investment of product models and prototypes.

## Intangible Benefits

Intangible benefits can be real, but they are usually more difficult to quantify.

### Improve Customer Relations

The need to improve customer relations can sometimes take a strange turn. Rainer and Cegielski (2011, p. 309) offer an interesting example. On January 15, 2009, U.S. Airways Flight 1549 crash-landed in the Hudson River after striking a flock of Canada geese on takeoff. Fortunately, all 155 passengers survived. Like all carriers, U.S. Airways had a playbook for such incidents. Their network of gate agents, reservation clerks, and other employees were dispatched to handle the survivors in the most customer-friendly way they could. They provided cash and credit cards for employees to buy medicines, toiletries, and personal items that passengers needed. They had cases of prepaid cell phones and sweatsuits for anyone who needed dry clothes. They escorted each passenger to either a new flight or a local hotel, where the company arranged for around-the-clock buffets. They arranged train tickets and rental cars for those who did not want to fly. They retained locksmiths to help passengers who lost their keys get back into their cars and homes. They also followed up after the crash with ticket refunds and a $5000 check to cover the costs of replacing their possessions. The case describes other measures the airline took with the result that at the time the case was written, over one-third of the passengers had flown again with U.S. Airways.

### Improve Decision Making

ERP integrates information across functional departments. Employees in different departments can see the same information and synchronize the updating process. Good information is essential to the performance measurement process. Information gathered from company operations and reports can be further refined for use on management dashboards or balanced scorecard formats for use at all levels of the organization.

### Streamline Administrative Processes

RFID automates the product identification process and increases information transparency. Because of the enhancement of product visibility, RFID can cut waste and streamline administrative processes. Software offers business process management solutions to help clients integrate their enterprise applications such as order fulfillment and warehouse management. Chevron Texaco streamlined the data flow from its existing ERP software. Major benefits include the increased visibility of spending transactions, improvement of invoice payment process, increase of employee productivity, and reduction of operational costs (Chevron 2008).

### Integrate Company Planning and Execution

ERP is enterprise-wide software designed to integrate functional information systems. Integration provides managers a broad global overview of enterprise operations. A manager can orchestrate and integrate business activities across the supply chain to increase the ability to sense, and respond, to external environment changes. ERP can increase the degree of alignment between strategic planning and operational execution. Basell Polyolefins, a Netherlands-based manufacturer, used an ERP application to improve scheduling productivity and efficiency. ERP applications empowered Coca-Cola to align its strategic planning with execution. Major benefits resulting from Coca-Cola's ERP adoption include the increase in shipment accuracy, a reduction in delivery cycle times, and an improvement in labor productivity (sap.com/ecosystem).

## Barriers to IT Adoption

Most ITs fall into the category of innovations. As with all innovations, they take time to gain acceptance and use in the marketplace. Rogers documented a number of instances in which potential recipients rejected obviously beneficial technologies (Rogers 2004).

An organization needs to recognize three major groups of obstacles that inhibit IT technology reaching general acceptance: technological, managerial, and societal. Technological obstacles encompass the technological capabilities, standards interoperability, as well as application scope. These issues contribute to the differences among companies in the perceived ease of use and perceived usefulness of the technology. We will use RFID as a representative technology in the following section. These obstacles are also applicable to adopting emerging supply chain technologies, such as videoconferencing SaaS and cloud computing.

### Technological Obstacles

RFID technology is expected to be superior to bar code technology at some point in the future, although it does not yet equal bar codes in read accuracy and reliability and is also more costly. RFID has the potential to be superior to bar codes in its technical capabilities. It can read multiple objects at the same time, has a higher tolerance of harsh conditions, and provides better security control. However, these promises are not yet reality.

#### *Standards Interoperability*

RFID needs three key components to operate: an RFID reader, a tag with antenna, and a middleware software. RFID readers and tags vary in their consistency at reading objects at different granularity levels and in different forms of objects. The standards variation has limited the scalability and application scopes. Standards interoperability issues occur in front end and back end. The incompatibility between readers and tags is front-end issues. The incompatibility between information stored in RFID stations and information stored in the existing system across supply chains is back-end issues.

#### *Back-End Interoperability*

RFID can generate considerably more data than obtained from bar codes. It is crucial to build a secure and reliable network ecosystem to process the data collected via RFID and move it across thousands of companies in an integrated supply chain. However, how to process the huge amount of data on a real-time basis to support business decisions between business partners can be a major challenge. Three key layers of applications account for the back-end interoperability: (1) middleware, (2) integration layer, and (3) enterprise applications.

#### *Scope of Applications*

Uses of RFID include toll collection, access authentication, toll ticketing, and automobile immobilization devices. RFID provides tracking features that help a company know the physical location of an object within the supply chain during its lifetime. Its tracing (pedigree) features provide historical information about the life of the object, such as the time spent at each location, record of ownership, packaging configuration, and environmental storage conditions for a particular object. This is an especially important feature for perishable goods and some pharmaceuticals.

### *Security*

Information has value in an organization. Data are transformed into usable information through the variety of technologies described in this chapter. As this transformation progresses, the information created increases in value. In some cases, it reaches the state of intellectual property or information that is unique to a particular company and that provides it with a competitive advantage. As companies engage in more connectivity with other companies, they must use care in the design and operation of their information systems to prevent loss of information to competitors or other undesired sources. Security in electronic information systems is a growing concern and requires continuing diligence by IT management.

## Managerial Obstacles

In addition to the technical obstacles, managerial issues including power structure, trust, and economic issues could influence the network effects of RFID adoption. The low degree of enthusiastic cooperation from Wal-Mart suppliers is a visible case to substantiate the importance of managerial issues. Interorganizational relationships include strategic alliance, partnership, joint venture, delegation, R&D consortium, and many networked organizations (Ring and Van de Van 1994).

To reduce cost or improve operational efficiency through sharing a greater volume of sensitive information, RFID applications require buyers and suppliers to form a stronger IOS than with bar code applications because of the increased amounts of shared information. This includes the willingness of both parties to share information and collaborate. For instance, how willingly and how much trust do the suppliers—P&G and its competitors—have in sharing their company's information with Wal-Mart? Or conversely, how much confidence do retailers have in sharing their sales and marketing plans with their suppliers?

### *Power Structure*

A set of mechanisms should be deployed, ranging from convincing power to compulsory power, to influence an organization to adopt an IOS. To realize the aforementioned benefits, RFID needs to be used as part of an IOS to support business operations. As such, these two kinds of power mechanisms are important for the consideration of RFID adoption (Hart and Saunders 1997).

The convincing power mechanism focuses on reward and incentive approaches to encourage an organization to adopt RFID; this approach is useful to maintain a long-term alliance relationship. In contrast, the compulsory power mechanism, like punishment, is often used when an organization has many partners and helps influence business partners with a relatively low bargaining power to implement RFID.

Firms with dominating power over its business partners—such as the ability to penalize business partners by reducing or canceling orders—are more likely to force the use of IOS. The more bargaining power a firm has over its business partners, the higher the possibility that an organization can obtain resources from its business partners. This unbalanced relationship will force companies to maintain a cooperative relationship and make them more receptive to the adoption of RFID.

### Trust

Trust is another important factor influencing the interorganizational relationship (Smith et al. 1995). Once an organization adopts an IOS, it decides to extend its use based on the degree of trust by business partners (Hart and Saunders 1997). That is, when the degree of integration between business partners increases, they can access information not previously available or inaccessible. Trust in business partners could encourage the willingness of an organization to be open for negotiation and share information (Nidumolu 1989). Trust is also an important factor for the success of RFID. Trusting organizations are more willing to invest in RFID and share information with their business partners. Moreover, trust can stop opportunist behavior from appearing. When the opportunists' behavior declines, the opportunity to share information with business partners improves.

### Economic Issues

There are hidden costs associated with the early adoption stage of RFID. These costs include not only the direct costs of tags and readers but also middleware, R&D, consulting fees, system changes, maintenance, training, testing, management support, and conversion from bar code to RFID, service by third-party service providers, and achieving integration across the supply chain. The concept of TCO can provide a clearer picture of the cost associated with RFID investment.

RFID spending falls into four categories: hardware, software, internal labor, and consulting services. Most of the successful applications of RFID occurred when the cost of the RFID tag was insignificant when compared with the cost of the tagged item, such as a truck or a locomotive. RFID tag cost becomes much more of a concern when the cost of the item being tagged is lower, such as in clothing or groceries.

### Employee Acceptance

The best designed IT can fail if the employees are not properly trained in its use. Improper or lack of training can cause damage to both customer relations and employee morale. Learning a new method of doing things is often difficult, and employees have to be convinced of its benefits before they are willing to accept it enthusiastically. Sometimes, companies learn the hard way when they overlook good employee orientation and training.

## Societal Obstacles

Despite RFID's potential benefits, many companies are interested in obtaining the information for commercial usage. For instance, an insurance company would be interested in knowing the health history of its client in order to adjust the insurance coverage for that client. Retailers may potentially misuse the information by tracking their customers' buying habits for marketing purposes. Surveys indicate a high percentage of consumers are concerned about invasion of privacy. Consumers believe that government is the most likely organization to abuse consumer privacy rights with RFID, followed by banks, insurance companies, and credit card companies (Stegeman 2004).

All of the areas described earlier—technological, managerial, and societal—have an effect on the rate of RFID adoption. Businesses will adopt RFID technology more quickly in some industries or applications where the obstacles are less formidable.

## Model of an Integrated Supply Chain Information System

Business partners often have trouble integrating their systems and sharing information in an effective manner after making independent IT investment decisions over time. Without the coordination of business partners in making IT investment decisions, supply chain strategy would not be able to align with the required architecture and infrastructure among business partners. The misalignment could increase operational costs and decrease flexibility of supply chains. Therefore, a long-term perspective of how to integrate IT across supply chain partners is equally important as the short-term IT investment decision.

An integrated supply chain information system needs to create a closed loop from supply chain strategy to information systems architecture and to infrastructure, as shown in Figure 14.3. Information systems are a means to achieve supply chain strategy, which can help develop business goals. Business requirements are associated with each business goal. It is important to design supply chain systems to meet various requirements and help accomplish corresponding business goals. The designer translates the high-level architecture into detailed technical design, including systems requirements, standards, data elements, and process flows. Information systems, including hardware, software, procedures, IT professionals, and databases, can be developed in-house or outsourced from vendors. To support the varying needs of business partners in the supply chain, these information system components should be integrated as a coherent infrastructure. The infrastructure can exist at several levels, encompassing the entire organization, and across the supply chain. The presence of an integrated information system infrastructure is indispensable for the success of supply chain strategy. Once the infrastructure is in place, the detailed system design is completed, implemented, and operated until the next round of supply chain strategies are considered.

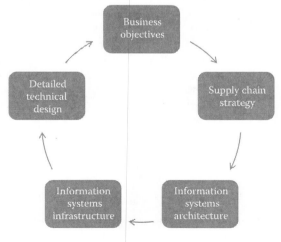

**FIGURE 14.3**
An integrated supply chain information system model.

## Summary

Information flow is essential for operating and managing a successful supply chain. The absence of information flow can jeopardize the flow of goods and funds. As a result, many business problems arise because of the inefficient and ineffective information flows—customer complaints about back orders or too much inventory in the warehouse. The delivery of goods and services is unpredictable. However, sharing information is not a natural trait for business partners of internal and external factors. Interorganizational trust and building bargaining power are internal factors. Unpredictable market dynamics are external factors. The presence of these factors demonstrates the urgency of sharing information across the supply chain to ensure uninterrupted and relevant information flows.

Information needed to support upstream, midstream, and downstream segments of a supply chain varies greatly. It is important to know what information needs to be created, distributed, and shared among business partners. A wide range of supply chain technologies are available to support business activities within each supply chain segment. Each supply chain technology has its strengths and weaknesses. Learning about them is critical to the success of adopting supply chain technology. A business also needs to learn how to measure its tangible and intangible costs and benefits of the adopted technology. Technological, managerial, economic, and social barriers can arise in the process of assimilating supply chain technology. An effective assimilation of supply chain technology can help enhance the competitiveness of a business and its supply chain.

Many supply chain technologies, such as videoconferencing, SOA, and cloud computing, are maturing. A manager needs to learn about these new technologies and assess their applicability to the existing supply chain infrastructure. A company can have the first-mover advantage if it can turn the identified emerging supply chain technologies into its competitive advantage. Adopting an existing or evolving supply chain technology is an IT project. Effective IT project management can help a company support its business strategies with the use of supply chain technology.

One of the current topics related to supply chain information flow is big data and data analytics. Persons knowledgeable about data analytics agree it has potential but requires top management support if it is to be effective in a company. According to McKinsey & Company, "Today, the power of data and analytics is profoundly altering the business landscape, and once again companies may need more top management muscle. It's becoming apparent that without extra executive horsepower, stoking the momentum of data analytics will be difficult for many organizations" (Brown et al. 2013).

We will describe those topics more fully in Chapter 18 in a discussion of the future of supply chains.

## Hot Topic

The company profile for Chapter 12 was about Boeing and its difficulties in integrating its 787 Dreamliner supply chain. It is also the basis for this chapter's hot topic that explores the need for good information flow.

## Hot Topics—The Boeing 787: A New Supply Chain Model in the Commercial Aircraft Industry—Part 1

The Boeing 787, also known as the Dreamliner, was introduced in Chapter 11 "Company Profile" of Boeing. In this series of hot topics on the Boeing 787 (hereafter referred to as the Dreamliner), we will expand on the implications of this aircraft within the supply chain. The Dreamliner is the most innovative commercial aircraft to date and represents a new approach to SCM for that industry. Despite the fact that the introduction was fraught with delays and a battery problem, the plane is now off the ground and deliveries are being made to the airlines. For students of supply chains, it is important to see how the Dreamliner is different from previous aircraft that Boeing has produced.

## HOW THE BOEING 787 SUPPLY CHAIN IS DIFFERENT

Models built prior to the Dreamliner used a build-to-print model whereby Boeing specified in detail the parts and assemblies to be made (Bryson and Rusten 2008). With the Dreamliner, Boeing moved the manufacturing and development (specifications) of key components to the supplier. This move helped to spread out the development cost of the aircraft by vesting tier-one suppliers into the project. The details of this new supply chain model are discussed in the following.

*The sourcing strategy has shifted to more outsourcing*: Approximately 70% of the Dreamliner is outsourced compared to 35%–50% for previous Boeing models (Tang and Zimmerman 2009). For the first time on an aircraft model, Boeing has outsourced the design and making of the wings (Bryson and Rusten 2008). This function is in the hands of three Japanese companies: Fuji, Mitsubishi, and Kawasaki.

*The number of suppliers is reduced*: Perhaps the most dramatic shift in supply chain strategy is the movement away from using thousands of suppliers for the making of an aircraft. Instead, for the Dreamlifter, approximately 50 tier-one suppliers are used (Tang and Zimmerman 2009).

*The overall responsibility of the suppliers has increased*: The 50 tier-one suppliers are now responsible for both the design and construction of a major subassembly to the aircraft. This is in sharp contrast to previous Boeing models whereby suppliers did not design parts and subassemblies, but instead built according to Boeing specifications (Bryson and Rusten 2008).

*Boeing has moved to the role of systems integrator*: Boeing's role in the plane-making process is more akin to that of a systems integrator (Hart-Smith 2001), an approach that manages the production of subassemblies made by the tier-one suppliers. The tier-one suppliers now take responsibility for the entire design and building of a subassembly and in managing their own suppliers (MacPherson and Pritchard 2007).

*Large subassemblies are transported via specially made Dreamlifters*: Modified Boeing 747s, called Dreamlifters, are used to transport larger subassemblies of the plane to its final assembly point in Everett, Washington. For example, after the wings are completed by the Japanese partners, they are loaded on Dreamlifters and flown to the United States. For this logistics function of the supply chain, Boeing shares a common practice with its rival, Airbus, which uses a similar approach by moving large sections of its A-380 via a modified Airbus A-300 called Beluga.

*Passenger input was obtained*: Unlike previous Boeing models, the company utilized several research methods to tap into the emotional and psychological feelings of passengers (Emery 2010). What was learned from this research was then incorporated into the design of the aircraft. Within the cabin, higher humidity levels and an enhanced air purification system add to passenger comfort. Outside the plane, innovations to the wing have improved the gust suppression abilities of the aircraft, enabling passengers to experience a smoother flight (Emery 2010).

*More composite materials are used*: The Dreamliner uses more composite materials than any existing commercial aircraft. Composites are mixtures of resins and high-strength carbon, graphite or glass fibers, and boron glass fibers (Bryson and Rusten 2008). One advantage of composites is better fuel efficiency since lighter materials are being used. Composites take the place of aluminum, which is heavier and tends to corrode. Since composites do not corrode, lower maintenance costs to the aircraft are achieved. Composites are also stronger than aluminum and require fewer fasteners, which results in a lower manufacturing cost (Tang and Zimmerman 2009).

*Tier-one suppliers are selectively clustered throughout the global supply chain*: Boeing is using supply chain partners worldwide but has clustered the making of the main airframe components in Japan and several European countries. The following table shows this arrangement.

| Subassembly or Module for the Airframe | Supplier | Country of Origin |
|---|---|---|
| *Wing* | | |
| Center wing box | Fugi | Japan |
| Wing | Mitsubishi | Japan |
| Fixed trailing edge | Kawasaki | Japan |
| Leading edge | Spirit | United States |
| Wing tips | KAL-ASD | South Korea |
| Movable trailing edge | Boeing | Australia |
| *Fuselage and doors* | | |
| Front forward fuselage | Spirit | United States |
| Forward fuselage | Kawasaki | Japan |
| Center fuselage | Alenia | Italy |
| Aft fuselage | Boeing/Vought | United States |
| Passenger entry doors | Latecoere | France |
| Cargo access doors | Saab | Sweden |
| *Tail assembly* | | |
| Tail fin | Boeing | United States |
| Horizontal stabilizer | Alenia | Italy |
| *Landing gear* | | |
| Wheel well | Kawasaki | Japan |
| Landing gear doors | Boeing | Canada |
| Landing gear | Messier-Dowty | England |

Source:   Tang, C. and Zimmerman, J., *Supply Chain Forum*, 10(2), 74, 2009.

## QUESTIONS FOR RESEARCH AND DISCUSSION

1. What potential limitations does this new supply chain arrangement pose?
2. Why do you think Boeing choose to source to companies in Europe and Japan?

## REFERENCES

Bryson, J. and Rusten, G., Transnational corporations and spatial divisions of 'service' expertise as a competitive advantage: The example of 3M and Boeing, *The Service Industries Journal*, 28(3), 307–323, 2008.

Emery, B., Innovation in commercial aircraft: The 787 Dreamliner cabin, *Research Technology Management*, 53(6), 24–29, 2010.

Hart-Smith, L., Out-sourced profits: The cornerstone of successful subcontracting, Paper presented at the *Boeing Third Annual Technical (TATE) Symposium*, St. Louis, MN, February 14–15. http://seattletimes.nwsource.com/ABPub/2011/02/04/2014130646.pdf, retrieved January 25, 2014.

MacPherson, A. and Pritchard, D., Boeing's diffusion of commercial aircraft technology to Japan: Surrendering the U.S. industry for foreign financial support, *Journal of Labor Research*, 28(3), 552–566, 2007.

Tang, C. and Zimmerman, J., Managing new product development and supply chain risks: The Boeing 787 case, *Supply Chain Forum*, 10(2), 74–86, 2009.

## Discussion Questions

1. Why is information flow so essential for running and managing a supply chain today?

2. Business partners transmit information across the supply chain. What types of information do business partners transmit in upstream, midstream, to downstream supply chains?

3. What supply chain technologies are available today? How can a business leverage these technologies to support supply chain activities?

4. Supply chain technologies deliver tangible and intangible benefits. How can a business measure these benefits?

5. Supply chain technology follows the technology adoption path. A company needs to move through the adoption path in order to assimilate supply chain technology into the operational process. However, many different barriers are ahead of the adoption path. What are these barriers?

6. What are emerging supply chain technologies?

7. Supply chain systems like ERP and EDI transcend organizational boundaries. How can a business assimilate supply chain systems in an effective manner?

8. What are some of the security issues that inhibit increased flow of information along the supply chain?

9. Why is trust an important consideration in designing and implementing IOS systems?

10. Will videoconferencing ever replace face-to-face meetings? Why or why not?

## References

Anonymous, Videoconferencing vulnerable, security expert says, *Information Management*, 46(3), 13, May/Jun 2012.

Ashford, W., Managing risks in the cloud, *Computer Weekly*, 8, December 2–8, 2008.

Banerjee, S. and Golhar, D.Y., Electronic data interchange: Characteristics of users and nonusers, *Information & Management*, 26(1), 65–74, 1994.

Blackstone, J., *APICS Dictionary*, 14th edn., APICS—The Association for Operations Management, Chicago, IL, 2013.

Botta-Genoulaz, V., Millet, V., and Grabot, B., A survey on the recent research on ERP systems, *Computers in Industry*, 56, 510, 2005.

Bozarth, C.C. and Handfield, R.B., *Introduction to Operations and Supply Chain Management*, Prentice Education Inc., Upper Saddle River, NJ, 2005.

Brandel, M., Social networking goes corporate, *ComputerWorld*, 25–27, August 11, 2008.

Brown, B., Court, D., and Willmott, P., Mobilizing your C-suite for big-data analytics, McKinsey Insights and Publications, www.mckinsey.com/Insights/Business_Technology/Mobilizing_your_C_suite_for_big_data_analytics, accessed November 14, 2013.

Casola, V., Mancini, E.P., Mazzocca, N., Rak, M., and Villano, U., Computer communications, *Computer Communications*, 31(18), 4312, 2008.

Cecere, L., Building the extended supply chain: If only it was like Legos, *Forbes*, January 27, 2014. http://www.forbes.com/sites/loracecere/2014/01/27/building-the-extended-supply-chain-if-only-it-was-like-legos/, accessed April 15, 2014.

Chevron, 2008 Annual report supplement. http://www.chevron.com/search/?k = ERP&text = erp&Header = FromHeader&ct = All%20Types, accessed August 1, 2013.

Davenport, T.H., Putting the enterprise into the enterprise systems, *Harvard Business Review*, 76(4), 121, 1998.

Eck, M., Advanced planning and scheduling: Is logistics everything? A research on the use (fullness) of advanced planning and scheduling systems, Technique Report, Vrije University, Amsterdam, Holland, 2003.

Fan, M., Kumar, S., and Whinston, A.B., Short-term and long-term competition between providers of shrink-wrap software and software as a service, *European Journal of Operational Research*, 196(2), 661, 2009.

Farb, B., Cloud expected to keep rising, SaaS revenue will surpass $21 billion by 2015, Customer Relationship Management, September 2011. www.destinationCRM.com.

Finnie, S., Peering behind the cloud, *Computerworld*, 42, 22, 2008.

Fontana, J., Big-picture planning key to SaaS success, *Network World*, 26(10), 1, 2009.

Frederick, J., Need to reinvent Rx supply chain is RFID conference focus, *Drug Store News*, 29(15), 24, 2007.

Gallego, G., Phillips, R., and Sahin, O., Strategic management of distressed inventory, *Production and Operations Management*, 17(4), 402, 2008.

Hall, M.E., SaaS surprises, *Computerworld*, 43(12), 18, 2009.

Harris, A., IT consultant, Personal interview, August 8, 2014.

Hart, P. and Saunders, C., Power and trust—Critical factors in the adoption and use of electronic data interchange, *Organization Science*, 8(1), 23, 1997.

Hugos, M., *Essentials of Supply Chain Management*, 2nd edn., John Wiley & Sons, Inc., Hoboken, NJ, 2006.

IBM, Service oriented architecture. http://www-01.ibm.com/software/solutions/soa/, retrieved April 16, 2009.

IBM, Chubu Electric speeds information sharing and decision making by creating a new, enterprise-wide collaboration system using IBM Lotus solutions. http://public.dhe.ibm.com/software/in/lotus/SWC14016USEN.pdf, accessed August 1, 2013.

Johnson, J.R., Retailers: Item level RFID could be ubiquitous by the end of 2012, RFID 24-7, January 1, 2011. www.rfid24-7.com/articles/011011.shtml, accessed February 15, 2014.

Kenward, M., Supply chain communication (surveys edition), *Financial Times*, 5, February 17, 2003.

Kwon, T.H. and Zmud, R.W., Unifying the fragmented models of information systems implementation, in R.J. Boland and R.A. Hirschheim (eds.), *Critical Issues in Information Systems Research*, Wiley, Chichester, U.K., pp. 227–252, 1987.

Lambert, D.M., *Supply Chain Management: Processes, Partnerships, Performance*, 3rd edn., Supply Chain Management Institute, Sarasota, FL, 2008.

Leonard, L.N.K. and Davis, C.C., Supply chain replenishment: Before-and-after EDI implementation, *Supply Chain Management*, 11(3), 225, 2006.

Lin, P.P., SaaS: What accountants need to know, *The CPA Journal*, 80(6), 68–72, 2010.

MacVittie, L., A primer on cloud computing, *Network World*, 25(49), 21, 2008.

Malone, R.A., *Chain Reaction: How Today's Best Companies Manage Their Supply Chains for Superior Performance*, Kaplan Publishing, New York, 2007.

Merono-Cerdan, A.L., Soto-Acosta, P., and Lopez-Nicolas, C., How do collaborative technologies affect innovation in SMEs? *International Journal of e-Collaboration*, 4(4), 33, 2008.

Mitra, K. and Layak, S., Live in the cloud, *Business Today*, 18(8), 76, 2009

Nidumolu, S.R., The impact of interorganizational systems on the form and climate of seller-buyer relationships: A structural equations modelling approach, *Proceedings of the 10th International Conference on Information Systems*, Boston, MA, 1989.

Parker, K., SAP senior execs talk about ERP, SOA, and the demise of three-letter acronyms, *Manufacturing Business Technology*, 26(1), 8, 2008.

Peters, L.R.R., Is EDI dead? The future of the Internet in supply chain management, *Hospital Material Management Quarterly*, 22(1), 42, 2000.

Premkumar, G., Ramamurthy, K., and Crum, M., Determinants of EDI adoption in the transportation industry, *European Journal of Information Systems*, 6(2), 107, 1997.

Rainer, R.K., Jr. and Cegielski, C.G., *Introduction to Information Systems*, 3rd edn., John Wiley & Sons, Inc., Hoboken, New Jersey, 2011.

RFID 24-7, RFID in retail: Midyear report, http://www.rfid24-7.com/article/rfid-in-retail-midyear-report/, July 17, 2014, accessed July 20, 2014.

Richmond, R., Microsoft to offer free trial of computer-security service, *The Wall Street Journal*, B2, November 30, 2005.

Ring, P.S. and Van de Van, A.H., Developmental processes of cooperative interorganizational relationships, *Academy of Management Review*, 19(1), 90, 1994.

Robey, D., Im, G., and Wareham, J.D., Theoretical foundations of empirical research on interorganizational systems: Assessing past contributions and guiding future directions, *Journal of the Association for Information Systems*, 9(9), 497, 2008.

Rogers, E.M., *Diffusion of Innovation*, 5th edn., Free Press, New York, 2004.

Rowell, A., Siemens PLM launches "industry catalysts" to accelerate PLM deployments—and other 2013 Siemens PLM event highlights, IDS Manufacturing Insights, November 2013. www.idc.com/downloads/MIcurrentnewsletter.html, accessed January 13, 2014.

SAP, Enterprise resource planning, http://www.sap.com/solutions/business-suite/erp/index.epx, accessed October 16, 2013.

SAP, Explore customer stories, http://www.sap.com/ecosystem/customers/customers/index.epx, accessed October 10, 2013.

Smith, K.G., Carroll, S.J., and Ashford, S.J., Intra- and interorganizational cooperation: Toward a research agenda, *Academy of Management Journal*, 38(1), 7, 1995.

Turban, E., Leidner, D., McLean, E., and Wetherbe, J., *Information Technology or Management: Transforming Organizations in the Digital Economy*, John Wiley & Sons, Inc., Hoboken, NJ, 2008.

Turbide, D., Advanced planning and scheduling (APS) systems, *Midrange ERP Magazine*, 1, 1998.

Walker, W.T., Supply chain consultant, Personal correspondence, May 4, 2009.

Weinberg, N., Nine hot technologies for '09, *Network World*, 26, 13, 2009.

# 15

## Funds Flow along the Supply Chain

### Learning Outcomes

After completing this chapter, you should be able to

- Describe the three types of flows along a supply chain
- Explain why cash flow is needed within a company
- Discuss the benefits and obstacles of funds flow in the supply chain
- Identify the various sources of external funds in a supply chain
- Explain why funds flow depends on product and information flows
- Explain why net income is different from net cash flow
- Discuss how inventory management affects cash flow
- Explain the importance of electronic funds transfer (EFT)
- Describe the cash-to-cash cycle
- Explain the relationship between the balance sheet, income statement, and cash flow statement

### Company Profile: Wells Fargo

In Chapter 14, we discussed how information flow is necessary to facilitate the flow of goods. Both goods and information have to flow before funds flow. Funds flow is about how customers, at any point in the supply chain, pay their suppliers for the goods and services received. Just as good inventory management is necessary for companies to succeed, good cash flow management is also necessary, even essential. Suppliers are often smaller than their customers and depend on their customers for prompt payment.

Funds are increasingly being transmitted electronically, another job for the information systems. As you read this company profile and the chapter in the book, think about these questions:

- What are the ways in which the operations function can help improve cash management?
- What is the cash-to-cash cycle time? What are its components?
- Why are companies increasing their use of electronic payment systems?
- Why is net cash generated not the same as net profit?

What additional questions do you think relate to funds, or cash, flow?

## Company Profile: Wells Fargo

### INTRODUCTION

Chapter 15 is about funds flow. In today's business world, the customers often are much larger than their suppliers. In addition to putting pressure on suppliers for lower prices, customers often request, even demand, longer payment terms for their purchases. Sometimes, it takes a third party to ease the tension between customer and supplier, especially when it relates to cash flow. The third party in this case is a bank, Wells Fargo. Founded in 1852 by Henry Wells and William Fargo to serve the Western frontier, the bank has grown steadily from a local bank to a global provider of extensive and diversified financial services. One area in which it has moved rapidly is in supply chain financing, a new form of supplier financing that will be described later in this profile.

Many companies, especially those growing rapidly, are discovering, or rediscovering, that profits are different from cash flow. Does participation in a supply chain help or hinder cash flow? Sometimes, it depends on where your company fits in the supply chain. Would companies like to improve their flow of funds? Almost universally, the answer is "Yes," especially small companies, who are often the most pressed for funds to pay their bills.

What do we mean by cash flow? One definition is that "the financial supply chain is the flow and uses of cash throughout the physical supply chain. Where products, services, and/or information are transferred, there is an accompanying flow of cash" (Roberts 2002).

There are three major flows in a supply chain—the physical flow of goods and services, information, and funds. All are necessary if a supply chain is to function and flourish. The flow of goods and services was described in earlier chapters, and information flow was covered in Chapter 14.

Funds flow, or the flow of money, is required in a supply chain. The money flows from the consumer upstream in a supply chain until all suppliers have received payment for the goods and services they provided. Although there are other funds flow in a company, such as for equipment purchases and payroll, we will focus on the flow along the supply chain, which affects the working capital of a company—its accounts receivable, inventory, and accounts payable. While the flow of funds is mandatory if a supply chain is to exist, it is still an uncoordinated and suboptimized flow in many supply chains.

## BENEFITS OF IMPROVED FUNDS FLOW

Apple Inc., which will be discussed in the Company Profile for Chapter 16, is quite unique.

## WAYS TO IMPROVE

What can companies do to improve the flow of funds along the supply chain? First, companies should work to improve the flow of goods and information, because funds flow depends on their correct and complete flow. They should align the funds flow more closely with the first two flows. As Bernabucci (2008) puts it, "Fixing a cash-flow problem requires companies to examine and improve the three key flows of commerce: goods, information and funds." If optimizing the physical supply chain hinders the financial supply chain, it is not an optimal method for the overall supply chain.

Companies should try to integrate more closely the processes dealing with funds flow. Eliminating the batch processing and the functional silos would be desirable. However, if this is not feasible or appropriate, at least, they should view the process as continuous and try to create a semblance of flow for funds.

One of the most promising measures of supply chain performance is the cash-to-cash cycle time. Practices such as lean manufacturing and sales and operations planning (S&OP) can improve (reduce) the cash-to-cash cycle time (Blanchard 2013).

The ways to reduce the cash-to-cash cycle time is to reduce accounts receivable, reduce inventory, or increase accounts payable. Assuming that reducing inventory is a result of lean manufacturing, the next areas to consider are accounts receivable and accounts payable. Supply chain finance (SCF) is a technique to do this.

## SUPPLY CHAIN FINANCE

SCF is a new concept that is beginning to get closer attention.

Banks are interested because it offers them an opportunity to get additional business. The customer gets extended payment terms from the lender—a bank or some other financing organization. The lender pays the supplier after discounting the payment, but at terms they would extend to the larger buyer, which are lower than the terms for the smaller supplier. "The result is that the supplier's working capital costs are reduced, even though its payment terms have been extended. It is then in a position to convert this cost reduction into price reductions to satisfy the buyer; the buyer gets the benefit of extended terms, lower prices, reduced capital costs and alignment of its procurement and finance interests; and the lender gets the benefit of a higher margin on the exposure to the buyer company" (Kerle 2007, p. 4).

Automation of the information is an important step in the SCF implementation process. One of the most important considerations when selecting SCF technology is the ease of interaction among the parties involved—buyers, suppliers, and third-party financial institution. Some of the processes that leading SCF users are automating include

- Electronic invoice presentment
- Trade-related document preparation and management
- Invoice matching/reconciliation (internal)
- Invoice approval process (internal)
- Purchase order management
- Electronic payment process

- Collaborative invoice discrepancy management
- Invoice discount management
- Charge-back management (e.g., invoice deductions)

The top challenges encountered by study participants during their deployment and implementation of their SCF technology included

- Internal integration issues
- Need to redesign business processes to fit the new solution
- Staff resistance
- Training time and costs
- Lack of internal information technology (IT) resources (Aberdeen 2007)

Collaboration is an active concept in the physical flow of supply chains. Collaboration is gaining greater acceptance in information flows. It follows that collaboration among supply chain participants will improve funds flow. Knechtges and Watts (2000) say this is especially important for the small business. They believe, "If the entire channel and the entire chain is working together to serve the customer's expectations to the highest levels, it can truly be a win-win partnership. The common goals go back to being able to have a positive impact on customer service, profit, cash flow, and transportation costs. Long-term, well-managed partnerships can help everyone in the supply chain achieve their goals."

## SUPPLY CHAIN FINANCE AT WELLS FARGO

Wells Fargo is a bank that is actively involved in SCF. SCF enables a bank to establish continuing relationships with both buyers and sellers. It is a program that benefits buyers, sellers, and the bank. The buyer can get extended payable terms while the seller reduces the days to collect their receivables. The bank benefits by getting additional revenues they have not had access to in the past. For the large company, often the buyer, it is a convenient way to earn income. However, for the seller, usually a smaller company, it is often essential to the welfare of the business.

The bank may also view SCF as another way in which they can serve their customers. Wells Fargo puts great emphasis on providing customer service. The chairman, CEO, and president, John G. Stumpf, starts his 2012 Letter to the stakeholders with this statement: "Each morning, across the company, our day starts with conversations – conversations about how best to serve our customers and help them succeed financially. We've been having those conversations at Wells Fargo for more than 160 years, and they are the cornerstone of our success" (Wells Fargo Annual Report 2013a, p. 2).

In addition to being a financially successful enterprise, Wells Fargo also prides itself on being a good corporate citizen. They prepare a Corporate Social Responsibility Interim Report, and, in their Annual Report, they outline some of the programs to which they devote resources.

- *Community investment*: We provide human and financial resources to help build strong communities. Their philanthropy includes community development, education, human services, arts and culture, civic, and environment.
- *Environmental stewardship*: We focus on integrating environmental mindfulness into our products, services, and operations. The company contributed $8.0 million to environmental grants.

- *Product and service responsibility*: We offer all customers responsible financial advice and solutions for now and the future. In this sector, they report lending small businesses $16 billion in 2012.
- *Team member engagement*: We support our team members professionally, financially, and personally. As part of this program, employees volunteered 1.5 million hours to nonprofit organizations in 2012.
- *Ethical business practices*: We ensure all business functions run responsibly and ethically. As a result of this emphasis, 99.96% of eligible team members completed the Code of Ethics and Business Conduct annual training in 2012 (Wells Fargo 2013 Annual Report, p. 26).

"Determining the optimal supply chain finance instrument enables both the buyer and supplier to minimize the cost of capital and develop a symbiotic business relationship. Newer financing options have succeeded in gaining the swiftness of the physical supply chain without creating administrative and cost burdens or added risk to either party. As technology has evolved, increased transparency into the physical-goods supply chain has enhanced information flow and lowered costs. The same is now becoming true for the financial supply chain. All of this equates to a new supply chain paradigm that reduces costs and risks, benefiting both the buyer and the seller" (Wells Fargo 2013b).

The objective of the Wells Fargo SCF program is to increase working capital and offer new financing options to their customers. Their channel finance program works with manufacturers and distributors to reduce credit exposure and shift risk off their balance sheets, provide growth capital to resellers (suppliers to the large companies) without increased credit risk, and outsource credit administration and receivables collection and management. The SCF program offers resellers the opportunity to obtain greater purchasing power with a larger credit facility, achieve greater liquidity with longer-term financing, and reduce costs with interest-free financing.

The bank also has a key accounts purchase program whereby large corporations that sell to domestic and international buyers can improve liquidity, increase sales and market share, reduce concentration risk, and mitigate critical business risks associated with selling internationally.

In the supplier finance program, large corporations with wide supplier networks can lower the cost of capital for domestic and international suppliers, meet target payable terms and cost-reduction objectives, and leverage their own creditworthiness to help solidify the financial well-being of their suppliers (Wells Fargo 2013c).

For a more comprehensive explanation of SCF from a banking perspective, see the paper by Bob Dyckman (2011), who, at the time he wrote the article, was the managing director and sales manager for the SCF team at Wells Fargo Capital Finance.

## SUMMARY

Successful supply chains manage three areas. First, they manage the physical flow of goods and services that provide value to the consumer. Second, they manage the information flow that documents and supports the physical flow. Finally, the information flow becomes an essential input to the funds flow from the consumer upstream in the supply chain until all suppliers receive compensation for their efforts. New technologies are making it easier for companies to improve the funds flow. As competition increases and supply chains lengthen around the world, the time required for reimbursement also grows longer. To maintain effective supply changes and to assure the effective flow of funds, the large companies and third-party providers will become important links in keeping the small, struggling suppliers alive and well.

## REFERENCES

Anonymous, Technology platforms for supply chain finance, Aberdeen Group, March 2007.
Bernabucci, R.J., Supply chain gains from integration, *Financial Executive*, 24(3), 46–48, 2008.
Blanchard, D., Measure what you're managing, *Industry Week*, 258(5), 44–46, May 2013.
Dyckman, B., Supply chain finance: Risk mitigation and revenue growth, *Journal of Corporate Treasury Management*, 4(2), 168–173, 2011.
Kerle, P., Steady supply, *Supply Chain Europe*, 16(4), 18, 4p., 2007.
Knechtges, J.P. and Watts, C.A., Supply chain management for small business—How to avoid being part of the food chain, *Hospital Material Management Quarterly*, 22(1), 29, 2000.
Wells Fargo, CEO letter to stakeholders, 2013a. https://www.wellsfargo.com/downloads/pdf/invest_relations/2012-annual-report.pdf, accessed October 24, 2013.
Wells Fargo, Treasury insights, The right supply chain finance instrument benefits buyers and suppliers, 2013b. https://treasuryinsights.wellsfargotreasury.com/?elqPURLPage=1624, accessed October 17, 2013.
Wells Fargo, Supply chain finance, 2013c. https://wellsfargo.com/com/international/businesses/supply-chain-finance, accessed October 17, 2013.

## Overview of the Flow of Funds

In Chapter 2, we explained that supply chains contain flows for three areas—goods and services, information, and funds or money. We described the flow of goods and services in Chapters 5 through 9 and information flow in Chapter 14. This chapter describes the flow of funds along the supply chain and the technology used to move funds electronically between supply chain partners.

Figure 15.1 illustrates the flow of goods and information along the supply chain. Goods and services flow downstream toward the ultimate consumer. Information flows both downstream from suppliers to customers and back upstream from customers to suppliers. Funds flow upstream, from the consumer to the origin of the supply chain.

The ultimate consumer is the primary source of funds for an entire supply chain. Once the consumer pays for the products or services purchased, the funds flow back upstream through the supply chain members.

The rate at which funds flow back through the supply chain varies with the practices of each entity in the supply chain. There is not yet a standard approach to the flow of funds. While a great deal of effort has been expended in smoothing and speeding the flow of products downstream to the consumer, the same amount of effort does not appear in the upstream flow of funds. While technology exists to speed the flow, there is not always the same sense of urgency among operations managers to improve the flow of funds.

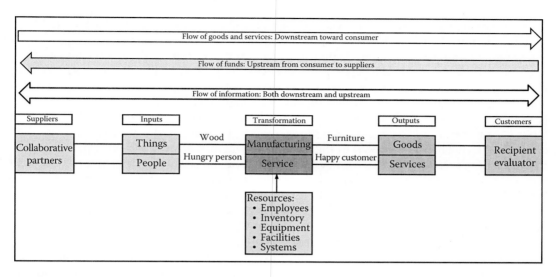

**FIGURE 15.1**
Supply chain flows—products, information, and funds.

## Need for Cash Flow within a Company

Before funds can flow smoothly along a supply chain, each participant in the supply chain must practice good cash management. In the short term, this usually involves management of working capital. In the long term, cash management involves selecting the best methods of using the cash generated from operations. It may mean investing in capital expenditures to grow or modernize the business, or it may mean reimbursing shareholders by paying dividends or buying back stock.

Good cash management within a company requires collaboration among all functions of an organization, especially among marketing, finance, and operations. At the end of this chapter, we include a comprehensive example of the relationships between the income statement (a report of how well the company performed) and the balance sheet (a report of the current financial condition of the company). We also show an example of how improvements in working capital management can significantly improve cash flow within a company.

An adequate cash flow is necessary for a business to survive. Profit does not equal cash flow, especially in small businesses. Most businesses use an accrual accounting system, as opposed to a cash-basis accounting system. Accrual accounting attempts to match the flow of expenses in the income statement with the flow of revenues. As a result, certain costs accrue and are not congruent with the cash expenditures.

One of the most obvious examples is the purchase of a capital asset, such as a piece of equipment. The equipment is expected to last several years, or well beyond interim, or even annual, financial reports. Rather than treat the entire cost of the equipment as an expense in the month in which it is purchased, the costs are amortized over the life of the equipment, for example, 5 years. Each month a portion of the total cost is charged to the income statement as depreciation, a cost of operation. In this case, the cash is expended in a particular month, but the costs are spread over a 5-year period.

Inventory is another example. When products are built, cash is expended for materials, labor, and overhead costs. When production exceeds sales, inventory increases. When inventory increases, some of the expenses associated with the inventory are stored as assets in the balance sheet and do not appear as costs in the income statement. In this case, cash expenditures exceed the expenses shown in the income statement, and the income looks better than if all the expenses were included. When inventory decreases, it has the opposite effect. The costs shown in the income statement exceed the cash expenses used in production; accordingly, income declines. This financial accounting practice has caused some companies to stop or even reverse their efforts to reduce inventories. Some managers incorrectly conclude that reducing inventories reduces income.

When businesses are growing, they need to acquire capital equipment and facilities, and they increase their accounts receivable and inventories. This means that they are probably showing a profit on their income statement, but may be desperate for cash to keep up with their growth. In this case, they have to look for funding from a lending institution, and this may be difficult because they may not have a credit history or a formal cash flow forecast.

## Supply Chain Funds Flow*

We will use *funds* to describe the flow of money along the supply chain because most companies do not use actual cash to satisfy their obligations. While the use of cash has decreased in business-to-business (B2B) transactions, it is still a significant factor with consumers. A comprehensive study of the supply chain describes how to keep cash flowing in the United States from the United States Mint through banks to ATMs to consumers (Rajamani et al. 2006).

### Flows in a Supply Chain

As previously described, there are three major flows in a supply chain—the physical flow of goods and services, information, and funds. All are necessary if a supply chain is to function and flourish.

The most visible flow is that of the goods and services. In manufactured products, they originate in the extraction (mining) industries and farms and flow toward the consumer through fabricators, assemblers, distributors, and retailers. Although not yet perfect, this flow is improving rapidly. For the most part, it receives major management attention and resource commitment. Of the three supply chain flows, the flow of goods and services is in the most advanced state of development.

Information flow is also receiving a great deal of attention. A supply chain requires only a minimal amount of required information, such as customer purchase orders, shipping notices, and invoices, to keep it operating. Most companies provide this information willingly and promptly. This minimal information is necessary to create the flow of funds along the supply chain. Additional information flows on a voluntary basis among supply

---

* This section has been adapted from Crandall (2008).

chain participants. This includes sales results, demand forecasts, and plans for special events, such as sales or product promotions. This additional information makes it possible for goods and services to flow faster and more smoothly. It also helps funds to flow more smoothly. Information flows in both directions—toward the consumer and from the consumer up the supply chain toward the suppliers. New technologies, and advances in collaboration among supply chain participants, are making information flow better, although it still encounters some turbulence in most supply chains.

Banks are an essential participant in electronic funds transfer (EFT). They act as the clearinghouse for the funds transfer. Usually, the customer has an account at a bank. After receiving the product from a supplier, the customer sends a message to the bank to release the funds needed to pay for the received shipment. The customer's bank then sends a wire transfer to the bank where the supplier has its account. The entire process takes only minutes instead of hours or days, as in the past.

## Benefits of Improved Funds Flow

The primary benefit of improved funds flow is to reduce the cash-to-cash cycle time. Most large retail companies operate with cash-to-cash cycle times in the 15–20-day range. Apple Inc., discussed in the Company Profile for Chapter 16, is quite unique with a negative cash-to-cash cycle time—they have minimal receivables and inventory and receive extended payable terms from their suppliers.

Improved funds flow improves customer–supplier relationships, and, conversely, improved customer–supplier relationships improve funds flow. As the flow of goods and services improves and information sharing advances, funds flow will improve. If payments were made more promptly and consistently, relationships would improve. The result would be a win–win situation throughout the supply chain.

Improved funds flow would also tend to reduce the imbalances among supply chain participants. Large retailers tend to demand more liberal payable terms from manufacturers. In turn, large manufacturing companies tend to demand more liberal payable terms from their smaller suppliers. If funds flow were aligned with product and information flows and integrated along the supply chain, it would tend to reduce the inequities resulting from company size or creditworthiness (Grealish 2005).

Blanchard (2013) points out that "Certain best practices, such as vendor-managed inventory, lean manufacturing and short-cycle demand planning, as well as a comprehensive sales and operations planning (S&OP) process, can reduce cash-to-cash cycle times."

## Obstacles

If improved funds flow is so desirable, what is keeping companies and their supply chain partners from achieving it? Researchers have found several reasons that appear to be the biggest obstacles.

*Technology*: Companies are not using the latest technology available. While few companies use cash anymore, many are still using checks. Checks are slower and less secure than EFT methods. In addition, companies have not adequately integrated funds flow with goods and information flows. Other companies have moved aggressively into web-based systems to economize international payments (Gentry 2006).

*Administrative processes*: Many companies have not yet streamlined their internal procedures. Although lean methodology is making great progress in manufacturing and distribution to create flow of physical goods, it is still waiting in the wings to make its entrance into most paperwork processes. As mentioned earlier, many administrative processes are still functionally separated into silo-like stations. Accordingly, batch processing of purchase orders, invoices, and checks is still more the norm than the exception.

*Errors*: Errors in the physical flow of goods and services, such as partial shipments or defective goods, cause delays in funds flow. Customers will not complete payment until their orders have been satisfied. Even if the physical flow is correct, there may be errors in the paperwork. Satisfactory reconciliation of a purchase order, a receiving report, and an invoice is a prerequisite to payment. Even small discrepancies can cause delays and expenditures of employee time to resolve. International transactions are especially risky because the buyer and seller may not have met or developed a trusting relationship. The seller will often insist on being paid up front or by an electronic letter of credit (Walker 2009).

*Adversarial attitude*: Many companies have not yet made the transition from an adversarial attitude toward their customers and suppliers to a collaborative one. It almost seems natural or inbred to say to customers: "Pay up or else." Credit managers develop an aura of being unyielding and unsympathetic. On the other hand, it is equally natural to push suppliers for extended credit terms because of *all the business we do with you*. This is especially true if the supplier is smaller and heavily dependent on *our business*. In addition, money is not something most companies want to talk about in mixed company (customers and suppliers). It is among the most confidential of topics. Companies may be willing to reveal their sales information or planned promotions, but they may not be willing to reveal their bad debt ratio or how well they are pressuring their suppliers for extended terms.

## External Sources of Funds

While members of the supply chain transfer funds among themselves, there are other entities that play important roles in the flow of funds along the supply chain. They are lending institutions, such as banks and third-party service providers, such as value-added networks (VANs).

### Banks and Other Lending Institutions

Companies, especially small companies, have always needed external sources of funds. Banks and other lending institutions have provided this funding in a variety of ways. Some of the most common methods of lending include the following:

- Short-term loans have been secured with working capital. These may be for weeks or a season—a farmer gets a loan to cover expenses until harvest time, or a manufacturer gets funds to build inventory for the Christmas season.
- Sometimes loans secured with working capital evolve into year-round *lines of credit* to finance the fluctuations during the year.
- Funds to purchase equipment are secured by the equipment itself. Equipment manufacturers, such as GE (airplane engines) or Caterpillar (earth-moving equipment), finance the purchase in order to sell the equipment.

- Another type of long-term debt is to cover expansion of a retailer's new stores. These can be secured directly by a lien on the store or based on the financial condition of the company.
- Sometimes customers provide financing to suppliers by lending money to the supplier to buy materials or add capacity to meet the customer's needs.

There are many variations of these. All of these methods existed before the modern-day supply chains were developed.

## Supply Chain Finance

Supply chain finance was described in the Company Profile at the beginning of this chapter.

In recent years, banks have also become more active in SCF. SCF is a new concept that is getting closer attention. Banks are interested because it offers them an opportunity to obtain additional business.

Timme and Wanberg (2011) describe a typical SCF cycle as follows:

*Step 1*: The supplier ships the goods, and the buyer matches the receipt to the invoice and approves it for payment on the due date.

*Step 2*: The buyer sends an electronic copy of the approved invoice to the bank verifying that it will pay the supplier the net amount on the due date via the bank.

*Step 3*: The bank notifies the supplier of the approved invoices, usually on an electronic platform web portal, and asks the supplier if it would like to finance the invoice to get paid early.

*Step 4*: If the supplier agrees, the bank discounts the payment at an agreed interest rate and purchases the invoice from the supplier. At this point, the supplier is cashed out of the transaction.

*Step 5*: The bank receives the full payment from the buyer upon the due date of the invoice.

The use of SCF is growing rapidly. Customers, often large retailers, or manufacturers, by getting extended payment terms, reduce their cash-to-cash cycle times and improve the return on their working capital investment (Hintze 2012).

One of the keys to implementing SCF is to have the technology to automate the process. This requires linking buyer with supplier through a third party, usually a bank. Third-party technology providers, such as PrimeRevenue, Syncada, Ariba, Bottomline Technologies, and Kyriba, are among a number of technology vendors. These vendors can also link multiple banks into a network to manage the SCF program for very large corporate customers that can have billions in payables outstanding (Hintze 2012).

Recently, SCF has been popular in Europe, where over 90% of the banks are experiencing strong growth in the demand for SCF. In the United States, there is also strong demand by large customers who want to ensure that key suppliers remain responsive and flexible to the buyers' changing needs in a dynamic business environment. According to Philip Kerle, "As the global economy builds up a new head of steam, SMEs – who make up the larger part of many essential supply chains – desperately need access to reasonably priced finance, but are currently unable to do so in their own right. By joining a supply chain finance programme, their cash flow can be eased through access to credit that is secured on their large corporate customer's credit rating – an extremely attractive proposition in the current tight lending market. No wonder, therefore, that financiers are reporting such interest in SCF" (Logistics Manager.com 2011).

SCF is one of the cash management techniques that involves multiple members of the supply chain. Where there is continuing collaboration, all parties appear to benefit. The buyer gets extended terms and lower costs, the supplier gets faster reimbursement, and the bank gets additional business. However, as in all new techniques that involve multiple parties, it takes time and coordination to link the parties together with compatible IT systems.

## Performance Measurement

One of the problems affecting funds flow is the performance measures used by different functions. The financial accounting function uses performance measures that differ from those used by management accounting.

### Financial Accounting Measures: Current Ratio

The financial accounting part of a business and its creditors look at working capital and current ratios. The total working capital is the sum of accounts receivable and inventory, minus accounts payable. The current ratio is calculated by dividing the sum of accounts receivable and inventory by accounts payables (in dollars or other monetary unit). From this perspective, high accounts receivable, high inventory, and low accounts payable are desirable because it indicates a greater capability to repay loans issued by the creditors. The traditional thinking has been that the higher the current ratio, the better. The general rule in many accounting books has been that the current ratio should be about 2.0 or better, although this guideline has been modified in recent editions as the effect of lean manufacturing has reduced the amount of inventory companies carry.

### Management Accounting Measures

Management accounting is more concerned with how well the company is using its resources. From this perspective, low accounts receivable, low inventory, and high accounts payable are desirable. The performance measures include accounts receivable (DRO), days of inventory (DIO), and days of account payable (DPO). The credit department is responsible for accounts receivable, the inventory managers for inventory, and the accounts payable department for extracting extended terms from their suppliers. In recent reports, Wal-Mart, Lowe's, Apple, and Hewlett–Packard have all reported current ratios well below the traditional standards, a result of their effective cash management processes.

In recent years, many businesses have been dealing with liquidity, or cash flow, problems. While the current ratio has been a long-term favorite among many financial people, some are now looking to a better measure of liquidity and have turned to the cash conversion cycle (CCC) or the cash-to-cash cycle time. The cash-to-cash cycle time, which is an integrated approach, is calculated by adding the days of accounts receivable (DRO) and days of inventory (DIO) and subtracting the days of accounts payable (DPO). The DRO is calculated by dividing the average accounts receivable by the net sales per day. The DIO is calculated by dividing the average inventory by the average cost of goods sold per day. The DPO is calculated by dividing the average accounts payable by the average costs of goods sold per day (Cagle et al. 2013). All of these calculations are in dollars or some appropriate monetary unit.

A lower cycle time is better, because it means that the company is using less cash to manage its business. Farris and Hutchison describe the cash-to-cash cycle and offer a number

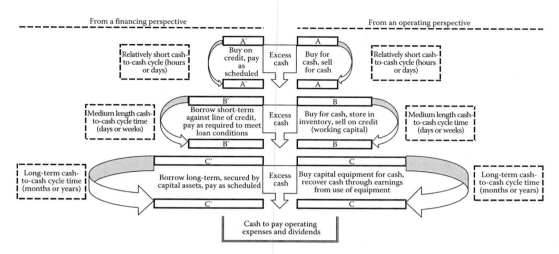

**FIGURE 15.2**
Cash flow cycles.

of examples of its effect in such companies as Dell (Farris and Hutchison 2002). Walker (2005) provides additional insight about the importance of this measure.

Figure 15.2 shows the different cash flow cycles that are common to most businesses. There are three major cash flow cycles. The first is when a business sells on a cash basis and buys for either cash or credit and repays their suppliers quickly. The second cycle takes longer, because the business invests their cash in the inventory and sells on credit. On the financing side, they may use a line of credit at a bank to support their medium-term purchasing requirements. The third cycle is the longest. This occurs when a business invests cash in long-term capital projects, such as equipment or buildings. To finance this, they may need to set up long-term repayments schedules with banks or third-party lenders.

A company that is known for its good cash management is Apple. Some of this has come from their extensive outsourcing of manufacturing processes to low-cost suppliers, thereby reducing their capital investment in equipment and facilities. A recent report suggests that Apple is spending more on robots and other manufacturing automation, perhaps signaling a desire to become more of a direct participant in their own manufacturing (Satariano 2013). This ties into an earlier announcement by Tim Cook, CEO of Apple, in revealing that Apple intended to begin manufacture some of its Mac computer in the United States (Satariano and Tyrangiel 2012). Obviously, Apple is taking a long-range view of cash management in considering these investments.

From the beginning of performance measurement methods, companies have tried to measure financial performance. Morgan provides a brief history of performance measurement evolution as consisting of the following phases:

- *Year 1450–present*: Dominated by transaction costs and profit determination
- *Year 1900–present*: Enhanced to include operations and value-adding perspectives
- *Year 1970–present*: Enhanced to include process, quality, and customer focus
- *Year 1990–present*: Enhanced to give a balanced view of the organization
- *Year 2000–present*: Enhanced to include the supply chain and interprocess activities (Morgan 2007)

In summary, goods and services flow relatively well while information flow is erratic but improving. Funds flow lags, because it depends on the flow of products and information. Funds flow is also disconnected and biased, sometimes for reasons unrelated to the flow of goods and information.

## Need for Finance and Operations to Collaborate

Up to this point, we have discussed the need for a smooth flow of funds along the supply chain and some of the programs used to effect this flow. But who has the responsibility for making this happen? It would be convenient to say that the finance function in an organization should be responsible; however, in almost every organization, large or small, the finance function cannot achieve effective cash flow without the help from the rest of the organization, especially the operations groups of purchasing and production.

Inventory is one of the working capital components that represents a significant investment for most companies and is also critical in providing good customer service. Crandall (2012) outlines a number of ways in which the operations groups can influence the flow of cash within a company and, subsequently, within the supply chain.

A recent survey of midsize firms found that managing working capital was a major priority. What the survey also uncovered was that lack of bargaining power was the most cited barrier to improving their cash position. But the survey respondents also revealed that organizational and cultural issues—lack of coordination, internal pressure to accept unfavorable terms, and lack of a shared organizational mandate for improvement—were also considered major issues. Organizational and cultural matters edge out process weakness and technological inadequacy by a narrow margin and far exceed financial matters like weakened credit ratings, credit constriction, or lack of access to affordable financial products and services (Surka 2012).

## Effect of Production Strategies on Funds Flow

Production strategies can have an effect on funds flow. In Chapter 7, we discussed the different strategies of make to stock (MTS), assemble to order (ATO), make to order (MTO), and engineer to order (ETO). Each of these strategies affects the positioning of inventory along the supply chain and the type of inventory carried. Figure 15.3 shows this effect. The horizontal axis shows the progression along the supply chain from raw materials to finished products available to the consumer. The vertical axis shows the level of customization for the product. The effect can be summarized as follows:

- *MTS*: Standard product—inventory is positioned as finished goods inventory.
- *ATO*: Some customization of product—inventory is positioned as modular subassemblies.
- *MTO*: Considerable customization of product—inventory is positioned as component parts.
- *ETO*: Ultimate customization of product—no specific inventory is required.

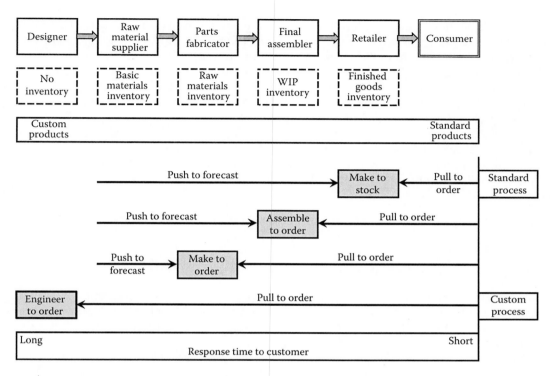

**FIGURE 15.3**
Effect of production strategies on inventory.

Figure 15.3 points out that keeping the inventory in an unfinished state reduces both the cost of inventory and the cash required to maintain it.

As shown in Figure 15.3, retailers maintain their inventories in the finished goods state to provide fast response times to their customers. If they do not have the product in stock, they build quick response systems to obtain product from their suppliers.

Manufacturers have more flexibility in the type of inventory they can carry. As long as they complete the order fast enough to satisfy their customer, they postpone adding the labor and overhead to the product.

## Effect of Outsourcing on Product Costs and Capital Requirements

Outsourcing has a significant effect on funds flow. First, the longer lead times to obtain goods from suppliers, especially those located in foreign countries, increase the level of inventory that must be maintained by the customer to sustain sales until replenishment orders are received. In addition, there may be greater volatility in delivery times and product quality that could cause lost sales and increased strains on funds flow.

Outsourcing has another effect that does not receive as much attention as the effect on direct costs. Outsourcing reduces the capital investment in production capacity (equipment and facilities) required by the outsourcing company. In the short term, this reduces cash requirements. In the long term, failures in the outsourcing efforts may make it more expensive to acquire the capacity needed to bring the production back into the company.

As described earlier, financial institutions are important participants in supply chains. Rajamani et al. (2006) describe the flow of cash throughout the world. They point out that cash is expensive to transport, handle, and store; therefore, it is usually the least automated of the major payment systems. Movement of cash requires careful handling to assure that it is received where it is needed on time and securely.

In Chapter 14, we described a number of information technologies used in supply chains. In this chapter, we describe two of those technologies in more depth because of their importance in funds flow. They are interorganizational systems (IOSs) and electronic data interchange (EDI).

## Interorganizational Systems*

What do supply chains, outsourcing, quick response systems, customer relationship management (CRM), supplier relationship management (SRM), globalization, and collaborative planning, forecasting and replenishment (CPFR) have in common? They all require an IOS.

Definitions for IOSs tend to be short. Kumar and van Dissel (1996) explain that IOSs can be considered as planned and managed cooperative ventures between otherwise independent agents. Turban et al. (2006) define IOS as involving information flow among two or more organizations.

Chi and Holsapple (2005) offer a more expansive definition. "In the broadest sense, an IOS consists of computer and communications infrastructure for managing interdependencies between firms. From a knowledge management perspective, this infrastructure enables and facilitates knowledge flows among organizations (and their participating representatives) such that the needed knowledge gets to the relevant participants on a timely basis in a suitable presentation(s) in an affordable way for accomplishing their collaborative work." IOS is a way for different organizations to communicate with each other to accomplish some mutually beneficial goals.

An IOS includes the following technology elements:

*EDI:* The electronic movement of business documents between business partners.

*Extranets:* Extended intranets that link business partners.

*Extensible Markup Language (XML):* An emerging B2B standard, promoted as a companion or even a replacement for EDI systems.

*Web services:* The emerging technology for integrating B2B and intrabusiness applications. One of the real difficulties in implementing IOSs is achieving compatibility between computer systems in different entities.

In addition to the technology, an IOS must include other necessary ingredients. The first is a worthwhile application. Companies must have a business reason to communicate with each other. While the needs may be obvious, often it is necessary to choose among

---

* This section has been adapted from Crandall (2007).

alternatives. Which processes and relationships should be worked on first? The fact that it is difficult to establish a smooth-running relationship means that there may be a finite number of relationships that can be successfully maintained. Consequently, it is important to choose those applications with the greatest payoff.

Once the applications have been determined, businesses must make sure they have accurate and timely information available to use in the IOS. The IOS is of limited value, and may even be detrimental, without good information to flow along the streamlined communication channels.

Finally, companies have to build relationships with one another. Relationships can be one to one (traditional EDI between one customer and one supplier), one to many (extranet from manufacturer to dealers), many to one (reverse auction), or many to many (electronic marketplace, such as an electronic hub in a supply chain configuration of suppliers and customers). However, relationships are more than a mechanical linking of computers in which a store's computer automatically orders replenishment stock. They also link individual persons, departments, and entire companies. This type of relationship takes conviction and perseverance.

## Benefits

IOSs were developed because of two business pressures: the desire to reduce costs and the need to improve the effectiveness and timeliness of business processes (Turban et al. 2006). Chi and Holsapple (2005) noted that IOSs can

- Become an important source of sustainable competitiveness
- Reduce the cost of communication while expanding its reach (time and distance)
- Increase the number and quality of alternatives while decreasing the cost of transactions
- Enable tight integration between firms, while reducing the cost of coordination
- Facilitate knowledge sharing and trust building
- Speed up expertise exploitation and knowledge application
- Enhance innovation and knowledge generation

Other motives for implementing IOSs include the following: to comply with mandates from regulatory agencies, to exert power over other organizations, to pursue common or mutually beneficial goals with other entities, to gain internal and interorganizational efficiencies, to increase agility and responsiveness, to promote innovation, to reduce environmental uncertainty, and to increase its legitimacy and reputation as a progressive member of its peers (Chi and Holsapple 2005). Regardless of the motives, companies have moved rapidly to implement IOSs.

## Obstacles

When companies enter into an IOS, they become more dependent on each other and must recognize interdependency; else, they may not realize the benefits (Clark and Lee 2000). Interdependency increases as the IOS moves from being a participant, to understanding a participant's business, to exploiting a participant's dependency (Johnston and Vitale 1988).

The PC industry has moved toward a build-to-order (BTO) approach. This increases the complexity of an IOS to the point that it must be customized between parties. The customization increases the cost of the IOS and limits the number of IOS relationships a company can economically sustain (Dedrick and Kraemer 2005).

Another issue in interdependencies is who has the power. Often, the initiator of the system is the customer and the supplier becomes the follower. Suppliers who adopt IOS technology at the insistence of their customer often avoid implementing the technology in a sophisticated way. This increases the risk and may hinder not only their ability to gain benefits but also the initiator's ability to realize many of the originally anticipated benefits.

Ownership of the IOS is another important issue. As companies share greater amounts of information, the risk of information exploitation increases. The owner of the IOS may expect to get information that is useful but is denied of that information because of the concern of other participants. While this has been an important topic in the literature, there is little guidance as to how to handle the situation (Han et al. 2004).

IOSs require the building of relationships. Relationships breed interdependence. It takes trust among the parties involved to assure comfortable and productive interdependent relationships. Building trust is a slow and tenuous process.

## Evolution

IOSs evolve through several phases. Companies ideally move from a competitive standoff to a collaborative embrace; however, there are some interim stages in the evolution of an IOS.

*Competition*: The beginning stage is the one in which customers and suppliers compete in a zero-sum game with the objective being "I win—you lose." Other descriptors of this condition include "cards close to the vest" and "You go first."

*Communication*: At some point, the icebreaker is for companies to begin to communicate in a way that suggests the need and willingness to do something for their mutual benefit. They may not be convinced yet, but they begin to see the possibilities.

*Coordination*: As the relationship grows, the companies begin to coordinate their activities so that both will benefit. This is still a somewhat formal stage of information exchange, but it does provide enough benefits that the companies are encouraged to keep progressing in their relationship building.

*Cooperation*: If the relationship grows into a more comfortable one, the parties involved can be more enthusiastic about the progress they are making and look for ways to make it more valuable for each other.

*Collaboration*: Collaboration implies working together. At this stage, the participants are convinced that they help each other and trust each other. This ideal state still awaits most companies.

## Future

IOSs are not a dream. They exist, although not all company relationships have reached the collaboration stage. This stage will come as the technology, infrastructures, and relationships of companies merge into alignment. Two examples describe the IOS potential.

Clark and Lee (2000) describe a continuous replenishment program (CRP) in the grocery industry. It is a vendor-managed inventory (VMI) program, where the manufacturer receives sales data from the retailer and generates the replenishment order through an automated computer system.

Dedrick and Kraemer (2005) speak of the changes in the PC industry because of the competitive pressures to reduce costs and the efficiencies of the Dell direct-sales/BTO strategy. They report these changes were driven by competitive and market conditions. However, IT has enabled particular forms of organizational restructuring, such as the shift from supply-driven to demand-driven production and the formation of different value chains to most effectively support demand-driven production processes. PC vendors moved to an MTO mode and outsourced functions to reduce costs and become more flexible. PC makers increasingly focused internal efforts on core activities such as marketing, sales, and product management. They coordinated other activities of product development, manufacturing, distribution, and customer service with external partners who included contract manufacturers (CMs), original design manufacturers (ODMs), logistics providers, distributors, and various service specialists. IT has played a critical role in supporting the growth of this flexible value network.

Herbert Simon, a highly respected writer on management topics, said this about IOS. "The main requirement in the design of organizational communication systems is not to reduce scarcity of information but to combat the glut of information, so that we may find time to attend to that information which is most relevant to our tasks— something that is possible only if we can find our way expeditiously through the morass of irrelevancies that our information systems contain" (Simon 1997).

IOSs are another means by which companies seek to gain a competitive advantage by fitting together all of the pieces to the IOS puzzle. Better IOSs provide for better supply chains.

---

## EDI*

Traditional EDI is the paperless (electronic) exchange of trading documents, such as purchase orders, shipment authorizations, advanced shipment notices, and invoices, using standardized document formats (Blackstone 2010). It became available to companies in the early 1970s as a means of transferring information electronically from one business to another. One of the more popular applications was in order processing. A business could place an order with a supplier, who would acknowledge the order and ship the order with an invoice. The receiving business could prepare a receiving report, match it with their purchase order and the invoice, and authorize payment through their bank, which would send the payment to the supplier's bank. This process eliminated paperwork because all the transactions and accompanying documents described earlier were electronic. EDI reduced costs of order processing, reduced errors, speeded up deliveries, and reduced the time required of employees to identify errors and track orders. The biggest problem was that it was expensive to implement and operate. Many of the companies that used EDI did so reluctantly because their

---

* This section has been adapted from Crandall (2005).

major customers demanded it. It was secure because it was confined to a one-to-one kind of communication. However, it required each customer–supplier relationship to be set up individually.

Large companies implemented EDI and found it worth the investment. Some of the major applications included global communications, financial funds transfers, health-care claims processing, and manufacturing and retailing transactions (Kalakota and Whinston 1996). Small companies could not achieve the volume necessary to make it a worthwhile investment. If they had to use EDI, they usually did it through VANs, which were entities that facilitated the transfer of information between customer and supplier. Using a VAN eliminated a portion of the initial investment cost, but did not change the high transaction costs. Sawabini (2001) reported that where EDI fails, it is because (1) it is cost prohibitive and too complex for smaller suppliers and (2) it offers few bottom-line benefits for suppliers.

In summary, EDI was good for a few but a burden for the masses. Today, competitive pressures make the need for rapid information transfer imperative and Internet EDI has largely replaced the earlier point-to-point versions of EDI.

## Benefits

Internet EDI is attractive because of the lower-cost potential. Both the initial investment and transactions costs are lower (Angeles 2000). In addition, Senn (1998) suggested the following reasons why the use of the Internet is attractive:

- Publicly accessible network with few geographical constraints
- Potential to reach the widest possible number of trading partners
- Provides powerful tools facilitating IOSs
- Consistent with the growing interest in increasing electronic services
- Complement or replaces current EDI strategies
- Leads to an electronic commerce strategy

The Internet also offers close to real-time transactions because there is no longer the need to go through VANs that use the batch-and-forward method of transmitting information.

With the ever-expanding global business community, it is no longer possible to use hard copy documents in most business transactions. The speed and efficiency of electronic information is compelling businesses to make the transition to paperless systems.

Internet EDI must overcome the problem that restricted the growth of traditional EDI—dominance by a few large companies in a supply chain. Smaller firms are often forced to adopt innovations in order to maintain supplier relationships with large customers, but without realizing any of the savings provided by them (Grossman 2004).

Any IOS is more than technology. It also includes organization and people. Trust between participants is necessary if the relationship is to be an effective one, supporting collaboration among entities. Even if the cost and security issues can be resolved, the trust issue may remain. Ruppel (2004) found that company culture and trust were significant issues when it came to the use of EDI in either the traditional or Internet format. In order to move from coordination to collaboration, trust is required.

## Funds Flow in the Reverse Supply Chain

Funds also flow in the reverse supply chain, described in Chapter 10. Developing a smooth flow of funds in the reverse supply chain is even more difficult than in the forward supply chain. Usually, the participants are different in the reverse supply chain, and the procedures are less regular and well established. A part of the regular process for retail stores is issuing credits for returned merchandise. This is a more prominent requirement for e-business sellers, where returns represent a higher percentage of sales. Accounting systems do not *run backwards* and must be supplemented to account for the flow of funds in the reverse supply chain (Walker 2009).

## Comprehensive Example*

Even successful companies can need to borrow money to pay the bills. Developing a cash flow plan, with significant input from the operations and supply chain management areas, is crucial for a company's long-term health. What factors increase cash availability? Why should those who work in the operations and supply chain management functions— purchasing, production planning, production, and distribution—care?

Availability of cash is important in all companies, especially for small and growing companies. Many small business managers have wondered how they can have such cash flow problems while sales and profits have increased year after year. The bottom line is that managing cash flow is not simply the responsibility of the financial groups. Supply chain managers and operations managers can have a significant impact on the amount of cash a company has on hand.

### Components of a Financial Statement

Cash is usually the first item listed on a company's balance sheet. Balance sheets often are far less interesting to operations managers than the income statement of a business. This is because the income statement describes the things with which they are most concerned. Sales (shipments), cost of sales (materials, direct labor, and overhead), and income (profit) are the results used as performance measures for operations managers. Consequently, these managers are naturally more interested in those measures than in the contents of the balance sheet. They may be interested in some of the balance sheet items, such as inventory, but they usually see this information in a selected format, rather than in the complete balance sheet format. However, understanding the balance sheets, or more specifically the changes in the balance sheet accounts, is essential to understanding cash flow.

In recent years, the accounting community has added a cash flow statement to the income statement and balance sheet as part of its regular financial statement package. Some formats used for the cash flow statement do not help explain the relationships between the income statement and the cash flow statement or between the balance sheet and the cash

---

* This section has been adapted from Crandall and Main (2002).

flow statement. Consequently, the cash flow statement has not received widespread acceptance among operations managers as a useful tool.

Figure 15.4 shows an example of a hypothetical set of financial statements for a small company. It shows the relationships among the balance sheet, the income statement, and the cash flow statement. A typical balance sheet format is on the left side. The numbers represent the changes in the balance sheet from the beginning of the fiscal year to the end of the fiscal year. The income statement on the right side shows the total results for last year. The cash flow statement is in the middle, also reflecting the actual results. The series of arrows on the left side shows the links between the balance sheet changes and the cash flow results, and the arrows on the right show the links between the income statement and the cash flow statement.

## Analyzing the Cash Flow Statement

The first item listed is the net income after taxes of $421 thousand. That comes from the net income line on the income statement. Next is depreciation (a noncash expense), also from the income statement. This identifies the income after cash expenses summarized as *cash from operations*. It is apparent that a positive net income increases cash, but that is only one of the factors affecting cash. It also appears that increased depreciation increases cash. However, depreciation is a noncash expense, and the impact on cash occurs when the asset (shown as a fixed asset in the balance sheet) to be depreciated is bought. Adding it back to the net income is just a convenient way of using the net income as reported and then calculating the net income before depreciation or any other noncash expense.

The next section shows the changes in the working capital accounts. Working capital equals current assets minus current liabilities and is usually viewed as the net amount of assets that can be quickly converted to cash. For accounts receivable, we show a negative $543 thousand. This means that accounts receivable increased by that amount during the year. If the amount that customers owe increases, then it represents cash owed but not collected, so it has a negative impact on cash flow. The inventory shows a negative $1023 thousand or an increase in the inventory of that amount. If the inventory increases, you have to invest in purchases and production of that inventory; therefore, it has a negative impact on cash. Both accounts receivable and inventory come from the asset part of the balance sheet. Accounts payable shows a positive of $198 thousand. This comes from the liability portion of the balance sheet and indicates an increase in the account (opposite effect from the asset accounts). This increases the amount owed to suppliers and has a positive impact on cash flow. Prepaid expenses, an asset, other current assets, and accrued expenses, a liability, all show an increase in cash. These three accounts are usually minor in relation to the other working capital accounts. The sum of the changes in the working capital accounts is shown as the *total from (to) working capital*.

The next section on the cash flow statement deals with transactions involving fixed assets, such as the purchase or sale of capital assets. It also includes the payment of the current portion of long-term debt, which can generally be associated with the regular payments for fixed assets that have previously been acquired. There are also *other* accounts that reflect adjustments in the value of assets or other accounts, such as goodwill. The key item for operations managers is the amount for capital expenditures, which is $653 thousand. This is the amount spent for new equipment or facilities and is a reduction in cash. Operations managers heavily influence expenditures for capital assets.

Links among financial statements

**ABC Enterprises — Changes in balance sheets — for the year ending December 31, XXXX — Dollars in thousands**

| | Inc (Dec) |
|---|---|
| Cash | -20 |
| Accounts receivable | 543 |
| Inventory | 1023 |
| Prepaid expenses | -72 |
| Other | -117 |
| Total current assets | 1357 |
| Land and Buildings | -71 |
| Production equipment | 494 |
| Leasehold improvements | 230 |
| Gross fixed assets | 653 |
| Less: Accumulated depreciation | 654 |
| Net fixed assets | -1 |
| Other | 192 |
| Total assets | 1548 |
| Accounts payable | 198 |
| Notes payable—bank | -384 |
| Current of long-term debt | -158 |
| Accrued expenses | 7 |
| Other (including FIT payable) | -160 |
| Total current liabilities | -497 |
| Long-term debt | 1624 |
| Total liabilities | 1127 |
| Common stock | 0 |
| Preferred stock | 0 |
| Paid-in capital | 0 |
| Retained earnings | 421 |
| Total stockholders' equity | 421 |
| Total—Liabilities and equity | 1548 |

**ABC Enterprises — Cash flow results — for the year ending December 31, XXXX — Dollars in thousands**

| | Total |
|---|---|
| Cash provided (required) | |
| Net income before taxes | 421 |
| Plus: Depreciation | 708 |
| Cash from operations | 1129 |
| Changes in working capital | |
| Accounts receivable | 543 |
| Inventory | -1023 |
| Accounts payable | 198 |
| Prepaid expenses | 72 |
| Accrued expenses | 7 |
| Other changes in current assets | 117 |
| Total from (to) working capital | -1172 |
| Other changes in fixed assets | -192 |
| Notes payable | -384 |
| Debt payments | -158 |
| Capital expenditures | -653 |
| Other disbursements (Credit) | -160 |
| Total from other changes | -1547 |
| Total provided (Required) | -1590 |
| Financed by | |
| Bank loan increase (decrease) | 1624 |
| Gain (Loss)—Sale of assets | -54 |
| Other (including equity adjust.) | 0 |
| Total financed | 1570 |
| Net cash increase (decrease) | -20 |

**ABC Enterprises — Income statement — for the year ending December 31, XXXX — Dollars in thousands**

| | Total |
|---|---|
| Net sales | 22,655 |
| Cost of sales | 15,082 |
| Gross profit | 7,573 |
| Operating expenses | |
| Payroll and fringes | 2,977 |
| Advertising | 754 |
| Rent and utilities | 585 |
| Telephone | 156 |
| Insurance | 180 |
| Repairs and maintenance | 194 |
| Shipping and delivery | 189 |
| Warehouse supplies | 191 |
| Professional fees | 263 |
| Travel and entertainment | 101 |
| Office supplies | 5 |
| Freight out | 0 |
| Depreciation | 708 |
| Other | 202 |
| Total expenses | 6,575 |
| Operating income | 998 |
| Other income | 240 |
| Other expense (interest) | 817 |
| Income before taxes | 421 |
| Provision for taxes | 0 |
| Net income | 421 |

**FIGURE 15.4**

Cash flow links between income statement and changes in balance sheets.

The final section of the cash flow statement, headed as "Financed By," shows the effect of financing transactions. In the example, the bank loan (a short-term note usually guaranteed by working capital) increased by $1624 thousand, thereby providing cash. In addition, the sale of assets resulted in a negative cash flow of $54 thousand. Had the company issued more stock, the increased cash available would have been shown in the *other (including equity adjustment)* line. The last entry shows a net decrease in the cash account of $20 thousand, a number that links to the same number in the balance sheet changes.

An overall picture for the company emerges. While it had $1129 thousand in cash from operations, it required $1172 thousand to finance the working capital requirements. The company also required $1547 thousand for capital expenditures, payment of the current portion of long-term debt, and other requirements for fixed assets. To finance these cash requirements, it had to increase debt by $1570 thousand. While the company was able to do this, it is obvious that management should strive to finance more of their cash requirements through internally generated funds, especially in its management of working capital. A financial plan focused on the cash shortcoming is needed.

## Looking at Alternatives

In preparing a plan for the next year, it is often useful to consider alternatives. In the following example, we considered alternatives in the management of working capital—accounts receivable, inventory, and accounts payable. What is the effect on net income and cash flow, if we were able to improve the utilization of working capital? Obviously, operations managers contribute to the utilization of these items.

Table 15.1 is a summary of key information selected from the plan's financial forecasts to show the effect of working capital alternatives. The first column represents a base forecast. This forecast assumes 50 DROs, 60 DIOs, and 30 DPOs.

**TABLE 15.1**

Net Work Capital Forecasts

| Net Working Capital Forecast (Dollars in Thousands) | Alternatives: Working Capital Targets | | | | | | |
|---|---|---|---|---|---|---|---|
| | 1 | 2 | 3 | 4 | 5 | 6 | 7 |
| Sales | 25,000 | 25,000 | 25,000 | 25,000 | 25,000 | 25,000 | 25,000 |
| Cost of goods sold | 16,750 | 16,750 | 16,750 | 16,750 | 16,750 | 16,750 | 16,750 |
| Days per year | 365 | 365 | 365 | 365 | 365 | 365 | 365 |
| DROs | 50 | 40 | 30 | 30 | 30 | 30 | 30 |
| Average accounts receivable | 3,425 | 2,740 | 2,055 | 2,055 | 2,055 | 2,055 | 2,055 |
| DIOs | 60 | 60 | 60 | 40 | 20 | 20 | 20 |
| Average inventory | 2,753 | 2,753 | 2,753 | 1,836 | 918 | 918 | 918 |
| DPOs | 30 | 30 | 30 | 30 | 30 | 45 | 60 |
| 0.6 Purchases at 60% of COGS | 10,050 | 10,050 | 10,050 | 10,050 | 10,050 | 10,050 | 10,050 |
| Average accounts payable | 826 | 826 | 826 | 826 | 826 | 1,239 | 1,652 |
| Net working capital | 5,352 | 4,667 | 3,982 | 3,064 | 2,147 | 1,734 | 1,321 |
| Current ratio | 7.5 | 6.7 | 5.8 | 4.7 | 3.6 | 2.4 | 1.8 |
| Changes in working capital | | | | | | | |
| Accounts receivable | | 685 | 1,370 | 1,370 | 1,370 | 1,370 | 1,370 |
| Inventory | | 0 | 0 | 918 | 1,836 | 1,836 | 1,836 |
| Accounts payable | | 0 | 0 | 0 | 0 | 413 | 826 |
| Total cash from (to) working capital | | 685 | 1,370 | 2,288 | 3,205 | 3,618 | 4,032 |

*Option 1: Reducing accounts receivable*: Suppose operations and supply chain managers improve processes to reduce partial shipments, ship correct quantities, ship defect-free products, and ship on time. These factors affect the accounts receivable because they delay customer payments. If the sample company can help reduce accounts receivable from 50 days (alternative 1 in Table 15.1) to 40 days (alternative 2), then the company's cash flow would increase by $685 thousand. A further improvement reducing accounts receivable from 40 to 30 days increases cash flow by another $685 thousand, or a total of $1370 thousand (alternative 3).

*Option 2: Reducing inventory*: Inventory reduction can have an even greater impact on the net income and cash flow, because it usually represents a higher percentage of total assets. Alternative 3 shows an average inventory of 60 days. If this can be reduced to an average of 40 days, as in alternative 4, while maintaining accounts receivable at 30 days and accounts payable at 60 days, the cash flow increases by $918 thousand. A further reduction to 20 DIOs increases cash flow by another $918 thousand, or a total of $1.836 million. This strategy results in an increase of total cash flow to $3.205 million (alternative 5).

*Option 3: Negotiating better payment terms*: Should the purchasing function be able to negotiate even better payment terms by the suppliers, alternatives 6 and 7 show the effect of these changes. In alternative 6, accounts receivable remain at 30 days and inventory at 20 days, but accounts payable are increased to an average of 45 days. This provides a $413 thousand increase in cash flow, with another $413 thousand possible if accounts payable are increased to 60 days. The total increase in cash from the changes in accounts receivable, inventory, and accounts payable amounts to $4.032 million. Table 15.2 lists a few of the ways in which the operations function, along with purchasing and accounting, can improve working capital management.

Each company must decide what level of working capital is feasible. This example is intended only to show that better management of working capital can result in significant increases in cash flow. A reduction in the amount borrowed from lenders reduces interest expense. A reduction in the inventory has many other benefits, such as reductions in obsolescence or damage and reduction of storage space required.

**TABLE 15.2**

Actions to Improve Working Capital Management

| Cash Flow Element | Action by Operations |
|---|---|
| Revenues | Fast, on-time, high-quality complete shipments increase revenues. Good sales and operations planning are the key to achieving this. |
| Direct product costs and operating expenses | Careful management keeps costs low. Programs such as lean manufacturing and Six Sigma are two current programs to do this. |
| Accounts receivable | Correct, complete, on-time, and defect-free shipments. Many suppliers want the complete order before they authorize payment. |
| Inventory | Have the needed quantities of the correct item, but no more. This is the never-ending balancing challenge for operations managers. |
| Accounts payable | Good supplier relationships can gain discounts and extended terms. Delaying payments does not always work over the long term. |
| Capital investment | Careful management of capacity and currency of facilities. Maintaining currency is an essential core competency. |
| Sale of capital assets | It does not pay to be too far behind the curve. Careful maintenance and operations increases disposal value. |

In all of the improvements in net income and cash flow projected from the changes in working capital, it is interesting to observe what happens to the current ratio. The current ratio—current assets divided by current liabilities—is one of the oldest ratios used by creditors to decide whether the borrower is creditworthy. In Table 15.1, as the management of working capital improves, the current ratio declines from a high of 7.5–1.8, which normally indicates a deterioration of the company's ability to repay loans. In this case, the creditors need to understand that a decrease in the current ratio does not necessarily indicate a decreased ability to repay loans. The company is actually getting into a better financial position because of its better management of working capital.

In practice, the goals and metrics used to measure a company's success link to the company's financial objectives. Developing strategies that create a positive cash flow for the company is a good start for companies struggling in this area. Those strategies will reach out into all facets of the organization, highlighting the significant role played by operations and supply chain managers.

## Summary

This chapter discussed the flow of funds in supply chains. While not as widely publicized, funds flow is equally important as product and information flows in supply chains. Without timely and accurate funds flow, product flows would soon slow and eventually stop because suppliers would stop selling to customers who did not pay for the goods and services delivered.

We described the importance of funds flow, along with the benefits and obstacles of effective implementation. We outlined some of the performance measures used to evaluate the effectiveness of the flow of funds and how technologies like IOS and EDI are essential to successful funds flow.

Finally, we provided a detailed example to show how a company can evaluate the effect of working capital management on funds flow.

## Hot Topic: Boeing 787 Dreamliner

The hot topic for this chapter is about the Boeing 787 Dreamliner. It took several years longer than expected to complete the design and get the plane into production. Think about the lost revenues as a result of this delay.

### Hot Topics in the Supply Chain: The Boeing 787, Problems in the Supply Chain (Part 2)

The supply chain for the Boeing 787 (the Dreamliner) represents a major shift from how Boeing has managed its suppliers in the past. In part 1 (Chapter 14), we overviewed the specifics of this new configuration. In part 2, we outline the problems that have resulted with this new arrangement.

## PROBLEMS IN THE SUPPLY CHAIN

Boeing has outsourced approximately 70% of the Dreamliner to 50 tier-one suppliers. These suppliers function as strategic partners in that they assume the development and financial responsibilities for key subassemblies of the aircraft (Tang and Zimmerman 2009). In addition, they take total control of managing the supply chains of their tier-one and tier-two suppliers. Boeing utilized this approach in hopes that it would reduce both the development costs and production time in building the Dreamliner. The model is similar to Toyota, which has learned to cut development time and costs in offering new vehicle models (Tang 1999). Like Boeing, Toyota has outsourced about 70% of its vehicles to a close knit group of suppliers (Denning 2013). However, a number of problems have surfaced with the new Boeing supply chain configuration and are outlined next.

*Development time and budget costs were higher than projected*: Ironically, the outsourcing strategy had the goals of reducing development time for a new aircraft model from the usual 6–4 years. Instead, the end result was a development period 3 years behind schedule. Budget costs were also hoped to be reduced from the estimated $10 to $6 million. Instead, the Dreamliner was fraught with cost overruns (Denning 2013).

*Tier-one suppliers had to be bailed out with on-site help*: Boeing had originally planned to outsource all subassembly development and cost management to its 50 tier-one suppliers. The goal was to work with just a handful of suppliers and simplify the development process. However, the opposite occurred in that Boeing had to send engineers to a number of its tier-one suppliers as well as to their tier-two and tier-three suppliers. Engineering expertise was needed at these levels that Boeing had not anticipated (Denning 2013).

One of the weakest links in the supply chain was with the tier-one supplier, Vought Aircraft Industries. Vought's role was to build the aft fuselage but was also considered the weakest link in the supply chain (Tang and Zimmerman 2009). Vought, a long-time Boeing supplier, was slow to deliver its part of the plane, complicated carbon fiber subassembly with over 6000 components (Lunsford 2007). Boeing eventually had to purchase two units of the company in order to regain control of that part of the supply chain.

*There were major problems with the lithium batteries*: In January 2013, about a year after the introduction of the Dreamliner for commercial service, all Boeing 787s were grounded because of concerns over the lithium batteries. Lithium batteries are an industry first on commercial aircraft. Their role is critical because they run the electronic flight equipment that replaced the heavier hydraulic systems, a move to increase the fuel efficiency of the aircraft (Venables 2013).

There are two batteries on each aircraft that start the auxiliary power unit and function as backups for electronic flight systems. Incidents involving low voltages, electrical arching, and other electrical issues had been occurring with the aircraft since its introduction. However, a January 7 battery fire and a January 16 battery failure prompted the FAA to ground all aircraft (Williard et al. 2013).

In regard to the supply chain, the battery system was outsourced to GS Yuasa, a Japanese firm. Speculation exists that the firm was under pressure to fast-track the approval process for the battery system so the Dreamliner could be delivered (Billones 2013). Ironically, the first airlines to receive the Dreamliner were Japanese airlines, All Nippon Airways (ANA), and Japan Airlines (JAL).

The Dreamliners were allowed to fly again in April 2013, after a change in the battery design. However, the root cause of the failures was never established; instead, the design change focused on containing a fire so smoke and fire would not spread throughout the aircraft (Williard et al. 2013).

*There were communication problems in the supply chain*: A problem in any type of sophisticated supplier relationship is an overreliance on computer technology for communication purposes. Boeing attempted to use an online communications tool, Exostar, to facilitate progress and information exchange along the supply chain. However, the system was not as effective as planned as suppliers did not consistently input necessary information (Denning 2013). In hindsight, more face-to-face communications were needed, rather than a web-based system.

## QUESTIONS FOR RESEARCH AND DISCUSSION

1. Why do you think this new supply chain configuration has done well for Toyota, but has been less effective for Boeing?
2. The analysis of problems earlier implies that such occurrences were unique for Boeing and its tier-one suppliers. However, all of the problems discussed could occur in other supply chains. What evidence have you seen of similar problems in your supply chains?

## REFERENCES

Billones, C., Japan relaxed safety standards to fast-track Boeing 787 rollout, *The Japan Daily News*, January 29, 2013. Retrieved February 1, 2014, from http://japandailypress.com/japan-relaxed-safety-standards-to-fast-track-boeing-787-rollout-2922325/.

Denning, S., What went wrong at Boeing? *Forbes*, January 21, 2013. Retrieved January 29, 2104, from http://www.forbes.com/sites/stevedenning/2013/01/21/what-went-wrong-at-boeing/.

Lunsford, J., Jet blues: Boeing scrambles to repair problems with new plane, *Wall Street Journal*, A1, December 7, 2007.

Tang, C., The supplier relationship map, *International Journal of Logistics*, 2(1), 39–56, 1999.

Tang, C. and Zimmerman, J., Managing new product development and supply chain risks: The Boeing 787 case, *Supply Chain Forum*, 10(2), 74–86, 2009.

Venables, M., Dreamliner becomes Boeing's nightmare as fleet grounded, *Engineering & Technology*, 8(2), 24–25, March 2013.

Williard, N., He, W., Hendricks, C., and Pecht, M., Lessons learned from the 787 Dreamliner issue on lithium-ion battery reliability, *Energies*, 6, 4682–4695, 2013.

## Discussion Questions

1. Explain how funds flow differs from product and information flows in a supply chain.
2. How does an IOS help companies improve funds flow?
3. What is the difference between traditional EDI and Internet EDI?
4. Describe how changes in working capital affect cash flow.
5. Why does funds flow depend on goods and information flows?
6. Describe cash-to-cash cycle time. How is it calculated?

7. Why is operations management involved in funds flow?

8. Using the example in Table 14.1, how would you respond to a creditor who is concerned about the company's declining current ratio?

9. In what ways can net working capital be managed more effectively?

10. Explain the difference between net income and net cash flow.

## References

Angeles, R., Revisiting the role of Internet-EDI in the current electronic commerce scene, *Information Management*, 13(1), 45, 2000.

Blackstone, J.H., *APICS Dictionary*, 13th edn., APICS—The Association for Operations Management, Chicago, IL, 2010.

Cagle, C.S., Campbell, S.N., and Jones, K.T., Analyzing liquidity: Using the cash conversion cycle, *Journal of Accountancy*, 215(5), 44–48, 8–9, 2013.

Chi, L. and Holsapple, C.W., Understanding computer-mediated interorganizational collaboration: A model and framework, *Journal of Knowledge Management*, 9(1), 53, 2005.

Crandall, R., Straight flush, streamline cash flow for a winning chain, *APICS Magazine*, 18(7), 26, 2008.

Crandall, R.E., Exploring internet EDI, *APICS Magazine*, 15(8), 18, 2005.

Crandall, R.E., Beating impossible deadlines, *APICS Magazine*, 16(6), 20, 2006.

Crandall, R.E., Let's talk, better communication through IOS, *APICS Magazine*, 17(6), 20, 2007.

Crandall, R.E. and Crandall, W.R., A cross-functional view of inventory management, Why collaboration among marketing, finance/accounting and operations management is necessary, *2012 SEINFORMS Proceedings*, pp. 16–26.

Crandall, R.E. and Main, K.T., Cash is king, *APICS—The Performance Advantage*, 12(1), 36, 2002.

Dedrick, J. and Kraemer, K.L., The impacts of IT on firm and industry structure: The personal computer industry, *California Management Review*, 47(3), 122, 2005.

Gentry, C.R., Moving money, *Chain Store Age*, 82(2), 59, 2006.

Grealish, A., Making the business case for the e-financial supply chain, *Commercial Lending Review*, 20(3), 40, 2005.

Grossman, M., The role of trust and collaboration in the internet-enabled supply chain, *Journal of American Academy of Business*, 5(½), 391, 2004.

Han, K., Kauffman, R.J., and Nault, B.R., Information exploitation and interorganizational systems ownership, *Journal of Management Information Systems*, 21(2), 109, 2004.

Hintze, J., Supplier finance advantage, *Treasury & Risk Magazine*, reprint May 1, 2012, April/May 2012.

Jessup, L.M. and Valacich, J.S., *Information Systems Today*, 2nd edn., Pearson Education, Inc., Prentice Hall, Upper Saddle River, NJ, 2006.

Johnston, H.R. and Vitale, M.R., Creating competitive advantage with interorganizational information, *MIS Quarterly*, 12(2), 153, 1988.

Kalakota, R. and Whinston, A.B., *Frontiers of Electronic Commerce*, Addison-Wesley Publishing Company, Inc., Reading, MA, 1996.

Kerle, P., Steady supply, *Supply Chain Europe*, 16(4), 18, 2007.

Krishnan, M. and Shulman, J., Reducing supply chain risk, *The McKinsey Quarterly*, 29(1), 10, 2007.

Logistics Manager.com, Strong growth for supply chain finance, January 6, 2011, http://www.logisticsmanager.com/Articles/15272/Strong+growth+for+supply+chain+finance+.html, accessed August 8, 2014.

Morgan, C., Supply network performance measurement: Future challenges? *International Journal of Logistics Management*, 18(2), 255, 2007.

Rajamani, D., Geismar, H.N., and Sriskandarajah, C., A framework to analyze cash supply chains, *Production and Operations Management*, 15(4), 544, 2006.

Roberts, J.S., The supply chain of dollars and cents, *Inside Supply Management*, 13(6), 38, 2002.

Ruppel, C., An information systems perspective of supply chain tool compatibility: The roles of technology fit and relationships, *Business Process Management Journal*, 10(3), 311, 2004.

Sadlovski, V. and Enslow, B., Supply chain finance benchmark report, *The New Opportunity to Improve Financial Metrics and Create a Cost-Advantaged Supply Chain*, Aberdeen Group, 2006.

Satariano, A., Apple's $10.5B on robots to lasers shores up supply chain, November 13, 2013. http://www.bloomberg.com/news/2013-11-13/apple-s-10-5b-on-robots-to-lasers-shores-up-supply-chain.html.

Satariano, A. and Josh, T., Apple to invest in manufacturing Macs in U.S., Cook says, *Bloomberg Business Week*, December 6, 2012. http://www.bloomberg.com/news/2012-12-06/apple-to-invest-in-manufacturing-macs-in-u-s-ceo-cook-says.html.

Sawabini, S., EDI and the internet, *Journal of Business Strategy*, 22(1), 41, 2001.

Senn, J.A., Expanding the reach of electronic commerce, *Information Systems Management*, 15(3), 7, 1998.

Simon, H.A., *Administrative Behavior: A Study of Decision-Making Processes in Administrative Organizations*, 5th edn., The Free Press, New York, 1997.

Surka, M., *Cash and Working-Capital Discipline*, CFO Research Services, CFO Publishing, Boston, MA, 2012.

Timme, S.G. and Wanberg, E., How supply chain finance can drive cash flow, *Supply Chain Management Review*, 15(1), 18–24, 2011.

Turban, E., Leidner, D., McLean, E., and Wethebe, J., *Information Technology for Management: Transforming Organizations in the Digital Economy*, John Wiley & Sons, New York, 2006.

Walker, W.T., *Supply Chain Architecture: A Blueprint for Networking the Flow of Material, Information, and Cash*, CRC Press, Boca Raton, FL, 2005.

Walker, W.T., Supply chain consultant, Personal correspondence, April 29, 2009.

# 16

## ROI for Supply Chains and Other Issues

### Learning Outcomes

After reading this chapter, you should be able to

- Describe the management programs that require close supply chain relationships
- Describe the value of determining return on investment (ROI) for the entire supply chains
- Explain why supply chain costs and benefits should be allocated among its participants
- Identify the costs associated with integrating the supply chain
- Identify the tangible and intangible benefits from integrating the supply chain
- Discuss the factors to consider when distributing the costs and benefits along the supply chain
- Identify the four types of supply chain governance categories
- Describe the role of the prime mover in a supply chain
- Identify the various changes that can occur with supply chain composition

One of the more difficult problems facing supply chain members is how to fairly divide responsibilities and proceeds from the supply chain operation. In the past, individual businesses could calculate the ROI for its own operation. In this age of supply chains, where the success of individual companies is dependent on the success of the supply chain in which they operate, there is a need to share the common pain and gain.

To date, there is not a lot of evidence to indicate that companies and supply chains are working to share the net proceeds of their combined efforts. However, we believe that as supply chains mature, more companies will search for ways to accomplish this. In order to share benefits and costs, supply chain members must develop an integrated supply chain. In Chapter 13, we described some of the actions necessary to achieve effective supply chain integration. Much of what has to be done involves supply chain learning. Unless the supply chain partners share their knowledge and understand its implications, there is little likelihood that they will agree on the positive proceeds from the supply chain and how to distribute them among the members.

In an attempt to "gain insight into supply chain competence and the factors that enhance its development," researchers Spekman et al. (2002) posed the following requirements for supply chain learning:

- There must be trust between partners and they must be committed to the concept of supply chain management (SCM).
- Partners must share their combined knowledge and understand its portent.

- Relationships between partners must be comingled to enhance knowledge transfer.
- Decision makers must be flexible, adaptive, and ready to learn.
- Cultures must be trusting and open to facilitate learning.
- Partners must believe in and support a win–win orientation.
- Knowledge sharing is a necessary prelude to financial sharing.

## Company Profile: Apple

Chapter 16 is an enigma. How does a company manage its supply chain when the supply chain is composed of different companies under different ownerships? In a project involving more than one company in a supply chain, how do they equitably share the costs and benefits of the project? If a retailer discovers one of their suppliers is using child labor in producing the product, how does the retailer resolve this issue? If a retailer wants suppliers to replace bar codes with radio-frequency identification (RFID) tags, how do they achieve their objective? As you read this company profile and the chapter in the book, think about these questions:

- Who manages the supply chain that provides the orange juice at your grocery store?
- Who has the greater stake in managing a supply chain—customer or supplier? Explain.
- Who, in an organization, has the responsibility for managing a supply chain?
- How do you resolve disputes, or problems, in a supply chain? Give an example.

What additional questions do you think relate to management of supply chains?

### Company Profile: Apple

### INTRODUCTION

Apple Inc., formerly known as Apple Computer, Inc., was started by Steve Jobs and Steve Wozniak in 1976. The two met in 1971 when Wozniak was 21 and Jobs 16. Wozniak, the technical talent of the pair, had developed a box that enabled a person to make long-distance calls at no cost. Jobs, the promoter, provided the drive to sell several hundred of the boxes and also pushed Wozniak to develop the Apple I computer, which became the first product of the company. Despite its limitations—no keyboard—it was a pioneering effort and earned Wozniak a reputation as a master designer (Wikipedia 2013).

The company sold enough Apple I computers to finance the development of Apple II in 1977, a major player in developing the home computer market. It was followed in 1980 with the Apple III, which had some overheating problems and was not a major success. Also in 1980, Apple became a public company. In 1983, the Lisa computer was developed to be Apple's first GUI computer. However, at a price of $9,995, it failed to gain market position because of its high price. In 1984, the Macintosh computer was introduced with a memorable Super Bowl ad. The Macintosh was an advance in technology but not a major marketing success.

After an internal power struggle with the new CEO, John Sculley, a marketing executive from PepsiCo, Steve Jobs left the company in 1985 and acquired Pixar, a visual effects house. He also founded NeXT Inc., a computer company that built machines with a futuristic design. While Jobs was engaged in his own pursuits, Apple Computers, Inc. continued under the leadership of CEOs Sculley, Spindler, and Amelio. During this period, the company survived but lost much of its image as an innovative product developer. In 1996, Apple purchased NeXT and Steve Jobs returned to Apple. In 1997, Jobs replaced Gil Amelio as interim CEO to begin restructuring the company's product line. In 1998, Apple introduced the iMac, which became a hit and reintroduced the company's new emphasis on the design and aesthetics of its products. Jobs continued as CEO until he resigned in August 2011, shortly before his death from pancreatic cancer in October 2011 (Apple Company History 2014).

Under Jobs, the success of Apple products is unparalleled. They include iTunes (2001), iPod (2002), iPhone (2007), and iPad (2010). During Jobs' second tenure as CEO, they also introduced supporting products—drives, printers, input devices, displays, and networking systems. For a complete history of Apple products, see the display under Company History at http://apple-history.com/h1.

Steve Jobs was succeeded by Tim Cook, who had been with Apple as an operations specialist for over a decade. Cook is almost the antithesis of Jobs in personality, and this difference is being reflected in the public's image of the company. Apple has been extremely successful over the past 5 years, including the last two under Cook. However, it is being viewed increasingly as an outstanding operating company but less imposing as an innovative product development company. In an article in Fortune, Adam Lashinsky Sr. (2012) describes it as Cook's leadership coming into focus. "Cook is maintaining, by words and actions, most of Apple's unique corporate culture. But shifts of behavior and tone are absolutely apparent; some of them affect the core of Apple's critical product-development process. In general, Apple has become slightly more open and considerably more corporate."

While the increased openness under Cook's leadership is welcomed by the investing public and stock analysts, it is no doubt creating waves in the culture of the company. While operations and supply chain managers may welcome the increased importance of their positions, engineers may be less enthusiastic about the loss of stature as the center of creative genius (Lashinsky 2012).

Table 16.1 compares the attitudes about supply chains between Steve Jobs and Tim Cook. The contrast between the two represents a challenge for Tim Cook as he manages

**TABLE 16.1**

Views of Supply Chains from Jobs and Cook

| Seven Supply Chain Lessons from Steve Jobs | Seven Traits of Great Supply Chain Leader—Tim Cook |
| --- | --- |
| 1. Customer comes first; cost cutting second. | 1. Seek first to understand. |
| 2. Set impossible target. | 2. Put first thing first. |
| 3. Prioritize actions based on importance. | 3. Be proactive. |
| 4. Adopt process view of organization. | 4. Avoid analysis paralysis. |
| 5. Simplify product and process. | 5. Keep calm. |
| 6. Make radical change when necessary. | 6. Be humble. |
| 7. Enhance relationship via face-to-face meeting. | 7. Be honest and learn from mistakes. |

*Source:* Adapted from SupplyChainOpz, 7 Supply chain lessons from Steve Jobs, 2013a. http://www.supplychainopz.com, accessed October 28, 2013; SupplyChainOpz, 7 Traits of great supply chain leader: A case study of Tim Cook, 2013b. http://www.supplychainopz.com, accessed October 28, 2013.

the transition from the leadership of Steve Jobs to himself. It is obvious that Jobs was results oriented, while Cook is more process oriented and has traits that make him more approachable by employees, customers, and investors.

## APPLE'S RANKING AS A SUPPLY CHAIN COMPANY

Apple Inc. has been ranked No. 1 for the past 6 years in the Gartner Top 25 Supply Chain Rankings. As the Gartner analysts explain it,

> Apple tops our ranking for a record-breaking sixth year in a row, continuing to outpace everyone else by a wide margin on all five measures we use. It's not just the stellar financials. Apple was ranked No. 1 again by the peer voters, capturing 75% of the highest possible points a company can get across the voting pool. At the same time, as a company traditionally known for its product innovation, Apple now faces formidable competition in the mobile device market. With Tim Cook at the helm, the company known for its focus on simplicity has expanded its product portfolio to a broader array of sizes and price points to address the competition, driving the need for more complexity management in its supply chain. The ex-operations chief has also fostered increased transparency on supplier responsibility, particularly with suppliers and manufacturing partners in China
>
> **Hofman et al. (2013)**

The Gartner ranking is based on the following factors with the indicated factor weights:

- Peer opinion (172 voters)—25%
- Gartner opinion (33 voters)—25%
- Three-year weighted return on assets (ROA)—25%
- Inventory turns—15%
- Three-year weighted revenue growth—10%

Apple ranked first in peer opinion, 3-year weighted ROA, and 3-year weighted revenue growth. They were second to McDonald's in inventory turns and were fifth in the Gartner opinion ranking. Their composite score of 9.51 (out of a possible 10) was well ahead of McDonald's second place score of 5.87 (Hofman et al. 2013).

## SUPPORTING EVIDENCE

Apple ranked high in the opinion categories of the Gartner ranking; however, they excelled in the financial measures. Table 16.2 summarizes some of Apple's key performance measures for the past 5 years. Starting with a revenue growth rate of over 50% per year, they added an amazing annualized inventory turn rate of close to 100 turns. Their ROA averaged over 22% annualized. Part of their success is that they have minimized their investment in capital equipment by outsourcing practically all of their manufacturing operations with relatively modest investments in distribution and retailing.

Although Gartner has not yet used the cash-to-cash cycle as a factor, they have indicated they will include it when they are able to get reliable data. Apple has done an outstanding job in cash management to go along with their revenue and income growth. As Table 16.2 indicates, Apple has a negative cash-to-cash cycle time, indicating they

**TABLE 16.2**

Apple Management Performance Measures

| Selected Key Performance Indicators | September 27, 2008 | September 26, 2009 | September 25, 2010 | September 24, 2011 | September 29, 2012 | % Increase (Decrease), 2008–2012 |
|---|---|---|---|---|---|---|
| Revenues | 32,479 | 36,537 | 65,225 | 108,249 | 156,508 | 382 |
| Cost of sales | 21,334 | 23,397 | 39,541 | 64,431 | 87,846 | 312 |
| Gross margin | 11,145 | 13,140 | 25,684 | 43,818 | 68,662 | 516 |
| Income (loss) before provision for taxes | 6,895 | 7,984 | 18,540 | 34,205 | 55,763 | 709 |
| Net income (loss) | 4,834 | 5,704 | 14,013 | 25,922 | 41,733 | 763 |
| Gross margin (%) | 34 | 36 | 39 | 40 | 44 | 28 |
| Income before taxes (%) | 21 | 22 | 28 | 32 | 36 | 68 |
| Net income (%) | 15 | 16 | 21 | 24 | 27 | 79 |
| Current assets | 34,690 | 36,265 | 41,678 | 44,988 | 57,653 | 66 |
| Current liabilities | 14,092 | 19,282 | 20,722 | 27,970 | 38,542 | 174 |
| Current ratio | 2.5 | 1.9 | 2.0 | 1.6 | 1.5 | −39 |
| Accounts receivable | 2,422 | 3,361 | 5,510 | 5,369 | 10,930 | 351 |
| Inventory | 509 | 455 | 1,051 | 776 | 791 | 55 |
| Accounts payable | 5,520 | 5,601 | 12,015 | 14,632 | 21,175 | 284 |
| Number of full-time employees | 32,000 | 34,300 | 46,600 | 60,400 | 72,800 | 128 |
| Sales per full-time employee (thousands of dollars) | 1,015 | 1,065 | 1,400 | 1,792 | 2,150 | 112 |
| Sales per day (thousands of dollars) | 88,984 | 100,101 | 178,699 | 296,573 | 428,789 | 382 |
| Cost of sales per day (thousands of dollars) | 58,449 | 64,101 | 108,332 | 176,523 | 240,674 | 312 |
| Days of accounts receivable | 27 | 34 | 31 | 18 | 25 | −6 |
| Days of inventory | 9 | 7 | 10 | 4 | 3 | −62 |
| Days of accounts payable | 94 | 87 | 111 | 83 | 88 | −7 |
| Cash-to-cash cycle time (days) | −59 | −47 | −70 | −60 | −59 | 1 |

*Source:* Adapted from Mergent Online, 2013, http://www.mergentonline.com, accessed October 15, 2013.
*Note:* Dollars are in million except where otherwise indicated.

collect funds from customers before they have to pay for their purchases. This is the result of obtaining extended accounts payable terms from suppliers and achieving very high inventory turns.

Another accolade to Apple's supply chain capabilities surrounded the debut of the iPhone 5 in 2012. In speaking of Tim Cook, one source reported, "Rather, it was the speed of the global launch that astounded, validating the new CEO's much-touted wizardry at the essential but unglamorous task of managing a supply chain." The writers go on to explain that the rollout included 100 countries and contracts with 240 carriers to deliver the phones to waiting customers (Henderson and Gupta 2012).

## SEGMENT SALES

Table 16.3 shows how Apple's sales were distributed among markets in 2011 and 2012. Although sales increased overall in 2012 over 2011 by 45%, the highest increase was in Japan and the lowest in Europe, perhaps reflecting the economic recovery in Japan and the continuing slump in Europe.

Table 16.4 shows sales by product segment. As in Table 16.3, the overall increase of 45% in dollars showed the highest increase in new products such as the iPhone and the iPad. In computer products, desktops showed a decline in dollars although the number of units stayed about the same. While iPod unit sales declined, the dollar amount

**TABLE 16.3**

Sales by Market Area

| Market Area | Apple Net Sales (Dollars in Millions) | | | |
|---|---|---|---|---|
| | **2011** | **2012** | **Increase (%)** | **Total, 2012 (%)** |
| Americas | $38,315 | $57,512 | 50 | 37 |
| Europe | 27,778 | 36,323 | 31 | 23 |
| Japan | 5,437 | 10,571 | 94 | 7 |
| Asia Pacific | 22,592 | 33,274 | 47 | 21 |
| Retail | 14,127 | 18,828 | 33 | 12 |
| Total | $108,249 | $156,508 | 45 | 100 |

*Source:* Adapted from Apple 10-K report, 2012, www.mergentonline.com, accessed October 24, 2013.

**TABLE 16.4**

Sales by Product Segment

| Product | Apple Net Sales (Dollars in Millions) | | | |
|---|---|---|---|---|
| | **2011** | **2012** | **Increase (%)** | **Total, 2012 (%)** |
| Desktops | $6,439 | $6,040 | −6 | 4 |
| Laptops | 15,344 | 17,181 | 12 | 11 |
| Total computer | 21,783 | 23,221 | 7 | 15 |
| iPod and related products | 13,767 | 14,149 | 3 | 9 |
| iPhone and related products | 47,057 | 80,477 | 71 | 51 |
| iPad and related products | 20,358 | 32,424 | 59 | 21 |
| Peripherals and other hardware | 2,330 | 2,778 | 19 | 2 |
| Retail | 2,954 | 3,459 | 17 | 2 |
| Total | $108,249 | $156,508 | 45 | 100 |

| Product | Apple Net Sales (Thousands of Units) | | | |
|---|---|---|---|---|
| | **2011** | **2012** | **Increase (%)** | **Price per Unit, 2012** |
| Desktops | 4,669 | 4,656 | 0 | $1297 |
| Laptops | 12,066 | 13,502 | 12 | 1272 |
| Total computer | 16,735 | 18,158 | 9 | 1279 |
| iPod and related products | 42,620 | 35,165 | −17 | 402 |
| iPhone and related products | 72,293 | 125,046 | 73 | 644 |
| iPad and related products | 32,394 | 58,310 | 80 | 556 |
| Total | 164,042 | 236,679 | 44 | $661 |

*Source:* Adapted from Apple 10-K report, 2012, www.mergentonline.com, accessed October 24, 2013.

increased slightly, reflecting the revenues from the iTunes Store, App Store, and iBookstore in addition to revenues from related services. It is also interesting that the percent increase in units is almost the same as the percent increase in dollars.

The distribution of revenues by product segment suggests that Apple Inc. needs new products to sustain its rapid growth. The modest growth in computers and iPods was more than offset by iPhones and iPads. Increasing competition and short product life cycles create demand for new products. As new products are introduced, it puts even greater pressure on the supply chains to keep up with the changes.

## WHY APPLE MUST MANAGE THEIR SUPPLY CHAINS

Apple does not manufacture its own products, as the company points out in their 2012 10-K report.

> Substantially all of the Company's hardware products are manufactured by outsourcing partners that are located primarily in Asia. A significant concentration of this manufacturing is currently performed by a small number of outsourcing partners, often in single locations. Certain of these outsourcing partners are the sole-sourced suppliers of components and manufacturers for many of the Company's products. Although the Company works closely with its outsourcing partners on manufacturing schedules, the Company's operating results could be adversely affected if its outsourcing partners were unable to meet their production commitments. The Company's purchase commitments typically cover its requirements for periods up to 150 days
>
> **Apple 10-K Report (2012, p. 7)**

The fact that Apple is so dependent on a few *sole-sourced* suppliers indicates the need for the company to manage those suppliers carefully on a continuing basis. Apple must adjust its demand requirements of suppliers regularly because of the changing demand for Apple products, both increasing and decreasing. New product introductions are especially important because of the spike in demand as new products are announced.

The heavy emphasis on outsourcing manufacturing, primarily to companies in Asia, has not been without problems. Foxconn, the largest assembler of Apple products, has been targeted as a major contract manufacturer guilty of unsafe working conditions and oppressive employee practices. Cook has investigated these allegations, even visiting a Foxconn factory personally. Apple also joined the Fair Labor Association, an industry-financed third-party monitoring group that has the ability to visit factories and report its findings independently. Even with the problems, Cook's Apple is increasing its dependence on manufacturing in China. They reported assets of $2.6 billion, primarily in China, most of it in material and equipment bought on behalf of its suppliers. Even though Apple owns much of the manufacturing capabilities, its partners will operate the facilities (Lashinsky 2012).

## OBSTACLES OR PROBLEMS

Not all observers are convinced that Apple has the greatest supply chain strategy. In speaking of the problems Apple experienced with suppliers, notably Foxconn, in early 2012, Krzykowski (2012) said, "In January, the world may have received the most convincing evidence yet of how difficult – and perhaps impossible – it is to know everything that occurs in a supply chain that spans the globe." As a result, Apple increased the number of supplier audits.

In 2013, as a result of its continued rapid growth and concerns about the employee practices at some suppliers, Apple increased its business with Pegatron, also a Chinese company. The company already manufactures IPad minis and some versions of the iPhone. This addition should provide Apple with more capacity and the ability to react faster to supply chain disruptions. This change also indicated Apple's willingness to continue its dependence on outsourcing manufacturing to China companies.

In speaking of the iPhone 5 launch, it was reported that Apple had overcommitted to FedEx and other carriers in their need for movement of product. This may have been the result of forecasting errors or delays in production of the products (Lennane 2013). Cohan (2012) blames supply chain problems for the lower-than-expected initial iPhone 5 sales.

While relying on its successful supply chain model, the company itself (Apple 10-K Report 2013) points out that its heavy dependence on offshore outsourcing exposes it to a number of risks, including the following:

- Global economy could affect the company.
- Some resellers may also distribute products from competing manufacturers.
- Inventories can become obsolete or exceed anticipated demand.
- Some components are currently obtained from single or limited sources.
- Some custom components are not common to the rest of the industries.
- The ability to obtain components in sufficient quantities is important.
- Supply chain disruption such as natural and man-made disasters can be serious.
- The company depends on logistical services provided by outsourcing partners.
- The company also relies on its partners to adhere to supplier code of conduct.

As the company continues to grow, the potential for disruptions in their supply chain will become a greater concern.

## CONCLUSIONS

That Apple Inc. is successful is evident. Its success has resulted from both innovative products and well-designed supply chains. Any concerns are more about the company's capability to sustain its growth and profitability, while dealing with the social and environmental problems encountered in its use of outsourcing as a prime strategy in manufacturing. With Tim Cook as CEO, most observers would agree that the supply chain issues will be handled. The jury is still out with respect to the company's ability to continue to develop innovative products.

## REFERENCES

Apple 10-K Report, 2012, www.mergentonline.com, accessed October 24, 2013.
Apple Company History, http://apple-history.com/h1, accessed August 8, 2014.
Cohan, P., Apple can't innovate or manage supply chain, *Forbes*, October 26, 2012. http://www.forbes.com/sites/petercohan/2012/10/26/apple-cant-innovate, accessed October 24, 2013.
Henderson, P. and Gupta, P. Apple supply chain mastery is key to CEO Tim Cook's success, *Reuters*, 2012. http://www.huffingtonpost.com/2012/09/22/apple-supply-chain-tim-cook_n_1905674.html, accessed October 25, 2013.
Hofman, D., Aronow, S., and Nilles, K., The Gartner supply chain top 25 for 2013, May 22, 2013. http://www.gartner.com/imagesrv/summits/docs/apac/supply-chain/the_gartner_supply_chain_top_252126.pdf, accessed October 25, 2013.

Krzykowski, B., Bruised, not beaten, *Quality Progress*, 45(3), 10, March 2012.

Lashinsky, A. Sr., How Tim Cook is changing Apple, *Fortune Magazine*, June 11, 2012.

Lennane, A., Apple supply chain slackens as it slips behind with deliveries of new iPhone, *The Loadstar*, September 26, 2013. http://theloadstar.co.uk/apple-supply-chain-slackens-as-it-slips-behind, accessed October 23, 2013.

Mergent Online, Apple, Inc., 2013. http://www.mergentonline.com, accessed October 15, 2013.

SupplyChainOpz, 7 Supply chain lessons from Steve Jobs, 2013a. http://www.supplychainopz.com, accessed October 28, 2013.

SupplyChainOpz, 7 Traits of great supply chain leader: A case study of Tim Cook, 2013b. http://www.supplychainopz.com, accessed October 28, 2013.

## Supply Chain Configurations

In Chapter 13, we described the evolution of supply chains from tightly coupled vertically integrated configurations to tightly coupled collaborative configurations. In the first configuration, one company owned or directly controlled the major portions of the supply chain and could allocate costs and benefits through transfer pricing mechanisms. In the second configuration, companies depended on each other to achieve a smooth flow of goods and services from their origin to the ultimate consumer. The second configuration could be termed virtual integration.

It is probably safe to say that most supply chains are somewhere in between the extremes described earlier. But where are they? How are companies dealing with the problem of moving from the security of vertical integration to the uncertainty of loosely coupled globally dispersed independent operations? Figure 16.1 shows a supply chain that moves toward the ultimate consumer through loosely coupled links that work at varying levels of effectiveness and efficiency. It gets the job done, and the individual companies may be operating at high performance levels. The opportunities for improvement are in the intercompany links. While these improvements are real, they are difficult to evaluate. It is even more difficult to assign credit for the individual contributors.

In this chapter, we first describe a process whereby supply chain members can attempt to determine the ROI in a supply chain. We then outline ways in which the supply chain members can divide the net proceeds among themselves. We also discuss the role of the prime mover company in a supply chain, because it is likely that a single company will exert the greatest influence and, as a leader, must take responsibilities that other participants need not assume. Inasmuch as supply chain relationships are dynamic and apt to

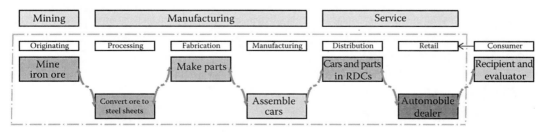

**FIGURE 16.1**
Loosely linked supply chains without direct route to ultimate consumer.

change over time, we describe ways in which these transition periods can be managed. Finally, we provide examples of approaches to the distribution of supply chain proceeds among its members.

## Programs Requiring Close Supply Chain Relationships*

Global competition is forcing companies to improve—or go out of business! Successful companies continue to search for some competitive advantage, whether it is in price, quality, flexibility, or, more recently, in response times. Just as they strive to reduce costs and improve quality, companies are trying to reduce the time it takes to get an order from the supplier to the customer and to get new products to the market. It is no longer sufficient to offer low cost and high quality; a company must be responsive and flexible, too.

Some programs have evolved over the past two decades to reduce the lead times demanded by customers, mostly retailers who have tipped the balance of power from the manufacturers to retailers, such as Wal-Mart, Home Depot, and Kroger. While these programs have individual identities, they are gradually being absorbed under the umbrella of SCM.

Some of the more prominent programs are

- Quick response program (QRS)
- Continuous replenishment program (CRP)
- Efficient consumer response (ECR)
- Vendor-managed inventory (VMI)
- Sales and operations planning (S&OP)
- Collaborative planning, forecasting, and replenishment (CPFR)

These programs were described in earlier chapters.

## Need to Evaluate Supply Chain ROI

How do companies quantify the ROI for an entire supply chain? This cannot likely be done by starting with the total income and investment of a company, as often reported in a company's annual report. This is a macro number that is the result of numerous variables, many well beyond the scope of the workings of the supply chain. Rather, the evaluation of the ROI for a supply chain requires a project analysis approach, where the effects of individual incremental actions are considered. This approach should consider both tangible and intangible benefits and costs.

### Tangible Benefits

It follows that the term *tangible* implies that dollar values are readily available (we will use dollars for our discussion; however, any unit of currency applies). Unfortunately,

---

* The next section has been adapted from Crandall (2006).

this is not the case. Even where there are tangible benefits, it may be difficult to assign a dollar value, even indirectly.

### Reduced Inventory

One of the most obvious benefits from integrated supply chains is reduced inventory. This is a prime objective of a supply chain, so it follows that there should be a reduction in the total inventory over time. Although none of the approaches are easy, it should be possible to establish the total inventory related to the supply chain operation at some point in time. Subsequently, the total inventory in the supply chain can be reported to determine how much the total inventory has decreased. By using a carrying cost percentage (one that is accepted by other supply chain members), it is possible to calculate a dollar savings.

### Reduced Cycle Times

Another objective is to determine how much the supply chain reduces the total cycle times for moving the product along the supply chain to the consumer. This time can be expressed in days or weeks, and when compared with a base period, a cycle time reduction can be calculated. Converting a reduction to cycle time into dollar savings requires assigning a value to the reduction in the response time. Does the reduction in cycle time result in increased sales or reduced costs? Are there other factors to consider? Does this calculation duplicate some of the savings for reduced inventory?

### Improved Customer Service

Improved customer service also has some tangible features, such as increased percentage of on-time deliveries, reduced stockouts, and reduced errors in invoicing. As with cycle times, customer service performance indicators are tangible but may also be difficult to convert to dollar increase in sales.

### Improved Quality

Improved quality means fewer defects, reduced inspections, reduced repairs or reworks, and fewer ruined products downstream in the supply chain. However, the costs of quality have been elusive and few accounting systems are set up to measure how these improvements relate to reduced costs or increased sales.

## Tangible Costs

In order to achieve some of the benefits described earlier, companies incur some tangible costs. Some of the more likely costs are described in the following section.

### Communications

One of the primary costs in setting up an integrated supply chain is in designing and implementing an interorganizational communications system (IOS). An IOS requires not only hardware and software but ongoing consulting and maintenance services as well. There may be a need for software upgrades and modifications. Often, IOS systems require years to fully implement and it is more difficult to accumulate project costs over a multiyear period. However, an IOS will be one of the major costs in setting up integrated supply chains.

### Retraining Internal Employees

Retraining employees in the use of new systems can be time consuming and frustrating for both the employee and the trainer. Formal training classes take time away from the job and may be considered nonproductive; however, they are necessary if the new system is ever to work at a fraction of the level for which it was designed.

### Restructure Supplier Network

Identifying and organizing the supplier network take time. Companies must identify the companies that they want to include in the supply chain they are designing. These are companies that make a significant contribution to the supply chain, and they should be rewarded for their participation. Not every supplier will be included, because some will not have a large enough role to make it worthwhile to include them. For those that are included, it will be necessary to identify the key representatives of the organization that will be involved in the eventual distribution process of the proceeds.

The costs involved in restructuring the supplier network includes those associated with compiling information, meeting with representatives from the supplier organizations to develop supplier agreements, time to monitor adherence to the agreements, and other costs that can be directly assigned to the restructuring effort.

### Design Customer Network

Designing a customer network will be at least as costly as designing the supplier network because customers may not be as willing to cooperate in developing a supply chain that shares information among its members. Today, many retailers are the largest entities in a supply chain, and they may believe that they should command the major share of the benefits, by receiving lower prices from their upstream suppliers. Any suggestions that suppliers should share in the benefits may not be well received by the retail establishment.

### Capital Investment

There may be a need for capital investment. For example, in implementing RFID, there is a need to buy equipment for attaching and reading tags. When the investment is required by the same company that gains the benefits, there may not be a problem. However, when the company that gains the benefit, such as a retailer, is not the same as the company that incurs the investment cost, such as a supplier, an imbalance results that needs correction. Even if chain-level benefits are obvious, an innovation may not go forward if a key participant does not believe that it is sharing the benefits appropriately (Wouters 2006).

## Indirect Costs

By definition, the indirect costs are those costs that are less obvious when trying to create an integrated supply chain. The following examples illustrate some of these indirect costs.

## Meetings Required to Organize Customer and Supplier Relationships

Under direct costs, we listed meetings, where the costs of transportation, lodging, and wages can be directly assigned. However, there will be other costs that are less direct, such as informal meetings with other functional areas, telephone conversations, or e-mail messages with supplier representatives and time spent thinking about building the supply chain relationship. Usually, companies do not try to record these costs and assign them to specific activities.

## Programs to Change Internal Culture

Time will be required to change the internal culture of an organization. While it may be possible to document the costs of formal orientation meetings of groups within the organization, many companies may not consider this necessary or even desirable. Informal meetings of small groups or individuals are even less likely to be documented; however, a considerable amount of time will be required to ensure that the internal employees are on board with the changes being made in the relationships with the customers and suppliers, especially if any employee has direct contact with these external entities.

## Changes in Organization Structure

Changing the organization structure may be easier than changing the culture, but it will take time to consider the changes required, design the revised organization, effect the changes, retrain the affected employees in their new responsibilities, and monitor the ongoing operations to ensure that the new organization structure is working effectively. It is unlikely that any company could, or even try to, accumulate an accurate record of the costs involved in this activity.

## Realignment of Roles of Supply Chain Participants

In addition to the realignment of roles within a company, it will also be necessary to realign the roles between the companies. A review of individual practices may uncover duplication of effort that can be reduced or omissions that have to be covered. It takes time to identify what needs to be done, negotiate the agreement of where the activity will be performed, and decide what changes will be required in the day-to-day operations.

The preceding discussion shows that quantifying the value of benefits and costs is difficult, if not impossible, in many cases. Some benefits and costs will occur over a multiyear period; this suggests the use of time value of money or discounted cash flow calculations. However, how do you track cash flow in a multiunit supply chain? This task would appear to be even more difficult than quantifying the individual benefits or costs. Where a company may be involved in multiple supply chains or have operations not directly connected with a known supply chain, the situation becomes even more muddled.

One of the problems may be that there are changes in both costs and capital requirements. This makes the redistribution task more difficult. For example, direct costs may decrease, but capital investment may increase.

An equally daunting task is to factor in changes and realignments of the supply chain over time. New customers enter, old customers leave, new suppliers are added, and old suppliers are discontinued. The start and stop dates do not necessarily correspond with known accounting or calendar periods. Consequently, it is unlikely that the effect of these changes can be considered at all in the total scheme of things.

We described the difficulty of quantifying indirect benefits and indirect costs. Even more difficult is trying to quantify what might be called intangible benefits and intangible costs.

## Intangible Benefits

A number of intangible benefits have been attributed to the successful integration of supply chains. We will describe several of these benefits to illustrate they exist and are elusive to quantify.

### Integrated Flow of Goods and Services

Achieving an integrated supply chain should mean that there is an improved flow of goods and services through the supply chain. But how does a company measure the benefits? Do they consider the benefit to be primarily in the reduced response time? Consider that the supplier is asked to increase the variety of products that they supply and, at the same time, reduce the response time. This added requirement may cause an increase in response time, but it is difficult to know the extent of the benefit if the average response time does not change.

They could also consider the benefit to be a reduction in suppliers. However, consider that they increase the volume and variety of products purchased, which would normally require more suppliers. If they end up with the same number of suppliers, does that suggest no progress?

### Faster Resolution of Problems

Integrated supply chains should make it easier and faster to identify and resolve problems that may occur between the customer and supplier. Is the benefit the reduced time required of the participants to solve the problem? On the other hand, is the benefit the result of the solved problem that makes it possible to get a shipment on time to complete and ship a critical order? How much of the value of the total shipment should be assigned to the solved problem?

There may be a more elusive situation. Consider the problem that arises from trying to decide how an operation that was previously performed by the customer, say, a retailer, is to be transferred to its supplier, a wholesaler. The problem would not have occurred if the companies were not trying to integrate the supply chain; therefore, should there be any benefit to solving a problem created in this fashion?

### Match Customer Wants with Products Provided

Most would agree that an integrated supply chain should make it possible to more closely match the products provided with what the customer wants. How do you measure this? Can you assume that an increase in the percentage of orders filled measures this improvement? If so, what about all the other actions that helped to increase the fill rate? How do you allocate the benefits and costs among all the factors involved?

### Reduced Excess Capacity along the Supply Chain

Another benefit of integrated supply chains is a reduction in the excess capacity along the supply chain, because of reduced demand variability. Consider that, simultaneously, the supply chain's effectiveness results in an increase in demand, so there is no change in the overall capacity requirements. The net result is two significant benefits, but measuring the net benefit of each is a daunting prospect.

### Increased Knowledge

As indicated earlier, most managers would agree that increased knowledge should lead to improved performance, including increased financial returns along the supply chain. However, few managers would be willing to quantify the exact relationship between increased knowledge and increased income. To make the waters even murkier, one study found that learning appears to have a positive impact on performance measures relating to end-customer satisfaction and being a more market-focused supply chain. Learning does not appear to affect supply chain performance related to cost. The authors stated that tacit knowledge (knowledge that resides in the minds of people) is more difficult to transfer than implicit knowledge (knowledge that is documented in policies and reports) (Spekman et al. 2002).

The challenges affecting the supply chain learning include

- The natural tension that exists among partnerships regarding cooperation versus competition
- The need to ensure that learning happens throughout the supply chain at the supply chain enterprise, the firm, and the individual
- The need to have a flexible structure to enhance learning

For supply chains to develop a sustainable competitive advantage, the partners must focus on the end-use customer. Learning does have a positive impact on the end-use customer-focused performance metrics (Spekman et al. 2002).

### Reduced Risk of Supply Chain Disruption

Risk of disruption is reduced when a supply chain becomes more tightly coupled and the participants operate more collaboratively. However, this improvement occurs over time and incremental improvements may be difficult to recognize by those involved in the change process.

## Intangible Costs

Determining the intangible costs of integrating a supply chain may be as elusive as determining the benefits.

### Loss of Confidential Information

One of the dangers of integrating supply chains and beginning to exchange information is the potential loss of confidential information. While there is also the opportunity to gain confidential information, it is difficult to assign a value to either type of transaction. If the information were sold, a value could be implicitly assigned; however, when the information exchange is informal, perhaps, not even known, no explicit value is quantifiable.

### Increased Awareness of Inequitable Treatment among Participants

Closer relationships in integrated supply chains may expose inequitable treatment among the participants. This situation may be in the form of discounts allowed, reduced inspection requirements, or the level of information sharing. Irrespective of the form and magnitude, someone may object and require mollification. Even if they do not object openly, they may silently worry about it to the point of providing diminished service.

### Discrepancy between Contribution and Payoff among Participants

One extension of the equitable treatment idea mentioned earlier may be that a member of the supply chain believes they are not getting a fair return on their contribution to the welfare of the supply chain. This may cause them to reconsider the desirability of continuing in the supply chain. If they are an important member, a disgruntled member may adversely affect other participants.

### Legal Actions

As supply chains become more geographically dispersed and complex, formal written agreements become more necessary to clarify the terms and conditions of exchange, especially when participants are in different countries. Globalization increases the need for documentation and, at the same time, increases the risk of legal actions when disagreements arise (den Butter and Linse 2008). While the costs of legal actions are tangible, the likelihood of it becoming a reality is nebulous. Sharing these costs along a supply chain could become a fascinating, but perhaps fruitless, exercise.

We have discussed a number of factors that make it difficult to determine the net benefits of an integrated supply chain and the relevant contribution of each member. This does not suggest that an attempt to determine the benefits and costs has no value; it does suggest that doing it will be difficult. Later, we will describe some cases where dyads of companies made a definite effort to determine the net benefits and found a way to distribute these benefits. However, successes in determining and allocating net benefits are still rare.

## Obstacles to Equitable Distribution among Members

There are a number of factors to consider when organizing to distribute benefits among the supply chain participants.

### How to Organize?

How do you organize supply chain members to develop a cost/benefit distribution system? A facetious answer would be "very carefully." However, it is a delicate question and one that many supply chains have not yet dealt with. Supply chains have organized carefully to physically move goods and services from the point of origin to the point of consumption. They have also organized to share information as well as move funds from the point of consumption upstream through the supply chain. However, few have attempted to organize in order to share the benefits.

Do all members participate equally? If not, how do you decide? Is it based on company size? Or amount of goods sold in the supply chain? Or the absence of problems? Or the level of innovation displayed? There are a number of measures that could be used; however, selecting and measuring them offer a challenge.

Should the participants elect or appoint a steering committee to establish policies and procedures to clarify and prevent, or at least reduce, disputes? Should large companies get more votes?

Is there a need for formal contracts among the participants? If the proceeds to be distributed are significant, it is unlikely that informal agreements will suffice. Consequently, there will be a need for formal contracts that spell out the accumulation and distribution process.

Is there a need for a third party to reconcile differences? Consider that the participants cannot agree or do not have written guidelines to follow. Maybe there is a need to have an arbitrator or mediator to work with the parties to come up with a reasonable way to proceed. If so, there will be a need to select the arbitrator. Will the supply chain representatives vote or, in some way, arrive at a consensus?

### How to Distribute?

If the supply chain participants can arrive at a consensus figure for the net benefit amount, how should these benefits be distributed? One approach would be to distribute the proceeds according to some predetermined weight, based on performance. But how do you measure performance, especially the relative performance of different participants? Is it based on effort or results? Most would vote for results, but now we are back to the difficulty of determining the identity and value of the results. There seems to be no end of difficulties in determining the financial amounts that may have accrued as the result of effective supply chain integration.

Once determined, how can the benefits be distributed? Should it be in the form of direct payments from one entity to another? How often should the payments be made, or should paper debits and credits be maintained indefinitely? Perhaps, participants should make payments into a central pool with distributions to be made at designated intervals.

There is no simple or objective answer. It is inevitable that the determination and distribution of the proceeds must include judgment and negotiation—judgment as to what should be included and the related amounts and negotiation as to how the proceeds should be distributed. If the participants try to develop a verifiable methodology, it may take an exhaustive amount of time and wipe out all the benefits that could be accrued from making the distribution.

### Supply Chain Governance Models

There are some attempts to develop models of what could be considered as *supply chain management* or *supply chain governance* (SCG). One model addressed the issue of ways in which sustainability practices could be introduced throughout a supply chain network. Table 16.5

**TABLE 16.5**

Supply Chain Governance Categories

| | | Centrality of the Focal Organization | |
|---|---|---|---|
| | | **Low** | **High** |
| **Supply Chain Density** | **Low** | *Transactional SCG*: Limited interaction with customers or supplier; short-term relationship; its role is that of a negotiator with the primary objective to maintain ongoing operations with no major changes. | *Prime Mover SCG*: Extensive interaction with supply chain participants, both downstream and upstream; long-term commitment; its role is one that of an orchestrator; its objective is to exercise control over the collaborative efforts. |
| | **High** | *Acquiescent SCG*: Limited interaction upstream and downstream; short-term perspective; full participant with a role of an executor; its objective is to maintain membership in the supply chain with access to markets and partners. | *Participative SCG*: Fully integrated both upstream and downstream, with long-term relationships; leader by example with a role of a compromiser to achieve results; flexible and adaptable to multiple parties and changing conditions. |

*Source:*  Adapted from Vurro, C. et al., *J. Bus. Ethics*, 90, 607, 2009.

summarizes the primary types of networks that could be considered. While some supply chains may operate without a dominant company present (low centrality of focal organization), most supply chains have a major retailer or manufacturer as a dominant company. We will use the Prime Mover SCG block in Table 16.5 to describe the role of a dominant party in a supply chain. We will use the participative SCG block in Table 16.5 to describe the typical role of a third-party provider in a supply chain (Vurro et al. 2009).

The *prime mover* in the supply chain assumes the pseudo position of manager of the supply chain and sets up the informal, but highly influential, management mechanism. Wal-Mart is an example of a prime mover. Another approach is what was designated as the *mini-maestro* model by Bitran et al. (2007), in which a third party assumes the management of a significant portion of a company's supply chain. Our description about the role of the prime mover covers some of the functions that could also be performed by a third party.

## Prime Mover in the Supply Chain

Most supply chains have a single company that is the prime mover of activity in that supply chain. It may be the retailer, such as Wal-Mart, that wants to assure itself of dependable supply, or a manufacturer, such as Procter & Gamble, who may want to assure itself of a dependable customer network.

## Organize

Often a supply chain is organized through the initiative of a single company. It is not likely that a group of companies gathered and decided to organize a supply chain. The company that takes the initiative does so, because they see an advantage in having a supply chain arrangement. They may not even realize that they are setting out on this noble mission;

it may start with simply building a relationship with a single customer or supplier and then building onto that beginning.

In recent years, retailers have often been the prime movers to establish working supply chains. Retailers in the clothing sector, especially those in style-oriented clothes such as women's dresses, as opposed to more stable designs, such as men's shirts, have moved toward quick replenishment programs. Typically, a retailer would try to forecast the demand for an assortment of clothes—sizes, styles, and colors—and order enough to last the entire selling season, because they would not be able to get replenishment orders before the season ends. With this arrangement, it is inevitable that the retailer ends up with an excess of slower-moving items and stockouts in faster-moving items. Under quick replenishment programs, the retailer could order an assortment to start the season and use replenishment orders for faster-moving items. QRSs required close coordination and collaboration between the retailer and manufacturer. Obviously, a QRS offers considerable benefits to the retailer, which often takes the initiative in organizing such a supply chain arrangement. The "Company Profile" in Chapter 2 describes the supply chain developed by Inditex (Zara) to achieve fast response to market changes.

Most businesses of any size will have multiple supply chains. We discussed this requirement in earlier chapters, but it is worth repeating that different product categories will probably require different supply chain arrangements. Consequently, a business that is the prime mover for one of its supply chains is only a participant in another of its supply chains.

### Select the Team

The organizer of the supply chain system is also apt to design the system to suit its needs. By design, we refer to the composition of the businesses that are in the supply chain. Ideally, a supply chain requires certain functions to be performed, and businesses are selected to perform all the required functions. Sometimes, it is necessary to have a degree of duplication or redundancy in functions, especially at the early stages, until the flow of goods and services is well established.

In addition to having businesses that perform the required functions, those businesses must be compatible in technology, organization, and culture. Interorganizational communication is essential. Businesses must be able to freely communicate, usually through electronic media, to exchange information about demand forecasts, actual sales, purchase orders, shipments, payments, and all the related normal flow of information. In addition, they need to communicate abnormal information, such as rush orders, impending weather disruptions, and quality problems.

Even with compatible communications technology, businesses may fail to communicate successfully if their organization structures are not compatible. It is more difficult, and takes longer, for a business with a rigid and vertical hierarchical line-and-staff organization to complete transactions with a business that has a flat, horizontal structure with empowered employees. The two organizations may want to move at different speeds in making and implementing decisions.

Organizational cultures also play a role. An organization with a long history of success may be less willing to consider new ways of operating. If they are the prime mover, they may not have to change much. However, if they are being asked to change in order to accommodate changes that the supply chain organizer wants to make, it may take a while for the cultures to mesh.

Organizing a supply chain is more than selecting the best player for each position. It involves getting those players to work together as a team.

## Monitor Ongoing Operations

Once the supply chain is working, it is important to evaluate its performance. While all members of the supply chain will be interested, the prime mover organization will be vitally interested. It will need to have some way to assess the progress being made in getting participants to cooperate, redundant functions eliminated, service gaps filled, and all the other problems, both small and large, that can arise in organizing and operating something as complex as a supply chain.

It is unlikely there will be a neat reporting system to provide this kind of information to the concerned parties. They will have to ask questions, make visits, and challenge delays to uncover the true causes of problems or deficiencies in the effective workings of the supply chain. Sometimes, fixes will be as simple as introducing a new purchasing manager to other companies. Often, the fixes will be difficult and of a long-term nature, such as replacing a major software application. As with many programs of this magnitude, monitoring the progress in creating an integrated supply chain is a journey, not a destination.

## Evaluate Performance

Even before the supply chain is fully operational, the question of performance measurement should be considered. Performance measurement of a supply chain is different from performance measurement for an individual company. While some measures may be the same—inventory turns or cycle times—they are not necessarily the sum or average of all the participants. Supply chain participants have to agree on the measures used, develop a way to accumulate information, and design a system to report the results to the appropriate recipients.

## Initiate Change

Regardless of how well a supply chain performs, there is bound to be the need for change at some point. When this happens, the prime mover may again have to take the initiative to introduce the change, sometimes, at the risk of causing major disruptions in the composition and effectiveness of the supply chain.

A classic example of this occurred when Wal-Mart introduced the idea of replacing bar codes with RFID chips. While executives at Wal-Mart recognized the magnitude of the change, they believed it to be necessary for the long-term vitality of their supply chains. There were a number of problems to be solved—the cost of RFID tags, lack of standards, investment required for suppliers, and uncertainty of the reliability of the technology.

This project has been a multiyear program. Some suppliers resisted, because they did not see a return on their investment. Other suppliers had to invest in equipment to make and position the tags on cases and pallets of Wal-Mart. Many suppliers felt they incurred the costs and Wal-Mart reaped the benefits. We will come back to this example later in this chapter when we discuss about the methods of allocating costs and benefits among the supply chain participants.

Another example that may require the prime mover to take a leading role is in the movement toward green supply chains. While individual companies can *go green*, it is usually difficult for them to do it alone, because of the need to coordinate changes in raw materials or packaging. Consequently, changes are facilitated when approached from a supply chain perspective.

## Third-Party Provider

In Chapter 9, we described the role of third-party logistics providers (3PLs) in helping move the goods along the supply chain. There is a growing awareness that logistics providers can assume even greater roles as supply chain facilitators, or even "a 'maestro' that can coordinate the entire network and align the incentives for all the participating players" (Bitran et al. 2007, p. 34). The authors concede that most large companies would not be interested, or willing, to turn over the entire supply chain to a third party; however, there are some who would be willing to outsource a substantial portion of the supply chain to a *mini-maestro*. Li & Fung is a Hong Kong–based company that serves private-label apparel firms and maintains a network of over 10,000 suppliers in 40 countries but does not own any of them. The company coordinates each process in the supply chain, from raw materials sourcing through manufacturing and onto final shipment of goods. Another firm that also offers product design, engineering, and manufacturing services that are vertically integrated is Flextronics.

One of the responsibilities of the *mini-maestro* is to "develop ground-level mechanisms for allocating and sharing the net costs and benefits of partnering, recognizing that business processes changes can impact some players more than others" (Bitran et al. 2007, p. 35). If a manufacturer postpones manufacturing in one location, it will mean that a distribution center will have to carry more inventories. How should the distribution center be compensated? If the supply chain participants improve the supply chain network through innovation, how should the participants be rewarded? The *mini-maestro* must find an equitable solution to these and also develop trust among the members of the supply chain and develop the systems and standards by which the members operate (Bitran et al. 2007).

## Changes in Supply Chain Composition

The changes described earlier focused on changes that affect existing members of the supply chain. However, what if the members of the supply chain change? Is this a simple decision to replace one supplier with another supplier, or is it a more involved decision?

Consider, as suggested earlier, that the supply chain has been carefully crafted by selecting individual businesses that perform designated functions well and, equally important, work with other members of the supply chain as a team member. As the supply chain members begin to acclimate and work together, they form additional links with one another because of being in the same supply chain. For example, suppose two suppliers (A and B) of different but complementary parts pool their deliveries to reduce costs and coordinate the deliveries for the convenience of their customer. Removing one of them as a supplier (A) has implications for both the customer and the remaining supplier (B). Does this mean that the customer should consult with or be influenced by the remaining supplier (B) before replacing (A)?

Today, this hypothetical situation may be highly theoretical, because most changes are still handled on an ad hoc basis without undue regard for the secondary consequences. However, as supply chains become more formal, such as in virtual organizations for building aircraft, there may be a need for a more interactive process in selecting or displacing members.

### Dictated by Prime Mover

In some cases, such as with Wal-Mart, the prime mover may be powerful enough to make changes on its own assessment of the situation. However, most retailers do not have the influence of a Wal-Mart and may have to consider the implications of membership changes. How do they proceed?

### Consensus of the Supply Chain Participants

One of the most obvious ways would be to gain a consensus from among the remaining members of the supply chain. However, if there are dozens of supply chain participants, should they all be consulted, or should there be something like an executive committee composed of the key members of the supply chain? If so, how should the executive committee be selected? Should the prime mover handpick them, or should the members of the supply chain, with fixed terms of office, vote them on? While this scenario may seem like a theoretical question now, it could become a real issue in the future.

### Consultation with an Outside Adviser

If it is impossible, or impractical, to gain a consensus within the supply chain, it may be possible to call in an external adviser to solve the dilemma. The business community has an array of advising consultants today, and it is clearly within the realm of possibility that another product line could be added to the existing supply chain consultants or become a specialty of organizational consultants.

### Mediation by Third Party

If outside advisers cannot resolve the situation, a further step in the process could be to agree on a mediator to decide on what should be done. Mediation is a normal practice in labor disputes; it could become a regular practice in supply chain member disputes as well.

### Legal Action

As supply chains become more tightly integrated and the financial stakes get higher, it may be more difficult to replace one supplier with another. It is conceivable that the day may come when legal action will be required to effect this change. Displaced suppliers may sue for damages, customers may countersue for nonperformance, and judges or juries may decide on real or punitive damages.

---

## Case Studies

In this chapter, we described why close relationships among supply chain members are necessary for the welfare of the entire supply chain. We also looked at how difficult it is to organize and operate integrated supply chains. It is probably even more difficult to design a way to distribute benefits and costs among the supply chain members. We do not believe that there is a generic way to handle this distribution in all types of businesses.

Neither is it possible to outline a general approach that can use macro measures of income and costs, such as net income before taxes or return on the invested assets. The approach must involve an analysis of incremental benefits and costs arising from specific actions or programs by the affected participants.

In this final section, we have included examples of benefit sharing among supply chain members. They represent the limitations of what is practical and realistic today; in the future, new methods may be developed that are more generalizable.

## Use of Accounting Records

One study looked at the use of accounting records to provide information used to decide on the distribution of profits from a specific project. Two companies, a retail supermarket chain and a supplier of hair colorant products, agreed to jointly implement an improvement program and share in the proceeds. Using a category management structure, they implemented a number of changes—improved display fixture location, increased space allocation, in-store promotions, and increased customer education. The program was a success for both organizations. They collaborated on joint forecasts and discovered that the gains in revenues were well above the industry averages. By using accounting records, they were able to agree on a way to share the proceeds from the project. They also learned the accounting records played an enabling role that opened up strategic possibilities for enhanced cooperation. The companies concluded that the accounting records, originally a control measure, provided an opportunity for learning and problem solving between category partners and contributed to a higher level of trust (Free 2007).

## RFID Implementation

RFID is a technology that is gaining acceptance in a number of applications. The transition to RFID tags from bar codes or from no visible identification is an obvious change. Therefore, it should be one of the changes where costs and benefits can be clearly documented. However, one of the barriers to its adoption is the difficulty in assessing the potential ROI to members of the supply chain. To date, it appears that retailers gain the largest benefits, while suppliers incur most of the costs. One group attempted to design a model that would capture the benefits in five areas: lower operating costs, increase in revenue, lower overhead costs, reduced inventory capital costs, and lead time reduction. While encouraged by the initial results in evaluating the results for one of the Wal-Mart's top 100 suppliers, they concluded, "The difficulty in assessing the potential benefits of RFID implementation is one of the key barriers to adoption of this promising technology" (Veeramani et al. 2008).

## Cost Reductions with Investment Requirements

Sometimes, improvement initiatives reduce costs while requiring capital investments. If the benefits are achieved in one part of the supply chain, while the capital investment is required in another part, determining the net ROI becomes more difficult. In such a situation, one empirical study investigated whether companies were able to negotiate a price adjustment that led to a positive economic result for both the companies. The study was further complicated by the introduction of a third party that assumed the capital investment role. The research concluded that the introduction of a capital investment reduced the potential for finding a solution to the distribution problem. The existence of benefits in the supply chain does not guarantee that a mutually acceptable solution can be found. There are several managerial implications from the analysis:

- Use of a net present value (NPV) is appropriate; however, participants should agree to a specific rate at the beginning of the project.
- Price adjustments alone may not be a feasible solution; therefore, managers should be creative in their consideration of redistribution mechanisms.
- Managers should be aware of the implementation costs, both within their own companies and in the supply chain partners.
- Managers considering the implementation of an innovation should consider the investment goals and negotiating stance of their supply chain partner.

The study concluded that the existence of supply chain benefits does not guarantee that an innovation will go forward (Wouters 2006).

## Supply Chain Finance

We described supply chain finance (SCF) in Chapter 15 as a means of improving cash flows along the supply chain. The customer—the buyer—receives extended payment terms from the lender, a bank or some other financing organization. The lender pays the supplier after discounting the payment, but at terms that they would extend to the larger buyer, which are lower than the terms for the smaller supplier. "The result is that the supplier's working capital costs are reduced, even though its payment terms have been extended. It is then in a position to convert this cost reduction into price reductions to satisfy the buyer; the buyer gets the benefit of extended terms, lower prices, reduced capital costs and alignment of its procurement and finance interests; and the lender gets the benefit of a higher margin on the exposure to the buyer company" (Kerle 2007, p. 4).

SCF has some tangible benefits to the buyers, primarily in reduced working capital needs and enhanced relationships with the suppliers. Suppliers gain from increased sales and reduced financing costs. Financiers gain from additional business and reduced lending risk (Kerle 2007). Are all these benefits equitable among the parties involved? If not, how are the benefits to be redistributed?

## Benefits of Supply Chain Collaboration

A theoretical model illustrates the increased difficulty of achieving mutual benefits from supply chain improvements. This model identified a requirement for four coordination modes that would lead to supply chain integration. These modes are

1. Logistics synchronization—the value creation loop
2. Information sharing—the facilitation loop
3. Incentive alignment—the motivation loop
4. Collective learning—the capability loop

The model designers believed the successful simultaneous implementation of these integrative loops will help to synchronize interdependent activities, increase visibility to match supply and demand, align the supply chain partners in actions that lead to supply chain profitability, and acquire new capabilities with the supply chain. As a result, all the participants will recognize a need for matching processes, information, incentives, and capabilities, in order to combine their efforts in activities that will benefit the entire supply chain (Simatupang et al. 2002).

## Summary

This chapter focused on sharing among the supply chain participants. Although a commendable objective, it remains a challenge for many companies and supply chains. We described some of the benefits and difficulties in creating effective supply chain sharing.

First, we described three phases in supply chain configuration—past (vertical integration), future (virtual integration), and present (loosely coupled). We examined some of the programs that make supply chain integration a potentially beneficial target. A number of these programs have progressed to a level that participants have benefited; however, almost all programs still have added benefits to be attained.

In order to share benefits, supply chains must first determine there are benefits and then figure out how to share these benefits among its members. There are tangible and intangible benefits as well as tangible and intangible costs. There are also capital investment requirements that may or may not coincide with the benefits and costs.

Before the benefits can be allocated, the supply chain needs a structure or configuration. At present, the configurations are informal and often not very transparent to the supply chain members. Consequently, any agreement to share benefits may end up being used on a project basis or ad hoc basis, instead of being an ongoing agreement.

Currently, there is little in the way of standard approaches to supply chain sharing. This is largely an open territory for companies to work on as they continue to improve their supply chain operations.

## Hot Topic

If there was ever a supply chain that needed management, it has been Boeing's 787 Dreamliner supply chain. This chapter's hot topic examines this issue.

### Hot Topics in the Supply Chain: The Boeing 787, Pushing the Limits of Outsourcing (Part 3)

In the hot topics for Chapters 14 and 15, we looked at Boeing's supply chain strategy for the Boeing 787 Dreamliner. In this last hot topic, we consider the long-term impact of this strategy. Recall that Boeing has built this plane by working closely with 50 tier-one suppliers. At the writing of this book, 117 Dreamliners had been delivered to 16 different airlines for a total of 67,000 flights (Boeing 2014). With this early evidence, and an order list numbering in the hundreds, it appears the Dreamliner is about to become a major success.

The business model that Boeing is using is that of a systems integrator (Hart-Smith 2001; MacPherson and Pritchard 2007). This role outsources the design of subassemblies to supply partners who share the risk and rewards of being part of a larger project, such as building an aircraft. However, there are voices on the horizon expressing caution as to the long-term effects of a supply chain that outsources excessively, including the design and manufacturing of major subassemblies, such as wings and fuselages. It should be acknowledged that Airbus has some similarities with Boeing in that it outsources over a wide geographic area, primarily four major countries in the European Union (EU). But it does not excessively outsource design like Boeing has on its Dreamliner (MacPherson and Pritchard 2007).

It should also be recognized that in the building of a large complex product such as a jetliner, there is some marketing sense in spreading production subassembly contracts over a group of countries—it generates sales from airlines in those countries. Boeing has long sourced to companies in Japan, and not surprisingly, many orders for aircraft go to the Japanese airlines. However, what is different in the supply chain for the Dreamliner is that Boeing has outsourced both design and production of major airframe subassemblies to other companies, wings to three companies in Japan. In return, Boeing has made these companies strategic partners that share both the risk and future financial rewards of the Dreamliner. It sounds like a win–win situation and hopefully, it will be, but two concerns have been raised with this type of agreement: the loss of intellectual capital and the concern that a supplier could later become a competitor.

## LOSS OF INTELLECTUAL CAPITAL

Former Boeing engineer Dr. L.J. Hart-Smith wrote a paper in 2001 outlining the dangers of outsourcing the development role to other companies. His main concern was that once you outsource the writing of specifications, the loss of intellectual capital will soon follow (Hart-Smith 2001). When this occurs, the company will eventually lose its ability to innovate. Hart-Smith cites the former Douglas Aircraft Company as an example of an overoutsourced enterprise that lost its ability to create aircraft. This note is important since Boeing is currently one of only two companies that make large commercial aircraft (Airbus is the other). To outsource engineering expertise to its supplier partners erodes the base that the company is built on.

For a commercial aircraft maker, it is important to regularly introduce new products on the market, as they generate early sales, and then an aftermarket business of variants (updated versions of the same model aircraft). Maintaining this business model of new products and variants keeps the aircraft maker flying, so to speak, insuring a sustainable source of revenue for the future. However, the link that keeps this company running smoothly is the intellectual capital that makes the aircraft maker innovative. Outsourcing design interrupts this innovation stream and can eventually cripple the aircraft maker (Hart-Smith 2001).

## FORMER SUPPLIERS MAY BECOME COMPETITORS

The other concern of outsourcing the design of airframe subassemblies is that the supply partners may eventually become competitors. MacPherson and Pritchard (2007) question the role of Japanese supply partners of the Dreamliner. They wonder if those same partners may use their tacit knowledge of airframe development to build their own aircraft and then compete with Boeing! An intermediate step would be to build a smaller composite-based plane that would compete in the regional jet market. From there, the next step up would be to build an aircraft to compete with the Boeing 737, the company's flagship plane that dominates in the short-haul market (MacPherson and Pritchard 2007).

The Boeing 737 market is crucial to watch in the future as China would also like to develop its own short-haul plane, the C919. While China is considerably further behind in its aviation programs than the United States, it has stated that desire to enter the short-haul market (Cendrowski 2013). Boeing's long-time Chinese supplier, Comac, has already been working on the C919 model as its answer to the Boeing 737. With China and Japan possibly advancing into the short-haul market, one wonders what the future will be for Boeing and Airbus, who currently share this strategic market niche.

## QUESTIONS FOR RESEARCH AND DISCUSSION

1. For the aircraft manufacturing industry, explain how the diffusion of engineering expertise could occur if a company outsources the writing of specifications to its suppliers.

2. For a company that outsources many of its operations, what key functional areas should it retain?

## REFERENCES

Boeing, 784 Dreamliner, 2014. Retrieved February 5, 2014, from http://www.newairplane.com/787/dreamliner-live/.

Cendrowski, S., China gets ready for takeoff, *Fortune Magazine*, pp. 90–94, 97, 99, November 18, 2013.

Hart-Smith, L., Out-sourced profits: The cornerstone of successful subcontracting, Paper presented at *the Boeing Third Annual Technical (TATE) Symposium*, St. Louis, MO, February 14–15, 2001. Retrieved January 25, 2014, from http://seattletimes.nwsource.com/ABPub/2011/02/04/2014130646.pdf.

MacPherson, A. and Pritchard, D., Boeing's diffusion of commercial aircraft technology to Japan: Surrendering the U.S. industry for foreign financial support, *Journal of Labor Research*, 28(3), 552–566, 2007.

## Discussion Questions

1. Which management programs require close collaboration among supply chain participants?

2. Why is it important for all members of the supply chain to share in the costs and benefits of operating the supply chain?

3. What are some of the financial benefits gained by operating an integrated supply chain?

4. Why is it difficult to determine the total financial benefits of operating a supply chain?

5. What are some of the intangible benefits gained by operating an integrated supply chain?

6. What are some ways in which members of a supply chain can divide the benefits equitably among the members?

7. What is the role of the prime mover in the supply chain?

8. In your opinion, which company has done a good job of managing their supply chain? Why?

9. In your opinion, which company has done a poor job of managing their supply chain? Why?

10. Discuss the differences between supply chain integration and SCM.

## References

Bitran, G.R., Gurumurthi, S., and Sam, S.L., The need for third-party coordination in supply chain governance, *MIT Sloan Management Review*, 48(3), 30, 2007.

Crandall, R.E., Beating impossible deadlines: A variety of methods can help, *APICS Magazine*, 16(6), 20, 2006.

den Butter, F.A.G. and Linse, K.A., Rethinking procurement in the era of globalization, *Sloan Management Review*, 50(1), 76, 2008.

Free, C., Supply chain accounting practices in the UK retail sector: Enabling or coercing cooperation? *Contemporary Accounting Research*, 24(3), 6, 2007.

Kerle, P., Steady supply, *Supply Chain Europe*, 16(4), 18, 2007.

Simatupang, T.M., Wright, A.C., and Sridharan, R., The knowledge of coordination for supply chain integration, *Business Process Management Journal*, 8(3), 289, 2002.

Spekman, R.E., Spear, J., and Kamauff, J., Supply chain competency: Learning as a key component, *Supply Chain Management*, 7(1), 41, 2002.

Veeramani, D., Tang, J., and Gutierrez, A., A framework for assessing the value of RFID implementation by tier-one suppliers to major retailers, *Journal of Theoretical and Applied Electronic Commerce Research*, 3(1), 55, 2008.

Vurro, C., Russo, A., and Perrini, F., Shaping sustainable value chains: Network determinants of supply chain governance models, *Journal of Business Ethics*, 90, 607–621, 2009.

Wouters, M., Implementation costs and redistribution mechanisms in the economic evaluation of supply chain management initiatives, *Supply Chain Management*, 11(6), 510, 2006.

# Section VI

# The Future

# 17

## Trends in Supply Chain Management

### Learning Outcomes

After reading this chapter, you should be able to

- Describe the evolution of critical success factors (CSFs) in the United States
- Identify the major drivers of change in supply chains
- Identify the major changes expected in supply chains in the future
- Explain what is meant by the vanishing boundary between manufacturing and services
- Explain the role of technology in future supply chains
- Explain the role of infrastructure in future supply chains
- Explain the role of culture and its impact on employees in future supply chains
- Discuss how relationships among supply chain participants will need to change in the future
- Explain why today's supply chains are risk prone
- Discuss the trend in sustainability and its impact on the supply chains

Changes in the business environment have triggered the need for integrated supply chains. Will changes in the future reinforce the need for supply chains or make them less beneficial?

### Company Profile: Amazon

This chapter gives you a chance to use your ability to identify the major changes that are taking place in supply chains. The future of your company may depend on how well you anticipate and manage through the transitions. Do you remember Montgomery Ward? Circuit City? Borders? Those companies failed to recognize the need to change. What is the impact of online retailing on traditional retailers? Will automation replace the need for production workers? Will the sustainability movement become an integral part of a company's strategic management program? As you read this company profile and the chapter in the book, think about these questions:

- What items do you buy online? What products do you not buy online? Why?
- What items will your children buy online? Not buy? Why?
- How will packages be delivered to your house in the future?
- Will offshore outsourcing increase or decrease in the future? Explain.

What other questions do you think relate to trends in supply chain management (SCM)?

### Company Profile: Amazon.com

Amazon started as an Internet bookseller. They were incorporated in 1994 and began operation in 1995. During the years since, they have expanded into a number of new businesses, including retailing, distribution (fulfillment), manufacturing, and services. They have aggressively built on their position as the leading Internet reseller to move into related businesses. At times, they have sacrificed profits to achieve revenue growth and market position as well as providing the infrastructure to support future growth.

### OVERVIEW

Amazon provides the following overview of the company: "Our primary source of revenue is the sale of a wide range of products and services to customers. The products offered on our consumer-facing websites primarily include merchandise and content we have purchased for resale from vendors and those offered by third-party sellers, and we also manufacture and sell Kindle devices. Generally, we recognize gross revenue from items we sell from our inventory as product sales and recognize our net share of revenue of items sold by other sellers as services sales. We also offer other services such as AWS, fulfillment, publishing, digital content subscriptions, advertising, and co-branded credit cards" (Amazon 2012a, p. 18).

Some of the recent financial results for Amazon are shown in the following tables. Table 17.1 includes the sales, income, and cash flows for the 7 years of 2007–2013. Sales have increased almost fivefold during this period. Income also increased proportionately during 2007–2010, but slipped in 2011, dipped to a loss in 2012, and regained slightly in 2013. The decline in recent years has resulted from aggressive increases in spending for supporting activities. This shows more clearly in Table 17.2, which has a breakdown of operating expenses, both in dollars and as a percent of net sales.

As shown in Table 17.2, the cost of sales decreased by 4.7% from 2009 until 2013 as the company expanded its product line into items with higher gross profit margins and increased the sale of services. This was offset by increases in fulfillment (3.2%), marketing (1.4%), and technology and content (3.8%). These increases reflect the aggressive expansion of fulfillment centers to assure a high level of customer service by increasing fill rates and reducing lead times on standard items. The company continues to expand to meet their anticipated growth in shipments of both their own products and those of third parties for whom they provide fulfillment services (Amazon 2012a, p. 28).

The company is also investing heavily in technology. They have added computer scientists, software engineers, and merchandising employees. Amazon has also invested in several areas of technology and content such as digital initiatives and expansion of new and existing physical and digital product categories and offerings as well as in technology infrastructure to enhance the customer experience and improve the company's process efficiencies (Amazon 2012a, p. 19).

**TABLE 17.1**

Amazon Sales, Income, and Cash Flow for 2007–2013

| | For the Years Ending December 31 (in Millions of Dollars) | | | | | | |
|---|---|---|---|---|---|---|---|
| | 2007 | 2008 | 2009 | 2010 | 2011 | 2012 | 2013 |
| *Income Statement* | | | | | | | |
| Net sales | $14,835 | 19,166 | 24,509 | 34,204 | 48,077 | 61,093 | 74,452 |
| Income from operations | 655 | 842 | 1,129 | 1,406 | 862 | 676 | 745 |
| Net income from operations (%) | 4.4 | 4.4 | 4.6 | 4.1 | 1.8 | 1.1 | 1.0 |
| Net income | 476 | 645 | 902 | 1,152 | 631 | (39) | 274 |
| Net income (% of sales) | 3.2 | 3.4 | 3.7 | 3.4 | 1.3 | −0.1 | 0.4 |
| *Cash flow* | 2007 | 2008 | 2009 | 2010 | 2011 | 2012 | 2013 |
| Net cash from operations | $1,405 | 1,697 | 3,293 | 3,495 | 3,903 | 4,180 | 5,475 |
| Purchase of fixed assets | 224 | 333 | 373 | 979 | 1,811 | 3,785 | 3,444 |
| Free cash flow | 1,181 | 1,364 | 2,920 | 2,516 | 2,092 | 395 | 2,031 |
| Cash flow from investing activities | 266 | −866 | −1,964 | −2,381 | −119 | 190 | −832 |
| Cash flow from financing activities | 50 | −198 | −280 | 181 | −482 | 2,259 | −539 |
| Foreign currency effect | 20 | −70 | −1 | 17 | 1 | −29 | −86 |
| Net cash provided (required) | 1,517 | 230 | 675 | 333 | 1,492 | 2,815 | 574 |
| Cash balance at end of year | $2,539 | 2,869 | 3,444 | 3,777 | 5,269 | 8,084 | 8,658 |

*Source:* Mergent Online, http://www.mergentonline.com.

**TABLE 17.2**

Operating Expenses by Type for 2009–2013

| Sales and Operating Expenses (Millions of Dollars) | 2009 | 2010 | 2011 | 2012 | 2013 | % Inc (December) 2009–2013 |
|---|---|---|---|---|---|---|
| Net sales | $24,509 | 34,204 | 48,077 | 61,093 | 74,452 | 204 |
| Cost of sales | 18,978 | 26,561 | 37,288 | 45,971 | 54,181 | 185 |
| Fulfillment | 2,052 | 2,898 | 4,576 | 6,419 | 8,585 | 318 |
| Marketing | 680 | 1,029 | 1,630 | 2,408 | 3,133 | 361 |
| Technology and content | 1,240 | 1,734 | 2,909 | 4,564 | 6,565 | 429 |
| General and administrative | 328 | 470 | 658 | 896 | 1,129 | 244 |
| Other | 102 | 106 | 154 | 159 | 114 | 12 |
| Total operating expenses | $23,380 | 32,798 | 47,215 | 60,417 | 73,707 | 215 |
| **Percent of Net Sales** | 2009 (%) | 2010 (%) | 2011 (%) | 2012 (%) | 2013 (%) | % Inc (December) 2009–2013 |
| Cost of sales | 77.4 | 77.7 | 77.6 | 75.2 | 72.8 | −4.7 |
| Fulfillment | 8.4 | 8.5 | 9.5 | 10.5 | 11.5 | 3.2 |
| Marketing | 2.8 | 3.0 | 3.4 | 3.9 | 4.2 | 1.4 |
| Technology and content | 5.1 | 5.1 | 6.1 | 7.5 | 8.8 | 3.8 |
| General and administrative | 1.3 | 1.4 | 1.4 | 1.5 | 1.5 | 0.2 |
| Other | 0.4 | 0.3 | 0.3 | 0.3 | 0.2 | −0.3 |
| Total operating expenses | 95.4 | 95.9 | 98.2 | 98.9 | 99.0 | 3.6 |

*Source:* Mergent Online, http://www.mergentonline.com.

Sales in North America represented 57% of total revenues in 2012, with the remaining 43% in the rest of the world. With supply chains becoming more widely dispersed, being able to operate in various countries throughout the world will be a definite advantage for Amazon. Merchandise sales are the largest and fastest growing area, representing 63% of sales in 2012. Media sales have increased, but at a slower rate than for merchandise sales and represented 33% of sales in 2012. *Other* sales, which were 4% of sales in 2012, represent the newest product category. This includes Amazon Web Services (AWS), which provides access to technology infrastructure that enables virtually any type of business to use the service. They gained in unit sales by reducing prices, selling faster growing categories such as electronics and other general merchandise, increasing in-stock inventory availability, and increasing selection of product offerings (Amazon 2012a, p. 28).

## RETAILING: PRODUCTS AND SERVICES

Amazon had its beginning as an online retailer and this continues to be the main portion of their business revenue. The company lists the following selection of services and merchandise for sale on their website (Amazon 2012b):

- Books & Audible
- Movies, Music & Games
- Electronics & Computers
- Home, Garden & Tools
- Beauty, Health & Grocery
- Toys, Kids & Baby
- Clothing, Shoes & Jewelry
- Sports & Outdoors
- Automobile & Industrial
- Unlimited Instant Videos
- MP3s & Cloud Player
- Amazon Cloud Drive
- Kindle Books & E-readers
- Kindle Fire Tablets
- Appstore for Android
- Digital Games & Software

Amazon employs a multilevel e-commerce strategy. They started by focusing on business-to-consumer (B2C) relationships between itself and its individual customers, while using business-to-business (B2B) relationships between itself and its suppliers. By adding customer reviews as part of the product descriptions, the company developed customer-to-business (C2B) transactions. It now also facilitates customer-to-customer (C2C) business by using the Amazon marketplace as an intermediary to facilitate consumer-to-consumer transactions.

Amazon places great emphasis on tracking the interests of individual customers or prospective customers. They greet returning customers by name "Welcome Richard" and do their best to recommend books or other products their software determines may be of interest to the customer. Their system is streamlined to make ordering and paying easy, such as the use of the *1-Click ordering* provision, which allows repeat customers to shortcut steps in the payment process.

Selling books in electronic form is being accelerated with the introduction of tablet readers such as Amazon's Kindle, Apple's iPad, and Barnes & Noble's Nook. Amazon has been especially aggressive in selling e-books to enhance the sales of their Kindle readers.

## DISTRIBUTION (FULFILLMENT CENTERS)

Amazon quickly discovered that sending individual customer orders to book publishers and expecting the publisher to ship individual books to consumers worked in theory but not in practice. As book sales increased during an early Christmas season in Amazon's history, publishers were overwhelmed and were unable to handle the large volumes of orders. Amazon decided they must establish their own distribution centers (later named fulfillment centers) to stock some of the most popular selling books in order to meet customer demand. Although originally designed for books, the fulfillment centers quickly became a distribution center for all types of products that Amazon sold through its online website.

Once they established their fulfillment centers, Amazon is capitalizing on these assets by expanding their use beyond Amazon's internal needs. Amazon provides distribution and shipping services for third parties. During the early part of 2012, these shipments represented over a third of the units shipped. Outside vendors that sell on the Amazon site pay Amazon a commission on sales, usually about 10%, as well as fees for storing and shipping the product in Amazon's fulfillment center. If an outside vendor does not sell on the Amazon website, they can still use Amazon's distribution service, designated as Fulfillment by Amazon (FBA), where Amazon stores and ships the products to the third party's customers (Kucers 2012).

Retailers who sell both through stores and online find the retail combination difficult; however, many are finding the distribution of the combination even more difficult. That is why some are turning to Amazon to handle the distribution function at their fulfillment centers. "Integrating online and brick-and-mortar supply chains raises tricky questions, such as how to manage transportation between various origins and destinations, how to handle store returns for online-only merchandise, and how to supply online orders from store inventories without turning salespeople into order pickers and packers" (Bonney 2012).

Amazon tries to operate its fulfillment centers as efficiently as possible. They use the latest technology in storing and selecting items, where employees called stowers and pickers perform these functions. They use random storage, as opposed to having products occupy a certain space in the center, to maximize the use of storage space. They also use robots for some of the stowing and picking. They acquired Kiva Systems Inc., in 2012, a robot manufacturer, to have the latest in robotics technology (Kucers 2012).

There are other technology applications. Automated scales compare the weight of each box against the known weight of the items in the order. An information technology (IT) system releases and routes orders to coordinate the picking operations with the outgoing truck schedule. Algorithms are used in Seattle to determine the best fulfillment center to handle any given order based on inventory availability, service requirements, and transportation cost. In addition to scaling each of the fulfillment centers through labor and shift adjustments to meet fluctuating demand, Amazon sets up fulfillment operations at supplier distribution centers to allow Amazon to ship directly from suppliers during peak seasons (Johnson 2011).

At the beginning of 2014, Amazon had approximately 100 fulfillment centers, about equally divided between North America and other countries, with over 40 million ft$^2$ of space in North America and almost 30 million ft$^2$ outside of

North America (MWPVL 2014). Amazon is extending its fulfillment centers by adding sortation centers that sort packages into zip code order for faster delivery by the U.S. Postal Service. As of mid-2014, they have seven sortation centers and plan additional ones to further support their fulfillment centers (Wulfraat 2014).

For Christmas, 2013, Amazon experienced delays in deliveries promised for Christmas, when the orders overwhelmed the capacities of the primary package carriers, UPS and FedEx. While the delayed deliveries represented a very small percentage of the total orders for the year, it was widely reported in the newspapers and other news media. The sortation centers will also move Amazon toward less dependence on the package carriers UPS and FedEx. They have structured an agreement with the U.S. Postal Service to deliver packages on Sunday. This arrangement also enables Amazon to increase their drive to reduce delivery times to next day or even same day deliveries (Greene 2014). In their obsession to expand their product offerings, Amazon has announced the launch of an online store for 3-D printed items to allow consumers to customize and personalize a variety of more than 200 items (France-Presse 2014). Amazon has also announced their desire to begin delivering packages by drones as soon as the Federal Aviation Administration (FAA) approves their application.

## MANUFACTURING (KINDLE)

Amazon introduced the first Kindle in 2007. It is a portable reader that wirelessly downloads books, magazines, newspapers, blogs, and personal documents to a high-resolution display. Kindle is now the best selling product in the history of Amazon. Kindle Touch and Kindle Touch 3G are new additions that feature a touch screen that makes it easier to turn pages, search, shop, and take notes. Kindle Fire provides access to movies, TV shows, music, books, magazines, apps, games, and web browsing with free storage in Amazon Cloud. It has a color touch screen and a dual-core processor (Overview 2012).

The market for e-books is increasing rapidly. Forrester Research estimates there are 40 million e-readers and 65 million tablets in use in the United States (Alter 2012). While Amazon does not directly manufacture the Kindle line, they are *virtual* manufacturers in that they outsource the production process to contract manufacturers. They have suppliers in China (flex circuit connectors, injection-molded case, controller board, and lithium polymer battery), Taiwan (electrophoretic display), and South Korea (wireless card), according to Steve Denning (2011). In this way, Amazon is assuming greater control over the Kindle supply chain. It makes the product, markets it, and supplies books, both print and electronic, for Kindle users.

## ORIGINATION (BOOK AUTHORS)

Amazon began their move into digital books as early as 2005, when they acquired BookSurge (print on demand) and Mobipocket (e-books). As one writer puts it: "It was a clear message to both the digital and traditional book publishing industry that Amazon was investing in their vision of the future of book publishing" (Treanor 2010). Intrigued by the possibilities, Treanor (2010) tracked a series of 18 acquisitions by Amazon, beginning as early as 1998–2008. These acquisitions were investments in technology infrastructure to support their book business and other industries as well.

In 2009, Amazon began its move into the publishing business by establishing AmazonEncore to allow writers to digitally self-publish their work and to republish books that were no longer in print. This was nonthreatening to the major book publishers, who also viewed Amazon as one of their major customers. In 2010, Amazon introduced AmazonCrossing to publish English translations of foreign-language books. It has since added several other imprints devoted to mysteries, thrillers, romance, and science fiction (Stone 2012).

Amazon brought their venture into publishing to a new level by hiring Laurence J. Kirshbaum to head up Amazon Publishing. Kirshbaum was once a top executive of Time Warner Book Group and a well-accepted member of the publishing community. His movement to Amazon Publishing has, in the eyes of the major publishers, transformed him into a traitor and bitter enemy. In addition, Amazon encourages writers to publish their writing with Amazon and has recently announced the addition of several well-known authors to Amazon's stable of writers (Stone 2012).

It appears that Amazon is intent on becoming a major factor in the publishing world, moving them further upstream the supply chain, from retailer to distributor to manufacturer and even to originator, as they encourage authors to publish for Amazon. The goal is to connect the customer with what they want, regardless of the medium used to move the content to them (Welly 2011).

## SERVICES

Amazon provides marketing services for their third-party sellers, those companies that sell products through the Amazon website. They also provide fulfillment services for those same companies.

Jeffrey Bezos, CEO of Amazon, Inc., has this to say about AWS: "Amazon Web Services has grown to have thirty different services and thousands of large and small businesses and individual developers as customers. All AWS services are pay-as-you-go and radically transform capital expense into a variable expense. AWS is self-service; you don't need to negotiate a contract or engage with a salesperson – you can just read the online documentation and get started. AWS services are elastic – they easily scale up and easily scale down" (Bezos 2012).

## SUMMARY

Amazon's business segments include the following:

1. *Books*: This is the main business in which Amazon started and is still the backbone of their company. As described earlier, they moved quickly from being purely an online seller of books to developing their own distribution (fulfillment) function.
2. *Other merchandise*: It soon became apparent that the Amazon website that sold books could also sell other merchandise, especially those products that were well known to consumers, such as clothes and small appliances. As consumers became more knowledgeable and trusting in Amazon, the variety of merchandise expanded rapidly. As with books, Amazon was able to use its fulfillment centers to stock and deliver the merchandise to consumers quickly and reliably.
3. *Kindle*: Once the books and merchandise supply chains were operating well, the company recognized that physical books would eventually be replaced, at least in part, by electronic books and that readers would need a device on which to download and store books for reading as they desired. The Kindle was first introduced as a reader and then was further developed into a tablet computer—the Kindle Fire. As described earlier, this also enabled Amazon to enter the manufacturing part of the book supply chain. Adding in-house authors moved Amazon back into the originating portion of the supply chain.

4. *E-books*: Acquiring and delivering physical products have enabled Amazon to grow into an international business with markets and facilities throughout the world. Their knowledge of electronic communications also enabled them to quickly become a major supplier of e-books to readers. Electronic downloads were also something that consumers were becoming more comfortable with, especially with the growth in music downloads. It was a small step to downloading text materials, and Amazon was quick to grasp the potential in this area.

5. *Cloud computing services*: Ever aware of potential new businesses, Amazon did not miss the interest in cloud computing services. The company is gearing up its portfolio of services and is expecting to become a major player in the game.

Amazon is expanding horizontally as a retailer and has become a competitor in a wide variety of industries. One of their more recent entries is in the home improvement field, where they are rapidly becoming a competitor to Home Depot, Lowe's, and Sears as a seller of tools and appliances. They are also venturing into the grocery home delivery business in selective test areas.

Amazon is clearly in the middle of several dominant trends in SCM—online selling, omnichannel marketing and distribution, individual customer service, and automation of distribution (fulfillment) centers. They are innovative in their quest to be able to sell to every person whatever they want to buy.

## REFERENCES

Alter, A., Your e-book is reading you; digital-book publishers and retailers now know more about their readers than ever before. How that's changing the experience of reading, *Wall Street Journal (Online)*, June 29, 2012.

Amazon, Annual report, 2012a, http://phx.corporate-ir.net/phoenix.zhtml?c=97664&p=irol-reportsAnnual, accessed January 15, 2014.

Amazon, Amazon Earth's biggest selection (Full line of products and services), 2012b. http://www.amazon.com/gp/site-directory/ref=sa_menu_fullstore, accessed July 27, 2012.

Bezos, J.P., Letter to the shareholders, 2011 Amazon annual report, 2012 , http://phx.corporate-ir.net/phoenix.zhtml?c=97664&p=irol-reportsAnnual, accessed January 15, 2014.

Bonney, J. and Editor, S., Amazon's supply chain: Delivering clicks and bricks, *Journal of Commerce*, pp. n/a, January 30, 2012.

Denning, S., Why Amazon can't make a Kindle in the USA, *Forbes*, August 12 2011. http://www.forbes.com/sites/stevedenning/2011/08/20/does-it-really-matter-that-amazon-cant-manufacture-a-kindle-in-the-usa/, accessed July 27, 2012.

France-Presse, A., Amazon launches 3-D printing store, http?//www.industryweek.com/print/technology/amazon-launches-3-d-printing-store?NL=IW-02&YM_RID, July 28, 2014, accessed August 7, 2014.

Greene, J., Amazon launches Sunday delivery from Kent warehouse, *Seattle Times*, http://seattletimes.com/html/businesstechnology/2024114127_amazonkentxml.html, July 20, 2014, accessed July 24, 2014.

Johnson, M.E., Virtualizing fulfillment operations—Physical clouds allow Amazon to scale, Center for Digital Strategies Blog, 2011. http://www.tuck.dartmouth.edu/digital/about/blog/detail/virtualizing-fulfillment-operations, accessed June 29, 2012.

Kucers, D., Amazon wrings profit from fulfillment as speaking soars: Tech, *Bloomberg Business Week*, May 21, 2012. http://www.bloomberg.com/news/2012-03-21/amazon-wrings-profit-from-fulfillment-as-spending-soars-tech.html, accessed July 25, 2012.

MWPVL, Amazon global fulfillment center network, 2014. www.mwpvl.com/html/amazon_com.html, accessed February 20, 2014.

Overview of Amazon, 2012. http://phx.corporate-ir.net/phoenix.zhtml?c=176060&p=irol-mediaKit, accessed June 29, 2012.

Stone, B., Amazon's hit man, *Business Week*, 1, January 30, 2012. , http://www.businessweek.com/magazine/amazons-hit-man-01252012.html, accessed August 15, 2013.

Treanor, T., Amazon: Love them? Hate them? Let's follow the money, *Publishing Research Quarterly*, 26(2), 119–128, 2010.

Welly, K., Amazon knows the medium doesn't matter. *EContent*, 34(8), 8–9, 12, 2011.

Wulfraat, M., Logistics comment: Amazon is building a new distribution network—Quickly and quietly!, *SupplyChainDigest*, July 23, 2014, www.scdigest.com/experts/wulfraat_14-07-23.php?cid=8309, accessed July 30, 2014.

In this chapter, we will try to identify some of the trends in supply chain management. This implies a *from* and *to* tendency. To set the stage, we will provide some background on the evolution of the growth of business and supply chains in the U.S.

## From the Past to the Present

In his book, *The Visible Hand*, Alfred Chandler described the evolution of the American business organization from Revolutionary War days until after the close of World War II. His basic theme was that the *invisible hand* of the marketplace, as described by Adam Smith, was replaced by the modern business enterprise in coordinating flows of goods through existing processes of production and distribution and of allocating funds and personnel for future production and distribution (Chandler 1977). In so doing, Chandler outlined the importance of supply chains and their management.

### From the American Revolution to World War II

In the latter part of the eighteenth century, around 1790, the typical American business was privately owned, often by one family and managed by the owners. Its market area was confined to the local town or community because there were no formal, effective means of distribution—limited by the speed of horse-drawn wagons. The retail establishments were small general stores that tried to carry a wide assortment of goods, from lace to animal feed. The business organization was simple with little in the way of formal titles or job descriptions.

From 1790 to about 1840, not much changed in business organizations, largely because the speed of distribution did not increase. While the population increased and the size of markets increased proportionately, business organizations did not increase in size; there were just more of them. One exception was in textiles. Here, larger companies began to appear in New England. They had a source of power—the rivers to provide water power to run the machinery—and had a number of employees, usually young women from the neighboring farm communities. This arrangement required they begin to develop supervisors, some of whom were not part of the owner's family.

During the industrial revolution period, from 1840 to about 1890, the situation changed. Steam power became available to drive equipment in the factory and railroads. The speed of distribution increased, from the slow pace of horses or canal boats to faster-moving trains. In addition to being faster, railroads were more dependable (Chandler 1977). The size of markets increased, both in volume and in distance covered. Manufacturing began to employ power-driven equipment to assist workers. Job specialization became an accepted practice.

Before the end of the century, the concept of interchangeable parts became an accepted objective and, in some cases, became the actual practice. Major industries began to emerge, such as in textiles, firearms, sewing machines, and clocks. Oil refining and coal mining became the fuels of choice to drive the railroad locomotives. The railroads pioneered the beginning of the multitier organization structure, because they needed a way to manage their widespread and diverse activities.

Although owners continued to be actively involved in the strategic decisions, such as in the acquisition and allocation of resources, they began to turn over the day-to-day management of the business to full-time managers. Some companies were involved in exporting cotton or tobacco or importing goods, such as metals and spices. These extended operations were beyond the scope of what the owners could manage; therefore, they began to work through a series of added service groups—brokers, agents, financiers, and the like. This was the beginnings of supply chains in the modern era.

During the period from 1890 until the beginning of World War II, the United States began to emerge as a major industrial power. The automobile industry became a major factor when Ford introduced what became known as mass production. Frederick Taylor began his work at Midvale Steel and created the scientific management approach. To a large extent, professional managers replaced owners as the business managers. The owners who remained active in management were more than owners; they were trained managers. Some of the companies recognized as progressive in their management systems included DuPont, General Electric, U.S. Steel, and General Motors.

The American industry faced new challenges during World War II. Major manufacturers, such as General Motors and Ford, shifted from producing consumer automobiles to producing tanks and other vehicles for use by the Allied military. They did this successfully and demonstrated the flexibility and effectiveness of producing large volumes of goods with largely inexperienced employees. Professional managers, working within the framework of multitier organization structures, were able to achieve new levels of output.

## From World War II until the Present

After World War II, the latter half of the twentieth century included a number of shifts in both organization structure and approaches to production. The emphasis shifted from internal production improvement to external business conglomerates or diversified assemblages of disparate businesses into one centrally financed, but locally managed, enterprise. Teledyne and IT&T led this parade of conglomerates, but many public companies were caught up in the drive to increase stock prices by capitalizing on the synergistic effect of acquiring companies with stock, regardless of their fit with existing businesses.

During this period of financial manipulation, foreign competitors began to enter the American market with new technologies and new products. Germany, with Volkswagen, and Japan, with Toyota and Honda, began to chip away in the domestic automobile market. Companies like Sony and Toshiba entered the consumer electronics market and soon had dominant positions, eliminating the lead held by RCA and Zenith, among others. Even the mighty steel manufacturers—U.S. Steel and Bethlehem—began to lose market share, as Japanese companies used their modern technologies to provide products of better quality and lower prices. Initially caught unprepared, American manufacturers had to react quickly to stem the loss of market share. In some industries, they are still recovering, as evidenced by the grave situations that confronted the *Big Three* automobile companies of General Motors, Ford, and Chrysler during the past decade.

With the increased availability of products and competitors, customers became dominant drivers of change. During this period, organizations became flatter, in order to reduce response times to customers. The hierarchical organization structure—designed to assure control of the lower levels—became unwieldy. Managers, and eventually hourly employees, were empowered to make decisions at lower levels of the organization. Companies worked feverishly to smooth the flow of goods and services through their internal organizations. Just-in-time (JIT) and lean manufacturing programs represented effective ways to achieve the needed improvements.

The need for faster response forced companies to look beyond their own organizations and toward their supply chains. It became apparent that companies who were part of an integrated supply chain would be better able to compete in the modern environment. SCM became the focus for leading companies.

As the world enters the new millennium, a number of challenges await American industry, some involving SCM. Thomas Friedman publicized the rise of global competition in his book *The World is Flat* (Friedman 2005). American businesses face the problem of an imbalance of imports over exports. To meet the pressures to reduce costs, a number of manufacturers have outsourced portions of their activities to foreign suppliers. Even service companies have joined the outsourcing movement, especially in the IT area. While offshore outsourcing by American companies may have slowed and some are announcing the shift to reshoring (bringing production back to the United States), the issue is still very much in transition.

In summary, the tightly vertically integrated corporation of the 1950s has given way to a loosely connected array of functions scattered throughout the world. In seeking lower costs, many companies are experiencing a loss of central control. Although there are many ideas about how to organize these activities into an integrated supply chain with its resultant smooth and dependable flow of products, not many companies have completely achieved this objective.

Rifkin (2014) also provides a comprehensive view of the evolution of industry, the rise of capitalism and the movement toward what he calls the *Collaborative Commons*.

## Evolution of Critical Success Factors in the United States*

In this section, we describe how global CSFs have evolved in the United States. CSFs vary over time, among industries, among companies in the same industry, and among managers within the same company. Ultimately, CSFs must be specific to the need at hand.

In order to compete effectively, manufacturing companies should include those CSFs that relate to improved customer service. Likewise, service companies have to include CSFs that include improvements in cost efficiencies and quality.

We will describe the effect of the competitive situation in the United States on CSFs; however, similar trends are occurring in other countries, particularly those engaged in global markets—the European Union and the Far East, such as Japan, China, and India. It is especially relevant to see CSFs in light of the economic impact of political revolutions, such as the transition of communist countries to market-driven economies.

---

* This section has been adapted from Crandall and Crandall (2014).

In the book *The Commanding Heights,* Daniel Yergin and Joseph Stanislaw provide a fascinating description of how individual countries are making this transition (Yergin and Stanislaw 2002).

The United States is an example of a country that has witnessed changes in CSFs throughout its history. Table 17.3 summarizes these changes in CSFs over time in the United States. In the table headings, we use the terms *order qualifiers* and *order winners* as introduced by Terry Hill in his book *Manufacturing Strategy* (Hill 1989). Order qualifiers are "those competitive characteristics that a firm must exhibit to be a viable competitor in the marketplace" and order winners are "those competitive characteristics that cause a firm's customers to choose that firm's goods and services over those of its competitors" (Blackstone 2013).

**TABLE 17.3**

Evolution of CSFs in the United States

| A Look at the CSFs in the United States | | | |
|---|---|---|---|
| **Time Period** | **Competitive Situation** | **Order Qualifiers** | **Order Winners** |
| 1840s–1890s, Industrial Revolution | Little competition: The United States was primarily an importer of finished goods and an exporter of raw materials, developing the *American system* of manufacturing. | Function | Function, availability |
| 1890s–1930s, Growth and recovery | Limited competition: The United States was beginning to become an industrialized nation, next to Europe; little competition from Far East countries. | Function, availability | Price (Ford), variety (General Motors) |
| 1940s–1950s, Mass production | Little competition after World War II (Germany and other European countries rebuilding; Japan not a major factor; the United States could sell all it could make). | Function, availability | Price, quality, delivery |
| 1960s–1970s, Arisings | Stirrings from Germany and Japan, in automobiles and basic industries (steel); the United States living off successes of the 1950s and acquisitions. | Availability, price | Quality, delivery, service |
| 1980s, Awakening | Inroads from Japan and other Far East countries, in consumer electronics and other basic industries; the United States not prepared to compete. | Availability, price, delivery (JIT) | Quality, service, product variety |
| 1990s, Globalization I | U.S. firms countered in some industries; Japan and other Far East countries faltered; Europe solid; rapid movement to global market. | Availability, price, delivery, quality (TQM) | Product variety, service, flexibility, integrated systems |
| 2000 plus, Globalization II | Global market; offshore outsourcing; strategic partners; financial integration; mass customization; integrated systems; lean manufacturing. | Availability, price, quality, delivery, flexibility; variety, integrated systems | Mass customization, paperless transactions, integrated supply chains, virtual corporations |

*Source:* Adapted from Crandall, R.E. and Crandall, W.R., *Vanishing Boundaries, How Integrating Manufacturing and Services Creates Customer Value,* CRC Press, Boca Raton, FL, 2014. With permission.

## Beginning (from First Settlements through 1800)

In the first stages of the country's industrial development, there was little internal competition among businesses. Most of the competition for finished goods and services came from other countries; however, without some industries, there was little money available for residents to buy the finished goods. Agriculture was the principal industry. Residents were largely self-sufficient and any goods they produced, such as furniture or fresh vegetables, were for their own use. If they produced more than they needed, they sold the excess or traded it to their neighbors. The order qualifiers were simply the ability to make something; order winners were having some additional goods available. In the United States, settlers cleared the land and exported timber, while farmers grew tobacco and cotton for export. In return, they imported tools, clothing, and equipment for use in developing increased industrial capability.

## Industrial Revolution (1840s–1890s)

During most of the nineteenth century, the United States developed its industrial capability to record levels and made the transition from a primarily agricultural economy to one that included a more significant industrial capability. The industrial revolution saw the beginnings of major industries such as the telegraph, telephones, railroads, textiles, oil, and steel. The country needed all of these basic industries to feed, clothe, and transport the rapidly expanding population. While there was still competition from other countries, primarily Europe, American companies were successful if they could produce goods (e.g., steel) or provide a service (e.g., railroads). The market was expanding so rapidly that it was difficult for an organization's available capacity to meet the demand for its products.

## Growth and Recovery (1890s–1930s)

During this period, the United States made the transition to an industrialized country and started to become a major player in the global marketplace. The automobile industry played a major role in this transition. Henry Ford set the stage by emphasizing his company's ability to mass produce a standard automobile and sell it at a low price. General Motors used a different strategy—that of offering an array of choices. The competition between the two companies created the need to meet new CSFs—that of lower prices and greater variety. It also provided for the transition of the United States from an agricultural to an industrial-based economy. This growth period also set the stage for the origins of the service economy.

## Mass Production (1940s–1950s)

During this period, the United States emerged from the Great Depression and became involved in World War II. As a result, the country increased in terms of industrial capability, especially in the areas of automobile and aircraft manufacturing as well as atomic energy. Immediately following the war, the United States was able to sell almost anything it could produce in excess of domestic demand. This feat was possible since other industrialized countries in Europe and Japan had their industrial capability destroyed during the war. Competition was almost nonexistent and U.S. companies did not have much to worry about, except making sure they had the capacity to meet demand. As a result, companies did not strive to make improvements in their technology or management methods. Indeed, life was good!

### Arisings (1960s–1970s)

As Europe and Japan recovered from World War II, they rebuilt their industrial capacity with the latest technology and looked for new markets to develop, once they satisfied their own domestic markets. Japan transformed itself from a maker of low-quality goods to a producer of high-quality products. Two areas of particular improvement were in the automobile and consumer electronics industries.

### Awakening (1980s)

The need for quality products reached a new level during this period and became a CSF recognized in other countries. However, because of the lack of competition, producing quality products was not a CSF many companies in the United States acknowledged as important. For example, the steel industry resisted modernization because it did not need additional capacity and could not justify new technology. Consequently, this was a period of financial wizardry, as conglomerates abounded and synergy was the magic word. Manufacturing lost its luster and automobile companies tried to compete on styling and reduced prices. The need for additional CSFs such as quality and response times became a reality.

### Globalization I (1990s)

While some degree of globalization had existed before this period, a movement to expand more aggressively was emerging. This movement progressed, despite the fact that some remnants of *Buy American* campaigns were still active. Wal-Mart became the leader in low-price retailing, but at the expense of buying many products overseas. Many other companies watched and waited before finally deciding that buying overseas was not a bad thing, especially since Americans wanted low prices.

### Globalization II and Mass Customization (2000 and after)

The United States is currently in a transition mode. Traditional manufacturing businesses are diminishing, as basic production operations are being outsourced to offshore companies. The major U.S. automobile manufacturers—General Motors, Ford—have recovered from the recent drop in demand but may still be losing market share to foreign competitors. The apparel industry is also disappearing as companies purchase more clothing from foreign countries that offer dramatically lower labor costs and generally acceptable quality. Understandably, different constituencies have mixed reactions to these trends. Some are pessimistic and want tighter controls on imports; others are optimistic and want industry leaders to help make the transition from traditional industries to futuristic industries. To make this transition requires a change not only in the type of products and services offered but also in the attitudes and knowledge of the workers.

New CSFs are imperative. In addition to low prices, high quality, and reduced response times, businesses must be agile, flexible, and virtual. The ante is going up; successful businesses have to offer a larger variety of products and services. However, offering a variety of products or services is not always the ultimate answer. The ideal situation, at least in the minds of some, is to treat each consumer as a market of one and design and deliver products and services unique to each person. While this scenario may appear overly idealistic, many see it as the direction of the future. What will this transition involve for the United States? It is a transition from one life cycle to a new life cycle. In fact, that transition is already taking place when a country abandons the former life cycle built around traditional manufacturing, with service support, to a new life cycle—one that is built around a service/manufacturing package of goods and services that is systems oriented and holistic with respect to the customers' needs.

## Major Drivers of Change in Supply Chains

Supply chains have been an area of major emphasis for at least the past three decades. When companies were vertically integrated, a good portion of the supply chain, especially on the supply side, was contained within the parent organization. In addition, much of the marketing focus was on domestic customers. Manufacturing companies were dominant and their strategy could be described as a *make and sell* approach in which they developed products they could make at low costs and for which they thought there would be demand. Only then did they develop a marketing approach to sell those products. By the last part of the twentieth century, retailers had become the dominant members of the supply chain, and product manufacturers found they needed to change to a *sense and respond* approach in which they would first determine what the consumer wanted. Then they would figure out how to make the product to meet consumer expectations (Haeckel 1999).

Today, supply chains are accepted as a vital part of doing business. There is no reason to believe they will lose that importance in the near future. However, they will have to continue to adapt to changes that are taking place. Several of the major changes that appear most imperative for today's supply chains include

- Global business perspective
- Balanced approach to offshore outsourcing
- Continuing advances in technology
- Evolution from transactions to processes
- Vanishing boundaries between manufacturing and services
- Infrastructure refinements
- Culture and employees
- Supply chain relationships
- Emergence of third-party logistics providers
- Risk management
- Sustainability
- Strategic employee plan

We describe each of these areas in the following sections. Some of these trends were discussed in earlier chapters when appropriate to that chapter subject.

## Global Business Perspective

Global competition forced companies to consider the merit of having partners who could help in reducing costs, increasing market scope, reducing response times, and developing new products. Competition also forced companies to consider how to make their supply chains effective and efficient. One of the goals was to develop integrated supply chains in which partners could collaborate to achieve perfection in customer service. Supply chain integration implied the need to reduce variation and differences among participants.

Globalization will continue to be a major driver of business strategy. Companies that do not consider the effect of both foreign competition and foreign opportunities in their planning are not likely to succeed. While this does not mean all companies must become a multinational, it does suggest all companies will be touched, in some way, with the globalization movement, as described by Thomas Friedman (2005).

Thomas Friedman writes about three periods of globalization. He considered Globalization 1.0 as the period from about 1492 until 1800, with the dynamic force being countries globalizing. Globalization 2.0 was from 1800 until 2000 and was about companies globalizing. Globalization 3.0 began about 2000 and he considers it as the period when individuals globalize. He describes this phenomenon as the flat-world platform. "The flat-world platform is the product of a convergence of the personal computer (which allowed every individual suddenly to become the author of his or her own content in digital form) with fiber-optic cable (which suddenly allowed all those individuals to access more and more digital content around the world for next to nothing) with the rise of work flow software (which enabled individuals all over the world to collaborate on that same digital content from anywhere, regardless of the distances between them). No one anticipated this convergence. It just happened—right around the year 2000."

As Friedman points out, individuals will become more knowledgeable about goods and services that are available throughout the world. E-commerce will make products available, even when the consumer is located a long distance from the supplier. Third-party logistics companies such as UPS and FedEx will become increasingly important as delivery services. In addition, they will also begin to perform additional auxiliary services to preclude multiple handling of goods, especially in the reverse logistics cycle.

Businesses will have to be selective in their choice of global initiatives. In some cases, they will be looking for low-value creation and will locate their routine, repetitive operations in low-cost countries. In other cases, they will be looking for high-value creation and will locate their high-technology operations in countries that are leaders, such as Japan in robotics. As an added consideration, they will be aware of the market potential, wherever they are located.

## Balanced Approach to Offshore Outsourcing

Outsourcing, especially offshore outsourcing, has introduced variations and differences among supply chain participants. Although companies have long evaluated make-or-buy alternatives, they were primarily considering domestic suppliers, and often in obvious noncore functions, such as food service, custodial services, and payroll. As foreign countries, especially those who were previously under state-controlled economies, appeared, the potential for major cost savings was too much for many, if not most, companies to resist. Offshore outsourcing began as a trickle, but soon became a torrent, as companies moved major portions of their business to external suppliers.

This complicated the life of supply chain managers who could no longer stroll down the street to their supplier's convenient location. They had to deal with suppliers in different countries, with different cultures and language, operating under different governments, and different attitudes, especially in ethics, worker treatment, and environmental concerns. For companies that had begun to approach the integrated, collaborative, supply chain relationships so desired, they now found these relationships disrupted and suppliers more remote than originally intended.

While the movement to offshore outsourcing continues, it has now assumed a more deliberate pace instead of the torrent of turbulence that existed for a few years. Companies are evaluating the trade-offs between lower costs and increased difficulties in achieving a smooth and consistent flow from their remote suppliers. In the next section, we will outline some of the steps necessary to complete this transition from a vertically integrated domestic producer to a virtually organized global enterprise.

Offshore outsourcing was discussed in some depth in earlier chapters. As we write this book, it is a *hot* topic, and many companies have already, or are considering, moving some part of their manufacturing or service operations to a foreign country. The primary driver of this movement is the desire to reduce product or service costs. A secondary objective is to reduce the investment requirements that would be required if the product were continued as an internal activity.

At the same time, some companies are rethinking their decisions to outsource, and some companies are considering bringing some manufacturing back to the United States. While this movement is still in its early stages, it will be interesting to see what the outcome will be.

## Total Cost of Ownership

Two factors are slowing the movement to offshore outsourcing and, in some cases, reversing it. While companies continue with this strategy, they are moving with more deliberation. One factor is the need to consider the total cost of ownership. Some companies have moved quickly but failed to consider the hidden costs of outsourcing. While some of these costs are tangible, such as the increased lead times required to get product, other costs are less tangible, such as the cost of lowered acceptance in the domestic marketplace, or the cost of lost proprietary knowledge.

Another factor that may become more significant in the future is the trend toward equalization between countries. Today, China and India have large competitive cost advantages. As the standard of living increases over a broader part of the population in these countries, and as wages rise, the cost differential will lessen. While wage rates in these countries may never equal those in the United States, the difference will not be enough to be readily attractive. If, and when, that occurs, there will be a reversal of the offshore outsourcing movement. We are already seeing this in the *reshoring* movement of recent years.

## Risk Management

"Risk is the expected outcome of an uncertain event, i.e., uncertain events lead to the existence of risks. We call these uncertain events 'risk events'" (Manuj and Mentzer 2008).

If there were no uncertainties, we would have no risks. Linking multiple companies together into a supply chain creates opportunities for uncertainty; therefore, SCM contains risks.

Risk can be classified into four categories: supply, demand, operational, and security risks. Supply risks relate to disruptions in the inbound supply of goods that adversely affect the ability of the receiving firm to meet customer demand in terms of quantity, quality, cost, and time. Operations risk results from variances in the production process that affect the ability to meet volume, quality, timeliness, and profitability goals. Demand risk occurs when customers unexpectedly vary their ordering patterns in volumes, timing, and mix. Security risk is the variety of events that could threaten human resources, operations, and information systems and could lead to undesirable outcomes, such as loss of proprietary information, vandalism, crime, and sabotage (Manuj and Mentzer 2008).

Risk management is closely allied with the total cost of ownership, although different performance measures are often used. Risk can be associated with almost every step in offshore outsourcing. Supplier dependability is often unknown in the early stages. Delivery times may vary, as well as the product quality. Process variance, in any form, causes the need for increased inventories and lowered customer service levels. Ultimately, it may be possible to assign a cost to these negative factors; however, in earlier stages, they represent negative results that are often not considered in the initial analysis or, if considered, are understated. We provide a fuller discussion of risk management later in this chapter.

## Other Issues

There are at least two other issues that will affect outsourcing. The first is the strategy employed by unions. The second is the government's position on immigration laws. To date, unions have opposed offshore outsourcing as a threat to union member employment, especially in manufacturing industries.

Immigration has become an issue primarily in agriculture and service industries. The federal government has the responsibility to establish some direction that will satisfy the needs of business for low-cost employees while appeasing the negative reaction from displaced domestic employees.

## Continuing Advances in Technology

Technology has long been a leading driver of change. Some of the technologies that could have the greatest effect on supply chains relate to retail operations, demand forecasting, transportation methods, and information processing.

## Retail Operations

The supply chain is dependent on collecting accurate and timely information at the retailer's operation and sharing that information back along the members of the supply chain, in the expectation that every participant will have a better insight into what the true demand for a product is. Most stores have point-of-sale (POS) terminals that record sales information in great detail. A number of suppliers are moving into the use of radio-frequency identification (RFID) to collect even more data and track the movement of products along the supply chain. One industry group reports accelerated adoption of RFID technology. Some retailers have gained confidence in the technology and plan to deploy it in the coming holiday season, even when introducing new technology is a risk. Zara is on a fast track to deploy RFID at all of its 6,000 stores by 2016. The big drivers for manufacturers include receiving accuracy, shipping accuracy, pick pack accuracy, and electronic proof of delivery. One of the big benefits is the reduction of retail claims about missing product shipments. For retailers, RFID offers the opportunity to collect more data about product movement and customer preferences (RFID 24-7 2014).

## Demand Forecasting

All businesses need a demand forecast to plan production and service levels, decide on resource requirements, and organize for the orderly and timely flow of goods and services. In the past, each company depended on its history of sales and fragments of information derived from a variety of sources, both internal and external. As a result, it was possible for individual companies to have widely varying estimates of future demand, even when the companies were directly connected as supplier and customer.

Today, supply chain organization makes it possible to share demand information among all its members. While the forecasts are still not foolproof, they tend to be more consistent. Retail customers can share POS information with their suppliers. They can also indicate future marketing events—sales promotions, new store openings, and seasonal events.

Advocates of the value of data analytics claim that more sophisticated analysis of the increasing availability of unstructured data will reveal clues to future demand. We described some of the changes in data collection and analysis in Chapter 14.

## Transportation and Distribution

In transportation, the changes have been incremental. While the speed of ships, trains, and trucks has not changed much in the past half-century, cargo handling has. The use of containerization has not only reduced loading costs, it has also increased the reliability of the transportation service. Containers are packed at the manufacturing or distribution site, and they arrive at the receiving site with the cargo untouched. For a more complete history of the role of containerization in supply chains, see *The Box* by Marc Levinson (2008).

In Chapter 6, we described the distribution function in some detail. As globalization increases, the distribution function will become even more of a factor in the design and operations of supply chains. In Chapter 9, we described the increased importance of the logistics functions in improving the flow of goods along the supply chain.

## Information Systems

On the information systems side, electronic communications has changed the way people buy. Internet purchases have increased. The electronic communications capability has reduced the cost of paperwork processing and increased the reliability of the information communicated.

In Chapter 14, we described information flows in the supply chains. Businesses continue to make dramatic improvements in the field of IT, and there is no reason to believe this will not continue. RFID and service-oriented architecture (SOA) are still in the growth stage of their life cycles; they appear to have sufficient value and momentum to gain a substantial share of the market in the next few years.

One of the most widely recognized technologies is the enterprise resource planning (ERP) system. Beginning with material requirements planning (MRP) in the 1970s, evolving to MRP II in the 1980s, and emerging as ERP in the 1990s, this integrated system evolved even more in the first decade of the twenty-first century to extend its reach beyond the enterprise to include other participants along the supply chain. In anticipation of the impact of the Internet, there are several desired capabilities of the interenterprise systems of the future. They include the following:

- Facilitates an integrated supply chain
- Provides smooth and consistent data transfer between users
- Enables flexibility to meet changing business environments
- Reflects evolving enterprise models and technology, like mobile wireless
- Includes databases to support transaction and query-intensive applications
- Accommodates characteristics like culture, language, technology level, standards, information flow, and partnering relationship changes
- Encourages global vendor alliances to meet the needs of clients throughout the world
- Adopts standards like XML, web service architecture, and evolving wireless standards
- Incorporates flexibility and interoperability of modules, systems, and enterprises

Whatever it is called, the enterprise system of the future will be found at the convergence of business need and technological change (McGaughey and Gunasekaran 2007).

## From Special-Purpose to General-Purpose Resources

The mass production era concentrated on how to make high volumes of standard products at low cost and high quality. One of the ways to achieve these objectives was to use higher levels of equipment automation that became more special purpose in design and operation. Companies designed jobs with limited content to achieve job specialization. Workers could be quickly trained and were productive, but with limited flexibility to move from job to job or to handle an increased scope of work. Special-purpose facilities were designed to make a limited variety of products. In the automobile industry, often a plant designed for one model could not switch over to manufacture another model without extensive conversion costs.

Over the past couple of decades, there has been an increasing interest in the concept of mass customization, or making high volumes of customized products. To move in the direction of mass customization requires that companies rethink their approach to designing resources used in the transformation process. Job designs have been enlarged and employees have been empowered to perform a greater variety of tasks and assume greater responsibility. Equipment is being designed to be more flexible—faster changeovers and setups and wider scope of products produced. Where equipment was once designed to operate separately from the employees or replace employees, more equipment and processes are being designed to blend employee and equipment skills to get the most from each resource. Even facilities are being designed with flexibility in mind. Automobile manufacturers are designing new plants that are capable of producing multiple models within the same plant, at the same time. This form of flexibility will make it possible to more closely match supply with demand, even when demand fluctuates in both volume and product mix.

The movement from mass production to mass customization, described in Chapter 7, is accelerating the need for supply chains to be flexible and agile. Manufacturers are using the postponement concept to introduce flexibility into existing supply chains. For the future, they are designing products that are more easily customized to meet consumer needs.

Lee highlighted the need for flexibility by declaring that the best supply chains are more than just fast and cost effective. They are also agile, adaptable, and aligned. He emphasized that companies do not need more technology and investment. What they need to change is their culture, especially among the managers who can make the transition happen (Lee 2004).

---

## Evolution from Transactions to Processes*

One of the changes advocated is that companies should change from a transaction orientation to a process orientation. We described this type of change in Chapter 7 and will only briefly summarize it in this chapter.

### Transactions versus Processes

First, what is a transaction? Transactions are individual events reported to the system, for example, issues, receipts, transfers, and adjustments (Blackstone 2013). Transactions are individual activities or operations that go on throughout a business every day as part of getting the total work of the business done. They are distinct and separated from one another by organization structure, location, functional specialty, and historical practice. They are internally focused and operated to be efficient. They are usually housed in separate departments, sometimes referred to as *functional silos* that every knowledgeable manager now tries to hide from peer and public scrutiny. Are transactions bad? No, because they are necessary; however, they may cause delays and increased costs unless properly managed.

The next question is what are processes? The *APICS Dictionary* defines a process as "a planned series of actions or operations (e.g., mechanical, electrical, chemical, inspection, test) that advances a material or procedure from one stage of completion to another" (Blackstone 2013). Simply stated, a process connects individual transactions into a sequence of activities that accomplish a desired result. A process orientation leads nicely into the

---

* This section has been adapted from Crandall (2007).

supply chain concept. Processes tend to be both internally focused toward operations and externally focused toward the customer or supplier, and while we want our processes to be efficient (doing things right) and of high quality, we want them especially to be customer focused and effective (doing the right thing) (Drucker 1974).

Consider a simple example—customer order processing. Think about all of the functional areas that an order could pass through before the customer receives the goods— order entry, credit, materials management, production, accounting, distribution, and shipping. Each function performs a transaction. When combined in a flow of activities, the customer experience is enhanced.

It gets even more interesting to think about linking processes together to form what could be considered as master processes. If we link the customer sales order process involving accounts receivable, with the stock replenishment process involving inventory, with the supplier purchase order process involving accounts payable, we can form a cash-to-cash cycle process that measures how well a business is using its cash (Farris and Hutchison 2002).

### Benefits of a Process Orientation

What are the benefits of a process orientation? Processes tend to be customer oriented and, as a result, should improve customer service. One of the ways this is accomplished is through reducing response times, whether it is in product development or normal order processing. A transaction orientation tends to be batch oriented, where in-process inventories build up to assure efficient processing, whether it is in manufacturing washing machines or approving supplier invoices. A process orientation is designed to achieve a smooth and regular flow of work through a series of workstations. This is desirable within an entity; it is even more desirable if the flow can be extended to external entities, such as customers and suppliers to form supply chains. A process perspective will not only enable a business to provide better customer service, it will also enable it to reduce processing costs and errors (Snee and Hoerl 2005).

Changing from a transaction orientation to a process orientation is one of the major challenges facing many businesses. It involves a blend of the old with the new.

### Vanishing Boundaries between Manufacturing and Services

The manufacturing/services boundary is vanishing. Because of increased global competition, businesses must find ways to compete on additional CSFs. This quest to compete better has created a somewhat ironic dilemma; manufacturers must become more customer oriented, while service companies must become more effective and efficient. Each side is learning from the other. While there are differences between manufacturing and service objectives, it is becoming more important to combine product-producing activities with service-providing activities to have a successful business.

Blending manufacturing and services requires the capability to combine different kinds of businesses. This blending raises the complexity of businesses and increases the need to continually improve.

The boundary between manufacturing and service businesses is vanishing and new composite businesses are emerging. These composite businesses have to continually improve their competitive advantage through the introduction, implementation, and assimilation of management improvement programs. The process of achieving this competitive advantage relies on the careful implementation of changes in technology, infrastructure, and organizational culture.

It is no longer sufficient to manufacture a low-cost, high-quality product. A complementary service package is necessary. The product is rapidly becoming only the order qualifier; the service that accompanies the product represents the order winner (Hill 1989). As a result, manufacturers must focus more attention on the service sectors of their businesses. When they do, they find some of the techniques used in manufacturing also work in the service sector, such as the organization of workflows or the need for capacity planning. However, they may also find that some functions are different, such as greater difficulty in measuring productivity or the less certain control of quality.

Conversely, for service businesses, they can no longer depend on providing a high level of personal service. They must strive for low costs, high quality, and fast response times. To achieve these objectives, they must adopt some of the techniques developed in the manufacturing industries.

## Infrastructure Refinements

Infrastructure includes organization structure, policies and procedures, and practices of an organization. The international consulting firm of Booz Allen Hamilton conducted a study of organization effectiveness. The study spanned 30 industries, both manufacturing and service. They concluded sustainable success in organizations is *a matter of execution*. They further found effective execution depends on execution ability and agility. They defined execution ability as a measure of conversion—the conversion of intention into action. It reflects how well companies can implement important strategic and operational decisions. Execution agility concerns how well companies can deal successfully with discontinuous change in its environment (Kletter and Harada 2008).

The study also concluded organizational survival results from clarity of decision-making responsibilities and the value of information, especially at the top management level. These two attributes are more important than organization restructuring (the blocks and lines on the organization chart) and motivators (the incentives for employees). As might be expected, the study also found that companies possessing higher levels of execution ability and execution agility outperformed those with lower levels.

### From Vertical to Horizontal Organizations*

Organization structures have changed from rigid, vertical hierarchies to flexible, horizontal structures. In order to be responsive to customers and suppliers, lower-level employees have been empowered to talk with representatives in other companies.

Organizations can be tall or flat. Tall organizations have many levels of management tiers. Tall organizations were typical of many businesses prior to the 1980s. Then, a wave of restructuring began that eliminated many managers from the middle of the company.

---

* This section has been adapted from Crandall and Crandall (2014).

The goal of reengineering was to reorganize the company so it would function more efficiently and, at the same time, better serve customers (Hammer and Champy 1993). No doubt, reengineering accomplished this feat in many companies that were heavy with too many managers and inefficient in responding quickly to customer needs. A number of stories relate how companies cut their number of employees and responded faster to customer demands. Some of these cuts in employee staff occurred among hourly employees. Other cuts were more far reaching as entire levels of management, usually in the mid areas of the company, were eliminated.

The impact of reengineering has also led to the creation of a new organizational structure, the horizontal organization. "Unlike traditional vertically oriented organizations with functional units such as production, marketing, and finance, horizontal organizations are flat and built around core processes aimed at satisfying customers" (Kreitner and Kinicki 2007, p. 551).

Cross-functional process teams are replacing the traditional functional departments. These teams are designed to better address the specific needs of the customer as opposed to running an internal process of the organization. Hopefully, the result is an expedited process transaction with the customer. For example, instead of taking weeks to process an insurance policy or a mortgage, the goal is to complete the transaction in a matter of days or even hours. Streamlining the transaction process requires a careful analysis of the job, a process human resource management specialists like to call job analysis.

Of course, a lesser form of the horizontal structure is simply a flatter organization. Delayering the organization compresses the hierarchy and widens the organization chart. The result is a flatter organization that is more flexible in responding to change. Translated, management is able to make decisions faster since there are not as many levels of approval needed. Because fewer personnel are on the payroll, particularly in middle management, there is also a savings in payroll costs.

### From Rigid Rules to Flexible Policies and Procedures

When businesses are small, they operate with informal policies and procedures. As companies grow larger, employees need more formal guidelines to follow. In keeping with job designs that emphasized job specialization, it followed that procedures would become more narrow and specialized. Within any single company, this was a manageable arrangement. However, as companies began to form partnerships within the supply chain, it became apparent that the narrowly focused procedures within one company did not necessarily fit the narrowly focused procedures in another company.

To resolve these differences, it meant one company had to change to use another company's procedure, or each company had to modify its procedures to be compatible. This usually meant the procedures had to become more flexible. This increased flexibility placed more responsibility and decision-making authority with employees at lower levels in the organization. This trend toward flexibility and empowered employees will likely continue in the future.

### From Tacit Knowledge to Implicit Knowledge

Traditionally, many companies prided themselves on retaining employees for a long time, especially managerial employees. This meant these employees built up a great deal of tacit knowledge about their jobs and company practices. Under these conditions, it was not as necessary to document practices and procedures because someone within the company knew about them.

Today, employee turnover, both enforced and voluntary, has increased to the point where few companies can depend on having the storehouse of tacit knowledge available. Consequently, they are finding it necessary to convert their tacit knowledge to implicit knowledge regularly and thoroughly. This becomes even more important because of the dynamic nature of most businesses. Today's good practice may become tomorrow's albatross, if not adapted to meet the changing conditions.

## From Financial Accounting to Management Accounting

In the move to integrated and agile supply chains, it is important that companies not overlook the need to adapt their accounting systems and performance measures to reflect the changed operating conditions. In order to manage better, companies must change from a primarily cost-based orientation to one that considers strategic business processes. Rising competition, globalization, and advances in technology are forcing enterprises all over the world to invest in more effective and extensive managerial accounting systems.

As supply chains change, accountants and accounting must also change. One of the most critical changes is that accountants will increasingly become an active partner in decision making at all levels of an organization. They will need to develop reporting systems that are useful to all functions in a business; they must convert data into relevant information. In many organizations, they will become part of cross-functional teams to identify and implement improvement programs.

These changes present a challenge not only to accountants but also to other functional areas who become partners in the efforts to make the accounting information more meaningful.

---

## Culture and Employees

Competition and technology are driving change in supply chains. Companies respond by designing products and processes to meet the changing conditions. They are also adapting their infrastructure to be more responsive to today's needs. However, if they do not adapt their company cultures and employees to embrace the new business environment, their progress will be slowed, even arrested by the resistance within their own company.

## From Passive or Obstructing Culture to Engaged and Receptive Culture

In companies that are progressing through the transition to supply chain integration, their company cultures are changing from being passive or obstructive to being engaged and receptive. This change does not come easily, but, when it does, it can be a powerful force in helping the company become a leader in the supply chain.

In the book *Vanishing Boundaries: How Integrating Manufacturing and Services Create Customer Value* (2014), the authors looked at organizational culture as a key component in addressing change in the company. It is the *elusive key to change* because of its more subtle and hidden nature. Unlike technology and infrastructure, which are more externally oriented, organizational culture resides *beneath the surface* in the form of embedded values and beliefs held by organizational members. Fortunately, culture can change, although it is a slower process than changing technology and infrastructure. Implementing major improvement programs almost always requires a change in the organization culture.

The key to changing organizational culture is to first address the values and beliefs of the employees. Fundamental values, held dearly by many, may have to be changed or at least challenged. Usually, the source of this value change will come from top management, most likely the president or CEO. In addition to changing beliefs, more external cultural elements, called artifacts, also need to be addressed. New slogans may need to be implemented, the physical structure of the facilities may need to be improved, stories will have to be told of employees who achieve outstanding goals, and reward systems will need to formally acknowledge excellent performance by organizational members.

In short, organizational culture is an elusive key to change, but only if it is not addressed. When culture is considered part of the equation for making organizational change, the rest of the change process has a better chance at being successful (Crandall and Crandall 2014).

## Employees: From Specialized to Empowered

The role of the employee is changing. New technologies have made it possible to redesign jobs to best utilize employee skills. However, it requires that employees learn new skills and adopt new attitudes toward work. Table 17.4 portrays the changes between twentieth and twenty-first century manufacturing (Handler and Healy 2008).

As manufacturing tasks change, the employees must also change. Table 17.5 lists some of the changes in skills and traits required of manufacturing employees (Handler and Healy 2008).

Changes in the supply chain will probably mean the work force will become even more mobile than in recent years. As continuous improvement programs continue to disrupt internal operations and outsourcing adds its effect, employee turnover will increase. Retraining programs will be necessary, and government–employer–union collaborations may finally become meaningful and effective.

**TABLE 17.4**

Changes in the Nature of Manufacturing Tasks

| Twentieth Century | Twenty-First Century |
|---|---|
| Fixed production roles | Job rotation and sharing |
| Permanent full-time staff | Combination of full- and part-time staff |
| Low autonomy to make decisions | Independent decision making |
| Emphasis on technical skills | Both technical and interpersonal skills important |
| Unchanging procedure; general safety guidelines | Detailed, dynamic procedures; evolving guidelines |
| Operation of low-tech machines | Computer and IT skills critical |

*Source:* Adapted from Handler, C.A. and Healy, M.C., *Hiring Manufacturing Staff in the 21st Century: A Fundamental Shift in Skills*, Rocket-Hire, Pearson Education, Inc., London, U.K., 2008. With permission.

**TABLE 17.5**

Changes in Required Skills for Manufacturing Personnel

| Old World | New World |
|---|---|
| Learning one or two specific technical roles | Mechanical reasoning, logic, spatial visualization |
| Physical strength and flexibility | Personal flexibility, communication, and cooperation |
| Ability to follow fixed, unchanging procedures | Initiative, persistence, and independence |
| Attention to production and safety procedures | Attention to detail, self-control, and dependability |
| Following orders | Making independent decisions |
| Operating and maintaining mechanical machinery | Using computers for a range of critical functions |

*Source:* Adapted from Handler, C.A. and Healy, M.C., *Hiring Manufacturing Staff in the 21st Century: A Fundamental Shift in Skills*, Rocket-Hire, Pearson Education, Inc., London, U.K., 2008. With permission.

## Supply Chain Relationships

Supply chains contain individual companies that perform a variety of functions, as described throughout this book. The supply chain also contains links between the companies. As described in Chapter 16, these links are loosely connected and the members of the supply chain depend on active and effective relationships with their supply chain partners for success.

### From Adversarial to Collaborative

How do companies avoid increasing the complexity and risk levels in their supply chains? The surest way is by building a trusting relationship with their supply chain partners. Building relationships requires three elements—the need for the relationship, the means by which the entities can communicate, and the willingness to enter into the relationship.

Being a partner in a supply chain usually provides the need to build closer relationships. Every customer has a supplier and every supplier has a customer. This interface can be enhanced by improving the casual or formal exchange of goods or services with a more mutually beneficial ongoing business relationship.

Improved communication capability is often the key to the means by which companies can improve their working relationships. Today, communications technologies abound. Companies can communicate informally or formally throughout the world. The means of communicating exist; it is up to the parties involved to use them.

The last requirement for a collaborative relationship is the willingness to enter into such a working arrangement. If there is a need and a means, the willingness often boils down to whether the parties involved trust each other. They may be able to coordinate activities, or even cooperate to a certain extent without trust, but they will never collaborate in an effective way without a level of mutual trust. However, trust is a complex topic and trusting relationships are difficult to develop. How do companies develop trust? Do they do it by having individuals begin to trust individuals in the other company until they reach a critical mass of individual trusting relationships and consider they now have a trusting relationship between the two companies? Or do they try to build an impersonal trusting relationship between the two companies?

## Trust and Distrust

Supply chains pose an interesting dilemma concerning trust. Companies can both trust and distrust the other organization in the relationship. Supply chain relationships can be multifaceted and multiplex; they are not single, simple relationships but rather involve contacts and transactions between multiple people and systems. Some of these contacts or transactions may build to a high level of confidence that the outcomes are consistent and predictable; other contacts or transactions may deteriorate to a high level of confidence that the outcomes are unpredictable and inconsistent. In such a situation, it is not reasonable to use an average as a measure of trust. Both trust and distrust are present and the management of the relationship must reflect this. Table 17.6 explains these relationships (Lewicki et al. 1998).

**TABLE 17.6**

Trust versus Distrust

| | | Level of Distrust | |
| --- | --- | --- | --- |
| | | Low | High |
| Level of Trust | High | **Phase 3.** The interactions between the companies go well. One company has a number of successful experiences; therefore, they begin to trust the other company and expect that future transactions will be favorable to the first company. Conversely, very few of the experiences may have had negative outcomes; consequently, the first company may have little reason to distrust the other company. This appears to be an ideal state, and companies may move easily to collaborate with one another. | **Phase 4.** This may be the eventual outcome for many relationships. As time goes on, one party learns more about the other party and finds both positives and negatives in the relationship. There are a number of situations that work well, and a high level of trust emerges. On the other hand, there are situations that do not work well, so a high level of distrust develops. This does not mean the parties cannot do business together or even collaborate in certain situations. It just means they should be discriminating in their choices. This is a relationship in which formal, contractual agreements will be useful. |
| | Low | **Phase 1.** This is a beginning state for many relationships. The initial contacts are casual, of limited importance, and with formal or professional contact. At this stage, there is no reason to either trust or distrust. Business can be conducted and the parties can even cooperate on certain transactions without the need to hesitate or monitor the other's actions or performance. For parties with little interdependence, their relationship may not progress beyond this stage. However, if there is a higher level of interdependence, the companies will need to move to a more dependent stage and begin to cultivate a deeper relationship with one another. | **Phase 2.** This state is a difficult one for sustained relationships. If one company can control the other, they may be able to carry on a business relationship, but it will never be a comfortable one. The tendency will be to suspect the worst consequences of business transactions, and companies will be looking to discontinue the existing relationship and try to find other partners that offer the opportunity to build a more productive relationship. |

*Source:* Adapted from Lewicki, R.J. et al., *Acad. Manage. Rev.*, 23(3), 438, 1998. With permission.

Lewicki warns benign and unconditional trust appears to be an extremely dangerous strategy for managing social relations. He suggests that while there is a need for trust, it must be tempered with the reality that relationships contain a myriad of opportunities that may require elements of distrust. However, companies do not have much experience in managing these ambivalent relationships. The result is that multiple-motive conditions challenge people's ability to manage the complexities of simultaneous trust and distrust (Lewicki et al. 1998).

Another consideration is that any relationship may be temporary or transient. People change and conditions change. These disruptions cause changes in the trust relationships. If the change is minor, such as an upgrade in software, it may not affect the relationship significantly. However, if the change is major, such as a new CEO who wants to put more pressure on purchasing costs, the shift in the trust relationship may change drastically.

Knowledge sharing is one way to increase the level of trust. Much has been written about learning organizations. Some research suggests that as the unit of competition changes from organization versus organization to supply chain versus supply chain, a learning organization will not be enough. Learning supply chains will be created to facilitate innovation and creativity necessary to survive and thrive in the unpredictable global business environment of today. SCM can ensure the flow of goods and services and foster a level of trust among partners. This, in turn, sets the stage for ever-increasing levels of knowledge transfer and eventually to the learning supply chain.

Factors necessary for building trust include ability, benevolence, and integrity. Ability is the capacity to perform the predicted action, but is not constrained to a physical or cognitive capacity but also includes the environment in which a trust challenge may be situated. Benevolence is being considerate of another's welfare and not wanting to do anything that would harm it. Integrity means a party acts in a predictable fashion, as demonstrated by its actions. Maqsood et al. (2007) make some summary statements about trust:

- Trust is a frame of mind; it requires challenges and conflict to be validated.
- Trust and distrust coexist.
- The nature of trust changes over the time that the relationship continues.
- Power imbalances and quality of information and knowledge exchange are tightly bound in the trust–distrust experience.

Trust, when moderated by risk evaluation, can lead to commitment by the trusting parties.

## Emergence of Third-Party Supply Chain Coordinators

The movement away from vertically integrated companies to supply chains has accelerated over the past few years. One of the drivers has been the search for lower costs through outsourcing. Another driver has been the desire to remove noncore activities and to concentrate on core activities in the search for higher productivity and more focused use of resources. A third driver has been the enhanced IT capabilities that have made it easier to manage remote operations or suppliers.

Regardless of the reason, many companies are now partners in multiple fragmented supply chains. As described in earlier chapters, this fragmented arrangement gives rise to a number of new problems. In addition to the need to coordinate and collaborate on physical flow problems, there is a need to organize and manage information flows, described in Chapter 14, and funds flow, described in Chapter 15. In addition, there is almost certainly the knotty problem of how to allocate benefits, costs, and problem resolution among the supply chain participants, described in Chapter 16. At present, most supply chains are still feeling their way toward a workable solution to all these issues. Sometimes, a company takes the lead and mandates a solution, such as Wal-Mart with RFID. However, there is not a universally accepted approach.

A somewhat novel approach has been suggested. The developers call it the *maestro model*. They explain, "Drawing from the set of capabilities we have outlined for logistics players (LP), we anticipate a brand-new role in the value network for a neutral third party—a 'maestro' that can coordinate the entire network and align the incentives for all the participating players" (Bitran et al. 2007, p. 34). In anticipation that some large companies may not be receptive to the idea of one entity managing the entire supply chain, they also suggest a *mini-maestro* model where a coordinator manages only a part of the supply chain and the principal company manages the remainder of the supply chain. The authors suggest that companies like UPS and FedEx are already acting in the maestro role for many supply chains. Some of the benefits expected from this type of arrangement include a better alignment of costs and benefits; increased mutual dependence and trust; greater uniformity of systems, standards, and IT; and a social contract that encourages individual players to reach their potential.

## Risk Management*

### Increased Complexity and Risk

As supply chains extend over longer distances and the number of separate entities increases, supply chains become more complex, and the risks associated with their operation increase. These supply chains involve risk in variable demand as well as less tangible risks in social, political, and environmental issues. One way to reduce these risks is by moving from a *production-centric* mode to a *service-centric* mode. Outsourcing can not only provide savings but also provide process improvement, increased agility, and strategic refocusing (Bretton 2008).

Multiple supply chains also increase the complexity and risk. It is likely that some supply chains will involve the same suppliers. In this case, the failure of one supplier will adversely affect more than one supply chain.

Mahuj and Mentzer developed a model of three major global supply chain risks, which they classified as demand risks, supply risks, and operational risks. They reported the managers in their study expect growth in outsourcing and offshoring, increasingly demanding customers, geographical dispersion of supply chains, access to markets in emerging economies, and unanticipated events such as terrorist acts and

---

* This section on risk management has been adapted from Crandall (2012).

natural disasters will lead to increased complexity, risks, and opportunities in global supply chains (Manuj and Mentzer 2008).

Risk management in supply chains has become a concern for all types of organizations. As businesses move to loosely coupled networks of customers and suppliers, spread over wide geographic areas and diverse business environments, the level of potential supply chain disruptions increases. Whether an end consumer or an originating supplier, a company faces risks that could have a significant negative effect on its well-being. Many organizations recognize they have a core responsibility to manage these risks to minimize the negative effects. However, as reported in a recent APICS study, "the survey results reveal that supply chain risk management is still at an early stage of maturity and that there are gaps at the organizational management level and the supply chain and operations management level" (APICS 2011, p. 2). The study found 72% of the respondent organizations did not have a risk management role or position, and almost a third of the organizations had practiced risk management for no more than 5 years.

Another survey found 85% of the companies surveyed said they had suffered at least one supply chain disruption during 2011, with the following major causes: adverse weather (51%), unplanned IT or telecommunications outage (41%), transport network disruption (21%), and earthquake or tsunami (21%) (Veysey 2011).

That risk management is becoming more important in most organizations is exemplified in an article in *Strategic Finance*, the journal for management accounting. The author summarized the difference between the traditional financial risks and the contemporary, as shown in Table 17.7. Involvement of management accountants in risk management should help in integrating the allocation of the necessary resources to prevent or mitigate risks with their identification and strategy formulation by operations management (Thomson 2010).

The variety of risk categories described earlier shows the wide range of risks in supply chains. There are also a variety of actions to mitigate or prevent risk disruptions, as outlined in the following text.

Figure 17.1 provides an overview of a proposed model. The horizontal axis shows a progression from internal causes on the left, to external causes in the center, to natural disasters on the right. The vertical axis shows a progression from low impact at the bottom, to medium impact in the center, to high impact at the top. Internal risks of disruption carry a high frequency of occurrence but a low potential impact. Natural disasters have a very low frequency of occurrence but a very high potential impact. Disruptions from external sources fall somewhere between internal and natural disaster disruptions in both frequency and impact.

**TABLE 17.7**

Comparison of Risk Management Paradigms

| Old Paradigms of *Risk* | New Paradigms of *Risk* |
|---|---|
| Ad hoc activity | Continuous activity imbued in culture |
| Treasury, audit, and controllership functions | All management, especially accountants |
| Risks hidden in silos | Risks discussed cross functionally |
| Risk management prevents bad things | Risk management creates opportunities |
| ERM is a consultant's program | ERM is a business imperative |
| No ROI in risk management | Positive ROI in risk management |

*Source:* Adapted from Thomson, J.C., *Strateg. Finance*, 43, December 2010.

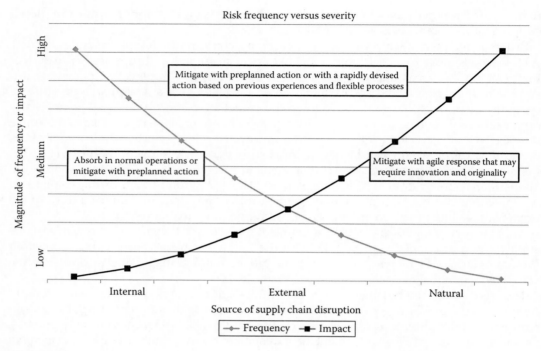

**FIGURE 17.1**
Risk frequency versus severity.

## Internal Risks

Internal risks can result from actions of customers, suppliers, or internal to any company within the supply chain, such as the manufacturer. They can also result from the relationships that have been established as the supply chain is assembled.

*Low-impact, high-frequency, expected, minor disruptions*: In normal operations, the more closely linked supply chains are designed to minimize disruptions; however, they may also be the most adversely affected, should a disruption occur. If lean production practices are used throughout, there is little buffering with inventory or excess capacity; therefore, a disruption will have a greater negative impact. Disruptions during normal operations can be identified and planned for; therefore, they should have limited effect on the supply chain's operation.

*Medium-impact, moderate-frequency, anticipated, moderate disruptions*: The introduction of new projects or major events into a supply chain partner, or along the supply chain, can introduce greater risk. The mandate by Wal-Mart to use RFID by suppliers introduced real, or potential, disruptions that probably could be anticipated, but which represented more deliberate planning to accommodate.

*High-impact, low-frequency, low predictability, major disruption*: A final category of risk in supply chains occurs when participants opt to remove themselves from the supply chain. A customer may find another supplier, or a supplier may go out of business. Product recalls or noncompliance with government regulations can also represent unanticipated, major disruptions.

## External (Open System Environment) Influences

Operating in an open system environment presents an array of supply chain disruptions. While some of these changes can be anticipated, their timing and magnitude often cannot. Their disruptive impact can range from minimal to major. While they represent uncertainty, a firm must consider their potential impact and develop flexible processes in order to cope with their eventuality.

*Competitors*: Competitors introduce new products, change prices, launch major advertising initiatives (free shipping for Christmas), and buy suppliers. A company should identify the most likely moves by a competitor and plan an appropriate response.

*Economy*: Recent fluctuations in the United States and global economies caused many companies to rethink their strategies about outsourcing, new product launches, investment in added capacity, and other resource-intensive decisions. These are *new normal* times; decisions that worked well during growth periods do not fit well in economic downturns.

*Technology*: Technology continues to be a source of progress; unfortunately, it also introduces disruptions in normal supply chain operations. IT can be especially disruptive when it introduces major changes in processes and interorganizational communications, such as electronic data interchange (EDI) on the Internet or cloud computing.

*Government*: The federal government can change tax incentives for environmentally friendly investment, strengthen the enforcement of product tracking, or require healthcare insurance for all employees. State or local governments can change the sales tax rate, restrict waste disposal, increase recycling requirements, or increase incentives for new business start-ups. Governments move slowly and in somewhat uncertain paths, but their impact can be significant.

*Environment*: At present, the future direction and timing of environmentally sustainable movements are uncertain. However, it appears that their impact will be a major opportunity or threat to many establishments.

*Society*: Cultural and generational differences abound. Buying habits, especially in the e-business age, are changing rapidly. Decisions about investments in bricks and mortar retail stores, and malls, are brain twisting in their variations. Executives can recognize the future will not be like the past, but are struggling to know how to use that knowledge in planning their organization's future.

## Natural Disasters

Yuva (2010) provides a summary of natural disasters occurring throughout the world during the last decade, using data from the Center for Research on Epidemiology of Disasters (CRED), as shown in Table 17.8.

The total indicates some type of natural disaster occurs, on the average, daily. While some have greater impact, any could be disruptive to a supply chain. The recent earthquake in Japan and flooding in Thailand and the Philippines are dramatic evidence these kinds of natural disasters can have a significant effect on supply chains. Firms must develop the agility to quickly adapt to these unpredictable, yet not unexpected, occurrences.

Designing and implementing an effective supply chain is difficult, even without the threat of disruptive risks. However, good risk management is now another requirement in this age of extended and complex supply chains.

**TABLE 17.8**

Number of Major Disasters

| Type of Disaster | Average 2000–2008 | Year 2009 |
|---|---|---|
| Flood | 178 | 147 |
| Storm | 108 | 84 |
| Mass movement, wet (avalanches) | 18 | 30 |
| Earthquake (including tsunami) | 30 | 22 |
| Extreme temperature | 22 | 22 |
| Drought | 17 | 10 |
| Wildfire | 15 | 9 |
| Volcano | 6 | 2 |
| Mass movement, dry (landslides) | 1 | 1 |
| Total | 392 | 327 |

*Source:* Adapted from Yuva, J., *Inside Supply Manag.*, 21(9), 28, 2010.

## Sustainability*

Sustainability is an equally *hot topic* in SCM. While special interest groups may have been the origin of the sustainability movement, it has now become a widespread topic among business and governmental leaders. Supply chain managers must consider more carefully their choice of materials to avoid or preserve endangered items. They must also consider more carefully the processes used in transporting products over long distances to reduce carbon emissions. Supply chains are becoming greener as companies work harder to implement changes that will reduce negative impacts on the Earth's environment but, at the same time, maintain a competitive advantage.

After a supply chain has done all of the ordinary things to succeed, it must begin to do some extraordinary things before their competing supply chains beat them to it. We described one of these at the beginning of this chapter—the triple bottom line (TBL). We will now expand on the concept of sustainability and its relationship to the TBL.

Many businesses and nonprofit organizations of all types and sizes want to *go green*. The number of articles written about environmental issues is increasing rapidly. There are even television ads about green products. Thomas Friedman follows his *The World is Flat* bestseller with *Hot, Flat and Crowded* (2008), a challenge for America to "replace our wasteful, inefficient energy practices with a strategy for clean energy, energy efficiency, and conservation." The attention is good, because most of us want to preserve the environment. As usual, there are some catches. A company that goes bankrupt trying to be green is not acceptable; we want businesses to be in the black, not in the red. A company that takes advantage of its employees or suppliers, or whose management engages in illegal activities, no matter how green they appear, is also undesirable.

Sustainability, for a business, is the ability to keep operating successfully. The Brundtland Commission report, *Our Common Future*, defined sustainable development as "development that meets the need of the present world without compromising the ability of the future generations to meet their own needs" (Anderson 2006).

Much of the business literature focuses on how to run a company to gain economic or financial success. A number of academic journals and trade publications actively solicit

---

* A portion of this section is adapted from Crandall (2009).

and publish articles about environmental and social issues, often with the emphasis on one or the other sectors. A new journal that attempts to look at the picture holistically is the *International Journal of Sustainable Strategic Management* (*IJSSM*) (Parnell 2008). It is in gaining a broader perspective that the sustainability concept becomes a strategic opportunity, not a compliance threat.

Technology can introduce negative factors into both environmental and societal environments. On the other hand, it can help to alleviate, or even eliminate, the negative impact from other sources. The same is true for the type of governance.

## Triple Bottom Line

The mission of business is to make a profit. That mandate has long been a *given* in the management literature and in the management of companies. Efforts to introduce social and environmental concerns were often viewed as obstacles, or at least distractions, by executives, especially at the highest levels. Just as victories count in measuring a coach's performance in football, profits were the measure of executive performance. Economic, or financial, performance is a tangible, short-term game in which only the strong survive. This is the first bottom line.

Social issues, such as providing employees with suitable working conditions, fair wages, and a host of other matters, were topics debated with unions during contract negotiations. Other matters, such as increased traffic flows, noise, and tax relief, were considered as unavoidable in bringing jobs into a community. Lead-laden toys, polluted milk, and side effects of drugs were too often considered the unfortunate consequences of the drive for business continuity. Companies often used human resource or public relations departments as buffers against the barrage of new social agendas. It is difficult to measure, report, and set goals for achieving social responsibility (Bulkin 2003). Consequently, it has been difficult for many businesses to come to grips with their obligations in the societal arena, the second bottom line.

Environmental issues have moved through a transition. Protecting the environment was originally viewed by most businesses as simply a need to comply with government regulations. Saving endangered species and trying to assess the effects of global warming were for special-interest groups and the government, not business. As businesses began to reluctantly comply with mandates to reduce polluted discharge water or exhaust gases or to substitute benign for hazardous materials or to reduce the waste going into landfills, they made an astounding discovery—the changes could actually save them money! Companies began to associate doing good—environmental improvement projects—with doing well or making a profit (Laszlo 2008). A separate, and important, topic is the role of reverse logistics and its contribution to improved resource utilization (Crandall 2006). Environmental projects can be planned and measured; they are often multiyear, and they represent the third bottom line.

Although the movement is still in the early stages, a number of companies are publicly stating they are now not only aware of but also accepting their responsibilities in the sustainability movement. One of the most obvious ways is registering with the Global Reporting Initiative (GRI), whose "mission is to make sustainability reporting by all organizations as routine and comparable as financial reporting." The U.S.-based nonprofit Ceres, formed the GRI for environmentally responsible economies and Tellus Institute, with the support of the United Nations Environment Programme (UNEP) in 1997. As of the middle of 2014, Ceres reports that over 1,800 companies use the GRI standard for corporate reporting on environmental, social, and economic performances (Ceres 2014).

## Beyond the Triple Bottom Line

Some writers are extending the scope of the TBL by pointing out the reporting responsibility is more than to shareholders. There are other stakeholders, such as financial stakeholders, supply chain stakeholders, regulatory stakeholders, political stakeholders, and social stakeholders. The conceptualization of sustainability beyond the TBL includes sustainable supply chain management (SSCM). SSCM incorporates strategy, organizational culture, reporting transparency, and risk management as key interfaces to merge economic, social, and environmental performance (Carter and Rogers 2008).

## Why Should Business Take the Lead?

For years, special-interest groups have been the prime movers in both the social and environmental movements. They have tried to get local, state, and national governments to enact legislation or form agencies to require businesses to comply. Although they have made some progress, their efforts have not reached the critical mass to make their programs a concerted and widespread realization. Special-interest groups are motivated and, in some cases, large enough to exert considerable influence. However, they lack the direct power to make the changes and must work through other agencies to achieve their objectives.

Governments respond to constituent pressures, but do so slowly and often only partially. In a democratic government, the mix of disparate interests usually prevents radical change. Consequently, governments do not often lead aggressively.

This leaves the business community with an option to take on the leadership in this area. They have the power to directly implement change but, in the past, have been more narrowly focused on the economic aspects of business, not the social or environmental aspects. To lead in the direction of the TBL, businesses will have to expand their missions and strategies.

## Need for Alliances

A number of companies are demonstrating progress in their sustainability efforts. These organizations obviously achieve economic success, and an increasing number are demonstrating environmental progress, at least in those projects that combine environmentalism with cost reduction. Social pressures forced some companies to review the employee practices and product quality of their suppliers. Hence, partnerships between businesses and stakeholders can be an effective sustainability risk control strategy (Anderson 2006).

There is also progress in forming alliances to achieve mutually desired results. Elkington observed effective, long-term partnerships will be crucial during the sustainability transition. Some will be between the public and private sectors, some between companies, and some between companies and groups campaigning for a broad range of TBL objectives. While he reported progress in environmental partnerships as well as in such areas as Third World development and human rights, he concludes with this:

> One emerging recognition is that, however much a single company may be able to do on the eco-efficiency front, int the end sustainability will depend on the progress of entire concentrations of industry, complete value chains, and whole economies. (Elkington 1999, p. 237)

An example of a special-interest group taking a step toward an alliance with business is the announcement by Greenpeace of their *practical blueprint* to not only improve the world's energy mix and stop climate change but also help end the global financial crisis (France-Presse 2008). If so, the transition from *us versus them* to *we* could be the harbinger of good things to come.

## Benefits and Obstacles

Moving toward the TBL could bring short-term benefits of cost savings and a more favorable public image for businesses. Over the long-term, it increases the likelihood of a prosperous survival and even a feeling of accomplishment among employees and managers. Sustainability strategies can decrease sustainability risk costs, augment competitive positions, protect company reputations, and improve bottom lines (Anderson 2006).

However, there are obstacles. Some projects will incur costs without immediate financial returns. There will be conflicting interests among the myriad of stakeholders. The reporting requirements will be new and uncertain for some time. It will be difficult to evaluate the potential value of some actions—the costs will be more tangible than the benefits, so making the decision to go ahead will be an act of faith for some executives. Conflicts may arise among members of the same supply chain. Supply chains will need reorganization, both inside a company and between the participants (de Brito et al. 2008).

It will be difficult to form a collaborative effort; not many companies want to be first. Major paradigm changes will be required in business, government, and society. The need to integrate economic, social, and environmental objectives while integrating supply chain participants will be a challenge of the first order.

## Sustainability in the Future

Individuals, societies, governments, and businesses are interested, and concerned, about the future. At present, the movement toward sustainability and its components seems to be gathering momentum. While there have been highs and lows in the history of sustainability, it now appears to be on an upward wave (Elkington 2006). The TBL concept appears to be an acceptable approach to integrate the somewhat different interests of a variety of groups.

Supply chain professionals will be among the most active in implementing sustainability practices. Some activities that can reduce costs while improving corporate reputations include reduced packaging, improved working conditions in warehouses, more fuel-efficient transportation, and supplier requirement to undertake environmental and social programs (Carter and Rogers 2008).

The APICS organization publishes a widely recognized magazine for its membership. As part of an effort to promote the sustainability movement in industry, it publishes its November–December issue each year as *green*, completely paperless with all of the articles written about sustainability issues (APICS 2014).

## Strategic Employee Plan

Global competition and extended supply chains have triggered a major challenge for companies in their pursuit of a strategic employee plan. They must confront questions as "What kind and how many employees will we need in the future? How will we find them? Who will provide the education or training? What kind of in-house training should we provide?"

At present, there is not a well-coordinated program to effectively match supply with demand. There is a good deal of attention being given to the reshoring initiative; however, there is less attention being given to providing the skilled workers that will be needed if and when those jobs return to the United States.

## Summary

This chapter summarizes the current state of supply chains and identifies the changes expected in future supply chains. Although much has been accomplished to create effective and efficient supply chains, much remains to be done, especially in managing the links, or relationships between companies, in the supply chains.

We described some of the forces causing companies to initiate, and then modify, their supply chains. Businesses recognized the value of supply chains when confronted with global competition. After forming their initial supply chains, many found the need for modification as they moved more actively into offshore outsourcing. Globalization and offshore outsourcing increased the complexity of supply chains and the risks inherent in their use.

In order to adapt, businesses needed to develop a strategic global business perspective. Many companies found the need to temper their enthusiasm for offshore outsourcing as they found the benefits were not as much as expected and the costs and risks were more than expected. There is a continuing need to use the latest technology, although choosing among the alternatives requires knowledgeable and objective analysis. The infrastructures and cultures in companies have to change—often a slow and arduous process.

Most of all, companies are finding they must manage their relationships with their supply chain partners in new and exploratory ways. While the objectives are clear, the process is not. All roads to integrated supply chains energized with collaborative relationships go through a simple five-letter word—trust—a subject for our final chapter.

## Hot Topic

One of the most controversial topics about supply chains is the discovery of sweatshops used by low-cost suppliers. This chapter's "Hot Topic" reviews some of the issues associated with this subject.

### Hot Topics in the Supply Chain: Finding Solutions to the Sweatshop Problem

There is a common perception that social activists enjoy exposing the foibles of large multinational corporations (MNCs). It can be a sport, because a large successful company is also a vulnerable target for criticism. Indeed, investigative journalists and even academics can make a career going after the bad guy, an MNC that is making megabucks and led by a well-compensated CEO. In regard to supply chains, MNCs have often been the target for criticism because of their strategy of contracting with sweatshops in developing countries. In fact, this one issue alone may be the top public relations crisis facing MNCs in regard to their supply chains.

## BACKGROUND

Retailers, particularly those in the discount industry, have received the bulk of such criticism. The problem is also seated in a lesser talked about source—us. That is right, as consumers we vote with our wallets, pocketbooks, and credit cards. Each time we walk into a Wal-Mart, Target, or Dollar General and make a purchase, we are voting for sweatshops. On the other hand, we LOVE to preach against sweatshops, because it makes us feel good inside, real good as a matter of fact.

However, staying away from discount retailers is not a foolproof way for consumers to avoid associations with sweatshops. Clothing, even nice clothing, has been linked to sweatshops. Likewise, laptops and iPhones are not immune to sweatshop association.

So, are sweatshops necessarily a bad thing? The answer of course is yes and no. Sweatshops do provide employment, and they help stimulate the local economy of the developing country they are in. They also keep prices competitive. Certainly, some goods made in developing countries are sold in those countries, so a market is created as well. But does that mean we have to live with the institution of the sweatshop? Can the lot of those who work in sweatshops be improved? Of course, it can, and a handful of companies are setting the path for the future in making sweatshops a thing of the past. In fact, the goal is to improve the workplace, so that what was once a sweatshop no longer needs to be called a sweatshop, but a viable place to work. If global trade is to remain healthy, sweatshops will need to be transitioned into good places to work. Two long-term trends must be acknowledged: (1) MNCs will need to take a long-term view toward their suppliers in developing countries and (2) resources will need to be committed to improve first-tier suppliers in developing countries.

## RECOMMENDATIONS

### MNCs Will Need to Take a Long-Term View toward Their Suppliers

A supplier to an MNC should be approached as a long-term relationship, not a casual fling. As we have mentioned throughout this book, retailers hold the power advantage over manufacturers. Large discount retailers can scan the world for the lowest-cost supplier, rejecting one after the other over a few cents per item (Shell 2009). This strategy can even be used to the retailer's advantage when it calls off a relationship with a supplier. Former plant inspector Frank (2008) recalls how some companies approach their foreign suppliers with intentions of using them only for a short term, knowing that such suppliers probably are not abiding by good labor standards anyway. Their strategy is to place orders with an unsuspecting supplier who has come in at a rock-bottom price. Then after orders have been coming in, and about the time they sense the supplier may be crooked, they send plant monitors to the factory, and then proudly reject that company from being a supply chain partner when it is "discovered" that they are not abiding by proper labor standards.

This contrived relationship makes the MNC look good, because it appears that they have rejected the supplier for nobler reasons. Such short-term relationships ensure a quick supply of a product at a low price, but do nothing to cement a potential relationship. A supplier relationship like that described earlier is not a sincere effort to build a good supply chain partner. Instead, it is a way to extract the lowest price possible for making a product. Frank (2008) recommends that companies presource their suppliers first before placing an order. Suppliers who cannot meet acceptable labor standards should not be selected, even if they do offer the lowest price.

## MNCs WILL NEED TO COMMIT RESOURCES TO IMPROVE THEIR FIRST-TIER SUPPLIERS IN DEVELOPING COUNTRIES

It is this recommendation that will make many managers uncomfortable. They might think, are we in the social welfare business too? The answer is, well yes, as a matter of fact, you are. We expect U.S. companies to take care of their employees. We also work extensively with our home-based suppliers. Why should we view our overseas partners as any different?

To understand this recommendation, we have to realize that, in developing countries, raising labor standards is not always an acceptable practice. To do so means that either the host government or the supplier themselves will have to invest funds in improving the business. Such funds from those two sources are rare in developing countries. Instead, foreign suppliers must compete by offering the lowest prices possible to entice business to use their factories. This often translates into offering low wages and substandard working conditions. Raising wages and work standards is just not possible under these circumstances (Chan et al. 2013).

Foreign direct investment (FDI) can help, but it is more aggressive in that the MNC will usually secure ownership of the factory, a model that Mattel has followed in some of its factories in Asia (Frank 2008). Another approach is to strategically partner with the supplier, a relationship that commits the companies on a long-term path working together. The auto industry in China is a good example of how car makers from the United States, Europe, and Japan have set up strategic partnerships with automakers in China. This model has worked well, but of course, cars are a different industry from clothing, durable goods, and toys, industries that are more sweatshop prone.

A third approach differs from FDI and strategic partnerships in terms of resource allotment. In this approach, the MNC provides funding to help its supplier achieve higher work standards. The Danish brand Select Sports uses this strategy with one of its suppliers in Sialkot, Anwar Khawaja Industries. Select Sports contributes 50% of the costs of running this supplier's social program (Lund-Thomsen 2008). It should be noted that this approach is still rare in supply chains, as most MNCs expect the supplier to fund its own programs.

## QUESTIONS FOR RESEARCH AND DISCUSSION

1. It was pointed out that MNCs may need to supply resources to help its suppliers achieve upgrades in the workplace. Would simply raising prices not achieve the same objective? Why or why not? Could this vary by industry?

2. Do you think consumers would share in the cost of funding upgrades for suppliers in developing countries by paying higher prices? If so, what industries might this work best in?

## REFERENCES

Chan, J., Pun, N., and Selden, M., The politics of global production: Apple, Foxconn and China's new working class, *New Technology, Work and Employment*, 28(2), 100–115, 2013.
Frank, T.A., Confessions of a sweatshop inspector, *The Washington Monthly*, pp. 34–37, 2008.

Lund-Thomsen, P., The global sourcing and codes of conduct debate: Five myths and five recommendations, *Development and Change*, 39(6), 1005–1018, 2008.

Shell, E., *Cheap: The High Cost of Discount Culture*, Penguin Books, New York, 2009.

## Discussion Questions

1. How have supply chains moved from tightly coupled to loosely coupled structures?
2. Discuss the evolution of CSFs in the United States over the past two centuries.
3. How does the globalization movement affect the design of supply chains?
4. How does offshore outsourcing affect the design of supply chains?
5. How does risk management affect the design of supply chains?
6. Discuss some of the new technologies available for supply chain design.
7. How will the infrastructure of companies change in the future?
8. Discuss the role of organizational cultures in future supply chains.
9. How can supply chain participants develop a higher level of trust with each other?
10. What is the role of sustainability in supply chain design and operation?

## References

Anderson, D.R., The critical importance of sustainability risk management, *Risk Management*, 53(4), 66, 2006.

APICS, 2014, www.apics.org, accessed August 11, 2014.

APICS, *Supply Chain Risk Challenges and Practices, 2011*, APICS The Association for Operations Management, Chicago, IL, 2011.

Becker, T., The business behind green, eliminating fear, uncertainty, and doubt, *APICS Magazine*, 18(2), 20–21, 2008.

Bitran, G.R., Gurumurthi, S., and Sam, S.L., The need for third-party coordination in supply chain governance, *MIT Sloan Management Review*, 48(3), 30, 2007.

Blackstone, J.H., *APICS Dictionary*, 14th edn., APICS—The Association for Operations Management, Chicago, IL, 2013.

Bretton, M., Getting back to your roots: How to manufacture a better future, *Supply Chain Europe*, 17(4), 6, 2008.

Bulkin, B.J., BP = bringing profits: In a socially responsible way, *Mid-American Journal of Business*, 18(1), 7, 2003.

Carter, C.R. and Dale, S.R., A framework of sustainable supply chain management: Moving toward new theory, *International Journal of Physical Distribution & Logistics Management*, 38(5), 360, 2008.

Ceres, History and vision, 2014, http://www.ceres.org/about-us/our-history/our-history, accessed August 11, 2014.

Chandler, A.D., Jr., *The Visible Hand: The Managerial Revolution in American Business*, Belknap Press of Harvard University Press, Cambridge, MA, 1977.

Crandall, R.E., Opportunity or threat? Uncovering the reality of reverse logistics, *APICS Magazine*, 16(4), 24–27, April 2006.

Crandall, R.E., Productive renovations, converting your business from a transaction-oriented to a process-oriented model, *APICS Magazine*, 17(2), 16, 2007.

Crandall, R.E., Riding the green wave: More businesspeople catch on to the sustainability movement, *APICS Magazine*, 19(1), 24–27, 2009.

Crandall, R.E., Perceptions of peril, Evaluating risk management in supply chains, *APICS Magazine*, 22(3), 21–25, 2012.

Crandall, R.E. and Crandall, W.R., *Vanishing Boundaries, How Integrating Manufacturing and Services Creates Customer Value*, CRC Press, a Taylor & Francis Group, Boca Raton, FL, 2014.

de Brito, M.P., Carbone, V., and Blanquart, C.M., Towards a sustainable fashion retail supply chain in Europe: Organisation and performance, *International Journal of Production Economics*, 114(2), 534, 2008.

Drucker, P., *Management: Tasks, Responsibilities, Practices*, Harper & Row, New York, 1974.

Elkington, J., *Cannibals with Forks: The Triple Bottom Line of 21st Century Business*, New Society Publishers, Gabriola Island, British Columbia, Canada, 1999.

Elkington, J., Governance for sustainability, *Corporate Governance: An International Review*, 14(6), 522, 2006.

Farris, M.T. II and Hutchison, P.D., Cash-to-cash: The new supply chain management metric, *International Journal of Physical Distribution & Logistics Management*, 32(3/4), 288, 2002.

France-Presse, A., Greenpeace offers "blueprint" for climate, economic crises, *Industry Week*, October 27, 2008. http://www.industryweek.com/PrintArticle.aspx?ArticleID=17635, accessed October 15, 2013.

Friedman, T.L., *The World Is Flat: A Brief History of the Twenty-First Century*, Farrar, Straus and Giroux, New York, 2005.

Friedman, T.L., *Hot, Flat, and Crowded, Why We Need a Green Revolution and How It Can Renew America*, Farrar. Straus and Giroux, New York, 2008.

Gupta, K.M. and Gunasekaran, A., Costing in new enterprise environment: A challenge for managerial accounting researchers and practitioners, *Managerial Auditing Journal*, 20(4), 337, 2005.

Haeckel, S.H., *Adaptive Enterprise: Creating and Leading Sense-and-Respond Organizations*, Harvard Business School Press, Boston, MA, 1999.

Hammer, M. and Champy, J., *Reengineering the Corporation: A Manifesto for Business Revolution*, Harper Collins, New York, 1993.

Handler, C.A. and Healy, M.C., *Hiring Manufacturing Staff in the 21st Century: A Fundamental Shift in Skills*, Rocket-Hire, Pearson Education, Inc., London, U.K., 2008.

Hill, T., *Manufacturing Strategy*, Irwin, Homewood, IL, 1989.

Kletter, D. and Harada, C., *The Dominant Genes: Organizational Survival of the Fittest*, Booz, Allen, Hamilton, New York, 2008.

Kreitner, R. and Kinicki, A., *Organizational Behavior*, McGraw-Hill, Irwin, NY, 2007.

Laszlo, C., *Sustainable Value, How the World's Leading Companies Are Doing Well by Doing Good*, Stanford Business Books, Stanford, CA, 2008.

Lee, H.L., The triple-A supply chain, *Harvard Business Review*, 82(10), 102, 2004.

Levinson, M., *The Box: How the Shipping Container Made the World Smaller and the World Economy Bigger*, Princeton University Press, Princeton, NJ, 2008.

Lewicki, R.J., McAllister, D.J., and Bies, R.J., Trust and distrust: New relationships and realities, *The Academy of Management Review*, 23(3), 438, 1998.

Manuj, I. and Mentzer, J.T., Global supply chain risk management strategies, *International Journal of Physical Distribution & Logistics Management*, 38(3), 192, 2008.

Maqsood, T., Walker, D., and Finegan, A., Extending the "knowledge advantage: Creating learning chains, *The Learning Organization*, 14(2), 23, 2007.

McGaughey, R.E. and Gunasekaran, A., Enterprise resource planning (ERP): Past, present and future, *International Journal of Enterprise Information Systems*, 3(3), 23, 2007.

Parnell, J.A., Sustainable strategic management: Construct, parameters, research directions, *International Journal of Sustainable Strategic Management*, 1(1), 35–45, 2008.

RFID 24-7, RFID in retail: Midyear report, July 17, 2014, http://www.rfid24-7.com/article/rfid-in-retail-midyear-report/, accessed July 20, 2014.

Rifkin, J., *The Zero Marginal Cost Society, The Internet of Things, the Collaborative Commons, and the Eclipse of Capitalism*, Palgrave Macmillan, New York, 2014.

Snee, R.D. and Hoerl, R.W., *Six Sigma beyond the Factory Floor*, Pearson Prentice Hall, Upper Saddle River, NJ, 2005.

Staff, Virtuality becoming reality, *International Journal of Physical Distribution and Logistics Management*, 26(7), 38, 1996.

Thomson, J.C., Staying on track in a turnaround, *Strategic Finance*, 92(6), 43–50, December 2010.

Veysey, S., Majority of companies suffered supply chain disruption in 2011: Survey, businessinsurance.com, accessed November 8, 2011.

Yergin, D. and Stanislaw, J., *The Commanding Heights: The Battle for the World Economy*, Touchstone, New York, 2002.

Yuva, J., Assess your vulnerability to natural disasters, *Inside Supply Management*, 21(9), 28–31, 2010.

# 18

## *Preparation for the Future*

### Learning Outcomes

After reading this chapter, you should be able to

- Explain why supply chains must continue to evolve
- Identify the major changes anticipated in supply chains
- Explain the triple bottom line (TBL) concept
- Discuss the role of governments in future supply chains
- Identify the roles of third parties in future supply chains
- Describe information technology changes in future supply chains
- Discuss potential performance measures for future supply chains
- Describe the changes in organizational structure as it relates to future supply chains
- Describe the expanding role of knowledge management in the future
- Discuss the role of data analytics in future supply chains
- Explain how chaos theory could be applied in a business setting

While the future will not be like the past, it is difficult, if not impossible, to forecast accurately all the variations that could arise in the years to come. Thus, a company should prepare to provide some structure, such as in a supply chain, but also be prepared to adapt to changing conditions. Some of the future conditions will require capabilities beyond those existing in today's supply chains. In Chapter 17, we described some of the areas that are changing. This chapter describes steps necessary for future supply chains to be successful. It is a chapter about change—some incremental and linear, some radical and nonlinear. It is about identifying and planning for the expected and successfully anticipating and reacting to the unexpected.

### Company Profile

That the future will be different from the present is a given. The question for supply chain managers is how to prepare for expected, and unexpected, changes. What is the risk of not preparing? Will the change be disruptive enough to destroy the company? Can it be prevented or at least mediated enough to survive? Most managers are so busy managing day-to-day operations, they do not have time to worry, or plan, for something that may, or may not, occur in the future. Questions are plentiful; answers are scarce. As you read this company profile and the chapter in the book, think about these questions:

- Who, in the organization, is responsible for forecasting the future? How do they do it?
- What is the most damaging event you can think of for a company?
- Which industry is most likely to change in the future? Explain.
- Identify one of the Fortune 50 companies that you think will not exist in 10 years. Why?

What other questions do you think relate to preparing supply chains for the future?

## Company Profile: Google

In Chapter 17, the company profile was about Amazon, a company that wants to be able to sell you any product you would like to buy. In this chapter, the company profile is about Google, a company that would like to be able to provide you with whatever information you need or want. Google has the most popular Internet search engine on the market.

> Google is a global technology leader focused on improving the ways people connect with information. We aspire to build products and provide services that improve the lives of billions of people globally. Our mission is to organize the world's information and make it universally accessible and useful. (2012 10-K Report, p. 2)

Google was not the first to develop an Internet search engine. The first was named Archie (archive without the "v") and developed by computer science students at McGill University in Montreal. At least six other search engines were identified before Larry Page and Sergey Brin introduced Google in 1998. An article by Seymour et al. (2011) traces the history of search engines in considerable detail.

The acquisition of Motorola Mobility in 2012 moved Google into the hardware business. The mobile segment of Motorola includes mobile wireless devices and related products and services. The home segment provides video entertainment services to consumers. The home segment was sold to Lenovo Group, Inc. in 2013.

It is hard to believe that Google has been around for only a little over 15 years. One study compared top companies on their 15th birthday. The results are shown in Table 18.1.

### REVENUE

Revenues for Google, including Motorola Mobility, were almost $60 billion in the year-ending December 31, 2013. Of that total, Google provided 92.8%, primarily from advertising, and Motorola 7.2%, from a combination of products and services (Mergent Online 2014). Of the advertising revenues, about two-thirds came from Google websites, with the other one-third from Google Network members' websites and other sources (2012 Google 10-K Report). Revenues for the past 3 years have averaged about 47% from the United States, 11% from the United Kingdom, and about 42% from the rest of the world.

**TABLE 18.1**

Comparison of Big Tech Companies at Age 15

| When They Were 15: Big Tech Companies | | | | |
|---|---|---|---|---|
| Company | Year They Reached 15 | Sales (Millions of Dollars) | Market Cap (Millions of Dollars) | Number of Employees |
| Google | 2013 | $56,594 | $290,371 | 44,777 |
| Amazon | 2009 | 20,508 | 36,141 | 23,300 |
| Yahoo | 2009 | 7,209 | 16,236 | 13,600 |
| Dell | 1999 | 23,636 | 102,359 | 36,500 |
| Oracle | 1992 | 1,058 | 1,539 | 7,466 |
| Apple | 1991 | 5,992 | 8,089 | 14,528 |
| Microsoft | 1990 | 1,654 | 12,300 | 5,635 |

*Source:* Adapted from Chemi, E., Google at 15: Living large with a $290 billion market value, *BusinessWeek*, September 27, 2013, http://0-www.businessweek.com.wncln.wncln.org/articles/2013-09-27/google-at-15-living-large-with-a-290-billion-market-value.

## FINANCIAL RESULTS

Google has been a profitable company from the very beginning. Table 18.2 summarizes the key information for the last 5 years, 2009–2013. Revenues have more than doubled during the last 5 years. Net income has increased at a slower rate than revenues because of higher cost of revenues (sales)—lower gross margins.

Table 18.3 shows a breakdown of expenses in more detail. The increase in cost of sales resulted from the addition of manufacturing and inventory-related costs and increases in traffic acquisition costs, data center costs, hardware product costs, and content acquisition costs (Google 10-K, p. 30). Marketing expenses have also increased significantly as a percent of sales from 8.7% to 12.1%, as the company increased its marketing staff in anticipation of continued growth. Research and development costs have also increased slightly, as a percent of sales, as well as general and administrative expenses. While these increases in expenses are significant, the company continues to generate attractive growth in revenues and income.

## MAJOR BUSINESS SEGMENTS (FROM THE 2012 10-K REPORT)

### GOOGLE SEARCH

The company has been built around this technology, and they continue to add features that include Flight Search, Search plus Your World, Google Now, and Google Knowledge Graph. They want to help searchers "find the right thing, get the best summary, and go deeper and broader" (2012 Annual 10-K Report, p. 3).

### ADVERTISING

The company's main source of revenue is through selling advertising, much of it based on a cost-per-click basis. They continue to improve their programs through the addition of features that enable users to better manage their advertising programs. Some of the specific segments of their advertising program include the following:

*Google display*: Provides advertisers' services related to the delivery of display advertising across publishers participating in the company's AdSense, DoubleClick Ad Exchange, YouTube, and Google Finance.

**TABLE 18.2**

Google Income and Cash Flow Results (2009–2013)

| | For the Year Ending December 31 (Dollars in Millions) | | | | | |
|---|---|---|---|---|---|---|
| | 2009 | 2010 | 2011 | 2012 | 2013 | % Increase (Decrease) from 2009 until 2013 |
| *Income Statement* | | | | | | |
| Google revenues | $22,890 | 29,321 | 37,905 | 46,039 | 55,519 | 143 |
| Motorola revenues | | | | 4,136 | 4,306 | |
| Total net revenues | 22,890 | 29,321 | 37,905 | 50,175 | 59,825 | 161 |
| Cost of Google revenues | 8,844 | 10,417 | 13,188 | 17,176 | 21,993 | 149 |
| Cost of Motorola revenues | | | | 3,458 | 3,865 | |
| Total cost of revenues | 8,844 | 10,417 | 13,188 | 20,634 | 25,858 | 192 |
| Gross margin | 14,046 | 18,904 | 24,717 | 29,541 | 33,967 | 142 |
| Income from operations | 8,312 | 10,381 | 11,742 | 12,760 | 13,966 | 68 |
| Net income | $6,520 | 8,505 | 9,737 | 10,737 | 12,920 | 98 |
| *As Percent of Revenues* | | | | | | |
| Gross margin (% of revenues) | 61.4 | 64.5 | 65.2 | 58.9 | 56.8 | −5 |
| Income from operations (%) | 36.3 | 35.4 | 31.0 | 25.4 | 23.3 | −13 |
| Net income (% of revenues) | 28.5 | 29.0 | 25.7 | 23.3 | 23.3 | −5 |
| *Cash Flow* | | | | | | |
| Net cash from operations | $9,316 | 11,081 | 14,565 | 16,619 | 18,659 | 100 |
| Purchase of fixed assets | (810) | (4,018) | (3,438) | (3,273) | (7,358) | 808 |
| Free cash flow | 8,506 | 7,063 | 11,127 | 13,346 | 11,301 | 33 |
| Cash flow from investing activities | (8,019) | (10,680) | (19,041) | (13,056) | (13,679) | 71 |
| Cash flow from financing activities | 233 | 3,050 | 807 | 1,229 | (857) | −468 |
| Foreign currency effect | 11 | (19) | 22 | 3 | (3) | −127 |
| Net cash provided (required) | 1,541 | 3,432 | (3,647) | 4,795 | 4,120 | 167 |
| Cash balance at the end of the year | $10,198 | 13,630 | 9,983 | 14,778 | 18,898 | 85 |

*Source:* Mergent Online, Google Inc., 2014, http://www.mergentonline.com, accessed February 17, 2014.

*Google mobile*: Extends the products and services by providing mobile-specific features to mobile devices. The technology includes search by voice, search by sight, and search by location.

*Google local*: Provides users with relevant local information for over 80 million places globally. Users can find addresses, phone numbers, hours of operation, directions, and more for shops, restaurants, parks, and landmarks.

## OPERATING SYSTEMS AND PLATFORMS

*Android*: Working closely with the Open Handset Alliance, a business alliance of more than 75 technology and mobile companies, the company developed Android, a free, fully open source mobile software platform that any developer can use to create applications that mobile and handset manufacturers can install on a device.

*Google Chrome OS and Google Chrome*: Google Chrome OS is an open source operating system with the Google Chrome web browser as its foundation.

*Google+*: A new way to share online, with tightened integration between Google+ and other Google properties, Gmail and YouTube.

**TABLE 18.3**

Comparison of Key Financial Results for Google (2009–2013)

| | For the Year Ending December 31 (Dollars in Millions) | | | | | |
|---|---|---|---|---|---|---|
| | 2009 | 2010 | 2011 | 2012 | 2013 | % Increase (Decrease) 2009–2013 |
| *Sales and Operating Expenses (Millions of Dollars)* | | | | | | |
| Net sales ($) | 22,890 | 29,321 | 37,905 | 50,175 | 59,825 | 161 |
| Cost of sales | 8,844 | 10,417 | 13,188 | 20,634 | 25,858 | 192 |
| Research and development | 2,843 | 3,762 | 5,162 | 6,793 | 7,952 | 180 |
| Marketing | 1,984 | 2,799 | 4,589 | 6,143 | 7,253 | 266 |
| General and administrative | 1,667 | 1,962 | 2,724 | 3,845 | 4,796 | 188 |
| Other | | | 500 | | | |
| Total operating expenses | 15,338 | 18,940 | 26,163 | 37,415 | 45,859 | 199 |
| *Percent of Net Sales* | | | | | | |
| Cost of sales (%) | 38.6 | 35.5 | 34.8 | 41.1 | 43.2 | 4.6 |
| Research and development (%) | 12.4 | 12.8 | 13.6 | 13.5 | 13.3 | 0.9 |
| Marketing (%) | 8.7 | 9.5 | 12.1 | 12.2 | 12.1 | 3.5 |
| General and administrative (%) | 7.3 | 6.7 | 7.2 | 7.7 | 8.0 | 0.7 |
| Other (%) | 0.0 | 0.0 | 1.3 | 0.0 | 0.0 | 0.0 |
| Total operating expenses (%) | 67.0 | 64.6 | 69.0 | 74.6 | 76.7 | 9.6 |

*Source:* Mergent Online, Google Inc., 2014, http://www.mergentonline.com, accessed February 17, 2014.

*Google Play*: An entirely cloud-based, digital entertainment destination with more than 700,000 apps and games.

*Google Drive*: A place where users can share and collaborate by using Google Docs to work with others in real time on documents, spreadsheets, and presentations.

*Google Wallet*: A place to store credit and debit cards and offers and reward cards. Users can pay in-store at over 200,000 merchants across the United States.

*Google TV*: A platform to experience television and the Internet on a single screen, with the ability to search for the content users want. It is based on the Android operating system and runs Google Chrome.

#### ENTERPRISE

These products provide Google technology for business settings. It uses Google Apps, which includes Gmail, Google Docs, Google Calendar, and Google Sites. It provides hosted, web-based applications for any device with a browser and an Internet connection.

#### MOTOROLA

Buying Motorola to gain access to making smartphones turned out to bring along some difficulties. While the intent was to put more emphasis on lower-cost phones using the Android system, it also put Google into competition with some of the other Android-using phone manufacturers. By selling to Lenovo, this reduces the likelihood that other Android users will move to other systems and also brings more balance to the Android ecosystem, where Samsung has been the dominant player. The purchase enabled Google to evaluate the need for a single vendor to create both hardware and software for the smartphone market, as Apple does. The divesture of Motorola indicates that Google has decided hardware and software vendors can be separated (Winkler 2014).

## SWOT ANALYSIS

MarketLine did a SWOT analysis on Google in 2013. They summarized the company's strengths as follows: (1) having search engine dominance; (2) Android's success is a growth driver in the mobile market; and (3) increasing cash from operations. The report estimated Google's search engine had a market share of more than 88% measured against global website traffic. The company also had a 52% of mobile ad revenue in 2012. On the weakness side, the report viewed Google as having a weaker strategy in social media advertising space, as compared with Facebook.

The opportunities identified in the report are as follows: (1) poised to benefit from the growing smartphone and tablet market because of the strength of Android and the movement of the market toward lower-cost phones, (2) display advertisement and mobile ad spend will be strong drivers as the use of mobile devices continues to increase, and (3) positive outlook for global online video market, with their position in YouTube.

Threats are also on the horizon. They include the following: (1) European Commission antitrust investigations could result in fines if not satisfactorily resolved; (2) PRISM electronic surveillance program could have a negative effect on brand image and market share; and (3) intense competition will continue as new disruptive technologies could emerge that would erode Google's present dominant position.

## GOOGLE LOCATIONS

Google moved into its headquarters in Mountain View, California—better known as the Googleplex—in 2004. Today, Google has more than 70 offices in more than 40 countries around the globe. As an indication of the environment the company wants to create, they describe their locations as follows:

> Though no two Google offices are the same, visitors to any office can expect to find a few common features: murals and decorations expressing local personality; Googlers sharing cubes, yurts and "huddles"; video games, pool tables and pianos; cafes and "microkitchens" stocked with healthy food; and good old fashioned whiteboards for spur-of-the-moment brainstorming. (Google Locations 2014)

## OTHER INTERESTS

The founders of Google, Larry Page and Sergey Brin, both turned 40 in 2013. Both are still active in the company, Page as CEO and Brin in charge of special projects, including Google X, the company's skunkworks. They met at Stanford and have remained close friends, as well as partners, and share a passion for confronting new challenges. Some of their current interests include Google Glass and driverless cars.

### GOOGLE GLASS

Google Glass is a computer that can be worn like a pair of glasses. It has a small screen above one eye and allows the wearer to record video, audio, and location data. It can also connect wirelessly with a smartphone so the wearer can send text messages, check e-mail, make phone calls, and use maps without reaching into a pocket. It is the latest in a growing field of wearable computers. The technology is exciting; however, there are major concerns about privacy. It appears likely that some limits will be necessary on the use of these glasses, self-imposed either by the sellers or by the government (Bloomberg View 2013).

### DRIVERLESS CAR

The idea of a driverless car excites some as an answer to traffic problems while others view it as technology gone mad. Regardless of your point of view, it is a technical accomplishment that is coming, perhaps within the next few years. Google has cars that have driven themselves over 200,000 miles around California. According to Ashley Vance of Business Week (2012), "Car companies have their own autonomous bits and bobs in the works, and we see manifestations of the technology today. Some vehicles assist with parking, others warn drivers about collisions. Some carmakers caution, though, that the introduction of truly autonomous vehicles will likely need to wait until pretty basic transportation problems are fixed."

There are other projects, some of which are still in the minds of the people who work at Google. The company tries to encourage innovation. The culture at Google is described as follows: "It's really the people that make Google the kind of company it is. We hire people who are smart and determined, and we favor ability over experience. Although Googlers share common goals and visions for the company, we hail from all walks of life and speak dozens of languages, reflecting the global audience that we serve. And when not at work, Googlers pursue interests ranging from cycling to beekeeping, from frisbee to foxtrot. We strive to maintain the open culture often associated with startups, in which everyone is a hands-on contributor and feels comfortable sharing ideas and opinions" (Google Inc. 2014).

### SUMMARY

Google is on the forefront of many projects that require good supply chains to deliver the products and services to the consumer. One of the more ambitious is Google Ventures. This is a venture capital firm started in 2009 with a $100 million investment by the parent firm, Google Inc. It is independently managed, but with a number of ex-Googlers as employees and partners and including some entrepreneurs from companies purchased by Google Ventures. It has a wide-ranging portfolio of investments, including company in clean energy, education, robotics, transportation, shopping, and medicine. Google Ventures contributes more than money; it also provides web design services, hosts workshops for portfolio company employees, and helps with marketing, recruiting, and engineering (Helft 2014).

> Times change, but Page and Brin still are the Google Guys, in pursuit of a more connected tomorrow. (Helft 2013)

### REFERENCES

Bloomberg View, Who's afraid of google glass? *Business Week*, June 6, 2013, http://www.businessweek.com/articles/2013-06-06/bloomberg-view-whos-afraid-of-google-glass, accessed February 26, 2014.

Chemi, E., Google at 15: Living large with a $290 billion market value, *BusinessWeek*, September 27, 2013.

Google Inc., Form 10-K, 2012, https://investor.google.com/pdf/2012_google_annual_report.pdf, accessed February 25, 2014.

Google Inc., Our culture, https://www.google.com/about/company/facts/culture/, accessed February 26, 2014.

Google Locations, 2014, http://www.google.com/about/company/facts/locations/, accessed February 25, 2014.

Helft, M., The Google guys, *Fortune*, 167(3), 16 February 25, 2013.

Helft, M., Where Google Ventures is pinning its hopes, *Fortune*, 169(1), 76–79, January 13, 2014.

MarketLine, Google Inc., SWOT analysis, MarketLine Advantage, September 2013, http://www.marketline.com.

Mergent Online, Google Inc., 2014, http://www.mergentonline.com, accessed February 17, 2014.

Seymour, T., Frantsvog, D., and Kumar, S., History of search engines, *International Journal of Management & Information Systems*, 15(4), 47–58, 2011.

Vance, A., Prepare for autonomous traffic jams, *Business Week*, March 21, 2012, http://www.businessweek.com/articles/2012-03-21/prepare-for-autonmous-traffic-jams, accessed February 26, 2014.

Winkler, R., Google's costly Motorola maneuver may pay dividends, *Wall Street Journal*, Eastern Edition, B. 6, January 2014.

## Recognize the Need to Adapt

The first requirement for dealing successfully with change is to recognize that the future business environment will be rife with change. Successful companies, and their supply chains, will view the coming changes as opportunities to more firmly establish themselves as leaders in their industries, while those who delay or avoid dealing with change will falter, even fail.

Change will take two forms: change coming from outside the supply chain (the open-system environment) and change from within the supply chain (often in reaction to the external changes). We will describe sources of external change and in the remainder of the chapter describe ways to deal with the external changes.

There are a number of external drivers of change. We will highlight six—globalization, competition, economy, technology, customers, and employees.

### Globalization

One of the areas of sweeping change is the globalization movement. Definitions abound. The *APICS Dictionary* defines globalization as "the interdependence of economies globally that results from the growing volume and variety of international transactions in goods, services, and capital, and also from the spread of new technology" (Blackstone 2013). Countries, communities, and cultures not only trade with one another, but they learn from each other and use this increased knowledge in both beneficial and detrimental ways. Friedman (2007) identified globalization as a key element in economic advancements in the world.

### Competition

A company's competition may come from anywhere in the world. Increased competition results from the ease with which individuals and companies communicate with one another. Another cause is the movement of a number of large countries away from state-controlled economies to market-driven economies. The breakup of communism has unleashed a torrent of competition from countries within the former Soviet bloc and China.

A major source of increased competition is the absence of worldwide wars. Since World War II, European countries, and Japan, have been able to rebuild their industrial capacity with new technologies and have become more competitive than before the war. Asian countries that did not have significant industrial capability during most of the twentieth century have begun to establish themselves as formidable suppliers of goods. Central and South American countries have also begun to emerge as recognizable forces in the global marketplace. Thomas Friedman, in his book *The World is Flat*, offers the Dell Theory of Conflict Prevention to illustrate the power of supply chains in linking countries in a common cause. In an earlier book, Friedman posed a similar theory he called the Golden Arches Theory of Conflict Prevention, featuring McDonalds. The Dell Theory stipulates that no two countries will ever fight a war against each other as long as they are part of the same global supply chain. Their goals will be to make just-in-time (JIT) deliveries and share in the rising standard of living that comes with economic prosperity (Friedman 2007).

In the United States, it is becoming increasingly obvious that the competition-free period immediately following World War II rapidly disappeared in the last quarter of the twentieth century. Successful companies need to be a part of successful supply chains in order to compete.

## Economy

While the economies of countries will still have cyclical ups and downs, the long-term trend is definitely upward. Globalization is making people all over the world hungry for the goods and services many have long been without. In addition, the population of the world is increasing and the demand for products grows along with the population. In addition to the gigantic potential in the large populations of China, India, and Russia, there are many smaller countries with needs and wants to satisfy.

This growth in market demand has both positive and negative implications. From a business perspective, it offers exciting opportunities for rapid growth. However, perhaps for the first time, it is forcing a widespread recognition the resources of the planet Earth are not inexhaustible. They must be used wisely to strike a balance between demand and supply.

## Technology

We describe a number of technologies throughout this book and summarize some with future implications in Chapter 17. New, and revolutionary, technologies continue to enable companies, and supply chains, to attain new levels of capability. They will also create new areas of concern as long as their consequences remain unknown.

Future supply chain members must recognize that the introduction of new technologies usually requires changes in the infrastructure and culture of the individual supply chain partners and in their relationships with one another.

## Customers

We described customers in Chapter 3. Customers in developed countries are becoming more selective, and the mass markets are becoming niche markets. Customers in developing countries do not necessarily represent a reemergence of previous mass markets.

Each developing country has different requirements. It may be as obvious as different tastes in food, but it may also be in less obvious areas as religious or political beliefs.

While the total demand for goods and services will increase, there will be a need for a wider selection of goods and services. The marketplace demand will be more complex; therefore, the products to supply that demand will also be more complex. There seems to be no doubt that supply chains will need to be versatile and adaptable.

### Employees

Just as in other areas, employees of the future will be different. Although an oversimplification, employees in developed countries will be more mobile and selective in their job choices. They will also tend to become knowledge workers, instead of just crafts persons. Developing countries will provide the bulk of the labor-intensive workers, at least in the short run.

In developed countries, several issues will be of major concern. Higher wages will be expected as workers attain higher skills and companies grow larger and more profitable. Employees will probably move from job to job more often than in the past. This will cause a need for improved training methods and a rethinking of such program areas as health-care benefits that can be carried from one company to another. Another major concern will be how to generate and retain knowledge within a company—how to convert the tacit knowledge of individuals into an implicit form of knowledge available company wide.

## Develop New Measures of Success

In the past, companies were measured, and its management rewarded, on the financial results of the company. Whether the measure was direct, as in profits and return on investment (ROI), or indirect, such as the stock price, financial results were the primary, if not sole, measure used to assess the contribution of firms. In the future, that will also change, and a number of writers are advocating a new measure of success—sustainability or the triple bottom line (TBL). This term includes three areas: financial, social, and environmental. Companies will have to satisfy all three criteria to be considered fully successful (Elkington 1999).

### Financial Success

Financial success will continue to be the primary and most visible measure of a company's success. Large, publicly traded companies will find it important not only to perform well but also to forecast how well they will perform in the future. The stock market is becoming future or expectation oriented. In the past, stock buyers considered the past performance of companies as an indicator of future earnings. However, because of the rapidly changing business environment, stock buyers are now becoming more interested in the *guidance* numbers they get from company executives.

The financial measures are also becoming more comprehensive. Income, income margins, and earnings per share were primary measures. Today, there is more attention to ROI, return on asset (ROA), and economic value–added (EVA) measures. Although financial measures have not been extended to the total supply chain, it will be a natural extension as the identity of supply chains becomes more evident.

## Social Responsibility

John Elkington (1999) defines social capital as a measure of the ability of people to work together for common purposes in groups and organizations. A key element of social capital is the sense of mutual trust.

In addition to profitability, what is the obligation of a business to society? Must it be a good corporate citizen as well as being a positive economic force in the community? Some of the questions that could be asked in the future include the following:

- Should a company participate in local government decisions such as traffic flow, infrastructure growth, public school financing and instructional content, and public safety issues?
- Should a company encourage employees to become members of governmental study groups and pay them for the time they spend on these projects?
- Should a company support local charities through direct contributions and by encouraging their employees to aid their support in time and money?
- Should a company endorse, and enforce, a code of ethics that would prevent actions that would bring embarrassment and potential economic hardship to a community?
- Should a company have a responsibility to employees, beyond severance pay and unemployment insurance, who lose their jobs because of company actions?
- Should a company be more open in communicating their plans, especially in areas that may pose problems for the local community?

These questions are just a sample of the issues that fall into the area of social responsibility. Even if there were agreement on the correct role of business in these areas, it is also apparent that it is extremely difficult, if not impossible, to measure the contribution that a company provides. In today's society, a company's participation could even be considered self-serving or a conflict of interests by some, resulting in a negative contribution score.

In view of the lack of accepted performance measurement standards, companies must proceed cautiously. The danger is that a lack of participation will be viewed negatively while an excess of participation may also be viewed negatively. The most positive position will probably be in the middle of the continuum from none to full participation. To further complicate decision making, each of the aforementioned questions represents just one issue or area of consideration. There are more, depending on the geographic area and its own stage of development.

## Environment

A third area that is rapidly demanding attention is the company's responsibility to environmental concerns. In speaking of studies to define what it would take to enable a global society to be sustainable, one of the early writers in this area posed the need for businesses to be involved. "We must hope that business people will be actively involved in shaping and implementing such projects. In contrast to the antiindustry, antiprofit, and antigrowth orientation of much early environmentalism, it has become increasingly clear that business must play a central role in achieving the goals of sustainable development strategies" (Elkington 1994). The same author also offers the following definition of sustainability from the World Commission on Environment and Development, which suggests that development is sustainable where it meets the needs of the present without compromising the ability of future generations to meet their own needs (Elkington 1999).

Just as with achieving social equity, businesses will have difficulty in measuring the benefits of moving forward with sustainability programs. It is a decision space filled with trade-offs. Will the increased use of ethanol in automobile fuels cause an increase in food costs and increase the starvation rate in underdeveloped countries? Will the use of *new* coal processes, which claim to have lower emission rates in generating electric power, extend the availability of fossil fuels for its optimal use? Will the development of shale gas prove to be a desired alternative to oil?

## Integrating Financial Results, Society Equity, and Sustainability

While it is possible to measure the level of emissions, at least with some reliability, how can they be converted to a number that defines the contribution of a business? Will companies someday be able to establish a correlation between the level of sustainability contribution to their gain in market share from appreciative consumers? Beyond that, will they be able to trace that acceptance to the financial results of a company? It is almost an overwhelming task. However, if that is the challenge, then surely the business community will be able to find ways to solve the problem in a satisfactory way. Businesses are becoming increasingly aware that they must consider all three areas in making their strategic decisions.

Finding a way to measure results depends on what must be measured. In the next section, we summarize strategies necessary for future supply chains.

---

## Identify What Needs to Be Done

A number of well-qualified organizations completed studies in which they forecast what they believe to be the challenges ahead in the area of supply chain management.

### APICS E&R Foundation Inc.

A study sponsored by the APICS Educational and Research Foundation Inc. was entitled *Supply Chain Management 2010 and Beyond: Mapping the Future of the Strategic Supply Chain*. The study is appropriate because of the changing scope of supply chains. Initially, the supply chain was viewed as primarily oriented to the supply, or upstream, side. By the mid-1990s, the scope broadened to include both the upstream and downstream, or customer focus, of the supply chain. Recently, there has been another enlargement—from operational supply chain management to strategic supply chain management.

The new study attempts to identify the differences expected between operational supply chain management and strategic supply chain management. Table 18.4 contains a list of the major gaps identified, along with a further breakdown of microgaps.

The APICS study offers a number of areas in which additional research should be beneficial. The authors offer the following concluding comments:

- Today's supply chain is evolving from one that is fundamentally order oriented, cost driven, and execution focused to one that is strategically focused, design oriented, dynamic, and driven by multiple objectives.

**TABLE 18.4**

Gaps in Implementing Global Supply Chain Integration

| Gaps | Microgaps |
|---|---|
| Strategic visibility and alignment | Make significant strategic supply chain investment and improvements. Develop business and intercultural skills to operate in a global environment. Make information visible and readily available throughout the supply chain. Focus on *cradle to cradle* beyond recycling and *cradle to grave*. |
| Talent management and leadership | Acquire and develop supply chain leaders. Design supply chains with little opportunity for further cost reduction. Blend industry and educational institutions to develop supply chain talent. |
| Supply chain models including optimization, risk, and cost | Total supply network optimization requires new models for solving problems. Develop processes and systems that are transparent and able to be monitored. Understand the balance between buffers, postponement, and optimal costs. |
| Process orientation including measures, information, and integration | Design network measures to achieve financial and strategic objectives. Work with supply chain partners to improve product design and time to market. Expand localized functions to company and intercompany processes. |
| Relationship and trust | Build trust between organizations to improve collaboration and relationships. Redefine boundaries of what people focus on to include collaboration. Reward long-term and effective collaboration. |
| Supply chain architecture and structure | Build supply chain layers—product, financial, information, and competency. Refine supply chain structures—locations, physical links, and types of nodes. |

*Source:* Adapted from Melnyk, S.A. et al., Supply chain management 2010 and beyond: Mapping the future of the strategic supply chain, Special Report, APICS Educational & Research Foundation Inc., Chicago, IL, 2008. With permission.

- When suggesting the supply chain is too preoccupied with costs, what is actually meant is that it is too preoccupied with direct or individual item or process costs. This is in contrast with the strategic supply chain management, which is concerned with total costs.

- Supply chain management is changing from being a system jointly managed by three traditional corporate functions (purchasing/sourcing, logistics/transportation, and operations management) to a system that must draw on the capabilities of all functions of the firm (accounting, finance, engineering, and marketing, as well as the three traditional functions).

- Supply chain management is increasingly forcing managers and researchers to think in terms of managing operations and processes across corporate boundaries.

- The focus of supply chain management is shifting from the upstream/supply side to the downstream/demand side. The customer drives the supply chain, not just the supply side.

- The focus of the supply chain management is shifting from management to supply chain design/redesign—of products, processes, and the entire supply chain.

- Supply chain management is now becoming a core competency—a skill set some firms are developing and turning into a vehicle for generating and maintaining a sustainable competitive advantage in their respective markets.

- The term *supply chain management* may be a misleading term. It implies a supply or upstream focus; it implies management, not design; it implies a linear chain. Several replacement terms were suggested in the study—supply network optimization, value network optimization, and value network systems.

- A new view of the supply chain is proposed—the adaptive supply chain. The supply chain must deal with and respond to challenges and changes taking place on both the supply and demand sides. As new conditions emerge, the supply chain must be able to quickly and efficiently realign itself to compensate and respond to these changes (Melnyk et al. 2008).

## McKinsey Study

The results from a McKinsey study are partially summarized in Table 18.5. The left side of the table shows the major challenges cited by the survey respondents, who were asked to list the top two challenges. The fact that 12 challenges were listed shows a range of challenges over the companies participating in the study. The challenges are ranked from highest to lowest in the "Next 5 Years" column. The right side of the table shows the level of preparedness of the companies, where about two-thirds answered with a definite "Prepared" or "Not Prepared." In general, companies indicated a higher level of preparedness for the higher-rated challenges.

**TABLE 18.5**

Challenges for Future Supply Chains

| Companies Were Asked to List Their Top Two Challenges | % of Companies Responding | | Changes from the Past to the Future (Considered "Significant" if More than 10% Change) | % Level of Preparedness (Does Not Include "neither Prepared nor Unprepared" or "Don't Know") | |
|---|---|---|---|---|---|
| Challenges | Past 3 Years | Next 5 Years | | Prepared | Not Prepared |
| Increasing pressure from global competition | 27 | 35 | Significant increase | 41 | 25 |
| Increasing consumer expectations about service/quality | 32 | 28 | Decrease | 47 | 16 |
| Increasingly complex patterns of customer demand | 24 | 27 | Increase | 36 | 25 |
| Increasing cost pressure in logistics and transportation | 30 | 25 | Significant decrease | 33 | 23 |
| Increasing volatility of commodity prices | 25 | 24 | Decrease | 28 | 37 |
| Increasing financial volatility in currency and inflation | 22 | 24 | Increase | 26 | 36 |
| Growing exposure to differing regulatory requirements | 14 | 24 | Significant increase | 37 | 37 |
| Increasing global markets for labor and talent, rising wages | 17 | 23 | Significant increase | 20 | 37 |
| Increasing volatility of customer demand | 37 | 21 | Significant decrease | 29 | 22 |
| Increasing environmental concerns | 12 | 21 | Significant increase | 35 | 38 |
| Increasing complexity in supplier landscape | 14 | 15 | Increase | 24 | 26 |
| Geopolitical instability | 2 | 7 | Significant increase | 5 | 67 |

*Source:* Adapted from Gyorey, T. et al., Challenges ahead for supply chains, McKinsey Global Survey, McKinsey & Company, New York, 2010.

## University of Tennessee

In a study sponsored by Ernst and Young, and Terra Technology, nine faculty members at the University of Tennessee identified the following 10 *game-changing* trends in supply chains:

1. Customer service to customer relationship management
2. Incremental change to a transformational agile strategy
3. Functional focus to process integration
4. Absolute value for the firm to relative value for the customer
5. Forecasting to demand management
6. Managerial to value-based accounting
7. Absolute to relative customer value
8. Vertical to virtual integration
9. Adversarial to collaborative relationships
10. Information hoarding to sharing

The study concludes that firms have made significant progress in all these areas; the last five areas offer the greatest opportunity for further improvement (Stank et al. 2013).

These three studies all highlight the need for companies to recognize customers are becoming more demanding, supply chains are more widely dispersed and complex, and there is a growing need for talent to manage the supply chains that are being established.

In facing these challenges, companies have to be prepared for two external forces: what are governments at all levels going to do? Is help available from third-party service providers?

---

## Adapt to Government Actions

How will governments, at all levels, adapt to the future? Will they be more involved in business or less? Will they help or restrict? How can businesses most effectively work with governments? What are the issues that will require government involvement? These and other questions offer worthy challenges to both government officials and business managers. Some of the areas that appear to be most likely to be on the agenda include the environment, business ethics, product safety, social equity, and fiscal responsibility. Enacting appropriate legislation and regulations will require bipartisan support, which in itself will require a change in the political climate in the U.S. legislative bodies and other governments throughout the world.

### Environment

Historically, businesses were considered successful if they operated *in the black*, instead of *in the red*. In the future, businesses will also have to worry about being *green*. Concerns about the environment are mounting, and the industrial complex of businesses is increasingly viewed as the source of many environmental problems.

This concern magnifies in the supply chain. Companies have been working hard to establish supply chains that move product and services from the sources to the consumer—the forward supply chain. Now, they must become equally concerned with the reverse supply chain—reverse logistics, to use a commonly accepted term—or the movement of goods from the consumer back through the reuse and recycle steps to the product's eventual disappearance. It is sometimes viewed as a *cradle-to-grave* cycle when the product ends up as a recycled product in a landfill. An even more optimistic perspective is called a *cradle-to-cradle* cycle when the product eventually ends up as a biodegradable component of a landfill that generates new products from the discarded waste and uses the methane gas given off by decomposing solids as a new energy source. Landfills have been designed as new forms of factories that offer amazing capabilities to convert what was once considered waste into useful products.

## Business Ethics

The financial meltdown on Wall Street in 2008 highlighted the result from business executives who were willing to take extreme risks to create profits that enhanced their own well-being. However, the results were disastrous for their stockholders and even more disastrous for society, both in general and for individuals who were adversely affected.

It is unlikely we will find a way to legislate ethics; individuals, and entities, will find a way to get around the most tightly written laws. Until society finds a way to motivate business executives to operate ethically, whatever that may eventually mean, we will face periodic examples of this malady (Hawken 2007).

Part of the responsibility belongs to the education system—at all levels, and especially at the level that purports to be training *managers* for the workplace. Henry Mintzberg has written a scathing criticism of the typical MBA program in America. He believes the present system is creating analysts, not managers, who perform poorly when placed in executive positions, as they often are, because of the still-favorable perception of the MBA education. If Mintzberg is even partially correct, the American system of business education requires a vigorous and rapid transformation, an objective requiring considerable change in today's academic environment (Mintzberg 2005). Christiansen also believes that higher education institutions must deal with radical changes needed to meet tomorrow's educational needs (Christiansen and Eyring 2011).

## Product Safety

Product safety problems abound. Although the quality movement has increased the level of awareness about product quality, recent events have made product safety front-page news. Some of the highest profile cases have involved tainted foods, medicines, and toys for small children, a defenseless consumer group. Many of the problems originated in offshore outsourcing situations where the products came from companies that were less stringently regulated or audited for safety compliance by the government. The result is a major negative factor for the outsourcing movement. No doubt, the threat of importing unsafe products will cause many companies to proceed more cautiously. We have tried to highlight some of these issues in our Hot Topics series at the end of each chapter.

One area of particular concern is in leading-edge medication for diseases that are mass killers, such as AIDS, heart attacks, and cancer. Pharmaceutical companies are rushing to bring new drugs to the market, some motivated by humanitarian needs, but some by *first-mover* profit motives. Regardless of the motivation, there is the constant need to balance speed to market with efficacy of product. Government oversight requires competent overseers.

## Social Equity

Will businesses of the future be of kinder, gentler natures that care as much for the needs of its stakeholders—employees, society, and the environment—as it does about its executives and shareholders? Will they change willingly or will government intervention be required to facilitate these changes? Can companies' management commit to the new challenge of the expanded obligations? Can management learn how to balance sometimes complementary, sometimes conflicting, demands? We know we can travel to the moon and build cell phones that can perform amazing functions, but can we learn to trust each other enough to collaborate on a mutually beneficial project?

One issue that needs a more rational analysis is the perceived disparity between CEO pay and benefits, as compared with lower-level workers' pay and benefits. Does the answer lie in government intervention or voluntary adjustments by business leaders?

Companies are beginning to see they will have to reuse and recycle products and materials in an environmentally sound way. Does it not also appear reasonable that businesses should be concerned with how their product is used? Consider a couple of examples. What would it take to design cell phones that would automatically shut off when the automobile engine is running? How do spirits manufacturers prevent users from operating any kind of powered equipment? Intelligent individuals may be able to solve some of these problems if motivated to do so.

## Infrastructure

Governments, at all levels, and in all countries, have a key role to play in building transportation systems—roads, airports, railroads, and ports—and supporting facilities. The transportation infrastructures are necessary to make the movement of goods along the supply chain efficient and effective. At the same time, supply chains are necessary to support the construction and maintenance of the infrastructure. Some countries, such as China, are helping other countries, such as in Africa, to build their infrastructure in exchange for access to natural resources (Walker 2009).

## Capitalize on Third-Party Skills

Third parties abound in supply chains. They perform a variety of functions, although most at the discretion of individual members of the supply chain, not of the supply chain as an integrated entity. Third-party providers, or 3PL, were described in Chapter 9 as being key movers in bringing about tighter integration of supply chains.

### Direct Support: Outsourcing

One of the most obvious, and active, role of third parties is to be an outsourced supplier. Today, individual companies outsource; tomorrow, some of the outsourcing could be done as the result of mutual agreement between two or more members of the supply chain. For example, suppose multiple companies agree to outsource all of their transportation activities to a third party such as UPS or FedEx. The third party could design and operate an integrated logistics system that would be of benefit to the entire supply chain, not just one company.

### Indirect Support: Financing and Insurance

Another area in which third parties participate is in providing financing and insurance. While it is normally provided for individual companies, it could also be done for the supply chain. A third party could take ownership of the products as they move along from supplier to customer and provide both financing and insurance. In lieu of taking ownership, the third party could lend money to each supplier and collect it as the product moves along the supply chain. Insurance could be provided as needed to cover the value of the in-process inventory.

### Advisory: Consulting and Training

Consulting companies already provide services and training to individual companies. They could extend their services to multiple entities in the supply chain. An example could be to design and help implement an information-sharing system for two, or more, participants. Another example would be to provide training to align the cultures of multiple companies to work together more effectively. The role of the third party could reduce the inbred resistance to change that exists in many organizations today.

### Analyst: Measure Performance and Identify Needs

In an earlier section, we described the difficulties in designing performance measures that can capture the effectiveness of a supply chain. Perhaps a third party can best do this. Independent government agencies measure productivity of industries; it is logical that outside entities could do this for supply chains.

In addition to the measurement activity, third parties could also analyze the information to identify opportunities for improvement, sometimes for individual companies, sometimes for the entire supply chain.

### Manager: Virtual Holding Company

It is difficult to see how an individual company can manage a supply chain. Even a company with the dominance of Wal-Mart does not manage a supply chain. It influences supply chain members, even exercising a level of control, but it does not manage in the usual sense of the word. Perhaps there is a need for a third party to perform the management function for a supply chain, much as a holding company does, or even as a board of directors. Larger, or more essential, members of the supply chain would have more representation in the third-party organization.

## Utilize Information Technology

A number of information technologies (ITs) will gain acceptance and use in future supply chains. We described them in Chapter 14 and will only briefly describe a few in this section. However, the reader should expect that more innovative and valuable technologies will be developed as supply chains continue to evolve.

## Enterprise Resource Planning Extension

ERP systems became popular in the 1990s, reaching a crescendo of implementations just before the end of the last century, as companies rushed to avoid the Y2K scare. ERP systems offered a way to consolidate information flows within a company.

Since then, ERP systems and supply chain management software have been on a collision course. After the surge in business, demand for ERP systems abated, and vendors had to look for new applications. They seized on the opportunity to focus their product development efforts on the needs in supply chain management. Some companies developed stand-alone software such as customer relationship management (CRM) and supplier relationship management (SRM). It was logical for ERP vendors to extend their systems beyond a specific organization, with applications that connected downstream with customers and upstream with suppliers. Some did it by developing their own applications, such as SAP, some by acquisition, such as Oracle.

ERP is a system that can integrate all applications within a company; the new applications are add-ons that link smoothly with the integrated system. While this feature sounds good, it has not always resulted in a smooth coupling.

## Service-Oriented Architecture

In order to facilitate the linking process, a conceptual approach evolved—service-oriented architecture (SOA). This is an approach to achieve an effective way to combine all of the desired software applications into a workable system. We described SOA in Chapter 14. It involves a combination of hard technology, astute and innovative judgment, and perseverance in order to prevail.

## Internet Processes

The arrival of the use of the Internet for commercial use in the 1980s introduced a number of new ways of doing business, and there are more changes to come. We will use *e-business* to be the umbrella term to cover all of the applications that involve business use of the Internet. Included under this umbrella are such applications as e-commerce (the marketing and selling of goods), e-procurement (the buying of goods and services), and all of the other "e" applications in between.

e-Business companies face a basic question: should they sell directly to the consumer, or should they go through resellers? At present, a number of combinations are being used:

- Sell directly by receiving the order from a consumer and forwarding it to a third party for shipment (Amazon for companies such as Target).
- Sell directly by receiving the order and having it shipped through the company's own distribution system (Amazon books).
- Sell through resellers. The e-commerce website takes the order and directs the consumer to a retail store for pickup (Wal-Mart).
- Sell through own retail system. Offer a choice of buying on the website or from the retail store (Barnes & Noble books).

Based on the developments in this area, it appears that most businesses will use websites to receive the order and more conventional distribution channels to move the product to the consumer. Crandall (2014) describes how retailers and supply chain managers are moving to the omnichannel concept.

### Interorganizational Systems

Future supply chains will be longer and more complex. They will require greater communications capabilities, both within individual companies and between companies. Companies that develop effective communications systems with their supply chain partners will achieve better financial results than those that do not (Blankley 2008). However, a good interorganizational system only makes it possible to share information. Unless supply chain partners are willing to share information, the results will be less than satisfactory.

## Take Advantage of Other Technologies

While there will certainly be new developments in computer-related technology, especially in the areas of communications and information handling, there will also be developments in technologies that address management concepts—organization, teams, project management, and people management.

### Organization and Teams

Organizations will be flatter and more flexible; companies will develop virtual organizations that change as needed to adapt to different situations or external stimuli. Some companies have moved to eliminate permanent offices for employees who can take the next available desk when they are in the office and require desk space. While it sounds like a chaotic situation, it represents a high level of flexibility among the employees.

Teams will continue to be a useful way of adapting available skills to a needed task. There will be fewer teams—companies will learn to use the concept as needed and not just as a demonstration of forward thinking. Teams will have a purpose; when that purpose is achieved, the team will disband and members will revert to a specific assignment or become a member of another team.

Companies have tried different approaches to people management, especially since the Hawthorne studies. In those studies, we learned people respond to a variety of stimuli, sometimes as individuals and sometimes as a group. Organizations use different ways to obtain responses from employees favorable to business, most often resulting in greater productivity. We continue to learn how to motivate people; however, in some ways, we are just beginning to recognize the value of that capability.

### Project Management

Project management will become a normal part of everyday operations. During the mass production era, production lines could be set up for long runs. As companies move closer to mass customization and other approaches requiring flexibility and adaptability, they encounter the need for ways to manage change, and that is a primary benefit of the project management technique. It offers a way to accomplish in the shortest possible time a series of related tasks performed intermittently, not continuously. The project may require an hour, such as the changeover of a production line, or 2 years, such as the building of a new distribution center. Projects will become a normal part of doing business.

## Process Technology

At the same time they are linking with supply chain partners, companies must continue to develop their internal processes with the latest in technologies such as design modeling, automated process control, laser technology nanotechnology, photonics, sensor technology, smart materials, energy technologies, knowledge-based engineering for automation, and simulated flows (Jain and Benyoucef 2008).

Additive manufacturing, or 3D printing, is a new technology that offers amazing possibilities to transform existing manufacturing processes. This is a process that could have disruptive impact on supply chains as the technology becomes more fully developed and incorporated in manufacturing companies (Crandall 2013a).

How do companies build these desired supply chains? We described this process in considerable detail in earlier chapters. At this point, we will describe some of the major steps in the continuing journey.

## Build Strategic and Operational Plans

Strategic and operational planning will continue to be necessary in a company's preparation for the future. Company plans will be coordinated with other members of the supply chain. However, planning must be flexible enough to enable the supply chain to react to unexpected events and to innovate when the opportunity presents itself.

There is a vast array of literature that describes strategic planning and the process by which strategic plans are designed and implemented. A full discussion of strategic planning is beyond the scope of this book. However, it is essential to the success of supply chain management in the future. An essential feature of a strategic plan is that it must link closely with the operational plans for employees to implement.

## Continue the Drive for Collaboration

The need for collaboration along the supply chain is not a new topic. However, results to date suggest that the majority of companies are not truly collaborating with other participants of the supply chain. They are coordinating and even cooperating, but they are not collaborating. This will be a continuing need for the foreseeable future, because companies will have to make substantial changes in their infrastructure and cultures if they are to achieve collaboration. One approach that could lead to enhanced collaboration includes the following stages:

*Stage I*: Create a commitment and an understanding among supply chain participants.

*Stage II*: Remove the resisting forces to supply chain collaboration through

- Information sharing and systems integration
- People management and development
- Supply chain performance measurement

- Rationalization and simplification
- Relationship management and trust building

*Stage III*: Continuously improve collaboration capability. This stage requires that supply chains be able to adapt to ever-changing environments.

Unfortunately, few of the subjects interviewed in the study felt their companies had arrived at Stage III. Rather, they found themselves in the midst of Stage II—the transformation journey (Fawcett et al. 2008).

## Develop Performance Measures for Supply Chain Management

Many companies are finding it difficult to establish an effective performance measurement system within their own companies. The trend is to move farther away from the traditional financial performance measures to those that can measure what is happening in the operations and projects that are actively being pursued. Consequently, while the measures may be more meaningful, they are increasingly difficult to relate to the financial results of a company.

If individual companies are finding it difficult to define and implement desired performance measurement systems, it follows that doing it for an entire supply chain will be considerably more difficult. In earlier chapters, we discussed the difficulties of establishing the physical supply chain, the accompanying information flow, and the subsequent funds flow. As companies move through supply chain implementation, they will also be moving through the transition of designing and testing performance measures meaningful to all members of the supply chain.

In Chapter 16, we described some of the attempts to develop equitable distribution of benefits and costs among participants in a supply chain. Just as companies use public accounting firms to assess the reliability of their financial reports, it may be necessary to use a third party to provide objective reports about the performance of the entire supply chain.

In the following text, we describe some of the measures that would appear to have the greatest applicability across the supply chain.

### Integrate Delivery Effectiveness Measures

The following measures are the same delivery measures that individual companies use, except they will have to be applied to entire supply chains. The supply chain will have to establish goals, or standards, for each of the measures for all members of the supply chain. It will then be possible to measure where the supply chain is not performing to the expectations of the combined membership.

*Response time from customer order to customer receipt*: If the supply chain sells stock items, the goal will be to have those items available at the retail store, or on the Internet, for immediate pickup or delivery to the consumer. If the product or service offers a level of customization, the response time will be longer.

*Fill rate*: The fill rate will be a useful measure for the entire supply chain, just as it is for individual companies. When the fill rate falls below a target level, the cause of the inadequate performance can be identified and corrective action taken.

*Inventory level*: One of the benefits of a supply chain is to reduce the total amount of inventory. It may also be useful to express the inventory as days of inventory. As a supply chain develops a target level of inventory, the performance report can identify the amount, and location, of excess inventory in the supply chain.

*Speed of funds flow*: An area of increasing interest is the speed with which funds flow through the supply chain. Today, it can be approximated by adding days of receivables to days of inventory and subtracting days of payables. In theory, these numbers could be accumulated for each member of the supply chain to arrive at a composite number for the entire supply chain.

## Integrate Cost and Quality Measures

*Product or service cost*: Businesses know the cost of a product or service now; it is just they know it only for their own businesses. The purchase price for the retailer is the total cumulative cost for the entire supply chain. However, the retailer has little visibility into the costs of upstream participants in the supply chain. Transfer prices from one supplier to its customer contain material, labor, and overhead cost elements, but these individual components become the purchase price, or material costs, for the next company in the supply chain. To make an analysis of the cost elements throughout the supply chain, it would be necessary to look at the material, labor, and overhead components for each participant in the supply chain. It is difficult enough for a single company to analyze its cost components, especially overhead costs; it is unlikely much progress will be made soon in performing a comprehensive cost analysis for a supply chain. The best way may be to look at the effect of incremental changes on costs.

*Product or service quality*: The final quality of a product or service can be determined; however, as with costs, it is more difficult to identify the contribution, either positive or negative, toward quality, by each participant in the supply chain. Rather than trying to develop an information system that compiles the quality record of a product at each stage of its development, it is more likely that specific quality problems recorded at the end of the supply chain will be traced back to the cause of the problem, where corrective action can be taken.

## Supplier Profitability

In Chapter 16, we discussed the desirability of equitably spreading the benefits and costs along the supply chain. At this point, while we believe the idea has some merit, it is probably not realistic to expect that it will happen anytime soon. Other performance measurements will provide a greater contribution to the members of the supply chain.

## Effectiveness of Supply Chain Integration

A number of factors affect the level of supply chain effectiveness. Most of them are not easy to measure, yet are considered important, even critical, to the success of the supply chain. One approach may be to conduct longitudinal surveys of participating members on a regular basis (quarterly or annually) to determine their best estimate of the success of supply chain members in achieving desired levels for each factor. Some factors worthy of measurement include the following:

*Level of supply chain collaboration*: Many advocate the need for increased collaboration among supply chain members. This suggests that there is a benefit from increased collaboration.

It would be desirable to be able to measure the level of collaboration and then relate that to the increased benefits realized. To do this would require a composite measure of the number of successful contacts between companies, a way to weight the importance of each type of contact and a Likert-type scale to judge the effectiveness of each contact. A weighted factor analysis approach could be used, if companies consider it worthwhile expending the resources to collect and compile the information.

*Level of information sharing*: It is conceivable a weighted factor analysis could also be used to judge the level of information sharing among companies—identify the types of information to be shared, assign a weight to each type, and have knowledgeable employees judge the effectiveness of the sharing. Although feasible, each company would have to assess the desirability of committing resources to carry out this type of analysis.

For the short-term future, it is likely that individual companies will develop metrics about supply chain performance only when they see a quantifiable benefit to doing so. They may assess incremental changes, such as the difference in the cost between two suppliers or the cost reduction achieved by a recycling of packing materials. It is unlikely that individual companies will try to evaluate the performance of the entire supply chain.

On the other hand, supply chain performance measurement may be an opportunity for a consulting company or a nonprofit agency to develop a measurement system that will be useful and practical. Programs such as activity-based costing, balanced scorecard, and product lifestyle management (PLM) are concepts that can be extended to supply chains.

## Structure the Organization to Manage Change

There is a need for companies to restructure their organizations to become more agile. Organizations should become more horizontally focused, rather than using the traditional vertical hierarchy. This means lower-level employees must be empowered with greater decision-making authority. It does not mean that everyone has the freedom, or responsibility, to talk with anyone in other companies. Of all the functions in an organization, the ones that will have to change the most are the purchasing, or procurement, function and the sales, or marketing, function.

### Purchasing

Purchasing talks with suppliers. They will have to expand their range of discussions to include more information gathering about materials improvements, capacity limitations, potential disruptions, environmental concerns, and cash flow considerations. Purchasing can be a conduit from its company to the suppliers that enables other functional areas, such as engineering or operations, to discuss, with their counterparts, such topics as the effect of product redesign or inventory management.

In reporting on a CAPS Research study, Roberta Duffy lists the following key attributes that many successful organizations have identified as necessary for their supply chain management programs:

- Commitment to total quality management (TQM)
- Commitment to JIT

- Commitment to total cycle time reduction
- A long-range strategic plan
- Meaningful supplier relationships
- Strategic cost management that includes a total cost of ownership (TCO) philosophy
- Information systems and automation
- Appropriate performance measurement
- Training and professional development
- Service excellence
- Corporate/social responsibility
- Learning organization
- Management and leadership

While the study identifies the attributes needed, it does not provide specific metrics to measure the level of performance. While each purchasing organization must develop its own metrics, the study does provide some guidelines to help:

- *Rule 1*: The measurements should follow what the organization itself values.
- *Rule 2*: Make sure measurements are compatible with those in other functional areas.
- *Rule 3*: Do not measure for the fun of measuring. Use them to improve.
- *Rule 4*: Be willing to dig for meaningful measures.
- *Rule 5*: Trends are more important than targets (Duffy 1999).

The position of chief purchasing officer (CPO) is a new title in many companies. The purchasing function was viewed as a necessary, but nonstrategic, function. Many companies did not consider it at the same level as finance, marketing, engineering, or operations. That perspective is changing as the supply chain concept, accompanied by outsourcing, is placing more responsibility on purchasing. While more companies are recognizing the CPO function, not all require that the CPO have a purchasing background. Almost 50% of the persons selected came from outside the purchasing function, although most of the candidates have experience in some supply chain function (Johnson and Leenders 2008).

## Integrate All Functions

While marketing and purchasing probably have the greatest transformations to make, all functions within an organization will have to adapt to the needs of future supply chains. Managing the supply chain requires the integration of all functions, including sales, marketing, logistics, finance, accounting, operations, and purchasing. In addition to the cross functional, supply chain management needs to include all interactions with customers and suppliers (Lambert et al. 2008).

## Educate the Work Force

Increased efforts in training and education will be necessary in the future, for at least two reasons. The first results from the redistribution of skills along the supply chain because of outsourcing. Employees in firms performing the outsourced functions will need training. In addition, employees remaining in the firms that outsourced will need retraining, as the composition of their jobs change, most likely to extended responsibilities. A secondary consideration is that the work force in the United States is aging, and increasing numbers of skilled workers are retiring (Katz 2008).

The second reason is that products and processes will be changing at a rapid rate. These changes require skills in the use of new materials and manufacturing processes as well as in managing the increased rate of goods and information flowing through the supply chain.

Professional certifications will become more important as an indicator of the knowledge base a person has. Organizations such as APICS (the association for operations management) and the Institute for Supply Management (ISM) have multiple certification programs, including supply chain management. These certifications are available to practitioners who are motivated to continue their learning processes.

## Increase Marketing Influence

Marketing talks with customers. They will have to expand their range of discussions to include more information gathering and obtaining participation of customers to improve product design, reduce non-value-added features or process steps, and reduce the variation in demand patterns.

Peter Drucker pointed out over a half-century ago that satisfying customers was the primary purpose of a business (Drucker 1954). In order to become more customer sensitive, companies should transform themselves from an *inside-out* orientation to an *outside-in* way of thinking. This means they should focus on what the customer needs first, then decide how they should best satisfy those needs (Bund 2006).

Marketing has to take the lead in transforming the company into a more customer-centric organization. Manufacturing companies have been criticized for taking a *make and sell* approach—they make what they know how to make and then try to figure out how to sell it. The opposite of this is to use a *sense and respond* approach—find out what the customer wants and then make it (Haeckel 1999). While the latter approach sounds appealing, it takes a change in the marketing culture first, even before the rest of the organization.

## Overcome Inertia

The need to change the organization structure is not always obvious, especially if the company has been operating successfully. However, as history proves on a regular basis, past success is not a guarantee of continued success. Many of the largest companies have declined in importance or gone out of business because of their failure to overcome their internal resistance to change. Examples include Montgomery Ward, Westinghouse, and the Pennsylvania Railroad. More recently, General Motors and Chrysler faced this problem. The question is not whether or not they need to change; the problem is deciding on how to change and doing it quickly enough to avoid disaster.

One researcher attributes the reluctance to inertia—things at rest tend to stay at rest. Organizations tend to stay the way they are unless there is a dramatic need to change. Automobile companies in the United States have seen their market share gradually erode because of the competition from Japanese and European competitors, who have even set up locations in the United States. However, the American companies apparently did not see the need to change, at least, until the economic crisis in 2008 forced them to the point of extinction without government financial aid. Many companies have reduced their work forces so much they may have lost much of their capability to change (Morgan 2008). It is interesting that the first of the *Big Three* American automobile manufacturers to make a comeback was Ford. It was also the first of the three to bring in outside top management to lead the transition to a new way of doing business.

Earlier, we described some of the attributes that supply chains need to be successful. If supply chains are to possess these characteristics, does it follow that individual members of the supply chain must also have these attributes first? In Chapter 13, we described the need to change the culture of the company as part of the progress toward an integrated and agile supply chain.

Changing cultures takes time and resources. A commitment to change the culture of even a single company is often a huge task; doing it for multiple companies working together is difficult to imagine. It is only worthwhile when the supply chain members are tightly coupled in a quasi-permanent relationship, with large future expectations. Even then, it will probably take a third party to design and coordinate a program to achieve tangible results.

## Expand Knowledge Management*

Another challenge for supply chain management will be knowledge management. In his book *Powershift*, Toffler (1991) outlined how power has shifted from military power to financial power and, finally, to the power of information. However, knowledge management has stages of development, as we will show in this section.

- Simple definitions for data and information are easy to find. Knowledge and wisdom are a different story. Data—a set of discrete, objective facts about events, in an organizational context and structured records of transactions.
- Information—a message, usually in the form of a document or an audible or visible communication. It has a sender and a receiver. Information is meant to change the way the receiver perceives something, to have an impact on his or her judgment and behavior. It must inform.
- Knowledge—broader, deeper, and richer than data or information. "Knowledge is a fluid mix of framed experience, values, contextual information, and expert insight that provides a framework for evaluating and incorporating new experiences and information. It originates and is applied in the minds of knowers. In organizations, it often becomes embedded not only in documents or repositories but also in organizational routines, processes, practices, and norms."
- Knowledge includes wisdom and insight (Davenport and Prusak 1998).

Word Thesaurus includes these synonyms:

- Wisdom—understanding, knowledge, insight, perception, astuteness, intelligence, acumen, good judgment, and penetration
- Clairvoyant—intuitive, psychic, telepathic, second-sighted, perceptive, and far sighted

The definitions imply that value is added as data progress to information and beyond to knowledge and wisdom.

---

* A portion of this section is adapted from Crandall (2007).

## From Data to Information

Figure 18.1 shows a series of transformation processes moving from left to right. Data are transformed into information, largely by organizing bits of data into meaningful clusters of information. An example would be to take individual daily sales by item, by customer, and by store and summarize this into stock replenishment orders by item, buying trends by customer, and revenue performance by store. Data are collected transaction by transaction and, in today's environment, by computers, such as point-of-sale terminals. Computers also do most of the data-organizing activities. IT converts tasks from nonroutine to routine (Dibrell and Miller 2002). At present, most organizations face the problem of having too much data that need conversion to meaningful information.

Most organizations are still learning to use their information effectively to make routine decisions in areas such as inventory management, CRM, and resource utilization. To do this, they need a system for disseminating the right information to the right people to make the right decisions. They want to send enough information to users so that they can make better decisions; however, they do not want to overburden users with excess information. It is a complex and continuous process to design an effective flow from the data collection point to the decision-making point. This is the challenge of the big data, or data analytics, movement—to design a system to collect, organize, analyze, and use both the structured and unstructured data that exist and are available to organizations. Today, the effectiveness of information systems ranges from leader companies with smooth and comprehensive flows to laggard companies who have undirected data collection and erratic information flow.

Figure 18.1 shows converging lines that move from a widely separated state at the data stage of the knowledge chain to more narrowly separated states on the right-hand side of the diagram. This illustrates the need for a selection process that separates and preserves the most important data into more concise elements of information.

## From Information to Knowledge

The next major transformation process converts information into knowledge. While converting data to information can be handled largely by computers, most authorities agree it

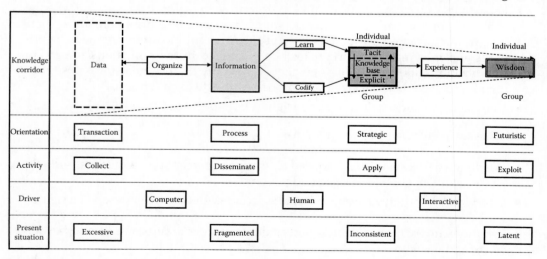

**FIGURE 18.1**
The knowledge corridor.

takes people to convert information to knowledge. This conversion has two major paths—learning and codifying. Learning is a process in which individuals convert information to tacit knowledge or knowledge lodged within their own minds. Several sources describe this learning process (Fahey and Prusak 1998; Girad 2006; Lester and Parnell 2007). Tacit knowledge remains with an individual until they share it with another individual or codify it to make it available to groups. From an organization's perspective, tacit knowledge is only valuable as long as that individual stays with the organization. It is to the organization's benefit to convert tacit knowledge to explicit knowledge.

Codifying information involves documenting in some formal process the rules, policies, and procedures of an organization into explicit knowledge. Explicit knowledge is often the result of a group effort and is available to a wide cross section of the organization. It extends beyond the tenure of any individual or groups of individuals; it is part of the organization's knowledge base.

Companies use knowledge for strategic planning and making decisions. Today, it appears there is a greater disparity among organizations in using knowledge than there is in using information. Information comes from systems that are standardized to facilitate intercompany communications. Knowledge, however, comes largely from individuals, and there is a great variation among people. Companies are often willing to share information; they are less willing to share knowledge. As with the movement from data to information, Figure 18.1 extends the converging lines to convey the concept that knowledge is extracted from information into more concentrated and focused resources.

**From Knowledge to Wisdom**

The final stage shown in Figure 18.1 is the transformation of knowledge into wisdom. In their book *Working Knowledge: How Organizations Manage What They Know*, Davenport and Prusak (1998) combine wisdom, or insight, into a broader category of knowledge. However, wisdom warrants a separate category, because of its tremendous potential for companies. This is unexplored territory for most companies and for individuals within those companies. Wisdom implies a level of understanding beyond that shown for knowledge. Wisdom is largely unique to individuals. It is hard to conceive of wisdom as coming from an inanimate organization. Individuals gain wisdom in a variety of ways, but largely through their own experiences. Some individuals gain wisdom while others, with similar experiences, do not. The expression "twenty years of experience" versus "one year of experience twenty times" seems to capture this distinction. One of the anomalies is that some people have it but most do not. Individuals with wisdom may have an uncanny knack of predicting outcomes or spotting problems or opportunities before others do. They may appear clairvoyant or having a form of second sight, a characteristic that would be a desirable capability for businesses to have in the future. When an organization discovers wisdom in an individual, they should exploit it for the good of both the organization and the individual.

Gaining wisdom is a learning process; exactly how that process works is still largely unknown. In fact, some individuals may have wisdom but are not aware of it because it is difficult to recognize and document. As a result, wisdom is an untapped resource in most organizations because it takes a rare combination of circumstances and individuals to expose wisdom. The converging lines in Figure 18.1 continue to narrow as the diagram reaches wisdom.

Moving along the knowledge corridor requires learning, both by individuals and by organizations. Lytras and Pouloudi (2006) point out the need for integrating knowledge management and learning activities, which they feel have been underrepresented in most companies, and illustrate how the two can be jointly supported by various knowledge management systems.

## Some Ways to Learn

One of the classic books on the subject of learning is Peter Senge's *The Fifth Discipline: The Art & Practice of the Learning Organization*. He calls systems thinking (a discipline for seeing patterns of change) the fifth discipline because it underlies all of his learning disciplines of personal mastery, mental models, shared vision, and team learning (Senge 1990).

Rapid change is driving the need to develop learning systems. The first requirement of learning in a computer-pervasive, networked world is speed of learning. The second requirement is to integrate learning across cross-functional units, and the third requirement is to facilitate learning in real time. A dynamic learning process consists of (1) learning *for* performance, (2) learning during performance, and (3) learning from performance. If an organization knows what is to be learned, they can *transfer* existing knowledge before performance, *prompt* learners during performance, and *evaluate* the results after the performance. If an organization does not know what people need to learn, they *focus* on providing basic knowledge before performance, stress *capture* of information during performance, and *interpret* (synthesize, digest) what was learned after performance (Baird and Griffin 2006).

Daniel Pink believes the rational, logical linear thinking that carried business to its present level is inadequate for the complexity and increasingly competitive business world of tomorrow. Instead, as he puts it, "Today, the defining skills of the previous era—the "left brain" capabilities that powered the Information Age—are necessary but no longer sufficient. And the capabilities we once disdained or thought frivolous—the "right-brain" qualities of inventiveness, empathy, joyfulness, and meaning—increasingly will determine who flourishes and who flounders. For individuals, families, and organizations, professional success and personal fulfillment now require a whole new mind" (Pink 2005, p. 3).

## Obstacles to Knowledge Transfer

While most would agree that acquiring and disseminating knowledge is important, knowledge transfer does not come easy, because of several obstacles:

1. *The organizational structure makes it harder to transfer knowledge*: Companies with rigid functional structures, that is, departments that may actually compete against each other, are at a disadvantage when it comes to sharing knowledge within the organization (Mohamed et al. 2004). It should come as no surprise then, which flatter structures, virtual organizations, and companies that use cross-functional teams do a better job at knowledge management.

2. *The right technology is not in place to share knowledge*: Moving knowledge throughout the organization should be a systematic process, not a random series of events. Achieving this will involve a new culture of openness among management and staff. It also helps to use the right technology to move knowledge through the company. Intranets are good vehicles for doing this, especially in larger companies where

employees are physically separated from each other. There are also knowledge management software (KM software) programs that can aid in this process.

Knowledge management technology may have already saved your life or at least kept you from becoming very sick. Pharmacists and doctors use knowledge management to make sure certain prescription drugs are not taken together, lest the patient have a serious reaction, or even die. Hackensack University Medical Center uses such a system to make sure patients do not receive dangerous combinations of medication. The medical center also utilizes a robot to help doctors make rounds of their patients. The device, called Mr. Rounder, can be operated from the doctor's laptop computer from home. Mr. Rounder can enter a hospital room and use a two-way video to talk to the patient about their condition. The robot even wears a white lab coat and stethoscope. Hackensack University Medical Center may be slightly ahead of the curve when it comes to using knowledge management techniques, and the results have been good. According to a recent *Business Week* article, patient mortality rates are down, while productivity and quality of care is up (Mullaney and Weintraub 2005).

3. *The culture may not be receptive to knowledge transfer*: Some managers and executives fear sharing knowledge will erode their power base. This mindset could actually be true in some businesses, especially those not committed to knowledge management. In fact, some consider the willingness to share, along with a supportive organizational culture, as the main factor needed for knowledge management to flourish.

Knowledge is not limited to specific problems such as the investigation of employee turnover or the effect of supplier variances on customer service. It can be anything that is of value to someone in the business. It might be tips on how to better perform a task, or it could be a series of workshops on how to prevent certain types of operational crises, such as machine failures or safety mishaps. Ernst and Young document best practices and then share them throughout their organization with a computer application called community of interest (COIN). Other companies actively involved in knowledge management include General Electric, Toyota, Hewlett–Packard, and Buckman Laboratories (Robbins and Coulter 2007).

Nonaka and Takeuchi (1995) explain the interactive process necessary for knowledge transfer is similar to how the ball moves spontaneously in a rugby match through intensive and laborious interaction among members of the team. They believe that "creating organizational knowledge is as much about bodily experience and trial and error as it is about mental modeling and learning from others. Similarly, it is as much about ideals as it is about ideas."

## Will Knowledge Replace "Things"?

Will knowledge, or the management of knowledge, ever replace things such as products or services? Perhaps not, at least in the immediate future. However, the question raises some interesting possibilities because the availability of knowledge could reduce the needs for some goods or services. Let us consider two situations. Suppose the knowledge, or experience, of visiting a tourist attraction, such as the Grand Canyon, is packaged into a virtual reality presentation that could be viewed (experienced) in the comfort of your home.

Would that not reduce the need to travel to that attraction, with the incumbent burden of driving a car or flying in an airplane and all of the services required along the way? (McConnon 2007) In the second situation, suppose that all of the knowledge required to qualify for a degree from a university were packaged and available for individuals to access at their convenience and ability to absorb. Would that not reduce the need for facilities (campus and buildings) and complementary services (faculty and staff) now required? Lest you write these off as fantasies, the technology to achieve both of these situations is being developed; it only remains to sell the ideas to the consumer, and, probably more difficult, to the present providers.

## Acquire Data Analytics Capabilities*

In the preceding section, we outlined the need to convert data to meaningful information. One of the current popular topics relating to supply chain information is the collection and conversion of unstructured data to information. This process is called data analytics, or big data.

In the APICS Operations Management Now—Joining the Big Data Revolution (October 26, 2012), Abe Eshkenazi highlighted the increased interest in the business community to participate in what some are calling an information revolution. He also called attention to the APICS 2012 Big Data Insights and Innovations report, which identified the following trends:

- Data overload and the abundance of trivial information are challenges many organizations face. There are a lot of data but also a lack of useful data.
- Important data are not reaching practitioners in efficient time frames.
- Despite current systems, data are still not always easily accessible.
- Current IT has not yet delivered optimal satisfaction in terms of what is easily measurable, reportable, or quantifiable data such as scheduling, inventory levels, and customer demand across the supply chain.
- Supply chain dataflow includes direct suppliers and customers, but there are gaps in a complete or true end-to-end supply chain dataflow model for most respondents (APICS 2012).

Is big data different? Why? The Leadership Council (2012) says that big data is not a precise term. "It describes data sets that are growing exponentially and that are too large, too raw or too unstructured for analysis using relational database techniques." Other definitions reinforce the idea that businesses are now collecting immense quantities of data in a myriad of ways. The result is that the traditional methods of data organization and analysis cannot keep pace. This does not mean you should throw away your spreadsheets, yet; however, you should look for some new ways to manage the onslaught that is headed your way. Forrester Research estimates that organizations effectively utilize less than 5% of their available data (Leadership Council 2012).

---

* This section is adapted from Crandall (2013).

## Background

Most of the early papers in the late 1990s and early 2000s used data analytics as the key descriptive term. Later, writers began to talk about *big data*, and, in the past 2–3 years, the references to both *big data* and data analytics have become commonplace.

How much data are being collected? One author estimates that more than 15 petabytes (a petabyte is one million gigabytes) of new information, or more than eight times the information in all U.S. libraries, is being collected each day (McKendrick 2010). Another source estimates that about 80% of the data are unstructured (Stevenson 2012).

Where do these data come from? A New York Times articles report that data are doubling every 2 years and coming not only from existing sources but also from entirely new streams, including industrial equipment, automobiles, electrical meters, and shipping crates. They measure and communicate location, movement, vibration, temperature, humidity, and even chemical changes in the air (Lohr 2012).

Edmund Schuster (2012) reported in a LinkedIn post that the Massachusetts Institute of Technology (MIT) initiated the study of big data in 2003 with the advent of the MIT Data Center Program. The researchers published some of their findings in an article "An Introduction to Semantic Modeling for Logistical Systems" (Brock et al. 2005).

One of the characteristics of the new data is that much of it is *unstructured*—tweets, product reviews, and Facebook *likes*—as opposed to more traditional data, which has been considered *structured* for use in spreadsheets and rational databases. The author points out that the need for the right infrastructure and a good idea of what to do with it (Provost 2012a).

IBM is a leading provider of big data systems. In a book for those wanting more, the authors define big data using four Vs:

- *Volume*: Big data can be measured in zettabytes (ZB), which are a million petabytes.
- *Variety:* Trying to capture all the forms of new unstructured data to complement the existing structured data.
- *Velocity:* The rate at which data arrive at the enterprise and are processed or well understood.
- *Veracity:* The quality or reliability of the data (Zikopoulos et al. 2013).

The authors dig deeply into the details of a big data system, with emphasis on IBM's capability in this area.

Called a new discipline, one group says *that* big data analytics is a workflow that distills terabytes of low-value data (e.g., every tweet) down to, in some cases, a single bit of high-value data (Fisher et al. 2012).

## Benefits

A joint survey by IndustryWeek and SAS solicited respondents as to their level of satisfaction with their SCM and ERP systems. Only about 12% were *very satisfied* with their existing systems because the general conclusion is that these systems report the past but do not help predict the future. For those companies that were very confident in their data, 43% saw increases in their revenues and gross margins while only 28% of those were less confident their data achieved increased in revenues and gross margins. Other benefits from the leading companies included improving key metrics, deriving more value from their assets, and increased inventory turns (IW and SAS 2012).

In analyzing ways to improve supplier performance, Seth (2012) cited these less tangible benefits. Gain visibility into lower tiers of your supply chain, so you can understand their performance trends and use this information to further improve performance:

- Benchmark your supplier's performance continuously against your peer group.
- Deliver qualified predictions about developments to be expected based on the trends in KPI scores in the aggregated information.
- Assess future risk within your supply base and proactively take corrective actions (Seth 2012).

One study reports that companies that incorporate big data and analytics into their operations show productivity rates and profitability that are 5%–6% higher than their peers (Barton and Court 2012). It is intuitively rational to believe that increased relevant information will lead to better decision making, which, in turn, will lead to increased well-being for the business. A recurring theme in many articles *is* that data analytics can help identify trends that facilitate the prediction of the future.

## Obstacles

Can data analytics really be that great? Remember, most of the early articles in trade journals are written from a positive perspective. The authors are reporting success stories and explaining how the concept or technique can be of value. It is only later when academics and other more penetrating writers begin to point out the actions necessary to realize those benefits. A group from Microsoft prepared a very comprehensive article about data analytics. They outlined a five-step approach as necessary:

1. *Acquiring data*: Data are available from a variety of sources, and users must decide which data will be useful.
2. *Choosing architecture on cost and performance*: This involves choosing a platform to organize the computation around a set of programming abstractions substantially different from those of the normal desktop environment. Most companies do not have the in-house capability and using cloud sources is expensive.
3. *Shaping the data to the architecture*: The analyst must ensure the data are uploaded in a way that is compatible with how the computation will be structured and distributed and partitioned appropriately. Moving back and forth between the cloud and a local machine is common but often a pain.
4. *Writing code*: Select a programming language, design the system, decide how to interface with the cloud, and be prepared for the rapidly changing environment. (Note: Most of this explanation was over the head of this writer.)
5. *Debugging and iteration*: This is a process of debugging and looking for errors (code rarely works the first time), changing code, and visualization at multiple scales (Fisher et al. 2012).

Another source reported obstacles can include wrong metrics, too much data, poor data quality, lack of know-how, and growing complexity in the marketplace (Robbins 2010).

One group reporting on their work with dozens of companies in six data-rich industries found companies needed three capabilities: (1) identify, combine, and manage multiple sources of data; (2) build advanced analytics models for predicting and optimizing outcomes; and (3) possess the muscle to transform its organization to yield better decisions (Barton and Court 2012).

Getting into big data in a meaningful way requires a hefty assortment of IT hardware and software, supported by *data scientists* (Fisher et al. 2012). It also requires building relationships with cloud companies, because most businesses would not have the resources required to go it alone.

Finally, managing change of any kind is always a challenge, and changing the work habits, or culture, of a company can be a major disruption to a program as ambitious as big data.

## Applications

Some of the better-known users of big data analytics include Google, Amazon, IBM, and Hewlett–Packard (Barton and Court 2012). Some of the companies most active have been working on their systems to collect and analyze for over a decade.

Some of the industries that use data analytics include airlines for determining what price it charges on each flight; banks, for deciding how to best provide customer service across its multiple offerings; and retailers, in deciding how much and where to place their inventories for optimum effectiveness (Court 2012).

Another study reports applications in identifying fraud in real time; evaluating patients for health risks, changes in consumer sentiment, and a need for service on a jet engine; and exploring network relationships in social media sites (Davenport et al. 2012).

IBM developed an analytics application for helping an online university identify students who may drop out of the program, thereby enabling the organization to improve their retention rates (IBM 2010).

For those who watched the 2012 presidential elections on television, it was obvious that the networks were digging into the data in an unprecedented way, not only in the level of granularity (state, county, and precinct) but also by age, gender, ethnic background, and other classifications.

## Techniques

There are a myriad of analysis techniques being used or considered that go beyond the traditional. They include more advanced statistical techniques, logic models, and pattern analysis. A structured approach suggested by a combined report by SAS and IndustryWeek could includes the following levels:

1. Standard reports
2. Ad hoc reports
3. Query drilldown (or OLAP)
4. Alerts
5. Statistical analysis
6. Forecasting
7. Predictive modeling
8. Optimization

Other techniques include artificial intelligence techniques such as natural-language processing, pattern recognition, and machine learning (Lohr 2012).

There is little doubt that the technology of data collection is driving organizations to confront the issue of what to do with it. Entities of all types are being forced to recognize they must deal with this issue or suffer the consequences of competitors who do.

## Conclusions

Although big data and data analytics are in the spotlight, many companies are unsure how to proceed. Top managers are hesitant to make large investments until they have a better idea of how to capitalize on the potential. Many are not convinced their organizations are ready or they are not effectively analyzing the data they already have (Barton and Court 2012).

A 2011 survey by the Economist Intelligence Unit of almost 600 senior executives found the following:

- There is a strong link between effective data management strategy and financial performance.
- Extracting value from big data remains elusive for many organizations.
- Many companies struggle with the most basic aspects of data management, such as cleaning, verifying, or reconciling data across the organization.
- Companies that are furthest along the data management competency continuum—strategic data managers—provide a useful model for how organizations will need to evolve if they are to extract and utilize valuable data-driven insights (Briody 2011).

Leaders in data analytics differ from traditional users by paying attention to data flows instead of data stocks in a data warehouse, relying on data scientists and product and process developers rather than data analysts, and moving analytics away from the IT function and into core business, operational, and production functions (Davenport et al. 2012).

One of the potential issues for many companies will be deciding doing analytics between in-house and in the cloud. Many companies will not have the in-house capabilities and will have to rely on third-party providers; however, they will need to select their applications wisely because of the high cost. Larger companies many elect to build their own analytics capability; they will need commitment and perseverance to succeed. They will not only need to understand the technology, they will have to change their infrastructure and culture to effect the new way of managing. Positive results will come only after a substantial investment in time and money to establish the base necessary for success.

Several papers indicate that data analytics does more than interpret the past; it identifies leading indicators of the future. While traditional SCM/ERP systems are not advanced enough for current dynamic economic conditions, analytics are needed for next-generation supply chains (IW and SAS 2012).

Data analytics can help in making the world a better place in which to live, by improving education systems and learning, identifying health risks and designing prevention systems, and improving agriculture processes to feed the growing population of the world (World Economic Forum 2012).

Big data can also be an environmental concern. All of that data are stored in vast data warehouses covering acres of land and consuming lots of energy. Emerson Network Power estimated there were over 500,000 data centers in the world, occupying nearly 300 million ft$^2$ of space and releasing tons of carbon emissions. However, someone is working on making data centers smarter and less energy wasteful. One observer puts it this way: "Kermit knew it wasn't easy being green, but as frogs vanish because of a host of ecological inputs, some manmade, it's getting harder than ever" (Provost 2012b).

The Big Data concept is here, but in the early stages of its life cycle. Data collection techniques are ahead of data conversion techniques; data conversion techniques are ahead of user comprehension; and user comprehension is ahead of general population acceptance.

## Integrate Manufacturing and Services

As we indicated in Chapter 17, the boundary between manufacturing and services is vanishing. Progressive companies are building a bridge over the chasm that separated *hard goods* from *soft services*. For years, programs that originated in the manufacturing area are being adapted for use in service applications. Authors suggest that MRP has application in academic institutions, JIT works in retail, banks use TQM, lean manufacturing revolutionizes distribution, and the health-care industry benefits from Six Sigma.

While most of the articles describe using manufacturing techniques in services, an increasing number report on how concepts and techniques that originated in service industries are adopted in manufacturing. While manufacturing industries concentrated on reducing costs and improving quality, service industries focused on improving customer relationships and service. Certainly, service businesses want to reduce costs and improve quality, that is why they adapt manufacturing techniques to fit their needs. Beyond that, they understand that businesses exist because of their ability to find customers and satisfy their needs. As Peter Drucker described it, "With respect to the definition of business purpose and business mission, there is only one such focus, one starting point. It is the customer. The customer defines the business. Businesses exist for one purpose only—to find customers and provide service to them" (Drucker 1974, p. 79). Companies must look at their business from the outside in—from the perspective of the customer. Bund builds on Drucker's basic ideas and includes a number of case histories of businesses that have succeeded, in part because they have taken the *outside-in* look at their companies. She uses service companies, such as FedEx, eBay, and Costco, to illustrate the customer-centric thinking. She also considers GE and Dell as outside-in thinkers, but only because they think of themselves as service businesses (Bund 2006). Earlier, Mathé and Shapiro (1993) pointed that low cost would not be enough for survival in the future, but that there was a need for manufacturing companies to integrate service strategies in their planning. Even such hard core manufacturing units as machine shops recognize the opportunity to *supersize* their products with services (Zelinski 2004).

Does this mean that manufacturing companies are not interested in their customers? Of course not; however, most manufacturing customers are other businesses, not individual consumers. The service industries deal with other businesses, but they also focus on individual consumers. This requires that they have a more diverse customer base and deal with a greater variety of needs and wants. There are no mass markets anymore.

Does it matter where the ideas originate? Most of the programs require a cross-functional approach. Whether in manufacturing or services, sales and marketing provide the customer contact; operations and finance provide the analysis and execution. The world is changing and competing successfully is becoming more difficult. Businesses need to take advantage of good ideas, regardless of its source. Manufacturing and services can learn from each other; many companies have already started (Crandall 2007).

## Apply Chaos Theory to Business

Chaos theory is a concept for future application, although perhaps the not-too-distant future. As we point out in virtually every chapter in this book, supply chains are complex and dynamic. Sometimes, the changes do not always seem to follow a logical systematic pattern, and that is the essence of chaos theory. It attempts to explain nonlinear behavior and the reasons for unexpected events in an otherwise predictable business. "Chaos theory is rooted in mathematics and the natural sciences. Chaos is a state whereby phenomena that appear to be unrelated actually follow an unknown or hidden pattern. Chaotic systems possess two characteristics: sensitive dependence to initial conditions and unpredictability in the long run" (Crandall et al. 2010).

One of the first applications of chaos theory was in weather forecasting. Edward Lorenz tried to develop equations he could use in forecasting the weather. At the time, weather forecasting was considered an art form, not a science, and Lorenz was trying to change this idea. He found his equations worked most of the time, but once in a while, something unexpected happened. Among other things, he discovered weather systems contained order—warmer in the summer, colder in the winter—but within the orderliness, there were unexpected disruptions. Patterns repeated, but were dependent on starting conditions and the existence of other variables. For example, all hurricanes in the Atlantic Ocean spawn off the coast of Africa and move westward toward the United States. However, they do not all have the same intensity or follow the same path. This idea of disorder within order can be extended beyond weather to other systems, including business.

Economists can observe the same patterns in trying to forecast business cycles. Everyone accepts the concept of the business cycle provides order, but with variations. When will the next business downturn occur? How long will it last? How severe will it be? When will the upturn begin?

Marketers experience it with new product introductions. The product life cycle will provide a pattern or a framework of order, but each product has its own idiosyncrasies. Will it be a success? How many will sell? How long will it be before a competing product comes along? How effective will the advertising program be?

Throughout this book, we have stressed the idea that successful businesses have to change to remain competitive. This idea suggests that the most successful businesses will be leaders of change, or innovators. Doing nothing suggests stability and orderliness, but with limited future prospects. Being innovative suggests turbulence, with higher probabilities of success or failure. The desired path is one of orderly innovation, or leading-edge successes amid noncatastrophic failures. Businesses can plan a general direction and begin their execution; however, they must be flexible enough to adjust their plans as unexpected disruptions occur.

The application of chaos theory to business requires a blend of scientific theory and practical knowledge. We do not have room to describe it in detail; we can only suggest that it will probably become one of the more important considerations as supply chains progress. Frederick provides a good explanation of the application of chaos theory to the business setting (Frederick 1998).

William T. Walker captures the essence of this need for flexibility. He poses this scenario. "Are we there yet in our ability to fulfill each single customer order through a different supply chain? Can the current state of information technology, and technology in general, support the exchange of goods and services plus information plus the flow of cash plus the measurement of end-to-end performance through a set of plug 'n play trading partner relationships that execute a single order perfectly and profitably? The next order will, of course, be executed through a different set of relationships" (Walker 2009).

## Summary

We began this chapter with a discussion of why there is a need for supply chains to adapt to changing conditions. Supply chains must respond to open-system requirements and to the growing movement in sustainability. We then identified a number of challenges compiled by the ISM, APICS, and the combined efforts of IBM and IndustryWeek.

As they build their supply chains, companies need to consider a number of factors, including the role of government, the increasing participation of third parties, and the impact of an array of technologies. After that, they need to take a number of steps to be able to adapt their supply chains to keep pace with the changes in the business environment. In particular, they must build collaborative supply chains. To do this, they will need to change their technology, their infrastructure, their culture, and their performance reporting systems.

In addition, businesses, and their supply chains, will need to recognize the need to include sustainability as a primary reason for their existence. Collaborative and sustainable supply chains require knowledge management and a continuing search for wisdom.

We introduced the idea that the future will be uncertain. Supply chains will need to understand how to cope with this uncertainty, most likely by using some of the knowledge generated in the study of complex systems or chaos theory. The business world needs critical thinkers—the ability to focus on the most relevant information, ask the right questions, and separate facts from false assumptions. Sadly, one study found there is a lack of critical thinkers coming along (Chartrand et al. 2009). However, there is hope. Another study of the human brain found that increased knowledge can tilt our vision toward the positive side and enable us to cope with the problems we may encounter (Sharot 2011). We hope this book has been helpful in adding to your understanding of supply chains.

## Hot Topic

One of the current topics of interest in today's food supply chains is transparency in the contents of food. This chapter's hot topic is about a food product that is struggling to survive as a result of negative publicity, even though it has never caused illness or death.

## Hot Topics in the Supply Chain: How Social Media Knocked Down the Lean Finely Textured Beef Industry

The future of supply chains will involve one relatively new trend, the intense scrutiny of stakeholders that use social media. A number of companies have recently seen their supply chains rocked by videos gone viral. One company, Beef Products Inc. (BPI), felt a sales plunge in 2012 after reports surfaced on how their product was made.

### BACKGROUND

For companies that process ground beef, there is a well-known product that has been used for decades that serves as an additive—lean finely textured beef (LFTB). By adding LFTB to regular ground beef, the fat content of the final product can be reduced, a process that has some obvious health benefits. The additive also serves to extend the batch of regular ground beef, which accrues cost benefits as well. If you have eaten a hamburger in the last 20 years, the odds are good that you have also consumed LFTB (Gruley and Campbell 2012). The LFTB has one other interesting benefit—its use saves the amount of cattle that needs to be raised each year by about 1.5 million head (Berman 2012).

The production of LFTB involves moving scraps of meat through a centrifugal force process that eliminates the fat. The remaining product is treated with ammonia hydroxide to kill bacteria such as *E. coli* and salmonella. LFTB is used as an additive to ground beef, low-fat hot dogs, pepperoni, frozen meat entrees like meatballs, and canned foods. McDonald's, Taco Bell, and Burger King used LFTB until 2012 (Bloomgarden-Smoke, 2012). The use of LFTB is deemed safe by the U.S. Department of Agriculture, and to date, there has never been a health-related crisis associated with its use (Gleason and Berry 2012).

However, the production and sale of LFTB decreased significantly in 2012. After two decades of successful sales, this key component in the supply chain for the meat industry came to an abrupt halt when somebody referred to it as *pink slime*. That term was thrown about the social media world, and within days, the future of LFTB was left in question.

BFI is the key company in the supply chain that manufactures LFTB. The company was founded by entrepreneur Eldon Roth, who invented the process of making LFTB. BFI sold its product to other companies that made beef patties and other meat products. High-profile companies that used LFTB in their products included McDonald's, Burger King, and Taco Bell. Grocery stores that sold products containing LFTB included Wal-Mart and Kroger (Gruley and Campbell 2012).

### TROUBLE BREWING

Despite the success of the product, trouble had been brewing in the background. In April 2011, celebrity chef, Jamie Oliver, drew attention to the ammonia process on his television show, Food Revolution (Gruley and Campbell 2012). In the episode, Oliver doused a mound of beef with ammonia, a replication that was not entirely accurate of the real process, but still visually disturbing. Viewers were led to question the link between ammonia and beef. But it was a follow-up image of the product that led to an even greater shock—critics called it *pink slime*. The blogosphere and video references to this newly renamed product were unfortunate. Now, the gross factor was involved, and LFTB maker BPI began to see a decline in sales as it was faced with a hopeless public relations disaster (Bloomgarden-Smoke 2012).

Food blogger Bettina Elias Siegel responded by launching an online petition to have it banned from the federal school lunch program (Gruley and Campbell 2012). *ABC News* with Diane Sawyer covered the story, and the blogosphere became active as various stakeholders weighed in on the product.

Within the supply chain, the impact was being felt widely. BPI, the maker of LFTB, had to close three of its four plants and eliminate 650 jobs. Meat processor, AFA Foods, Inc., had to file for bankruptcy as a direct result of the pink slime controversy (Gleason and Berry, 2012).

One of the ironies of the pink slime story is that it did not involve outsourcing the production of a product to low-wage workers in Asia. Instead, the production of the product occurred in the United States with American jobs at stake when the downfall began. Furthermore, the concern was not founded on the actual occurrence of a foodborne illness. Instead, it was based on the perceptions of stakeholders—that LFTB is, in some way, bad.

## REFERENCES

Berman, R., Deflate food-purity claims before they explode, *Nation's Restaurant News*, 46(13), 48, June 25, 2012.

Bloomgarden-Smoke, K., "Pink slime": Health crisis or misunderstood meat product? *Christian Science Monitor*, March 27, 2012, from http://www.csmonitor.com/USA/2012/0327/Pink-slime-Health-crisis-or-misunderstood-meat-product, retrieved June 21, 2013.

Gleason, S. and Berry, I., Beef processor falters amid "slime." *Wall Street Journal*, p. B2, April 3, 2012.

Gruley, B. and Campbell, E., Slimed: Was a food innovator unfairly targeted? *Bloomberg Businessweek*, pp. 18–20, April 16, 2012.

## Questions for Discussion

1. The use of LFTB helps to reduce the cost of ground beef products, particularly in school lunch programs. Conduct an Internet or library search, and find out what the current status is on the purchase of meat with LFTB within school lunch programs.

2. The use of ammonia hydroxide is not unique to just LFTB. What other food products involve the use of this process?

## Discussion Questions

1. What are some of the factors that will cause supply chains to change?

2. What are the implications of the sustainability movement for businesses?

3. What is the TBL concept? What are its implications for the business community?

4. Discuss the responsibility of business in building social capital.

5. Discuss the responsibility of business in addressing environmental concerns.

6. What is knowledge management? How do businesses capitalize on knowledge?

7.  What is chaos theory? Does it have a place in future business planning?

8.  Research a company that has failed to change and, as a result, lost much of its market position.

9.  Research a company that you believe has been successful because of its willingness and capability to adapt to changing conditions.

10. Describe one technology that you think will be a major driver of change in the future. Why did you select it?

## References

APICS 2012 Big data insights and innovations, Discovering emerging data practices in supply chain and operations management, Chicago, IL, 2012.

Baird, L. and Griffin, D., Adaptability and responsiveness: The case for dynamic learning, *Organizational Dynamics*, 35(4), 372, 2006.

Barton, D. and Court, D., Making advanced analytics work for you, *Harvard Business Review*, October 2012, http://hbr.org/2012/10/making-advanced-analytics-work-for-you/ar/1.

Blackstone, J.H., Jr., *APICS Dictionary*, 14th edn., APICS, Chicago, IL, 2013.

Blankley, A., A conceptual model for evaluating the financial impact of supply chain management technology investments, *International Journal of Logistics Management*, 19(2), 155, 2008.

Briody, D., Big data, Harnessing a game-changing asset, Economist Intelligence Unit Limited, London, U.K., 2011.

Brock, D.L., Schuster, E.W., Allen, S.J., and Kar, P., An introduction to semantic modeling for logistical systems, *Journal of Business Logistics*, 26(2), 97–118, 2005.

Bund, B.E., *The Outside-In Corporation: How to Build a Customer-Centric Organization for Breakthrough Results*, McGraw-Hill, New York, 2006.

Carter, P.L., Carter, J.R., Monczka, R.M., Slaight, T., Blascovich, J.D., and Markham, W.J., Succeeding in a dynamic world, CAPS Research, A.T. Kearney and Institute of Supply Management (ISM) Research Study, Tempe, AZ, 2008.

Chartrand, J., Ishikawa, H., and Flander, S., *Critical Thinking Means Business: Learn to Apply and Develop the New #1 Workplace Skill*, Pearson TalentLens, Pearson Education, Inc., Upper Saddle River, NJ, 2009.

Christiansen, C.M. and Eyring, H.J., *The Innovative University: Changing the DNA of Higher Education from the Inside Out*, Jossey-Bass, Hoboken, NJ, 2011.

Court, D., Putting big data and advanced analytics to work, McKinsey, New York, NY, 2012, http://www.mckinsey.com/features/advanced_analytics (includes video)

Crandall, R.E., From data to wisdom, Is knowledge management the key to the future? *APICS Magazine*, 17(9), 30, 2007.

Crandall, R.E., Where will additive manufacturing take us? *APICS Magazine*, 23(1), 20–23, 2013a.

Crandall, R.E., The big data revolution, Investigating recent developments in collection and analytics, *APICS Magazine*, 23(2), 20–23, 2013b.

Crandall, R.E., Bricks and clicks, how retailers and supply chain managers are embracing omnichannel, *APICS Magazine*, 24(2), 26–29, 2014.

Crandall, W.R., Parnell, J.A., and Spillan, J.E., *Crisis Management in the New Strategy Landscape*, SAGE, Thousand Oaks, CA, p. 221, 2010.

Davenport, T.H., Barth, P., and Bean, R., How "big data" is different, *MIT Sloan Management Review*, 54(1), 43–46, Fall 2012.

Davenport, T.H. and Prusak, L., *Working Knowledge: How Organizations Manage What They Know*, Harvard Business School Press, Boston, MA, 1998.

Dibrell, C.C. and Miller, T.R., Organization design: The continuing influence of information technology, *Management Decision*, 40(5/6), 620, 2002.

Drucker, P.F., *Management: Tasks, Responsibilities, Practices*, Harper & Row, New York, 1974.

Drucker, P.F., *The Practice of Management*, Harper & Row, New York, 1954.

Duffy, R.J., Strategic moves across the board, *Purchasing Today*, 10(8), 36, 1999.

Elkington, J., Towards the sustainable corporation: Win-win-win business strategies for sustainable development, *California Management Review*, 36(2), 90, 1994.

Elkington, J., *Cannibals with Forks: The Triple Bottom Line of 21st Century Business*, New Society Publishers, Gabriola Island, British Columbia, Canada, 1999.

Elkington, J., Governance for sustainability, *Corporate Governance: An International Review*, 14(6), 522, 2006.

Fahey, L. and Prusak, L., The eleven deadliest sins of knowledge management, *California Management Review*, 40(3), 265, 1998.

Fawcett, S.E., Magnan, G.M., and McCarter, M.W., A three-stage implementation model for supply chain collaboration, *Journal of Business Logistics*, 29(1), 93, 2008.

Fisher, D., DeLine, R., Czerwinski, M., and Drucker, S., Interactions with big data analytics. *Interactions*, 19(3), 50, 2012.

Frederick, W.C., Creatures, corporations, communities, chaos, complexity, *Business and Society*, 37(4), 358, 1998.

Friedman, T.L., *The World is Flat: A Brief History of the Twenty-First Century*, Picador/Farrar, Straus and Giroux, New York, 2007.

Girad, J., Where is the knowledge we have lost in managers? *Journal of Knowledge Management*, 10(6), 22, 2006.

Gyorey, T., Jochim, M., and Norton, S., Challenges ahead for supply chains, McKinsey Global Survey, McKinsey & Company, New York, NY, 2010.

Haeckel, S.H., *Adaptive Enterprise: Creating and Leading Sense-and-Respond Organizations*, Harvard Business School Press, Boston, MA, 1999.

Hawken, P., *Blessed Unrest, How the Largest Movement in the World Came into Being and Why No One Saw It Coming*, Penguin, London, U.K., 2007.

IBM, Keeping students on track, IBM Corporation, Somers, NY, 2010.

IBM and Industry Week, Managing supply chains for growth and efficiency, IBM and Industry Week Custom Research, 2008.

Industry Week and SAS, Supply-chain analytics: Beyond ERP & SCM, IW custom research and SAS, 2012, www.industryweek.com and www.sas.com, accessed November 5, 2012.

Jain, V. and Benyoucef, L., Managing long supply chain networks: Some emerging issues and challenges, *Journal of Manufacturing Technology Management*, 19(4), 469, 2008.

Johnson, P.F. and Leenders, M.R., Building a corporate supply function, *Journal of Supply Chain Management*, 44(3), 39, 2008.

Katz, J., Manufacturing in the classroom, *Industry Week*, September 1, 2008. http://www.industryweek.com/PrintArticle.aspx?ArticleID = 17040

Lambert, D.M., Garcia-Dastugue, S.J., and Croxton, K.L., The role of logistics managers in the cross-functional implementation of supply chain management, *Journal of Business Logistics*, 29(1), 113, 2008.

Leadership Council for Information Advantage, Big data: Big opportunities to create business value, 2012, http://www.emc.com/microsites/cio/articles/big-data-big-opportunities/LCIA-Big Data-Opportunities-Value.pdf.

Lester, D. and Parnell, J., *Organizational Theory: A Strategic Perspective*, Atomic Dog Publishing, Mason, OH, 2007.

Lohr, S., The age of big data, *The New York Times*, February 11, 2012, http://www.nytimes.com/2012/02/12/sunday-review/big-datas-impact-in-the-world.html?_r = 1&pagewanted = all][11/6/2012 7:15:32 AM], accessed November 5, 2012.

Lytras, M.D. and Pouloudi, A., Towards the development of a novel taxonomy of knowledge management systems from a learning perspective: an integrated approach to learning and knowledge infrastructures, *Journal of Knowledge Management*, 10(6), 64, 2006.

Mathé, H. and Shapiro, R.D., *Integrating Service Strategy in the Manufacturing Company*, Chapman & Hall, London, U.K., 1993.

McConnon, A., Just ahead: The web as a virtual world, *Business Week*, August 13, 2007, p. 62.

McKendrick, J., Big data, big issues, database trends and applications, E-Edition, December 2010, www.dbta.com.

Melnyk, S.A., Lummus, R., Vokurka, R.J., and Sandor, J., Supply chain management 2010 and beyond: Mapping the future of the strategic supply chain, Special Report, APICS Educational & Research Foundation Inc., Chicago, IL, 2008.

Mintzberg, H., *Managers Not MBAs: A Hard Look at the Soft Practice of Managing and Management Development*, Berrett-Koehler Publishers, Inc., San Francisco, CA, 2005.

Mohamed, M., Stankosky, M., and Murray, A., Applying knowledge management principles to enhance cross-functional team performance, *Journal of Knowledge Management*, 8(3), 127, 2004.

Morgan, M., Unmasking transformation, *Industrial Engineer*, 40(9), 26, 2008.

Mullaney, T. and Weintraub, A., The digital hospital, *Business Week*, March 28, 2005, p. 77.

Nonaka, I. and Takeuchi, H., *The Knowledge-Creating Company: How Japanese Companies Create the Dynamics of Innovation*, Oxford University Press, New York, 1995.

Pink, D.H., *A Whole New Mind: Moving from the Information Age to the Conceptual Age, Riverhead*, Penguin Group, New York, 2005.

Provost, T., No fad, big data is real deal, CFO, 2012a, http://www3.cfo.com/article/2012/11/technology_sloan-summit-big-data-predictive-analytics-netsuite-kpmg-infrastructure?_mid=99002&_rid=99002.51300.10225

Provost, T., Will 2013 be the year of green data? CFO, 2012b, http://www3.cfo.com/blogs/technology/technology-blog/2012/12/Will-2013-Be-the-Year-of-Green-Data?mid = 107204&rid = 107204.51300.3785[12/21/2012 8:34:12 AM]

Robbins, J., Using more analytics can help industrial manufacturers, *Industry Week*, http://www.industryweek.com/articles/using_more_analytics_can_help_industrial_manufacturers_23248.aspx, November 22, 2010, accessed January 15, 2014.

Robbins, S. and Coulter, M., *Management*, 9th edn., Pearson-Prentice Hall, Upper Saddle River, NJ, 2007.

Schuster, E., Big data is a big reality, 2012, http://ingehygd.blogspot.com/2012/02/big-data-is-big-reality.html#!/2012/02/big-data-is-big-reality.html

Senge, P.M., *The Fifth Discipline: The Art & Practice of the Learning Organization*, Currency Doubleday, New York, 1990.

Seth, V., The next frontier of competitive wars in supplier management, *Industrial Distribution*, November/December 2012, pp. 49–51.

Sharot, T., The optimism bias, *Time Magazine*, http://content.time.com/time/health/article/0,8599,2074067,00.html, May 28, 2011, accessed January 15, 2014.

Stank, T., Dittmann, J.P., Tate, W., Autry, C., Moon, M., Peterson, K., Bell, J., and Bradley, R., Game changing trends in supply chain, parts I–V, *Supply Chain Management Review*, March 15, 2013, www.scmr.com, accessed August 14, 2013.

Stevenson, T., Why big data offers big opportunity, *Investment Week*, 2012, pp. 43–43.

Toffler, A., *Powershift: Knowledge, Wealth, and Violence at the Edge of the 21st Century*, Bantam, New York, 1991.

Walker, W.T., Supply chain consultant, personal correspondence, May 2, 2009.

World Economic Forum, Big data, big impact: New possibilities for international development, The World Economic Forum, 2012, www.weforum.org/mfs, accessed November 5, 2012.

Zelinski, P.C., Manufacturing in a service economy, *Modern Machine Shop*, 77, 1, 14, 2004.

Zikopoulos, P., deRoos, D., Parasuraman, K., Deutsch, T., Corrigan, D., and Giles, J., *Harness the Power of Big Data, The IBM Big Data Platform*, McGraw-Hill, New York, 2013.

# *Index*